Operator Theory: Advances and Applications
Vol. 149

Current Trends in Operator Theory and its Applications

Joseph A. Ball
J. William Helton
Martin Klaus
Leiba Rodman
Editors

Springer Basel AG

Editors:

Joseph A. Ball
Department of Mathematics
Virginia Tech
Blacksburg, VA 24061
USA
ball@math.vt.edu

J. William Helton
Department of Mathematics
University of California
La Jolla, CA 92093
USA
helton@math.ucsd.edu

Martin Klaus
Department of Mathematics
Virginia Tech
Blacksburg, VA 24061
USA
klaus@math.vt.edu

Leiba Rodman
Department of Mathematics
College of William and Mary
Williamsburg, VA 23187-8795
USA
lxrodm@math.wm.edu

2000 Mathematics Subject Classification Primary: 47xx; Secondary: 32xx, 42xx, 93xx, 94xx

A CIP catalogue record for this book is available from the
Library of Congress, Washington D.C., USA

Bibliographic information published by Die Deutsche Bibliothek
Die Deutsche Bibliothek lists this publication in the Deutsche Nationalbibliografie; detailed bibliographic data is available in the Internet at <http://dnb.ddb.de>.

ISBN 978-3-0348-9608-5 ISBN 978-3-0348-7881-4 (eBook)

DOI 10.1007/978-3-0348-7881-4

Originally published by Birkhäuser Verlag, Basel – Boston – Berlin in 2004
Softcover reprint of the hardcover 1st edition 2004

Printed on acid-free paper produced from chlorine-free pulp. TCF ∞
Cover design: Heinz Hiltbrunner, Basel

ISBN 978-3-0348-9608-5

9 8 7 6 5 4 3 2 1

www.birkhasuer-science.c

Contents

Preface

This volume is a collection of original and review articles which reflects many developments on the cutting edge of research in operator theory and its applications, and related areas of mathematics: multivariable operator theory and its connections with multi-dimensional systems and the function theory of several complex variables, operator theory in indefinite inner product spaces, distributed parameter systems, mathematical physics (inverse scattering, Dirac-type equations, Schrödinger operators), wavelets (including connections with the representation theory for operator algebras), and numerical analysis. Review articles include a report on recent achievements and future directions of research in the area of operator theory and its diverse applications.

The collection appears as proceedings of the International Workshop on Operator Theory and its Applications (IWOTA) which was held at Virginia Tech, August 2002, and reflects lectures, talks, and discussions which took place at the workshop.

We gratefully acknowledge the following agencies for their financial support of IWOTA 2002:

NSF (National Science Foundation)
Department of Mathematics, Virginia Tech
College of Arts and Sciences, Virginia Tech
Research Division, Virginia Tech

Joseph A. Ball, J. William Helton, Martin Klaus, and Leiba Rodman
Organizers of IWOTA 2002 and editors of the volume

Operator Theory:
Advances and Applications, Vol. 149, ix–xxii

IWOTA 2002 and Recent Achievements and New Directions in Operator Theory and Applications

Joseph A. Ball, J. William Helton, Martin Klaus, and Leiba Rodman

Abstract. This is a report on IWOTA 2002 and recent achievements, challenges and directions for future research in operator theory and its applications, as reflected at the International Workshop on Operator Theory and its Applications (IWOTA), held at Virginia Tech, Blacksburg, Virginia, from August 6 till August 9, 2002. The report incorporates input received from workshop participants and others whose fields of research are included in the IWOTA 2002 themes. We thank all who helped us to prepare this report.

1. The workshop IWOTA 2002

The primary objective of IWOTAs, which have met biannually since 1981 and more frequently in recent years, is to bring together major researchers in the area of operator theory and related fields, including applications in engineering and mathematical physics. IWOTAs emphasize cross-disciplinary interaction between mathematicians, engineers, physicists, and scientists in other fields in which results and methods of operator theory have significant applications. IWOTA 2002 had 14 plenary lectures; the plenary speakers and the titles of their presentations were as follows:

Damir Arov, *The class of strongly regular J-inner matrix functions*
Vladimir Bolotnikov, *On interpolation problems for contractive-valued functions on certain domains in $\mathbb{C}^n$*
Stephen Boyd, *Optimization over linear matrix inequalities*
Sarah Ferguson, *A new look at H^2 of the bidisk*

Received by the editors August 8, 2003.
Mathematics Subject Classification (2000). 47-99, 93-99.
Key words and phrases. Interpolation, commutant lifting, multivarible operator theory, multidimensional systems, indefinite metric, harmonic analysis, wavelets, structured matrices, factorization, matrix functions, linear matrix inequalities, distributed parameter systems, direct scattering, inverse scattering.
Work of all four authors supported in part by an NSF Grant.

Tryphon Georgiou, *High resolution spectral analysis: advances and applications*
Israel Gohberg, *Orthogonal systems and convolution equations*
Palle E.T. Jorgensen, *Factorizations and decompositions for operators motivated by wavelet analysis*
Marinus A. Kaashoek, *Two new additions to the commutant lifting theorem*
Vadim Olshevsky, *Efficient algorithms for structured matrices*
James Rovnyak, *Spectral problems for some indefinite generalizations of canonical differential and difference equations*
Lev Sakhnovich, *Matrix finite-zone Dirac type equations*
Victor Vinnikov, *Systems on Lie groups and nonselfadjoint representations of Lie algebras*
Alexander Volberg, *Electric intensity capacity and operator theory*
Ricardo Weder, *The time-dependent approach to inverse scattering*

In addition there were 73 talks grouped by special sessions: Interpolation theory (V. Bolotnikov, organizer); Scattering and Inverse Scattering (M. Klaus, organizer); Operator Theory and Applications (H. Langer, organizer); Wavelets (D. Larson, organizer); Structured Matrices (V. Olshevsky, organizer); Distributed Parameter Systems/Operator Theory (O. Staffans, organizer); Multivariable Operator Theory (V. Vinnikov, organizer). The speakers and titles of their presentations were as follows:

Interpolation Theory:

Yury Arlinskiĭ, *On von Neumann's problem in the extension theory of nonnegative operators*
Animikh Biswas, *Intertwining lifting of operators*
Tiberiu Constantinescu, *Szegö kernels and orthogonal polynomials in several noncommuting variables*
Aad Dijksma, *Nudel'man's problem in an indefinite setting*
Harry Dym, *Application of Riccati equations to singular interpolation problems*
Gilbert J. Groenewald, *J-spectral factorization for arbitrary rational matrix functions*
H. Turgay Kaptanoğlu, *Radial differential operators and Bergman projections on diagonal Besov spaces on the ball*
Alexander Kheifets, *Defect and equality in boundary interpolation*
Linda Patton, *Pick interpolation on the boundary of the bidisc*
Dirk Pik, *The Kalman-Yakubovich-Popov inequality and infinite-dimensional discrete-time systems*
André C.M. Ran, *A fixed point theorem in partially ordered sets and some applications to matrix equations*
Eduard Tsekanovskiĭ, *Realization theory of Herglotz-Nevanlinna matrix-valued functions, interpolation problems and explicit system solutions*

Scattering and Inverse Scattering:

Tuncay Aktosun, *Inverse scattering with partial data*
Jared Bronski, *Eigenvalue asymptotics, passive scalar intermittency and small ball estimates for fractional Brownian motions*
Ira Herbst, *Quantum scattering for a class of long-range magnetic fields at high energies*
Peter Miller, *Scattering theory for nonselfadjoint Zakharov-Shabat operators in the semiclassical limit: some facts and open questions*
Alexander Rybkin, *Higher order WKB-type phase for the Schrödinger equation in 1D and scattering on long-range potentials*
Ken Shaw, *Eigenvalues and pulse shapes for Zakharov-Shabat systems*
Alexey Tikhonov, *Functional model for operators with spectrum on a curve*
Vadim Tkachenko, *1-d periodic differential operator of order 4*
Alexander Tovis, *Semiclassical limit of a focusing nonlinear Schrödinger equation and related scattering problems*
Cor van der Mee, *State space methods for solving inverse scattering problems on the line*
Rudi Weikard, *On the inverse resonance problem*
Peter Yuditskiĭ, *Functional models for almost periodic Jacobi matrices*

Operator Theory and Applications:

Miron Bekker, *Helson-Szegö theorem for operator-valued weights*
Aad Dijksma, *The Schur algorithm for generalized Schur functions and J-unitary matrix polynomials*
Seppo Hassi, *A reproducing kernel space model for N_κ-functions*
Lázló Kérchy, *Generalized Toeplitz operators*
Heinz Langer, *Self-adjoint block operator matrices with non-separated diagonal entries and their Schur complements*
Matthias Langer, *Variational principles for eigenvalues of pencils of unbounded operators*
Victor Lomonosov, *The Bishop-Phelps theorem fails for uniform non-self-adjoint dual operator algebras*
Annemarie Luger, *A characterization of generalized poles*
Konstantin A. Makarov, *Variation of spectral subspaces under off-diagonal perturbations*
Dmitry J. Nicolsky, *Approximation of the spectrum of a quasi-periodic function sampled at a finite number of points distributed uniformly*
Artem D. Pulemyotov, *On the generalized joint eigenvector expansion for commuting normal operators*
Peter Stollman, *Quasicrystals and aperiodic order: ergodic properties and operator theoretic consequences*
Christiane Tretter, *On some quadratic operator pencils associated with damped dynamical systems*

Wavelets:

Radu Balan, *The C^*-algebra of nonexpansive operators and application to a frame redundancy result*
Yuanan Diao, *The s-elementary frame wavelets are path connected*
Eugen J. Ionascu, *Wavelet sets and their induced isomorphisms on* $[0,1)$
David Larson, *Wavelets, frames and operator theory*
Kerri Kornelson, *Ellipsoidal frames*
Steen Pedersen, *Exponential bases for Cantor measures*
David Walnut, *Sampling and the local Pompeiu problem*
Eric Weber, *Geometry of affine frames*

Structured Matrices:

Albrecht Böttcher, *The product of a structured matrix and a random vector*
Tiberiu Constantinescu, *Displacement structure, quantum channels, and tensor algebras*
Mark Embree, *Pseudospectra of piecewise continuous Toeplitz matrices*
Dario Fasino, *Fast orthogonalization of Cauchy-like matrices and rational functions*
Georg Heinig, *Fast direct algorithms for Toeplitz least squares problems*
Christian Mehl, *Normal matrices in indefinite inner products*
Karla Rost, *Toeplitz-plus-Hankel matrices*
Michael Stewart, *Orthogonalization of structured matrices using inner products*

Distributed Parameter Systems:

Birgit Jacob, *When is an observation operator admissible?*
A. Karelin, *On a reduction of singular integral operators with a linear-fractional involution to matrix characteristic operators*
Gabriel Prajitura, *Classes of operators with fixed local spectral properties*
David Russell, *Some wave and buckling phenomena in nonlinear elastic beams with supporting constraints*
Olof J. Staffans, *Conservative continuous time systems out of thin air*
Ping Wang, *The application of the Moore-Penrose generalized inverse to an operator equation*

Multivariable Operator Theory:

Raúl Curto, *Algebraic varieties arising in truncated complex moment problems*
Ronald G. Douglas, *Quotient modules and nilpotent actions*
Michael Dritschel, *A completely positive approach to Andô's theorem*
Chen Dubi, *Realization of rational functions in* $\mathbb{C}^N$
Xiang Fang, *Dimension, multiplicity, and some other invariants for operator tuples*
Lawrence Fialkow, *Truncated planar moment problems: the algebraic variety*

D. Kalyuzhniy-Verbovetskiĭ, *On the Bessmertnyĭ class of homogeneous positive analytic functions of several variables*
Mircea Martin, *Uniform approximation by solutions of elliptic equations and seminormality in higher dimensions*
John E. McCarthy, *Operator theory on varieties*
Scott McCullough, *Noncommutative polynomials and Linear-Matrix-Inequality representations*
Gelu Popescu, *Multivariable Nehari problem and interpolation*
Mihai Putinar, *Privilege on strictly convex domains of* $\mathbb{C}^n$
Jim Solazzo, *Reflexivity, factorization, and Hankel operators*
Tavan T. Trent, *Corona theorem for Dirichlet space*

2. Interpolation and commutant lifting theory

Recent achievements. Interpolation theory played a defining role in the development of H^∞-control theory beginning in the 1980s. As part of these developments, we include the Adamjan-Arov-Kreĭn extension of the Nehari theorem (characterizing infinity-norm approximation of a given L^∞ function on the unit circle by a meromorphic function with prescribed number of poles inside the disk) and its later implementation in state-space form, as well as the development of an array of approaches (via commutant lifting, projective geometry/indefinite Beurling-Lax theorem, reproducing kernel Hilbert spaces, unitary extensions of isometry, "fundamental matrix identities" related to positive kernels, state space implementation via directional zero/pole and/or factorization analysis) to the solution of matricial Nevanlinna-Pick interpolation problems. A more recent advance has been the parametrization of the set of all solutions of a Nevanlinna-Pick interpolation problem also meeting a low McMillan-degree constraint, originally through application of techniques from nonlinear functional analysis (degree theory); this result has important implications in speech synthesis applications. More recently there have been extensions of the matricial Nevanlinna-Pick interpolation theory to a context of time-varying systems (both in discrete and continuous time) and there have been various refinements of the commutant lifting theorem (more flexibility in the formulation of the norm constraint and in the intertwining hypothesis) which have led to new types of applications. In another direction, interpolation theory and related problems have been studied in classes of functions whose Carathéodory-Pick matrices have a bounded number of negative eigenvalues (with no a priori regularity hypotheses).

Future directions. A newer emerging area for applications of interpolation problems is signal processing; specifically, problems arising in the construction of antennae arrays call for the solution of some nonstandard interpolation problems. A general theme for future research is the characterization of interpolants to certain non-standard interpolation problems. In this context, as well as in the context of more traditional analytic interpolation theory, it is important to quantify the size of families of interpolants as it relates to modeling uncertainty and resolution in engineering applications, as well as to study the effect of interpolation nodes/constraints (which relate to, e.g., antennae array geometry and sensor dynamics) to the size of such families of interpolants.

3. Multivariable operator theory, several complex variables and multi-dimensional system theory

Recent achievements. There has been a recent flurry of activity extending the interpolation and commutant lifting theory described in the previous section to multivariable settings. We now have a theory of Nevanlinna-Pick interpolation (including the case of interpolation at points on the boundary of the domain) analogous to the classical theory in several multivariable settings, including the polydisk, the ball and the general setting of a reproducing kernel Hilbert space with "Pick kernel". Commutant lifting and Toeplitz corona theorems (including some progress on the solution of the Carleson corona theorem on domains beyond the disk case) have now been obtained for both the ball and polydisk setting. More broadly, this work includes connections with multi-dimensional system theory, model theory for tuples of commuting operators in various classes and more general multivariable settings for Lax-Phillips scattering. The solution of the interpolation problem (or the characteristic operator function or the scattering operator, depending on the context) is usually realized explicitly as the transfer function of some sort of conservative, multi-dimensional linear system. There now has appeared some work which extends these ideas beyond a specific domain (such as the polydisk or the ball) to a general domain in $\mathbb{C}^n$ defined as the points of contractivity for a given matrix multivariable polynomial; the classical Cartan domains of the first three types arise as special cases. Related work deals with a *noncommutative* version of Schur-class of functions (the unit ball of the "Cuntz-Toeplitz algebra"), the elements of which are represented as formal power series in some number of noncommuting indeterminants, multiplication by one of which gives rise to a contraction operator on the Fock space commuting with the right creation operators. Results in the noncommutative theory can be symmetrized to produce the corresponding results for the multiplier class of functions on the ball studied by Arveson. More abstract approaches are also being pursued for the commutative case which use the language of modules over function rings and homological algebra. There have also been recent breakthroughs in the understanding of the various types of multivariable classes of functions of bounded mean oscillation (BMO) and in estimating the norms of big and little multivariable Hankel operators, as well as continuing progress on various sorts of multivariable moment problems (which have important applications in signal processing).

Future directions. There already exist a number of older results (due to Cartan and others) on holomorphic extension of a given complex-valued function defined on a subvariety to a larger domain which were obtained by homological-algebra methods. An important direction for future research is the assimilation of these older results based on homology methodology into the operator/system theory framework which has led to the recent analogues of Nevanlinna-Pick interpolation discussed above.

There already exists a now relatively old result due to Andô on the existence of a commuting unitary dilation for a commuting pair of dilations. We now appear to be on the verge (from a number of independent directions) of getting an explicit understanding of the geometric structure of this dilation.

We expect areas of continuing activity to be the development of classical interpolation theory on nonclassical domains. More broadly, we expect to see the integration of operator model theory and system theory in more general contexts. Currently, we have such an integration for polydisks and interpolation in the Schur-Agler class (a class somewhat smaller than the Schur-class of functions mapping the polydisk in $\mathbb{C}^d$ $(d > 2)$ to the unit disk in $\mathbb{C}$), for balls and the associated Arveson class of contractive multipliers with respect to the kernel $k_d(z, w) = 1/(1-\langle z, w\rangle)$, and for the class of contractive bundle maps between two Hardy spaces consisting of holomorphic sections of parahermitian vector bundles over a real Riemann surface. There have been generalized settings of operator model theory due to Curto, Vasilescu and others for which the connections with system theory and Lax-Phillips scattering theory are not clear.

By taking the Riemann surface to be the double of a finitely-connected planar domain, this last setting includes the case of Hardy spaces over planar domains studied by Abrahamse and Douglas in the 1970s. Since the original work of Abrahamse on the Nevanlinna-Pick interpolation problem on a finitely-connected planar domain, there has been some but limited progress in understanding the structure of the set of all solutions, or in obtaining the result via system-theory ideas (whereby one arrives at the solution realized as the transfer function of some linear system). A general setting for these issues is the Nevanlinna-Pick interpolation problem for contractive multipliers on a reproducing kernel Hilbert space, where the kernel is not a "Pick kernel" for which the Nevanlinna-Pick interpolation theorem holds in essentially the classical form. We expect future work eventually to unveil the mysteries currently surrounding this area.

While there has been much work in the multivariable-operator-theory literature on commuting contractions or commuting nonselfadjoint operators close to selfadjoint, this work has not made any connections with the extensive literature on hyponormal operators, which until now has used different techniques and dealt with different issues. The theory of hyponormal operators somehow lies deeper – more work needs to be done here to make connections with the recent developments in other parts of multivariable operator theory.

We now have a couple of isolated instances of multivariable commutant lifting – this should be understood in a broader, more fundamental setting; in this connection, the Taylor functional calculus for operator tuples needs to be better known and more broadly used. This general commutant lifting theorem appears now to be a key piece (along with the $\overline{\partial}$-technique) toward an operator-theoretic simplification and proof of the Carleson corona theorem on more general domains.

We also expect a more complete theory of multivariable moment problems to emerge.

For all these directions we expect to see a continuing use of commutative algebra and algebraic geometry, as well as operator algebra and C^*-algebra techniques.

4. Operator theory in spaces with an indefinite metric

Recent achievements. Indefinite inner products have arisen in several natural ways over the last couple of decades: J-inner matrix functions arising in the parametrization of all solutions of a matricial Nevanlinna-Pick interpolation or of a Nehari problem, solution of the inverse problem of characterizing which J-inner functions parametrize all solutions of an interpolation problem, the Adamjan-Arov-Kreĭn generalization of Nehari's theorem to handle the case of approximation of an L^∞-function by a function in the class H^∞_κ (bounded on the unit circle and meromorphic on the unit disk with at most κ poles), extensions of the concept of reproducing kernel Hilbert space to the notion of reproducing kernel Pontryagin space. A fundamental result now relatively old is the modified spectral theorem for an operator selfadjoint in a Pontryagin space. Recent achievements in this area represented at IWOTA2002 include: interpolation problems involving Pick matrices having some number κ of negative eigenvalues, spectral theory for special cases of matrices normal in an indefinite inner product, spectral properties of polynomials orthogonal in an indefinite inner product, spectral theory of canonical differential equations involving an indefinite inner product, Darlington representations in an indefinite-inner-product setting, an analysis of the Schur algorithm for a function in the generalized Schur class $\mathcal{S}_\kappa$, new approaches to and understanding of the Kreĭn-Langer factorization for functions in the class $\mathcal{S}_\kappa$, and continuing exploitation of Kreĭn-space operator theory techniques for the analysis of assorted generalized boundary-value problems for differential operators coming up in engineering and mathematical-physics contexts.

Future directions. There are many concrete problems in differential equations or mathematical physics where the theory of operators in indefinite metric spaces is relevant. Examples include the problem of the motion of an ideal or viscous liquid around a solid body, the theory of linear systems having Pontryagin state spaces, and direct and inverse problems for indefinite cases of canonical differential equations. We expect to see continuing interest in and work on such applications of Kreĭn-space operator theory. On the more theoretical side, development of a structure theory for various new classes of operators in indefinite metric spaces, even in finite-dimensional spaces, remains a challenge.

5. Operator theory in harmonic analysis

Recent achievements. Recent achievements include the introduction of operator-theoretic techniques, in particular, the optimality principle for the principle of dynamic programming in control theory, for new treatments and new solutions of problems in harmonic analysis (e.g., estimates for Carleson measures and Hunt-Muckenhoupt-Wheeden weights, including new results for the case of matrix- or operator-valued weights).

Recently X. Tolsa solved the Painlevé and Vitushkin problems on analytic capacity. The major contribution of Vitushkin was the reduction of problems in rational approximation to analytic capacity estimates, which then made verification of subadditivity conditions crucial. Now it is possible to give an equivalent, more operator-theoretic formulation of analytic capacity, thereby making subadditivity easier to prove.

Future directions. The full operator-theoretic reformulations and implications of the recent solution of Vitushkin's problem remain to be explored. There should follow more explicit results in rational approximation and harmonic analysis in which operator theory will play a prominent role.

6. Factorization of matrix functions

Recent achievements. Factorization of matrix functions was developed in the first half of the last century as a tool for studying and solving systems of singular integral equations or, equivalently, certain boundary value problems for analytic functions. In the 70s and 80s its applications to inverse scattering theory were found and intensively developed. Recently researchers have discovered several more rather unexpected applications, including the theory of random matrices, asymptotics of orthogonal polynomials, completion and extension problems for (almost) periodic functions of several variables. As a result, some new approaches were developed in the factorization theory per se, such as the steepest descent method in the asymptotical analysis of parametrized factorization problems. The cornerstones of the intrinsic development of the factorization theory also include the complete treatment of the piecewise continuous case on Carleson curves with Muckenhoupt weights, and the factorization theory of semi-almost periodic matrix functions.

Future directions. Several classical problems of factorization theory still have appeal. We mention here the problem of explicit factorization of new classes of matrix functions, and we expect further progress to be achieved here with more systematic use of special functions and the Riemann boundary problem on Riemann surfaces. Computation of partial indices and, in particular, finding new verifiable conditions for their stability, remains an important open problem. Substantial progress in this direction was achieved lately using the current ideas of numerical methods, but more remains to be done.

New problems, arising from the needs of the latest developments (both theoretical and applied), include the systematic treatment of matrix functions defined on curves with intersections and the asymptotic behavior of the factorization in case of parameterized families of matrices more general than piecewise continuous (in particular, having discontinuities of semi-almost periodic type). Though the properties of the almost periodic factorization, when it exists, are rather well understood by now, the intrinsic nature of its existence remains a mystery and calls for further research. Work on explicit factorization of concrete almost periodic matrix functions, arising from convolution type equations on finite intervals, is also in order. The recently discovered connections with the Corona problem might prove to be useful in this regard.

Yet another natural direction of development is the factorization theory of matrix functions defined on abstract ordered abelian groups. Such a theory, when constructed, would embrace the classical results on Wiener-Hopf factorization of continuous matrix functions on the unit circle, on the one hand, and (already mentioned) more recent almost periodic factorization, on the other. One should anticipate completely new phenomena here, and a need to develop radically new ideas and approaches; indeed, even the almost periodic factorization is far from being analogous to the classical case of functions defined on the unit circle.

7. Integration of operator theory with wavelets

Recent achievements. While the theory of wavelets and its applications to data compression and image reconstruction has been undergoing intense development in the past couple of decades, the understanding of its connections with operator theory (and operator algebra) ideas is more recent (in the work of Larson, Jorgensen and their collaborators), and the interaction (between wavelets and operator theory) goes both ways. Here are a few samples of the connections:

The theory of wavelets has led to a generalization of Halmos's idea of a wandering vector for a (unitary or isometric) shift operator on a Hilbert space to the notion of wandering vector for a unitary system (with generators satisfying some sort of intertwining condition rather than commuting). This operator-theory point of view has then given further insight and new results in the theory of wavelets.

A fundamental operator for many wavelet considerations has turned out to be the so-called *transfer operator*. Indeed, this operator determines the regularity properties of wavelets, e.g., the strength of the approximations, and additional fine structure of direct practical significance. The spectral theory of this operator, including application of an infinite-dimensional extension of Perron-Frobenius theory, has led to better understanding of wavelet constructions. The transfer operator has applications in statistics, mathematical physics, and dynamical systems. Within pure mathematics, it is studied in analytic number theory and harmonic analysis in connection with zeta functions and trace formulas. The spectral theory of the operator varies with the context of the function spaces: these are the regularity spaces which, at the same time, play a crucial role in questions around discretization of PDE problems (e.g., finite element methods). The operator theoretic aspects of wavelet analysis are covered strongly at a number of centers, in the US, Europe, and Asia. The wavelet algorithm is now known to have a very fast polynomial quantum version, while the classical one is not known to be polynomial. This is an area with a lot of interest in the physics community.

Future directions. We expect to see more interaction between the wavelet and operator theory communities. Wavelet analysis has led to a number of problems in areas which have traditionally been in the domain of the operator theory community, e.g., spectral theory of matrix polynomials, factorization of unitary-operator-valued functions, Fredholm index and related homotopy questions, spectral and index theory for concrete operators on a function space. To this point, researchers in the wavelet community have generally come up with their own ad hoc solutions rather than using the systematic theories developed in the operator-theory community. On the other hand, the problems in wavelet theory are often slightly different than anything which has already been done in the operator-theory community. For example, the transfer operator mentioned above does not fit into established trends in operator theory – the closest, but not exact, fit is perhaps the Cowen-Douglas class of operators. These various connections give a fruitful area for continuing cross-fertilization.

8. Structured matrices

Recent achievements. Structured matrices, such as Toeplitz, Hankel, Vandermonde, Cauchy, Pick, Bezout, separately and in combinations, such as Toeplitz-plus-Hankel, or more generally, matrices with "displacement structure", have been extensively studied for a long time, from diverse and seemingly unrelated points of view, in mathematics, computer science, and engineering. This includes also the point of view of operator theory: for example, Toeplitz matrices have been studied using methods of reproducing kernel Hilbert space and lifting of commutants. A relatively recent achievement is the development of fast solution algorithms for Toeplitz, Vandermonde and Cauchy matrices, solutions of Lyapunov/Riccati equations, types of Pick matrices. One approach to these fast solution algorithms is based on the theory of rational matrix interpolation. In particular, there now exist superfast algorithms for solving passive tangential interpolation problems of Nevanlinna-Pick type, and a new algorithm for list decoding of Reed-Solomon codes. New ideas entering the mainstream are the notion of pseudospectrum and applications of spectral theory for various classes of infinite-dimensional operators, and of random matrices to understanding conditioning and asymptotics issues for various Galerkin approximations.

Future directions. There remain matrices for which fast algorithms should be developed: circulant matrices, multi-dimensional block Toeplitz (existing algorithms do not use the complete special structure). We expect to see more interaction between numerical analysis, the theory of random matrices, and the spectral theory of infinite-dimensional operators.

9. Linear matrix inequalities

Recent achievements. The recent development of efficient interior-point algorithms for convex optimization problems involving linear matrix inequalities (LMIs) has spurred research in a wide variety of application fields, including control system analysis and synthesis, combinatorial optimization, circuit design, structural optimization, experiment design, and geometric problems involving ellipsoidal bounding and approximation. The interior point methods for convex optimization were reviewed in a plenary lecture at IWOTA2002.

Future directions. Linear system problems convert readily to systems of matrix inequalities; then the trick (often impossible) is to convert them to "nice" matrix inequalities. There are only a few tricks known, the main one being to use "Schur complements". Despite the attention of hundreds of engineers, essentially nothing general or systematic is known. Matrix inequalities fall directly within the expertise of the operator-theory community; recently engineering-type inequalities are getting scrutiny in this community. A major issue central to numerical solution is which problems are convex matrix inequalities and which convert to linear matrix inequalities. There are beginning to be results on such questions and the pace is quickening.

10. Distributed-parameter system theory

Recent achievements. Axiomatization of the integral form of an input-state-output system (involving bounded operators between input, state and output signals over each time interval rather than between instantaneous input, state and output) in the notion of "well-posed system" has led to formulation of a useful rigged setup for an operator colligation with unbounded infinitesimal input, state dynamics, and output operators. Such colligations (including extensions corresponding to systems which are not well-posed) have led to new realization theorems for functions which are positive-real, Herglotz-Nevanlinna, or bounded-real over a half-plane. This foundation in turn is leading to better understanding of state-space methods for feedback stabilization, J-spectral factorization, and optimal and H^∞-control for distributed parameter systems. There have been also interesting connections between function theory (e.g., the corona problem) and well-posedness questions for a given dynamics and unbounded input or output operator.

Future directions. Application of the theory to new concrete examples and applications, including such situations as quantum systems and linearized fluid flow, is an important future direction for the theory. Also integration of the more general behavior-theoretic/implicit-system viewpoint can be expected.

11. Direct and inverse scattering and spectral problems

Recent achievements. Riemann-Hilbert problems/Wiener-Hopf factorization: Inverse scattering problems arise in wave propagation and quantum mechanics: one seeks to recover the potential function or the differential operator from knowledge of the scattering data or spectral measure. The direct problem is to compute the scattering function and spectral measure from the operator. Well-known inverse spectral methods are via solving Riemann-Hilbert problems and related integral equations and Wiener-Hopf factorization problems, or by making use of the theory of Riemann surfaces and theta functions. A more recent method borrows the so-called state-space method from system theory: one assumes that the scattering function is rational and then solves the inverse scattering problem in terms of the matrices arising from the realization of the reflection coefficient as the transfer function of a linear system. The connection with interpolation problems is exploited to obtain solutions of bitangential inverse scattering and spectral problems for canonical integral and differential systems. Formulas are obtained via reproducing kernel Hilbert space methods with kernels based on strongly regular J-inner matrix-valued functions.

Nonlinear PDEs: Operator theory helps to analyze various linear and nonlinear differential equations describing important physical phenomena. One important application is the inverse scattering transform; specifically, solutions of certain nonlinear partial differential equations (nonlinear evolution equations) are obtained by solving the inverse scattering problems of related linear ordinary differential equations. Operator-theory techniques to solve inverse scattering problems thus play an important role in solving various nonlinear partial differential equations. Some special solutions of such nonlinear partial differential equations, known as soliton solutions, have led to advances in communications via fiber optics. The analysis of the inverse scattering transformation leads to questions in the spectral theory of nonselfadjoint systems of differential equations, like the Zakharov-Shabat system or the Manakov system. Research in this area has been of interest for both mathematicians and engineers.

Ergodic and almost ergodic operators: Functional models of operators on Riemann surfaces which were introduced in operator theory are now being used to develop spectral and scattering theories of ergodic and almost periodic difference/differential operators.

Future directions. Perturbations of selfadjoint (and normal) operators which are not, say, selfadjoint or even accretive: Applications include spectral problems arising in the theory of pulse propagation in nonlinear optical fibers.

Selfadjoint perturbations in Schatten-von Neumann ideals larger than the trace ideal, and Schrödinger operators with long range non-smooth potentials: The two issues mentioned here are connected and actually related to an old question of M.G. Kreĭn raised at the Moscow Mathematical Congress (1966): *find spectral interpretations of M.M. Djrbashian's factorization of meromorphic functions from generalized Nevanlinna classes.* The connection between the two is as follows. Schrödinger operators with long-range potentials can be viewed as non-trace class perturbations of free Hamiltonians. Consequently, many analytic functions (like the perturbation determinant) appearing in the trace class perturbation theory need suitable modifications. The problem is that Nevanlinna functions do not belong to this setting. There are strong indications to conjecture that the relevant functions are from the so-called Djrbashian class of analytic functions of generalized bounded type but it appears to be a very difficult issue. Thus, important developments in this direction should also address the Kreĭn problem.

A proper solution of questions arising in the spectral theory of almost periodic difference/differential operators alluded to above requires a real breakthrough in the subject, namely, functional models on Riemann surfaces of infinite genus going beyond the Widom class.

Joseph A. Ball
Department of Mathematics
Virginia Tech
Blacksburg, VA 24061-0123 USA
e-mail: `ball@math.vt.edu`

J. William Helton
Department of Mathematics
University of California at San Diego
La Jolla, CA 92093-0112 USA
e-mail: `helton@osiris.ucsd.edu`

Martin Klaus
Department of Mathematics
Virginia Tech
Blacksburg, VA 24061-0123 USA
e-mail: `klaus@math.vt.edu`

Leiba Rodman
Department of Mathematics
College of William and Mary
Williamsburg, VA 23187-8795 USA
e-mail: `lxrodm@math.wm.edu`

Operator Theory:
Advances and Applications, Vol. 149, 1–22

Inverse Scattering Transform, KdV, and Solitons

Tuncay Aktosun

Abstract. In this review paper, the Korteweg-de Vries equation (KdV) is considered, and it is derived by using the Lax method and the AKNS method. An outline of the inverse scattering problem and of its solution is presented for the associated Schrödinger equation on the line. The inverse scattering transform is described in order to solve the initial-value problem for the KdV, and the time evolution of the corresponding scattering data is obtained. Soliton solutions to the KdV are derived in several ways.

1. Introduction

The *Korteweg-de Vries equation* (KdV, for short) is used to model propagation of water waves in long, narrow, and shallow canals. It was first formulated [1] in 1895 by the Dutch mathematicians Diederik Johannes Korteweg and Gustav de Vries. Korteweg was a well known mathematician of his time, and de Vries wrote a doctoral thesis on the subject under Korteweg.

After some scaling, it is customary to write the KdV in the form

$$\frac{\partial u}{\partial t} - 6u\frac{\partial u}{\partial x} + \frac{\partial^3 u}{\partial x^3} = 0, \qquad x \in \mathbf{R}, \quad t > 0, \tag{1.1}$$

where $-u(x,t)$ corresponds to the vertical displacement of the water from the equilibrium at the location x at time t. Replacing u by $-u$ amounts to replacing -6 by $+6$ in (1.1). Also, by scaling x, t, and u, i.e., by multiplying them with some positive constants, it is possible to change the constants in front of each of the three terms on the left-hand side of (1.1) at will.

Note that the KdV is a *nonlinear* partial differential equation (PDE, for short) due to the presence of the uu_x term, where we use a subscript to denote the partial derivative. The u_{xxx} term makes it *dispersive*, i.e., in general an initial

---

Received by the editors April 7, 2003; revised November 6, 2003. Submitted by M. Klaus.
Mathematics Subject Classification (2000). Primary 35Q53; Secondary 35Q51, 81U40.
Key words and phrases. Korteweg-de Vries equation, Inverse scattering transform, Soliton.
The research leading to this article was supported in part by the National Science Foundation under grants DMS-0296022 and DMS-0204437 and the Department of Energy under grant DE-FG02-01ER45951.

wave $u(x, 0)$ will broaden in space as time progresses. In addition to its solutions showing behavior of nonlinearity and dispersiveness, the KdV possesses certain special solutions, known as *solitary wave solutions,* which would not be expected from a nonlinear and dispersive PDE. A single solitary wave solution to a PDE retains its shape in time and moves either to the left or right. It turns out that such a solution to (1.1) only moves to the right, and any general solitary wave solution asymptotically resembles a train of single solitary wave solutions.

It was a Scottish engineer named John Scott Russell who first observed a solitary water wave. This happened in 1834 on the Edinburgh-to-Glasgow canal, some 60 years before the formulation of the KdV as a model for water waves. Russell reported [2] his observation to the British Association for the Advancement of Science in September 1844. The details of Russell's observation, including the rivalry between him and George Airy who did not believe in the existence of solitary water waves, can be found in [3–6].

As outlined in the famous preprint [7], which was never published as a journal article, Enrico Fermi in his summer visits to Los Alamos, jointly with J. Pasta and S. Ulam, studied a one-dimensional (1-D, for short) dynamical system of 64 particles with forces between neighbors containing nonlinear terms. These computational studies were carried out on the Los Alamos computer named Maniac I. The primary aim of the study was to determine the rate of approach to the equipartition of energy among various degrees of freedom. Contrary to expectations, Fermi, Pasta, and Ulam observed very little, if any, tendency towards the equipartition of energy, but instead the almost ongoing recurrence to the initial state, which was puzzling. After Fermi's death in November 1954, Pasta and Ulam completed their last few computational examples. Their preprint appears in Fermi's Collected Papers [8] and is also available on the internet [9].

The importance of the KdV arose in 1965, when Zabusky and Kruskal [10] were able to explain the Fermi-Pasta-Ulam puzzle in terms of solitary-wave solutions to the KdV. In their analysis of numerical solutions to the KdV, Zabusky and Kruskal observed *solitary-wave pulses,* named such pulses *solitons* because of their particle-like behavior, and noted that such pulses interact with each other nonlinearly but come out of their interaction virtually unaffected in size or shape. Such unusual nonlinear interactions among soliton solutions to the KdV created a lot of excitement, but at that time no one knew how to solve such a nonlinear PDE, except numerically.

In their celebrated paper [11] of 1967, Gardner, Greene, Kruskal, and Miura presented a method, now known as the *inverse scattering transform,* to solve the initial-value problem for the KdV, assuming that the initial value $u(x, 0)$ approaches a constant sufficiently rapidly as $x \to \pm\infty$. There is no loss of generality in choosing that constant as zero. They showed that $u(x, t)$ can be obtained from $u(x, 0)$ with the help of the solution to the *inverse scattering problem* for the 1-D Schrödinger equation with the *time-evolved scattering data.* They also explained that soliton solutions to the KdV correspond to the case of zero reflection coefficient in the scattering data. They observed from various numerical studies of the

KdV that, for large t, $u(x,t)$ in general consists of a finite train of solitons traveling in the positive x direction and an oscillatory train spreading in the opposite direction.

In our paper, we present an elementary review of the inverse scattering transform for the KdV. We consider the time-evolved Schrödinger equation, where $V(x)$ in (2.1) is replaced by $u(x,t)$; namely, we deal with

$$\frac{d^2\psi(k,x;t)}{dx^2} + k^2\,\psi(k,x;t) = u(x,t)\,\psi(k,x;t), \qquad x \in \mathbf{R}, \tag{1.2}$$

where $t > 0$ is a parameter that is usually interpreted as time. Thus, we view $V(x)$ as the initial value $u(x,0)$ of the potential $u(x,t)$ and look at $\psi(k,x;t)$ as the time evolution of $\psi(k,x)$ of (2.1) from the initial time $t = 0$. The *scattering coefficients* $T(k;t)$, $R(k;t)$, and $L(k;t)$ associated with (1.2) are viewed as evolving from the corresponding coefficients $T(k)$, $R(k)$, and $L(k)$ of (2.1), respectively, from $t = 0$. Thus, our notation is such that

$$V(x) = u(x,0), \quad \psi(k,x) = \psi(k,x;0),$$
$$T(k) = T(k;0), \quad R(k) = R(k;0), \quad L(k) = L(k;0).$$

Our review paper is organized as follows. In Section 2 we consider *scattering solutions* and *bound state solutions* to the Schrödinger equation (2.1) and introduce the scattering coefficients, *bound-state norming constants,* and *dependency constants* corresponding to a potential in the so-called *Faddeev class.* In Section 3 we present an outline of the inverse scattering problem for (2.1) and review some solution methods based on solving an associated *Riemann-Hilbert problem.* In Section 4 we consider the time evolution of the scattering coefficients, bound-state norming constants, and dependency constants when the potential evolves from $u(x,0)$ to $u(x,t)$. We also introduce the *Lax pair* associated with the KdV and derive the KdV by using the *Lax method.* In Section 5 we study the *AKNS method* and derive the KdV via that method. In Section 6 we present some of the methods to solve the initial value problem for the KdV. In Section 7 we concentrate on soliton solutions to the KdV and obtain various representations of the N-soliton solution. Finally, in Section 8 we provide certain remarks on the *Bäcklund transformation,* the *conserved quantities,* and some other aspects related to the KdV.

2. Schrödinger equation and the scattering data

Consider the Schrödinger equation

$$\frac{d^2\psi(k,x)}{dx^2} + k^2\,\psi(k,x) = V(x)\,\psi(k,x), \qquad x \in \mathbf{R}, \tag{2.1}$$

where V is real-valued and belongs to $L^1_1(\mathbf{R})$. Here, $L^1_n(\mathbf{R})$ denotes the class of measurable potentials such that $\int_{-\infty}^{\infty} dx\,(1+|x|^n)\,|V(x)|$ is finite. The class of real-valued potentials in $L^1_1(\mathbf{R})$ is sometimes called the Faddeev class, after Ludwig Faddeev's analysis [12] of the inverse scattering problem for (2.1) within that class of potentials. In appropriate units, (2.1) describes the quantum mechanical behavior of a particle of total energy k^2 under the influence of the potential V. The inverse scattering problem for (2.1) consists of the determination of V from an appropriate set of scattering data.

There are two types of solutions to (2.1). The scattering solutions consist of linear combinations of e^{ikx} and e^{-ikx} as $x \to \pm\infty$, and they occur for $k \in \mathbf{R} \setminus \{0\}$. Real k values correspond to positive energies of the particle, and a particle of positive energy can be visualized as being capable of escaping to $x = \pm\infty$ as a result of scattering from V. Heuristically, since $V(x)$ vanishes at $x = \pm\infty$, the particle will still have some kinetic energy at infinity and hence is allowed to be at infinity. On the other hand, a bound state of (2.1) is a solution that belongs to $L^2(\mathbf{R})$ in the x variable. It turns out that, when V belongs to the Faddeev class, the bound-state solutions to (2.1) decay exponentially as $x \to \pm\infty$, and they can occur only at certain k-values on the imaginary axis in $\mathbf{C}^+$. We use $\mathbf{C}^+$ to denote the upper-half complex plane and put $\overline{\mathbf{C}^+} := \mathbf{C}^+ \cup \mathbf{R}$. Each bound state corresponds to a negative total energy of the particle, and as a result the particle is bound by the potential and does not have sufficient kinetic energy to escape to infinity. We will use N to denote the number of bound states, which is known to be finite when V is in the Faddeev class, and suppose that the bound states occur at $k = i\kappa_j$ with the ordering $0 < \kappa_1 < \cdots < \kappa_N$.

Among the scattering solutions to (2.1) are the *Jost solution from the left,* f_{l}, and the *Jost solution from the right,* f_{r}, satisfying the respective boundary conditions

$$e^{-ikx} f_{\mathrm{l}}(k,x) = 1 + o(1), \quad e^{-ikx} f_{\mathrm{l}}'(k,x) = ik + o(1), \qquad x \to +\infty, \tag{2.2}$$

$$e^{ikx} f_{\mathrm{r}}(k,x) = 1 + o(1), \quad e^{ikx} f_{\mathrm{r}}'(k,x) = -ik + o(1), \qquad x \to -\infty, \tag{2.3}$$

where the prime is used for the derivative with respect to the spatial coordinate x. From the spatial asymptotics

$$f_{\mathrm{l}}(k,x) = \frac{e^{ikx}}{T(k)} + \frac{L(k)\, e^{-ikx}}{T(k)} + o(1), \qquad x \to -\infty, \tag{2.4}$$

$$f_{\mathrm{r}}(k,x) = \frac{e^{-ikx}}{T(k)} + \frac{R(k)\, e^{ikx}}{T(k)} + o(1), \qquad x \to +\infty, \tag{2.5}$$

we obtain the scattering coefficients, namely, the *transmission coefficient* T, and the *reflection coefficients* L and R, *from the left and right,* respectively. It is also possible to express the scattering coefficients in terms of certain Wronskians [12–16] involving f_{l} and f_{r}. We have

$$T(k) = \frac{2ik}{[f_{\mathrm{r}}(k,x); f_{\mathrm{l}}(k,x)]}, \tag{2.6}$$

$$L(k) = \frac{[f_{\mathrm{l}}(k,x); f_{\mathrm{r}}(-k,x)]}{[f_{\mathrm{r}}(k,x); f_{\mathrm{l}}(k,x)]}, \quad R(k) = \frac{[f_{\mathrm{l}}(-k,x); f_{\mathrm{r}}(k,x)]}{[f_{\mathrm{r}}(k,x); f_{\mathrm{l}}(k,x)]},$$

where the Wronskian is defined as $[F; G] := FG' - F'G$.

It is known [12–16] that, for each fixed $x \in \mathbf{R}$, the Jost solutions $f_{\mathrm{l}}(\cdot, x)$ and $f_{\mathrm{r}}(\cdot, x)$ have analytic extensions in k to $\mathbf{C}^+$. Moreover,

$$f_{\mathrm{l}}(-k^*, x) = f_{\mathrm{l}}(k, x)^*, \quad f_{\mathrm{r}}(-k^*, x) = f_{\mathrm{r}}(k, x)^*, \qquad k \in \overline{\mathbf{C}^+},$$

$$T(-k) = T(k)^*, \quad R(-k) = R(k)^*, \quad L(-k) = L(k)^*, \qquad k \in \mathbf{R},$$

where the asterisk denotes complex conjugation. We also have

$$R(k)\, T(k)^* = -L(k)^*\, T(k), \qquad k \in \mathbf{R}, \tag{2.7}$$

$$|T(k)|^2 + |L(k)|^2 = 1 = |T(k)|^2 + |R(k)|^2, \qquad k \in \mathbf{R}. \tag{2.8}$$

Thus, the scattering coefficients cannot exceed one in absolute value for real k. Furthermore, $T(k) \neq 0$ if $k \in \mathbf{R} \setminus \{0\}$, and hence the reflection coefficients are strictly less than one in absolute value when $k \in \mathbf{R} \setminus \{0\}$. In general, R and L are defined only for real k values, but T has a meromorphic extension to $\mathbf{C}^+$. For large k one has

$$T(k) = 1 + O(1/k), \qquad k \to \infty \text{ in } \overline{\mathbf{C}^+},$$

$$R(k) = o(1/k), \quad L(k) = o(1/k), \qquad k \to \pm\infty.$$

Each bound state corresponds to a pole of T in $\mathbf{C}^+$ and vice versa. It is known that the bound states are simple, i.e., at each $k = i\kappa_j$ there exists only one linearly independent solution to (2.1) belonging to $L^2(\mathbf{R})$. The bound-state norming constants $c_{\mathrm{l}j}$ and $c_{\mathrm{r}j}$ are defined as

$$c_{\mathrm{l}j} := \left[\int_{-\infty}^{\infty} dx\, f_{\mathrm{l}}(i\kappa_j, x)^2\right]^{-1/2}, \quad c_{\mathrm{r}j} := \left[\int_{-\infty}^{\infty} dx\, f_{\mathrm{r}}(i\kappa_j, x)^2\right]^{-1/2},$$

and they are related to each other via the residues of T as

$$\mathrm{Res}\,(T, i\kappa_j) = i\, c_{\mathrm{l}j}^2\, \gamma_j = i\, \frac{c_{\mathrm{r}j}^2}{\gamma_j}, \tag{2.9}$$

where γ_j is the dependency constant given by

$$\gamma_j := \frac{f_{\mathrm{l}}(i\kappa_j, x)}{f_{\mathrm{r}}(i\kappa_j, x)}. \tag{2.10}$$

The sign of γ_j is the same as that of $(-1)^{N-j}$ and hence

$$c_{\mathrm{r}j} = (-1)^{N-j} \gamma_j c_{\mathrm{l}j}.$$

The normalized bound-state solution $\varphi_j(x)$ at $k = i\kappa_j$ is defined as

$$\varphi_j(x) := c_{\mathrm{l}j}\, f_{\mathrm{l}}(i\kappa_j, x) = (-1)^{N-j} c_{\mathrm{r}j}\, f_{\mathrm{r}}(i\kappa_j, x).$$

The *scattering matrix* associated with (2.1) is given by

$$\mathbf{S}(k) := \begin{bmatrix} T(k) & R(k) \\ L(k) & T(k) \end{bmatrix}, \qquad k \in \mathbf{R},$$

and it can be constructed in terms of the bound-state energies and either one of the reflection coefficients R and L. Given $R(k)$ for $k \in \mathbf{R}$ and the bound-state poles $k = i\kappa_j$, one can construct T as

$$T(k) = \left(\prod_{j=1}^{N} \frac{k + i\kappa_j}{k - i\kappa_j} \right) \exp\left(\frac{1}{2\pi i} \int_{-\infty}^{\infty} ds \, \frac{\log(1 - |R(s)|^2)}{s - k - i0^+} \right), \qquad k \in \overline{\mathbf{C}^+}, \quad (2.11)$$

and use (2.7) to construct $L(k)$ for $k \in \mathbf{R}$. Similarly, given $L(k)$ for $k \in \mathbf{R}$ and the bound-state poles $k = i\kappa_j$, one can construct T [cf. (2.8) and (2.11)] as

$$T(k) = \left(\prod_{j=1}^{N} \frac{k + i\kappa_j}{k - i\kappa_j} \right) \exp\left(\frac{1}{2\pi i} \int_{-\infty}^{\infty} ds \, \frac{\log(1 - |L(s)|^2)}{s - k - i0^+} \right), \qquad k \in \overline{\mathbf{C}^+},$$

and obtain $R(k)$ for $k \in \mathbf{R}$ via (2.7).

3. Inverse scattering problem

When there are no bound states, either one of the reflection coefficients R and L uniquely determines the corresponding potential in the Faddeev class. However, when there are bound states, for the unique determination of V, in addition to one reflection coefficient and the bound-state energies, one must also specify a bound-state norming constant or, equivalently, the dependency constant for each bound state. To recover V uniquely, as our scattering data we may use either the *left scattering data* $\{R, \{\kappa_j\}, \{c_{\mathrm{l}j}\}\}$ or the *right scattering data* $\{L, \{\kappa_j\}, \{c_{\mathrm{r}j}\}\}$; these two are equivalent to each other, and each is also equivalent to $\{\mathbf{S}, \{\gamma_j\}\}$.

A *characterization* for a specific class of potentials consists of specifying some necessary and sufficient conditions on the scattering data which guarantee that there exists a corresponding unique potential in that class. Such conditions are usually obtained by using the *Faddeev-Marchenko method* [12–19]; this method is also known as the *Marchenko method,* and sometimes in the literature it is referred to as the *Gel'fand-Levitan-Marchenko method* even though this is a misnomer [20]. The characterization conditions can be stated for the left scattering data, for the right scattering data, or for the combination of both. For a characterization in the class of real-valued potentials belonging to $L^1_2(\mathbf{R})$, the reader is referred to [14]. Various characterizations in the Faddeev class can be found in [13,17–19].

Since k appears as k^2 in (2.1), the functions $f_{\mathrm{l}}(-k,x)$ and $f_{\mathrm{r}}(-k,x)$ are also solutions to (2.1) and they can be expressed as linear combinations of the Jost solutions $f_{\mathrm{l}}(k,x)$ and $f_{\mathrm{r}}(k,x)$ as

$$f_{\mathrm{l}}(-k,x) = T(k)\, f_{\mathrm{r}}(k,x) - R(k)\, f_{\mathrm{l}}(k,x), \qquad k \in \mathbf{R},$$

$$f_{\mathrm{r}}(-k,x) = T(k)\, f_{\mathrm{l}}(k,x) - L(k)\, f_{\mathrm{r}}(k,x), \qquad k \in \mathbf{R},$$

or equivalently as

$$m_{\mathrm{l}}(-k,x) = T(k)\, m_{\mathrm{r}}(k,x) - R(k)\, e^{2ikx}\, m_{\mathrm{l}}(k,x), \qquad k \in \mathbf{R}, \quad (3.1)$$

$$m_{\mathrm{r}}(-k,x) = T(k)\, m_{\mathrm{l}}(k,x) - L(k)\, e^{-2ikx}\, m_{\mathrm{r}}(k,x), \qquad k \in \mathbf{R}, \tag{3.2}$$

where m_{l} and m_{r} are the *Faddeev functions* defined as

$$m_{\mathrm{l}}(k,x) := e^{-ikx} f_{\mathrm{l}}(k,x), \quad m_{\mathrm{r}}(k,x) := e^{ikx} f_{\mathrm{r}}(k,x). \tag{3.3}$$

Each of (3.1) and (3.2) can be viewed as a Riemann-Hilbert problem [6,15], where, knowing the scattering coefficients for $k \in \mathbf{R}$, the aim is to construct m_{l} and m_{r} such that, for each $x \in \mathbf{R}$, $m_{\mathrm{l}}(\cdot,x)$ and $m_{\mathrm{r}}(\cdot,x)$ are analytic in $\mathbf{C}^+$, continuous in $\overline{\mathbf{C}^+}$, and behave like $1 + O(1/k)$ as $k \to \infty$ in $\overline{\mathbf{C}^+}$. Once $m_{\mathrm{l}}(k,x)$ or $m_{\mathrm{r}}(k,x)$ is constructed, the potential can be obtained with the help of (2.1) and (3.3), namely, by using

$$V(x) = \frac{m_{\mathrm{l}}''(k,x)}{m_{\mathrm{l}}(k,x)} + 2ik\frac{m_{\mathrm{l}}'(k,x)}{m_{\mathrm{l}}(k,x)}, \quad V(x) = \frac{m_{\mathrm{r}}''(k,x)}{m_{\mathrm{r}}(k,x)} - 2ik\frac{m_{\mathrm{r}}'(k,x)}{m_{\mathrm{r}}(k,x)}, \tag{3.4}$$

where the right-hand sides can be evaluated at any particular value of $k \in \overline{\mathbf{C}^+}$.

Alternatively, the potential can be constructed by the Faddeev-Marchenko method; namely, V can be obtained from the left scattering data $\{R, \{\kappa_j\}, \{c_{\mathrm{l}j}\}\}$ by solving the *left Marchenko integral equation* or from the right scattering data $\{L, \{\kappa_j\}, \{c_{\mathrm{r}j}\}\}$ by solving the *right Marchenko integral equation.*

The left Marchenko equation using the left scattering data as the input is given by

$$B_{\mathrm{l}}(x,y) + \Omega_{\mathrm{l}}(2x+y) + \int_0^\infty dz\, \Omega_{\mathrm{l}}(2x+y+z)\, B_{\mathrm{l}}(x,z) = 0, \qquad y > 0, \tag{3.5}$$

where

$$\Omega_{\mathrm{l}}(y) := \frac{1}{2\pi} \int_{-\infty}^{\infty} dk\, R(k)\, e^{iky} + \sum_{j=1}^{N} c_{\mathrm{l}j}^2\, e^{-\kappa_j y}.$$

One can obtain (3.5) from (3.1) via a Fourier transformation. Once (3.5) is solved and $B_{\mathrm{l}}(x,y)$ is obtained, the potential is recovered as

$$V(x) = -2\,\frac{dB_{\mathrm{l}}(x,0^+)}{dx}, \tag{3.6}$$

and the Faddeev function from the left is constructed as

$$m_{\mathrm{l}}(k,x) = 1 + \int_0^\infty dy\, B_{\mathrm{l}}(x,y)\, e^{iky}. \tag{3.7}$$

Similarly, via a Fourier transformation on (3.2), using the right scattering data as the input one obtains the right Marchenko equation

$$B_{\mathrm{r}}(x,y) + \Omega_{\mathrm{r}}(-2x+y) + \int_0^\infty dz\, \Omega_{\mathrm{r}}(-2x+y+z)\, B_{\mathrm{r}}(x,z) = 0, \qquad y > 0, \tag{3.8}$$

where

$$\Omega_{\mathrm{r}}(y) := \frac{1}{2\pi} \int_{-\infty}^{\infty} dk\, L(k)\, e^{iky} + \sum_{j=1}^{N} c_{\mathrm{r}j}^2\, e^{-\kappa_j y}.$$

Once (3.8) is solved, the potential is recovered by using

$$V(x) = 2\,\frac{dB_{\rm r}(x,0^+)}{dx}, \tag{3.9}$$

and the Faddeev function from the right is constructed as

$$m_{\rm r}(k,x) = 1 + \int_0^\infty dy\, B_{\rm r}(x,y)\, e^{iky}. \tag{3.10}$$

When the characterization conditions on the scattering data corresponding to potentials in the Faddeev class are satisfied, both the left and right Marchenko equations are uniquely solvable, and the right-hand sides of (3.6) and (3.9) are equal to each other and belong to $L_1^1(\mathbf{R})$. Thus, V can be obtained from either (3.6) or (3.9). There are various other methods to recover V from an appropriate set of scattering data. We refer the reader to [16,19] for a review of some of those methods.

4. Lax method and evolution of the scattering data

Soon after Gardner, Green, Kruskal, and Miura demonstrated [11] that the initial value problem for the KdV can be solved by the inverse scattering transform, Peter Lax gave [21] a criterion to show that the KdV can be viewed as a *compatibility condition* related to the time evolution of solutions to (1.2). Since Lax's criterion is applicable to other nonlinear PDEs solvable by an inverse scattering transform (e.g., the initial-value problem for the nonlinear Schrödinger equation can be solved [4,6,22] via the inverse scattering transform for the Zakharov-Shabat system), we first outline the general idea behind the Lax method and next demonstrate its application on the KdV.

Given a linear operator $\mathcal{L}$ with $\mathcal{L}\psi = \lambda\psi$, we are interested in finding another operator $\mathcal{A}$ (the operators $\mathcal{A}$ and $\mathcal{L}$ are said to form a Lax pair) such that:

(i) The spectral parameter λ does not change in time.

(ii) The quantity $\psi_t - \mathcal{A}\psi$ must remain a solution to $\mathcal{L}\psi = \lambda\psi$.

(iii) The quantity $\mathcal{L}_t + \mathcal{L}\mathcal{A} - \mathcal{A}\mathcal{L}$ must be a multiplication operator.

As the following argument shows, for compatibility, we are forced to have

$$\mathcal{L}_t + \mathcal{L}\mathcal{A} - \mathcal{A}\mathcal{L} = 0, \tag{4.1}$$

which is interpreted as an integrable PDE and in general is nonlinear. From condition (ii) above we see that

$$\mathcal{L}\left(\psi_t - \mathcal{A}\psi\right) = \lambda\left(\psi_t - \mathcal{A}\psi\right),$$

or equivalently,

$$\begin{aligned} \mathcal{L}\psi_t - \mathcal{L}\mathcal{A}\psi &= \lambda\psi_t - \mathcal{A}\left(\lambda\psi\right) \\ &= \partial_t\left(\lambda\psi\right) - \mathcal{A}\mathcal{L}\psi \\ &= \partial_t\left(\mathcal{L}\psi\right) - \mathcal{A}\mathcal{L}\psi \\ &= \mathcal{L}_t\psi + \mathcal{L}\psi_t - \mathcal{A}\mathcal{L}\psi, \end{aligned} \tag{4.2}$$

where we have used $\mathcal{L}\psi = \lambda\psi$ and $\lambda_t = 0$. After canceling the $\mathcal{L}\psi_t$ terms in (4.2), we get

$$(\mathcal{L}_t + \mathcal{L}\mathcal{A} - \mathcal{A}\mathcal{L})\,\psi = 0. \tag{4.3}$$

Because of condition (iii) listed above, from (4.3) we obtain the compatibility condition (4.1).

Let us write the Schrödinger equation (1.2) as

$$\mathcal{L}\psi = \lambda\psi, \quad \mathcal{L} := -\partial_x^2 + u(x,t), \tag{4.4}$$

and try

$$\mathcal{A} = \alpha\partial_x^3 + \beta\partial_x^2 + \xi\partial_x + \eta, \tag{4.5}$$

where the coefficients α, β, ξ, and η may depend on x and t, but not on λ. Note that $\mathcal{L}_t = u_t$ and λ is the same as k^2. Using (4.4) and (4.5) we can write (4.1) explicitly as

$$(\ )\partial_x^5 + (\ )\partial_x^4 + (\ )\partial_x^3 + (\ )\partial_x^2 + (\ )\partial_x + (\ ) = 0, \tag{4.6}$$

where each ( ) in (4.6) denotes the appropriate coefficient. The coefficient of ∂_x^5 automatically vanishes. Setting the coefficients of ∂_x^j equal to zero for $j = 4, 3, 2, 1$, we obtain

$$\alpha = c_1, \quad \beta = c_2, \quad \xi = c_3 - \frac{3}{2}c_1 u, \quad \eta = c_4 - \frac{3}{4}c_1 u_x - c_2 u,$$

where c_1, c_2, c_3, and c_4 are arbitrary constants. Using $c_1 = -4$ and $c_3 = 0$ in the last term on the left-hand side of (4.6) and setting that last term equal to zero, we obtain the KdV equation given in (1.1). Further, by using $c_2 = c_4 = 0$, we obtain the operator $\mathcal{A}$ associated with $\mathcal{L}$ as

$$\mathcal{A} = -4\partial_x^3 + 6u\partial_x + 3u_x. \tag{4.7}$$

Let us remark that condition (ii) stated above allows us to determine the time evolution of any solution to the Schrödinger equation (1.2) as the initial potential $u(x,0)$ evolves to $u(x,t)$. For example, let us find the time evolution of $f_l(k,x;t)$, the Jost solution from the left. By using (2.2) and (2.5) with $T(k)$ and $R(k)$ replaced by $T(k;t)$ and $R(k;t)$, respectively, we see that

$$e^{-ikx} f_l(k,x;t) = 1 + o(1), \quad e^{-ikx} f_l'(k,x;t) = ik + o(1), \qquad x \to +\infty, \tag{4.8}$$

$$f_r(k,x;t) = \frac{e^{-ikx}}{T(k;t)} + \frac{R(k;t)\,e^{ikx}}{T(k;t)} + o(1), \qquad x \to +\infty. \tag{4.9}$$

From condition (ii) of the Lax method and (4.7) we obtain

$$\begin{aligned}\partial_t f_l(k,x;t) - \left(-4\partial_x^3 + 6u\partial_x + 3u_x\right) f_l(k,x;t)\\ = p(k,t)\, f_l(k,x;t) + q(k,t)\, f_r(k,x;t),\end{aligned} \tag{4.10}$$

where we have used the fact that the quantity $\partial_t f_l - \mathcal{A} f_l$ remains a solution to (1.2) and hence can be expressed as a linear combination of the two linearly independent Jost solutions $f_l(k,x;t)$ and $f_r(k,x;t)$ with coefficients $p(k,t)$ and $q(k,t)$, respectively. For each fixed t, assuming that $u(x,t) = o(1)$ and $u_x(x,t) = o(1)$ as

$x \to +\infty$, the coefficients $p(k,t)$ and $q(k,t)$ can be evaluated by letting $x \to +\infty$ in (4.10). Using (4.8) and (4.9) in (4.10), we get

$$\partial_t e^{ikx} + 4\partial_x^3 e^{ikx} = p(k,t)\, e^{ikx} + q(k,t) \left[\frac{1}{T(k;t)}\, e^{-ikx} + \frac{R(k;t)}{T(k;t)}\, e^{ikx}\right]. \tag{4.11}$$

From (4.11), by comparing the coefficients of e^{ikx} and e^{-ikx} on both sides, we obtain

$$q(k,t) = 0, \quad p(k,t) = -4ik^3. \tag{4.12}$$

Thus, the time evolution of $f_l(k,x;t)$ is determined by the linear third-order PDE

$$\partial_t f_l(k,x;t) - \mathcal{A} f_l(k,x;t) = -4ik^3 f_l(k,x;t). \tag{4.13}$$

Similarly, letting $x \to -\infty$ in (4.10), with the help of (2.3), (2.4), and (4.12) we obtain

$$\begin{aligned}\partial_t &\left[\frac{1}{T(k;t)}\, e^{ikx} + \frac{L(k;t)}{T(k;t)}\, e^{-ikx}\right] \\ &= \left[-4\partial_x^3 - 4ik^3\right] \left[\frac{1}{T(k;t)}\, e^{ikx} + \frac{L(k;t)}{T(k;t)}\, e^{-ikx}\right],\end{aligned} \tag{4.14}$$

where we have also used that $u(x,t) = o(1)$ and $u_x(x,t) = o(1)$ as $x \to -\infty$. From (4.14), comparing the coefficients of e^{ikx} and e^{-ikx} on both sides, we obtain

$$\partial_t T(k;t) = 0, \quad \partial_t L(k;t) = -8ik^3 L(k;t),$$

and hence

$$T(k;t) = T(k;0) = T(k), \quad L(k;t) = L(k;0)\, e^{-8ik^3 t} = L(k)\, e^{-8ik^3 t}. \tag{4.15}$$

Thus, the transmission coefficient remains unchanged and the reflection coefficient from the left undergoes a simple phase change as t progresses.

Proceeding in a similar manner, we can obtain the time evolution of the Jost solution $f_r(k,x;t)$ and the right reflection coefficient $R(k;t)$. With $\mathcal{A}$ as in (4.7), we get

$$\partial_t f_r(k,x;t) - \mathcal{A} f_r(k,x;t) = 4ik^3 f_r(k,x;t), \tag{4.16}$$

$$R(k;t) = R(k;0)\, e^{8ik^3 t} = R(k)\, e^{8ik^3 t}. \tag{4.17}$$

In order to evaluate the time evolution of the dependency constants $\gamma_j(t)$, we can substitute $\gamma_j(t) f_r(i\kappa_j,x;t)$ for $f_l(i\kappa_j,x;t)$ [cf. (2.10)] and evaluate (4.13) at $k = i\kappa_j$. We get

$$\begin{aligned}f_r(i\kappa_j,x;t)\, \partial_t \gamma_j(t) + \gamma_j(t)\, \partial_t f_r(i\kappa_j,x;t) - \gamma_j(t) \mathcal{A} f_r(i\kappa_j,x;t) \\ = -4\kappa_j^3 \gamma_j(t) f_r(i\kappa_j,x;t).\end{aligned} \tag{4.18}$$

On the other hand, from (4.16) at $k = i\kappa_j$, we obtain

$$\gamma_j(t)\, \partial_t f_r(i\kappa_j,x;t) - \gamma_j(t)\, \mathcal{A} f_r(i\kappa_j,x;t) = 4\kappa_j^3 \gamma_j(t)\, f_r(i\kappa_j,x;t). \tag{4.19}$$

Subtracting (4.19) from (4.18) we conclude that $\partial_t \gamma_j(t) = -8\kappa_j^3 \gamma_j(t)$, which leads to

$$\gamma_j(t) = \gamma_j(0)\, e^{-8\kappa_j^3 t} = \gamma_j\, e^{-8\kappa_j^3 t}. \tag{4.20}$$

Then, from (2.9) and (4.15) we obtain the time evolution of the norming constants $c_{lj}(t)$ and $c_{rj}(t)$ as

$$c_{lj}(t) = c_{lj}(0)\, e^{4\kappa_j^3 t} = c_{lj} e^{4\kappa_j^3 t}, \quad c_{rj}(t) = c_{rj}(0)\, e^{-4\kappa_j^3 t} = c_{rj} e^{-4\kappa_j^3 t}.$$

5. AKNS method to derive the KdV

In the previous section we have seen that the KdV arises as a compatibility condition in the Lax method. There are other methods to derive nonlinear PDEs that can be solved by the inverse scattering transform, i.e., by solving the inverse problem with the time-evolved scattering data for a corresponding linear differential equation. One of these methods was developed by Ablowitz, Kaup, Newel, and Segur, and it was first applied to the Sine-Gordon equation [23]. Here we outline the basic idea behind the method of Ablowitz, Kaup, Newel, and Segur (AKNS method, for short) and use it to derive the KdV equation.

Given a linear operator $\mathcal{X}$ associated with the system $v_x = \mathcal{X}v$, we are interested in finding another operator $\mathcal{T}$ (the operators $\mathcal{X}$ and $\mathcal{T}$ are said to form an *AKNS pair*) such that:

(i) The spectral parameter λ does not change in time.

(ii) The quantity $v_t - \mathcal{T}v$ must remain a solution to $v_x = \mathcal{X}v$.

(iii) The quantity $\mathcal{X}_t - \mathcal{T}_x + \mathcal{X}\mathcal{T} - \mathcal{T}\mathcal{X}$ must be a (matrix) multiplication operator.

Note that in general $\mathcal{X}$ contains the spectral parameter λ, and hence $\mathcal{T}$ also depends on λ as well. Usually, $\mathcal{X}$ and $\mathcal{T}$ are matrix-valued with entries depending on x, t, and λ. As the operator $\mathcal{A}$ in the Lax method determines the time evolution of solutions to $\mathcal{L}\psi = \lambda\psi$, in the AKNS method the operator $\mathcal{T}$ determines the time evolution of solutions to $v_x = \mathcal{X}v$ according to condition (ii) listed above.

As the following argument shows, for compatibility, we are forced to have

$$\mathcal{X}_t - \mathcal{T}_x + \mathcal{X}\mathcal{T} - \mathcal{T}\mathcal{X} = 0, \tag{5.1}$$

which leads to an integrable PDE and is in general nonlinear. From condition (ii) above we see that

$$(v_t - \mathcal{T}v)_x = \mathcal{X}(v_t - \mathcal{T}v),$$

or equivalently,

$$\begin{aligned} v_{tx} - \mathcal{T}_x v - \mathcal{T} v_x &= \mathcal{X} v_t - \mathcal{X}\mathcal{T}v \\ &= (\mathcal{X}v)_t - \mathcal{X}_t v - \mathcal{X}\mathcal{T}v \\ &= (v_x)_t - \mathcal{X}_t v - \mathcal{X}\mathcal{T}v \\ &= v_{xt} - \mathcal{X}_t v - \mathcal{X}\mathcal{T}v. \end{aligned} \tag{5.2}$$

We expect v to be smooth enough so that $v_{tx} = v_{xt}$. Let us replace $\mathcal{T}v_x$ by $\mathcal{T}\mathcal{X}v$ on the left-hand side in (5.2), from which we get $(\mathcal{X}_t - \mathcal{T}_x + \mathcal{X}\mathcal{T} - \mathcal{T}\mathcal{X})v = 0$, which in turn, as a result of condition (iii) listed above, gives us the compatibility condition (5.1).

Let us write the Schrödinger equation (1.2), by replacing the spectral parameter k^2 by λ, in the form of the first-order linear system $v_x = \mathcal{X} v$ by choosing

$$v = \begin{bmatrix} \psi_x \\ \psi \end{bmatrix}, \quad \mathcal{X} = \begin{bmatrix} 0 & u(x,t) - \lambda \\ 1 & 0 \end{bmatrix}.$$

We will construct $\mathcal{T}$ so that $\mathcal{T}$ and $\mathcal{X}$ will form an AKNS pair. Let us try

$$\mathcal{T} = \begin{bmatrix} \alpha & \beta \\ \xi & \eta \end{bmatrix},$$

where the entries α, β, ξ, and η may depend on x, t, and λ. The compatibility condition (5.1) leads to

$$\begin{bmatrix} 0 & u_t \\ 0 & 0 \end{bmatrix} - \begin{bmatrix} \alpha_x & \beta_x \\ \xi_x & \eta_x \end{bmatrix} + \begin{bmatrix} 0 & u-\lambda \\ 1 & 0 \end{bmatrix} \begin{bmatrix} \alpha & \beta \\ \xi & \eta \end{bmatrix} - \begin{bmatrix} \alpha & \beta \\ \xi & \eta \end{bmatrix} \begin{bmatrix} 0 & u-\lambda \\ 1 & 0 \end{bmatrix} = \begin{bmatrix} 0 & 0 \\ 0 & 0 \end{bmatrix},$$

or equivalently

$$\begin{bmatrix} -\alpha_x - \beta + \xi(u-\lambda) & u_t - \beta_x + \eta(u-\lambda) - \alpha(u-\lambda) \\ -\xi_x + \alpha - \eta & -\eta_x + \beta - \xi(u-\lambda) \end{bmatrix} = \begin{bmatrix} 0 & 0 \\ 0 & 0 \end{bmatrix}. \tag{5.3}$$

From the $(1,1)$, $(2,1)$, and $(2,2)$ entries of the above matrix equation, we obtain

$$\beta = -\alpha_x + (u-\lambda)\xi, \quad \eta = \alpha - \xi_x, \quad \eta_x = -\alpha_x. \tag{5.4}$$

Then the $(1,2)$ entry in (5.3) is given by

$$u_t + \frac{1}{2}\xi_{xxx} - u_x \xi - 2\xi_x(u-\lambda) = 0. \tag{5.5}$$

Letting $\xi = \lambda\zeta + \mu$ in (5.5), where λ is the spectral parameter, we obtain

$$2\zeta_x\lambda^2 + \left(\frac{1}{2}\zeta_{xxx} - 2\zeta_x u + 2\mu_x - u_x\zeta\right)\lambda + \left(u_t + \frac{1}{2}\mu_{xxx} - 2\mu_x u - u_x\mu\right) = 0.$$

Equating the coefficients of each power of λ to zero, we get

$$\zeta = c_1, \quad \mu = \frac{1}{2}c_1 u + c_2, \quad u_t - \frac{3}{2}c_1 u u_x - c_2 u_x + \frac{1}{4}c_1 u_{xxx} = 0, \tag{5.6}$$

where c_1 and c_2 are arbitrary constants. Using $c_1 = 4$ and $c_2 = 0$, from (5.6) we obtain the KdV given in (1.1). Moreover, with the help of (5.4) we get

$$\alpha = u_x + c_3, \quad \beta = -4\lambda^2 + 2\lambda u + 2u^2 - u_{xx}, \quad \xi = 4\lambda + 2u, \quad \eta = c_3 - u_x,$$

where c_3 is an arbitrary constant. Letting $c_3 = 0$, we obtain

$$\mathcal{T} = \begin{bmatrix} u_x & -4\lambda^2 + 2\lambda u + 2u^2 - u_{xx} \\ 4\lambda + 2u & -u_x \end{bmatrix}. \tag{5.7}$$

It is possible to obtain the time evolution of the Jost solutions and of the scattering data via the AKNS method. Note that condition (ii) of the AKNS method is satisfied if we have

$$v_t - \mathcal{T} v = p(t,\lambda)\, v, \tag{5.8}$$

for some (scalar) coefficient $p(t, \lambda)$. For example, if we choose

$$v = \begin{bmatrix} f_l'(k,x;t) \\ f_l(k,x;t) \end{bmatrix},$$

by letting $x \to +\infty$ in (5.8) and by using (2.2), we obtain the time evolution of $f_l(k, x; t)$ given in (4.13). By using (2.4) with $T(x)$ replaced by $T(k; t)$ and with $L(k)$ by $L(k; t)$, and by letting $x \to -\infty$ in (5.8) and using (2.2), we get the time evolutions given in (4.15). In a similar way, with the help of (2.3), (4.9), and (5.7), by choosing

$$v = \begin{bmatrix} f_r'(k,x;t) \\ f_r(k,x;t) \end{bmatrix},$$

and by letting $x \to \pm\infty$ in (5.8), we obtain (4.16) and (4.17).

6. Solution to the Cauchy problem for the KdV

The Cauchy problem (initial-value problem) for the KdV consists of finding $u(x,t)$ when $u(x, 0)$ is known. As shown by Gardner, Greene, Kruskal, and Miura [11], this problem can be solved by using the inverse scattering transform. Let $\mathcal{D}(t)$ denote any of the equivalents of the scattering data for the Schrödinger equation with the time-evolved potential $u(x,t)$. In other words, we have

$$\mathcal{D}(t) := \{R(k;t), L(k;t), T(k;t), \{\kappa_j(t)\}, \{\gamma_j(t)\}, \{c_{lj}(t)\}, \{c_{rj}(t)\}\}, \tag{6.1}$$

or equivalently,

$$\mathcal{D}(t) = \{R(k)\, e^{8ik^3t}, L(k)\, e^{-8ik^3t}, T(k), \{\kappa_j\}, \{\gamma_j\, e^{-8\kappa_j^3 t}\}, \{c_{lj}\, e^{4\kappa_j^3 t}\}, \{c_{rj}\, e^{-4\kappa_j^3 t}\}\}.$$

Note that $\mathcal{D}(0)$ corresponds to the initial scattering data associated with the potential $V(x)$, where $u(x, 0) = V(x)$.

Below we outline the solution to the Cauchy problem for the KdV.

(i) Given $u(x, 0)$, determine the corresponding scattering data $\mathcal{D}(0)$. This is done by solving the *direct scattering problem* $V(x) \mapsto \mathcal{D}(0)$. The solution to this problem is essentially equivalent to solving (2.1) and obtaining the Jost solutions $f_l(k,x)$ and $f_r(k,x)$, from which $\mathcal{D}(0)$ can be constructed.

(ii) Evolve in time the scattering data as $\mathcal{D}(0) \mapsto \mathcal{D}(t)$ in accordance with (6.1). Note that the time evolution of the scattering data is really simple. On the other hand, the time evolution of the potential $u(x, 0) \mapsto u(x,t)$ will be much more complicated. Similarly, we expect that the time evolution of the Jost solutions $f_l(k,x) \mapsto f_l(k,x;t)$ and $f_r(k,x) \mapsto f_r(k,x;t)$, governed by the PDEs (4.13) and (4.16), respectively, will be complicated.

(iii) Having obtained the time-evolved scattering data $\mathcal{D}(t)$, solve the corresponding inverse scattering problem $\mathcal{D}(t) \mapsto u(x,t)$ for (1.2). This problem is known to be uniquely solvable [13] when the initial potential V belongs to the Faddeev class.

As indicated in Section 3, in step (iii) above one can obtain the solution to the inverse scattering problem by solving, for example, the time-evolved Riemann-Hilbert problem [cf. (3.1)]

$$m_{\rm l}(-k,x;t)=T(k)\,m_{\rm r}(k,x;t)-R(k)\,e^{2ikx+8ik^3t}\,m_{\rm l}(k,x;t),\qquad k\in\mathbf{R},\tag{6.2}$$

and recover $u(x,t)$ by using [cf. (3.4)]

$$u(x,t)=\frac{m_{\rm l}''(k,x;t)}{m_{\rm l}(k,x;t)}+2ik\frac{m_{\rm l}'(k,x;t)}{m_{\rm l}(k,x;t)},\qquad k\in\overline{\mathbf{C}^+},\tag{6.3}$$

where the right-hand side can be evaluated at any k value, including $k=0$ and $k=\pm\infty$. Equivalently, one can solve the time-evolved Riemann-Hilbert problem [cf. (3.2)]

$$m_{\rm r}(-k,x;t)=T(k)\,m_{\rm l}(k,x;t)-L(k)\,e^{-2ikx-8ik^3t}\,m_{\rm r}(k,x;t),\qquad k\in\mathbf{R},$$

and use [cf. (3.4)]

$$u(x,t)=\frac{m_{\rm r}''(k,x;t)}{m_{\rm r}(k,x;t)}-2ik\frac{m_{\rm r}'(k,x;t)}{m_{\rm r}(k,x;t)},\qquad k\in\overline{\mathbf{C}^+},$$

where the right-hand side can be evaluated at any k value in $\overline{\mathbf{C}^+}$.

Alternatively, one can solve the time-evolved left Marchenko equation [cf. (3.5)]

$$B_{\rm l}(x,y;t)+\Omega_{\rm l}(2x+y;t)+\int_0^\infty dz\,\Omega_{\rm l}(2x+y+z;t)\,B_{\rm l}(x,z;t)=0,\qquad y>0,\tag{6.4}$$

with

$$\Omega_{\rm l}(y;t):=\frac{1}{2\pi}\int_{-\infty}^\infty dk\,R(k)\,e^{8ik^3t+iky}+\sum_{j=1}^N c_{{\rm l}j}^2\,e^{8\kappa_j^3t-\kappa_j y},$$

and recover $u(x,t)$ by using [cf. (3.6)]

$$u(x,t)=-2\,\frac{\partial B_{\rm l}(x,0^+;t)}{\partial x}.\tag{6.5}$$

Equivalently, one can solve the time-evolved right Marchenko equation [cf. (3.8)]

$$B_{\rm r}(x,y;t)+\Omega_{\rm r}(-2x+y;t)+\int_0^\infty dz\,\Omega_{\rm r}(-2x+y+z;t)\,B_{\rm r}(x,z;t)=0,\qquad y>0,\tag{6.6}$$

with

$$\Omega_{\rm r}(\alpha;t):=\frac{1}{2\pi}\int_{-\infty}^\infty dk\,L(k)\,e^{-8ik^3t+iky}+\sum_{j=1}^N c_{{\rm r}j}^2\,e^{-8\kappa_j^3t-\kappa_j y},$$

and obtain $u(x,t)$ by using [cf. (3.9)]

$$u(x,t)=2\,\frac{\partial B_{\rm r}(x,0^+;t)}{\partial x}.\tag{6.7}$$

7. Soliton solutions to the KdV

Consider the Cauchy problem for the KdV corresponding to the initial scattering data with zero reflection coefficients, N bound states at $k = i\kappa_j$, and dependency constant γ_j, where we have the ordering $0 < \kappa_1 < \cdots < \kappa_N$. When $R \equiv 0$, from (2.11) we see that the transmission coefficient is given by

$$T(k) = \prod_{j=1}^{N} \frac{k + i\kappa_j}{k - i\kappa_j}. \tag{7.1}$$

In this case, the inverse scattering problem $\mathcal{D}(t) \mapsto u(x,t)$ can be solved algebraically in a closed form, and the resulting solution $u(x,t)$ to the KdV is known as the N-soliton solution. Using (7.1) we can write (6.2) as

$$m_{\mathrm{l}}(-k,x;t) \prod_{j=1}^{N} (k - i\kappa_j) = m_{\mathrm{r}}(k,x;t) \prod_{j=1}^{N} (k + i\kappa_j), \qquad k \in \mathbf{R}. \tag{7.2}$$

From the analyticity properties of m_{l} and m_{r} it follows that each side in (7.2) is entire in k with a polynomial growth of leading term k^N as $k \to \infty$ in the complex plane $\mathbf{C}$. Further, since k appears as ik in the Faddeev functions, both sides in (7.2) must be a polynomial of the form

$$k^N + ik^{N-1} a_{N-1}(x,t) + \cdots + i^N a_0(x,t), \tag{7.3}$$

where $a_j(x,t)$ are real valued and to be determined by using [cf. (2.10), (3.3), and (4.20)]

$$\gamma_j \, e^{2\kappa_j x - 8\kappa_j^3 t} = \frac{m_{\mathrm{l}}(i\kappa_j,x;t)}{m_{\mathrm{r}}(i\kappa_j,x;t)}, \qquad j = 1,\ldots,N. \tag{7.4}$$

From (7.2) and (7.3) we get

$$m_{\mathrm{l}}(k,x;t) = \frac{k^N - ik^{N-1} a_{N-1}(x,t) + \cdots + (-i)^N a_0(x,t)}{\prod_{j=1}^{N}(k + i\kappa_j)}, \tag{7.5}$$

$$m_{\mathrm{r}}(k,x;t) = \frac{k^N + ik^{N-1} a_{N-1}(x,t) + \cdots + i^N a_0(x,t)}{\prod_{j=1}^{N}(k + i\kappa_j)}. \tag{7.6}$$

Let us define

$$\omega_j := \gamma_j \, e^{2\kappa_j x - 8\kappa_j^3 t}, \qquad j = 1,\ldots,N, \tag{7.7}$$

where we recall that the sign of ω_j is the same [cf. (2.10)] as that of $(-1)^{N-j}$. Using (7.5)-(7.7) in (7.4) we get a system of linear algebraic equations for the N unknowns $a_j(x,t)$, namely

$$\omega_j = \frac{\kappa_j^N - \kappa_j^{N-1} a_{N-1}(x,t) + \cdots + (-1)^N a_0(x,t)}{\kappa_j^N + \kappa_j^{N-1} a_{N-1}(x,t) + \cdots + a_0(x,t)}, \qquad j = 1,\ldots,N,$$

which can be written as

$$\begin{bmatrix} \kappa_1^{N-1}(\omega_1+1) & \kappa_1^{N-2}(\omega_1-1) & \cdots & \omega_1-(-1)^N \\ \kappa_2^{N-1}(\omega_2+1) & \kappa_2^{N-2}(\omega_2-1) & \cdots & \omega_2-(-1)^N \\ \vdots & \vdots & \ddots & \vdots \\ \kappa_N^{N-1}(\omega_N+1) & \kappa_N^{N-2}(\omega_N-1) & \cdots & \omega_N-(-1)^N \end{bmatrix} \begin{bmatrix} a_{N-1}(x,t) \\ a_{N-2}(x,t) \\ \vdots \\ a_0(x,t) \end{bmatrix} = \begin{bmatrix} \kappa_1^N(1-\omega_1) \\ \kappa_2^N(1-\omega_2) \\ \vdots \\ \kappa_N^N(1-\omega_N) \end{bmatrix}. \tag{7.8}$$

Let $M^{(N)}$ denote the coefficient matrix in (7.8) whose (j,n) entry is given as

$$M_{jn}^{(N)} := \kappa_j^{N-n}\left[\omega_j-(-1)^n\right], \qquad j,n=1,\dots,N. \tag{7.9}$$

From (7.8), via Cramer's rule, it is possible to extract $a_{N-j}(x,t)$ explicitly as

$$a_{N-j}(x,t) = -\frac{\det P^{(N-j)}}{\det M^{(N)}}, \qquad j=1,\dots,N,$$

where the $(N+1)\times(N+1)$ matrix $P^{(N-j)}$ is given by

$$P^{(N-j)} := \begin{bmatrix} 0 & Y^{(N,j)} \\ Q^{(N)} & M^{(N)} \end{bmatrix},$$

with $Q^{(N)}$ being the $N\times 1$ matrix whose jth row contains the entry $\kappa_j^N(1-\omega_j)$ for $j=1,\dots,N$; and $M^{(N)}$ being the $N\times N$ matrix given in (7.9); and $Y^{(N,j)}$ being the $1\times N$ matrix whose jth column contains the entry 1 and all the remaining entries being zero. For example, we have

$$P^{(N-1)} := \begin{bmatrix} 0 & 1 & 0 & \cdots & 0 \\ \kappa_1^N(1-\omega_1) & \kappa_1^{N-1}(\omega_1+1) & \kappa_1^{N-2}(\omega_1-1) & \cdots & \omega_1-(-1)^N \\ \kappa_2^N(1-\omega_2) & \kappa_2^{N-1}(\omega_2+1) & \kappa_2^{N-2}(\omega_2-1) & \cdots & \omega_2-(-1)^N \\ \vdots & \vdots & \vdots & \ddots & \vdots \\ \kappa_N^N(1-\omega_N) & \kappa_N^{N-1}(\omega_N+1) & \kappa_N^{N-2}(\omega_N-1) & \cdots & \omega_N-(-1)^N \end{bmatrix}.$$

We note that $u(x,t)$ can be constructed from $a_{N-1}(x,t)$ alone or from $a_0(x,t)$ alone. For example, using (6.3) and (7.5) in the limit $k\to\pm\infty$, we see that

$$u(x,t) = 2\,\frac{\partial a_{N-1}(x,t)}{\partial x}, \tag{7.10}$$

or using (6.3) and (7.5) at $k=0$ we see that

$$u(x,t) = \frac{1}{a_0(x,t)}\,\frac{\partial^2 a_0(x,t)}{\partial x^2}. \tag{7.11}$$

In fact, using (7.5) in (6.3) we see that one can recover $u(x,t)$ by using any one of the following $(N+1)$ equations:

$$u(x,t) = \frac{1}{a_{N-j}(x,t)}\left[\frac{\partial^2 a_{N-j}(x,t)}{\partial x^2} + 2\frac{\partial a_{N-j-1}(x,t)}{\partial x}\right], \qquad j = 0, 1, \ldots, N,$$

where we have defined $a_{-1}(x,t) := 0$ and $a_N(x,t) := 1$.

Alternatively, we can obtain the N-soliton solution $u(x,t)$ via (6.5) by solving, in a closed form, the left Marchenko equation (6.4), which has a degenerate kernel thanks to the fact that $R \equiv 0$. Similarly, $u(x,t)$ can be obtained via (6.7) by solving, in a closed form, the right Marchenko equation (6.6), which has a degenerate kernel.

It is also possible to obtain the solutions to the Marchenko equations (6.4) and (6.6) in an algebraic manner, without really solving the integral equations themselves. Below we illustrate this for the recovery of $B_\text{l}(x,y;t)$ and obtain $u(x,t)$ via (6.5). Using $R \equiv 0$ we can write (6.2) as

$$m_\text{l}(-k,x;t) - 1 = [T(k) - 1]\, m_\text{r}(k,x;t) + [m_\text{r}(k,x;t) - 1], \qquad k \in \mathbf{R}.$$

Taking the Fourier transform of both sides with $\frac{1}{2\pi}\int_{-\infty}^{\infty} dk\, e^{iky}$ and exploiting the analyticity properties of m_l and m_r, we obtain [cf. (3.7), (3.10)]

$$B_\text{l}(x,y;t) = \frac{1}{2\pi}\int_{-\infty}^{\infty} dk\, [T(k) - 1]\, m_\text{r}(k,x;t)\, e^{iky}, \qquad y > 0. \tag{7.12}$$

The right-hand side of (7.12) can be evaluated as a residue integral along the semicircular arc that is the boundary of $\mathbf{C}^+$. Using (7.1) in (7.12), we get

$$B_\text{l}(x,y;t) = i\sum_{j=1}^{N} [\text{Res}\,(T, i\kappa_j)]\, m_\text{r}(i\kappa_j,x;t)\, e^{-\kappa_j y}, \qquad y > 0. \tag{7.13}$$

Using (2.9) and (7.4) in (7.13), we obtain

$$B_\text{l}(x,y;t) = -\sum_{j=1}^{N} \varepsilon_j\, m_\text{l}(i\kappa_j,x;t)\, e^{-\kappa_j y}, \tag{7.14}$$

where we have defined

$$\varepsilon_j := c_{\text{l}j}^2 e^{-2\kappa_j x + 8\kappa_j^3 t}.$$

Using the time-evolved version of (3.7), from (7.14) we get

$$m_\text{l}(k,x;t) = 1 - \sum_{j=1}^{N} \frac{i\varepsilon_j}{k + i\kappa_j} m_\text{l}(i\kappa_j,x;t)\, e^{-\kappa_j y}. \tag{7.15}$$

Putting $k = i\kappa_n$ in (7.15) for $n = 1, \dots, N$, it is possible to recover the $m_l(i\kappa_j, x; t)$ by solving the linear algebraic system

$$\begin{bmatrix} 1 + \dfrac{\varepsilon_1}{\kappa_1 + \kappa_1} & \dfrac{\varepsilon_2}{\kappa_1 + \kappa_2} & \cdots & \dfrac{\varepsilon_N}{\kappa_1 + \kappa_N} \\ \dfrac{\varepsilon_1}{\kappa_2 + \kappa_1} & 1 + \dfrac{\varepsilon_2}{\kappa_2 + \kappa_2} & \cdots & \dfrac{\varepsilon_N}{\kappa_2 + \kappa_N} \\ \vdots & \vdots & \ddots & \vdots \\ \dfrac{\varepsilon_1}{\kappa_N + \kappa_1} & \dfrac{\varepsilon_2}{\kappa_N + \kappa_2} & \cdots & 1 + \dfrac{\varepsilon_N}{\kappa_N + \kappa_N} \end{bmatrix} \begin{bmatrix} m_l(i\kappa_1, x; t) \\ m_l(i\kappa_2, x; t) \\ \vdots \\ m_l(i\kappa_N, x; t) \end{bmatrix} = \begin{bmatrix} 1 \\ 1 \\ \vdots \\ 1 \end{bmatrix}. \tag{7.16}$$

Let Γ denote the coefficient matrix in (7.16), namely, let its (n, j) entry be given by

$$\Gamma_{nj} = \delta_{nj} + \frac{\varepsilon_j}{\kappa_n + \kappa_j}, \qquad n, j = 1, \dots, N,$$

where δ_{nj} denotes the Kronecker delta. Then, using (7.14) and (7.16), we obtain $B_l(x, 0^+; t)$ as the ratio of two determinants as

$$B_l(x, 0^+; t) = \frac{\det Z}{\det \Gamma},$$

where Z is the matrix defined as

$$Z := \begin{bmatrix} 0 & \varepsilon_1 & \varepsilon_2 & \cdots & \varepsilon_N \\ 1 & 1 + \dfrac{\varepsilon_1}{\kappa_1 + \kappa_1} & \dfrac{\varepsilon_2}{\kappa_1 + \kappa_2} & \cdots & \dfrac{\varepsilon_N}{\kappa_1 + \kappa_N} \\ 1 & \dfrac{\varepsilon_1}{\kappa_2 + \kappa_1} & 1 + \dfrac{\varepsilon_2}{\kappa_2 + \kappa_2} & \cdots & \dfrac{\varepsilon_N}{\kappa_2 + \kappa_N} \\ \vdots & \vdots & \vdots & \ddots & \vdots \\ 1 & \dfrac{\varepsilon_1}{\kappa_N + \kappa_1} & \dfrac{\varepsilon_2}{\kappa_N + \kappa_2} & \cdots & 1 + \dfrac{\varepsilon_N}{\kappa_N + \kappa_N} \end{bmatrix}.$$

It can be shown that $\det Z = \partial_x(\det \Gamma)$ and hence with the help of (6.5), we get

$$u(x, t) = -2 \frac{\partial}{\partial x} \left[\frac{1}{\det \Gamma} \frac{\partial(\det \Gamma)}{\partial x} \right]. \tag{7.17}$$

Some Mathematica notebooks based on (7.10), (7.11), and (7.17) for the evaluation of N-soliton solutions to the KdV and their animations are available at the author's web page [24].

We can use (7.10), (7.11), or (7.17) to analyze properties of solitons of the KdV. For example, when $N = 1$, from (7.10) or (7.11) we get the single soliton solution to (1.1) as

$$u(x, t) = -2\kappa_1^2 \operatorname{sech}^2 \left(\kappa_1 x - 4\kappa_1^3 t + \sqrt{\ln \gamma_1} \right). \tag{7.18}$$

It is seen that the amplitude of this wave is $2\kappa_1^2$, it moves in the positive x direction with speed $4\kappa_1^2$, and the dependency constant γ_1 plays a role in the initial location of the soliton. The width of the soliton is inversely proportional to κ_1, which can be seen, e.g., by using the fact that $\int_{-\infty}^{\infty} dx\,\sqrt{-u(x,t)}$ is equal to $\sqrt{2}\pi$.

By exploiting the properties of one-soliton solutions to the KdV, one can show that as $t \to +\infty$, the N-soliton solution to the KdV resembles a train of N separate solitons each traveling with speed $4\kappa_j^2$. In this case, the KdV can be considered for all $t \in \mathbf{R}$ and it can be shown that each soliton emerges from the nonlinear interaction by experiencing only a change in the phase. For details, the reader is referred to [3,4,6,25].

8. Conclusion

In this section we will comment on three aspects of the KdV; namely, the time evolution stated in condition (ii) of the Lax method outlined in Section 4, the conserved quantities, and the Bäcklund transformation.

Some references carelessly state the time evolution associated with the Lax method. For example, in (1.2.10) of [3, page 6] it is stated that the evolution of the solutions to (1.2) is given by $\partial_t\psi = \mathcal{A}\psi$ with $\mathcal{A}$ as in (4.7), instead of the correct statement (ii) of Section 4. The incorrectness of $\partial_t\psi = \mathcal{A}\psi$ can be demonstrated explicitly by an elementary example. Consider the 1-soliton solution to the KdV given in (7.18). Let us choose $\gamma_1 = \kappa_1 = 1$. A solution to (1.2) is obtained as $f_l(i,x;t)$. Let us call that solution ψ. We have

$$\psi(x,t) = \frac{1}{2}\,e^{-4t}\,\mathrm{sech}(x-4t), \quad u(x,t) = -2\,\mathrm{sech}^2(x-4t).$$

It can directly be verified that ψ satisfies (1.2) with $k = i$, and $(\partial_t - \mathcal{A})\psi = -4\psi$, as indicated by the correct time evolvement (4.13) with $k = i$; hence $\partial_t\psi \neq \mathcal{A}\psi$.

In (1.7.6) of [6, page 25] and (4) of [5, page 81] it is carelessly stated that the evolution of the solutions to (1.2) is given by $\partial_t\psi - \mathcal{A}\psi = c\psi$ with $\mathcal{A}$ as in (4.7) and c is an arbitrary constant or a function of t. The incorrectness of this can be demonstrated, for example, by choosing ψ as $f_l(k,x;t) + f_r(k,x;t)$, i.e., the sum of the Jost solutions to (1.2). From (4.13) and (4.16) it follows that, with this choice of ψ, the equation $\partial_t\psi - \mathcal{A}\psi = c\psi$ would hold if and only if $(4ik^3 + c)f_r(k,x;t) - (4ik^3 - c)f_l(k,x;t) = 0$ for $x \in \mathbf{R}$ and $t > 0$, which is impossible due to the linear independence of $f_l(k,x;t)$ and $f_r(k,x;t)$ on $x \in \mathbf{R}$. Note that (4.4) has two linearly independent (generalized) eigenfunctions for each $\lambda > 0$ and hence $\partial_t\psi - \mathcal{A}\psi$ in general is not expected to be a constant multiple of ψ.

Let us note that a potential in the Faddeev class need not even be continuous. On the other hand, from (1.1) we see that classical solutions to the KdV are thrice differentiable with respect to x. Informally speaking, the discontinuities that may be present in the initial value $u(x,0)$ disappear and $u(x,t)$ becomes smoother for $t > 0$. On the other hand, even though $u(x,t)$ changes as t increases, certain integrals involving $u(x,t)$ with respect to x remain unchanged in time.

Such quantities are known as conserved quantities for the KdV. They can either be obtained directly from (1.1) or from the expansion of $T(k)$ in powers of $1/k$ as $k \to \pm\infty$ by using the fact that $T(k;t)$ does not change in time [cf. (4.15)]. When $u(x,t)$ is smooth, with the help of (2.6) we obtain

$$T(k;t) = 1 + \frac{C_1}{2ik} - \frac{C_1^2}{8k^2} - \frac{C_1^3}{48ik^3} + \frac{C_2}{8ik^3} + O(1/k^4), \qquad k \to \infty \text{ in } \overline{\mathbf{C}^+},$$

where we have defined

$$C_j := \int_{-\infty}^{\infty} dx\, u(x,t)^j, \qquad j = 1,2.$$

Thus, we have identified two of the infinite number of conserved quantities; i.e., C_1 and C_2 are independent of time and are equal to their values at $t = 0$. The time independence of C_1 and C_2 can also be obtained directly from (1.1). We can write (1.1) as $u_t = (3u^2 - u_{xx})_x$, and hence

$$\frac{dC_1}{dt} = \frac{d}{dt} \int_{-\infty}^{\infty} dx\, u(x,t) = 3u(x,t)^2 - u_{xx}(x,t)\Big|_{x=-\infty}^{\infty} = 0,$$

where we have used $u(x,t) = o(1)$ and $u_{xx}(x,t) = o(1)$ as $x \to \pm\infty$. Similarly, after multiplying (1.1) with $u(x,t)$, we can write the resulting equation as

$$(u^2)_t = (4u^3 - 2uu_{xx} + u_x^2)_x. \tag{8.1}$$

Integrating both sides of (8.1) on $x \in \mathbf{R}$, we get

$$\frac{dC_2}{dt} = \frac{d}{dt} \int_{-\infty}^{\infty} dx\, u(x,t)^2 = 4u(x,t)^3 - 2u(x,t)\, u_{xx}(x,t) + u_x(x,t)^2\Big|_{x=-\infty}^{\infty} = 0,$$

verifying the time independence of C_2.

Can we characterize the set of nonlinear PDEs solvable by an inverse scattering transform? In other words, can we find a set of necessary and sufficient conditions that guarantee an initial-value problem for a nonlinear PDE to be solvable via an inverse scattering transform related to a linear problem? There does not yet seem to be a satisfactory solution to this characterization problem. On the other hand, nonlinear PDEs solvable by an inverse scattering transform seem to have some common characteristic features such as the Lax pair, the AKNS pair, soliton solutions, an infinite number of conserved quantities, a Hamiltonian formalism, the Painlevé property, and the Bäcklund transformation. Here, we only briefly explain the last feature and refer the reader to [6] for details and other features.

A Bäcklund transformation is a means to produce another integrable nonlinear PDE from a given one. The basic idea is as follows. Assume v satisfies the integrable nonlinear PDE $\mathcal{M}(v) = 0$, and u satisfies another integrable nonlinear PDE, say $\mathcal{Q}(u) = 0$. A relationship $\mathcal{P}(u,v) = 0$, which is called a Bäcklund transformation, involving v, u, and their derivatives allows us to obtain $\mathcal{Q}(u) = 0$ from $\mathcal{M}(v) = 0$. A Bäcklund transformation can also be used on the same nonlinear PDE to produce another solution from a given solution.

As an example, assume that v satisfies the *modified KdV* given by

$$v_t - 6v^2 v_x + v_{xxx} = 0, \qquad x \in \mathbf{R}, \quad t > 0. \tag{8.2}$$

Then, choosing

$$u = v_x + v^2, \tag{8.3}$$

one can show that

$$u_t - 6uu_x + u_{xxx} = (\partial_x + 2v)(v_t - 6v^2 v_x + v_{xxx}), \qquad x \in \mathbf{R}, \quad t > 0.$$

Thus, (8.2) and (8.3) imply (1.1). The Bäcklund transformation given in (8.3) is known as Miura's transformation [26]. For a Bäcklund transformation applied on the KdV to produce other solutions from a given solution, we refer the reader to [6,27].

Another interesting question is the determination of the linear problem associated with the inverse scattering transform. In other words, given a nonlinear PDE that is known to be solvable by an inverse scattering transform, can we determine the corresponding linear problem? There does not yet seem to be a completely satisfactory answer to this question. We mention that Wahlquist and Estabrook [28] developed the so-called *prolongation method* to derive the linear scattering problem associated with the KdV and refer the reader to [6] for details.

References

[1] D.J. Korteweg and G. de Vries, *On the change of form of long waves advancing in a rectangular channel, and a new type of long stationary waves,* Phil. Mag. **39** (1895) 422–443.

[2] J.S. Russell, *Report on waves,* Report of the 14th meeting of the British Association for the Advancement of Science, John Murray, London, 1845, pp. 311–390.

[3] G.L. Lamb, Jr., *Elements of soliton theory,* Wiley, New York, 1980.

[4] M.J. Ablowitz and H. Segur, *Solitons and the inverse scattering transform,* SIAM, Philadelphia, 1981.

[5] P.G. Drazin, *Solitons,* Cambridge University Press, Cambridge, 1986.

[6] M.J. Ablowitz and P.A. Clarkson, *Solitons, nonlinear evolution equations and inverse scattering,* Cambridge University Press, Cambridge, 1991.

[7] E. Fermi, J. Pasta, and S. Ulam, *Studies of non linear problems,* Document LA-1940, Los Alamos National Laboratory, May 1955.

[8] E. Fermi, *Collected papers, Vol. II: United States, 1939–1954,* University of Chicago Press, Chicago, 1965.

[9] http://www.osti.gov/accomplishments/pdf/A80037041/A80037041.pdf

[10] N.J. Zabusky and M.D. Kruskal, *Interaction of "solitons" in a collisionless plasma and the recurrence of initial states,* Phys. Rev. Lett. **15** (1965) 240–243.

[11] C.S. Gardner, J.M. Greene, M.D. Kruskal and R.M. Miura, *Method for solving the Korteweg-de Vries equation,* Phys. Rev. Lett. **19** (1967) 1095–1097.

[12] L.D. Faddeev, *Properties of the S-matrix of the one-dimensional Schrödinger equation,* Am. Math. Soc. Transl. (Ser. 2) **65** (1967) 139–166.

[13] V.A. Marchenko, *Sturm-Liouville operators and applications,* Birkhäuser, Basel, 1986.

[14] P. Deift and E. Trubowitz, *Inverse scattering on the line,* Comm. Pure Appl. Math. **32** (1979) 121–251.

[15] R.G. Newton, *The Marchenko and Gel'fand-Levitan methods in the inverse scattering problem in one and three dimensions,* In: J. B. Bednar et al. (eds), *Conference on inverse scattering: theory and application,* SIAM, Philadelphia, 1983, pp. 1–74.

[16] K. Chadan and P.C. Sabatier, *Inverse problems in quantum scattering theory,* 2nd ed., Springer, New York, 1989.

[17] A. Melin, *Operator methods for inverse scattering on the real line,* Comm. Partial Differential Equations **10** (1985) 677–766.

[18] T. Aktosun and M. Klaus, *Small-energy asymptotics for the Schrödinger equation on the line,* Inverse Problems **17** (2001) 619–632.

[19] T. Aktosun and M. Klaus, *Chapter 2.2.4, Inverse theory: problem on the line,* In: E. R. Pike and P. C. Sabatier (eds), *Scattering,* Academic Press, London, 2001, pp. 770–785.

[20] R.G. Newton, *Inverse scattering. I. One dimension,* J. Math. Phys. **21** (1980) 493–505.

[21] P.D. Lax, *Integrals of nonlinear equations of evolution and solitary waves,* Comm. Pure Appl. Math. **21** (1968) 467–490.

[22] V.E. Zakharov and A.B. Shabat, *Exact theory of two-dimensional self-focusing and one-dimensional self-modulation of waves in nonlinear media,* Soviet Phys. JETP **34** (1972) 62–69.

[23] M.J. Ablowitz, D.J. Kaup, A.C. Newell, and H. Segur, *Method for solving the sine-Gordon equation,* Phys. Rev. Lett. **30** (1973) 1262–1264.

[24] http://www.msstate.edu/~aktosun/kdv

[25] G.B. Whitham, *Linear and nonlinear waves,* Wiley, New York, 1974.

[26] R.M. Miura, *Korteweg-de Vries equation and generalization. I. A remarkable explicit nonlinear transformation,* J. Math. Phys. **9** (1968) 1202–1204.

[27] H.D. Wahlquist and F.B. Estabrook, *Bäcklund transformation for solutions of the Korteweg-de Vries equation,* Phys. Rev. Lett. **31** (1973) 1386–1390.

[28] H.D. Wahlquist and F.B. Estabrook, *Prolongation structures and nonlinear evolution equations,* J. Math. Phys. **16** (1975) 1–7.

Tuncay Aktosun
Department of Mathematics and Statistics
Mississippi State University
Mississippi State, MS 39762, USA
e-mail: aktosun@math.msstate.edu

Operator Theory:
Advances and Applications, Vol. 149, 23–45

The Schur Algorithm for Generalized Schur Functions IV: Unitary Realizations

D. Alpay, T.Ya. Azizov, A. Dijksma, H. Langer, and G. Wanjala

Abstract. The generalized Schur transform as defined in [13] (see also [2]–[6]) is applied to the class $\mathbf{A}^0$ of all complex-valued functions, which are holomorphic at $z = 0$. Each such function has a coisometric and a unitary realization in some Krein space. We study the effect of this generalized Schur transform to the unitary realization; in [2], [3] we studied similar questions for the coisometric realizations. The main difference with [2], [3] is that a certain one-sidedness is replaced by a two-sidedness, comparable to the difference between the unilateral shift on one-sided sequences and the shift on two-sided sequences. We follow a direct approach in line with [2, 3, 6].

1. Introduction

1. Realizations. We denote by $\mathbf{A}^0$ the class of functions which are holomorphic in a neighborhood of 0 in the open unit disk $\mathbb{D}$. Each function $s(z) \in \mathbf{A}^0$ admits a *unitary realization*, that is, it can be written in the form

$$s(z) = \gamma + z\langle (I_{\mathcal{K}} - zT)^{-1}u, v\rangle_{\mathcal{K}}. \tag{1.1}$$

Here $\mathcal{K}$ is a Krein space with indefinite inner product $\langle\,\cdot\,,\,\cdot\,\rangle_{\mathcal{K}}$, $I_{\mathcal{K}}$ is the identity operator and T is a bounded operator on $\mathcal{K}$, u and v are elements in $\mathcal{K}$, and γ is a complex number: $\gamma = s(0)$, such that the matrix operator

$$U := \begin{pmatrix} T & u \\ \langle\,\cdot\,, v\rangle_{\mathcal{K}} & \gamma \end{pmatrix} : \begin{pmatrix} \mathcal{K} \\ \mathbb{C} \end{pmatrix} \to \begin{pmatrix} \mathcal{K} \\ \mathbb{C} \end{pmatrix}, \quad \begin{pmatrix} \mathcal{K} \\ \mathbb{C} \end{pmatrix} = \mathcal{K} \oplus \mathbb{C}, \tag{1.2}$$

is unitary: $U^* = U^{-1}$. We recall the terminology: The matrix U is identified with the *colligation* $\{\mathcal{K}, \mathbb{C}, T, u, v, \gamma\}$, $\mathcal{K}$ is called the *state space*, T is the *main operator* and the function on the right-hand side of (1.1) is called the *characteristic function* of the colligation. Evidently, the characteristic function of a unitary colligation belongs to the class $\mathbf{A}^0$.

Received by the editors February 12, 2003; revised June 2, 2003. Submitted by J. A. Ball.
Mathematics Subject Classification (2000). Primary 47A48, 47A57, 47B32, 47B50.
Key words and phrases. Generalized Schur function, minimal unitary colligation, unitary realization, Pontryagin spaces, generalized Schur transform.
The research for this paper was supported by the Netherlands Organization of Scientific Research NWO (grant 047-008-008), by the Russian Foundation for Basic Research RFBR (grant 02-01-00353), and by the Research Training Network HPRN-CT-2000-00116 of the European Union.

That the colligation is unitary can be expressed by the relations

$$
\begin{aligned}
TT^* + \langle\,\cdot\,, u\rangle_{\mathcal{K}} u &= I_{\mathcal{K}}, &(1.3)\\
\langle v, v\rangle_{\mathcal{K}} + |\gamma|^2 &= 1, &(1.4)\\
Tv + \gamma^* u &= 0, &(1.5)\\
T^*T + \langle\,\cdot\,, v\rangle_{\mathcal{K}} v &= I_{\mathcal{K}}, &(1.6)\\
\langle u, u\rangle_{\mathcal{K}} + |\gamma|^2 &= 1, &(1.7)\\
T^* u + \gamma v &= 0. &(1.8)
\end{aligned}
$$

The unitary colligation (1.2) can be chosen *minimal* which means that

$$\overline{\text{span}}\,\{T^n u,\, T^{*m} v \mid n, m = 0, 1, 2, \ldots\} = \mathcal{K}.$$

If this holds then (1.1) is called a *minimal realization* of $s(z)$. A minimal unitary realization of a function $s(z) \in \mathbf{A}^0$ is essentially unique: If besides (1.1) also

$$s(z) = \gamma + z\langle (I_{\widetilde{\mathcal{K}}} - z\widetilde{T})^{-1}\widetilde{u}, \widetilde{v}\rangle_{\widetilde{\mathcal{K}}}$$

is a minimal unitary realization of $s(z)$, then there is a *weak isomorphism* W from $\mathcal{K}$ to $\widetilde{\mathcal{K}}$ (by definition, an operator mapping a dense linear subset of $\mathcal{K}$ isometrically onto a dense linear subset of $\widetilde{\mathcal{K}}$) such that

$$WT\,|_{\operatorname{dom} W} = \widetilde{T}W, \quad Wu = \widetilde{u}, \quad Wv = \widetilde{v}.$$

In this case we have that if $\mathcal{K} = \mathcal{K}_+ \oplus \mathcal{K}_-$ and $\widetilde{\mathcal{K}} = \widetilde{\mathcal{K}}_+ \oplus \widetilde{\mathcal{K}}_-$ are any fundamental decompositions of $\mathcal{K}$ and $\widetilde{\mathcal{K}}$ respectively, then $\dim \mathcal{K}_\pm = \dim \widetilde{\mathcal{K}}_\pm$, in the sense that either both numbers are infinite or both are finite and equal. If at least one of these numbers is finite then W is bounded and can be extended by continuity to a unitary mapping from $\mathcal{K}$ onto $\widetilde{\mathcal{K}}$ and $WT = \widetilde{T}W$. These facts follow from, for example, [10, Section 5.3] and [14, Theorem 2.2]. We assume the reader is familiar with Krein and Pontryagin spaces and the theory of operators on these spaces such as in [10], [17], and [15].

2. Generalized Schur functions. Of the many equivalent definitions of a generalized Schur function from the class $\mathbf{S}^0$ (see [18] and also [7]) we select the one directly related to the preceeding subsection: $s(z)$ belongs to the class $\mathbf{S}^0$ if $s(z) \in \mathbf{A}^0$ and the state space in the minimal unitary realization of $s(z)$ is a Pontryagin space, that is, a Krein space $\mathcal{K}$ for which, in the notation above, $\dim \mathcal{K}_- < \infty$. The subclass of functions $s(z) \in \mathbf{S}^0$ for which $\dim \mathcal{K}_- = \kappa < \infty$, will be denoted by $\mathbf{S}^0_\kappa$. If $\kappa = 0$, the state space is a Hilbert space and $s(z)$ belongs to $\mathbf{S}^0_0$ if and only if it can be extended to a Schur function, that is, a holomorphic function defined on all of $\mathbb{D}$ which is bounded by 1; we write $\mathbf{S}_0$ for $\mathbf{S}^0_0$. Finally, we mention that $s(z) \in \mathbf{S}^0_\kappa$ if and only if it admits the representation

$$s(z) = \prod_{j=1}^{\kappa} \frac{1 - z z_j^*}{z - z_j}\, s_0(z)$$

with points $z_j \in \mathbb{D}$, $z_j \neq 0$, and a function $s_0(z) \in \mathbf{S}_0$, $s_0(z_j) \neq 0$, $j = 1, 2, \ldots, \kappa$.

3. The generalized Schur transform. In the sequel we write the Taylor expansion of $s(z) \in \mathbf{A}^0$ at $z = 0$ as

$$s(z) = \sigma_0 + \sigma_1 z + \sigma_2 z^2 + \cdots + \sigma_n z^n + \cdots, \quad \sigma_n = s^{(n)}(0)/n!, \ n = 0, 1, \ldots. \tag{1.9}$$

The *generalized Schur transform* $\widetilde{s}(z)$ of a function $s(z)$ is defined via a fractional linear transformation which depends on whether $|\sigma_0| < 1$, $|\sigma_0| > 1$, or $|\sigma_0| = 1$:

$$\widetilde{s}(z) = \begin{cases} \dfrac{1}{z}\dfrac{s(z) - \sigma_0}{1 - \sigma_0^* s(z)} & \text{if } |\sigma_0| < 1, \\ z^k \dfrac{1 - \sigma_0^* s(z)}{s(z) - \sigma_0} & \text{if } |\sigma_0| > 1, \\ z^q \dfrac{(Q(z) - z^k)s(z) - \sigma_0 Q(z)}{\sigma_0^* Q(z)s(z) - (Q(z) + z^k)} & \text{if } |\sigma_0| = 1 \end{cases} \tag{1.10}$$

and this equality holds for z in a neighborhood of 0 in $\mathbb{D}$. The Schur transform is defined for those functions $s(z) \in \mathbf{A}^0$ which satisfy

(i) $s(z) \not\equiv \sigma_0$ with $|\sigma_0| \geq 1$, and

(ii) $\sigma_0^* Q(z)s(z) - (Q(z) + z^k) \not\equiv 0$ if $|\sigma_0| = 1$.

In this definition k is the smallest integer ≥ 1 for which $\sigma_k \neq 0$. The condition (i) implies that such a k exists. The polynomial $Q(z)$ of degree $2k$ and the integer $q \geq 0$ in case $|\sigma_0| = 1$ are defined as follows. Introduce the complex numbers c_n, $n = 0, 1, \ldots$, by the relation

$$(s(z) - \sigma_0)(c_0 + c_1 z + \cdots + c_n z^n + \cdots) \equiv \sigma_0 z^k,$$

then $Q(z)$ is defined by

$$Q(z) = P(z) - z^{2k} P(1/z^*)^*, \quad P(z) := c_0 + c_1 z + \cdots + c_{k-1} z^{k-1}.$$

It is easy to see that $(Q(z) - z^k)s(z) - \sigma_0 Q(z)$ has a zero at $z = 0$ of order $2k$ and that $\sigma_0^* Q(z)s(z) - (Q(z) + z^k)$ has a zero at $z = 0$ of order at least $2k$. On account of condition (ii) the quotient of these functions has a pole at $z = 0$ of finite order. In the definition of the generalized Schur transform (1.10) this order is denoted by q.

It can be shown (see [16], [11], and also [2]) that a generalized Schur function which is not identically equal to a unimodular constant satisfies (i) and (ii) and that the generalized Schur transform of such a function is a generalized Schur function.

The Schur transform goes back to [19] and [20] in connection with the study of interpolation and moment problems. The generalized Schur transform was first developed by the authors of [12] and [16] in order to generate certain Pisot numbers. Later the transform was perfected to the present definition; for details see [11]. In [13] it was studied in a system theoretic context. The authors of [8] and [9] also investigated the Schur transform in the nonpositive case. They considered matrix-valued functions (in this connection see also [1]) and used reproducing kernel methods. They "avoided" the case where $|\sigma_0| = 1$ by applying the Schur transform in points of $\mathbb{D}$ different from the origin.

4. Contents of the paper. In this paper we investigate how to obtain the minimal unitary colligation of the Schur transform $\widetilde{s}(z)$ of $s(z)$ from the minimal unitary colligation of $s(z)$. In Section 2 we formulate the theorems: Theorem 2.1 for the case $|\sigma_0| < 1$, Theorem 2.3 for the case $|\sigma_0| > 1$, and Theorems 2.4 and 2.5 for the case $|\sigma_0| = 1$, the first one when $q = 0$ and the second one when $q \neq 0$. The final Theorem 2.6 concerns the minimal unitary realization of the function which is obtained by applying the generalized Schur transform finitely many times to $s(z)$. In Section 3 we provide proofs, for Theorems 2.1 and 2.3 rather briefly, for the Theorems 2.4 and 2.5 in detail.

The results are similar to the results in [2] and [3] where we studied the effect of the Schur transform on the *minimal coisometric realizations*: $s(z)$ has the representation (1.1) but now the colligation U in (1.2) is (i) *coisometric*, that is, $UU^* = I_{\mathcal{K}}$ or, equivalently, the relations (1.3), (1.4), and (1.5) hold, and (ii) *minimal*, that is,

$$\overline{\operatorname{span}}\,\{T^{*m}v \mid m = 0, 1, 2, \ldots\} = \mathcal{K}.$$

In [2], [3] a model for the main operator in the minimal *coisometric* colligation was obtained by considering the sequence

$$v,\ T^*v,\ T^{*2}v, \ldots, T^{*m}v, \ldots$$

as a basis for $\mathcal{K}$. The elements $\sum \alpha_n T^{*n}v$ in the span of this basis were identified with the one-sided sequences $(\alpha_n)_{n=0}^{\infty}$ with finitely many nonzero terms; for example, v was identified with $(1, 0, 0, \ldots)$. Then T^* is just the right shift on the sequence space. The inner product on $\mathcal{K}$ was expressed in terms of the inner product on the space ℓ^2:

$$\left\langle \sum \alpha_n T^{*n}v, \sum \beta_n T^{*n}v \right\rangle_{\mathcal{K}} = \left(\mathcal{G}(\alpha_n)_0^{\infty}, (\beta_n)_0^{\infty}\right)_{\ell^2},$$

where the Gram matrix is given by $\mathcal{G} = \mathcal{I} - \mathcal{F}\mathcal{F}^*$ with indentity matrix $\mathcal{I}$ and

$$\mathcal{F} = \begin{pmatrix} \sigma_0 & 0 & 0 & \cdots \\ \sigma_1 & \sigma_0 & 0 & \cdots \\ \sigma_2 & \sigma_1 & \sigma_0 & \cdots \\ \vdots & \vdots & \vdots & \ddots \end{pmatrix}.$$

In this note as a basis for the state space of the minimal *unitary* colligation U we choose the two sided sequence

$$\ldots, T^{*k}v, \ldots, T^{*2}v, T^*v, v \mid u, Tu, T^2u, \ldots, T^{\ell}u, \ldots.$$

The elements in its span can be identified with two-sided sequences with finitely many nonzero terms. For example,

$$u = (\cdots, 0, 0 \mid 1, 0 \cdots), \quad v = (\cdots, 0, 1 \mid 0, 0, \cdots).$$

The model for the main operator T is the matrix

$$\left(\begin{array}{cccc|cccc}
\ddots & & & & & & & \\
 & 1 & 0 & 0 & 0 & 0 & 0 & \\
\cdots & 0 & 1 & 0 & 0 & 0 & 0 & \cdots \\ \hline
\cdots & -\sigma_2^* & -\sigma_1^* & -\sigma_0^* & 0 & 0 & 0 & \cdots \\
 & 0 & 0 & 0 & 1 & 0 & 0 & \\
 & 0 & 0 & 0 & 0 & 1 & 0 & \\
 & 0 & 0 & 0 & 0 & 0 & 1 & \\
 & & & & & & & \ddots
\end{array}\right),$$

where $\sigma_0 = s(0)$ and $\sigma_{j+1} = \langle T^j u, v\rangle_{\mathcal{K}}$ is the coefficient of z^{j+1} in the Taylor series expansion of $s(z)$, $j = 0, 1, 2, \ldots$. By the relations (1.3)–(1.8) (see Lemma 3.1 in Section 3), the Gram matrix corresponding to this two-sided sequence is given by

$$\mathcal{G} = \left(\begin{array}{c|c} \mathcal{I} - \mathcal{A}\mathcal{A}^* & \mathcal{B} \\ \hline \mathcal{C} & \mathcal{I} - \mathcal{D}^*\mathcal{D} \end{array}\right)$$

with the infinite Hankel matrices

$$\mathcal{A} = \begin{pmatrix} \ddots & \vdots & \vdots & \vdots \\ \cdots & \sigma_0 & \sigma_1 & \sigma_2 \\ \cdots & 0 & \sigma_0 & \sigma_1 \\ \cdots & 0 & 0 & \sigma_0 \end{pmatrix}, \quad \mathcal{B} = \begin{pmatrix} \vdots & \vdots & \vdots & \\ \sigma_3 & \sigma_4 & \sigma_5 & \cdots \\ \sigma_2 & \sigma_3 & \sigma_4 & \cdots \\ \sigma_1 & \sigma_2 & \sigma_3 & \cdots \end{pmatrix},$$

$$\mathcal{C} = \begin{pmatrix} \cdots & \sigma_3^* & \sigma_2^* & \sigma_1^* \\ \cdots & \sigma_4* & \sigma_3^* & \sigma_2^* \\ \cdots & \sigma_5^* & \sigma_4^* & \sigma_3^* \\ & \vdots & \vdots & \vdots \end{pmatrix}, \quad \mathcal{D} = \begin{pmatrix} \sigma_0 & \sigma_1 & \sigma_2 & \cdots \\ 0 & \sigma_0 & \sigma_1 & \cdots \\ 0 & 0 & \sigma_0 & \cdots \\ \vdots & \vdots & \vdots & \ddots \end{pmatrix}.$$

Thus, unlike in the coisometric model, in the unitary model the roles of the elements u and v are similar. This is reflected in the theorems that follow.

2. The theorems

In the following theorems we consider a function $s(z) \in \mathbf{A}^0$ with Taylor expansion (1.9):

$$s(z) = \sigma_0 + \sigma_1 z + \sigma_2 z^2 + \cdots + \sigma_n z^n + \cdots, \quad \sigma_n = s^{(n)}(0)/n!,\ n = 0, 1, \ldots,$$

and assume

$$U = \begin{pmatrix} T & u \\ \langle\,\cdot\,, v\rangle_{\mathcal{K}} & \gamma \end{pmatrix} : \begin{pmatrix} \mathcal{K} \\ \mathbb{C} \end{pmatrix} \to \begin{pmatrix} \mathcal{K} \\ \mathbb{C} \end{pmatrix}, \tag{2.1}$$

is the minimal unitary colligation, whose characteristic function coincides with $s(z)$, that is, in a neighborhood of $z = 0$ in $\mathbb{D}$ we have

$$s(z) = \gamma + z\langle (I_{\mathcal{K}} - zT)^{-1} u, v\rangle_{\mathcal{K}}, \quad \gamma = s(0) = \sigma_0.$$

Here the state space $\mathcal{K}$ of U is a Krein space.

1. Case $|\sigma_0| < 1$. In this case, let $k \in \{1, 2, \ldots\}$ and $\sigma_1 = \cdots = \sigma_{k-1} = 0$ if $k \geq 2$. Unlike in the next case we do not require that $\sigma_k \neq 0$. If we apply the Schur transform k times to $s(z)$ then we obtain the function

$$\widetilde{s}(z) = \frac{1}{z^k} \frac{s(z) - \sigma_0}{1 - \sigma_0^* s(z)}. \tag{2.2}$$

We express the colligation of this function in terms of that of $s(z)$.

Theorem 2.1. *Assume* $s(z) \in \mathbf{A}^0$ *and* $|\sigma_0| < 1$. *Let* k *be an integer* ≥ 1 *such that* $\sigma_1 = \ldots = \sigma_{k-1} = 0$ *if* $k \geq 2$. *Let* U *in* (2.1) *be the minimal unitary colligation whose characteristic function coincides with* $s(z)$. *Then*

$$\widetilde{\mathcal{L}} = \operatorname{span}\{v, T^*v, \ldots, T^{*(k-1)}v\}, \quad \widehat{\mathcal{L}} = \operatorname{span}\{u, Tu, \ldots, T^{(k-1)}u\},$$

are k*-dimensional positive subspaces of* $\mathcal{K}$ *and* $\widetilde{s}(z)$ *defined by* (2.2) *is the characteristic function of the minimal unitary colligations* $\widetilde{U}$ *and* $\widehat{U}$:

$$\widetilde{U} = \begin{pmatrix} \widetilde{T} & \widetilde{u} \\ \langle \cdot, \widetilde{v} \rangle_{\widetilde{\mathcal{K}}} & \widetilde{\gamma} \end{pmatrix} : \begin{pmatrix} \widetilde{\mathcal{K}} \\ \mathbb{C} \end{pmatrix} \to \begin{pmatrix} \widetilde{\mathcal{K}} \\ \mathbb{C} \end{pmatrix}$$

with

$$\begin{aligned} \widetilde{T} &= PTP, & \widetilde{u} &= \frac{1}{\sqrt{1 - |\sigma_0|^2}} Pu, \\ \widetilde{v} &= \frac{1}{\sqrt{1 - |\sigma_0|^2}} PT^{*k}v, \quad & \widetilde{\gamma} &= \frac{\sigma_k}{1 - |\sigma_0|^2}, \end{aligned}$$

where P *is the orthogonal projection in* $\mathcal{K}$ *onto the subspace* $\widetilde{\mathcal{K}} := \mathcal{K} \ominus \widetilde{\mathcal{L}}$, *and*

$$\widehat{U} = \begin{pmatrix} \widehat{T} & \widehat{u} \\ \langle \cdot, \widehat{v} \rangle_{\widehat{\mathcal{K}}} & \widehat{\gamma} \end{pmatrix} : \begin{pmatrix} \widehat{\mathcal{K}} \\ \mathbb{C} \end{pmatrix} \to \begin{pmatrix} \widehat{\mathcal{K}} \\ \mathbb{C} \end{pmatrix}$$

with

$$\begin{aligned} \widehat{T} &= QTQ, & \widehat{u} &= \frac{1}{\sqrt{1 - |\sigma_0|^2}} QT^k u, \\ \widehat{v} &= \frac{1}{\sqrt{1 - |\sigma_0|^2}} Qv, \quad & \widehat{\gamma} &= \frac{\sigma_k}{1 - |\sigma_0|^2}, \end{aligned}$$

where Q *is the orthogonal projection in* $\mathcal{K}$ *onto the subspace* $\widehat{\mathcal{K}} := \mathcal{K} \ominus \widehat{\mathcal{L}}$.

Remark 2.2. The linear map W from

$$\operatorname{span}\{\widetilde{T}^n \widetilde{u}, \widetilde{T}^{*m} \widetilde{v} \mid n, m = 0, 1, 2, \ldots\} \subset \widetilde{\mathcal{K}}$$

onto

$$\operatorname{span}\{\widehat{T}^n \widehat{u}, \widehat{T}^{*m} \widehat{v} \mid n, m = 0, 1, 2, \ldots\} \subset \widehat{\mathcal{K}}$$

defined by $W\widetilde{T}^n \widetilde{u} = \widehat{T}^n \widehat{u}$, $W\widetilde{T}^{*m} \widetilde{v} = \widehat{T}^{*m} \widehat{v}$, $n, m = 0, 1, 2, \ldots$, is a weak isomorphism from $\widetilde{\mathcal{K}}$ to $\widehat{\mathcal{K}}$ such that

$$W\widetilde{T}\,|_{\operatorname{dom} W} = \widehat{T}W, \quad W\widetilde{u} = \widehat{u}, \quad W\widetilde{v} = \widehat{v}.$$

This remark also applies to Theorems 2.3 and 2.5 below. A proof is given at the end of Section 3.

The theorem implies that the state space $\mathcal{K}$ loses k positive squares. Thus if $s(z)$ belongs to $\mathbf{S}_\kappa^0$ and satisfies the conditions of the theorem, then $\widetilde{s}(z)$ defined by (2.2) also belongs to $\mathbf{S}_\kappa^0$.

2. Case $|\sigma_0| > 1$. We assume $s(z)$ is not identically equal to σ_0. Then (and only then) there is a smallest integer $k \geq 1$ such that $\sigma_k \neq 0$. Thus $k \in \{1, 2, \ldots\}$ and (a) $\sigma_1 \neq 0$ if $k = 1$ and (b) $\sigma_1 = \ldots \sigma_{k-1} = 0$ and $\sigma_k \neq 0$ if $k \geq 2$. The Schur transform of $s(z)$ is given by

$$\widetilde{s}(z) = z^k \frac{1 - \sigma_0^* s(z)}{s(z) - \sigma_0}. \tag{2.3}$$

We express the colligation of $\widetilde{s}(z)$ in terms of that of $s(z)$.

Theorem 2.3. *Assume $s(z) \in \mathbf{A}^0$, $|\sigma_0| > 1$, and $s(z) \not\equiv \sigma_0$. Let k be the smallest integer ≥ 1 such that $\sigma_k \neq 0$. Let U in (2.1) be the minimal unitary colligation whose characteristic function coincides with $s(z)$. Then*

$$\widetilde{\mathcal{L}} = \operatorname{span}\{v, T^*v, \ldots, T^{*(k-1)}v\}, \quad \widehat{\mathcal{L}} = \operatorname{span}\{u, Tu, \ldots, T^{(k-1)}u\},$$

are k-dimensional negative subspaces of $\mathcal{K}$ and $\widetilde{s}(z)$ defined by (2.3) is the characteristic function of the minimal unitary colligations $\widetilde{U}$ and $\widehat{U}$:

$$\widetilde{U} = \begin{pmatrix} \widetilde{T} & \widetilde{u} \\ \langle \cdot, \widetilde{v} \rangle_{\widetilde{\mathcal{K}}} & \widetilde{\gamma} \end{pmatrix} : \begin{pmatrix} \widetilde{\mathcal{K}} \\ \mathbb{C} \end{pmatrix} \to \begin{pmatrix} \widetilde{\mathcal{K}} \\ \mathbb{C} \end{pmatrix}$$

with

$$\begin{aligned} \widetilde{T} &= PTP - \frac{\langle \cdot, PT^{*k}v \rangle_{\widetilde{\mathcal{K}}}}{\sigma_k} Pu, & \widetilde{u} &= \frac{\sqrt{|\sigma_0|^2 - 1}}{\sigma_k} Pu, \\ \widetilde{v} &= \frac{\sqrt{|\sigma_0|^2 - 1}}{\sigma_k^*} PT^{*k}v, & \widetilde{\gamma} &= \frac{1 - |\sigma_0|^2}{\sigma_k}, \end{aligned}$$

where P is the orthogonal projection in $\mathcal{K}$ onto the subspace $\widetilde{\mathcal{K}} := \mathcal{K} \ominus \widetilde{\mathcal{L}}$, and

$$\widehat{U} = \begin{pmatrix} \widehat{T} & \widehat{u} \\ \langle \cdot, \widehat{v} \rangle_{\widehat{\mathcal{K}}} & \widehat{\gamma} \end{pmatrix} : \begin{pmatrix} \widehat{\mathcal{K}} \\ \mathbb{C} \end{pmatrix} \to \begin{pmatrix} \widehat{\mathcal{K}} \\ \mathbb{C} \end{pmatrix}$$

with

$$\begin{aligned} \widehat{T} &= QTQ - \frac{\langle \cdot, Qv \rangle_{\widehat{\mathcal{K}}}}{\sigma_k} QT^k u, & \widehat{u} &= \frac{\sqrt{|\sigma_0|^2 - 1}}{\sigma_k} QT^k u, \\ \widehat{v} &= \frac{\sqrt{|\sigma_0|^2 - 1}}{\sigma_k^*} Qv, & \widehat{\gamma} &= \frac{1 - |\sigma_0|^2}{\sigma_k}, \end{aligned}$$

where Q is the orthogonal projection in $\mathcal{K}$ onto the subspace $\widehat{\mathcal{K}} := \mathcal{K} \ominus \widehat{\mathcal{L}}$.

The theorem implies that the state space $\mathcal{K}$ loses k negative squares. Thus if $s(z)$ belongs to $\mathbf{S}_\kappa^0$ and satisfies the conditions of the theorem, then $1 \leq k \leq \kappa$ and $\widetilde{s}(z)$ defined by (2.3) belongs to $\mathbf{S}_{\kappa-k}^0$.

3. Case $|\sigma_0| = 1$. As in the previous case we assume that $s(z)$ is not identically equal to σ_0 and denote by k the smallest positive integer such that $\sigma_k \neq 0$. Let $c_0, c_1, \dots, c_n, \dots$ be the complex numbers satisfying

$$(s(z) - \sigma_0)(c_0 + c_1 z + \cdots + c_n z^n + \cdots) \equiv \sigma_0 z^k, \tag{2.4}$$

and set

$$Q(z) := P(z) - z^{2k} P(1/z^*)^*, \quad P(z) := c_0 + c_1 z + \cdots + c_{k-1} z^{k-1}. \tag{2.5}$$

Note that $c_0 \neq 0$, so that $P(0) \neq 0$ and $Q(z)$ is a polynomial of degree $2k$. We also assume that $s(z)$ is such that $\sigma_0^* Q(z)s(z) - (Q(z) + z^k)$ is not identically equal to 0. Then the Schur transform of $s(z)$ is defined by

$$\widetilde{s}(z) = z^q \frac{(Q(z) - z^k)s(z) - \sigma_0 Q(z)}{\sigma_0^* Q(z)s(z) - (Q(z) + z^k)}, \tag{2.6}$$

where q is the order of the pole of the quotient on the right-hand side; hence q is a nonnegative integer. At $z = 0$ the numerator has a zero of order $2k$ and the denominator has a zero of order $2k + q$. We write the Taylor expansion at $z = 0$ of the latter as

$$\sigma_0^* Q(z)s(z) - (Q(z) + z^k) = t_{2k+q} z^{2k+q} + \mathcal{O}(z^{2k+q+1}), \quad t_{2k+q} \neq 0. \tag{2.7}$$

In the following two theorems we express the colligation of $\widetilde{s}(z)$ in terms of that of $s(z)$. The first theorem concerns the case $q = 0$, which means that

$$t_{2k} = \frac{1}{\sigma_0}(c_0 \sigma_{2k} + c_1 \sigma_{2k-1} + \cdots + c_{k-1} \sigma_{k+1}) \neq 0. \tag{2.8}$$

The second theorem deals with the case $q > 0$.

Theorem 2.4. *Assume $s(z) \in \mathbf{A}^0$, $|\sigma_0| = 1$, and $s(z) \not\equiv \sigma_0$. Let k be the smallest integer ≥ 1 such that $\sigma_k \neq 0$ and assume (2.8) holds. Let U in (2.1) be the minimal unitary colligation whose characteristic function coincides with $s(z)$. Then*

$$\mathcal{L}_0 = \operatorname{span}\{T^{*(k-1)}v, \dots, T^* v, v, u, Tu, \dots T^{k-1}u\}$$

is a $(2k)$-dimensional subspace with k positive and k negative squares, and $\widetilde{s}(z)$ defined by (2.6) with $q = 0$ is the characteristic function of the minimal unitary colligation

$$\widetilde{U} = \begin{pmatrix} \widetilde{T} & \widetilde{u} \\ \langle \cdot, \widetilde{v} \rangle_{\widetilde{\mathcal{K}}} & \widetilde{\gamma} \end{pmatrix} : \begin{pmatrix} \widetilde{\mathcal{K}} \\ \mathbb{C} \end{pmatrix} \to \begin{pmatrix} \widetilde{\mathcal{K}} \\ \mathbb{C} \end{pmatrix}$$

with

$$\widetilde{T} = PTP - \frac{\langle \cdot, PT^{*k}v \rangle_{\mathcal{K}}}{\sigma_k t_{2k}} PT^k u, \quad \widetilde{u} = \frac{1}{t_{2k}} PT^k u,$$

$$\widetilde{v} = \frac{1}{t_{2k}^*} PT^{*k} v, \quad \widetilde{\gamma} = \sigma_0 - \frac{\sigma_k}{t_{2k}},$$

where P is the orthogonal projection in $\mathcal{K}$ onto the subspace $\widetilde{\mathcal{K}} := \mathcal{K} \ominus \mathcal{L}_0$.

Theorem 2.5. *Assume $s(z) \in \mathbf{A}^0$, $|\sigma_0| = 1$, and $s(z) \not\equiv \sigma_0$. Let k be the smallest integer ≥ 1 such that $\sigma_k \neq 0$ and assume that for some integer $q > 0$*

$$\sigma_0^* Q(z)s(z) - (Q(z) + z^k) = t_{2k+q}z^{2k+q} + t_{2k+q+1}z^{2k+q+1} + \cdots, \quad t_{2k+q} \neq 0.$$

Let U in (2.1) be the minimal unitary colligation whose characteristic function coincides with $s(z)$. Then

$$\widetilde{\mathcal{L}}_q = \operatorname{span}\{T^{*(k+q-1)}v, \ldots, T^*v, v, u, Tu, \ldots T^{k-1}u\}$$

and

$$\widehat{\mathcal{L}}_q = \operatorname{span}\{T^{*(k-1)}v, \ldots, T^*v, v, u, Tu, \ldots T^{k+q-1}u\}$$

are $(2k+q)$-dimensional subspaces of $\mathcal{K}$ with k positive and $k+q$ negative squares, and $\widetilde{s}(z)$ defined by (2.6) is the characteristic function of the minimal unitary colligations $\widetilde{U}$ and $\widehat{U}$:

$$\widetilde{U} = \begin{pmatrix} \widetilde{T} & \widetilde{u} \\ \langle \cdot, \widetilde{v}\rangle_{\widetilde{\mathcal{K}}} & \widetilde{\gamma} \end{pmatrix} : \begin{pmatrix} \widetilde{\mathcal{K}} \\ \mathbb{C} \end{pmatrix} \to \begin{pmatrix} \widetilde{\mathcal{K}} \\ \mathbb{C} \end{pmatrix}$$

with

$$\begin{aligned} \widetilde{T} &= PTP - \frac{\langle \cdot, PT^{*(k+q)}v\rangle_{\widetilde{\mathcal{K}}}}{t_{2k+q}} PT^k u, & \widetilde{u} &= \frac{1}{t_{2k+q}} PT^k u, \\ \widetilde{v} &= \frac{1}{t_{2k+q}^*} PT^{*(k+q)}v, & \widetilde{\gamma} &= -\frac{\sigma_k}{t_{2k+q}}, \end{aligned}$$

where P is the orthogonal projection in $\mathcal{K}$ onto the subspace $\widetilde{\mathcal{K}} := \mathcal{K} \ominus \widetilde{\mathcal{L}}_q$, and

$$\widehat{U} = \begin{pmatrix} \widehat{T} & \widehat{u} \\ \langle \cdot, \widehat{v}\rangle_{\widehat{\mathcal{K}}} & \widehat{\gamma} \end{pmatrix} : \begin{pmatrix} \widehat{\mathcal{K}} \\ \mathbb{C} \end{pmatrix} \to \begin{pmatrix} \widehat{\mathcal{K}} \\ \mathbb{C} \end{pmatrix}$$

with

$$\begin{aligned} \widehat{T} &= QTQ - \frac{\langle \cdot, PT^{*k}v\rangle_{\widehat{\mathcal{K}}}}{t_{2k+q}} QT^{k+q} u, & \widehat{u} &= \frac{1}{t_{2k+q}} QT^{k+q} u, \\ \widehat{v} &= \frac{1}{t_{2k+q}^*} QT^{*k}v, & \widehat{\gamma} &= -\frac{\sigma_k}{t_{2k+q}}, \end{aligned}$$

where Q is the orthogonal projection in $\mathcal{K}$ onto the subspace $\widehat{\mathcal{K}} := \mathcal{K} \ominus \widehat{\mathcal{L}}_q$.

The two subspaces

$$\operatorname{span}\{T^{*(k-1)}v, \ldots, T^*v, v\}, \quad \operatorname{span}\{u, Tu, \ldots T^{k-1}u\}$$

of $\mathcal{L}_0$ in Theorem 2.4 and of $\widetilde{\mathcal{L}}_q$ and $\widehat{\mathcal{L}}_q$ in Theorem 2.5 are neutral and skewly linked; see Lemma 3.4 below.

The theorems imply that the state space $\mathcal{K}$ of U loses k positive and $k+q$ negative squares. Thus if $s(z)$ belongs to $\mathbf{S}_\kappa^0$ and the case $|\sigma_0| = 1$ applies then $1 \leq k + q \leq \kappa$ and the Schur transform $\widetilde{s}(z)$ defined by (2.6) belongs to $\mathbf{S}_{\kappa-k-q}^0$.

4. The generalized Schur algorithm. Assume $s(z) \in \mathbf{A}^0$ satisfies the conditions (see Subsection 1.3):

(i) $s(z) \not\equiv \sigma_0$ with $|\sigma_0| \geq 1$, and
(ii) $\sigma_0^* Q(z)s(z) - (Q(z) + z^k) \not\equiv 0$ if $|\sigma_0| = 1$.

Repeated application of the Schur transform to $s(z)$ leads to a sequence of functions:

$$s_0(z) := s(z),\ s_1(z) := \widetilde{s}_0(z), \dots, s_j(z) := \widetilde{s}_{j-1}(z), \dots,$$

This is called the *generalized Schur algorithm.* The algorithm stops with the function $s_j(z)$ when this function no longer satisfies the conditions (i) and (ii). If $s(z)$ belongs to the class $\mathbf{S}^0$ then this is the case when $s_j(z)$ is identically equal to a constant of modulus 1.

Theorem 2.6. *Assume $s(z) \in \mathbf{A}^0$ satisfies the conditions* (i) *and* (ii) *and let U in* (2.1) *be the minimal unitary colligation whose characteristic function coincides with $s(z)$. Let $\widetilde{s}(z)$ be one of the functions appearing in the Schur algorithm of $s(z)$. Then there exist integers $\ell, r \geq 0$ and complex numbers $a \neq 0, b \neq 0$, and c, which may be* 0, *such that*

$$\mathcal{L} = \operatorname{span}\{T^{*(\ell-1)}v, \dots, T^*v, v, u, Tu, \dots T^{r-1}u\}$$

is a Pontryagin subspace of $\mathcal{K}$, and $\widetilde{s}(z)$ is the characteristic function of the minimal unitary colligation

$$\widetilde{U} = \begin{pmatrix} \widetilde{T} & \widetilde{u} \\ \langle\,\cdot\,, \widetilde{v}\rangle_{\widetilde{\mathcal{K}}} & \widetilde{\gamma} \end{pmatrix} : \begin{pmatrix} \widetilde{\mathcal{K}} \\ \mathbb{C} \end{pmatrix} \to \begin{pmatrix} \widetilde{\mathcal{K}} \\ \mathbb{C} \end{pmatrix}$$

with

$$\begin{aligned} \widetilde{T} &= PTP - c\langle\,\cdot\,, PT^{*\ell}v\rangle_{\mathcal{K}} PT^r u, & \widetilde{u} &= aPT^r u, \\ \widetilde{v} &= bPT^{*\ell}v, & \widetilde{\gamma} &= \widetilde{s}(0), \end{aligned}$$

where P is the orthogonal projection in $\mathcal{K}$ onto the subspace $\widetilde{\mathcal{K}} := \mathcal{K} \ominus \mathcal{L}$.

The proof of Theorem 2.6 is by induction on the steps of the generalized Schur algorithm. According to Theorems 2.1, 2.3, 2.4, and 2.5, this theorem is valid if $\widetilde{s}(z)$ is the immediate successor of $s(z)$. Assume that $\widetilde{s}(z)$ is as in the theorem and that the Schur transform can be applied to it. Then by invoking the theorems just mentioned one can show that the Schur transform of $\widetilde{s}(z)$ has a realization similar to that of $\widetilde{s}(z)$. We omit the details.

3. Proofs of the theorems and the remark

In this section we give the proofs of the theorems in the previous section. Since a unitary colligation is also coisometric, in our proofs we shall frequently refer to the theorems and the calculations given in [2] in which the minimal coisometric versions of these theorems are treated. We shall give complete proofs of Theorems 2.4 and 2.5 as they have no analogs in the coisometric setting.

Consider $s(z) \in \mathbf{A}^0$ and assume that it has the minimal unitary realization (1.1) and the Taylor expansion (1.9) around $z = 0$:

$$s(z) = \gamma + z\langle (I_{\mathcal{K}} - zT)^{-1}u, v\rangle_{\mathcal{K}} = \sigma_0 + \sigma_1 z + \cdots + \sigma_j z^j + \cdots,$$

whence $\sigma_0 = \gamma$ and $\sigma_{j+1} = \langle T^j u, v\rangle_{\mathcal{K}}$, $j = 0, 1, 2, \ldots$. The relations (1.3)–(1.8) imply the following lemma.

Lemma 3.1. *For $i, j = 0, 1, 2, \ldots$ it holds that $\langle T^i u, T^{*j} v\rangle_{\mathcal{K}} = \sigma_{i+j+1}$ and*

$$\langle T^i u, T^j u\rangle_{\mathcal{K}} = \langle T^{*j} v, T^{*i} v\rangle_{\mathcal{K}} = \\ = \begin{cases} (1 - |\sigma_0|^2 - |\sigma_1|^2 - \ldots - |\sigma_i|^2), & i = j \geq 0, \\ -(\sigma_0^* \sigma_{i-j} + \sigma_1^* \sigma_{i-j+1} + \ldots + \sigma_j^* \sigma_i), & i > j \geq 0. \end{cases}$$

For the last equality in the lemma see [2, Formula (4.7)]). From this lemma one can derive the formulas for the Gram matrices in Subsection 1.4.

1. Proofs of Theorems 2.1 and 2.3. In the proof of Theorem 2.1 we use the following corollary of Lemma 3.1.

Lemma 3.2. *If $\sigma_1 = \ldots = \sigma_{k-1} = 0$ then for $k \geq 1$, the Gram matrices associated with the finite sequences*

$$v, T^* v, \ldots, T^{*(k-1)} v \quad \text{and} \quad u, Tu, \ldots, T^{k-1} u$$

in $\mathcal{K}$ coincide and are equal to $(1 - |\sigma_0|^2) I_k$, where I_k is the $k \times k$ identity matrix, and $\langle T^{*k} v, T^{*k} v\rangle_{\mathcal{K}} = \langle T^k u, T^k u\rangle_{\mathcal{K}} = 1 - |\sigma_0|^2 - |\sigma_k|^2$.

Proof of Theorem 2.1. The statements concerning the colligation $\widehat{U}$ follow from those concerning the colligation $\widetilde{U}$, by applying the latter to the function $s^*(z) := s(z^*)^*$: $s^*(z)$ is the characteristic function of the minimal unitary colligation U^* and the Schur transform of $s^*(z)$ is given by $(\widetilde{s})^*(z)$. (Note that the roles of u and v are interchanged.)

As to the first part of the theorem we first observe that by Lemma 3.2, $\mathcal{L}$ is a k-dimensional positive subspace. That $\widetilde{U}$ is coisometric and that its characteristic function coincides with $\widetilde{s}(z)$ follows from [2, Theorem 5.1] for the case $k = 1$ and [2, Corollary 5.2] for the case $k > 1$, proved by induction on k. The relations (1.3) and (1.5) for $\widetilde{U}$ are proved explicitly there. The remaining relations can be shown in a similar way. We show this for (1.7): Using the same relation but then for U we obtain

$$\begin{aligned} \langle \widetilde{u}, \widetilde{u}\rangle_{\widetilde{\mathcal{K}}} + |\widetilde{\gamma}|^2 &= \frac{1}{1 - |\sigma_0|2} \langle Pu, u\rangle_{\mathcal{K}} + |\widetilde{\gamma}|^2 \\ &= \frac{1}{1 - |\sigma_0|^2} \left[\langle u, u\rangle_{\mathcal{K}} - \frac{|\langle v, u\rangle_{\mathcal{K}}|^2}{\langle v, v\rangle_{\mathcal{K}}} \right] + |\widetilde{\gamma}|^2 \\ &= \frac{1}{1 - |\sigma_0|^2} \left[1 - |\sigma_0|2 - \frac{|\sigma_1|^2}{1 - |\sigma_0|^2} \right] + |\widetilde{\gamma}|^2 = 1. \end{aligned}$$

Thus $\widetilde{U}$ is a unitary colligation. It remains to show its minimality. For that we first note that there are constants α_{mj} and β_{mj} such that for $m = 0, 1, 2, \ldots$,

$$(PT^*)^m v = T^{*m} v + \sum_{j=0}^{m-1} \alpha_{mj} T^{*k} v, \quad (TP)^m u = T^m u + \sum_{j=0}^{m-1} \beta_{mj} T^k u.$$

This can be seen by induction and the use of (1.3)–(1.8). Thus for $x \in \widetilde{\mathcal{K}}$ we have

$$\langle \widetilde{T}^{*m} \widetilde{v}, x \rangle_{\widetilde{\mathcal{K}}} = \langle T^* (PT^*)^m v, x \rangle_{\mathcal{K}} = \langle T^{*(m+1)} v, x \rangle_{\mathcal{K}} + \sum_{j=0}^{m-1} \alpha_{mj} \langle T^{*(j+1)} v, x \rangle_{\mathcal{K}}$$

and

$$\langle \widetilde{T}^m \widetilde{u}, x \rangle_{\widetilde{\mathcal{K}}} = \langle (PT)^m u, x \rangle_{\mathcal{K}} = \langle T^m u, x \rangle_{\mathcal{K}} + \sum_{j=0}^{m-1} \beta_{mj} \langle T^j u, x \rangle_{\mathcal{K}}.$$

Induction arguments again show that for $x \in \widetilde{\mathcal{K}}$,

$$\langle \widetilde{T}^{*m} \widetilde{v}, x \rangle_{\widetilde{\mathcal{K}}} = 0,\ m = 0, 1, 2, \ldots \quad \Rightarrow \quad \langle T^{*m} v, x \rangle_{\mathcal{K}} = 0,\ m = 1, 2, \ldots,$$

$$\langle \widetilde{T}^m \widetilde{u}, x \rangle_{\widetilde{\mathcal{K}}} = 0,\ m = 0, 1, 2, \ldots \quad \Rightarrow \quad \langle T^m u, x \rangle_{\mathcal{K}} = 0,\ m = 0, 1, 2, \ldots.$$

These implications in turn imply that the minimality of $\widetilde{U}$ follows from that of U. □

In the proof of Theorem 2.3 we follow the proof of its coisometric analog [2, Theorem 6.1]. We use the following lemma; see [2, Lemma 4.3] for its proof and references to earlier results. By the anti-space of a Krein space $(\mathcal{K}, \langle \cdot\,, \cdot \rangle)$ we mean the Krein space $(\mathcal{K}', \langle \cdot\,, \cdot \rangle')$ where $\mathcal{K}' = \mathcal{K}$ as linear spaces and $\langle x, y \rangle' = -\langle x, y \rangle$, $x, y \in \mathcal{K}'$. Evidently, $(\mathcal{K}')' = \mathcal{K}$ as Krein spaces.

Lemma 3.3. *Let $f(z)$ be the characteristic function of a minimal unitary colligation U on $\mathcal{K} \oplus \mathbb{C}$ of the form (1.2) and assume $\gamma = f(0) \neq 0$. Then the reciprocal function $f_r(z) := f(z)^{-1}$ is the characteristic function of the minimal unitary colligation*

$$U_r = \begin{pmatrix} T_r & u_r \\ \langle \cdot\,, v_r \rangle' & \gamma_r \end{pmatrix}$$

on $\mathcal{K}' \oplus \mathbb{C}$, where $\mathcal{K}'$ is the anti-space of the Krein space $\mathcal{K}$, with entries

$$T_r = T - \frac{\langle \cdot\,, v \rangle}{\gamma} u, \quad u_r = \frac{u}{\gamma},$$

$$v_r = \frac{v}{\gamma^*}, \quad \gamma_r = \frac{1}{\gamma},$$

that is, $f_r(z) = \gamma_r + z \left\langle (I_{\mathcal{K}} - zT_r)^{-1} u_r, v_r \right\rangle'$.

Proof of Theorem 2.3. We rewrite the formula (2.3) in the form

$$\frac{1}{\widetilde{s}(z)} = \frac{\sigma_0}{\sigma_0^*}\frac{1}{z^k}\frac{\dfrac{1}{s(z)} - \dfrac{1}{\sigma_0}}{1 - \dfrac{1}{\sigma_0^*}\dfrac{1}{s(z)}} \tag{3.1}$$

and note that $|1/\sigma_0| < 1$ and that

$$\frac{1}{s(z)} - \frac{1}{\sigma_0} = -\frac{\sigma_k}{\sigma_0^2}z^k + \mathcal{O}(z^{k+1}),$$

in particular, this difference has a zero at $z = 0$ of order k. We apply Lemma 3.3 to $s(z)$ and the corresponding colligation U and then we apply Theorem 2.1 to $1/s(z)$. The formula (3.1) implies that $\sigma_0^*/(\sigma_0\widetilde{s}(z))$ is the generalized Schur transform of $1/s(z)$; see (2.2). If $(\mathcal{K}', T, u, v, \gamma)$ is the minimal colligation for $\sigma_0^*/(\sigma_0\widetilde{s}(z))$, then

$$(\mathcal{K}', T, -\frac{|\sigma_0|}{\sigma_0^*}u, -\frac{|\sigma_0|}{\sigma_0}v, \frac{\sigma_0}{\sigma_0^*}\gamma)$$

is a minimal colligation for $1/\widetilde{s}(z)$. Finally we apply Lemma 3.3 to $1/\widetilde{s}(z)$ and this colligation and obtain the realizations mentioned in the theorem. □

2. Proof of Theorem 2.4. We begin with the following consequence of Lemma 3.1. For k complex numbers $z_1, z_2, \ldots z_k$ we define the $k \times k$ Toeplitz matrix

$$\Delta(z_1, z_2, \ldots, z_k) = \begin{pmatrix} z_1 & z_2 & \cdots & z_{k-1} & z_k \\ 0 & z_1 & & z_{k-2} & z_{k-1} \\ \vdots & \ddots & \ddots & \ddots & \vdots \\ 0 & 0 & & z_1 & z_2 \\ 0 & 0 & \cdots & 0 & z_1 \end{pmatrix}.$$

Lemma 3.4. *The following three statements are equivalent:*

(1) $|\sigma_0| = 1$ *and* $\sigma_1 = \ldots = \sigma_{k-1} = 0$.
(2) $\operatorname{span}\{T^{*(k-1)}v, \ldots, T^*v, v\}$ *is a k-dimensional neutral subspace of* $\mathcal{K}$.
(3) $\operatorname{span}\{u, Tu, \ldots, T^{k-1}u\}$ *is a k-dimensional neutral subspace of* $\mathcal{K}$.

If (1)–(3) *hold then the spaces in* (2) *and* (3) *are skewly linked if and only if* $\sigma_k \neq 0$. *In this case, the Gram matrix* $\mathcal{G}$ *for the sequence*

$$T^{*(k-1)}v, \ldots, T^*v, v \mid u, Tu, \ldots, T^{k-1}u$$

is the $2k \times 2k$ *matrix*

$$\mathcal{G} = \begin{pmatrix} 0 & \Delta \\ \Delta^* & 0 \end{pmatrix}, \quad \Delta = \Delta(\sigma_k, \sigma_{k+1}, \ldots, \sigma_{2k-1});$$

in particular, the space

$$\operatorname{span}\{T^{*(k-1)}v, \ldots, T^*v, v, u, Tu, \ldots, T^{k-1}u\}$$

is a $(2k)$*-dimensional Pontryagin space with positive index* k *and negative index* k.

Thus the space $\mathcal{L}_0$ in Theorem 2.4 is nondegenerate and the projection P in $\mathcal{K}$ onto $\widetilde{\mathcal{K}} = \mathcal{K} \ominus \mathcal{L}_0$ is well defined.

The coefficients $c_0, c_1, \ldots, c_{k-1}$ that make up the polynomial $Q(z)$, see (2.4) and (2.5), are determined by the relation

$$(\sigma_k + \sigma_{k+1} z + \cdots + \sigma_{2k-1} z^{k-1})(c_0 + c_1 z + \cdots + c_{k-1} z^{k-1}) = \sigma_0 + \mathcal{O}(z^k), \quad (3.2)$$

or, equivalently, by the relation

$$\Delta(\sigma_k, \sigma_{k+1}, \ldots, \sigma_{2k-1}) \begin{pmatrix} c_{k-1} & c_{k-2} & \cdots & c_0 \end{pmatrix}^t = \begin{pmatrix} 0 & 0 & \cdots & \sigma_0 \end{pmatrix}^t,$$

where t stands for the transpose of the vector. We define the complex numbers $s_0, s_1, \ldots, s_{k-1}$ and $r_0, r_1, \ldots, r_{k-1}$ in a similar way by

$$\Delta(\sigma_k, \sigma_{k+1}, \ldots, \sigma_{2k-1}) \begin{pmatrix} s_0 & s_1 & \cdots & s_{k-1} \end{pmatrix}^t = \begin{pmatrix} \sigma_{2k} & \sigma_{2k-1} & \cdots & \sigma_{k+1} \end{pmatrix}^t$$

and

$$\Delta(\sigma_k, \sigma_{k+1}, \ldots, \sigma_{2k-1})^* \begin{pmatrix} r_{k-1} & r_{k-2} & \cdots & r_0 \end{pmatrix}^t = \begin{pmatrix} -\sigma_0^* \sigma_k & 0 & \cdots & 0 \end{pmatrix}^t.$$

By Cramer's rule, s_0 can be expressed in terms of the σ_j's: $s_0 = n_0/d_0$ with $d_0 = \det \Delta(\sigma_k, \sigma_{k+1}, \ldots, \sigma_{2k-1})$ and

$$n_0 = \det \begin{pmatrix} \sigma_{2k} & \sigma_{k+1} & \cdots & \sigma_{2k-2} & \sigma_{2k-1} \\ \sigma_{2k-1} & \sigma_k & & \sigma_{2k-3} & \sigma_{2k-2} \\ \vdots & & \ddots & \ddots & \vdots \\ \sigma_{k+2} & 0 & & \sigma_k & \sigma_{k+1} \\ \sigma_{k+1} & 0 & \cdots & 0 & \sigma_k \end{pmatrix}.$$

Lemma 3.5. *Assume* $|\sigma_0| = 1$, $k \geq 1$, $\sigma_1 = \ldots = \sigma_{k-1} = 0$, *and* $\sigma_k \neq 0$. *Then*

$$s_0 = t_{2k}, \quad s_j = -\sigma_0^* \sigma_k c_{k-j}, \; j = 1, \ldots, k-1,$$

and

$$r_j = -\sigma_k c_{k-j-1}^*, \; j = 0, \ldots, k-1.$$

Moreover,

(1) $PTx = Tx - \dfrac{\langle T^k x, v\rangle}{\sigma_k} u, \quad x \in \widetilde{\mathcal{K}},$

(2) $PT^*x = T^*x - \dfrac{\langle T^{*k} x, u\rangle}{\sigma_k^*} v, \quad x \in \widetilde{\mathcal{K}},$

(3) $PT^k u = T^k u - \sum_{j=0}^{k-1} r_j T^{*j} v - \sum_{j=0}^{k-1} s_j T^j u,$

(4) $PT^{*k} v = T^{*k} v - \sum_{j=0}^{k-1} s_j^* T^{*j} v - \sum_{j=0}^{k-1} r_j^* T^j u,$

(5) $PT^* PT^k u = -r_{k-1} PT^{*k} v,$

(6) $PTPT^{*k} v = -r_{k-1}^* PT^k u,$

(7) $\langle PT^k u, PT^k u\rangle_{\widetilde{\mathcal{K}}} = \langle PT^{*k} v, PT^{*k} v\rangle_{\widetilde{\mathcal{K}}} = s_0 \sigma_0 \sigma_k^* + s_0^* \sigma_0{}^* \sigma_k - |\sigma_k|^2.$

Proof. The defining relations for $s_0, s_1, \ldots, s_{k-1}$ and $r_0, r_1, \ldots, r_{k-1}$, like (3.2) for the c_j's, can also be written as

$$\begin{aligned}(\sigma_k + \sigma_{k+1}z + \cdots + \sigma_{2k-1}z^{k-1})&(s_{k-1} + s_{k-2}z + \cdots + s_0 z^{k-1})\\ &= \sigma_{k+1} + \sigma_{k+2}z + \cdots + \sigma_{2k}z^{k-1} + \mathcal{O}(z^k)\end{aligned} \tag{3.3}$$

and

$$\begin{aligned}(\sigma_k + \sigma_{k+1}z + \cdots + \sigma_{2k-1}z^{k-1})&(r_{k-1}^* + r_{k-2}^* z + \cdots + r_0^* z^{k-1})\\ &= -\sigma_0\sigma_k^* + \mathcal{O}(z^k).\end{aligned} \tag{3.4}$$

If we multiply both sides of (3.3) by z and rearrange terms we get

$$\begin{aligned}(\sigma_k + \sigma_{k+1}z + \cdots + \sigma_{2k-1}z^{k-1})&(-1 + s_{k-1}z + s_{k-2}z^2 + \cdots + s_1 z^{k-1})\\ &= -\sigma_k - (\sigma_k + \sigma_{k+1}z + \cdots + \sigma_{2k-1}z^{k-1})s_0 z^k + \sigma_{2k}z^k + \mathcal{O}(z^{k+1})\\ &= -\sigma_k + \mathcal{O}(z^k).\end{aligned} \tag{3.5}$$

From this formula, (3.2), and (3.4) we conclude

$$\frac{1}{\sigma_k} = \frac{c_0}{\sigma_0} = -\frac{r_{k-1}^*}{\sigma_0\sigma_k^*}, \qquad -\frac{s_{k-j}}{\sigma_k} = \frac{c_j}{\sigma_0} = -\frac{r_{k-j-1}^*}{\sigma_0\sigma_k^*}, \qquad j = 1, 2 \ldots, k-1.$$

These relations readily imply those in the lemma, except for s_0. The formula (2.5) for $Q(z)$ can now be expressed in terms of the s_j's and r_j's:

$$Q(z) = \frac{\sigma_0}{\sigma_k}(1 - \sum_{j=1}^{k-1} s_j z^{k-j} + \sigma_0^* \sum_{j=0}^{k-1} r_j z^{k+j+1}). \tag{3.6}$$

Using this and (3.5), we obtain

$$\begin{aligned}\sigma_0{}^*Q(z)s(z) - (Q(z) + z^k) &= (s(z) - \sigma_0)\sigma_0{}^*Q(z) - z^k\\ &= \frac{1}{\sigma_k}(\sigma_k z^k + \sigma_{k+1}z^{k+1} + \cdots + \sigma_{2k}z^{2k} + \sigma_{2k+1}z^{2k+1} + \cdots) \times\\ &\quad \times(1 - s_{k-1}z - \cdots - s_1 z^{k-1} + \sigma_0{}^* r_0 z^{k+1} + \cdots + \sigma_0{}^* r_{k-1}z^{2k}) - z^k\\ &= \frac{1}{\sigma_k}(\sigma_{2k} - \sigma_{2k-1}s_{k-1} - \cdots - s_1\sigma_{k+1})z^{2k} + \mathcal{O}(z^{2k+1})\\ &= \frac{1}{\sigma_k}(\sigma_{2k} - \sigma_{2k} + \sigma_k s_0)z^{2k} + \mathcal{O}(z^{2k+1}) = s_0 z^{2k} + \mathcal{O}(z^{2k+1}),\end{aligned}$$

which shows that $s_0 = t_{2k}$.

The proofs of (1)–(4) involve (1.3)–(1.8) and the last part of Lemma 3.4 and are very much alike, so we only prove (1): We have

$$PTx = Tx - \sum_{j=0}^{k-1} \alpha_j T^{*j}v - \sum_{j=0}^{k-1} \beta_j T^j u,$$

where the α_j's and β_j's are determined by the relations

$$\langle PTx, T^{*\ell}v\rangle = 0, \quad \langle PTx, T^{\ell}u\rangle = 0, \quad \ell = 0,1,\ldots,k-1.$$

Since $x \in \widetilde{\mathcal{K}}$, we have

$$\langle Tx, T^{*\ell}v\rangle_{\mathcal{K}} = \langle x, T^{*(\ell+1)}v\rangle_{\mathcal{K}} = 0, \quad \ell = 0,1,\ldots,k-2,$$

and so by Lemma 3.4,

$$\sum_{j=0}^{k-1} \beta_j \sigma_{j+\ell+1} = 0, \quad \ell = 0,1,\ldots,k-2,$$

and, corresponding to $\ell = k-1$,

$$\sum_{j=0}^{k-1} \beta_j \sigma_{j+k} = \langle T^k x, v\rangle_{\mathcal{K}}.$$

Writing out these equations in detail and using that $\sigma_1 = \ldots = \sigma_{k-1} = 0$, we get

$$\beta_0 = \frac{1}{\sigma_k}\langle T^k x, v\rangle_{\mathcal{K}}, \qquad \beta_j = 0, \quad j = 1,2,\ldots,k-1.$$

Similarly, because $x \in \widetilde{\mathcal{K}}$ and by (1.6) and (1.8),

$$\langle Tx, T^{\ell}u\rangle_{\mathcal{K}} = \begin{cases} \langle x, (T^*T)T^{\ell-1}u\rangle_{\mathcal{K}} = \langle x, T^{\ell-1}u\rangle_{\mathcal{K}} - \sigma_\ell^* \langle x, v\rangle_{\mathcal{K}} = 0, \\ \hfill \ell = 1,\ldots,k-1, \\ \langle x, T^*u\rangle_{\mathcal{K}} = -\sigma_0^* \langle x, v\rangle_{\mathcal{K}} = 0, \; \ell = 0, \end{cases}$$

and hence

$$0 = \sum_{j=0}^{k-1} \alpha_j \langle T^{*j}v, T^{\ell}u\rangle_{\mathcal{K}} = \sum_{j=0}^{k-1} \alpha_j \sigma_{j+\ell+1}^*, \quad \ell = 0,1,\ldots,k-1.$$

Writing this out in detail we find that all α_j's are zero. This proves (1).

Formula (5) follows from formula (3) and (1.6).

Formula (6) follows from formula (4) and (1.3).

Part of (7) follows from Lemma 3.1, the last formula in Lemma 3.2 with $|\sigma_0| = 1$, $s_0 = t_{2k}$, and

$$\begin{aligned} \langle PT^k u, PT^k u\rangle_{\widetilde{\mathcal{K}}} &= \langle T^k u, T^k u - \sum_{j=0}^{k-1} r_j T^{*j}v - \sum_{j=0}^{k-1} s_j T^j u\rangle_{\mathcal{K}} \text{ by } (3) \\ &= \langle T^k u, T^k u\rangle_{\mathcal{K}} - \sum_{j=0}^{k-1} r_j^* \langle T^{k+j}u, v\rangle_{\mathcal{K}} - \sum_{j=0}^{k-1} s_j^* \langle T^k u, T^j u\rangle_{\mathcal{K}} \\ &= -|\sigma_k|^2 + \sigma_k^* \sum_{j=0}^{k-1} c_{k-j-1}\sigma_{k+j+1} + s_0^*\sigma_0^*\sigma_k = -|\sigma_k|^2 + \sigma_k^*\sigma_0 t_{2k} + s_0^*\sigma_0^*\sigma_k \text{ by } (2.8) \\ &= -|\sigma_k|^2 + s_0\sigma_0\sigma_k^* + s_0^*\sigma_0^*\sigma_k; \end{aligned}$$

the remaining part can be proved similarly. □

Proof of Theorem 2.4. In $\widetilde{\gamma} + z\langle(I_{\widetilde{\mathcal{K}}} - z\widetilde{T})^{-1}\widetilde{u}, \widetilde{v}\rangle_{\widetilde{\mathcal{K}}}$, where $\widetilde{T}$, $\widetilde{u}$, $\widetilde{v}$ and $\widetilde{\gamma}$ are as defined in the theorem, we set $x = (I_{\widetilde{\mathcal{K}}} - z\widetilde{T})^{-1}\widetilde{u}$. Then by Lemma 3.5 (1),

$$x = (I_{\mathcal{K}} - zT)^{-1}\widetilde{u} - z\frac{\langle T^k x, v\rangle_{\mathcal{K}}}{\sigma_k}(I_{\mathcal{K}} - zT)^{-1}u - z\frac{\langle T^k x, v\rangle_{\mathcal{K}}}{\sigma_k t_{2k}}(I_{\mathcal{K}} - zT)^{-1}PT^k u.$$

We apply T^k to both sides, take the inner product with v, and solve for $\langle T^k x, v\rangle_{\mathcal{K}}$:

$$\langle T^k x, v\rangle_{\mathcal{K}} = \frac{N(z)}{D(z)},$$

where

$$\begin{aligned} N(z) &= \quad \langle (I_{\mathcal{K}} - zT)^{-1}T^k\widetilde{u}, v\rangle_{\mathcal{K}}, \\ D(z) &= \quad 1 + z\frac{\langle (I_{\mathcal{K}} - zT)^{-1}T^k u, v\rangle_{\mathcal{K}}}{\sigma_k} + z\frac{\langle (I_{\mathcal{K}} - zT)^{-1}T^k PT^k u, v\rangle}{\sigma_k t_{2k}}. \end{aligned}$$

We express the denominator $D(z)$ in terms of $s(z) - \sigma_0$ and powers of z, using the power series expansion of $z(I_{\mathcal{K}} - zT)^{-1}$. The first two summands in the formula for $D(z)$ add up to

$$1 + z\frac{\langle (I_{\mathcal{K}} - zT)^{-1}u, v\rangle_{\mathcal{K}}}{\sigma_k} = 1 + \frac{1}{\sigma_k}\sum_{n=0}^{\infty} z^{n+1}\sigma_{n+k+1} = \frac{1}{\sigma_k z^k}(s(z) - \sigma_0).$$

The third summand, by Lemma 3.5 (3), is itself the sum of three terms:

$$\begin{aligned} &\frac{z}{\sigma_k t_{2k}}\langle (I_{\mathcal{K}} - zT)^{-1}T^{2k}u, v\rangle_{\mathcal{K}} \\ &\quad = \frac{1}{\sigma_k t_{2k} z^{2k}}(s(z) - \sigma_0 - (\sigma_k z^k + \ldots + \sigma_{2k}z^{2k})) \\ &\quad = \frac{s(z) - \sigma_0 - \sigma_k z^k}{\sigma_k t_{2k} z^{2k}} - \frac{\sigma_{k+1}z^{k+1} + \ldots + \sigma_{2k}z^{2k}}{\sigma_k t_{2k} z^{2k}}, \end{aligned}$$

$$\begin{aligned} &-\frac{z}{\sigma_k t_{2k}}\sum_{j=0}^{k-1} r_j \langle (I_{\mathcal{K}} - zT)^{-1}T^k T^{*j}v, v\rangle_{\mathcal{K}} \\ &\quad = -\frac{z}{\sigma_k t_{2k}}\sum_{j=0}^{k-1} r_j \langle (I_{\mathcal{K}} - zT)^{-1}T^{k-j}v, v\rangle_{\mathcal{K}} \text{ by } (1.3) \\ &\quad = \frac{\sigma_0^* z}{\sigma_k t_{2k}}\sum_{j=0}^{k-1} r_j \langle (I_{\mathcal{K}} - zT)^{-1}T^{k-j-1}u, v\rangle_{\mathcal{K}} \text{ by } (1.5) \\ &\quad = \sigma_0^*\frac{s(z) - \sigma_0}{\sigma_k t_{2k} z^{2k}}\sum_{j=0}^{k-1} r_j z^{k+j+1}, \end{aligned}$$

and

$$
\begin{aligned}
&-\frac{z}{\sigma_k t_{2k}}\sum_{j=0}^{k-1} s_j\langle (I_{\mathcal K}-zT)^{-1}T^{k+j}u,v\rangle_{\mathcal K}\\
&\quad=-\frac{1}{\sigma_k t_{2k}}\sum_{j=0}^{k-1}\frac{s_j}{z^{k+j}}(s(z)-\sigma_0-(\sigma_k z^k+\cdots+\sigma_{k+j}z^{k+j}))\\
&\quad=-\frac{s(z)-\sigma_0}{\sigma_k t_{2k}}\sum_{j=0}^{k-1}\frac{s_j}{z^{k+j}}+\frac{1}{\sigma_k t_{2k}}\sum_{j=0}^{k-1}\frac{s_j}{z^j}(\sigma_k+\cdots+\sigma_{k+j}z^j)\\
&\quad=-\frac{s(z)-\sigma_0}{\sigma_k t_{2k}z^{2k}}\sum_{j=0}^{k-1}s_j z^{k-j}+\frac{1}{\sigma_k t_{2k}}\sum_{j=0}^{k-1}\frac{1}{z^j}\sum_{n=0}^{k-j-1}s_{j+n}\sigma_{k+n}\\
&\quad=-\frac{s(z)-\sigma_0}{\sigma_k t_{2k}z^{2k}}\sum_{j=0}^{k-1}s_j z^{k-j}+\frac{\sigma_{k+1}z^{k+1}+\cdots+\sigma_{2k}z^{2k}}{\sigma_k t_{2k}z^{2k}},
\end{aligned}
$$

where we have used that $\sum_{n=0}^{k-j-1} s_{j+n}\sigma_{k+n}=\sigma_{2k-j}$, $j=0,1,\ldots,k-1$, according to the definition of the numbers s_j. Adding these terms and using (3.6) and $s_0=t_{2k}$, we get

$$
\begin{aligned}
D(z)&=\frac{1}{\sigma_k t_{2k}z^{2k}}[(s(z)-\sigma_0)(t_{2k}z^k+1+\sigma_0^*\sum_{j=0}^{k-1}r_j z^{k+j+1}-\sum_{j=0}^{k-1}s_j z^{k-j})-\sigma_k z^k]\\
&=\frac{1}{t_{2k}z^{2k}}((s(z)-\sigma_0)\sigma_0^*Q(z)-z^k)=\frac{1}{t_{2k}z^{2k}}(\sigma_0^*Q(z)s(z)-(Q(z)+z^k)).
\end{aligned}
$$

Similarly, by Lemma 3.5 (3),

$$
\begin{aligned}
N(z)&=\langle (I_{\mathcal K}-zT)^{-1}T^k\widetilde u,v\rangle_{\mathcal K}=\frac{1}{t_{2k}}\langle (I_{\mathcal K}-zT)^{-1}T^kPT^ku,v\rangle_{\mathcal K}\\
&=\frac{1}{t_{2k}z^{2k+1}}[(s(z)-\sigma_0)(\sigma_k\sigma_0^*Q(z)-t_{2k}z^k)-\sigma_k z^k].
\end{aligned}
$$

It is now easy to check that

$$
\widetilde\gamma+z\langle (I_{\widetilde{\mathcal K}}-z\widetilde T)^{-1}\widetilde u,\widetilde v\rangle_{\widetilde{\mathcal K}}=\sigma_0-\frac{\sigma_k}{t_{2k}}+\frac{z}{t_{2k}}\langle T^kx,v\rangle_{\mathcal K}=\sigma_0-\frac{\sigma_k}{t_{2k}}+\frac{z}{t_{2k}}\frac{N(z)}{D(z)}
$$

equals the expression on the right-hand side of (2.6) with $q=0$, and hence $\widetilde s(z)$ coincides with the characteristic function of $\widetilde U$.

To prove the unitarity of the colligation we must show the analogs of (1.3)–(1.8). This can be done using the formulas in Lemma 3.5. For example, the analog of (1.6) is proved as follows. For $x \in \widetilde{\mathcal{K}}$,

$$\widetilde{T}^*\widetilde{T}x = PT^*PTx - \frac{\langle x, T^*PT^k u\rangle_{\mathcal{K}}}{\sigma_k^* t_{2k}^*} PT^{*k}v$$
$$- \frac{\langle x, T^{*k}v\rangle_{\mathcal{K}}}{\sigma_k t_{2k}} PT^*PT^k u + \frac{\langle x, T^{*k}v\rangle_{\mathcal{K}}}{|\sigma_k|^2|t_{2k}|^2}\langle PT^k u, PT^k u\rangle_{\widetilde{\mathcal{K}}} PT^{*k}v.$$

The first summand on the right-hand side equals x by Lemma 3.5 (1), the second term is equal to

$$\frac{r_{k-1}^*}{\sigma_k^* t_{2k}^*}\langle x, T^{*k}v\rangle_{\mathcal{K}} PT^{*k}v = -\frac{\sigma_0}{\sigma_k t_{2k}^*}\langle x, T^{*k}v\rangle_{\mathcal{K}} PT^{*k}v$$

on account of Lemma 3.5 (5), the third is equal to

$$\frac{r_{k-1}}{\sigma_k t_{2k}}\langle x, T^{*k}v\rangle_{\mathcal{K}} PT^{*k}v = -\frac{\sigma_0^*}{\sigma_k^* t_{2k}}\langle x, T^{*k}v\rangle_{\mathcal{K}} PT^{*k}v$$

also because of Lemma 3.5 (5), and finally the fourth term equals

$$\left(\frac{\sigma_0}{\sigma_k t_{2k}^*} + \frac{\sigma_0^*}{\sigma_k^* t_{2k}} - \frac{1}{|t_{2k}|^2}\right)\langle x, T^{*k}v\rangle_{\mathcal{K}} PT^{*k}v$$

by Lemma 3.5 (7) and $s_0 = t_{2k}$. These four terms add up to

$$x - \frac{1}{|t_{2k}|^2}\langle x, T^{*k}v\rangle_{\mathcal{K}} PT^{*k}v = x - \langle x, \widetilde{v}\rangle_{\widetilde{\mathcal{K}}}\widetilde{v},$$

that is,

$$\widetilde{T}^*\widetilde{T}x + \langle x, \widetilde{v}\rangle_{\widetilde{\mathcal{K}}}\widetilde{v} = x, \quad x \in \widetilde{\mathcal{K}}.$$

The analogs of the other relations in (1.3)–(1.8) can be verified in a similar way.

It remains to prove the minimality statement: By induction there are constants α_{mj}, β_{mj}, γ_{mj}, and δ_{mj} such that for all $m = 0, 1, 2, \ldots$,

$$\widetilde{T}^{*m}\widetilde{v} = \frac{1}{t_{2k}^*}T^{*(k+m)}v + \sum_{j=0}^{k+m-1}\alpha_{mj}T^{*j}v + \sum_{j=0}^{k+m-1}\beta_{mj}T^j u,$$

$$\widetilde{T}^m\widetilde{u} = \frac{1}{t_{2k}}T^{k+m}u + \sum_{j=0}^{k+m-1}\gamma_{mj}T^{*j}v + \sum_{j=0}^{k+m-1}\delta_{mj}T^j u,$$

and hence if $x \in \widetilde{\mathcal{K}}$ is orthogonal to all elements of the two sided sequence

$$\ldots, \widetilde{T}^{*2}\widetilde{v}, \widetilde{T}^*\widetilde{v}, \widetilde{v} \mid \widetilde{u}, \widetilde{T}\widetilde{u}, \widetilde{T}^2\widetilde{u}, \ldots,$$

then it is also orthogonal to all elements

$$\ldots, T^{*(k+2)}v, T^{*(k+1)}v, T^{*k}v \mid T^k u, T^{k+1}u, T^{k+2}u, \ldots.$$

Hence the minimality of the colligation for $\widetilde{s}(z)$ follows from that of the colligation for $s(z)$. □

3. Proof of Theorem 2.5. The proof is similar to that of Theorem 2.3. There we applied Theorem 2.1 and Lemma 3.3; here besides these results we also invoke Theorem 2.4.

Proof of Theorem 2.5. We will show that the Schur transform of $1/s(z)$ is equal to $z^q/\widetilde{s}(z)$ and write down in terms of the minimal unitary colligation U for $s(z)$ given by (1.2) the minimal unitary colligations for

$s_1(z) := 1/s(z)$ with the help of Lemma 3.3,
$s_2(z) := \widetilde{s}_1(z) = z^q/\widetilde{s}(z)$ with the help of Theorem 2.4,
$s_3(z) := 1/\widetilde{s}(z)$ with the help of Theorem 2.1 with $k = q$,
and finally for $\widetilde{s}(z)$ again with the help of Lemma 3.3.

By Lemma 3.3, the minimal unitary colligation U_1 with characteristic function $s_1(z)$ is given by

$$\begin{array}{rclrcl} T_1 &=& T - \sigma_0^*\langle\,\cdot\,,v\rangle_{\mathcal{K}}u, & u_1 &=& \sigma_0^* u, \\ v_1 &=& \sigma_0 v, & \gamma_1 &=& \sigma_0^*. \end{array}$$

The state space of U_1 is $\mathcal{K}_1 = \mathcal{K}'$, the anti-space of $\mathcal{K}$.

We calculate the generalized Schur transform of $s_1(z)$. Note that $|s_1(0)| = 1$. We show that we can apply Theorem 2.4. Define σ_{1j}, k_1, c_{1j}, $Q_1(z)$, q_1, and $t_{1(2k_1)}$ for $s_1(z)$ as σ_j, k, c_j, $Q(z)$, q, and t_{2k} are defined for $s(z)$. As in the proof of [2, Theorem 8.1] by calculating the Taylor expansion of $s_1(z)$ for z near 0, we find that

$$\sigma_{1j} = \begin{cases} \sigma_0^*, & j = 0, \\ 0, & j = 1, \dots, k-1, \\ -\sigma_0^{*2}\sigma_j, & j = k, \dots, 2k-1, \\ -\sigma_0^{*2}(\sigma_{2k} - \sigma_0^*\sigma_k^2), & j = 2k, \end{cases}$$

and consequently that $k_1 = k$, $c_{1j} = -c_j$, $Q_1(z) = -Q(z)$, $t_{1(2k)} = t_{2k} - \sigma_0^*\sigma_k = -\sigma_0^*\sigma_k$ (since $q > 0$, $t_{2k} = 0$), and $q_1 = 0$. Finally, a simple calculation yields

$$\begin{aligned} s_2(z) = \widetilde{s}_1(z) &= \frac{(Q_1(z) - z^k)s_1(z) - \sigma_{10}Q_1(z)}{\sigma_{10}^* Q_1(z)s_1(z) - (Q_1(z) + z^k)} \\ &= \frac{\sigma_0^* Q(z)s(z) - (Q(z) + z^k)}{(Q(z) - z^k)s(z) - \sigma_0 Q(z)} = -\frac{t_{2k+q}}{\sigma_k} z^q + \mathcal{O}(z^{q+1}) = \frac{z^q}{\widetilde{s}(z)}. \end{aligned}$$

Hence we may invoke Theorem 2.4 to describe the minimal unitary colligation U_2 with characteristic function $s_2(z)$:

$$\begin{array}{rclrcl} T_2 &=& P_1T_1P_1 - \dfrac{\langle\,\cdot\,, P_1T_1^{*k}v_1\rangle'_{\mathcal{K}}}{\sigma_{1k}t_{1(2k)}} P_1T_1^k u_1, & u_2 &=& \dfrac{1}{t_{1(2k)}} P_1T_1^k u_1, \\ v_2 &=& \dfrac{1}{t_{1(2k)}^*} P_1T_1^{*k}v_1, & \gamma_2 &=& \sigma_{10} - \dfrac{\sigma_{1k}}{t_{1(2k)}}. \end{array}$$

Here P_1 is the projection in the space $\mathcal{K}$ onto the subspace $\mathcal{K}_2 := \mathcal{K} \ominus \mathcal{L}_0$ with

$$\begin{aligned} \mathcal{L}_0 &= \operatorname{span}\{T_1^{*(k-1)}v_1, \dots, v_1, u_1, \dots, T_1^{k-1}u\} \\ &= \operatorname{span}\{T^{*(k-1)}v, \dots, v, u, \dots, T^{k-1}u\}. \end{aligned}$$

Note that the state space of U_2 is the space $\mathcal{K}_2'$. Combining the formulas for U_1 and U_2 we get

$$T_2 = P_1TP_1 + \frac{\langle \cdot, P_1T^{*k}v\rangle_{\mathcal{K}}}{\sigma_0^*\sigma_k^2}P_1T^ku, \quad u_2 = -\frac{1}{\sigma_k}P_1T^ku, \quad v_2 = -\frac{1}{\sigma_k^*}P_1T^{*k}v, \quad \gamma_2 = 0.$$

Let $s_3(z)$ be the generalized Schur transform of $s_2(z) = z^q/\widetilde{s}(z)$. A minimal unitary colligation U_3 whose characteristic function coincides with $s_3(z)$ is given by Theorem 2.1 with $k = q$. We choose the first colligation:

$$T_3 = P_2T_2P_2, \quad u_3 = P_2u_2, \quad v_3 = P_2T_2^{*q}v_2, \quad \gamma_3 = \frac{1}{\widetilde{s}(0)} = -\frac{t_{2k+q}}{\sigma_k},$$

where P_2 is the orthogonal projection in the space $\mathcal{K}_2$ onto the subspace $\mathcal{K}_3 := \mathcal{K}_2 \ominus \mathcal{M}$ with

$$\mathcal{M} = \operatorname{span}\{T_2^{*(q-1)}v_2, \ldots, v_2\}.$$

The state space of U_3 is $\mathcal{K}_3'$. By induction on m and by (1.6) and (1.8), we find that there exist constants α_{mj}, β_{mj} such that for $m = 0, 1, \ldots$,

$$T_2^{*m}v_2 = -\frac{1}{\sigma_k^*}T^{*(k+m)}v + \sum_{j=0}^{k+m-1}\alpha_{mj}T^{*j}v + \sum_{j=0}^{k-1}\beta_{mj}T^ju.$$

It follows that $\mathcal{K}_3 = \mathcal{K}_2 \ominus \mathcal{M} = \mathcal{K} \ominus \widetilde{\mathcal{L}}$ with $\widetilde{\mathcal{L}}$ as described in the theorem, and that $P := P_2P_1$ is the orthogonal projection in $\mathcal{K}$ onto $\widetilde{\mathcal{K}} = \mathcal{K} \ominus \widetilde{\mathcal{L}}$ just as in the theorem. Combining the formulas for U_2 and U_3, we obtain

$$\begin{aligned} T_3 &= PTP + \frac{\langle \cdot, PT^{*k}v\rangle_{\mathcal{K}}}{\sigma_0^*\sigma_k^2}PT^ku = PTP, & u_3 &= -\frac{1}{\sigma_k}PT^ku, \\ v_3 &= -\frac{1}{\sigma_k^*}PT^{*(k+q)}v, & \gamma_3 &= -\frac{t_{2k+q}}{\sigma_k}. \end{aligned}$$

We now apply Lemma 3.3 to $s_3(z) = 1/\widetilde{s}(z)$ and the corresponding colligation U_3 with state space $\widetilde{\mathcal{K}}'$ and obtain the minimal unitary colligation for $\widetilde{s}(z)$ described in the first part of the theorem.

To prove the second part of the theorem one can use the same argument as in the proof of Theorem 2.1. □

4. Proof of Remark 2.2. The following proof holds for each of the Theorems 2.1, 2.3, and 2.5. Denote the Taylor coefficients of $\widetilde{s}(z)$ at $z = 0$ by $\widetilde{\sigma}_j$:

$$\widetilde{s}(z) = \widetilde{\sigma}_0 + \widetilde{\sigma}_1z + \widetilde{\sigma}_2z^2 + \cdots + \widetilde{\sigma}_nz^n + \cdots, \quad \widetilde{\sigma}_n = s^{(n)}(0)/n!, \ n = 0, 1, \ldots.$$

Then $\langle \widetilde{T}^n\widetilde{u}, \widetilde{v}\rangle_{\widetilde{\mathcal{K}}} = \widetilde{\sigma}_{n+1} = \langle \widehat{T}^n\widehat{u}, \widehat{v}\rangle_{\widehat{\mathcal{K}}}$, $n = 0, 1, \ldots$. Hence for $p, q = 0, 1, \ldots$,

$$\langle \widetilde{T}^p\widetilde{u}, \widetilde{T}^{*q}\widetilde{v}\rangle_{\widetilde{\mathcal{K}}} = \langle \widehat{T}^p\widehat{u}, \widehat{T}^{*q}\widehat{v}\rangle_{\widehat{\mathcal{K}}}$$

and, on account of Lemma 3.1,

$$\langle \widetilde{T}^p\widetilde{u}, \widetilde{T}^q\widetilde{u}\rangle_{\widetilde{\mathcal{K}}} = \langle \widehat{T}^p\widehat{u}, \widehat{T}^q\widehat{u}\rangle_{\widehat{\mathcal{K}}}, \quad \langle \widetilde{T}^{*p}\widetilde{v}, \widetilde{T}^{*q}\widetilde{v}\rangle_{\widetilde{\mathcal{K}}} = \langle \widehat{T}^{*p}\widehat{v}, \widehat{T}^{*q}\widehat{v}\rangle_{\widehat{\mathcal{K}}}.$$

The properties of W follow directly from these formulas and (1.3)–(1.8).

References

[1] R. Ackner, H. Lev-Ari, and T. Kailath, *The Schur algorithm for matrix-valued meromorphic functions*, Siam J. Matrix Anal. Appl. **15**(1) (1994), 140–150.

[2] D. Alpay, T. Azizov, A. Dijksma, and H. Langer, *The Schur algorithm for generalized Schur functions, I: Coisometric realizations*, Operator Theory: Adv. Appl., **129**, Birkhäuser Verlag, Basel, 2001, 1–36.

[3] D. Alpay, T. Azizov, A. Dijksma, and H. Langer, *The Schur algorithm for generalized Schur functions, II: Jordan chains and transformations of characteristic functions*, Monatshefte für Mathematik **138** (2003), 1–29.

[4] D. Alpay, T. Azizov, A. Dijksma, and H. Langer, *The Schur algorithm for generalized Schur functions, III: J-unitary matrix polynomials on the cirle*, Linear Algebra Appl. **369** (2003), 113–144.

[5] D. Alpay, T. Azizov, A. Dijksma, H. Langer, and G. Wanjala, *A basic interpolation problem for generalized Schur functions and coisometric realizations*, Operator Theory: Adv. Appl. **143**, Birkhäuser, Basel, 2003, 39–76.

[6] D. Alpay, T. Azizov, A. Dijksma, H. Langer, and G. Wanjala, *Characteristic functions of coisometric colligations and fractional linear transformations arising from a basic interpolation problem*, Research report IWI preprint 2002-2-01, Department of Mathematics, University of Groningen, Groningen, The Netherlands.

[7] D. Alpay, A. Dijksma, J. Rovnyak, and H. de Snoo, *Schur functions, operator colligations, and reproducing kernel Pontryagin spaces*, Operator Theory: Adv. Appl., **96**, Birkhäuser Verlag, Basel, 1997.

[8] D. Alpay and H. Dym, *On applications of reproducing kernel spaces to the Schur algorithm and rational J-unitary factorization*, Operator Theory: Adv. Appl., **18**, Birkhäuser Verlag, Basel, 1986, 89–159.

[9] D. Alpay and H. Dym, *On reproducing kernel spaces, the Schur algorithm, and interpolation in a general class of domains*, Operator Theory: Adv. Appl., **59**, Birkhäuser Verlag, Basel, 1992, 30–77.

[10] T.Ya. Azizov, I.S. Iokhvidov, *Linear operators in spaces with an indefinite metric*, Wiley, New York, 1989.

[11] M.J. Bertin, A. Decomps-Guilloux, M. Grandet-Hugot, M. Pathiaux-Delfosse, and J.P. Schreiber, *Pisot and Salem numbers*, Birkhäuser Verlag, Basel, 1992.

[12] C. Chamfy, *Fonctions méromorphes dans le cercle-unité et leurs séries de Taylor*, Ann. Inst. Fourier **8** (1958), 211–261.

[13] P. Delsarte, Y. Genin, and Y. Kamp, *Pseudo-Carathéodory functions and hermitian Toeplitz matrices*, Philips J. Res. **41**(1) (1986), 1–54.

[14] A. Dijksma, H. Langer, and H.S.V. de Snoo, *Unitary colligations in Krein spaces and their role in the extension theory of isometries and symmetric linear relations in Hilbert spaces*, Functional Analysis II, Proceedings Dubrovnik 1985, Lecture Notes in Mathematics 1242, Springer–Verlag, Berlin, 1987, 1–42.

[15] M. A. Dritschel and J. Rovnyak, *Operators on indefinite inner product spaces*, Lectures on operator theory and its applications (Waterloo, ON, 1994), Amer. Math. Soc., Providence, RI, 1996, 141–232.

[16] J. Dufresnoy, *Le problème des coefficients pour certaines fonctions méromorphes dans le cercle unité*, Ann. Acad. Sc. Fenn. Ser. A.I, **250**(9) (1958), 1–7.

[17] I.S. Iokhvidov, M.G. Krein, and H. Langer, *Introduction to the spectral theory of operators in spaces with an indefinite metric*, Akademie-Verlag, Berlin, 1982.

[18] M.G. Krein and H. Langer, *Über einige Fortsetzungsprobleme, die eng mit der Theorie hermitescher Operatoren in Raume Π_κ zusammenhängen, Teil I: Einige Funktionenklassen und ihre Darstellungen*, Math. Nachr. **77** (1977), 187–236.

[19] I. Schur, *Über die Potenzreihen, die im Innern des Einheitskreises beschränkt sind*, J. Reine Angew. Math. **147** (1917), 205–232; English translation in: *I. Schur methods in operator theory and signal processing*, Operator Theory: Adv. Appl., **18**, Birkhäuser Verlag, Basel, 1989, 31–59.

[20] I. Schur, *Über die Potenzreihen, die im Innern des Einheitskreises beschränkt sind; Fortsetzung*, J. Reine Angew. Math. **148** (1918), 122–145; English translation in: *I. Schur methods in operator theory and signal processing*, Operator Theory: Adv. Appl., **18**, Birkhäuser Verlag, Basel, 1989, 61–88.

D. Alpay
Department of Mathematics, Ben-Gurion University of the Negev
P.O. Box 653, 84105 Beer-Sheva, Israel
e-mail: `dany@math.bgu.ac.il`

T.Ya. Azizov
Department of Mathematics, Voronezh State University
394693 Voronezh, Russia
e-mail: `azizov@tom.vsu.ru`

A. Dijksma
Department of Mathematics, University of Groningen
P.O. Box 800, NL-9700 AV Groningen, The Netherlands
e-mail: `dijksma@math.rug.nl`

H. Langer
Institute of Analysis and Technical Mathematics
Vienna University of Technology
Wiedner Hauptstrasse 8–10, A-1040 Vienna, Austria
e-mail: `hlanger@mail.zserv.tuwien.ac.at`

G. Wanjala
Department of Mathematics, University of Groningen
P.O. Box 800, NL-9700 AV Groningen, The Netherlands
e-mail: `gerald@math.rug.nl`

Operator Theory:
Advances and Applications, Vol. 149, 47–77

Linear Systems with Schrödinger Operators and Their Transfer Functions

Yu. Arlinskiĭ and E. Tsekanovskiĭ

Dedicated with great pleasure and respect to Israel Gohberg on his 75th birthday

Abstract. We study linear, conservative, stationary dynamic systems (rigged operator colligations) and their transfer functions. The main operator of such systems is an extension with exit in a rigged Hilbert space of a Schrödinger operator on a half-line with non-self-adjoint boundary conditions.The description of all systems with accretive main operator in terms of transfer functions and their linear-fractional transformations and connections with Stieltjes and inverse Stieltjes functions are obtained. Using the bi-extension theory of symmetric operators in rigged Hilbert spaces and the system theory approach, we establish new properties of spectral functions of distributions (and corresponding Stieltjes integrals) of nonnegative self-adjoint extensions of a nonnegative Schrödinger operators on a half-line, new relations between spectral functions of the Friedrichs and Kreĭn-von Neumann extremal, nonnegative self-adjoint extensions, as well as new sharp inequalities involving pointwise functionals and extremal nonnegative extensions.

1. Introduction

In this paper we give a description and parametrization of all, so called, $(*)$-extensions of ordinary differential operators on a half-line. In particular, we consider one-dimensional non-self-adjoint Schrödinger operators and systems (rigged operator colligations) with the main operator a $(*)$-extension of the above-mentioned non-self-adjoint Schrödinger operator. We give a description of such type of systems when the linear fractional transformation of the transfer function (characteristic function) is a Stieltjes or inverse Stieltjes function and provide exact computations in the case of Bessel potentials. Using the system theory (operator colligation) approach, we establish new properties of spectral functions of distributions (and corresponding Stieltjes integrals) of nonnegative self-adjoint extensions

Received by the editors January 28, 2003; revised August 30, 2003. Submitted by J. A. Ball.
Mathematics Subject Classification (2000). Primary 47A63, 47B25; Secondary 47B65.
Key words and phrases. Linear systems, transfer functions, Schrödinger operators, spectral functions, Friedrichs and Krein-von Neumann extremal nonnegative extensions.

of a nonnegative Schrödinger operator on a half-line, the new relations between spectral functions of the Friedrichs and Kreĭn-von Neumann extremal, nonnegative, self-adjoint extensions as well as new sharp inequalities involving pointwise functionals and extremal nonnegative extensions.

2. Bi-extensions of ordinary differential operators

Let A be a closed, densely defined, symmetric operator in a Hilbert space $\mathfrak{H}$ with inner product $(f,g), f,g \in \mathfrak{H}$. Consider the rigged Hilbert space

$$\mathfrak{H}_+ \subset \mathfrak{H} \subset \mathfrak{H}_-,$$

where $\mathfrak{H}_+ = D(A^*)$ and

$$(f,g)_+ = (f,g) + (A^*f, A^*g), \quad f,g \in D(A^*).$$

Let $[\mathfrak{H}_+, \mathfrak{H}_-]$ be the set of all bounded linear operators from $\mathfrak{H}_+$ into $\mathfrak{H}_-$. Note that identifying the space conjugate to $\mathfrak{H}_\pm$ with $\mathfrak{H}_\mp$, we get that if $\mathbb{A} \in [\mathfrak{H}_+, \mathfrak{H}_-]$ then $\mathbb{A}^* \in [\mathfrak{H}_+, \mathfrak{H}_-]$.

Definition 1. *An operator $\mathbb{A} \in [\mathfrak{H}_+, \mathfrak{H}_-]$ is called a self-adjoint bi-extension of a symmetric operator A if $\mathbb{A} = \mathbb{A}^*$, $\mathbb{A} \supset A$, and the operator*

$$\widehat{A}f = \mathbb{A}f, \; f \in D(\widehat{A}) = \{f \in \mathfrak{H}_+ : \mathbb{A}f \in \mathfrak{H}\}$$

is self-adjoint in $\mathfrak{H}$.

Definition 2. *An operator $\mathbb{A} \in [\mathfrak{H}_+, \mathfrak{H}_-]$ is called a $(*)$-extension (or correct bi-extension) of an operator T (with non-empty set of regular points) if*

$$\mathbb{A} \supset T \supset A, \mathbb{A}^* \supset T^* \supset A$$

and the operator $\mathbb{A}_R = \frac{1}{2}(\mathbb{A} + \mathbb{A}^*)$ is a self-adjoint bi-extension of an operator A.

The existence, description and analog of von Neumann's formulas for self-adjoint bi-extensions and $(*)$-extensions were established in [14] (see also [5], [3], [6]). For instance, if Φ is an isometric operator from the defect subspace $\mathfrak{N}_i$ of the symmetric operator A onto the defect subspace $\mathfrak{N}_{-i}$, then the formulas below establish a one-to one correspondence between $(*)$-extensions of an operator T and Φ

$$\mathbb{A}f = A^*f + iR(\Phi - I)x, \; \mathbb{A}^*f = A^*f + iR(\Phi - I)y, \tag{2.1}$$

where $x, y \in \mathfrak{N}_i$ are uniquely determined from the conditions

$$f - (\Phi + I)x \in D(T), \; f - (\Phi + I)y \in D(T^*)$$

and R is the Riesz - Berezanskii operator of the triplet $\mathfrak{H}_+ \subset \mathfrak{H} \subset \mathfrak{H}_-$ that maps $\mathfrak{H}_+$ isometrically onto $\mathfrak{H}_-$ (see [14]). If the symmetric operator A has deficiency indices (n,n), then formulas (2.1) can be rewritten in the following form

$$\mathbb{A}f = A^*f + \sum_{k=1}^{n} \Delta_k(f)V_k, \mathbb{A}^*f = A^*f + \sum_{k=1}^{n} \delta_k(f)V_k, \tag{2.2}$$

where $\{V_j\}_1^n \in \mathfrak{H}_-$ is a basis in the subspace $R(\Phi - I)\mathfrak{N}_i$, and $\{\Delta_k\}_1^n$, $\{\delta_k\}_1^n$, are bounded linear functionals on $\mathfrak{H}_+$ with the properties

$$\Delta_k(f) = 0 \;\; \forall f \in D(T), \delta_k(f) = 0 \;\; \forall f \in D(T^*). \tag{2.3}$$

We will give a description of all $(*)$-extensions of a non-symmetric differential operator on half-line (see also [4]). Consider the following self-adjoint quasi-differential expression on $L_2[a, +\infty]$

$$l(y) = (-1)^n (p_0 y^{(n)})^{(n)} + (-1)^{n-1} (p_1 y^{(n-1)})^{(n-1)} + \cdots + p_n y,$$

where $\frac{1}{p_0(x)}, p_1(x), \ldots, p_n(x)$ are locally summable functions on $[a, +\infty)$. Denote by D^* the set of functions $y \in L_2[a, +\infty)$ for which the quasiderivatives $y^{[k]}$ $(k = 0, 1, \ldots, 2n-1)$ are locally absolutely continuous and $y^{[2n]} = l(y)$ belongs to $L_2[a, +\infty)$. Consider the symmetric operator

$$\begin{cases} Ay = l(y) \\ y^{[k-1]}(a) = 0 \end{cases}, \quad y \in D^*, k = 1, 2, \ldots, 2n$$

and suppose that this operator has deficiency indices (n, n). Consider the operators

$$\begin{cases} Ty = l(y), \\ y \in D^*, \\ \sum\limits_{k=1}^{2n} U_{jk} y^{[k-1]}(a) = 0 \end{cases}, \quad \begin{cases} T^*y = l(y), \\ y \in D^*, \\ \sum\limits_{k=1}^{2n} U_{*jk} y^{[k-1]}(a) = 0, \end{cases} \tag{2.4}$$
$$j = 1, 2, \ldots, n.$$

We will suppose also that $D(A) = D(T) \cap D(T^*)$ and that the operator T has a nonempty set of regular points. It is well known that

$$D(A^*) = D^*, \;\; A^*y = l(y), \;\; y \in D^*.$$

Consider $\mathfrak{H}_+ = D(A^*) = D^*$ with the scalar product

$$(y, z)_+ = (y, z)_{L_2[a, +\infty)} + (l(y), l(y))_{L_2[a, +\infty)}$$

and construct the triplet

$$\mathfrak{H}_+ \subset L_2[a, +\infty) \subset \mathfrak{H}_-.$$

Let $\{U_j\}_i^n$ and $\{U_{*j}\}_i^n$ be the set of elements in $\mathfrak{H}_-$ generating the functionals

$$\begin{aligned} (f, U_j) &= \sum_{k=1}^{2n} U_{jk} f^{[k-1]}(a) \\ (f, U_{*j}) &= \sum_{k=1}^{2n} U_{*jk} f^{[k-1]}(a) \end{aligned}, \quad j = 1, 2, \ldots, n.$$

Let $\{V_j\}_i^n \in \mathfrak{H}_-$ be the system of elements linearly independent with $\{U_j\}_i^n$ and generating the functionals

$$(f, V_j) = \sum_{k=1}^{2n} V_{jk} f^{[k-1]}(a), \;\; j = 1, 2, \ldots, n.$$

We also suppose that the operator

$$\begin{cases} \widehat{A}y = l(y) \\ (y, V_j) = 0 \end{cases}, \ j = 1, 2, \ldots, n$$

is self-adjoint in $L_2[a, +\infty)$.

Theorem 1. *Let T be a differential operator of the form (2.4) and $\rho(T) \neq \emptyset$. Then the formulas*

$$\begin{aligned} \mathbb{A}y &= l(y) + \sum_{k,j=1}^{n} (y, U_j) c_{kj} V_k, \\ \mathbb{A}^* y &= l(y) + \sum_{k,j=1}^{n} (y, U_{*j}) d_{kj} V_k \end{aligned} \tag{2.5}$$

establish a one-to-one correspondence between the set of $(*)$-extensions of an operator T and the set of matrices $\mathcal{C} = \|c_{jk}\|$ and $\mathcal{D} = \|d_{kj}\|$ uniquely determined by the matrices $\mathcal{U} = \|U_{jk}\|$, $\mathcal{U}_* = \|U_{*jk}\|$, $\mathcal{V}_* = \|v_{jk}\|$ by the following relation

$$\mathcal{U}_*^* \mathcal{D}^* \mathcal{V} - \mathcal{V}^* \mathcal{C} \mathcal{U} = \begin{bmatrix} 0 & I_{\nearrow} \\ -I_{\nearrow} & 0 \end{bmatrix} \left(\pm I_{\nearrow} := \left[\begin{matrix} 0 \cdots \pm 1 \\ \pm 1 \cdots 0 \end{matrix} \right] \right). \tag{2.6}$$

Proof. Let $\mathbb{A}$ be $(*)$-extension of an operator T of the form (2.4). Then, as we mentioned above, $\mathbb{A}$ has the form (2.2) with the property (2.3). Therefore

$$\Delta_k(f) = \sum_{j=1}^{n} (f, U_j) c_{kj},$$

$$\delta_k(f) = \sum_{j=1}^{n} (f, U_{kj}) d_{kj},$$

and the matrices $C = \|c_{kj}\|$ and $D = \|d_{kj}\|$ are invertible. From (2.2) it follows that

$$(A^* f, g) - (f, A^* g) = \sum_{k,j=1}^{n} (U_{kj}, g) \overline{d_{kj}} (f, V_k) - \sum_{k,j=1}^{n} (f, U_j) c_{kj} (V_k, g). \tag{2.7}$$

Applying the Lagrange's formula for differential operators and taking into account that the deficiency indices of the operator A are (n, n), we get from (2.7) for $g = f$:

$$\begin{aligned} &-\sum_{k} [f^{[k-1]}(a) f^{[2n-k]}(a) - f^{[2n-k]}(a) f^{[k-1]}(a)] = \\ &= \sum_{l,m} \Big(\sum_{k,j} \overline{d_{kj} U_{kjl} V_{km}} \Big) \overline{f^{[l-1]}(a)} f^{[m-1]}(a) - \\ &- \sum_{l,m} \Big(\sum_{k,j} c_{kj} U_{jl} \overline{V_{km}} \Big) f^{[l-1]}(a) \overline{f^{[m-1]}(a)}. \end{aligned} \tag{2.8}$$

Hence, we obtain (2.6). Going back to (2.6) we get (2.8), and as a consequence, (2.5) determines a $(*)$-extension of an operator T of the form (2.4). □

Now we will study Theorem 1 in the context of a non-self-adjoint Schrödinger operator. Let $\mathfrak{H} = L_2[a,+\infty)$ and $l(y) = -y'' + q(x)y$ where q is a real locally summable function. Suppose that the symmetric operator

$$\begin{cases} Ay = -y'' + q(x)y \\ y(a) = y'(a) = 0 \end{cases} \tag{2.9}$$

has deficiency indices (1,1). Let D^* be the set of functions locally absolutely continuous together with their first derivatives such that $l(y) \in L_2[a,+\infty)$. Consider $\mathfrak{H}_+ = D(A^*) = D^*$ with the scalar product

$$(y,z)_+ = \int_a^\infty \left(y(x)\overline{z(x)} + l(y)\overline{l(z)}\right) dx, \quad y,\ z \in D^*.$$

Let

$$\mathfrak{H}_+ \subset L_2[a,+\infty) \subset \mathfrak{H}_-$$

be the corresponding triplet of Hilbert spaces. Consider elements $U, U_*, V \in \mathfrak{H}_-$, and generating functionals

$$\begin{aligned}
(f,V) &= \frac{(\operatorname{Im} h)^{\frac{1}{2}}}{|\mu - h|}[\mu f(a) - f'(a)], \quad \operatorname{Im} h > 0,\ \operatorname{Im} \mu = 0,\\
(f,U) &= hf(a) - f'(a),\\
(f,U_*) &= \bar{h} f(a) - f'(a),
\end{aligned}$$

and operators

$$\begin{cases} Ty = l(y) = -y'' + q(x)y \\ (y,U) = hy(a) - y'(a) = 0 \end{cases}, \quad \begin{cases} T^*y = l(y) = -y'' + q(x)y \\ (y,U_*) - \bar{h}y(a) - y'(a) = 0 \end{cases}, \tag{2.10}$$

$$\begin{cases} \widehat{A}y = l(y) = -y'' + q(x)y \\ (y,V) = \mu y(a) - y'(a) = 0 \end{cases}, \quad \operatorname{Im} \mu = 0.$$

It is well known [1] that $\widehat{A} = \widehat{A^*}$. By Theorem 1, the $(*)$-extension $\mathbb{A}$ of an operator T of the form (2.10) can be represented as

$$\begin{aligned}
\mathbb{A}y &= -y'' + q(x)y + c(y,U)V,\\
\mathbb{A}y &= -y'' + q(x)y + d(y,U_*)V,
\end{aligned}$$

where the numbers c and d satisfy the relation (2.6). According to (2.4) and (2.10), the matrices $\mathcal{U}$, $\mathcal{U}_*$ and $\mathcal{V}$ have the form

$$\mathcal{U} = \begin{bmatrix} h & -1 \end{bmatrix}, \ \mathcal{U}_* = \begin{bmatrix} \bar{h} & -1 \end{bmatrix}, \ \mathcal{V} = \begin{bmatrix} \frac{(\operatorname{Im} h)^{1/2}}{|\mu-h|}\mu & \frac{-(\operatorname{Im} h)^{1/2}}{|\mu-h|} \end{bmatrix}.$$

Therefore,

$$\bar{d}\,\mathcal{U}_*^*\,\mathcal{V} - c\mathcal{V}^*\mathcal{U} = \bar{d}\begin{bmatrix} h \\ -1 \end{bmatrix} \begin{bmatrix} \frac{(\operatorname{Im} h)^{1/2}}{|\mu-h|}\mu & -\frac{(\operatorname{Im} h)^{1/2}}{|\mu-h|} \end{bmatrix} -$$

$$- c\begin{bmatrix} \frac{(\operatorname{Im} h)^{1/2}}{|\mu-h|}\mu \\ -\frac{(\operatorname{Im} h)^{1/2}}{|\mu-h|} \end{bmatrix} \begin{bmatrix} h & -1 \end{bmatrix} = \begin{bmatrix} 0 & 1 \\ -1 & 0 \end{bmatrix}.$$

Solving this matrix equation with respect to c and d we get

$$c = \frac{|\mu - h|}{(\mu - h)(\operatorname{Im} h)^{\frac{1}{2}}}, \quad d = \frac{|\mu - h|}{(\mu - \bar{h})(\operatorname{Im} h)^{\frac{1}{2}}}.$$

Consider now the elements $\mu\delta(x-a) + \delta'(x-a)$ and $\bar{h}\delta(x-a) + \delta'(x-a)$ in $\mathfrak{H}_-$, and generating functionals

$$(f, \mu\delta(x-a) + \delta'(x-a)) = \mu f(a) - f'(a),$$
$$(f, \bar{h}\delta(x-a) + \delta'(x-a)) = hf(a) - f'(a),$$

where $\delta(x-a)$ and $\delta'(x-a)$ are the delta-function and the derivative of the delta-function at the point a. Thus, we proved

Theorem 2. *The set of all $(*)$-extensions of a non-self-adjoint Schrödinger operator T_h of the form* (2.10) *in $L_2[a, +\infty)$ can be represented in the form*

$$\begin{aligned} \mathbb{A}y &= -y'' + q(x)y - \frac{1}{\mu - h}\,[y'(a) - hy(a)]\,[\mu\delta(x-a) + \delta'(x-a)], \\ \mathbb{A}y &= -y'' + q(x)y - \frac{1}{\mu - \bar{h}}\,[y'(a) - \bar{h}y(a)]\,\mu\delta(x-a) + \delta'(x-a)]. \end{aligned} \tag{2.11}$$

In addition, the formulas (2.11) *establish a one-to-one correspondence between the set of all $(*)$-extensions of a Schrödinger operator T_h of the form* (2.10) *and all real numbers $\mu \in [-\infty, +\infty]$.*

Definition 3. *An operator T with the domain $D(T)$ acting on a Hilbert space $\mathfrak{H}$ is called accretive if*

$$\operatorname{Re}(Tf, f) \geq 0 \ \ \forall f \in D(T).$$

Consider the symmetric operator A of the form (2.9) with defect indices (1,1), generated by the differential operation $l(y) = -y'' + q(x)y$. Let $\varphi_k(x, \lambda)(k = 1, 2)$ be the solutions of the following Cauchy problems:

$$\begin{cases} l(\varphi_1) = \lambda\varphi_1 \\ \varphi_1(a, \lambda) = 0 \\ \varphi_1'(a, \lambda) = 1 \end{cases}, \quad \begin{cases} l(\varphi_2) = \lambda\varphi_2 \\ \varphi_2(a, \lambda) = -1 \\ \varphi_2'(a, \lambda) = 0 \end{cases}.$$

It is well known [1] that there exists a function $m_\infty(\lambda)$ (called the Weyl-Titchmarsh function) for which

$$\varphi(x, \lambda) = \varphi_2(x, \lambda) + m_\infty(\lambda)\varphi_1(x, \lambda)$$

belongs to $L_2[a, +\infty)$.

Suppose that the symmetric operator A of the form (2.9) with deficiency indices (1,1) is nonnegative, i.e., $(Af, f) \geq 0$ for all $f \in D(A)$). It was shown in [15] that the Schrödinger operator T_h of the form (2.10) is accretive if and only if

$$\operatorname{Re} h \geq -m_\infty(-0). \tag{2.12}$$

For real h such that $h \geq -m_\infty(-0)$ we get a description of all nonnegative self-adjoint extensions of an operator A. For $h = -m_\infty(-0)$ the corresponding operator

$$\begin{cases} A^M y = -y'' + q(x)y \\ y'(a) + m_\infty(-0)y(a) = 0 \end{cases} \tag{2.13}$$

is the M. Kreĭn-von Neumann extension of A and for $h = +\infty$ the corresponding operator

$$\begin{cases} A^\mu y = -y'' + q(x)y \\ y(a) = 0 \end{cases} \tag{2.14}$$

is the Friedrichs extension of A. From (2.12), (2.13) and (2.14) it follows that a nonnegative operator A of the form (2.10) admits non-self-adjoint accretive extensions if and only if $m_\infty(-0) < \infty$.

3. Linear systems in the form of Brodskiĭ-Livsic rigged operator colligations

Let $\mathbb{A}$ be $(*)$-extension of an operator T, i.e.,

$$\mathbb{A} \supset T \supset A, \mathbb{A}^* \supset T^* \supset A$$

where A is a symmetric operator with deficiency indices (n,n) and $D(A) = D(T) \cap D(T^*)$.

Definition 4. *The array*

$$\Theta = \begin{pmatrix} & \mathbb{A} & K & J \\ \mathfrak{H}_+ \subset & \mathfrak{H} \subset \mathfrak{H} & & \mathfrak{E} \end{pmatrix} \quad (\dim \mathfrak{E} < \infty)$$

is called a linear, conservative, stationary dynamic system (l.c.s.d.s) or a Brodskiĭ-Livsic rigged operator colligation if:

1) $\mathfrak{E}$ *is a Hilbert space with scalar product* $(,)_\mathfrak{E}$ *and the operator* J *in this space satisfies the conditions* $J = J^* = J^{-1}$,
2) $K \in [\mathfrak{E}, \mathfrak{H}_-]$,
3) $\operatorname{Im} \mathbb{A} = KJK^*$, *where* $K^* \in [\mathfrak{H}_+, \mathfrak{E}]$ *is the adjoint of* K.

This array represents the following system Θ

$$\begin{cases} (\mathbb{A} - zI)x = KJ\varphi_- \\ \varphi_+ = \varphi_- - 2iK^*x \end{cases},$$

where $\varphi_- \in \mathfrak{E}$ is an input vector, $\varphi_+ \in \mathfrak{E}$ is an output vector, and x is a state space vector in $\mathfrak{H}$.

An operator $\mathbb{A}$ is said to be a main operator of the system Θ, J is said to be a direction operator, and K is said to be a channel operator. The operator-valued function

$$W_\Theta(\lambda) = I - 2iK^*(\mathbb{A} - \lambda I)^{-1}KJ \tag{3.1}$$

defined on the set $\rho(T)$ of regular points of an operator T is called the transfer operator-valued function (characteristic operator-valued function) of the system

Θ, i.e., $\varphi_+ = W_\Theta(\lambda)\varphi_-$. It was shown in [15] that any $(*)$-extension $\mathbb{A}$ of an operator $T(A^* \supset T \supset A)$, where A is a symmetric operator with deficiency indices (n,n) $(n < \infty)$, $D(A) = D(T) \cap D(T^*)$, can be included as a main operator of some Brodskiĭ-Livsic rigged operator colligation with $\dim E < \infty$ and invertible channel operator K.

It was also established in [15] that

$$V_\Theta(\lambda) = K^*(\operatorname{Re}\mathbb{A} - \lambda I)^{-1}K \tag{3.2}$$

is a Herglotz-Nevanlinna operator-valued function acting on a Hilbert space $\mathfrak{E}$, satisfying the following relation

$$V_\Theta(\lambda) = i[W_\Theta(\lambda) - I][W_\Theta(\lambda) + I]^{-1}J,\ \lambda \in \rho(T),\ \operatorname{Im}\lambda \neq 0. \tag{3.3}$$

It is easy to check that the $(*)$-extension

$$\mathbb{A}y = -y'' + q(x)y - \frac{1}{\mu - h}[y'(a) - hy(a)]\,[\mu\delta(x-a) + \delta'(x-a)],\quad \operatorname{Im}h > 0$$

of the non-self-adjoint Schrödinger operator T_h of the form (2.10) satisfies the condition

$$\operatorname{Im}\mathbb{A} = \frac{\mathbb{A} - \mathbb{A}^*}{2i} = (.,g)g, \tag{3.4}$$

where

$$g = \frac{(\operatorname{Im}h)^{\frac{1}{2}}}{|\mu - h|}\,[\mu\delta(x-a) + \delta'(x-a)] \tag{3.5}$$

and $\delta(x-a), \delta'(x)$ are the delta-function and its derivative at the point a. Moreover,

$$(y,g) = \frac{(\operatorname{Im}h)^{\frac{1}{2}}}{|\mu - h|}\,[\mu y(a) - y'(a)], \tag{3.6}$$

where

$$y \in \mathfrak{H}_+, g \in \mathfrak{H}_-, \mathfrak{H}_+ \subset L_2(a,+\infty) \subset \mathfrak{H}_-$$

and the triplet of Hilbert spaces is as discussed in Theorem 2. Let $\mathfrak{E} = \mathbb{C}$, $Kc = cg$ $(c \in \mathbb{C})$. It is clear that

$$K^*y = (y,g)y \in \mathfrak{H}_+ \tag{3.7}$$

and $\operatorname{Im}\mathbb{A} = KK^*$. Therefore, the array

$$\Theta = \begin{pmatrix} \mathbb{A} & K & 1 \\ \mathfrak{H}_+ \subset L_2[a,+\infty) \subset \mathfrak{H}_- & & \mathbb{C} \end{pmatrix} \tag{3.8}$$

is the Brodskiĭ - Livsic rigged operator colligation or l.c.s.d.s with the main operator $\mathbb{A}$ of the form (2.11), the direction operator $J = 1$ and the channel operator K of the form (3.7). Direct computations show that

$$W_\Theta(\lambda) = \frac{\mu - h}{\mu - \overline{h}}\,\frac{m_\infty(\lambda) + \overline{h}}{m_\infty(\lambda) + h}, \tag{3.9}$$

$$V_\Theta(\lambda) = \frac{(m_\infty(\lambda) + \mu)\operatorname{Im}h}{(\mu - \operatorname{Re}h)\,m_\infty(\lambda) + \mu\operatorname{Re}h - |h|^2} \tag{3.10}$$

and

$$\operatorname{Im} V_\Theta(i) = -\operatorname{Im} h \, \frac{|\mu - h|^2}{(\mu - \operatorname{Re} h)^2} \frac{\operatorname{Im} m_\infty(i)}{\left| m_\infty(i) + \frac{\mu \operatorname{Re} h - |h|^2}{\mu - \operatorname{Re} h} \right|^2}. \tag{3.11}$$

Remark 3. *The expression* (3.9) *of the characteristic function was obtained earlier by B.S. Pavlov* [13] *using a different approach based on the method of self-adjoint dilations of dissipative operators.*

Theorem 4. *Let* T_h $(\operatorname{Im} h > 0)$ *be an accretive Schrödinger operator of the form* (2.10)*. Then for all real* μ *satisfying the following inequality*

$$\mu \geq \frac{(\operatorname{Im} h)^2}{m_\infty(-0) + \operatorname{Re} h} + \operatorname{Re} h, \tag{3.12}$$

the operators

$$\begin{aligned} \mathbb{A} y &= -y'' + q(x)y + \frac{1}{\mu - h} [hy(a) - y'(a)] [\mu \delta'(x-a)], \\ \mathbb{A}^* y &= -y'' + q(x)y + \frac{1}{\mu - \overline{h}} [\overline{h} y(a) - y'(a)] [\mu \delta'(x-a)] \end{aligned} \tag{3.13}$$

define the set of all accretive $(*)$*-extensions* $\mathbb{A}$ *of the operator* T_h*. An operator* T_h *has a unique accretive* $(*)$*-extension* $\mathbb{A}$ *if and only if*

$$\operatorname{Re} h = -m_\infty(0).$$

In this case this unique $(*)$*-extension has the form*

$$\begin{aligned} \mathbb{A} y &= -y'' + q(x)y + [hy(a) - y'(a)] \, \delta(x-a), \\ \mathbb{A}^* y &= -y'' + q(x)y + [\overline{h} y(a) - y'(a)] \, \delta(x-a). \end{aligned} \tag{3.14}$$

Proof. From Theorem 2 it follows that the set of all $(*)$-extensions of the Schrödinger operator $T_h(\operatorname{Im} h > 0)$ of the form (2.10) can be described by the formula (3.13). Suppose that $\mathbb{A}$ of the form (3.13) is an accretive $(*)$-extension of the accretive Schrödinger operator T_h of the form (2.10). Then it follows from [15] that

$$\mu + m_\infty(-0) \geq 0. \tag{3.15}$$

Let $\xi < 0$. By the direct computations using (3.9) it follows that

$$\begin{aligned} \operatorname{Im} W_\Theta(\xi) = &\frac{-2[\mu + m_\infty(\xi) \operatorname{Im} h]}{|\mu - h|^2 |m_\infty(\xi) + h|^2} \times \\ &\times [-(\operatorname{Re} h)^2 + (\mu - m_\infty(\xi)) \operatorname{Re} h + \mu m_\infty(\xi) - (\operatorname{Im} h)^2]. \end{aligned} \tag{3.16}$$

Taking into account that $m_\infty(\xi) \to +\infty$ monotonically as $\xi \to -\infty$, we get that

$$\mu + m_\infty(\xi) > 0, \ \xi < 0. \tag{3.17}$$

From (3.2) and(3.3) we obtain

$$V_\Theta(\lambda) = i[W_\Theta(\lambda) + 1]^{-1} [W_\Theta(\lambda) - 1] = i - 2i[W_\Theta(\lambda) + 1]^{-1}. \tag{3.18}$$

Therefore,

$$\operatorname{Re} V_\Theta(\lambda) = -2[W_\Theta^*(\lambda)+1]^{-1}\,\frac{W_\Theta(\lambda)-W_\Theta^*(\lambda)}{2i}\,[W_\Theta(\lambda)+1]^{-1} \tag{3.19}$$

and for $\xi < 0$,

$$V_\Theta(\xi) = -2[W_\Theta^*(\xi)+1]^{-1}\operatorname{Im} W_\Theta(\xi)\,[W_\Theta(\xi)+1]^{-1}. \tag{3.20}$$

Comparing (3.20) and (3.2) we obtain

$$\operatorname{Im} W_\Theta(\xi) \le 0,\ \xi < 0.$$

From (3.16) it follows that

$$-(\operatorname{Re} h)^2 - (\mu - m_\infty(\xi))\operatorname{Re} h + \mu m_\infty(\xi) - (\operatorname{Im} h)^2 \ge 0$$

for all $\xi < 0$. Taking the limit $\xi \to -0$ we obtain

$$-(\operatorname{Re} h)^2 - (\mu - m_\infty(-0))\operatorname{Re} h + \mu m_\infty(-0) - (\operatorname{Im} h)^2 \ge 0,$$

which is equivalent to (3.12).

Suppose now that the inequality (3.12) holds. We will show that the operator $\mathbb{A}$ of the form (3.13) is accretive. Because $m_\infty(-0) < m_\infty(\xi)$ for all $\xi < 0$, it is easy to verify that the inequality

$$\begin{aligned} &-(\operatorname{Re} h)^2 - (\mu - m_\infty(-0))\operatorname{Re} h + \mu m_\infty(-0) - (\operatorname{Im} h)^2 \\ &\le -(\operatorname{Re} h)^2 - (\mu - m_\infty(\xi))\operatorname{Re} h + \mu m_\infty(\xi) - (\operatorname{Im} h)^2 \end{aligned} \tag{3.21}$$

holds for

$$\mu \ge \operatorname{Re} h. \tag{3.22}$$

By (3.12) one concludes (3.22) and the inequality (3.12) is then equivalent to the left-hand side of (3.21) being non-negative, i.e.,

$$\operatorname{Im} W_\Theta(\xi) \le 0,\ \xi < 0.$$

Therefore, applying (3.2) we get

$$V_\Theta(\xi) \ge 0,\ \xi < 0. \tag{3.23}$$

Because $V_\Theta(\lambda)$ is a Herglotz-Nevanlinna function, and it is nonnegative on the negative axis, then it follows from [9] that $V_\Theta(\lambda)$ is a Stieltjes function (i.e., $V_\Theta(\lambda)$ is a Herglotz-Nevanlinna function as well as $\lambda V_\Theta(\lambda)$). Therefore, from [9] it follows that $\mathbb{A}$ is accretive. When $\operatorname{Re} h = -m_\infty(-0)$ then (3.12) implies that $\mu = \infty$. Taking the limit $\mu \to +\infty$ in (3.13) we get that T_h has only one accretive (∗)-extension $\mathbb{A}$ if and only if $\operatorname{Re} h = -m_\infty(-0)$. This unique (∗)-extension $\mathbb{A}$ has the form (3.14). □

4. Some relations between spectral functions of the Friedrichs and Kreĭn-von Neumann extensions

Definition 5. *A linear, closed operator T acting on a Hilbert space $\mathfrak{H}$ with dense domain $D(T)$ is said to be prime if there is no reducing, nontrivial invariant subspace on which T induces a self-adjoint operator.*

It is well known [14], [15] that if the operator T satisfies the conditions

$$T \supset A, \qquad T^* \supset A, \tag{4.1}$$
$$D(A) = D(T) \cap D(T^*), \tag{4.2}$$

where A is a symmetric operator with dense domain $D(A)$, then T is prime if and only if the symmetric part A of T is prime. It is easy to verify that a symmetric operator A is prime if and only if

$$\underset{\lambda \neq \overline{\lambda}}{Span}\, \mathfrak{N}_\lambda = \mathfrak{H},$$

where $\mathfrak{N}_\lambda$ is the defect subspace of A. Moreover, if $A \geq 0$ and T is accretive then T is prime if and only if

$$\underset{\lambda \in \rho(T)}{Span}\, \mathfrak{N}_\lambda = \mathfrak{H},$$

where $\rho(T)$ is the set of all regular points of T.

Theorem 5. *Let* $\Theta_1 = \begin{pmatrix} \mathbb{A}_1 & K_1 & J \\ \mathfrak{H}_{+1} \subset \mathfrak{H}_1 \subset \mathfrak{H}_{-1} & & \mathfrak{E} \end{pmatrix}$ *and*
$\Theta_2 = \begin{pmatrix} \mathbb{A}_2 & K_2 & J \\ \mathfrak{H}_{+2} \subset \mathfrak{H}_2 \subset \mathfrak{H}_{-2} & & \mathfrak{E} \end{pmatrix}$ *be a l.c.s.d.s (Brodskiĭ-Livsic rigged operator colligations), where*

$$\begin{aligned} &\mathbb{A}_1 \supset T_1 \supset A_1, \mathbb{A}_1^* \supset T_1^* \supset A_1,\\ &\mathbb{A}_2 \supset T_2 \supset A_2, \mathbb{A}_2^* \supset T_2^* \supset A_2,\\ &D(A_1) = D(T_1) \cap D(T_1^*), D(A_2) = D(T_2) \cap D(T_2^*) \end{aligned} \tag{4.3}$$

and the operators T_1 and T_2 are accretive, prime, and A_1 and A_2 have finite and equal defect indices.

If

$$W_{\Theta_1}(\lambda) = W_{\Theta_2}(\lambda), \tag{4.4}$$

then there exists an isometric operator U from $\mathfrak{H}_1$ onto $\mathfrak{H}_2$ such that $U_+ = U|_{\mathfrak{H}_{+1}}$ is an isometry from $\mathfrak{H}_{+1}$ onto $\mathfrak{H}_{+2}$, $U_-^* = U_+^*$ is an isometry from $\mathfrak{H}_{-1}$ onto $\mathfrak{H}_{-2}$, and

$$UT_1 = T_2 U, U_- K_1 = K_2. \tag{4.5}$$

Proof. From (3.1) and (4.4) it follows

$$K_1^*(\mathbb{A}_1 - \lambda I)^{-1} K_1 = K_2^*(\mathbb{A}_2 - \lambda I)^{-1} K_2. \tag{4.6}$$

Since

$$\begin{aligned}&(\mathbb{A}_j - \lambda I)^{-1} - (\mathbb{A}_j^* - \overline{\lambda} I)^{-1}\\ &= (\mathbb{A}_j^* - \overline{\lambda} I)^{-1}[(\mathbb{A}_j^* - \overline{\lambda} I) - (\mathbb{A}_j - \lambda I)](\mathbb{A}_j - \lambda I)^{-1}\\ &= (\lambda - \overline{\mu})(\mathbb{A}_j^* - \overline{\lambda} I)^{-1}(\mathbb{A}_j - \lambda I)^{-1} - 2i(\mathbb{A}_j^* - \overline{\lambda} I)^{-1} K_j J K_j^* (\mathbb{A}_j - \lambda I)^{-1}\\ &(j = 1, 2)\end{aligned}$$

we get

$$\begin{aligned}&(\lambda - \overline{\mu}) K_1^* (\mathbb{A}_1^* - \overline{\lambda} I)^{-1}(\mathbb{A}_1 - \lambda I)^{-1} K_1\\ &= K_1^*(\mathbb{A}_1 - \lambda I)^{-1} K_1 - K_1^*(\mathbb{A}_1^* - \overline{\lambda} I)^{-1} K_1\\ &+2iK_1 * (\mathbb{A}_1^* - \overline{\lambda} I)^{-1} K_1 J K_1^* (\mathbb{A}_1^* - \lambda I)^{-1} K_1\\ &= K_2^*(\mathbb{A}_2 - \lambda I)^{-1} K_2 - K_2^*(\mathbb{A}_2^* - \overline{\lambda} I)^{-1} K_2\\ &+2iK_2^*(\mathbb{A}_2^* - \overline{\lambda} I)^{-1} K_2 J K_2^* (\mathbb{A}_2^* - \lambda I)^{-1} K_2\\ &= (\lambda - \overline{\mu}) K_2^*(\mathbb{A}_2^* - \overline{\lambda} I)^{-1}(\mathbb{A}_2 - \lambda I)^{-1} K_2.\end{aligned}$$

Therefore,

$$K_1^*(\mathbb{A}_1 - \overline{\lambda} I)^{-1}(\mathbb{A}_1 - \lambda I)^{-1} K_1 = K_2^*(\mathbb{A}_2^* - \overline{\lambda} I)^{-1}(\mathbb{A}_2 - \lambda I)^{-1} K_2. \tag{4.7}$$

Taking into account Hilbert's identity for resolvents

$$\begin{aligned}&(\mathbb{A}_j - \lambda I)^{-1} - (\mathbb{A}_j - \xi I)^{-1} = (\lambda - \xi)(\mathbb{A}_j - \xi I)^{-1}(\mathbb{A}_j - \lambda I)^{-1}\\ &(j = 1, 2)\end{aligned}$$

we get from (4.7)

$$\begin{aligned}&K_1^*(\mathbb{A}_1 - \overline{\mu} I)^{-m}(\mathbb{A}_1 - \lambda I)^{-n} K_1\\ &= K_2^*(\mathbb{A}_2^* - \overline{\mu} I)^{-m}(\mathbb{A}_2 - \lambda I)^{-n} K_1\\ &(m, n = 1, 2, \dots)\end{aligned} \tag{4.8}$$

and

$$\begin{aligned}&((\mathbb{A}_1 - \lambda I)^{-n} K_1 x, (\mathbb{A}_1 - \mu I)^{-m} K_1 y)_{\mathfrak{H}_1}\\ &= ((\mathbb{A}_2 - \lambda I)^{-n} K_2 x, (\mathbb{A}_2 - \mu I)^{-m} K_2 y)_{\mathfrak{H}_2}\\ &(x, y \in \mathfrak{E}).\end{aligned} \tag{4.9}$$

It is well known [14] that

$$\begin{aligned}(\mathbb{A}_1 - \lambda I)^{-1} K \mathfrak{E} &= \mathfrak{N}_\lambda^{(1)},\\ (\mathbb{A}_2 - \lambda I)^{-1} K \mathfrak{E} &= \mathfrak{N}_\lambda^{(2)}.\end{aligned} \tag{4.10}$$

Because the operators T_1 and T_2 are prime, then

$$\underset{\lambda \neq \overline{\lambda}}{Span}\, \mathfrak{N}_\lambda^{(1)} = \mathfrak{H}_1, \underset{\lambda \neq \overline{\lambda}}{Span}\, \mathfrak{N}_\lambda^{(2)} = \mathfrak{H}_2. \tag{4.11}$$

From (4.9) it follows that there exists an isometric operator $U : \mathfrak{H}_1 \to \mathfrak{H}_2$ such that

$$U(\mathbb{A}_1 - \lambda I)^{-n} K_1 = (\mathbb{A}_2 - \lambda I)^{-n} K_2, \;\; (n = 1, 2, \cdots).$$

Hence it follows that

$$\begin{aligned} &U(\mathbb{A}_1-\lambda I)^{-1}(\mathbb{A}_1-\lambda I)^{-1}K_1 \\ &=(\mathbb{A}_2^*-\lambda I)^{-1}(\mathbb{A}_2-\lambda I)^{-1}K_2 \\ &=(\mathbb{A}_2-\lambda I)^{-1}U(\mathbb{A}_1-\lambda I)^{-1}K_1. \end{aligned}$$

Taking into account (4.10), we get

$$U(T_1-\lambda I)^{-1}=(T_2-\lambda I)^{-1}U \tag{4.12}$$

and

$$UT_1=T_2U. \tag{4.13}$$

Therefore, $UA_1=A_2U$, $UA_1^*=A_2^*U$. Because

$$\begin{aligned} (x,y)_{\mathfrak{H}_{+1}} &= (x,y)_{\mathfrak{H}_1}+(A_1^*x,A_1^*y)_{\mathfrak{H}_1}, \\ (x,y)_{\mathfrak{H}_{+2}} &= (x,y)_{\mathfrak{H}_2}+(A_2^*x,A_2^*y)_{\mathfrak{H}_2} \end{aligned}$$

the operator U is an isometry of $\mathfrak{H}_{+1}$ onto $\mathfrak{H}_{+2}$. Since $\operatorname{Re}\mathbb{A}_j=\frac{1}{2}(\mathbb{A}_j+\mathbb{A}_j^*)$ $(j=1,2)$ is a self-adjoint bi-extension of the symmetric operator A_j, the operators

$$\begin{aligned} &\widehat{A}_j f=\operatorname{Re}\mathbb{A}_j f, \\ &D(\widehat{A}_j)=\{f\in\mathfrak{H}_{+j}:\operatorname{Re}\mathbb{A}_j f\in\mathfrak{H}_j\},\ j=1,2 \end{aligned} \tag{4.14}$$

are self-adjoint in $\mathfrak{H}_j$ $(j=1,2)$. These operators are self-adjoint extensions of A_j $(j=1,2)$. Since $U^*:\mathfrak{H}_{+2}\to\mathfrak{H}_{+1}$ is an isometric operator, the equality

$$(U^*x_2,\alpha_1)_{\mathfrak{H}_1}=(x_2,(U^*)^*\alpha_1)_{\mathfrak{H}_2},(x_2\in\mathfrak{H}_{+2},\alpha_1\in\mathfrak{H}_{-1}) \tag{4.15}$$

uniquely defines the operator $(U^*)^*:\mathfrak{H}_{-1}\to\mathfrak{H}_{-2}$, which is an adjoint of $U^*:\mathfrak{H}_{+2}\to\mathfrak{H}_{+1}$. The operator $(U^*)^*$ is an isometric operator from $\mathfrak{H}_{-1}$ onto $\mathfrak{H}_{-2}$. From (4.15) it follows

$$(U^*)^*\alpha_1=U\alpha_1\quad(\alpha_1\in\mathfrak{H}_1).$$

Thus, $(U^*)^*$ is an extension of U onto $\mathfrak{H}_{-1}$. Denote

$$U_-=(U^*)^*,\ U_+=U|_{\mathfrak{H}_{+1}}.$$

We obtained a triplet of operators (U_+,U,U_-) that maps isometrically the triplet

$$(\mathfrak{H}_{+1},\mathfrak{H}_1,\mathfrak{H}_{-1})$$

onto the triplet $(\mathfrak{H}_{+2},\mathfrak{H}_2,\mathfrak{H}_{-2})$. From (4.12) it follows that $U_+(\mathbb{A}_1-\lambda I)^{-1}=(\mathbb{A}_2-\lambda I)^{-1}U_-$. Taking into account that $U_-|_{\mathfrak{H}_{+1}}=U_+$, we have $(\mathbb{A}_2-\lambda I)U_+=U_-(\mathbb{A}_1-\lambda I)$. Finally, $\mathbb{A}_2=U_-\mathbb{A}_1U_+^{-1}$ and

$$\begin{aligned} \operatorname{Re}\mathbb{A}_2 &= U_-\operatorname{Re}\mathbb{A}_1U_+^{-1}, \\ \operatorname{Im}\mathbb{A}_2 &= U_-\operatorname{Im}\mathbb{A}_1U_+^{-1}. \end{aligned}$$

From (4.14) and the description of self-adjoint bi-extensions of the symmetric operators one concludes $\widehat{A}_2=U\widehat{A}_1U^{-1}$. This means that the operators $\widehat{A}_2$ and $\widehat{A}_1$ are unitarily equivalent. □

Next, consider the Brodskiĭ-Livsic rigged operator colligation (l.c.s.d.s) of the form (3.8)

$$\Theta = \begin{pmatrix} \mathbb{A} & K & 1 \\ \mathfrak{H}_+ \subset L_2[a,+\infty) \subset \mathfrak{H}_- & & \mathbb{C} \end{pmatrix},$$

where

$$\begin{aligned} \mathbb{A}y &= -y'' + q(x)y - \frac{1}{\mu - h}[y'(a) - hy(a)][\mu\delta(x-a) + \delta'(x-a)], \\ \mathbb{A}^*y &= -y'' + q(x)y - \frac{1}{\mu - \overline{h}}[y'(a) - \overline{h}y(a)][\mu\delta(x-a) + \delta'(x-a)] \end{aligned} \tag{4.16}$$

and suppose that the operator

$$\begin{cases} T_h y = -y'' + q(x)y \\ y'(a) = hy(a) \end{cases} \tag{4.17}$$

is accretive and the symmetric operator

$$\begin{cases} Ay = -y'' + q(x)y \\ y'(a) = y(a) = 0 \end{cases} \tag{4.18}$$

in $L_2[a,+\infty)$ has defect indices $(1,1)$. It is known [15] that the operator T_h is accretive if and only if $\operatorname{Re} h \geq -m_\infty(-0)$. Assume that $\operatorname{Re} h = -m_\infty(-0)$. Then, from Theorem 4 it follows that the $(*)$-extension (4.16) of the operator $T_h(\operatorname{Re} h = -m_\infty(-0))$ is accretive if and only if $\mu = \infty$ (see (3.12)). From the relations (4.16) we get for $\mu = \infty$

$$\begin{aligned} \mathbb{A}y &= -y'' + q(x)y + [hy(a) - y'(a)]\delta(x-a), \\ \mathbb{A}^*y &= -y'' + q(x)y + [\overline{h}y(a) - y'(a)]\delta(x-a) \end{aligned} \tag{4.19}$$

and

$$\begin{aligned} \operatorname{Re}\mathbb{A}y &= -y'' + q(x)y + [m_\infty(-0)y(a) + y'(a)]\delta(x-a), \\ \operatorname{Im}\mathbb{A}y &= (\operatorname{Im} h)y(a)\delta(x-a) = (y,g)g, \ \operatorname{Im} h > 0. \end{aligned} \tag{4.20}$$

Here

$$g = (\operatorname{Im} h)^{\frac{1}{2}}\delta(x-a) \tag{4.21}$$

and the operator $\operatorname{Re}\mathbb{A}$ of the form (4.20) is nonnegative, i.e., $(\operatorname{Re}\mathbb{A}y, y) \geq 0$ for all $y \in \mathfrak{H}_+$. Moreover, as was shown in [9]

$$V_\Theta(\lambda)\phi = K^*(\operatorname{Re}\mathbb{A} - \lambda I)^{-1}K\phi = \int_0^\infty \frac{d\sigma(t)}{t-\lambda}\phi, \ \phi \in \mathbb{C}.$$

So, we can associate $V_\Theta(\lambda)$ with the Stieltjes function

$$V_\Theta(\lambda) = \int_0^\infty \frac{d\sigma(t)}{t-\lambda}, \tag{4.22}$$

where

$$\int_0^\infty d\sigma(t) = \infty, \quad \int_0^\infty \frac{d\sigma(t)}{1+t} < \infty. \tag{4.23}$$

Because $\operatorname{Re}\mathbb{A} \supset A^M$ (see (4.20)), where

$$\begin{cases} A^M y = -y'' + q(x)y \\ y'(a) + m_\infty(-0)y(a) = 0 \end{cases} \tag{4.24}$$

is the Kreĭn-von Neumann extension of a nonnegative operator A of the form (4.18), $\sigma(t)$ is the spectral function of distribution of the Kreĭn-von Neumann boundary value problem of the form (4.24).

Theorem 6. *Let* $\Theta = \begin{pmatrix} \mathbb{A} & K & 1 \\ \mathfrak{H}_+ \subset L_2[a,+\infty) \subset \mathfrak{H}_- & & \mathbb{C} \end{pmatrix}$ *be the l.c.s.d.s (Brodskiĭ-Livsic rigged operator colligation)* (3.8)*, where* $\mathbb{A}$ *is a* $(*)$*-extension of the form* (4.19) *of the accretive Schrödinger operator* T_h *with* $\operatorname{Re} h = -m_\infty(-0)$*. Then the spectral measure in the representation*

$$V_\Theta(\lambda) = \int_0^\infty \frac{d\sigma(t)}{t-\lambda} \tag{4.25}$$

satisfies the following relation

$$\int_0^\infty \frac{d\sigma(t)}{1+t^2} = \operatorname{Im} h \left(\sup_{y \in D(A^M)} \frac{|y(a)|}{\left(\int_a^\infty \left(|y(x)|^2 + |l(y)|^2 \right) dx \right)^{\frac{1}{2}}} \right)^2. \tag{4.26}$$

Proof. Consider the Hilbert space $L_2^\sigma[0,+\infty)$, where $\sigma(t)$ is the spectral function of distribution from the parametric representation (4.22) and the operator

$$\vec{A}x = tx(t),$$

$$D(\vec{A}) = \left\{ x \in L_2^\sigma[0,\infty) : \ tx(t) \in L_2^\sigma[0,\infty) \quad \text{and} \quad \int_0^\infty x(t) d\sigma(t) = 0 \right\}.$$

It was proved in [2], [14] that operator $\vec{A}$ has defect indices (1,1) and that

$$\widehat{\vec{A}}x = tx(t),$$

$$D(\widehat{\vec{A}}) = \{ x \in L_2^\sigma[0,\infty) : tx(t) \in L_2^\sigma[0,\infty) \}$$

is a self-adjoint extension of $\vec{A}$. It was shown in [9] that the Stieltjes function (4.25) can be realized in the form (3.3) as a linear-fractional transformation of the transfer function (characteristic function) $W_{\vec{\Theta}}(\lambda)$ of some system (Brodskiĭ-Livsic rigged operator colligation) of the form

$$\vec{\Theta}=\begin{pmatrix} \vec{\mathbb{A}} & \vec{K} & 1\\ \vec{\mathfrak{H}}_+\subset L_2[a,+\infty)\subset\vec{\mathfrak{H}}_- & & \mathbb{C}\end{pmatrix},$$

where

$$\vec{\mathbb{A}}=\operatorname{Re}\vec{\mathbb{A}}+i(.,\vec{I})\vec{I},(\vec{I}\in\vec{\mathfrak{H}}_-) \tag{4.27}$$

and $\operatorname{Re}\vec{\mathbb{A}}\supset\vec{A}$. Therefore,

$$V_\Theta(\lambda)=\int\limits_0^\infty\frac{d\sigma(t)}{t-\lambda}=\vec{K}^*(\operatorname{Re}\vec{\mathbb{A}}-\lambda I)^{-1}K$$
$$=\left((\operatorname{Re}\vec{\mathbb{A}}-\lambda I)^{-1}\vec{I},\vec{I}\right)=V_{\vec{\Theta}}(\lambda).$$

From (3.1), (3.2) and (3.3) it follows that

$$W_\Theta(\lambda)=[I-iV_\Theta(\lambda)]\,[I+iV_\Theta(\lambda)]^{-1}$$
$$=[I-iV_{\vec{\Theta}}(\lambda)]\,[I+iV_{\vec{\Theta}}(\lambda)]^{-1}=W_{\vec{\Theta}}(\lambda).$$

By Theorem 5 there exists a triple of isometric operators (U_+,U,U_-) that maps the triplet $\mathfrak{H}_+\subset L_2[a,+\infty)\subset\mathfrak{H}_-$ isometrically onto $\vec{\mathfrak{H}}_+\subset L_2^\sigma[a,+\infty)\subset\vec{\mathfrak{H}}_-$ with $\vec{\mathbb{A}}=U_-\mathbb{A}U_+^{-1}$. Therefore,

$$\operatorname{Im}\vec{\mathbb{A}}=U_-\operatorname{Im}\mathbb{A}U_+^{-1}.$$

From (4.21) and (4.27) it follows that $U_-g=\vec{I}$, $g=(\operatorname{Im}h)^{\frac{1}{2}}\delta(x-a)$ and $\|g\|^2_{\mathfrak{H}_-}=\left\|\vec{I}\right\|^2_{\vec{\mathfrak{H}}_-}$. On the other hand, because $D(A^M)$ is a subspace in $\mathfrak{H}_+$ and dense in $L_2[a,+\infty)$, we can construct a new triple of Hilbert spaces, denoting $D(A^M)=\mathfrak{H}_+^M$

$$\mathfrak{H}_+^M\subset\mathfrak{H}_+\subset L_2[a,+\infty)\begin{matrix}\subset\mathfrak{H}_-^M\\ \subset\mathfrak{H}_-.\end{matrix}$$

In the same manner, denoting $D(\widehat{\vec{A}})=\widehat{\vec{\mathfrak{H}}_+}$, we get

$$\widehat{\vec{\mathfrak{H}}_+}\subset\vec{\mathfrak{H}}_+\subset L_2^\sigma[a,+\infty)\begin{matrix}\subset\widehat{\vec{\mathfrak{H}}_-}\\ \subset\vec{\mathfrak{H}}_-\end{matrix}. \tag{4.28}$$

Because the operators A^M and $\widehat{\vec{A}}$ are unitarily equivalent and $U_+|_{\mathfrak{H}_+^M}=U_+^M$ maps $\mathfrak{H}_+^M$ onto $\widehat{\vec{\mathfrak{H}}_+}$, there exists a triple of operators (U_+^M,U,U_-^M) that maps the triplet of Hilbert spaces

$$\mathfrak{H}_+^M\subset L_2[a,+\infty)\subset\mathfrak{H}_-^M$$

onto the triplet of Hilbert spaces

$$\widehat{\vec{\mathfrak{H}}_+}\subset L_2^\sigma[0,+\infty)\subset\widehat{\vec{\mathfrak{H}}_-}. \tag{4.29}$$

It was shown in [6] that

$$D(\widehat{\vec{A}}) = D(\vec{A}) + \frac{c}{1+t^2},\ c \in \mathbb{C}.$$

Consider the linear, continuous functional of the form $(x, \vec{I})$ $(x \in \vec{\mathfrak{H}}_+)$ on $\vec{\mathfrak{H}}_+$. Due to the embedding (4.28), we can consider this functional on $\widehat{\vec{\mathfrak{H}}}_+$ and it can be represented in the form

$$(x, \vec{I}) = (x, \widehat{\vec{I}}),\ x \in \widehat{\vec{\mathfrak{H}}}_+,\ \widehat{\vec{I}} \in \widehat{\vec{\mathfrak{H}}}_-,\ \vec{I} \in \vec{\mathfrak{H}}_-. \tag{4.30}$$

Let $\widehat{R}$ be the Riesz-Berezanskiĭ operator in the triple of Hilbert spaces (4.29) which maps $\widehat{\vec{\mathfrak{H}}}_+$ isometrically onto $\widehat{\vec{\mathfrak{H}}}_-$. As was shown before (see [2], [14])

$$\widehat{R}^{-1}c = \frac{c}{1+t^2},\ c \in \mathbb{C}.$$

Therefore,

$$\widehat{R}^{-1}1 = \frac{1}{1+t^2}$$

and 1 belongs to $\widehat{\vec{\mathfrak{H}}}_-$. We will now show that

$$(x, \vec{I}) = (x, 1),\ x \in \widehat{\vec{\mathfrak{H}}}_+,\ 1 \in \widehat{\vec{\mathfrak{H}}}_-,\ \vec{I} \in \vec{\mathfrak{H}}_-. \tag{4.31}$$

Taking into account (4.30) and properties of the resolvent of self-adjoint bi-extensions of a symmetric operator [14], we get

$$\begin{aligned}
&\left((\widehat{\vec{A}} - \bar{\lambda}I)^{-1}z, \vec{I}\right) = \left((\widehat{\vec{A}} - \bar{\lambda}I)^{-1}z, \widehat{\vec{I}}\right),\\
&\left(z, (\operatorname{Re}\mathbb{A} - \bar{\lambda}I)^{-1}\vec{I}\right) = \left(z, (\widehat{\vec{A}} - \lambda I)^{-1}\widehat{\vec{I}}\right),\\
&\left(z, \frac{1}{t-\lambda}\right) = \left(z, (\widehat{\vec{A}} - \lambda I)^{-1}c\right),\\
&\left(z, \frac{1}{t-\lambda}\right) = \bar{c}\left(z, \frac{1}{t-\lambda}\right),\ z \in \vec{\mathfrak{H}} = L_2^\sigma[0, +\infty)),\ c \in \mathbb{C}.
\end{aligned}$$

Therefore, $c = 1$ and $\widehat{\vec{I}} = 1$. It follows from (4.31) that

$$\begin{aligned}
&\sup_{x \in \widehat{\vec{\mathfrak{H}}}_+} \frac{|(x, \vec{I})|}{\|x\|_{\widehat{\vec{\mathfrak{H}}}_+}} = \sup_{x \in \widehat{\vec{\mathfrak{H}}}_+} \frac{|(x, 1)|}{\|x\|_{\widehat{\vec{\mathfrak{H}}}_+}} = \|1\|_{\widehat{\vec{\mathfrak{H}}}_-} = (\|1\|^2_{\widehat{\vec{\mathfrak{H}}}_-})^{\frac{1}{2}}\\
&= ((1,1)_{\widehat{\vec{\mathfrak{H}}}_-})^{\frac{1}{2}} = \left((\widehat{R}^{-1}1, 1)\right)^{\frac{1}{2}} = \left((\frac{1}{1+t^2}, 1)\right)^{\frac{1}{2}}\\
&= \left(\int_0^\infty \frac{d\sigma}{1+t^2}\right)^{\frac{1}{2}}.
\end{aligned}$$

On the other hand,

$$(y,g) = (U^*Uy,g) = (U_+^M y, (U^*)^* g) = (U_+^M y, U_- g) = (U_+^M y, \vec{I}),$$
$$y \in \mathfrak{H}_+^M,\ g \in \mathfrak{H}_-.$$

Therefore,

$$\sup_{y\in\mathfrak{H}_+^M} \frac{|(y,g)|}{\|y\|_{\mathfrak{H}_+^M}} = \sup_{y\in\mathfrak{H}_+^M} \frac{|(U_+^M y,\vec{I})|}{\|y\|_{\mathfrak{H}_+^M}}$$
$$= \sup_{y\in\mathfrak{H}_+^M} \frac{|(U_+^M y,\vec{I})|}{\|U_+^M y\|_{\widehat{\mathfrak{H}}_+}} = \sup_{x\in\widehat{\mathfrak{H}}_+} \frac{|(x,\vec{I})|}{\|x\|_{\widehat{\mathfrak{H}}_+}} = \left(\int_0^\infty \frac{d\sigma}{1+t^2}\right)^{\frac{1}{2}}.$$

Taking into account (4.21) we get (4.26). □

Consider the system (colligation) of the form (3.8), with $\mathbb{A}$ and $\mathbb{A}^*$ given by (4.16) and let $\mu = \operatorname{Re} h = -m_\infty(-0)$. Then we obtain the following system (colligation):

$$\Theta = \begin{pmatrix} \mathbb{A} & K & 1 \\ \mathfrak{H}_+ \subset L_2[a,+\infty) \subset \mathfrak{H}_- & & \mathbb{C} \end{pmatrix}, \tag{4.32}$$

where

$$\begin{aligned} \mathbb{A}y &= -y'' + q(x)y - \frac{i}{\operatorname{Im} h}\,[y'(a) - hy(a)]\,[\delta'(x-a) - m_\infty(-0)\delta(x-a)], \\ \mathbb{A}^*y &= -y'' + q(x)y - \frac{i}{\operatorname{Im} h}\,[y'(a) - \overline{h}y(a)]\,[\delta'(x-a) - m_\infty(-0)\delta(x-a)], \end{aligned} \tag{4.33}$$

and

$$\begin{aligned} \operatorname{Re}\mathbb{A}y &= -y'' + q(x)y - y(a)[\delta'(x-a) - m_\infty(-0)\delta(x-a)], \\ \operatorname{Im}\mathbb{A}y &= \frac{1}{\operatorname{Im} h}\,[-y'(a) - m_\infty(-0)y(a)]\,[\delta'(x-a) - m_\infty(-0)\delta(x-a)] \\ &= \frac{1}{\operatorname{Im} h}(y,g)g, \end{aligned} \tag{4.34}$$

where

$$g = \frac{1}{(\operatorname{Im} h)^{\frac{1}{2}}}[\delta'(x-a) - m_\infty(-0)\delta(x-a)].$$

By relation (3.9) we get that the transfer (characteristic) function $W_\Theta(\lambda)$ of the system (Brodskiĭ-Livsic operator colligation) of the form (4.32) has the following representation

$$W_\Theta(\lambda) = I - 2iK^*(\mathbb{A} - \lambda I)^{-1}K = -\frac{m_\infty(\lambda) + \overline{h}}{m_\infty(\lambda) + h}. \tag{4.35}$$

Denote by Θ_M the system (operator colligation) of the form (3.8), (4.16) and by Θ_μ the system (operator colligation) of the form (4.32), (4.33). Then it follows from (4.35) that

$$W_{\Theta_\mu}(\lambda) = -W_{\Theta_M}(\lambda).$$

Moreover(see (3.3)),

$$\begin{aligned} V_{\Theta_\mu}(\lambda) &= i[W_{\Theta_\mu}(\lambda) - I][W_{\Theta_\mu}(\lambda) + I]^{-1} \\ &= -i[W_{\Theta_M}(\lambda) + I][-W_{\Theta_M}(\lambda) + I]^{-1} \\ &= i[W_{\Theta_M}(\lambda) + I][W_{\Theta_M}(\lambda) - I]^{-1} \\ &= -\{i[W_{\Theta_M}(\lambda) - I][W_{\Theta_M}(\lambda) + I]^{-1}\}^{-1} = -V_{\Theta_M}^{-1}(\lambda). \end{aligned}$$

Since $V_{\Theta_M}(\lambda)$ is a Stieltjes function with representation (see (4.22))

$$V_{\Theta_M}(\lambda) = \int_0^\infty \frac{d\sigma(t)}{t-\lambda},$$

$-V_{\Theta_M}^{-1}(\lambda)$ is the inverse Stieltjes function [7] and has the following representation

$$V_{\Theta_M}(\lambda) = -V_{\Theta_M}^{-1}(\lambda) = \alpha + \int_0^\infty \left(\frac{1}{t-\lambda} - \frac{1}{t}\right) d\tau(t),$$

where $\alpha \le 0$, $\tau(t)$ is a nondecreasing function satisfying the relations

$$\int_0^\infty d\tau(t) = \infty, \quad \int_0^\infty \frac{d\tau(t)}{t^2+t} < \infty.$$

From (4.34) it follows that $\operatorname{Re}\mathbb{A} \supset A^\mu$, where

$$\begin{cases} A^\mu y = -y'' + q(x)y, \\ y(a) = 0 \end{cases}$$

is the Friedrichs extension of a nonnegative operator A of the form (3.1). Taking into account calculations similar to those in connection with Theorem 6, we get

Theorem 7. *Let* $\Theta = \begin{pmatrix} \mathbb{A} & K & 1 \\ \mathfrak{H}_+ \subset L_2[a,+\infty) \subset \mathfrak{H}_- & & \mathbb{C} \end{pmatrix}$ *be the l.c.s.d.s (Brodskiĭ-Livsic rigged operator colligation), where* $\mathbb{A}$ *is a* $(*)$*-extension of the form* (4.33) *of the accretive Schrödinger operator* T_h ($\operatorname{Re} h = -m_\infty(0)$)*. Then the spectral measure in the representation*

$$V_{\Theta_\mu}(\lambda) = V_\Theta(\lambda) = \alpha + \int_0^\infty \left(\frac{1}{t-\lambda} - \frac{1}{t}\right) d\tau(t) \tag{4.36}$$

satisfies the following relation

$$\int_0^\infty \frac{d\tau(t)}{1+t^2} = \frac{1}{\operatorname{Im} h}\left(\sup_{y\in D(A^\mu)} \frac{|y'(a)|}{\left(\int_a^\infty (|y(x)|^2 + |l(y)|^2)\,dx\right)^{\frac12}}\right)^2. \tag{4.37}$$

As a consequence of Theorems 6 and 7 it follows that the spectral measures $\sigma(t)$ and $\tau(t)$ in the representations (4.25) and (4.36), which correspond to the Kreĭn-von Neumann extension A^M and the Friedrichs extension A^μ, satisfy the following relation

$$\int_0^\infty \frac{d\sigma(t)}{1+t^2} \int_0^\infty \frac{d\tau(t)}{1+t^2} =$$

$$= \left(\sup_{y\in D(A^M)} \frac{|y(a)|}{\left(\int_a^\infty (|y(x)|^2 + |l(y)|^2)\, dx \right)^{\frac{1}{2}}} \right)^2$$

$$\times \left(\sup_{y\in D(A^\mu)} \frac{|y'(a)|}{\left(\int_a^\infty (|y(x)|^2 + |l(y)|^2)\, dx \right)^{\frac{1}{2}}} \right)^2$$

Remark 8. *The relations* (4.26) *and* (4.37) *can be rewritten as follows*

$$\int_a^\infty \left(|y(x)|^2 + |l(y)|^2\right) dx \geq \frac{\operatorname{Im} h\, |y(a)|^2}{\int_0^\infty \frac{d\sigma(t)}{1+t^2}}, \quad y'(a) + m_\infty(-0)y(a) = 0,$$

$$\int_a^\infty \left(|y(x)|^2 + |l(y)|^2\right) dx \geq \frac{|y'(a)|^2}{\operatorname{Im} h \int_0^\infty \frac{d\sigma(t)}{1+t^2}}, \quad y(a) = 0.$$

Theorem 9. *Let* $\Theta = \begin{pmatrix} \mathbb{A} & K & 1 \\ \mathfrak{H}_+ \subset L_2[a,+\infty) \subset \mathfrak{H}_- & & \mathbb{C} \end{pmatrix}$ *be the l.c.s.d.s (Brodskiĭ-Livsic rigged operator colligation), where* $\mathbb{A}$ *is a* $(*)$*-extension of the form* (4.16) *of the accretive Schrödinger operator* T_h *of the form* (4.17). *Then*

$$V_\Theta(\lambda) = K^*(\operatorname{Re}\mathbb{A} - \lambda I)^{-1}K = i[W_\Theta(\lambda) - I]\,[W_\Theta(\lambda) + I]^{-1} \tag{4.38}$$

is a Stieltjes function if and only if

$$\begin{cases} \mu \geq \dfrac{(\operatorname{Im} h)^2}{m_\infty(-0) + \operatorname{Re} h} + \operatorname{Re} h, \\ \operatorname{Re} h \geq -m_\infty(-0). \end{cases} \tag{4.39}$$

Proof. The proof of this theorem follows from Theorem 4 and from the fact [15] that T_h is accretive if and only if $\operatorname{Re} h \geq -m_\infty(-0)$, as well as from the statement [9] that $V_\Theta(\lambda)$ is a Stieltjes function if and only if $\mathbb{A}$ is accretive. □

It is well known that any Stieltjes function $V_\Theta(\lambda)$ has a parametric representation

$$V_\Theta(\lambda) = \gamma + \int_0^\infty \frac{d\sigma(t)}{t-\lambda}, \ (\gamma \geq 0), \tag{4.40}$$

where $\sigma(t)$ is a nondecreasing function for which

$$\int_0^\infty \frac{d\sigma(t)}{t+1} < \infty. \tag{4.41}$$

In addition to (4.41), since the Hermitian part of T_h has dense domain in $L_2[a,+\infty)$, one concludes that

$$\int_0^\infty d\sigma(t) = \infty. \tag{4.42}$$

Thus, the function $V_\Theta(\lambda)$ of the form (4.38) has the parametric representation (4.40), (4.41), (4.42) if and only if the inequalities (4.39) hold.

Theorem 10. *Let*

$$\Theta = \begin{pmatrix} \mathbb{A} & K & 1 \\ \mathfrak{H}_+ \subset L_2[a,+\infty) \subset \mathfrak{H}_- & & \mathbb{C} \end{pmatrix}$$

be the l.c.s.d.s (Brodskiĭ-Livsic rigged operator colligation), where $\mathbb{A}$ *is a* $(*)$*-extension of the form* (4.16) *of an accretive Schrödinger operator* T_h *of the form* (4.17). *Then*

$$V_\Theta(\lambda) = K^*(Re\mathbb{A} - \lambda I)^{-1}K = i[W_\Theta(\lambda) - I]\,[W_\Theta(\lambda) + I]^{-1}$$

is an inverse Stieltjes function [7] *if and only if*

$$-m_\infty(-0) < \mu \leq \operatorname{Re} h. \tag{4.43}$$

Proof. Consider the following $(*)$-extensions of the accretive Schrödinger operator T_h

$$\begin{aligned} \mathbb{A}_\mu y &= -y'' + qy - \frac{1}{\mu - h}[y'(a) - hy(a)]\,[\mu\delta(x-a) + \delta'(x-a)], \\ \mathbb{A}_\mu^* y &= -y'' + qy - \frac{1}{\mu - \overline{h}}[y'(a) - \overline{h}y(a)]\,[\mu\delta(x-a) - \delta'(x-a)] \end{aligned} \tag{4.44}$$

and

$$\begin{aligned} \mathbb{A}_\xi y &= -y'' + qy - \frac{1}{\xi - h}[y'(a) - hy(a)]\,[\xi\delta(x-a) + \delta'(x-a)], \\ \mathbb{A}_\xi^* y &= -y'' + qy - \frac{1}{\xi - \overline{h}}[y'(a) - \overline{h}y(a)]\,[\xi\delta(x-a) - \delta'(x-a)] \end{aligned} \tag{4.45}$$

with

$$\xi = \frac{\mu \operatorname{Re} h - |h|^2}{\mu - \operatorname{Re} h}.$$

Let

$$\Theta_\mu = \begin{pmatrix} \mathbb{A}_\mu & K_\mu & 1 \\ \mathfrak{H}_+ \subset L_2[a,+\infty) \subset \mathfrak{H}_- & & \mathbb{C} \end{pmatrix}$$

and

$$\Theta_\xi = \begin{pmatrix} & \mathbb{A}_\xi & K_\xi & 1 \\ \mathfrak{H}_+ \subset L_2[a,+\infty) \subset \mathfrak{H}_- & & & \mathbb{C} \end{pmatrix}$$

be the corresponding systems (Brodskiĭ-Livsic rigged operator colligations). From relation (3.9) it follows that

$$\begin{aligned} W_{\Theta_\mu}(\lambda) &= \frac{\mu - h}{\mu - \bar{h}} \frac{m_\infty(\lambda) + \bar{h}}{m_\infty(\lambda) + h}, \\ W_{\Theta_\xi}(\lambda) &= \frac{\xi - h}{\xi - \bar{h}} \frac{m_\infty(\lambda) + \bar{h}}{m_\infty(\lambda) + h}. \end{aligned} \tag{4.46}$$

Because

$$\xi = \frac{\mu \mathrm{Re}\, h - |h|^2}{\mu - \mathrm{Re}\, h},$$

one obtaines

$$\begin{aligned} \frac{\xi - h}{\xi - \bar{h}} &= \frac{\frac{\mu \mathrm{Re}\, h - |h|^2}{\mu - \mathrm{Re}\, h} - h}{\frac{\mu \mathrm{Re}\, h - |h|^2}{\mu - \mathrm{Re}\, h} - \bar{h}} = \frac{\mu \mathrm{Re}\, h - |h|^2 - \mu h + \mathrm{Re}\, h h}{\mu \mathrm{Re}\, h - |h|^2 - \mu \bar{h} + \mathrm{Re}\, h \bar{h}} = \\ &= \frac{\mu(\mathrm{Re}\, h - h) - h(\bar{h} - \mathrm{Re}\, h)}{\mu(\mathrm{Re}\, h - \bar{h}) - \bar{h}(\bar{h} - \mathrm{Re}\, h)} = \frac{-i\mu \mathrm{Im}\, h + ih \mathrm{Im}\, h}{i\mu \mathrm{Im}\, h - i\bar{h} \mathrm{Im}\, h} = \\ &= \frac{-(\mu - h)\mathrm{Im}\, h}{(\mu - \bar{h})\mathrm{Im}\, h} = -\frac{\mu - h}{\mu - \bar{h}}. \end{aligned}$$

Therefore, from (4.46) we get

$$W_{\Theta_\mu}(\lambda) = -W_{\Theta_\xi}(\lambda). \tag{4.47}$$

The relations (4.47) and (3.3) imply

$$\begin{aligned} V_{\Theta_\mu}(\lambda) &= K_\mu^*(\mathrm{Re}\, \mathbb{A}_\mu - \lambda I)^{-1} K_\mu = i[W_{\Theta_\mu}(\lambda) - I]\, [W_{\Theta_\mu}(\lambda) + I]^{-1} = \\ &= i[-W_{\Theta_\xi}(\lambda) - I][-W_{\Theta_\xi}(\lambda) + I]^{-1} = -V_{\Theta_\xi}^{-1}(\lambda). \end{aligned}$$

Finally,

$$V_{\Theta_\mu}(\lambda) = -V_{\Theta_\xi}^{-1}(\lambda). \tag{4.48}$$

As was shown in Theorem 9, the function $V_{\Theta_\xi}(\lambda)$ is a Stieltjes function if and only if

$$\xi \geq \frac{(\mathrm{Im}\, h)^2}{m_\infty(-0) + \mathrm{Re}\, h} + \mathrm{Re}\, h.$$

Therefore,

$$\xi = \frac{\mu \mathrm{Re}\, h - |h|^2}{\mu - \mathrm{Re}\, h} \geq \frac{(\mathrm{Im}\, h)^2}{m_\infty(-0) + \mathrm{Re}\, h} + \mathrm{Re}\, h. \tag{4.49}$$

Furthermore,

$$\frac{\mu \mathrm{Re}\, h - |h|^2}{\mu - \mathrm{Re}\, h} = \frac{\mu \mathrm{Re}\, h - (\mathrm{Im}\, h)^2 - (\mathrm{Re}\, h)^2}{\mu - \mathrm{Re}\, h} = -\frac{(\mathrm{Im}\, h)^2}{\mu - \mathrm{Re}\, h} + \mathrm{Re}\, h.$$

From (4.49) it follows that

$$-\frac{(\operatorname{Im} h)^2}{\mu - \operatorname{Re} h} + \operatorname{Re} h \geq \frac{(\operatorname{Im} h)^2}{m_\infty(-0) + \operatorname{Re} h} + \operatorname{Re} h$$

and

$$\frac{1}{\operatorname{Re} h - \mu} - \frac{1}{m_\infty(-0) + \operatorname{Re} h} \geq 0.$$

The last inequality is equivalent to (4.43). When μ satisfies (4.43), then $V_{\Theta_\xi}(\lambda)$ is a Stieltjes function and therefore $V_{\Theta_\mu}(\lambda)$ is an inverse Stieltjes function. □

Theorem 11. *Let*

$$\Theta = \begin{pmatrix} \mathbb{A}_\mu & K_\mu & 1 \\ \mathfrak{H}_+ \subset L_2[a, +\infty) \subset \mathfrak{H}_- & & \mathbb{C} \end{pmatrix}$$

be the l.c.s.d.s (Brodskiĭ-Livsic rigged operator colligation), where $\mathbb{A}_\mu$ is a (*)-extension of the form (4.44) of the accretive Schrödinger operator T_h and

$$\mu \geq \frac{(\operatorname{Im} h)^2}{m_\infty(-0) + \operatorname{Re} h} + \operatorname{Re} h.$$

Then the spectral measure in the representation

$$V_{\Theta_\mu}(\lambda) = \gamma + \int_0^\infty \frac{d\sigma_\xi(t)}{t - \lambda}, \quad \xi = \frac{\mu \operatorname{Re} h - |h|^2}{\mu - \operatorname{Re} h} \tag{4.50}$$

satisfies the following relation

$$\int_0^\infty \frac{d\sigma_\xi(t)}{1+t^2} = \frac{\operatorname{Im} h}{|\mu - h|^2} \left(\sup_{y \in D(\widehat{A}_\xi)} \frac{|\mu y(a) - y'(a)|}{\left(\int_a^\infty (|y(x)|^2 + |l(y)|^2)\, dx \right)^{\frac{1}{2}}} \right)^2, \tag{4.51}$$

where

$$\begin{cases} \widehat{A}_\xi y = -y'' + qy \\ y'(a) = \xi y(a) \end{cases}, \quad \xi = \frac{\mu \operatorname{Re} h - |h|^2}{\mu - \operatorname{Re} h} \tag{4.52}$$

is a nonnegative self-adjoint extension of the nonnegative operator A of the form (4.18).

Proof. It is easy to check that when $\mu \subset \left[\frac{(\operatorname{Im} h)^2}{m_\infty(-0) + \operatorname{Re} h} + \operatorname{Re} h, +\infty\right)$ then

$$\xi = \frac{\mu \operatorname{Re} h - |h|^2}{\mu - \operatorname{Re} h} \in [-m_\infty(-0), \operatorname{Re} h).$$

Therefore, $\widehat{A}_\xi$ of the form (4.52) is a nonnegative self-adjoint extension of the nonnegative minimal operator A of the form (4.18). From (4.43) it follows that

$$\operatorname{Re} \mathbb{A}_\mu = -y'' + qy + \left[\frac{\mu \operatorname{Re} h - |h|^2}{|\mu - h|^2} y(a) - \frac{\mu - \operatorname{Re} h}{|\mu - h|^2} y'(a) \right] \times$$
$$\times \left[\mu \delta(x - a) + \delta'(x - a) \right]$$

and that

$$\operatorname{Re}\mathbb{A}_\mu \supset \widehat{A}_\xi,\ \xi = \frac{\mu \operatorname{Re} h - |h|^2}{\mu - \operatorname{Re} h}.$$

Taking into account that $\operatorname{Im}\mathbb{A}_\mu = (\cdot, g)g$, where

$$g = \frac{(\operatorname{Im} h)^{\frac{1}{2}}}{|\mu - h|}[\mu\delta(x-a) + \delta'(x-a)],$$

and following the proof of Theorem 6, one obtains relation (4.51). □

Theorem 12. *Let*

$$\Theta_\xi = \begin{pmatrix} \mathbb{A}_\xi & K_\xi & 1 \\ \mathfrak{H}_+ \subset L_2[a,+\infty) \subset \mathfrak{H}_- & & \mathbb{C} \end{pmatrix}$$

be the l.c.s.d.s (Brodskiĭ-Livsic rigged operator colligation), where $\mathbb{A}_\xi$ is a $(*)$-extension of the form (4.45) of the accretive Schrödinger operator T_h and

$$-m_\infty(-0) \le \xi \le \operatorname{Re} h.$$

Then the spectral measure in the representation

$$V_{\Theta_\xi}(\lambda) = \alpha + \int_0^\infty \left(\frac{1}{t-\lambda} - \frac{1}{t}\right) d\tau_\mu(t)\ (\alpha \le 0)$$

satisfies the following relation:

$$\int_0^\infty \frac{d\tau_\mu(t)}{1+t^2} = \frac{\operatorname{Im} h}{|\xi - h|^2}\left(\sup_{y \in D(\widehat{A}_\mu)} \frac{|\xi y(a) - y'(a)|}{\left(\int_a^\infty (|y(x)|^2 + |l(y)|^2)\,dx\right)^{\frac{1}{2}}}\right)^2, \tag{4.53}$$

where

$$\xi = \frac{\mu \operatorname{Re} h - |h|^2}{\mu - \operatorname{Re} h},\ \mu > \operatorname{Re} h \tag{4.54}$$

and

$$\begin{cases} \widehat{A}_\mu y = -y'' + qy \\ y'(a) = \mu y(a) \end{cases}.$$

is a nonnegative self-adjoint extension of the nonnegative operator A of the form (4.18).

Proof. It was established in the proof of Theorem 10 that the inequalities

$$\mu \ge \frac{(\operatorname{Im} h)^2}{m_\infty(-0) + \operatorname{Re} h} + \operatorname{Re} h \tag{4.55}$$

and

$$-m_\infty(-0) \le \xi \le \operatorname{Re} h, \mu > \operatorname{Re} h,$$

where

$$\xi = \frac{\mu \operatorname{Re} h - |h|^2}{\mu - \operatorname{Re} h},$$

are equivalent. Consider two $(*)$-extensions $\mathbb{A}_\mu$ and $\mathbb{A}_\xi$ of the accretive Schrödinger operator T_h. As μ satisfies (4.55), the operator-valued function $V_{\Theta_\xi}(\lambda)$ is an inverse Stieltjes function and $V_{\Theta_\mu}(\lambda)$ is a Stieltjes function by Theorem 10, and

$$V_{\Theta_\mu}(\lambda) = -V_{\Theta_\xi}^{-1}(\lambda).$$

From (4.45) it follows that

$$\begin{aligned}\operatorname{Re}\mathbb{A}_\xi &= -y'' + qy \\ &+ \left[\frac{\xi \operatorname{Re} h - |h|^2}{|\xi - h|^2} y(a) - \frac{\xi - \operatorname{Re} h}{|\xi - h|^2} y'(a)\right] [\xi\delta(x-a) + \delta'(x-a)].\end{aligned}$$

Using (4.54) we get

$$\mu = \frac{\xi \operatorname{Re} h - |h|^2}{\xi - \operatorname{Re} h}.$$

Therefore,

$$\operatorname{Re}\mathbb{A}_\xi \supset \widehat{A}\mu.$$

Taking into account that $\operatorname{Im}\mathbb{A}_\xi = (\cdot, g)g$, where

$$g = \frac{(\operatorname{Im} h)^{\frac{1}{2}}}{|\xi - h|}[\xi\delta(x-a) + \delta'(x_a)]$$

and following the proof of Theorem 6, one will obtain relation (4.53). □

As a consequence of Theorems 11 and 12 we get the following: Let

$$\Theta_\mu = \begin{pmatrix} \mathbb{A}_\mu & K_\mu & 1 \\ \mathfrak{H}_+ \subset L_2[a,+\infty) \subset \mathfrak{H}_- & & \mathbb{C} \end{pmatrix},$$

$$\Theta_\xi = \begin{pmatrix} \mathbb{A}_\xi & K_\xi & 1 \\ \mathfrak{H}_+ \subset L_2[a,+\infty) \subset \mathfrak{H}_- & & \mathbb{C} \end{pmatrix},$$

where

$$\xi = \frac{\mu \operatorname{Re} h - |h|^2}{\mu - \operatorname{Re} h},$$

be the l.c.s.d.s (Brodskiĭ-Livsic rigged operator colligations), where $\mathbb{A}_\mu$ and $\mathbb{A}_\xi$ are $(*)$-extensions of the form (4.44), (4.45) of the accretive Schrödinger operator T_h and

$$\mu \geq \frac{(\operatorname{Im} h)^2}{m_\infty(-0) + \operatorname{Re} h} + \operatorname{Re} h.$$

Then the spectral measures in the representations

$$V_{\Theta_\mu}(\lambda) = \gamma + \int_0^\infty \frac{d\sigma_\xi(t)}{t - \lambda}$$

and

$$V_{\Theta_\xi}(\lambda) = \alpha + \int_0^\infty \left(\frac{1}{t-\lambda} - \frac{1}{t}\right) d\tau_\mu(t)$$

satisfy the following relation

$$\int\limits_0^\infty \frac{d\sigma_\xi(t)}{1+t^2}\int\limits_0^\infty \frac{d\tau_\mu(t)}{1+t^2} =$$
$$= \left(\sup_{y\in D(\widehat{A}_\xi)} \frac{\left|\frac{\mu}{\mu-h}y(a) - \frac{1}{\mu-h}y'(a)\right|}{\left(\int\limits_a^\infty (|y(x)|^2+|l(y)|^2)\,dx\right)^{\frac{1}{2}}}\right)^2 \tag{4.56}$$
$$\times \left(\sup_{y\in D(\widehat{A}_\mu)} \frac{\left|\frac{\mu\mathrm{Re}\,h-h^2}{\mu-h}y(a) - \frac{\mu-\mathrm{Re}\,h}{\mu-h}y'(a)\right|}{\left(\int\limits_a^\infty (|y(x)|^2+|l(y)|^2)\,dx\right)^{\frac{1}{2}}}\right)^2 .$$

The proof of relation (4.56) follows from Theorems 11 and 12 by multiplying both parts of the relations (4.50) and (4.53), and from the fact that

$$\xi = \frac{\mu\mathrm{Re}\,h - |h|^2}{\mu - h}$$

and

$$\begin{aligned}\xi - h &= \frac{\mu\mathrm{Re}\,h - |h|^2}{\mu - \mathrm{Re}\,h} - h = \frac{\mu\mathrm{Re}\,h - |h|^2 - \mu h - h\mathrm{Re}\,h}{\mu - \mathrm{Re}\,h} = \\ &= \frac{\mu(\mathrm{Re}\,h - h) - h(\overline{h} - \mathrm{Re}\,h)}{\mu - \mathrm{Re}\,h} = \frac{-i\mu\mathrm{Im}\,h + ih\mathrm{Im}\,h}{\mu - \mathrm{Re}\,h} = \\ &= i\mathrm{Im}\,h\frac{h-\mu}{\mu - \mathrm{Re}\,h}.\end{aligned}$$

Remark 13. *Using* (3.11), *we can rewrite relations* (4.51) *and* (4.37) *in the following way*

$$\begin{aligned}\sup_{y'(a)=\xi y(a)} \frac{|y(a)|^2}{\int_a^\infty (|y(x)|^2+|l(y)|^2)\,dx} &= -\frac{\mathrm{Im}\,m_\infty(i)}{|m_\infty(i)+\xi|^2}, \\ \sup_{y(a)=0} \frac{|y'(a)|^2}{\int_a^\infty (|y(x)|^2+|l(y)|^2)\,dx} &= -\mathrm{Im}\,m_\infty(i).\end{aligned} \tag{4.57}$$

Remark 14. *One can prove that*

$$\sup_{\sin\alpha y'(a)+\cos\alpha y(a)=0} \frac{|y'(a)|^2}{\int_a^\infty (|y(x)|^2+|l(y)|^2)\,dx} = \frac{1}{l(y_0)|_{x=a} + \tan\alpha\,(l(y_0))'\,|_{x=a}}, \tag{4.58}$$

where $y_0(x)$ is the unique solution of the boundary value problem

$$\begin{cases} l\left(l(y)\right) + y = 0, \\ y(a) = -\tan\alpha, \\ y'(a) = 1, \\ y,\, l(y) \in D^*. \end{cases}$$

5. Schrödinger operator with Bessel potentials

Consider the differential expression with the Bessel potential

$$l_\nu = -\frac{d^2}{dx^2} + \frac{\nu^2 - 1/4}{x^2}, \quad x \in [1, \infty)$$

of order $\nu > 0$ in the Hilbert space $\mathfrak{H} = L^2[1, \infty)$.

The minimal symmetric operator $Ay = l_\nu y$ generated by this expression and boundary conditions $y(1) = y'(1) = 0$ has defect numbers $(1, 1)$ [1]. If $\nu \geq 0$ the operator A is nonnegative. However, in this paper we will only consider the case $\nu \geq 1/2$. Let I_ν and Y_ν be the Bessel functions of the first and the second kind,

$$I_\nu(\lambda) = \sum_{k=0}^{\infty} \frac{(-1)^k (\lambda/2)^{\nu+2k}}{k!\Gamma(\nu + k + 1)}, \quad Y_\nu(\lambda) = \frac{I_\nu(\lambda)\cos\pi\nu - I_{-\nu}(\lambda)}{\sin\pi\nu}.$$

Consider the functions

$$u(x, \lambda) = \frac{\pi}{2}\sqrt{x}\left\{I_\nu(x\sqrt{\lambda})Y_\nu(\sqrt{\lambda}) - Y_\nu(x\sqrt{\lambda})I_\nu(\sqrt{\lambda})\right\},$$
$$v(x, \lambda) = \frac{\pi}{2}\sqrt{x\lambda}\left\{I_\nu(x\sqrt{\lambda})Y'_\nu(\sqrt{\lambda}) - Y_\nu(x\sqrt{\lambda})I'_\nu(\sqrt{\lambda})\right\}.$$

According to [1]

$$v(x, \lambda) + m_\nu(\lambda)u(x, \lambda) \in L_2[1, \infty).$$

Let

$$a(\lambda) = \frac{\pi}{2}\sqrt{\lambda}\left\{Y_\nu(\sqrt{\lambda})I'_\nu(\sqrt{\lambda}) - I_\nu(\sqrt{\lambda})Y'_\nu(\sqrt{\lambda})\right\}.$$

Then

$$u(1, \lambda) = 0, \ u'(1, \lambda) = a(\lambda),$$
$$v(1, \lambda) = -a(\lambda), \ v'(1, \lambda) = -\frac{a(\lambda)}{2}.$$

Let

$$\varphi_1(x, \lambda) = \frac{1}{a(\lambda)} u(x, \lambda),$$
$$\varphi_2(x, \lambda) = \frac{1}{2\,a(\lambda)}\left(u(x, \lambda) + 2v(x, \lambda)\right).$$

We have

$$\varphi_1(1, \lambda) = 0, \ \varphi'_1(1, \lambda) = 1,$$
$$\varphi_2(1, \lambda) = -1, \ \varphi'_2(1, \lambda) = 0.$$

Since

$$\begin{aligned}&\varphi_2(x,\lambda)+m_{\infty,\nu}(\lambda)\varphi_1(x,\lambda)=\\&=\frac{1}{2a(\lambda)}\left((1+2m_{\infty,\nu}(\lambda))u(x,\lambda)+2v(x\lambda)\right)\in L_2[1,\infty),\end{aligned}$$

we get the equality

$$m_{\infty,\nu}(\lambda)=m_\nu(\lambda)-\frac{1}{2}.$$

Using the definitions of $I_\nu(\lambda)$ and $Y_\nu(\lambda)$ it is easy to find the limiting value $m_\nu(-0)$:

$$m_\nu(-0)=\nu.$$

It follows that

$$m_{\infty,\nu}(-0)=\nu-\frac{1}{2}.$$

Let $H_\nu^{(1)}(\lambda)=I_\nu(\lambda)+iY_\nu(\lambda)$ be the Hankel functions of the first kind, then

$$\begin{aligned}m_\nu(\lambda)&=-\sqrt{\lambda}\,\frac{H_\nu^{'(1)}(\sqrt{\lambda})}{H_\nu^{(1)}(\sqrt{\lambda})},\\m_{\infty,\nu}(\lambda)&=-\frac{2\sqrt{\lambda}\,H_\nu^{'(1)}(\sqrt{\lambda})+H_\nu^{(1)}(\sqrt{\lambda})}{2H_\nu^{(1)}(\sqrt{\lambda})}.\end{aligned}$$

The Friedrichs and Kreĭn-von Neumann extensions of A are determined by the following boundary conditions

$$D(A^\mu)=\{y(1)=0\}\,,\;D(A^M)=\left\{y'(1)+\left(\nu-\frac{1}{2}\right)y(1)=0\right\}.$$

To perform the calculations for the cases $\nu=k+1/2$ we can use the following expressions

$$\begin{aligned}&I_{1/2}(\lambda)=\sqrt{\frac{2}{\pi}}\frac{\sin\lambda}{\sqrt{\lambda}},\;I_{3/2}(\lambda)=\sqrt{\frac{2}{\pi}}\left(\frac{\sin\lambda}{\lambda\sqrt{\lambda}}-\frac{\cos\lambda}{\sqrt{\lambda}}\right),\\&I_{k+1/2}(\lambda)=\sqrt{\frac{2}{\pi}}\,\lambda^{k+1/2}\left(-\frac{1}{\lambda}\frac{d}{d\lambda}\right)^k\frac{\sin\lambda}{\lambda},\;k=1,2,\ldots,\\&H_{k+1/2}^{(1)}(\lambda)=\sqrt{\frac{2}{\pi\lambda}}e^{-\frac{i\pi}{2}(k+1)}e^{i\lambda}\sum_{j=0}^{k}(-1)^j\frac{(k+j)!}{j!(k-j)!}\frac{1}{(2i\lambda)^j}.\end{aligned}$$

In particular,

$$\begin{aligned}&H_{1/2}^{(1)}(\lambda)=-i\sqrt{\frac{2}{\pi\lambda}}e^{i\lambda},\;H_{3/2}^{(1)}(\lambda)=-\sqrt{\frac{2}{\pi}}\frac{i+\lambda}{\lambda^{3/2}}e^{i\lambda},\\&m_{\infty,1/2}(\lambda)=-i\sqrt{\lambda},\;m_{\infty,3/2}(\lambda)=-\frac{i\lambda-\frac{3}{2}\sqrt{\lambda}-\frac{3}{2}i}{\sqrt{\lambda}+i}-\frac{1}{2}\end{aligned}$$

and

$$m_{\infty,1/2}(i) = \frac{1+\sqrt{2}-i\sqrt{2}}{2}, \; m_{\infty,3/2}(i) = 1 + \frac{1}{\sqrt{2}} - \frac{i}{2}.$$

Let $Ty = l_\nu y$ be the non-self-adjoint and maximal dissipative extension of A which is determined by the boundary condition

$$D(T) = \{y'(a) = hy(a)\}, \; \mathrm{Im}\, h > 0.$$

Then by relations (3.9) and (3.10) the transfer function $W_\Theta(\lambda)$ of the corresponding system Θ with $(*)$-extension of the Schrödinger operator T (with the Bessel potential) as main operator in the system and their fractional linear transformation take on the forms

$$W_\Theta(\lambda) = \frac{\mu - h}{\mu - \bar{h}} \frac{2\bar{h}H_\nu^{(1)}(\sqrt{\lambda}) - \left(2\sqrt{\lambda}\, H_\nu'^{(1)}(\sqrt{\lambda}) + H_\nu^{(1)}(\sqrt{\lambda})\right)}{2hH_\nu^{(1)}(\sqrt{\lambda}) - \left(2\sqrt{\lambda}\, H_\nu'^{(1)}(\sqrt{\lambda}) + H_\nu^{(1)}(\sqrt{\lambda})\right)},$$

$$V_\Theta(\lambda) = \frac{\mathrm{Im}\, h}{\mu - \mathrm{Re}\, h} \frac{2\mu H_\nu^{(1)}(\sqrt{\lambda}) - \left(2\sqrt{\lambda}\, H_\nu'^{(1)}(\sqrt{\lambda}) + H_\nu^{(1)}(\sqrt{\lambda})\right)}{2\xi H_\nu^{(1)}(\sqrt{\lambda}) - \left(2\sqrt{\lambda}\, H_\nu'^{(1)}(\sqrt{\lambda}) + H_\nu^{(1)}(\sqrt{\lambda})\right)},$$

where μ is a real number, and

$$\xi = \frac{\mu \mathrm{Rc}\, h \quad |h|^2}{\mu - \mathrm{Re}\, h}.$$

The operator T is accretive if and only if $\mathrm{Re}\, h \geq 1/2 - \nu$ and according to Theorem 4 the function $V(\lambda)$ belongs to the Stieltjes class if and only if

$$\mu \geq \frac{|h|^2 + (\nu - 1/2)\mathrm{Re}\, h}{\nu - 1/2 + \mathrm{Re}\, h},$$

and to the inverse Stieltjes class if and only if

$$\frac{1}{2} - \nu \leq \mu \leq \mathrm{Re}\, h.$$

Taking into account the relations (4.57), we get the following *sharp inequalities*:

$$y'(1) = \xi y(1) \Longrightarrow$$

$$\int_1^\infty \left(|y(x)|^2 + |y''(x)|^2\right) dx \geq \sqrt{2}\left(\left(\frac{1+\sqrt{2}}{2} + \xi\right)^2 + \frac{1}{2}\right) |y(1)|^2,$$

$$y(1) = 0 \Longrightarrow \int_1^\infty \left(|y(x)|^2 + |y''(x)|^2\right) dx \geq \sqrt{2}|y'(1)|^2,$$

$$y'(1) = \xi y(1) \Longrightarrow$$

$$\int_1^\infty \left(|y(x)|^2 + \left|y''(x) - \frac{2}{x^2}\right|^2\right) dx \geq 2\left(\left(1 + \frac{1}{\sqrt{2}} + \xi\right)^2 + \frac{1}{4}\right)|y(1)|^2,$$

$$y(1) = 0 \Longrightarrow$$

$$\int_1^\infty \left(|y(x)|^2 + \left|y''(x) - \frac{2}{x^2}\right|^2\right) dx \geq 2|y'(1)|^2.$$

The first, third and fourth inequalities are *new*. The second inequality was obtained for the first time using different approach in the paper [10] by F.Gesztesy, N.Kalton, K.Makarov and one of the authors.

Acknowledgment

We are very grateful to referee for valuable remarks.

References

[1] N.I. Akhiezer and I.M. Glazman. *Theory of linear operators.* Pitman Advanced Publishing Program, 1981.

[2] Yu.M. Arlinskiĭ. *On inverse problem of the theory of characteristic functions of unbounded operator colligations.* Dopovidi Akad.Nauk Ukrin.RSR, 2, 105–109, 1976.

[3] Yu.M. Arlinskiĭ. *Regular (∗)-extensions of quasiHermitian operators in rigged Hilbert spaces.* Izvestia Akad. Nauk Armyanskoĭ SSR, v.14, No.4, 297–312, 1979.

[4] Yu.M. Arlinskiĭ. *On regular (∗)-extensions and characteristic matrix-valued functions of ordinary differential operators.* Boundary value problems for differential operators, Kiev, 3–13, 1980.

[5] Yu. Arlinskiĭ and E. Tsekanovskiĭ. *Regular (∗)-extension of unbounded operators, characteristic operator-functions and realization problems of transfer functions of linear systems.* Preprint, VINITI, Dep.-2867, 72p., 1979.

[6] S.V. Belyĭ and E.R. Tsekanovskiĭ. *Realization theorems for operator-valued R-functions.* In I. Gohberg and Yu. Lyubich, editors, New results in operator theory and its applications, Operator Theory: Advances and Applications, 98, 55–91. Birkhäuser Verlag, Basel, 1997.

[7] S. Belyĭ, S. Hassi, H. de Snoo, and E. Tsekanovksiĭ. *On the realization of inverse Stieltjes functions.* Proceedings of MTNS 2002, University of Notre Dame, CD-ROM, 11p. 2002.

[8] M.S. Brodskiĭ. *Treugol'nye i zhordanovy predstavlenija lineinykh operatorov.* Nauka, Moskow, 1969. English translation: Triangular and Jordan representation of linear operators, American mathematical society, Providence, 1971.

[9] I. Dovzhenko and E. Tsekanovskiĭ. *Classes of Stieltjes operator-valued functions and their conservative realizations.* Soviet Math.Dokl., v.41, no.2, 1990, 201-204.

[10] F. Gesztesy, N. Kalton, K. Makarov and E. Tsekanovskiĭ. *Some Applications of Herglotz operator-valued functions.* Operator Theory, Advances and Applications, 123, 2001, 271–321.

[11] W. Helton. *Systems with infinite-dimensional state space: the Hilbert space approach.* Proceedings of IEEE, 64, 1, 145–160, 1976.

[12] M.S. Livšic. *Operator colligations, waves, open systems.* Transl. Math. Monog. AMS. AMS, Providence, R.I., 1973.

[13] B.S. Pavlov. *Dilation theory and spectral analysis of non-self-adjoint differential operators,* Proc. of VII Winter School on Mathemathical Programming and Related Topics, 3–69, 1976.

[14] E. Tsekanovskiĭ and Ju. Šmuljan. *The theory of biextensions of operators in rigged Hilbert spaces. Unbounded operator colligations and characteristic functions.* Uspehi. Mat. Nauk., 32, 5, 69–124, 1977.

[15] E. Tsekanovskiĭ. *Accretive extensions and problems on the Stieltjes operator-valued functions realization.* Operator Theory Advances and Applications, 59, 1992, 328–347.

Yu. Arlinskiĭ
Department of Mathematics
East Ukrainian National University
Lugansk, 91034, Ukraine
e-mail: yma@snu.edu.ua

E. Tsekanovskiĭ
Department of Mathematics
Niagara University, P.O.Box 2044
NY 14109 USA
e-mail: tsekanov@niagara.edu

Operator Theory:
Advances and Applications, Vol. 149, 79–106

Strongly Regular J-Inner Matrix Functions and Related Problems

Damir Z. Arov and Harry Dym

Abstract. A number of characterizations of the class of strongly regular J-inner matrix-valued functions and descriptions of the corresponding reproducing kernel Hilbert spaces and formulas for the reproducing kernels of these spaces are reviewed. Applications to bitangential interpolation problems, bitangential inverse problems for canonical integral and differential systems, J-unitary nodes and Livsic-Brodskii J-nodes are surveyed. Most of the furnished information is adapted from the papers [ArD1]–[ArD11]. However, in the last two sections, some new results are presented.

1. Introduction

The class $\mathcal{U}_{sR}(J) = \mathcal{U}_{sR}(J, \Omega_+)$ of strongly regular J-inner in Ω_+ mvf's (matrix-valued functions) was introduced, investigated and applied to bitangential inverse problems for canonical integral and differential systems in [ArD1]–[ArD8]. A survey of some of our results will be given below and some other related results on operator nodes with characteristic matrix functions of the class $\mathcal{U}_{sR}(J)$ will be formulated.

Here and below J is a fixed $m \times m$ signature matrix: $J^* = J^{-1} = J$; Ω_+ is one of the domains $\mathbb{D}_+$ or $\mathbb{C}_+$ in the complex plane $\mathbb{C}$, where

$$\mathbb{D}_+ = \{\lambda \in \mathbb{C} : |\lambda| < 1\}, \qquad \mathbb{C}_+ = \{\lambda \in \mathbb{C} : \Im\lambda > 0\}.$$

The matrices

$$j_{pq} = \begin{bmatrix} I_p & 0 \\ 0 & -I_q \end{bmatrix}, \qquad J_p = \begin{bmatrix} 0 & -I_p \\ -I_p & 0 \end{bmatrix}, \qquad \mathcal{J}_p = \begin{bmatrix} 0 & -iI_p \\ iI_p & 0 \end{bmatrix} \tag{1.1}$$

are often considered as a signature matrix J.

The class $\mathcal{U}_{sR}(j_{pq})$ is formed by the resolvent matrices (in the terminology of M.G. Krein) of generalized bitangential interpolation problems in the Schur

Received by the editors March 5, 2003. Submitted by L. Rodman.
Mathematics Subject Classification (2000). 30E05, 30D99, 34A55, 34L40, 47A56, 47A57.
Key words and phrases. J-inner matrix-valued functions, reproducing kernel Hilbert spaces, canonical systems, interpolation, inverse problems, operator nodes, Muckenhoupt condition, strongly regular.
Both authors thank the Mittag Leffler Institute (where a rough draft of this paper was extensively revised and extended) for hospitality and support. The first author also thanks the US CRDF and the Ukrainian government for partial support under grant UM 1-25670003. The second author also thanks Renee and Jay Weiss for endowing the Chair that supports his research.

class $\mathcal{S}^{p\times q}$ of $p \times q$ mvf's that are holomorphic and contractive in Ω_+ in the so called strictly completely indeterminate case. There is an analogous connection between the class $\mathcal{U}_{sR}(J_p)$ and generalized bitangential interpolation problems in the Caratheodory class $\mathcal{C}^{p\times p}$ of $p \times p$-mvf's that are holomorphic with positive real part in Ω_+. These connections were considered in [ArD4] and [ArD6]. They will be discussed in Section 3.

In Section 4, criteria for a mvf in the class $\mathcal{U}(J) = \mathcal{U}(J, \Omega_+)$ of J-inner mvf,s with respect to Ω_+ to belong to $\mathcal{U}_{sR}(J)$ that are taken from [ArD11] will be presented. They are formulated in terms of the Treil-Volberg matrix version of the Muckenhoupt (A_2)-condition on a mvf $\Delta(\zeta)$ that is nonnegative a.e. on the boundary $\partial\Omega_+$ of Ω_+ and is defined by the given mvf $U \in \mathcal{U}(J)$. Earlier in [ArD10] we obtained other criteria that was also formulated in terms of an (A_2)-condition but for other mvf's $\Delta(\zeta)$ and with an extra condition.

To every mvf $U \in \mathcal{U}(J)$ there corresponds a RKHS (reproducing kernel Hilbert space) $\mathcal{H}(U)$, the theory of which was developed and extensively studied by L. de Branges [Br1]–[Br3] and then applied to assorted problems of analysis by L. de Branges himself, partially in collaboration with J. Rovnyak [BrR], and by others, see, e.g., [Al], [ADRS], [AlD1]–[AlD3], [An], [Ba], [BaC], [Dy4], [DI], [Re], [Ro], [Sak], [Wi].

In Section 5 the characteristic properties of the space $\mathcal{H}(U)$ that correspond to $U \in \mathcal{U}_{sR}(J)$ are formulated and a description of the space $\mathcal{H}(U)$ for $U \in \mathcal{U}_{sR}(J_p)$ is given. These results were obtained in our papers [ArD4], [ArD7]. In these papers, we also obtained useful formulas for the RK's (reproducing kernels) $K_\omega^U(\lambda)$ for $J = j_{pq}$ and $J = J_p$. These expressions were then applied to obtain formulas for the solutions of bitangential inverse problems for canonical integral and differential systems with matrizants $U_t(\lambda) = U(t, \lambda)$, $0 \le t < d$, that satisfy the extra condition

$$U_t \in \mathcal{U}_{sR}(J), \qquad 0 \le t < d, \tag{1.2}$$

in [ArD2], [ArD4], [ArD7].

In our investigations, following a stategy that originates with M.G. Krein, we interpret the entire mvf's $U_t(\lambda) \in \mathcal{U}_{sR}(J)$ as the resolvent matrix of a generalized bitangential Krein extension problem for helical mvf in the so called strictly completely indeterminate case. Such GKEP's were investigated in [ArD9]. Some of our results on the bitangential inverse problems will be discussed in Section 6.

We wish to emphasize, that in our formulation of bitangential inverse problems, the frequency characteristic of the canonical system (such as a spectral function, an input impedance matrix, an input scattering matrix or a monodromy matrix) that is usually considered as the given data is augmented by a normalized monotonic continuous chain of pairs of entire inner mvf's, that is associated in a certain sense with the matrizant $U_t(\lambda)$ of the underlying system.

In Section 7, we present formulas for the solution of the bitangential inverse impedance problem that are adapted from [ArD7].

In Section 8, we consider a class of RKHS's $\mathcal{B}(\mathfrak{E})$ that are based on a $p \times 2p$ mvf $\mathfrak{E}(\lambda)$ that is meromorphic in $\mathbb{C}_+$. If $\mathfrak{E}(\lambda)$ is entire, then $\mathcal{B}(\mathfrak{E})$ is a RKHS of entire $p \times 1$ vvf's. The theory of such spaces of entire functions was introduced and extensively developed by L. de Branges [Br2], [Br3] and the references cited therein. These spaces play a useful role in the spectral theory of differential and integral equations and a number of other areas of analysis; see, e.g., [Dy1]–[Dy6], [DI], [DK], [DMc], [Dyu], [Gu], [KW1], [KW2], [Sak], [Wi], [Wo], [Yu]. We characterize those spaces $\mathcal{B}(\mathfrak{E})$ of meromorphic vvf's that are invariant under the backwards shift operator and introduce the notion of left and right strongly regular de Branges spaces in terms of two generalized Fourier transforms. We then present some characterizations of these two classes of strongly regular de Branges spaces that are related to the corresponding characterizations of left and right strongly regular J-inner mvf's.

There is a well-known two-sided connection between the theory of mvf's that are J-contractive in Ω_+ and the theory of characteristic mvf's of operator nodes. In particular, it is known that a mvf $U \in \mathcal{U}(J)$ that is holomorphic at the point $\lambda = 0$ may be considered as the characteristic mvf of a simple operator node: of a J-unitary node if $\Omega_+ = \mathbb{D}_+$ or of an LB (Livsic-Brodskii) J-node if $\Omega_+ = \mathbb{C}_+$ and $U(0) = I_m$. Moreover, there exist functional models of these nodes, in which the main operators are modelled as backward shift operators in the space $\mathcal{H}(U)$. On the basis of these models and our results on the spaces $\mathcal{H}(U)$ for $U \in \mathcal{U}_{sR}(J)$ the characteristic properties of simple operator nodes with characteristic mvf's $U(\lambda)$ of the class $\mathcal{U}_{sR}(J)$ were obtained by Z.D. Arova in [Ara1]–[Ara4]. Other characterizations of these classes of operator nodes are considered in [Ar] and Section 9 of this survey.

2. Definition of the class $\mathcal{U}_{sR}(J)$

We recall that an $m \times m$ mvf $U(\lambda)$ that is meromorphic in Ω_+ belongs to the class $\mathcal{U}(J) = \mathcal{U}(J, \Omega_+)$ of J-inner mvf's with respect to Ω_+ if it has J-contractive values $U(\lambda)$ at the points of holomorphy in Ω_+ and J-unitary boundary values $U(\zeta)$ a.e. on $\partial\Omega_+$.

Let $J \neq \pm I_m$ be a signature matrix and let

$$p = \text{rank } (I_m + J), \qquad q = \text{rank } (I_m - J). \tag{2.1}$$

Then there exists a unitary $m \times m$ matrix V such that

$$J = V^* j_{pq} V \qquad (V^*V = VV^* = I_m). \tag{2.2}$$

Moreover,

$$U \in \mathcal{U}(J) \Longleftrightarrow W(\lambda) = VU(\lambda)V^* \text{ belongs to the class } \mathcal{U}(j_{pq}). \tag{2.3}$$

It is well known that the linear fractional transformation

$$T_W[\varepsilon] = (w_{11}\varepsilon + w_{12})(w_{21}\varepsilon + w_{22})^{-1} \tag{2.4}$$

that is based on the four-block decomposition

$$W = \begin{bmatrix} w_{11} & w_{12} \\ w_{21} & w_{22} \end{bmatrix} \tag{2.5}$$

with blocks w_{11} and w_{22} of sizes $p \times p$ and $q \times q$ is well defined on $\mathcal{S}^{p\times q}$ when $W \in \mathcal{U}(j_{pq})$, i.e., $\mathcal{S}^{p\times q} \subset \mathcal{D}(T_W)$, where

$$\mathcal{D}(T_W) = \{\, \varepsilon : \det(w_{21}\varepsilon + w_{22}) \not\equiv 0 \,\}. \tag{2.6}$$

Moreover,

$$T_W[\mathcal{S}^{p\times q}] \subset \mathcal{S}^{p\times q}, \quad \text{and} \quad T_W[\mathcal{S}_{in}^{p\times q}] \subset \mathcal{S}_{in}^{p\times q}. \tag{2.7}$$

Here we use notations:

$$T_W[E] = \{\, T_W(\varepsilon) : \varepsilon \in E \,\}, \tag{2.8}$$

$\mathcal{S}_{in}^{p\times q}$ is the class of $p \times q$ mvf's that are inner with respect to Ω_+ if $p \geq q$ and are $*$-inner with respect to Ω_+ if $p < q$. By a theorem of L.A. Simakova [Si], the property (2.7) is characteristic for $W \in \mathcal{U}(j_{pq})$: if W is an $m \times m$ mvf that is meromorphic in Ω_+ with $\det W(\lambda) \not\equiv 0$ such that (2.7) holds, then $\rho(\lambda)W(\lambda) \in \mathcal{U}(j_{pq})$ for some scalar function $\rho(\lambda)$.

Let

$$\overset{\circ}{\mathcal{S}}{}^{p\times q} = \{\, s \in \mathcal{S}^{p\times q} : \|s\|_\infty < 1 \,\}. \tag{2.9}$$

Definition 2.1. A mvf $W \in \mathcal{U}(j_{pq})$ is said to be *strongly regular* if

$$T_W[\mathcal{S}^{p\times q}] \cap \overset{\circ}{\mathcal{S}}{}^{p\times q} \neq \emptyset. \tag{2.10}$$

Let $J \neq \pm I_m$ and let V be a unitary matrix such that (2.2) holds.

Definition 2.2. A mvf $U \in \mathcal{U}(J)$ is said to be a strongly regular J-inner mvf with respect to Ω_+ if the mvf $W(\lambda) = VU(\lambda)V^*$ is a strongly regular j_{pq}-inner mvf with respect to Ω_+.

The class of strongly regular J-inner mvf,s with respect to Ω_+ will be denoted by the symbol $\mathcal{U}_{sR}(J, \Omega_+)$, or $\mathcal{U}_{sR}(J)$ for short, when the domain is either clear from the context, or not important.

If $J = J_p$, then $q = p$ and

$$J_p = \mathfrak{V}^* j_p \mathfrak{V}, \text{ where } \quad j_p = j_{pp} \quad \text{and} \quad \mathfrak{V} = \frac{1}{\sqrt{2}} \begin{bmatrix} -I_p & I_p \\ I_p & I_p \end{bmatrix}. \tag{2.11}$$

Let V and $\mathfrak{V}$ be unitary matrices that are considered in the relations (2.2) and (2.11) and, for an $m \times m$ mvf $U(\lambda)$, let

$$W(\lambda) = VU(\lambda)V^* \quad \text{and, for } q = p, \text{ let} \quad A(\lambda) = \mathfrak{V}VU(\lambda)V^*\mathfrak{V}^*. \tag{2.12}$$

Then

$$U \in \mathcal{U}(J) \Longleftrightarrow W \in \mathcal{U}(j_{pq}) \quad \text{and, if } q = p, \quad U \in \mathcal{U}(J) \Longleftrightarrow A \in \mathcal{U}(J_p)$$
$$U \in \mathcal{U}_{sR}(J) \Longleftrightarrow W \in \mathcal{U}_{sR}(j_{pq}) \quad \text{and, if } q = p, \quad U \in \mathcal{U}_{sR}(J) \Longleftrightarrow A \in \mathcal{U}_{sR}(J_p).$$

Linear fractional transformations based on the four-block decomposition of mvf's $A(\lambda) \in \mathcal{U}(J_p)$

$$A(\lambda) = \begin{bmatrix} a_{11}(\lambda) & a_{12}(\lambda) \\ a_{21}(\lambda) & a_{22}(\lambda) \end{bmatrix} \tag{2.13}$$

with blocks $a_{ij}(\lambda)$ that are $p \times p$ mvf's map $\mathcal{C}^{p\times p} \cap \mathcal{D}(T_A)$ into $\mathcal{C}^{p\times p}$. However, it turns out to be more useful to consider the linear fractional transformations based on the mvf's

$$B(\lambda) = \begin{bmatrix} b_{11}(\lambda) & b_{12}(\lambda) \\ b_{21}(\lambda) & b_{22}(\lambda) \end{bmatrix} = A(\lambda)\mathfrak{V}, \tag{2.14}$$

since the set

$$\mathcal{C}(A) = T_B[\mathcal{S}^{p\times p} \cap \mathcal{D}(T_B)] \tag{2.15}$$

is a larger subset of $\mathcal{C}^{p\times p}$ than $T_A[\mathcal{C}^{p\times p} \cap \mathcal{D}(T_A)]$. We remark that

$$\mathcal{C}^{p\times p} = T_{\mathfrak{V}}[\mathcal{S}^{p\times p} \cap \mathcal{D}(T_{\mathfrak{V}})] \tag{2.16}$$

and define

$$\overset{o}{\mathcal{C}}{}^{p\times p} = T_{\mathfrak{V}}[\overset{o}{\mathcal{S}}{}^{p\times p}]. \tag{2.17}$$

Then

$$\overset{o}{\mathcal{C}}{}^{p\times p} = \{\, c \in \mathcal{C}^{p\times p} \,:\, c \in H_\infty^{p\times p} \text{ and } (\Re c(\zeta))^{-1} \in L_\infty^{p\times p} \,\}. \tag{2.18}$$

A mvf $A \in \mathcal{U}(J_p)$ belongs to the class $\mathcal{U}_{sR}(J_p)$ if and only if

$$\mathcal{C}(A) \cap \overset{o}{\mathcal{C}}{}^{p\times p} \neq \emptyset. \tag{2.19}$$

In the sequel we shall present a number of different characterizations of the class $\mathcal{U}_{sR}(J)$ and, somewhat later in the development, shall rename the class $\mathcal{U}_{sR}(J)$ to the class $\mathcal{U}_{rsR}(J)$ of right strongly regular J-inner mvf's and shall introduce a second class $\mathcal{U}_{lsR}(J)$ of left strongly regular J-inner mvf's.

3. Connection with generalized bitangential Nevanlinna-Pick problems

The generalized Schur Interpolation Problem GSIP $(b_1, b_2; s^o)$ based on a given set of mvf's $b_1 \in \mathcal{S}_{in}^{p\times p}$, $b_2 \in \mathcal{S}_{in}^{q\times q}$ and $s^o \in \mathcal{S}^{p\times q}$ is to describe the set

$$\mathcal{S}(b_1, b_2; s^o) = \{\, s \in \mathcal{S}^{p\times q} \,:\, b_1^{-1}(s - s^o)b_2^{-1} \in H_\infty^{p\times q} \,\}. \tag{3.1}$$

This problem is said to be *strictly completely indeterminate* if

$$\mathcal{S}(b_1, b_2; s^o) \cap \overset{o}{\mathcal{S}}{}^{p\times q} \neq \emptyset. \tag{3.2}$$

There is a two-sided connection between such problems and the class $\mathcal{U}_{sR}(j_{pq})$. To explain this, recall that the Potapov-Ginzburg transform $\mathcal{S} = PG(W)$ of a mvf $W \in \mathcal{U}(j_{pq})$ is defined by the formula

$$\mathcal{S} = \begin{bmatrix} s_{11} & s_{12} \\ s_{21} & s_{22} \end{bmatrix} = \begin{bmatrix} (w_{11}^{\#})^{-1} & w_{12}w_{22}^{-1} \\ -w_{22}^{-1}w_{21} & w_{22}^{-1} \end{bmatrix}, \tag{3.3}$$

where $f^{\#}(\lambda) = f(\lambda^{\sim})^*$,

$$\lambda^{\sim} = \begin{cases} 1/\overline{\lambda}, & \text{if} \quad \Omega_+ = \mathbb{D}_+, \\ \overline{\lambda}, & \text{if} \quad \Omega_+ = \mathbb{C}_+, \end{cases} \tag{3.4}$$

and the pseudocontinuation of a mvf $U \in \mathcal{U}(J)$ into the domain $\Omega_- = \text{Ext}\,\Omega_+$ is defined by the J-symmetry principle:

$$U(\lambda) = J(U^{\#}(\lambda))^{-1}J. \tag{3.5}$$

The diagonal blocks blocks of $\mathcal{S}(\lambda)$ satisfy the conditions $s_{11} \in \mathcal{S}^{p\times p}, \det s_{11} \not\equiv 0$ and $s_{22} \in \mathcal{S}^{q\times q}, \det s_{22} \not\equiv 0$ and hence admit inner-outer and outer-inner factorizations:

$$s_{11} = b_1\varphi_1, \qquad b_1 \in \mathcal{S}_{in}^{p\times p}, \quad \varphi_1 \in \mathcal{S}_{out}^{p\times p}, \tag{3.6}$$

$$s_{22} = \varphi_2 b_2, \qquad b_2 \in \mathcal{S}_{in}^{q\times q}, \quad \varphi_2 \in \mathcal{S}_{out}^{q\times q}. \tag{3.7}$$

The pair $\{b_1, b_2\}$ of inner mvf's considered in (3.6), (3.7) is called an *associated pair* of the j_{pq}-inner mvf W and we denote such a pair by writing $\{b_1, b_2\} \in \text{ap}\ (W)$. From this definition and general results on GSIP's we obtained the following two theorems.

Theorem 3.1. [ArD4]. *Let the GSIP $(b_1, b_2; s^o)$ be strictly completely indeterminate. Then there exists a mvf $W \in \mathcal{U}(j_{pq})$ such that*

$$\mathcal{S}(b_1, b_2; s^o) = T_W[\mathcal{S}^{p\times q}]. \tag{3.8}$$

Moreover, every such mvf $W(\lambda)$ belongs to the class $\mathcal{U}_{sR}(j_{pq})$ and, in the set of these mvf's, there exists a $W \in \mathcal{U}_{sR}(j_{pq})$ such that

$$\{b_1, b_2\} \in ap\ (W). \tag{3.9}$$

A mvf $W \in \mathcal{U}_{sR}(j_{pq})$ for which (3.8) *and* (3.9) *hold is defined up to a constant j_{pq}-unitary right multiplier. If $b_1(\lambda)$ and $b_2(\lambda)$ are both entire inner mvf's, then $W(\lambda)$ is also an entire mvf and may be uniquely specified by the conditions* (3.8), (3.9) *and the normalization condition $W(0) = I_m$.*

Theorem 3.2. [ArD4]. *Let $W \in \mathcal{U}_{sR}(j_{pq})$. Let*

$$\{b_1, b_2\} \in ap\ (W) \quad and \quad s^o \in T_W[\mathcal{S}^{p\times q}]. \tag{3.10}$$

Then the GSIP $(b_1, b_2; s^o)$ is strictly completely indeterminate and the formula (3.8) *holds. If $W(\lambda)$ is entire, then $b_1(\lambda)$ and $b_2(\lambda)$ are both entire.*

There exist analogous results on GCIP's (Generalized Caratheodory Interpolation Problems) and the class $\mathcal{U}_{sR}(J_p)$. To formulate them for a mvf $A(\lambda) \in \mathcal{U}(J_p)$ we consider the corresponding mvf $B(\lambda)$, defined by formula (2.14).

The mvf's $b_{21}^{\# \, -1}$ and b_{22}^{-1} have inner-outer and outer-inner factorizations in the Smirnov class [1] $\mathcal{N}_+^{p\times p}$ in Ω_+:

$$b_{21}^{\# \, -1} = b_3\varphi_3, \quad \text{where} \quad b_3 \in \mathcal{S}_{in}^{p\times p}, \quad \varphi_3 \in \mathcal{N}_{out}^{p\times p}, \tag{3.11}$$

$$b_{22}^{-1} = \varphi_4 b_4, \quad \text{where} \quad b_4 \in \mathcal{S}_{in}^{p\times p}, \quad \varphi_4 \in \mathcal{N}_{out}^{p\times p}. \tag{3.12}$$

The pair $\{b_3, b_4\}$ is called an *associated pair of the second kind* of the mvf $A(\lambda)$ and we write $\{b_3, b_4\} \in \text{ap}_{\text{II}}(A)$. (If $W(\lambda) = \mathfrak{V}A(\lambda)\mathfrak{V}$ and $\{b_1, b_2\} \in \text{ap}\,(W)$, then $\{b_1, b_2\}$ is called an *associated pair of the first kind* of $A(\lambda)$ and we write $\{b_1, b_2\} \in \text{ap}_{\text{I}}(A)$.)

Let $b_3 \in \mathcal{S}_{in}^{p\times p}$, $b_4 \in \mathcal{S}_{in}^{p\times p}$ and $c^o \in \mathcal{C}^{p\times p}$ be given mvf's. Then GCIP $(b_3, b_4; c^o)$ is to describe the set

$$\mathcal{C}(b_3, b_4; c^o) = \{\, c \in \mathcal{C}^{p\times p} \,:\, b_3^{-1}(c - c^o)b_4^{-1} \in \mathcal{N}_+^{p\times p}\}. \tag{3.13}$$

This problem is said to be strictly completely indeterminate, if

$$\mathcal{C}(b_3, b_4; c^o) \cap \overset{o}{\mathcal{C}}{}^{p\times p} \neq \emptyset. \tag{3.14}$$

Theorem 3.3. [ArD6]. *Let the GCIP $(b_3, b_4; c^o)$ be strictly completely indeterminate. Then there exists a mvf $A(\lambda) \in \mathcal{U}(J_p)$ such that*

$$\mathcal{C}(b_3, b_4; c^o) = \mathcal{C}(A). \tag{3.15}$$

Moreover every such mvf $A(\lambda)$ belongs to the class $\mathcal{U}_{sR}(J_p)$ and, in the set of these mvf's, there exists an $A \in \mathcal{U}_{sR}(J_p)$ such that

$$\{b_3, b_4\} \in ap_{\text{II}}(A). \tag{3.16}$$

A mvf $A \in \mathcal{U}_{sR}(J_p)$ for which (3.15) *and* (3.16) *hold is defined up to a constant J_p-unitary right multiplier. If $b_3(\lambda)$ and $b_4(\lambda)$ are both entire inner mvf's, then $A(\lambda)$ is also an entire mvf and may be uniquely specified by the conditions* (3.15), (3.16) *and the normalization condition $A(0) = I_m$.*

Theorem 3.4. [ArD6]. *Let $A \in \mathcal{U}_{sR}(J_p)$. Let*

$$\{b_3, b_4\} \in ap_{\text{II}}(A) \quad \text{and} \quad c^o \in \mathcal{C}(A). \tag{3.17}$$

Then the GCIP $(b_3, b_4; c^o)$ is strictly completely indeterminate and formula (3.15) *holds. If $A(\lambda)$ is entire, then $b_3(\lambda)$ and $b_4(\lambda)$ are both entire.*

A GCIP $(b_3, b_4; c^o)$ that is based on entire mvf's $b_3(\lambda)$ and $b_4(\lambda)$, is equivalent to a generalized bitangential Krein extension problem for the so called helical mvf's $g(t)$ that correspond to mvf's $c(\lambda) \in \mathcal{C}^{p\times p}$ by the formula

$$c(\lambda) = \lambda^2 \int_0^\infty e^{i\lambda t} g(t)\, dt, \quad g(0) \leq 0. \tag{3.18}$$

This GKEP was considered in [ArD9].

---

[1] $\mathcal{N}_+^{p\times p} = \{g/h : g \in \mathcal{S}^{p\times p} \text{ and } h \in \mathcal{S}_{out}^{1\times 1}\}$; $\mathcal{N}_{out}^{p\times p} = \{g/h : g \in \mathcal{S}_{out}^{p\times p} \text{ and } h \in \mathcal{S}_{out}^{1\times 1}\}$.

4. Criteria for $U \in \mathcal{U}_{sR}(J)$ in terms of Muckenhoupt conditions

In [ArD10] we obtained a necessary and sufficient condition on $U \in \mathcal{U}(J)$ under which $U \in \mathcal{U}_{sR}(J)$. In this formulation there are two conditions, one of which is the Treil-Volberg matrix version of the Muckenhoupt (A_2)-condition

$$\sup_I \|(\Delta_I)^{1/2}((\Delta^{-1})_I)^{1/2}\| < \infty, \tag{4.1}$$

where

$$\Delta_I = \frac{1}{|I|}\int_I \Delta(\zeta)\,|d\zeta|, \tag{4.2}$$

$\Delta(\zeta)$ is a mvf that is defined in terms of $U(\lambda)$ (and J) and is nonnegative a.e. on $\partial\Omega_+$, I is a subarc of $\mathbb{T} = \{\zeta : |\zeta| = 1\}$ if $\Omega_+ = \mathbb{D}_+$ and I is a finite subinterval of $\mathbb{R}$, if $\Omega_+ = \mathbb{C}_+$. In both settings, $|I|$ denotes the length of I.

Remark 4.1. The condition (4.1) is equivalent to the determinant condition

$$\sup_I((\det \Delta_I)\cdot(\det((\Delta^{-1})_I))) < \infty; \tag{4.3}$$

see [ArD11].

In [ArD11] we obtained new and simpler criteria for a mvf $U \in \mathcal{U}(J)$ to be strongly regular that is based on a different choice of the mvf $\Delta(\zeta)$ than was considered in [ArD10] and is summarized below. In fact, in both [ArD10] and [ArD11], we considered two notions of strong regularity: left and right, and introduced the classes $\mathcal{U}_{lsR}(J)$ and $\mathcal{U}_{rsR}(J)$ of left and right strongly regular J-inner mvf's. The latter coincides with the class $\mathcal{U}_{sR}(J)$ that was considered earlier. These two classes are connected by the relation

$$U(\lambda) \in \mathcal{U}_{lsR}(J) \Longleftrightarrow U^{\sim}(\lambda) \in \mathcal{U}_{rsR}(J),$$

where

$$U^{\sim}(\lambda) = JU^{\#}(-\lambda)J.$$

Let

$$P = \frac{1}{2}(I_m + J), \quad Q = \frac{1}{2}(I_m - J) \tag{4.4}$$

and, for a mvf $U \in \mathcal{U}(J)$ with $J \neq \pm I_m$, let

$$G_r(\zeta) = P + U(\zeta)^*QU(\zeta) \quad \text{and} \quad G_l(\zeta) = P + U(\zeta)QU(\zeta)^* \quad \text{a.e.} \quad \zeta \in \partial\Omega_+. \tag{4.5}$$

Let

$$W(\lambda) = VU(\lambda)V^* = \begin{bmatrix} w_{11}(\lambda) & w_{12}(\lambda) \\ w_{21}(\lambda) & w_{22}(\lambda) \end{bmatrix}, \tag{4.6}$$

for $U \in \mathcal{U}(J)$, where $J \neq \pm I_m$ and V satisfies (2.2). Let

$$s_{21}(\lambda) = -w_{22}^{-1}(\lambda)w_{21}(\lambda) \text{ and } s_{12}(\lambda) = w_{12}(\lambda)w_{22}(\lambda)^{-1}. \tag{4.7}$$

Then

$$V^*(G_r(\zeta))^{-1}V = \begin{bmatrix} I_p & s_{21}(\zeta)^* \\ s_{21}(\zeta) & I_q \end{bmatrix} =: \Delta_r(\zeta), \tag{4.8}$$

$$V^*(G_l(\zeta))^{-1}V = \begin{bmatrix} I_p & -s_{12}(\zeta) \\ -s_{12}(\zeta)^* & I_q \end{bmatrix} =: \Delta_l(\zeta) \tag{4.9}$$

and we have the following conclusions:

Theorem 4.2. [ArD11]. *Let $U \in \mathcal{U}(J)$. Then:*

(a) $U \in \mathcal{U}_{rsR}(J) \iff \Delta(\zeta) = \Delta_r(\zeta)$ *satisfies the* (A_2)*-condition* (4.1)
$\iff \Delta(\zeta) = G_r(\zeta)$ *satisfies the* (A_2)*-condition* (4.1).

(b) $U \in \mathcal{U}_{lsR}(J) \iff \Delta(\zeta) = \Delta_l(\zeta)$ *satisfies the* (A_2)*-condition* (4.1)
$\iff \Delta(\zeta) = G_l(\zeta)$ *satisfies the* (A_2)*-condition* (4.1).

5. The de Branges spaces $\mathcal{H}(U)$ for $U \in \mathcal{U}_{sR}(J)$

The RK $K_\omega^U(\lambda)$ of the RKHS $\mathcal{H}(U)$ based on a mvf $U \in \mathcal{U}(J)$ is defined by the formula

$$K_\omega^U(\lambda) = (\rho_\omega(\lambda))^{-1}(J - U(\lambda)JU(\omega)^*), \qquad \omega \neq \lambda^\sim, \tag{5.1}$$

where

$$\rho_\omega(\lambda) = \begin{cases} 1 - \lambda\overline{\omega}, & \text{if} \quad \Omega_+ = \mathbb{D}_+, \\ -2\pi i(\lambda - \overline{\omega}), & \text{if} \quad \Omega_+ = \mathbb{C}_+. \end{cases} \tag{5.2}$$

A mvf $U \in \mathcal{U}(J)$ and the vvf's $g \in \mathcal{H}(U)$ have pseudocontinuations into the extended complex plane $\mathbb{C} \cup \{\infty\}$ and we consider $U(\lambda)$ and $g(\lambda)$ on the sets of analyticity $\mathfrak{h}_U$ and $\mathfrak{h}_g$ of these extended functions in $\mathbb{C} \cup \{\infty\}$. The RK K_ω^U may be considered on $\mathfrak{h}_U \times \mathfrak{h}_U$. The vvf's $g(\lambda) \in \mathcal{H}(U)$ have nontangential boundary values $g(\zeta)$ at a.e. point $\zeta \in \partial\Omega_+$, when λ tends to ζ nontangentially (with respect to $\partial\Omega_+$) from $\mathbb{C} \setminus \partial\Omega_+$. We write $\mathcal{H}(U) \subset L_2^m$ if $g(\zeta) \in L_2^m(\partial\Omega_+)$ for every $g \in \mathcal{H}(U)$ and shall let $\Pi^{m\times k}$ denote the subclass of mvf's $f \in N^{m\times k}$ that admit a pseudocontinuation f_- to Ω_- such that $f_-^\# \in N^{k\times m}$.

Theorem 5.1. [ArD1]. *Let $U \in \mathcal{U}(J)$, where $J \neq \pm I_m$. Then $U \in \mathcal{U}_{sR}(J)$ if and only if one of the following equivalent properties holds:*

1) $\mathcal{H}(U) \subset L_2^m$.
2) $\mathcal{H}(U)$ *is a (closed) subspace of* L_2^m.
3)
$$\gamma_1\|g\|_{L_2^m} \le \|g\|_{\mathcal{H}(U)} \le \gamma_2\|g\|_{L_2^m} \tag{5.3}$$
for every $g \in \mathcal{H}(U)$, where γ_1 and γ_2 are finite constants such that $0 < \gamma_1 \le \gamma_2$.

Remark 5.2. The preceding theorem implies that $\mathcal{U}(J) \cap L_\infty^{m\times m} \subset \mathcal{U}_{sR}(J)$. Moreover, this inclusion is proper. Examples that illustrate this are furnished in Subection 7.6 of [ArD10].

Let $H_2^m(\Omega_\pm)$ be the Hardy spaces of $m \times 1$ vvf's $g(\lambda)$ that are holomorphic in $\Omega_\pm$ and have finite H_2-norm. (If $\Omega_- = \mathbb{D}_- = \operatorname{Ext} \mathbb{D}_+$ and $g \in H_2^m(\Omega_-)$ then $g(\infty) = 0$.)

Upon identifying the vvf's $g(\lambda)$ with their boundary values $g(\zeta)$ we have:

$$K_2^m \oplus H_2^m = L_2^m, \quad \text{where} \quad H_2^m = H_2^m(\Omega_+) \quad \text{and} \quad K_2^m = H_2^m(\Omega_-). \tag{5.4}$$

Let

$$\mathcal{H}_\pm(U) = \mathcal{H}(U) \cap H_2^m(\Omega_\pm) \tag{5.5}$$

for $U \in \mathcal{U}(J)$.

Theorem 5.3. [ArD1]. *Let $U \in \mathcal{U}_{sR}(J)$, where $J \neq \pm I_m$. Then*

$$\mathcal{H}(U) = \mathcal{H}_-(U) \dot{+} \mathcal{H}_+(U). \tag{5.6}$$

Let M_c be the operator in $L_2^p = L_2^p(\mathbb{R})$ of multiplication by a mvf $c \in L_\infty^{p\times p}$ and, for $b \in \mathcal{S}_{in}^{p\times p}$, let

$$\mathcal{H}(b) = H_2^p \ominus bH_2^p, \qquad \mathcal{H}_*(b) = K_2^p \ominus b^{-1}K_2^p. \tag{5.7}$$

These two spaces are RKHS's with respect to the standard inner product in L_2^p with RK's

$$k_\omega^b(\lambda) = \frac{I_p - b(\lambda)b(\omega)^*}{\rho_\omega(\lambda)} \text{ and } l_\omega^b(\lambda) = \frac{b(\lambda)^{-1}b(\omega)^{-*} - I_p}{\rho_\omega(\lambda)},$$

respectively. The following description of the space $\mathcal{H}(A)$ for $A \in \mathcal{U}_{sR}(J, \mathbb{C}_+)$ is given in [ArD7] in terms of the three mvf's

$$\{b_3, b_4\} \in \mathrm{ap}_{\mathrm{II}}(A) \tag{5.8}$$

and

$$c \in \mathcal{C}(A) \cap H_\infty^{p\times p}. \tag{5.9}$$

$$\Phi_{11} = \Pi_{\mathcal{H}(b_3)} M_c\big|_{H_2^p}, \quad \Phi_{12} = \Pi_{\mathcal{H}(b_3)} M_c\big|_{\mathcal{H}_*(b_4)}, \quad \Phi_{22} = \Pi_{K_2^p} M_c\big|_{\mathcal{H}_*(b_4)}, \tag{5.10}$$

where Π_L is the orthoprojection onto L.

Theorem 5.4. [ArD7]. *Let $A \in \mathcal{U}_{sR}(J, \mathbb{C}_+)$. Then the operator L_A that is defined by the rule*

$$L_A : \begin{bmatrix} g \\ h \end{bmatrix} \longrightarrow \begin{bmatrix} -\Phi_{11}^* g + \Phi_{22} h \\ g + h \end{bmatrix} \tag{5.11}$$

is a bounded linear map from $\mathcal{H}(b_3) \oplus \mathcal{H}_*(b_4)$ onto $\mathcal{H}(A)$:

$$\mathcal{H}(A) = \left\{ \begin{bmatrix} -\Phi_{11}^* g + +\Phi_{22} h \\ g + h \end{bmatrix} : g \in \mathcal{H}(b_3) \;\; \text{and} \;\; h \in \mathcal{H}_*(b_4) \right\}. \tag{5.12}$$

Moreover, if

$$f = L_A \begin{bmatrix} g \\ h \end{bmatrix} = \begin{bmatrix} -\Phi_{11}^* g + +\Phi_{22} h \\ g + h \end{bmatrix}$$

for some choice of $g \in \mathcal{H}(b_3)$ *and* $h \in \mathcal{H}_*(b_4)$, *and*

$$\Delta_A = 2\Re \begin{bmatrix} \Phi_{11}|_{\mathcal{H}(b_3)} & \Phi_{12} \\ 0 & \Pi_{\mathcal{H}(b_4)}\Phi_{22} \end{bmatrix},$$

then

$$\begin{aligned} \|f\|^2_{\mathcal{H}(A)} &= \langle \Delta_A \begin{bmatrix} g \\ h \end{bmatrix}, \begin{bmatrix} g \\ h \end{bmatrix} \rangle_{L_2^m} \\ &= \langle (c+c^*)(g+h), (g+h) \rangle_{L_2^p}. \end{aligned} \tag{5.13}$$

In [ArD7] we also obtained a formula for the RK K_ω^A in terms of the solution of a linear system of equations with coefficients $2\Re\Phi_{11}$, Φ_{12}, Φ_{12}^* and $2\Re\Phi_{22}$ and a right-hand side that is defined by the mvf's c, b_3 and b_4 and a dual pair $\dot{b}_3, \dot{b}_4$ of inner $p \times p$ inner mvf's that are defined essentially uniquely by the relations

$$\Phi_{11}^* \mathcal{H}(b_3) = \mathcal{H}(\dot{b}_3) \text{ and } \Phi_{22}^* \mathcal{H}_*(b_4) = \mathcal{H}_*(\dot{b}_4).$$

In applications to inverse problems and operator nodes, the case $0 \in \mathfrak{h}_A$ and formulas for $K_0^A(\lambda)$ are of particular interest; see Section 7 for a sample. In this case, the formulas referred to above yield the following result:

Theorem 5.5. [ArD7]. *Let* $A \in \mathcal{U}_{sR}(J, \mathbb{C}_+)$, *let* $0 \in \mathfrak{h}_A$ *and let* b_5, b_6 *be any pair of inner* $p \times p$ *mvf's such that*

$$(\dot{b}_3)^{-1} b_5 \in \mathcal{S}_{in}^{p\times p} \text{ and } b_6 (\dot{b}_4)^{-1} \in \mathcal{S}_{in}^{p\times p}.$$

Then

$$K_0^A = L_A \begin{bmatrix} \widehat{u}_{11} & \widehat{u}_{12} \\ \widehat{u}_{21} & \widehat{u}_{22} \end{bmatrix}, \tag{5.14}$$

where the $\widehat{u}_{ij} = \widehat{u}_{ij}(\lambda)$ *are* $p \times p$ *mvf's that are obtained as the solutions of the system of equations*

$$\Delta_A \begin{bmatrix} \widehat{u}_{11} & \widehat{u}_{12} \\ \widehat{u}_{21} & \widehat{u}_{22} \end{bmatrix} = \begin{bmatrix} -\Phi_{11} k_0^{b_5} & k_0^{b_3} \\ \Phi_{22}^* l_0^{b_6} & l_0^{b_4} \end{bmatrix} \tag{5.15}$$

and the operators in formulas (5.14) *and* (5.15) *act on the indicated matrix arrays column by column. In particular, the columns of* $\widehat{u}_{11}(\lambda)$ *and* $\widehat{u}_{12}(\lambda)$ *belong to* $\mathcal{H}(b_3)$ *and the columns of* $\widehat{u}_{21}(\lambda)$ *and* $\widehat{u}_{22}(\lambda)$ *belong to* $\mathcal{H}_*(b_4)$.

Earlier analogous results on the description of the RKHS $\mathcal{H}(W)$ and a formula for the RK $K_\omega^W(\lambda)$ were obtained for $W \in \mathcal{U}_{sR}(j_{pq})$ in [ArD4].

6. Canonical systems with matrizants of the class $\mathcal{U}_{sR}(J)$

In our papers [ArD2], [ArD4], [ArD7], we obtained formulas for the solutions of bitangential inverse problems for canonical integral systems of the form

$$u(t,\lambda) = u(0,\lambda) + i\lambda \int_0^t u(s,\lambda)\, dM(s) J, \quad 0 \le t < d, \tag{6.1}$$

and the corresponding canonical differential systems of the form

$$\frac{du}{dt} = i\lambda u(t,\lambda)H(t)J, \qquad 0 \le t < d \tag{6.2}$$

with $J = j_{pq}$ and $J = J_p$ from the formulas for $K_0^W(\lambda)$ and $K_0^A(\lambda)$.

In our analysis, we assume that the mvf M(t) that appears in the system (6.1) is a continuous nondecreasing $m \times m$ mvf on the interval $[0,d)$ with $M(0) = 0$ and that $u(t,\lambda)$ is a continuous $k \times m$ mvf on $[0,d)$. In the system (6.2) $H(t)$ is assumed to be locally summable $m \times m$ mvf that is nonnegative a.e. on $[0,d)$ and $u(t,\lambda)$ is a $k \times m$ mvf that is absolutely continuous on every closed subinterval of $[0,d)$. The system (6.2) is equivalent to the system (6.1) with

$$\mathrm{M}(t) = \int_0^t H(s)\,ds, \quad 0 \le t < d. \tag{6.3}$$

The matrizant $U_t(\lambda) = U(t,\lambda)$ of the system (6.1) (or (6.2)) is the solution of this system with initial condition $U(0,\lambda) = I_m$. It belongs to the class $\mathcal{E} \cap \mathcal{U}(J)$ of entire J-inner mvf's and $U_t(0) = I_m$ for every $t \in [0,d)$. Moreover

$$(U_{t_1})^{-1}U_{t_2} \in \mathcal{E} \cap \mathcal{U}(J), \quad \text{if} \quad 0 \le t_1 \le t_2 < d. \tag{6.4}$$

Our next objective is to introduce the notions of input scattering matrices and input impedance matrices for systems (6.1) and (6.2 with arbitrary signature matrices $J \ne \pm I_m$. To this end, we set

$$W_t(\lambda) = VU_t(\lambda)V^* \quad \text{and, if } q = p, \quad A_t(\lambda) = \mathfrak{V}W_t(\lambda)\mathfrak{V}^*, \quad 0 \le t < d, \tag{6.5}$$

where V and $\mathfrak{V}$ are defined by (2.2) and (2.11), and let

$$\mathcal{S}_{scat} = \bigcap_{0 \le t < d} T_{W_t}[\mathcal{S}^{p\times q}] \quad \text{and} \quad \mathcal{C}_{imp} = \bigcap_{0 \le t < d} \mathcal{C}(A_t). \tag{6.6}$$

The set $\mathcal{S}_{scat}$ of *input scattering matrices* is not empty because

$$T_{W_{t_2}} \subset T_{W_{t_1}} \subset \mathcal{S}^{p\times q} \quad \text{for} \quad t_1 \le t_2 < d$$

and $\mathcal{S}^{p\times q}$ is sequentially compact. We shall call a mvf $s \in \mathcal{S}_{scat}$ an input scattering matrix of the considered system. The sets $\mathcal{C}(A_t)$ are monotone non decreasing nonempty subsets of $\mathcal{C}^{p\times p}$. However, the set $\mathcal{C}_{imp}$ of *input impedance matrices* may be empty unless additional restrictions are imposed because $\mathcal{C}^{p\times p}$ is not sequentially compact. We shall call a mvf $c \in \mathcal{C}_{imp}$ an input impedance matrix of the considered system.

The mvf's $b_j^t(\lambda)$ in the chains of associated pairs

$$\{b_1^t, b_2^t\} \in \mathrm{ap}\,(W_t), \qquad 0 \le t < d \tag{6.7}$$

and

$$\{b_3^t, b_4^t\} \in \mathrm{ap}_{\mathrm{II}}\,(A_t), \qquad 0 \le t < d \tag{6.8}$$

of inner mvf's are entire and they may be normalized by the condition $b_j^t(0) = I$, $1 \le j \le 4$. Moreover, these two chains are monotonic in the following sense:

$$(b_1^{t_1})^{-1} b_1^{t_2} \in \mathcal{E} \cap \mathcal{S}_{in}^{p\times p}, \qquad b_2^{t_2}(b_2^{t_1})^{-1} \in \mathcal{E} \cap \mathcal{S}_{in}^{q\times q}, \tag{6.9}$$

$$(b_3^{t_1})^{-1} b_3^{t_2} \in \mathcal{E} \cap \mathcal{S}_{in}^{p\times p}, \qquad b_4^{t_2}(b_4^{t_1})^{-1} \in \mathcal{E} \cap \mathcal{S}_{in}^{p\times p}, \tag{6.10}$$

if $0 \le t_1 \le t_2 < d$. This property follows from (6.4).

In [ArD1]–[ArD8] we investigated canonical systems of the form (6.1) with property (1.2). We proved that for such systems the chains of normalized associated pairs are continuous in the sense that the mvf's $b_j^t(\lambda)$ are continuous with respect to t on the interval $[0, d)$ for every fixed $\lambda \in \mathbb{C}$. In our considerations, the given data for a bitangential inverse problem is not just a frequency characteristic of the system such as an input scattering matrix $s(\lambda)$, or an input impedance matrix $c(\lambda)$, or a spectral function or a monodromy matrix, but also a normalized monotonic continuous chain of pairs of entire mvf's $\{b_1^t, b_2^t\}$, $0 \le t < d$ or $\{b_3^t, b_4^t\}$, $0 \le t < d$, that relate to the system via (6.7) or (6.8), respectively.

The next theorem, which guarantees the existence and uniqueness of a solution to the bitangential inverse impedance problem with given data $(c, b_3^t, b_4^t, 0 \le t < d)$ is established in [ArD6]. The proof of this theorem is based on Theorem 3.3. This theorem is applied to obtain similar conclusions for the bitangential inverse spectral problem in [ArD8]. Analogous results were obtained earlier for the bitangential inverse monodromy problem in [ArD2] and for the bitangential inverse input scattering problem in [ArD4].

Theorem 6.1. *Let $c \in \mathcal{C}^{p\times p}$, let $\{b_3^t(\lambda), b_4^t(\lambda)\}, 0 \le t < d$, be a normalized monotonic continuous chain of entire inner $p \times p$ mvf's and assume that*

$$\mathcal{C}(b_3^t, b_4^t; c) \cap \mathring{\mathcal{C}}^{p\times p} \neq \emptyset \text{ for every } t \in [0, d)\ . \tag{6.11}$$

Then there exists exactly one canonical integral system (6.1) with matrizant $A_t(\lambda) = A(t, \lambda)$, $0 \le t < d$, that satisfies the following conditions:

(1) $c \in \mathcal{C}_{imp}$.
(2) $\{b_3^t, b_4^t\} \in ap_{II}(A_t)$ *for every* $t \in [0, d)$.
(3) $A_t \in \mathcal{U}_{sR}(J_p)$ *for every* $t \in [0, d)$.

In particular, the condition (6.11) is satisfied if $c \in \mathring{\mathcal{C}}^{p\times p}$.

7. Formulas for the solution $M(t)$ of the bitangential inverse impedance problem and for the corresponding matrizant

Under assumption (6.11), there exists a mvf

$$c^t \in \mathcal{C}(A_t) \cap H_\infty^{p\times p} \tag{7.1}$$

for every $t \in [0, d)$ and hence, upon writing Φ_{ij}^t for the operators Φ_{ij} defined in formula (5.10) with $b_3^t(\lambda)$ in place of $b_3(\lambda)$, $b_4^t(\lambda)$ in place of $b_4(\lambda)$ and $c^t(\lambda)$ in place of $c(\lambda)$, we obtain

$$\Phi_{11}^t = \Pi_{\mathcal{H}(b_3^t)} M_{c^t}\Big|_{H_2^p}\,,\ \Phi_{22}^t = \Pi_- M_{c^t}\Big|_{\mathcal{H}_*(b_4^t)} \text{ and } \Phi_{12}^t = \Pi_{\mathcal{H}(b_3^t)} M_{c^t}\Big|_{\mathcal{H}_*(b_4^t)}\,. \tag{7.2}$$

Correspondingly, let

$$Y_1^t = \Pi_{\mathcal{H}(b_3^t)}\left\{M_{c^t} + (M_{c^t})^*\right\}\Big|_{\mathcal{H}(b_3^t)} = 2\Re\left(\Phi_{11}^t\Big|_{\mathcal{H}(b_3^t)}\right)\,, \tag{7.3}$$

$$Y_2^t = \Pi_{\mathcal{H}_*(b_4^t)}\left\{M_{c^t} + (M_{c^t})^*\right\}\Big|_{\mathcal{H}_*(b_4^t)} = 2\Re\left(\Pi_{\mathcal{H}_*(b_4^t)}\Phi_{22}^t\right)\,, \tag{7.4}$$

$$\tau_3(t) = \overline{\lim}_{\nu\uparrow\infty}\frac{\log\|b_3^t(-i\nu)\|}{\nu} \text{ and } \tau_4(t) = \overline{\lim}_{\nu\uparrow\infty}\frac{\log\|b_4^t(-i\nu)\|}{\nu}\,. \tag{7.5}$$

Then, the entire inner $p\times p$ mvf $b_j^t(\lambda)$ is of exponential type $\tau_j(t)$, $\mathcal{H}(b_3^t)\subset\mathcal{H}(e_{\tau_3(t)}I_p)$, $\mathcal{H}_*(b_4^t)\subset\mathcal{H}(e_{\tau_4(t)}I_p)$ and, as the type of $\dot{b}_j^t$ is equal to the type of b_j^t,

$$\Phi_{11}^*\mathcal{H}(b_3^t)\subseteq\mathcal{H}(e_{\tau_3(t)}I_p) \text{ and } \Phi_{22}\ \mathcal{H}_*(b_4^t)\subseteq\mathcal{H}_*(e_{\tau_4(t)}I_p)\,.$$

Therefore, we can choose

$$b_5(\lambda) = b_5^t(\lambda) = e^{i\lambda\tau_3(t)}I_p \text{ and } b_6(\lambda) = b_6^t(\lambda) = e^{i\lambda\tau_4(t)}I_p$$

in Theorem 5.5.

In order to keep the notation relatively simple, an operator T that acts in the space of $p\times 1$ vvf's will be applied to $p\times p$ mvf's with columns $f_1,\dots,f_p$ column by column: $T[f_1\cdots f_p] = [Tf_1\cdots Tf_p]$.

Theorem 7.1. *In the setting of the last theorem, let $\tau_3(t)$ and $\tau_4(t)$ be defined by formula (7.5), and recalling that $(R_0f)(\lambda) = \{f(\lambda)-f(0)\}/\lambda$, let*

$$\widehat{y}_{11}^t(\lambda) = -\frac{1}{i}\left(\Phi_{11}^t(R_0e_{\tau_3(t)}I_p)\right)(\lambda)\,,\ \widehat{y}_{12}^t(\lambda) = \frac{1}{i}(R_0b_3^t)(\lambda) \tag{7.6}$$

$$\widehat{y}_{21}^t(\lambda) = \frac{-1}{i}\left((\Phi_{22}^t)^*(R_0e_{-\tau_4(t)}I_p)\right)(\lambda)\,,\ \widehat{y}_{22}^t(\lambda) = -\frac{1}{i}(R_0(b_4^t)^{-1})(\lambda) \tag{7.7}$$

Then the RK $K_\omega^t(\lambda)$ of the RKHS $\mathcal{H}(A_t)$ evaluated at $\omega=0$ is given by the formula

$$K_0^t(\lambda) = \frac{1}{2\pi}\begin{bmatrix}\widehat{x}_{11}^t(\lambda)+\widehat{x}_{21}^t(\lambda) & \widehat{x}_{12}^t(\lambda)+\widehat{x}_{22}^t(\lambda)\\ \widehat{u}_{11}^t(\lambda)+\widehat{u}_{21}^t(\lambda) & \widehat{u}_{12}^t(\lambda)+\widehat{u}_{22}^t(\lambda)\end{bmatrix}\,, \tag{7.8}$$

where:

(1) *The $\widehat{u}_{ij}^t(\lambda)$ are $p\times p$ mvf's such that the columns of $\widehat{u}_{1j}^t(\lambda)$ and $\widehat{u}_{12}^t(\lambda)$ belong to $\mathcal{H}(b_3^t)$ and the columns of $\widehat{u}_{21}^t(\lambda)$ and $\widehat{u}_{22}^t(\lambda)$ belong to $\mathcal{H}_*(b_4^t)$. The $\widehat{u}_{ij}^t(\lambda)$ may be defined as the solutions of the systems of equations:*

$$Y_1^t\widehat{u}_{1j}^t + \Phi_{12}^t\widehat{u}_{2j}^t = \widehat{y}_{1j}^t(\lambda) \tag{7.9}$$

$$(\Phi_{12}^t)^*\widehat{u}_{1j}^t + Y_2^t\widehat{u}_{2j}^t = \widehat{y}_{2j}^t(\lambda)\,,\ j=1,2\,. \tag{7.10}$$

(2) *The mvf's* $\widehat{x}^t_{ij}(\lambda)$ *are defined by the formulas*

$$\begin{aligned}\widehat{x}^t_{1j}(\lambda) &= -(\Phi^t_{11})^*\widehat{u}^t_{1j} \ , \\ \widehat{x}^t_{2j}(\lambda) &= \Phi^t_{22}\widehat{u}^t_{2j} \ , \ j = 1,2 \ .\end{aligned} \tag{7.11}$$

Remark 7.2. In the one-sided cases, when either $b^t_4(\lambda) = I_p$ or $b^t_3(\lambda) = I_p$, the formulas for recovering $M(t)$ are simpler. For example, if $b^t_4(\lambda) = I_p$, then $\tau_4(t) = 0$ and $\mathcal{H}_*(b^t_4) = \{0\}$ and hence $\widehat{u}^t_{2j} = 0, \widehat{x}^t_{2j} = 0$, and the equation for $\widehat{u}^t_{1j}$ and the formula for $\widehat{x}^t_{1j}$ simplify to

$$Y^t_1\widehat{u}^t_{1j} = \widehat{y}^t_{1j}(\lambda) \ , \ j = 1,2, \tag{7.12}$$

and

$$\widehat{x}^t_{1j} = -(\Phi^t_{11})^*\widehat{u}^t_{1j} \ , \ j = 1,2, \tag{7.13}$$

Theorem 7.3. *Let* $\{c(\lambda); b^t_3(\lambda), b^t_4(\lambda),\ 0 \le t < d\}$ *be given, where* $c \in C^{p\times p}$ *and* $\{b^t_3(\lambda), b^t_4(\lambda)\}$, $0 \le t < d$, *is a normalized monotonic continuous chain of pairs of entire inner* $p \times p$ *mvf's and let assumption* (6.11) *be in force. Then the unique solution* $M(t)$ *of the inverse input impedance problem considered in Theorem* 6.1 *is given by the formula*

$$M(t) = 2\pi K^t_0(0) = \int_0^{\tau_3(t)} \begin{bmatrix} x^t_{11}(a) & x^t_{12}(a) \\ u^t_{11}(a) & u^t_{12}(a)\end{bmatrix} da + \int_{-\tau_4(t)}^{0} \begin{bmatrix} x^t_{21}(a) & x^t_{22}(a) \\ u^t_{21}(a) & u^t_{22}(a)\end{bmatrix} da \tag{7.14}$$

and the corresponding matrizant may be defined by the formula

$$A_t(\lambda) = I_m + 2\pi i K^t_0(\lambda) J_p \ , \tag{7.15}$$

where $K^t_0(\lambda)$ *is specified by formula* (7.8) *and* $x^t_{ij}(a)$ *and* $u^t_{ij}(a)$ *designate the inverse Fourier transforms of* $\widehat{x}^t_{ij}(\lambda)$ *and* $\widehat{u}^t_{ij}(\lambda)$, *respectively.*

Remark 7.4. Upon writing L_t for L_{A_t}, Δ_t for Δ_{A_t}, and $K^t_\omega(\lambda)$ for $K^{A_t}_\omega(\lambda)$, the formula for the RK $K^t_0(\lambda)u = K^{A_t}_0(\lambda)u$ can be expressed in the form

$$K^t_0(\lambda)u = L_t\Delta_t^{-1}G_tu, \text{ where } G_t = \begin{bmatrix} -\Phi^t_{11}k^{b^t_5}_0 & k^{b^t_3}_0 \\ (\Phi^t_{22})^* l^{b^t_6}_0 & l^{b^t_4}_0 \end{bmatrix},$$

for every $u \in \mathbb{C}^m$ and hence

$$\begin{aligned} u^*M(t)u = 2\pi\langle K^t_0u, K^t_0u\rangle_{\mathcal{H}(A_t)} &= 2\pi\langle L_t\Delta_t^{-1}G_tu, L_t\Delta_t^{-1}G_tu\rangle_{\mathcal{H}(A_t)} \\ &= 2\pi\langle \Delta_t^{-1}G_tu, G_tu\rangle_{st}. \end{aligned}$$

8. The de Branges spaces $\mathcal{B}(\mathfrak{E})$

We turn next to a class of RKHS's $\mathcal{B}(\mathfrak{E})$ that were introduced and exploited to study spectral problems for integral and differential systems by L. de Branges [Br1], [Br2]; for subsequent developments and applications, see also the references cited in the first section. In the spectral theory of integral and differential systems, the spaces $\mathcal{B}(\mathfrak{E})$ are spaces of entire vvf's. However, in other problems of analysis it is useful to consider spaces $\mathcal{B}(\mathfrak{E})$ of vvf's that are meromorphic in Ω_+. In this more general setting, the space $\mathcal{B}(\mathfrak{E})$ may be defined in terms of a $p \times 2p$ mvf

$$\mathfrak{E}(\lambda) = [E_-(\lambda) \quad E_+(\lambda)], \tag{8.1}$$

where $\{E_-(\lambda), E_+(\lambda)\}$ is a pair of $p \times p$ mvf's that are meromorphic in $\mathbb{C}_+$ and are such that

$$\det E_+(\lambda) \not\equiv 0 \quad \text{and} \quad \chi = E_+^{-1}E_- \text{ belongs to the class } \mathcal{S}_{in}^{p\times p}. \tag{8.2}$$

We shall call a pair of mvf's that meet this property a de Branges pair and shall refer to the corresponding $p \times 2p$ mvf $\mathfrak{E}(\lambda)$ as a de Branges function. With each such de Branges function, there is an associated RKHS $\mathcal{B}(\mathfrak{E})$ of $p \times 1$ vvf's with RK

$$K_\omega^{\mathfrak{E}}(\lambda) = -\frac{\mathfrak{E}(\lambda)j_p\mathfrak{E}(\omega)^*}{\rho_\omega(\lambda)} = \frac{E_+(\lambda)E_+(\omega)^* - E_-(\lambda)E_-(\omega)^*}{\rho_\omega(\lambda)}. \tag{8.3}$$

The formula

$$f \longrightarrow E_+^{-1}f$$

defines a unitary map from $\mathcal{B}(\mathfrak{E})$ onto $\mathcal{H}(\chi)$.

For the rest of this section we shall restrict attention to the case $\Omega_+ = \mathbb{C}_+$. In this setting, if $E_+(\lambda)$ tends to a limit $E_+(\mu)$ as λ tends non tangentially to $\mu \in \mathbb{R}$ at a.e. point $\mu \in \mathbb{R}$ and $\det E_+(\mu) \neq 0$ a.e., then $f(\lambda)$ has nontangential limits $f(\mu)$ at a.e point $\mu \in \mathbb{R}$ for every $f \in \mathcal{B}(\mathfrak{E})$ and

$$\|f\|^2_{\mathcal{B}(\mathfrak{E})} = \|E_+^{-1}f\|^2_{st} = \int_{-\infty}^{\infty} f(\mu)^*\Delta_{\mathfrak{E}}(\mu)f(\mu)d\mu,$$

where

$$\Delta_{\mathfrak{E}}(\mu) = E_+^{-1}(\mu)^*E_+^{-1}(\mu). \tag{8.4}$$

Moreover,

$$\begin{aligned} f \in \mathcal{B}(\mathfrak{E}) \quad &\Longleftrightarrow \quad E_+^{-1}f \in \mathcal{H}(\chi) \\ &\Longleftrightarrow \quad E_-^{-1}f \in \mathcal{H}_*(\chi) \\ &\Longleftrightarrow \quad E_+^{-1}f \in H_2^p \text{ and } E_-^{-1}f \in K_2^p \end{aligned}$$

We turn next to de Branges spaces $\mathcal{B}(\mathfrak{E})$ based on mvf's $\mathfrak{E}(\lambda)$ that meet the following additional constraints:

$$0 \in \mathfrak{h}_{\mathfrak{E}}, E_+(0) = I_p, E_-(0) = I_p \text{ and } -i\chi'(0) > 0. \tag{8.5}$$

The significance of this extra assumption rests on the fact that

$$-i\chi'(0) > 0 \Longleftrightarrow K_0^{\mathfrak{E}}(0) > 0 \Longleftrightarrow \{f(0) : f \in \mathcal{B}(\mathfrak{E})\} = \mathbb{C}^p; \tag{8.6}$$

see [ArD7] for the justification. It is convenient to set

$$G_+(\lambda) = \frac{E_+(\lambda) - I_p}{\lambda} \quad \text{and} \quad G_-(\lambda) = \frac{E_-(\lambda) - I_p}{\lambda} \tag{8.7}$$

when these constraints are in effect.

Theorem 8.1. *Let $\{E_+, E_-\}$ be a de Branges pair that satisfies the extra constraints (8.5). Then the following conditions are equivalent:*

1. *The RKHS $\mathcal{B}(\mathfrak{E})$ is invariant under the action of the backwards shift operator*

$$R_0 : f \longrightarrow \frac{f(\lambda) - f(0)}{\lambda}. \tag{8.8}$$

2. *$G_+u \in \mathcal{B}(\mathfrak{E})$ and $G_-u \in \mathcal{B}(\mathfrak{E})$ for every $u \in \mathbb{C}^p$.*

3. $$E_+ \in \Pi^{p\times p},\ E_+^{-1}G_+ \in H_2^{p\times p}, E_- \in \Pi^{p\times p} \text{ and } E_-^{-1}G_- \in K_2^{p\times p}. \tag{8.9}$$

4. *There exists an $m \times m$ mvf $A \in \mathcal{U}(J_p, \mathbb{C}_+)$ such that*

$$0 \in \mathfrak{h}_A, A(0) = I_m \text{ and } \mathfrak{E}(\lambda) = \sqrt{2}[0 \quad I_p]A(\lambda)\mathfrak{V}. \tag{8.10}$$

Proof. Suppose first that (1) is in force and that $g \in \mathcal{B}(\mathfrak{E})$. Then the vvf's $g(\lambda)$ and $h(\lambda) = (E_+^{-1}g)(\lambda)$ are both holomorphic at the point $\lambda = 0$ and the identity

$$\frac{h(\lambda) - h(0)}{\lambda} = E_+^{-1}(\lambda)(R_0 g)(\lambda) + \frac{E_+^{-1}(\lambda) - I_p}{\lambda} g(0)$$

implies that the second term on the right belongs to $\mathcal{H}(\chi)$. Moreover, in view of the constraint (8.5) and the observation (8.6), it follows that

$$\frac{E_+^{-1}(\lambda) - I_p}{\lambda} u \in \mathcal{H}(\chi)$$

for every choice of $u \in \mathbb{C}^p$.

Next, since $\chi^{-1}\mathcal{H}(\chi) = \mathcal{H}_*(\chi)$ and $\lambda^{-1}(\chi^{-1}(\lambda) - I_p)u \in \mathcal{H}_*(\chi)$, the identity

$$\chi^{-1}(\lambda)\left\{\frac{E_+^{-1}(\lambda) - I_p}{\lambda}\right\}u = \frac{E_-^{-1}(\lambda) - I_p}{\lambda}u - \frac{\chi^{-1}(\lambda) - I_p}{\lambda}u$$

leads immediately to the conclusion that

$$\frac{E_+^{-1}(\lambda) - I_p}{\lambda} u \in \mathcal{H}(\chi) \iff \frac{E_-^{-1}(\lambda) - I_p}{\lambda} u \in \mathcal{H}_*(\chi).$$

Therefore, since

$$\frac{E_+^{-1}(\lambda) - I_p}{\lambda} u \in \mathcal{H}(\chi) \iff G_+u \in \mathcal{B}(\mathfrak{E})$$

and

$$\frac{E_-^{-1}(\lambda) - I_p}{\lambda} u \in \mathcal{H}_*(\chi) \iff G_-u \in \mathcal{B}(\mathfrak{E})$$

for every $u \in \mathbb{C}^p$ it is readily seen that (1) implies (2) and (2) implies (3).

Next, with the help of the evaluation

$$\langle \frac{h(\mu) - h(0)}{\mu}, f(\mu)\rangle_{st} = \langle h(\mu), \frac{f(\mu) - \chi(\mu)f(0)}{\mu}\rangle_{st},$$

which is valid for every choice of h and f in $\mathcal{H}(\chi)$, it may be shown that the adjoint of the backwards shift operator R_0 acting in $\mathcal{B}(\mathfrak{E})$ is given by the formula

$$\begin{aligned}(R_0^*g)(\lambda) &= (R_0g)(\lambda)\\ &- i\left\{G_-(\lambda)\frac{1}{2\pi}\int_{-\infty}^{\infty} G_-(\mu)^*\Delta_{\mathfrak{E}}(\mu)g(\mu)d\mu\right.\\ &- \left.G_+(\lambda)\frac{1}{2\pi}\int_{-\infty}^{\infty} G_+(\mu)^*\Delta_{\mathfrak{E}}(\mu)g(\mu)d\mu\right\}.\end{aligned}$$

Let

$$C = \begin{bmatrix} C_1 \\ C_2 \end{bmatrix}, \tag{8.11}$$

where

$$C_1 : g \in \mathcal{B}(\mathfrak{E}) \longrightarrow \frac{\sqrt{\pi}}{2\pi i}\int_{-\infty}^{\infty} \left(G_+(\mu) + G_-(\mu)\right)^* \Delta_{\mathfrak{E}}(\mu)g(\mu)d\mu \tag{8.12}$$

$$\begin{aligned}C_2 : g \in \mathcal{B}(\mathfrak{E}) \longrightarrow &\frac{\sqrt{\pi}}{2\pi i}\int_{-\infty}^{\infty} \left(G_+(\mu) - G_-(\mu)\right)^* \Delta_{\mathfrak{E}}(\mu)g(\mu)d\mu \\ &= \sqrt{\pi}g(0).\end{aligned} \tag{8.13}$$

Then

$$R_0 - R_0^* = iC^*J_pC \tag{8.14}$$

and hence the $m \times m$ mvf

$$A_{\mathfrak{E}}(\lambda) = I_m + i\lambda C(I - \lambda R_0)^{-1}C^*J_p \tag{8.15}$$

satisfies the identity

$$J_p - A_{\mathfrak{E}}(\lambda)J_pA_{\mathfrak{E}}(\omega)^* = -i(\lambda - \overline{\omega})C(I - \lambda R_0)^{-1}(I - \overline{\omega}R_0^*)^{-1}C^*. \tag{8.16}$$

Moreover, $A_{\mathfrak{E}}(\lambda)$ belongs to the class $\mathcal{U}(J_p, \mathbb{C}_+)$ and meets the constraints (8.10). This completes the proof that (3) implies (4).

It remains only to prove that (4) implies (1), but that follows from Theorem 2.4 of [ArD7]. □

Remark 8.2. The mvf $A(\lambda) = A_{\mathfrak{E}}(\lambda)$ that was constructed in the proof of the last theorem satisfies the condition

$$T_A[I_p] = i\alpha + \frac{1}{i\pi}\int_{-\infty}^{\infty}\left(\frac{1}{\lambda - \mu} - \frac{\mu}{1+\mu^2}\right)\Delta_{\mathfrak{E}}(\mu)d\mu \tag{8.17}$$

for some Hermitian matrix $\alpha \in \mathbb{C}^{p\times p}$. It is in fact the only mvf in the class $\mathcal{U}(J_p, \mathbb{C}_+)$ that meets both (8.10) and this last condition.

We shall say that a de Branges function $\mathfrak{E}(\lambda)$ is *regular* if it meets both of the conditions (8.5) and (8.9). Correspondingly, we shall say that the de Branges space $\mathcal{B}(\mathfrak{E})$ is *regular* if $\mathfrak{E}(\lambda)$ meets the constraint (8.5) and $\mathcal{B}(\mathfrak{E})$ is invariant under the backwards shift operator R_0. In view of the last theorem, the $p \times 2p$ mvf $\mathfrak{E}(\lambda)$ is regular if and only if the space $\mathcal{B}(\mathfrak{E})$ is regular. The class of regular de Branges spaces $\mathcal{B}(\mathfrak{E})$ of entire vvf's play a significant role in the spectral theory of integral and differential systems.

We now introduce a pair of generalized Fourier transforms on regular de Branges spaces $\mathcal{B}(\mathfrak{E})$ by the formulas

$$\begin{aligned}\mathcal{F}_r(\lambda) &= \frac{1}{\sqrt{2\pi}}C(I-\lambda R_0)^{-1}g \\ &= \frac{1}{\sqrt{2}}\begin{bmatrix} \frac{1}{2\pi i}\int_{-\infty}^{\infty}(G_+(\mu)+G_-(\mu))^*\,\Delta_{\mathfrak{E}}(\mu)\frac{\lambda g(\lambda)-\mu g(\mu)}{\lambda-\mu}d\mu \\ g(\lambda)\end{bmatrix}\end{aligned}$$

and

$$\begin{aligned}\mathcal{F}_l(\lambda) &= \frac{1}{\sqrt{2\pi}}C(I+\lambda R_0^*)^{-1}g \\ &= A_{\mathfrak{E}}^{\sim}(\lambda)\mathcal{F}_r(-\lambda).\end{aligned}$$

The identities (8.16) and

$$J_p - A_{\mathfrak{E}}^{\sim}(\lambda)J_pA_{\mathfrak{E}}^{\sim}(\omega)^* = -i(\lambda-\overline{\omega})C(I-\lambda R_0)^{-1}(I-\overline{\omega}R_0^*)^{-1}C^*. \tag{8.18}$$

imply that $\mathcal{F}_r$ is a unitary map from $\mathcal{B}(\mathfrak{E})$ onto $\mathcal{H}(A_{\mathfrak{E}})$ and that $\mathcal{F}_l$ is a unitary map from $\mathcal{B}(\mathfrak{E})$ onto $\mathcal{H}(A_{\mathfrak{E}}^{\sim})$. Consequently, $\mathcal{F}_r g \in \Pi^m$ and $\mathcal{F}_l g \in \Pi^m$ for every $g \in \mathcal{B}(\mathfrak{E})$. Therefore, both transforms have boundary values for a.e. $\mu \in \mathbb{R}$.

We shall say that a regular de Branges function $\mathfrak{E}(\lambda)$ is *right strongly regular* if $\mathcal{F}_r g \in L_2^m$ for every $g \in \mathcal{B}(\mathfrak{E})$ and that it is *left strongly regular* if $\mathcal{F}_l g \in L_2^m(\mathbb{R})$ for every $g \in \mathcal{B}(\mathfrak{E})$. Correspondingly, we shall say that the space $\mathcal{B}(\mathfrak{E})$ is left or right strongly regular according as $\mathfrak{E}(\lambda)$ is left or right strongly regular.

Theorem 8.3. *Let $\mathfrak{E}(\lambda)$ be a regular de Branges function. Then the following statements are equivalent:*

1. *$\mathfrak{E}(\lambda)$ is right strongly regular.*
2. *$\{\mathcal{F}_r g : g \in \mathcal{B}(\mathfrak{E})$ is a closed subspace of $L_2^m(\mathbb{R})$.*
3. *There exist a pair of positive constants γ_1 and γ_2 such that*

$$\gamma_1\|\mathcal{F}_r g\|_{st} \le \|g\|_{\mathcal{B}(\mathfrak{E})} \le \gamma_2\|\mathcal{F}_r g\|_{st}.$$

4. *At least one of the functions*

$$\langle g, (I-\lambda R_0^*)^{-1}G_+u\rangle_{\mathcal{B}(\mathfrak{E})}, \quad \langle g, (I-\lambda R_0^*)^{-1}G_-u\rangle_{\mathcal{B}(\mathfrak{E})}$$

has nontangential boundary limits that belong to $L_2(\mathbb{R})$ for every choice of $g \in \mathcal{B}(\mathfrak{E})$ and $u \in \mathbb{C}^p$.

Proof. In the given setting, the generalized transform $\mathcal{F}_r$ is a unitary operator from $\mathcal{B}(\mathfrak{E})$ onto $\mathcal{H}(A_{\mathfrak{E}})$ and $\mathfrak{E}(\lambda)$ is right strongly regular if and only if $A_{\mathfrak{E}} \in \mathcal{U}_{sR}(J_p)$. Thus the theorem is equivalent to Theorem 5.1. □

If $\mathfrak{E}(\lambda)$ is right strongly regular, then, as follows readily from the definition of $\mathcal{F}_r g$ and a well-known theorem of Banach, there exist a pair of positive constants α_1 and α_2 such that

$$\alpha_1\|g\|_{st} \le \|g\|_{\mathcal{B}(\mathfrak{E})} \le \alpha_2\|g\|_{st}.$$

Thus, in this setting, $\mathcal{B}(\mathfrak{E})$ is a closed subspace of $L_2^p(\mathbb{R})$ (with respect to Lebesgue measure). In fact,

$$\mathcal{B}(\mathfrak{E}) = \mathcal{H}(b_3) \oplus \mathcal{H}_*(b_4), \tag{8.19}$$

with scalar product

$$\langle g, h\rangle_{\mathcal{B}(\mathfrak{E})} = \int_{-\infty}^{\infty} h(\mu)^* \Delta_{\mathfrak{E}}(\mu) f(\mu) d\mu, \tag{8.20}$$

where the pair $\{b_3, b_4\}$ of inner $p \times p$ mvf's may be obtained from the inner-outer and outer-inner factorizations

$$(E_-^{\#})^{-1} = b_3\varphi_3 \quad \text{and} \quad E_+^{-1} = \varphi_4 b_4,$$

respectively. The equivalence of the second and fourth assertions follows from Lemma 6.1 of [ArD1].

We remark that if $\mathfrak{E}(\lambda)$ is a regular de Branges function, then the generalized transform $\mathcal{F}_l$ is a unitary operator from $\mathcal{B}(\mathfrak{E})$ onto $\mathcal{H}(A_{\mathfrak{E}}^{\sim})$ and

$$\mathfrak{E}(\lambda) \text{ is left strongly regular} \Longleftrightarrow A_{\mathfrak{E}} \in \mathcal{U}_{lsR}(J_p) \Longleftrightarrow A_{\mathfrak{E}}^{\sim} \in \mathcal{U}_{rsR}(J_p).$$

Thus, an application of Theorem 5.1 to $A_{\mathfrak{E}}^{\sim}$ yields the following result:

Theorem 8.4. *Let $\mathfrak{E}(\lambda)$ be a regular de Branges function. Then the following statements are equivalent:*

1. *$\mathfrak{E}(\lambda)$ is left strongly regular.*
2. *$\{\mathcal{F}_l g : g \in \mathcal{B}(\mathfrak{E})$ is a closed subspace of $L_2^m(\mathbb{R})$.*
3. *There exist a pair of positive constants γ_1 and γ_2 such that*

$$\gamma_1 \|\mathcal{F}_l g\|_{st} \le \|g\|_{\mathcal{B}(\mathfrak{E})} \le \gamma_2 \|\mathcal{F}_l g\|_{st}.$$

4. *At least one of the functions*

$$\langle g, (I - \lambda R_0)^{-1} G_+ u\rangle_{\mathcal{B}(\mathfrak{E})}, \quad \langle g, (I - \lambda R_0)^{-1} G_- u\rangle_{\mathcal{B}(\mathfrak{E})}$$

has nontangential boundary limits that belong to $L_2(\mathbb{R})$ for every choice of $g \in \mathcal{B}(\mathfrak{E})$ and $u \in \mathbb{C}^p$.

Theorem 8.5. *Let $\mathfrak{E}(\lambda)$ be a regular de Branges function. Then $\mathfrak{E}(\lambda)$ is left strongly regular if and only if the density $\Delta_{\mathfrak{E}}(\mu)$ satisfies the Treil-Volberg matrix version of the Muckenhoupt (A_2) condition.*

Proof. This is immediate from Theorem 7.5 in [ArD10]. □

9. Operator nodes with characteristic matrix functions of the class $\mathcal{U}_{sR}(J)$

Let $\mathcal{U}_0(J) = \mathcal{U}_0(J, \Omega_+)$ be the subclass of mvf's $U(\lambda)$ from $\mathcal{U}(J, \Omega_+)$ that are holomorphic at the point $\lambda = 0$ and,in addition, are normalized by the condition $U(0) = I_m$ when $\Omega_+ = \mathbb{C}_+$. It is known that every mvf $U(\lambda)$ of the class $\mathcal{U}_0(J)$ may be represented as the characteristic matrix function $U_\Sigma(\lambda)$ of a simple operator node Σ that is a J-unitary node if $\Omega_+ = \mathbb{D}_+$ and is an LB (Livsic-Brodskii) J-node, if $\Omega_+ = \mathbb{C}_+$.

We recall that a colligation $\Sigma = (A, C; X, Y)$ is called an LB J-node if $J \in \mathcal{L}(Y)$ is a signature operator, $A \in \mathcal{L}(X)$, $C \in \mathcal{L}(X, Y)$ and

$$A - A^* = iC^*JC. \tag{9.1}$$

Such a node Σ is called simple if

$$\bigcap_{n\geq 0} \text{Ker } CA^n = \{0\}. \tag{9.2}$$

The function

$$U_\Sigma(\lambda) = I + i\lambda C(I - \lambda A)^{-1}C^*J \tag{9.3}$$

is called the *characteristic function* of the LB J-node Σ. Two LB J-nodes $\Sigma_i = (A_i, C_i; X_i, Y_i)$ $(i = 1, 2)$ are said to be *unitarily equivalent* if

$$A_1 = R^{-1}A_2R \quad \text{and} \quad C_1 = C_2R, \tag{9.4}$$

where $R \in \mathcal{L}(X_1, X_2)$ is a unitary operator. It is known that if the characteristic functions $U_{\Sigma_1}(\lambda)$ and $U_{\Sigma_2}(\lambda)$ of two simple LB J-nodes Σ_1 and Σ_2 coincide in a neighbourhood of zero, then these nodes are unitarily equivalent. If $m = \dim Y < \infty$, then the operators from $\mathcal{L}(Y)$ may be defined in terms of an orthonormal basis in Y by $m \times m$ matrices and then the corresponding mvf $U_\Sigma(\lambda)$ is called the *characteristic matrix function* of Σ. In this case $\mathbb{C}^m$ may be considered instead of Y and then J and $U_\Sigma(\lambda)$ denote $m \times m$ matrices that define the corresponding operators from $\mathcal{L}(\mathbb{C}^m)$ in the standard basis.

An LB J-node is said to be *dissipative (accumulative)*, if $J = I$ $(J = -I$, respectively).

From now on we restrict attention to simple LB-J nodes Σ with characteristic function $U_\Sigma \in \mathcal{U}_0(J, \mathbb{C}_+)$.

The supplementary identities

$$K_\omega^{U_\Sigma}(\lambda) = \frac{1}{2\pi}C(I - \lambda A)^{-1}(I - \overline{\omega}A^*)^{-1}C^* \tag{9.5}$$

and

$$K_\omega^{U_\Sigma^\sim}(\lambda) = \frac{1}{2\pi}C(I + \lambda A^*)^{-1}(I + \overline{\omega}A)^{-1}C^* \tag{9.6}$$

follow easily from the formulas (9.1) and (9.4). These identities imply that the two generalized Fourier transforms

$$(\mathcal{F}_r x)(\lambda) = \frac{1}{\sqrt{2\pi}}C(I - \lambda A)^{-1}x \quad \text{and} \quad (\mathcal{F}_l x)(\lambda) = \frac{1}{\sqrt{2\pi}}C(I + \lambda A^*)^{-1}x \tag{9.7}$$

are unitary maps from X onto $\mathcal{H}(U_\Sigma)$ and from X onto $\mathcal{H}(U_\Sigma^\sim)$, respectively. Moreover, the simple LB-J node $\Sigma = (A, C; X, \mathbb{C}^m)$ with characteristic function

$U_\Sigma(\lambda) \in \mathcal{U}(J, \mathbb{C}_+)$ and the functional model $\overset{o}{\Sigma} = (R_0, C_0; \mathcal{H}(U_\Sigma), \mathbb{C}^m)$ based on the operators

$$R_0 : f \in \mathcal{H}(U_\Sigma) \longrightarrow \frac{f(\lambda) - f(0)}{\lambda} \tag{9.8}$$

$$C_0 : f \in \mathcal{H}(U_\Sigma) \longrightarrow \sqrt{2\pi} f(0) \in \mathbb{C}^m : \tag{9.9}$$

are unitarily equivalent:

$$A = (\mathcal{F}_r)^{-1} R_0 \mathcal{F}_r \quad \text{and} \quad C = C_0 \mathcal{F}_r.$$

In much the same way the transform $\mathcal{F}_l$ defines a unitary equivalence between the simple LB-J node $\widetilde{\Sigma} = (-A^*, C; X, \mathbb{C}^m)$ with characteristic function $U_{\widetilde{\Sigma}} \in \mathcal{U}(J, \mathbb{C}_+)$ and the functional model based on $\mathcal{H}(U_{\widetilde{\Sigma}})$.

If $J = J_p$ and the characteristic function $A_\Sigma(\lambda)$ of the simple LB-J_p node $\Sigma = (A, C; X, \mathbb{C}^m)$ belongs to the class $\mathcal{U}_0(J_p, \mathbb{C}_+)$ and is such that the mvf

$$c_0(\lambda) = (T_{A_\Sigma}[I_p])(\lambda) \tag{9.10}$$

admits a representation of the form

$$c_0(\lambda) = i\alpha + \frac{1}{\pi i} \int_{-\infty}^{\infty} \left(\frac{1}{\mu - \lambda} - \frac{\mu}{1 + \mu^2} \right) \Delta_{\mathfrak{E}}(\mu) d\mu \tag{9.11}$$

for some $p \times p$ Hermitian matrix α, then

$$\mathfrak{E}(\lambda) = \sqrt{2}[I_p \quad 0] A_\Sigma(\lambda)$$

is a regular de Branges matrix and

$$A_\Sigma(\lambda) = A_{\mathfrak{E}}(\lambda),$$

where $A_{\mathfrak{E}}(\lambda)$ is the mvf that was constructed in the proof of Theorem 8.1. This in turn provides another functional model $\Sigma_{\mathfrak{E}} = (R_0, C; \mathcal{B}(\mathfrak{E}), \mathbb{C}^m)$ of the simple LB-J_p node Σ, where R_0 now designates the backwards shift operator acting in $\mathcal{B}(\mathfrak{E})$ and the operator C is defined by formulas (8.11)–(8.13). This functional model is easily obtained from the model $\overset{o}{\Sigma}$ since the map

$$f \longrightarrow \sqrt{2}[0 \quad I_p] f$$

is a unitary operator that carries the backwards shift R_0 in $\mathcal{H}(A_\Sigma)$ into the shift acting in $\mathcal{B}(\mathfrak{E})$. If the mvf $c_0(\lambda)$ that is defined in (9.10) meets the condition (9.11), then the generalized Fourier transform

$$\mathcal{F}_r^{(2)} x = \frac{1}{\sqrt{2\pi}} [0 \quad I_p] C (I - \lambda A)^{-1} x$$

is a unitary operator from X onto $\mathcal{B}(\mathfrak{E})$. This serves to establish the unitary equivalence of the simple LB-J_p node Σ with the functional model $\Sigma_{\mathfrak{E}}$.

We remark that

$$(9.11) \text{ holds} \iff \bigcap_{n=0}^{\infty} [0 \quad I_p] C \ker A^n = \{0\}; \tag{9.12}$$

see [ArD8].

Theorem 9.1. *Let $\Sigma = (A, C; X, \mathbb{C}^m)$ be a simple LB J_p-node with a characteristic matrix function $U_\Sigma(\lambda)$ that is a right strongly regular J_p-inner mvf with respect to $\mathbb{C}_+$. Let*

$$\{b_3, b_4\} \in ap_{\rm II}(U_\Sigma) \quad \text{and} \quad b_2(0) = b_4(0) = I_p. \tag{9.13}$$

Let $\overset{o}{\Sigma}_+ = (\overset{o}{A}_+, \overset{o}{C}_+; \overset{o}{X}_+, \mathbb{C}^p)$ and $\overset{o}{\Sigma}_- = (\overset{o}{A}_-, \overset{o}{C}_-; \overset{o}{X}_-, \mathbb{C}^p)$ be simple dissipative and simple accumulative LB nodes with characteristic matrix functions $b_3(\lambda)$ and $b_4^{-1}(\lambda)$, respectively. Then

$$A = R^{-1}\begin{bmatrix} \overset{o}{A}_+ & 0 \\ 0 & \overset{o}{A}_- \end{bmatrix} R, \qquad [0_{p\times p}\ I_p]C = [\overset{o}{C}_+\ \ \overset{o}{C}_-]R, \tag{9.14}$$

for some R that satisfies the conditions

$$R \in \mathcal{L}(X, \overset{o}{X}_+ \oplus \overset{o}{X}_-), \qquad R^{-1} \in \mathcal{L}(\overset{o}{X}_+ \oplus \overset{o}{X}_-, X). \tag{9.15}$$

Moreover,

$$\overset{o}{X}_- = \{0\} \Longleftrightarrow b_4(\lambda) \equiv I_p \Longleftrightarrow U_\Sigma \in \mathcal{N}_+^{m\times m}$$

and

$$\overset{o}{X}_+ = \{0\} \Longleftrightarrow b_3(\lambda) \equiv I_p \Longleftrightarrow U_\Sigma \in \mathcal{N}_-^{m\times m}.$$

Proof. The given simple LB nodes $\Sigma, \overset{o}{\Sigma}_+, \overset{o}{\Sigma}_-$ can be replaced by their functional models. In these models, $A, \overset{o}{A}_-, \overset{o}{A}_+$, are backwards shifts in the spaces $X = \mathcal{H}(U_\Sigma), \overset{o}{X}_+ = \mathcal{H}(b_3), \overset{o}{X}_- = \mathcal{H}_*(b_4)$, respectively and the operators $C, \overset{o}{C}_+, \overset{o}{C}_-$ map a vvf g from their respective domains $X, \overset{o}{X}_+, \overset{o}{X}_-$ into $\sqrt{2\pi}g(0)$. In the given setting, Theorem 5.4 is applicable to the mvf $A(\lambda) = U_\Sigma(\lambda)$ and the operator $R = (L_A)^{-1}$ satisfies the stated assertions. □

Theorem 9.2. [Ar]. *Let $\Sigma = (A, C; X, \mathbb{C}^m)$ be an LB j_{pq}-node with characteristic matrix function $U_\Sigma(\lambda) = W(\lambda)$ that is j_{pq}-inner with respect to $\mathbb{C}_+$. Let*

$$\{b_1, b_2\} \in ap\,(W), \quad b_1(0) = I_p, \quad \text{and} \quad b_2(0) = I_q. \tag{9.16}$$

1. *Let Σ be a simple node and $W \in \mathcal{U}_{sR}(j_{pq})$. Let $\overset{o}{\Sigma}_+ = (\overset{o}{A}_+, \overset{o}{C}_+; \overset{o}{X}_+, \mathbb{C}^p)$ be a simple dissipative LB node with characteristic matrix function $b_1(\lambda)$ and $\overset{o}{\Sigma}_- = (\overset{o}{A}_-, \overset{o}{C}_-; \overset{o}{X}_-, \mathbb{C}^q)$ be a simple accumulative LB node with characteristic matrix function $b_2(\lambda)^{-1}$. Then*

$$A = R^{-1}\begin{pmatrix} \overset{o}{A}_+ & 0 \\ 0 & \overset{o}{A}_- \end{pmatrix} R, \qquad \overset{o}{C}_\pm = P_\pm C R^{-1}|\, \overset{o}{X}_\pm \tag{9.17}$$

for some R such that

$$R \in \mathcal{L}(X, \overset{o}{X}_+ \oplus \overset{o}{X}_-), \qquad R^{-1} \in \mathcal{L}(\overset{o}{X}_+ \oplus \overset{o}{X}_-, X), \tag{9.18}$$

where $P_+ = \operatorname{diag}\{I_p, 0_{q\times q}\}$ and $P_- = \operatorname{diag}\{0_{p\times p}, I_q\}$. Moreover,

$$\overset{o}{X}_- = \{0\} \Longleftrightarrow b_2(\lambda) \equiv I_q \Longleftrightarrow W \in \mathcal{N}_+^{m\times m}(\mathbb{C}_+)$$

and

$$\overset{o}{X}_+ = \{0\} \Longleftrightarrow b_1(\lambda) \equiv I_p \Longleftrightarrow W \in \mathcal{N}_-^{m\times m}(\mathbb{C}_-).$$

2. *Conversely, if* $\overset{o}{\Sigma}_+ = (\overset{o}{A}_+, \overset{o}{C}_+; \overset{o}{X}_+, \mathbb{C}^p)$ *is a dissipative LB node and if* $\overset{o}{\Sigma}_- = (\overset{o}{A}_-, \overset{o}{C}_-; \overset{o}{X}_-, \mathbb{C}^q)$ *is an accumulative LB node such that* (9.17) *and* (9.18) *hold, then* $W \in \mathcal{U}_{sR}(j_{pq})$.

Proof. A proof of the first statement may be based on the description of the space $\mathcal{H}(W)$ for $W \in \mathcal{U}_{sR}(j_{pq})$ that is furnished in [ArD4], in much the same way that the last theorem was verified.

The verification of the second assertion rests on the fact that the generalized right Fourier transforms based on the nodes $\overset{o}{\Sigma}_+$ and $\overset{o}{\Sigma}_-$ map into L_2^p and L_2^q, respectively and the fact that if

$$f = \begin{bmatrix} g \\ h \end{bmatrix} \in \mathcal{H}(W) \text{ and } W \in \mathcal{U}(j_{pq}),$$

then

$$f \in L_2^m \Longleftrightarrow g \in L_2^p \Longleftrightarrow h \in L_2^q. \qquad \square$$

If $U_\Sigma \in \mathcal{E} \cap \mathcal{U}(J)$, i.e., if the simple node Σ is a Volterra node, then the $b_j(\lambda)$ are entire inner functions and consequently the nodes $\overset{o}{\Sigma}_+$ and $\overset{o}{\Sigma}_-$ considered in the preceding two theorems are Volterra. If $m = 2$, then $b_1(\lambda) = b_3(\lambda) = e^{i\lambda\tau_3}$ and $b_2(\lambda) = b_4(\lambda) = e^{i\lambda\tau_4}$, where τ_3 and τ_4 are nonnegative numbers that may be computed by formula (7.5) with $b_j(\lambda)$ in place of $b_j^t(\lambda)$ for $j = 3$ and $j = 4$. The functional models based on the backwards shift acting in $\mathcal{H}(b_3)$ and $\mathcal{H}_*(b_4)$ are of course still applicable. However, since $\mathcal{H}(b_3) = L_2([0,\tau_3])^\wedge$ and $\mathcal{H}_*(b_4) = L_2([-\tau_4, 0])^\wedge$, the identities

$$h(\lambda) = \int_0^{\tau_3} e^{i\lambda a} h^\vee(a)da \Longrightarrow (R_0 h)(\lambda) = i\int_0^{\tau_3} e^{i\lambda b}\left(\int_b^{\tau_3} h^\vee(a)da\right)db$$

and

$$h(\lambda) = \int_{-\tau_4}^{0} e^{i\lambda a} h^\vee(a)da \Longrightarrow (R_0 h)(\lambda) = i\int_{-\tau_4}^{0} e^{i\lambda b}\left(\int_{-\tau_4}^{b} h^\vee(a)da\right)db$$

lead to well-known functional models based on integration operators acting in the indicated subspaces of L_2. In particular, the functional model of a simple dissipative Volterra node with characteristic function $e^{i\lambda\tau_3}$ may be chosen equal to $\overset{o}{\Sigma}_+ = (\overset{o}{A}_+, \overset{o}{C}_+; L_2([0,\tau_3]), \mathbb{C})$, where

$$(\overset{o}{A}_+ u)(t) = i\int_t^{\tau_3} u(a)da \text{ and } \overset{o}{C}_+ u = \int_0^{\tau_3} u(a)da \text{ for } u \in L_2([0,\tau_3]).$$

In much the same way, the functional model of a simple accumulative Volterra node with characteristic function $e^{-i\lambda\tau_4}$ may be chosen equal to the node $\overset{o}{\Sigma}_- =$

$(\overset{o}{A}_-, \overset{o}{C}_-; L_2([-\tau_4, 0])), \mathbb{C})$, where

$$(\overset{o}{A}_- u)(t) = i\int_{-\tau_4}^{t} u(a)da \text{ and } \overset{o}{C}_- u = \int_{-\tau_4}^{0} u(a)da \text{ for } u \in L_2([-\tau_4, 0])).$$

Correspondingly, the operator R considered in the preceding two theorems acts from X onto $L_2([0, \tau_3]) \oplus L_2([-\tau_4, 0]$.

There exist analogues of the preceding two theorems for the case $U_\Sigma \in \mathcal{U}_{lsR}(J)$ that may be obtained by applying the preceding results to $U_\Sigma^\sim$ and recalling that

$$U_\Sigma \in \mathcal{U}_{lsR}(J) \Longleftrightarrow U_\Sigma^\sim \in \mathcal{U}_{rsR}(J).$$

Analogs of Theorem 9.2 (see [Ar]) and of Theorem 9.1 may also be obtained for J-unitary nodes with characteristic matrix functions of the class $\mathcal{U}_{sR}(J, \mathbb{D}_+)$.

The properties of left strongly regular spaces $\mathcal{B}(\mathfrak{E})$ and of operators in these spaces related to the backwards shift R_0 were studied by other methods in the case that $E_+(\lambda)$ and $E_-(\lambda)$ are scalar entire functions by G.M. Gubreev [Gu], as an application of his theory of regular quasiexponentials. In particular, he noted the connection between the class $\mathcal{U}_{lsR}(J_p)$ and the class of left strongly regular de Branges spaces $\mathcal{B}(\mathfrak{E})$ when $p = 1$ and $\mathfrak{E}(\lambda)$ is entire. Some of his results may be obtained from the analysis in the last two sections.

Analogues of the last two theorems on LB J-nodes and J-unitary nodes with strongly regular J-inner characteristic matrix functions were obtained by Z. Arova in her PhD thesis, see also [Ara2]. Her characterizations of the class of simple operator nodes Σ with characteristic matrix function $U_\Sigma \in \mathcal{U}_{rsR}(J)$ used somewhat different nodes than were exhibited here. Thus, for example, in place of the relations (9.17), she used the relations

$$A = R^{-1}\begin{pmatrix} \overset{o}{A}_+ & 0 \\ 0 & \overset{o}{A}_- \end{pmatrix} R, \qquad \overset{o}{C}_\pm = CR^{-1}|\, \overset{o}{X}_\pm, \tag{9.19}$$

where $\overset{o}{\Sigma}_+ = (\overset{o}{A}_+, \overset{o}{C}_+; \overset{o}{X}_+, \mathbb{C}^m)$ and $\overset{o}{\Sigma}_- = (\overset{o}{A}_-, \overset{o}{C}_-; \overset{o}{X}_-, \mathbb{C}^m)$ are simple dissipative and simple accumulative LB nodes, respectively.

References

[Al] D. Alpay, *The Schur Algorithm, Reproducing Kernel Spaces and System Theory,* SMF/AMS TEXTS and Monographs, Amer. Math. Soc., Providence, RI, 2001.

[ADRS] D. Alpay, A. Dijksma, J. Rovnyak and H. de Snoo, *Schur functions, operator colligations and reproducing kernel Pontryagin spaces,* Oper. Theor. Adv. Appl., **96**, Birkhäuser Verlag, Basel-Boston, 1997.

[AlD1] D. Alpay and H. Dym, *Hilbert spaces of analytic functions, inverse scattering and operator models, I,* Integral Equations Operator Theory, **7** (1984), 589–741.

[AlD2] D. Alpay and H. Dym, *Hilbert spaces of analytic functions, inverse scattering and operator models, II,* Integral Equations Operator Theory, **8** (1985), 145–180.

[AlD3] D. Alpay and H. Dym, *On a new class of structured reproducing kernel spaces*, J. Funct. Anal., **111** (1993), 1–28.

[An] T. Ando, *de Branges Spaces and Analytic Operator Functions*, Division of Applied Mathematics, Research Institute of Applied Electricity, Hokkaido University, Sapporo, 1990.

[Ar] D.Z. Arov, *Conservative linear time-invariant lossless systems with strongly regular J-inner transfer functions,* Proceedings, MTNS-2000, Perpignon, France, 2000.

[ArD1] D.Z. Arov and H. Dym, *J-inner matrix functions, interpolation and inverse problems for canonical systems, I: Foundations,* Integral Equations Operator Theory, **29**, (1997), 373–454.

[ArD2] D. Arov and H. Dym, *J-inner matrix functions, interpolation and inverse problems for canonical systems, II: The inverse monodromy problem,* Integral Equations Operator Theory, **36** (2000), 11–70.

[ArD3] D. Arov and H. Dym, *J-inner matrix functions, interpolation and inverse problems for canonical systems, III: More on the inverse monodromy problem,* Integral Equations Operator Theory, **36** (2000), 127–181.

[ArD4] D.Z. Arov and H. Dym, *J-inner matrix functions, interpolation and inverse problems for canonical systems, IV: Direct and inverse bitangential input scattering problem,* Integral Equations Operator Theory, **43** (2002), 1–67.

[ArD5] D.Z. Arov and H. Dym, *J-inner matrix functions, interpolation and inverse problems for canonical systems, V: The inverse input scattering problem for Wiener class and rational* $p \times q$ *input scattering matrices,* Integral Equations Operator Theory, **43** (2002), 68–129.

[ArD6] D.Z. Arov and H. Dym, *The bitangential inverse input impendance problem for canonical systems, I: Weyl-Titchmash classification, and existence and uniqueness theorems*, Integral Equations Operator Theory, **47** (2003), 3–49.

[ArD7] D.Z. Arov and H. Dym, *The bitangential inverse input impendance problem for canonical systems, II: Formulas and examples*, Integral Equations Operator Theory, in press.

[ArD8] D.Z. Arov and H. Dym, *The bitangential inverse spectral problem for canonical systems*, J. Funct. Anal., in press.

[ArD9] D.Z. Arov and H. Dym, *On the three Krein extension problems and some generalizations,* Integral Equations Operator Theory, **31** (1998), 1–91.

[ArD10] D.Z. Arov and H. Dym, *Matricial Nehari problems, J-inner matrix functions and the Muckenhoupt condition,* J. Funct. Anal. **181** (2001), 227–299.

[ArD11] D.Z. Arov and H. Dym, *Criteria for the strongly regularity of J-inner functions and* γ*-generating matrices*, J. Math. Anal. Appl., **280** (2003), 387–399.

[Ara1] Z.D. Arova, *The functional model of J-unitary node with a given J-inner characteristic matrix function,* Integral Equations Operator Theory, **28** (1997), 1–16.

[Ara2] Z.D. Arova, *J-unitary nodes with strongly regular J-inner characteristic matrix functions,* Methods of Func. analysis and Topology, **6** (2000), no 3, 9–23.

[Ara3] Z.D. Arova, *On J-unitary regular J-inner characteristic functions in the Hardy class* $H_2^{n\times n}$, in: Operator Theoretical Methods, 12th International Conference on Operator Theory, Timisoara (Romania), 2000, 29–38.

[Ara4] Z.D. Arova, *On Livšic-Brodskii nodes with strongly regular J-inner characteristic matrix functions in the Hardy class*, in: *Recent advances in operator theory and related topics,* **OT127**, Birkhäuser Verlag, Basel, 2001, 83–97.

[Ba] J.A. Ball, *Models for noncontractions,* J. Math. Anal. Appl. **52** (1975), 235–254.

[BaC] J.A. Ball and N. Cohen, *de Branges-Rovnyak operator models and systems theory: a survey,* in: *Topics in Matrix and Operator Theory,* Oper. Theory Adv. Appl., **OT50**, Birkhäuser, Basel, 1991, pp.93–136.

[Br1] L. de Branges, *Some Hilbert spaces of analytic functions I,* Trans. Amer. Math. Soc., **106** (1963), 445–668.

[Br2] L. de Branges, *Hilbert Spaces of Entire Functions,* Prentice-Hall, Englewood Cliffs, 1968.

[Br3] L. de Branges, *The expansion theorem for Hilbert spaces of entire functions,* in: *Entire Functions and Related Parts of Analysis* Amer. Math. Soc., Providence, 1968.

[BrR] L. de Branges and J. Rovnyak, *Canonical models in quantum scattering theory,* in: *Perturbation Theory and its Applications in Quantum Mechanics* (C. Wilcox, ed.) Wiley, New York, 1966, pp. 295–392.

[Dy1] H. Dym, *An introduction to de Branges spaces of entire functions with applications to differential equations of the Sturm-Liouville type,* Adv.Math., **5** (1970), 395–471.

[Dy2] H. Dym, *Trace formulas for a class of Toeplitz-like operators* Israel J. Math., **27** (1977), 21–48.

[Dy3] H. Dym, *Hermitian block Toeplitz matrices, orthogonal polynomials, reproducing kernel Pontryagin spaces, interpolation and extensions,* in *Orthogonal Matrix-Valued Polynomials and Applications,* (I. Gohberg, ed.) Oper. Theory: Adv. Appl., **OT34**, Birkhäuser, Basel, 1988, pp. 79–135.

[Dy4] H. Dym, *J-contractive matrix functions, reproducing Kernel Hilbert spaces and interpolation,* CBMS Regional Conference series, N71, AMS, Providence, R.I., 1989.

[Dy5] H. Dym, *On reproducing kernels and the continuous covariance extension problem,* in: *Analysis and Partial Differential Equations: A Collection of Papers Dedicated to Mischa Cotlar,* Marcel Dekker, New York, 1990, pp. 427–482.

[Dy6] H. Dym, *On the zeros of some continuous analogues of matrix orthogonal polynomials and a related extension problem with negative squares,* Comm. Pure Appl. Math., **47** (1994), 207–256.

[DI] H. Dym and A. Iacob, *Positive definite extensions, canonical equations and inverse problems,* in: *Topics in Operator Theory, Systems and Networks* (H. Dym and I. Gohberg, eds.), *Oper. Theory Adv. Appl.* **12**, Birkhäuser, Basel, 1984, pp. 141–240.

[DK] H. Dym and N. Kravitsky, *On the inverse spectral problem for the string equation,* Integral Equations Operator Theory, **1/2** (1978), 270–277.

[DMc] H. Dym and H.P. McKean, *Gaussian Processes, Function Theory, and the Inverse Spectral Problem,* Academic Press, New York, 1976.

[Dyu] Yu. M. Dyukarev, The Stieltjes matrix moment problem and de Branges spaces associated with them, Proceedings of the Mathematics Conference (Birzeit/Nablus, 1998), 79–88, World Sci. Publishing, River Edge, NJ, 2000.

[Gu] G. M. Gubreev, *Spectral theory of regular and regular B-representative vector functions (projection method: 20 years after)*, Algebra and Anal. **12** (6), (2000), 1–97.

[KW1] M. Kaltenbäck and H.Woracek, *Pontryagin spaces of entire functions, I*, Integral Equations Operator Theory **33** (1999), no. 1, 34–97.

[KW2] M. Kaltenbäck and H.Woracek, *Pontryagin spaces of entire functions, II*. Integral Equations Operator Theory **33** (1999), no. 3, 305–380.

[Re] C. Remling, *Schrödinger operators and de Branges spaces*, J. Funct. Anal., **196** (2002), 323–394.

[Ro] J. Rovnyak, *Characterizations of spaces* $\mathbf{K}(M)$, unpublished manuscript, 1968.

[Sak] A.L. Sakhnovich, *Spectral functions of a canonical system of order* $2n$, Math. USSR Sbornik, **71** (1992), No.2, 355–369.

[Sar] D. Sarason, *Sub-Hardy Hilbert spaces in the unit disk*, Wiley, New York, 1994.

[Si] L.A. Simakova, *On meromorphic plus-matrix functions*, Mat. Issled., **10** (1975), no. 1, 287–292.

[Wi] H. Winkler, *The inverse spectral problem for canonical systems*, Integral Equations Operator Theory, **22** (1995), 360–376.

[Wo] H. Woracek, *de Branges spaces of entire functions closed under forming difference quotients*, Integral Equations Operator Theory, **37** (2000), 238–249.

[Yu] P. Yuditskii, *A special case of de Branges' theorem on the inverse monodromy problem*, Integral Equations Operator Theory, **39** (2001), 229–252. 238–249.

Damir Z. Arov
Department of Mathematics
South-Ukrainian Pedagogical University
65020 Odessa, Ukraine
e-mail: aspect@farlep.net

Harry Dym
Department of Mathematics
The Weizmann Institute of Science
Rehovot 76100, Israel
e-mail: harry.dym@weizmann.ac.il

Operator Theory:
Advances and Applications, Vol. 149, 107–132

Boundary Interpolation for Contractive-valued Functions on Circular Domains in $\mathbb{C}^n$

Joseph A. Ball and Vladimir Bolotnikov

Abstract. We consider a boundary interpolation problem for operator-valued functions defined on a class of bounded complete circular domains in $\mathbb{C}^n$ (including as particular cases, Cartan domains of types I, II and III) which satisfy the von Neumann's inequality. The solvability criterion is obtained and the set of all solutions is parametrized in terms of a family of Redheffer linear fractional transformations.

1. Introduction

The classical Schur class $\mathcal{S}$ consisting of complex-valued analytic functions mapping the unit disk $\mathbb{D}$ into the closed unit disk $\overline{\mathbb{D}}$ has been a source of much study and inspiration and has served as a proving ground for new methods for over a century now, beginning with the seminal work of Schur (for the original paper of Schur and a survey of some of the impact and applications in signal processing, see [28]). A major development has been the interpolation and realization theories for Schur functions and more recently, for their operator-valued analogues. The operator-valued Schur class $\mathcal{S}(\mathcal{E}, \mathcal{E}_*)$ consisting of analytic functions S on the unit disk with values $S(z)$ equal to contraction operators between two Hilbert spaces $\mathcal{E}$ and $\mathcal{E}_*$ has played a prominent role in both engineering and operator-theoretic applications (see, e.g. [20, 21, 33, 29, 34]). We mention in particular that any such function $S(z)$ can be realized in the form

$$S(z) = D + zC(I - zA)^{-1}B \tag{1.1}$$

where the connecting operator (or *colligation*)

$$U = \begin{bmatrix} A & B \\ C & D \end{bmatrix} : \begin{bmatrix} \mathcal{H} \\ \mathcal{E} \end{bmatrix} \mapsto \begin{bmatrix} \mathcal{H} \\ \mathcal{E}_* \end{bmatrix}$$

is unitary, and where $\mathcal{H}$ is some auxiliary Hilbert space (the *internal space* for the colligation). It is also well known that the Schur class of functions satisfies a von

Received by the editors January 28, 2003; revised May 12, 2003. Submitted by J. W. Helton.
Mathematics Subject Classification (2000). 47A57, 47A20, 47A48.
Key words and phrases. boundary interpolation, unitary extensions, characteristic function of a unitary colligation.

Neumann inequality: *if $S \in \mathcal{S}(\mathcal{E}, \mathcal{E}_*)$ and T is a contraction operator on a Hilbert space $\mathcal{H}'$, then $\|F(rT)\| \le 1$ for all $r < 1$.* Here $S(rT)$ can be defined, e.g., by

$$S(rT) = \sum_{n=0}^{\infty} r^n F_n \otimes T^n \in \mathcal{L}(\mathcal{E} \otimes \mathcal{H}', \mathcal{E}_* \otimes \mathcal{H}') \quad \text{if} \quad F(z) = \sum_{n=0}^{\infty} F_n z^n.$$

A related result is the Nevanlinna-Pick interpolation theorem, here stated for the scalar case for simplicity:

Given points $z_1, \dots, z_n \in \mathbb{D}$ and $w_1, \dots, w_n \in \mathbb{C}$, there exists $S \in \mathcal{S}$ with $S(z_j) = w_j$ for $j = 1, \dots, n$ if and only if the associated Pick matrix $\Lambda = \left[\frac{1 - w_i\overline{w_j}}{1 - z_i\overline{z_j}}\right]_{i,j=1}^n$ is positive semidefinite.

By the classical Fatou's lemma, the boundary radial limits $S(t) = \lim_{r\to 1} S(rt)$ exist at almost every point t on the unit circle $\mathbb{T}$ (for operator-valued functions such limits exist almost everywhere in the strong sense; see [33, Chapter 5]) and do not exceed one in modulus (are contractive in the operator-valued case). Another classical result by Julia-Carathéodory asserts that the boundedness of the quantity $d(z) = \frac{1-|S(z)|^2}{1-|z|^2}$ in some radial neighborhood of a point $t \in \mathbb{T}$ guarantees the existence of the limit $\lim_{r\to 1} S(rt)$ (which is unimodular) and $\lim_{r\to 1} d(rt)$ (which is nonnegative). The related interpolation result (stated again for the scalar case) is:

Given points $z_1, \dots, z_n \in \mathbb{T}$ and numbers $w_1, \dots, w_n$ and $\gamma_1, \dots, \gamma_n$ such that

$$|w_j| = 1 \quad \textit{and} \quad \gamma_j \ge 0 \quad \textit{for} \quad i = 1, \dots, n, \tag{1.2}$$

there exists $S \in \mathcal{S}$ with

$$\lim_{r\to 1} S(rz_j) = w_j \quad \textit{and} \quad \lim_{r\to 1} \frac{1 - |S(rz_j)|^2}{1 - |r|^2} \le \gamma_j \quad \textit{for} \quad i = 1, \dots, n \tag{1.3}$$

if and only if the associated Pick matrix $\Lambda = [\Lambda_{ij}]_{i,j=1}^n$ with the entries

$$\Lambda_{ij} = \begin{cases} \dfrac{1 - w_i\overline{w_j}}{1 - z_i\overline{z_j}}, & i \ne j \\ \gamma_i, & i = j, \end{cases}$$

is positive semidefinite.

Note that assumptions (1.2) are not restrictive as follows from the Julia-Carathéodory theorem.

The boundary Nevanlinna-Pick interpolation problem was worked out using quite different approaches: an indefinite modification of the Sz.-Nagy-Koranyi method [9], the method of fundamental matrix inequalities [31], the recursive Schur algorithm [25], the Grassmannian approach [14], via realization theory [13], and via unitary extensions of partially defined isometries [30]. Note also that a similar problem with equality sign in the second series of conditions in (1.3) was considered in [37, 27, 19].

In fact most of the above papers handled a more general tangential version of the boundary Nevanlinna-Pick problem for matrix and operator-valued Schur functions. The corresponding result is:

Given Hilbert spaces $\mathcal{E}$, $\mathcal{E}_*$ and $\mathcal{E}_L$, given points $z_1, \dots, z_n \in \mathbb{T}$ and operators $a_1, \dots, a_n \in \mathcal{L}(\mathcal{E}_L, \mathcal{E}_*)$, $c_1, \dots, c_n \in \mathcal{L}(\mathcal{E}_L, \mathcal{E})$, $\gamma_1, \dots, \gamma_n \in \mathcal{L}(\mathcal{E}_L)$ such that

$$a_j^* a_j = c_j^* c_j \quad \text{and} \quad \gamma_j \geq 0 \quad \text{for} \quad i = 1, \dots, n, \tag{1.4}$$

there exists $S \in \mathcal{S}(\mathcal{E}, \mathcal{E}_*)$ with

$$\lim_{r \to 1} S(rz_j)^* a_j = c_j \tag{1.5}$$

and

$$\lim_{r \to 1} a_j^* \frac{I_{\mathcal{E}_*} - S(rz_j)S(rz_j)^*}{1 - |r|^2} a_j \leq \gamma_j \quad \text{for} \quad i = 1, \dots, n \tag{1.6}$$

if and only if the associated Pick matrix $\Lambda = [\Lambda_{ij}]_{i,j=1}^n$ with the entries

$$\Lambda_{ij} = \begin{cases} \dfrac{a_i^* a_j - c_i^* c_j}{1 - z_i \overline{z_j}}, & i \neq j \\ \gamma_i, & i = j, \end{cases}$$

is positive semidefinite.

Interpolation theory for Schur functions has been extended recently to multivariable settings in several ways. Parametrizations of the set of all solutions of nonboundary Nevanlinna-Pick problems were obtained in [16] for the polydisk $\mathbb{D}^n \subset \mathbb{C}^n$ setting and in [17] for the unit ball $\mathbb{B}^n \subset \mathbb{C}^n$ setting. Note that solvability criteria for these problems were established earlier in [2, 1, 15, 3] for the case of the polydisk, in [36, 4, 32, 24, 35, 18] for the unit ball setting (including more general reproducing kernel Hilbert space and noncommutative Toeplitz-operator settings), and in [38] for Cartesian products of unit balls of arbitrary dimensions.

In [12] we considered nonboundary Nevanlinna-Pick interpolation problem for a class of contractive-valued functions analytic on a more general class of domains introduced in [5] and which we now recall.

We start with a polynomial $p \times q$ matrix-valued function

$$\mathbf{P}(z) = \begin{bmatrix} \mathbf{p}_{11}(z) & \dots & \mathbf{p}_{1q}(z) \\ \vdots & & \vdots \\ \mathbf{p}_{p1}(z) & \dots & \mathbf{p}_{pq}(z) \end{bmatrix} : \mathbb{C}^n \to \mathbb{C}^{p \times q} \tag{1.7}$$

and we define the domain $\mathcal{D}_{\mathbf{P}} \in \mathbb{C}^n$ by

$$\mathcal{D}_{\mathbf{P}} = \{z \in \mathbb{C}^n : \|\mathbf{P}(z)\|_{\mathbb{C}^{p \times q}} < 1\}.$$

Let $\mathcal{E}$ and $\mathcal{E}_*$ be two separable Hilbert spaces and $\mathcal{L}(\mathcal{E}, \mathcal{E}_*)$ be the space of bounded linear operators from $\mathcal{E}$ into $\mathcal{E}_*$. We denote by $\mathcal{SA}_{\mathbf{P}}(\mathcal{E}, \mathcal{E}_*)$ the Schur-Agler class of $\mathcal{L}(\mathcal{E}, \mathcal{E}_*)$-valued functions $S(z) = S(z_1, \dots, z_n)$ which are analytic on $\mathcal{D}_{\mathbf{P}}$ and such that

$$\|S(T_1, \dots, T_n)\| \leq 1$$

for any collection of n commuting operators $(T_1,\dots,T_n)$ on a Hilbert space $\mathcal{K}$, subject to

$$\|\mathbf{P}(T_1,\dots,T_n)\| < 1.$$

Domains $\mathcal{D}_{\mathbf{P}}$ and classes $\mathcal{SA}_{\mathbf{P}}(\mathcal{E},\mathcal{E}_*)$ (for $\mathcal{E}=\mathcal{E}_*=\mathbb{C}$) have been introduced in [5]. It was shown that the Taylor joint spectrum of the commuting n-tuple $(T_1,\dots,T_n)$ is contained in $\mathcal{D}_{\mathbf{P}}$ whenever $\|\mathbf{P}(T_1,\dots,T_n)\|<1$, and hence the operator $S(T_1,\dots,T_n)$ is well defined for any $\mathcal{L}(\mathcal{E},\mathcal{E}_*)$-valued function S which is analytic on $\mathcal{D}_{\mathbf{P}}$ by the Taylor functional calculus (see [23]). Operator-valued Schur-Agler classes were studied in [12]. The proof of the following theorem can be found in [12] and in [5] (for the case of scalar-valued functions); we state only the parts needed for our purposes here. In the statement and in the sequel, we often abbreviate expressions of the sort $\mathbf{P}(z)\otimes I_{\mathcal{H}}$ to simply $\mathbf{P}(z)$ without comment.

Theorem 1.1. *Let S be a $\mathcal{L}(\mathcal{E},\mathcal{E}_*)$-valued function. The following statements are equivalent:*

1. *S belongs to $\mathcal{SA}_{\mathbf{P}}(\mathcal{E},\mathcal{E}_*)$.*
2. *There exist an auxiliary Hilbert space $\mathcal{H}$ and an analytic function*
$$H(z)=\begin{bmatrix}H_1(z) & \dots & H_p(z)\end{bmatrix} \tag{1.8}$$
defined on $\mathcal{D}_{\mathbf{P}}$ with values in $\mathcal{L}(\mathbb{C}^p\otimes\mathcal{H},\mathcal{E}_*)$ so that
$$I_{\mathcal{E}_*}-S(z)S(w)^*=H(z)\left(I_{\mathbb{C}^p\otimes\mathcal{H}}-\mathbf{P}(z)\mathbf{P}(w)^*\right)H(w)^*, \tag{1.9}$$
or equivalently, there exists a positive kernel
$$\mathbb{K}=\begin{bmatrix}\mathbb{K}_{1,1} & \dots & \mathbb{K}_{1,p}\\ \vdots & & \vdots\\ \mathbb{K}_{p,1} & \dots & \mathbb{K}_{p,p}\end{bmatrix}:\Omega\times\Omega\mapsto\mathcal{L}(\mathbb{C}^p\otimes\mathcal{E}_*) \tag{1.10}$$
such that
$$I_{\mathcal{E}_*}-S(z)S(w)^*=\sum_{k=1}^{p}\mathbb{K}_{k,k}(z,w)-\sum_{k=1}^{p}\sum_{i,\ell=1}^{q}\mathbf{p}_{ik}(z)\overline{\mathbf{p}_{\ell k}(w)}\mathbb{K}_{i,\ell}(z,w)$$
$$\text{for all } z,w\in\mathcal{D}_{\mathbf{P}}. \tag{1.11}$$
3. *There is a unitary operator*
$$\mathbf{U}=\begin{bmatrix}A & B\\ C & D\end{bmatrix}:\begin{bmatrix}\mathbb{C}^p\otimes\mathcal{H}\\ \mathcal{E}\end{bmatrix}\to\begin{bmatrix}\mathbb{C}^q\otimes\mathcal{H}\\ \mathcal{E}_*\end{bmatrix} \tag{1.12}$$
such that
$$S(z)=D+C\left(I_{\mathbb{C}^p\otimes\mathcal{H}}-\mathbf{P}(z)A\right)^{-1}\mathbf{P}(z)B. \tag{1.13}$$

For S of the form (1.13) it holds that

$$I_{\mathcal{E}_*}-S(z)S(w)^*=C\left(I-\mathbf{P}(z)A\right)^{-1}\left(I-\mathbf{P}(z)\mathbf{P}(w)^*\right)\left(I-A^*\mathbf{P}(w)^*\right)^{-1}C^* \tag{1.14}$$

and therefore, the representation (1.9) is valid for

$$H(z)=C\left(I-\mathbf{P}(z)A\right)^{-1}. \tag{1.15}$$

The representation (1.13) is called *a unitary realization* of $S \in \mathcal{SA}_{\mathbf{P}}(\mathcal{E}, \mathcal{E}_*)$. For the case of the unit disk ($\mathbf{P}(z) = z$ and $\mathcal{D}_{\mathbf{P}} = \mathbb{D}$), the formula (1.13) presents a unitary realization for a clasdscial Schur function S, i.e., coincides with the realization formula (1.1) for the classical case.

Domains $\mathcal{D}_{\mathbf{P}}$ and classes $\mathcal{SA}_{\mathbf{P}}(\mathcal{E}, \mathcal{E}_*)$ enable us to consider in a unified way a wide variety of domains in $\mathbb{C}^n$ (we refer to [5, 12] for examples). In [12] we considered the bitangential Nevanlinna-Pick interpolation problem for Schur-Agler functions with all interpolation nodes inside $\mathcal{D}_{\mathbf{P}}$. Ambrozie and Eschmeier in [6] showed how to obtain results on interpolation as an application of a commutant lifting theorem for this setting. In the present paper we shall study the boundary Nevanlinna-Pick problem when prescribed (directional) values of an unknown interpolant are replaced by preassigned (directional) values of radial boundary limits. The techniques and results parallel those appearing in [11] for the special case of the polydisk and in [10] for the special case of the ball; the general case considered here, however, does present new features which must be understood before arriving at a complete solution. The foundation for the present method is the realization formula (1.13) for Schur-Agler functions in Theorem 1.1.

We assume that $\mathcal{D}_{\mathbf{P}}$ is bounded and completely circular, i.e., that *for every* $\zeta \in \mathbb{C}$, $|\zeta| \leq 1$, *and* $z \in \mathcal{D}_{\mathbf{P}}$, *it holds that*

$$\zeta z = (\zeta z_1, \dots, \zeta z_n) \in \mathcal{D}_{\mathbf{P}}.$$

We assume furthermore that the distinguished boundary of $\mathcal{D}_{\mathbf{P}}$ defined as

$$\partial\mathcal{D}_{\mathbf{P}} = \{z \in \mathbb{C}^n : \mathbf{P}(z)\mathbf{P}(z)^* = I_p\},$$

is nonempty. The latter condition implies immediately that $p \leq q$, which will be assumed throughout the paper.

In particular, Cartan domains of the first three types (more precisely, such domains with $p \leq q$) and their direct products and intersections are included in this scheme.

Let Ω be a subset of $\partial\mathcal{D}_{\mathbf{P}}$. The data set for the interpolation problem is as follows. We are given an auxiliary Hilbert space $\mathcal{E}_L$ and three operator-valued functions

$$\begin{aligned} &\mathbf{a}\colon \Omega \mapsto \mathcal{L}(\mathcal{E}_L, \mathcal{E}_*), \qquad \mathbf{c}\colon \Omega \mapsto \mathcal{L}(\mathcal{E}_L, \mathcal{E}), \\ &\Psi\colon \Omega \mapsto \mathcal{L}(\mathcal{E}_L) \text{ with } \Psi(\xi) \geq 0 \text{ for } \xi \in \Omega. \end{aligned} \tag{1.16}$$

The interpolation problem to be studied here is the following:

Problem 1.2. *Find all functions* $S \in \mathcal{SA}_{\mathbf{P}}(\mathcal{E}, \mathcal{E}_*)$ *such that*

$$\lim_{r\to 1} S(r\xi)^*\mathbf{a}(\xi) = \mathbf{c}(\xi) \quad \text{for all } \xi \in \Omega \tag{1.17}$$

and

$$\lim_{r\to 1} \mathbf{a}(\xi)^* \frac{I_{\mathcal{E}_*} - S(r\xi)S(r\xi)^*}{1-r^2} \mathbf{a}(\xi) \leq \Psi(\xi) \quad \text{for all } \xi \in \Omega, \tag{1.18}$$

where the limits in (1.17) *and* (1.18) *are understood in the strong and in the weak sense, respectively.*

Note that in case when Ω consists of finitely many points $z_1, \dots, z_n$, conditions (1.17) and (1.18) take the form (1.5) and (1.6), respectively, with $a_j = \mathbf{a}(z_j)$, $c_j = \mathbf{c}(z_j)$ and $\gamma_j = \Psi(z_j)$.

Condition (1.17) is called the *left-sided* interpolation condition for S. It follows by a multivariable operator-valued analogue of the classical Julia-Carathéodory theorem (see Lemma 2.1 below) that if the limit in (1.17) exists and equals $\mathbf{c}(\xi)$, then a necessary condition for the limits in (1.18) to exist and to be finite is

$$\mathbf{a}(\xi)^*\mathbf{a}(\xi) = \mathbf{c}(\xi)^*\mathbf{c}(\xi) \qquad (\xi \in \Omega). \tag{1.19}$$

It follows again by (the third assertion of) Lemma 2.1, that S satisfies also the right-sided interpolation condition

$$\lim_{r \to 1} S(r\xi)\mathbf{c}(\xi) = \mathbf{a}(\xi) \qquad (\xi \in \Omega).$$

Thus, Problem 1.2 is in fact a two-sided interpolation problem and conditions (1.19) are necessary for this problem to have a solution.

The paper is organized as follows. Upon completion of the present Introduction, in Section 2 we develop the Julia-Carathéodory theory for operator-valued functions on $\mathcal{D}_\mathbf{P}$ which we need for our purposes here, formulate the solution criterion (the existence of a positive kernel satisfying a Stein equation together with an inequality constraint) for existence of solutions to Problem 1.2 and derive the necessity part of this condition. In Section 3, under the assumption that this necessary condition holds, we show that solutions of Problem 1.2 are equal to characteristic functions of unitary colligations obtained as unitary extensions of a certain partial isometry constructed from any positive kernel meeting the conditions of the existence criterion. In Section 4 we adapt the techniques of Arov-Grossman to obtain the linear-fractional parametrization for the set of all characteristic functions of colligations equal to a unitary extension of one such partial isometry. When the partial isometry is taken to be that constructed from a particular positive kernel satisfying the conditions of the solution criterion, we have a parametrization for a particular subset of the set of all solutions of Problem 1.2. Taking the union of these over all positive kernels meeting the conditions of the solution criterion then gives us the set of all solutions. In particular, from this description of the set of all solutions we arrive at the sufficiency of the existence criterion.

2. The solvability criterion

In this section we establish the solvability criterion of Problem 1.2. We start with some auxiliary results. The assumption that $\mathcal{D}_\mathbf{P}$ is a bounded completely circular domain allows us to translate easily the classical Julia-Carathéodory theorem from the unit disk $\mathbb{D}$ to $\mathcal{D}_\mathbf{P}$ using slice-functions.

Lemma 2.1. *Let* $S \in \mathcal{SA}_\mathbf{P}(\mathcal{E}, \mathcal{E}_*)$, $\beta \in \partial\mathcal{D}_\mathbf{P}$, $\mathbf{x} \in \mathcal{E}_*$ *and let* H_j $(j = 1, \dots, p)$ *be* $\mathcal{L}(\mathcal{H}, \mathcal{E}_*)$*-valued functions from the representation* (1.9). *Then:*

I. The following three statements are equivalent:

1. *S is subject to* $\mathbf{L} := \sup_{0\le r<1} \mathbf{x}^* \frac{I_{\mathcal{E}_*} - S(r\beta)S(r\beta)^*}{1-r^2}\mathbf{x} < \infty.$
2. *The radial limit* $L := \lim_{r\to 1} \mathbf{x}^* \frac{I_{\mathcal{E}_*} - S(r\beta)S(r\beta)^*}{1-r^2}\mathbf{x}$ *exists.*
3. *The radial limit*

$$\lim_{r\to 1} S(r\beta)^*\mathbf{x} = \mathbf{y} \tag{2.1}$$

exists in the strong sense and serves to define the vector $\mathbf{y}\in\mathcal{E}$. *Furthermore,*

$$\lim_{r\to 1} S(r\beta)\mathbf{y} = \mathbf{x}, \qquad \|\mathbf{y}\|_{\mathcal{E}} = \|\mathbf{x}\|_{\mathcal{E}_*}, \tag{2.2}$$

(the limit is understood in the weak sense) and the radial limit

$$\widetilde{L} = \lim_{r\to 1} \frac{\mathbf{y}^*\mathbf{y} - \mathbf{x}^*S(r\beta)\mathbf{y}}{1-r} \tag{2.3}$$

exists.

II. Any two of the three equalities in (2.1) *and* (2.2) *imply the third.*

III. If any of the three statements in part I holds true, then the radial limits

$$T_j = \lim_{r\to 1} H_j(r\beta)^*\mathbf{x} \quad (j = 1,\dots,p) \tag{2.4}$$

exist in the strong sense and the block operator

$$T = \begin{bmatrix} T_1 \\ \vdots \\ T_p \end{bmatrix} \tag{2.5}$$

satisfies

$$T^*\Lambda_{\mathbf{P}}(\beta)T = L = \widetilde{L} \le \mathbf{L}, \tag{2.6}$$

where $\Lambda_{\mathbf{P}}(\beta)\in\mathcal{L}(\mathbb{C}^p\otimes\mathcal{H})$ *is defined via the limit*

$$\Lambda_{\mathbf{P}}(\beta) = \lim_{r\to 1} \frac{I_p - \mathbf{P}(\beta r)\mathbf{P}(\beta r)^*}{1-r^2}. \tag{2.7}$$

Proof. For the proof of all the statements for the single-variable case ($d = 1$) and $\mathbf{P}(z) = z$ (i.e., for the case of the unit disk) see [10, Lemma 2.3] (all the statements but those related to T_j's and for finite-dimensional $\mathcal{E}$ and $\mathcal{E}_*$ are contained in [26, Lemma 8.3, Lemma 8.4, and Theorem 8.5]). Note that in this case $\Lambda_{\mathbf{P}}(\beta) = I_p$. These proofs rely on reproducing kernel Hilbert space methods; for a more classical treatment (including for the case of the ball) and a survey of the various proofs of the Julia-Carathéodory theorem, we refer to Section 2.3 of [22].

For the general case, let us introduce the slice-function

$$S_\beta(\zeta) := S(\beta\zeta), \qquad (\zeta\in\mathbb{D}), \tag{2.8}$$

which clearly belongs to the classical Schur class $\mathcal{S}(\mathcal{E}, \mathcal{E}_*)$ (indeed, it follows from (1.9) that S takes contractive values at each point of $\mathcal{D}_{\mathbf{P}}$ and in particular at each point $z \in \mathcal{D}_{\mathbf{P}}$ of the form $z = \zeta\beta$ ($\zeta \in \mathbb{D}$)).

The two first statements of the lemma concern the boundary behavior of the function S_β near a boundary point $\zeta = 1$. Applying one-variable results to the slice-function S_β and returning to the original function S, we obtain the two first assertions. It is clear that the vectors $\mathbf{x} \in \mathcal{E}_*$ and $\mathbf{y} \in \mathcal{E}$ can be replaced by bounded linear operators from an auxiliary Hilbert space $\mathcal{E}_L$ into $\mathcal{E}_*$ and $\mathcal{E}$, respectively, with the limits then taken in the strong sense.

To prove the third statement, we first show that the limit in (2.7) exists. To this end we introduce the slice function

$$\mathbf{P}_\beta(\zeta) = \mathbf{P}(\beta\zeta), \qquad (\zeta \in \mathbb{D}). \tag{2.9}$$

Since $\mathcal{D}_{\mathbf{P}}$ is a bounded completely circular domain, it holds for every point $\beta \in \partial\mathcal{D}_{\mathbf{P}}$ and $\zeta \in \mathbb{D}$, that

$$\mathbf{P}_\beta(\zeta)\mathbf{P}_\beta(\zeta)^* = \mathbf{P}(\zeta\beta)\mathbf{P}(\zeta\beta)^* \le I_p$$

and thus, $\mathbf{P}_\beta$ belongs to the classical Schur class $\mathcal{S}(\mathbb{C}^q, \mathbb{C}^p)$. Since it takes a coisometric value at $\zeta = 1$, one can apply the one-variable result (statements I and II) of the lemma to $\mathbf{x} = I_p$ and $\mathbf{y} = S(\beta)^*$ to conclude that

$$\lim_{r\to 1} \frac{I_p - \mathbf{P}_\beta(r)\mathbf{P}_\beta(r)^*}{1-r^2} = \lim_{r\to 1} \frac{\mathbf{P}_\beta(1) - \mathbf{P}_\beta(r)}{1-r}\mathbf{P}_\beta(1)^* \tag{2.10}$$

and since $\mathbf{P}_\beta$ is a polynomial, the second limit in (2.10) exists and equals $\mathbf{P}'_\beta(1)\mathbf{P}_\beta(1)^*$. Therefore, the limit in (2.7) exists.

Since $\mathbf{P}_\beta(\zeta)$ is a Schur function, there is an auxiliary Hilbert space $\mathcal{G}_\beta$ and a unitary operator

$$\mathbf{U}_1 = \begin{bmatrix} A_1 & B_1 \\ C_1 & D_1 \end{bmatrix} : \begin{bmatrix} \mathcal{G}_\beta \\ \mathbb{C}^q \end{bmatrix} \to \begin{bmatrix} \mathcal{G}_\beta \\ \mathbb{C}^p \end{bmatrix} \tag{2.11}$$

such that

$$\mathbf{P}_\beta(\zeta) = D_1 + \zeta C_1\left(I_{\mathcal{G}_\beta} - \zeta A_1\right)^{-1} B_1. \tag{2.12}$$

On the other hand, upon setting $z = \zeta\beta$ in (1.13) we get

$$S_\beta(\zeta) = D + C\left(I_{\mathbb{C}^p\otimes\mathcal{H}} - \mathbf{P}_\beta(\zeta)A\right)^{-1}\mathbf{P}_\beta(\zeta)B \tag{2.13}$$

(in this realization we have in fact $\mathbf{P}_\beta(\zeta) \otimes I_{\mathcal{H}}$, which (by the above convention) is still denoted by $\mathbf{P}_\beta(\zeta)$) and similarly, we conclude from (1.9) that

$$\begin{aligned} L &= \lim_{r\to 1} \mathbf{x}^* \frac{I_{\mathcal{E}_*} - S(r\beta)S(r\beta)^*}{1-r^2}\mathbf{x} \\ &= \lim_{r\to 1} \mathbf{x}^* H(r\beta)\frac{I_{\mathbb{C}^p\otimes\mathcal{H}} - \mathbf{P}_\beta(r)\mathbf{P}_\beta(r)^*}{1-r^2}H(r\beta)^*\mathbf{x}. \end{aligned} \tag{2.14}$$

Now we combine realization formulas (2.12) and (2.13) to get a unitary realization for the Schur function S_β:

$$S_\beta(\zeta) = D_2 + \zeta C_2 \left(I - \zeta A_2\right)^{-1} B_2, \tag{2.15}$$

where the operator

$$\mathbf{U}_2 = \begin{bmatrix} A_2 & B_2 \\ C_2 & D_2 \end{bmatrix} : \begin{bmatrix} \mathcal{G}_\beta \\ \mathcal{E} \end{bmatrix} \to \begin{bmatrix} \mathcal{G}_\beta \\ \mathcal{E}_* \end{bmatrix} \tag{2.16}$$

given by

$$\begin{aligned} A_2 &= A_1 + B_1(I - AD_1)^{-1}AC_1, & (2.17)\\ B_2 &= B_1(I - AD_1)^{-1}B, & (2.18)\\ C_2 &= C(I - D_1A)^{-1}C_1, & \\ D_2 &= D + C(I - D_1A)^{-1}D_1B, & \end{aligned}$$

is unitary. Indeed, by (1.13) and (2.12),

$$\begin{aligned} H(\zeta\beta) &= C(I - \mathbf{P}_\beta(\zeta)A)^{-1} \\ &= C(I - D_1A - \zeta C_1\left(I - \zeta A_1\right)^{-1} B_1A)^{-1} \\ &= C(I - D_1A)^{-1}\left[I - \zeta C_1(I - \zeta A_1)^{-1}B_1A(I - D_1A)^{-1}\right]^{-1} \\ &= C(I - D_1A)^{-1} + \zeta C_2(I - \zeta A_2)^{-1}B_1A(I - D_1A)^{-1}, \qquad (2.19) \end{aligned}$$

and taking into account that

$$\zeta(I - \zeta A_2)^{-1}B_1A(I - D_1A)^{-1}C_1(I - \zeta A_1)^{-1} = (I - \zeta A_2)^{-1} - (I - \zeta A_1)^{-1},$$

we obtain

$$\begin{aligned} S_\beta(\zeta) &= D + H(\zeta\beta)\mathbf{P}_\beta(\zeta)B \\ &= D + \left[C(I - D_1A)^{-1} + \zeta C_2(I - \zeta A_2)^{-1}B_1A(I - D_1A)^{-1}\right] \\ &\qquad \times \left[D_1 + \zeta C_1\left(I - \zeta A_1\right)^{-1} B_1\right] B \\ &= D + C(I - D_1A)^{-1}D_1B + \zeta C_2\left(I - \zeta A_1\right)^{-1} B_1B \\ &\quad + \zeta C_2(I - \zeta A_2)^{-1}B_1A(I - D_1A)^{-1}D_1B \\ &\quad + \zeta C_2\left[(I - \zeta A_2)^{-1} - (I - \zeta A_1)^{-1}\right] B_1B \\ &= D_2 + \zeta C_2(I - \zeta A_2)^{-1}B_1\left[I + A(I - D_1A)^{-1}D_1\right] B \\ &= D_2 + \zeta C_2(I - \zeta A_2)^{-1}B_1(I - AD_1)^{-1}B \\ &= D_2 + \zeta C_2(I - \zeta A_2)^{-1}B_2. \end{aligned}$$

The verification of the fact that the operator $\mathbf{U}_2$ is unitary is straightforward; it involves the explicit formulas for the blocks A_2, B_2, C_2, D_2 and the fact that the operators $\mathbf{U}$ and $\mathbf{U}_1$ are unitary. Alternatively, one can pursue the following higher level path which perhaps is more informative.

The starting point is the observation that the colligation

$$\mathbf{U}_2 = \begin{bmatrix} A_2 & B_2 \\ C_2 & D_2 \end{bmatrix} : \begin{bmatrix} \mathcal{G}_\beta \\ \mathcal{E} \end{bmatrix} \to \begin{bmatrix} \mathcal{G}_\beta \\ \mathcal{E}_* \end{bmatrix}$$

arises as the feedback coupling of $\mathbf{U}_1 \otimes (I_{\mathbb{C}} \oplus I_{\mathcal{H}})$ with $\mathbf{U}$:

$$\mathbf{U}_2\colon \begin{bmatrix} x_1 \\ u \end{bmatrix} \mapsto \begin{bmatrix} \widetilde{x}_1 \\ y \end{bmatrix} \text{ if and only if } \mathbf{U}\colon \begin{bmatrix} x \\ u \end{bmatrix} \mapsto \begin{bmatrix} \widetilde{x} \\ y \end{bmatrix} \text{ and } \mathbf{U}_1 \otimes (I_{\mathbb{C}} \otimes I_{\mathcal{H}})\colon \begin{bmatrix} x_1 \\ \widetilde{x} \end{bmatrix} \mapsto \begin{bmatrix} \widetilde{x}_1 \\ x \end{bmatrix}.$$

By the general principle that a feedback coupling of 2×2 block unitary operators is again unitary, we see that the colligation $\mathbf{U}_2$ is unitary.

It follows from the unitarity of the realization (2.15) that

$$\frac{I_{\mathcal{E}_*} - S_\beta(\zeta)S_\beta(\omega)^*}{1 - \zeta\overline{\omega}} = C_2(I - \zeta A_2)^{-1}(I - \overline{\omega}A_2^*)^{-1}C_2^*$$

and by the one-variable result, there exists the strong limit

$$\widetilde{T} := \lim_{r \to 1}(I - rA_2^*)^{-1}C_2^*\mathbf{x}. \tag{2.20}$$

Setting $\zeta = r$ in (2.19) and letting $r \to 1$, we come to the existence of the following strong limit

$$T := \lim_{r \to 1} H(r\beta)^*\mathbf{x} = (I - A^*D_1^*)^{-1}\left(C^*\mathbf{x} + A^*B_1^*\widetilde{T}\right). \tag{2.21}$$

On account of the block structure (1.8) of H, the strong limits T_j's in (2.4) exist.

It remains to show that T satisfies relations (2.6). But now this follows from (2.14), (2.7) and (2.21):

$$L = \lim_{r \to 1} \mathbf{x}^*H(r\beta)\frac{I_{\mathbb{C}^p \otimes \mathcal{H}} - \mathbf{P}_\beta(r)\mathbf{P}_\beta(r)^*}{1 - r^2}H(r\beta)^*\mathbf{x} = T^*\Lambda_{\mathbf{P}}(\beta)T,$$

which completes the proof. □

To state the solvability criterion of Problem 1.2 we introduce some notations. Let

$$E_1 = \begin{bmatrix} I_{\mathcal{E}_L} \\ 0 \\ \vdots \\ 0 \end{bmatrix}, \quad E_2 = \begin{bmatrix} 0 \\ I_{\mathcal{E}_L} \\ \vdots \\ 0 \end{bmatrix}, \quad \ldots, \quad E_p = \begin{bmatrix} 0 \\ \vdots \\ 0 \\ I_{\mathcal{E}_L} \end{bmatrix} \in \mathcal{L}(\mathcal{E}_L, \mathbb{C}^p \otimes \mathcal{E}_L), \tag{2.22}$$

let $N_1, \ldots, N_q : \Omega \to \mathcal{L}(\mathcal{E}, \mathbb{C}^p \otimes \mathcal{E}_L)$ be the functions given by

$$N_k(\xi) = \begin{bmatrix} \overline{\mathbf{p}_{1,k}(\xi)}I_{\mathcal{E}_L} \\ \vdots \\ \overline{\mathbf{p}_{p,k}(\xi)}I_{\mathcal{E}_L} \end{bmatrix} \qquad (k = 1, \ldots, q) \tag{2.23}$$

and let $\Lambda_{\mathbf{P}} : \Omega \to \mathbb{C}^{p \times p}$ be the function defined via the limit

$$\Lambda_{\mathbf{P}}(\xi) = [\Lambda_{ij}(\xi)]_{i,j=1}^p = \lim_{r \to 1}\frac{I_p - \mathbf{P}(r\xi)\mathbf{P}(r\xi)^*}{1 - r^2}. \tag{2.24}$$

By the preceding analysis, $\Lambda_{\mathbf{P}}$ is well defined at each point $\xi \in \partial\mathcal{D}_{\mathbf{P}}$.

Theorem 2.2. *Problem* 1.2 *has a solution if and only if there exist* p^2 $\mathcal{L}(\mathcal{E})$*-valued kernels* $K_{ij}(\xi,\mu)$ *on* $\Omega\times\Omega$ *subject to the Stein identity*

$$\sum_{j=1}^{p} K_{jj}(\xi,\mu)-\sum_{k=1}^{q}\left(\sum_{i,\ell=1}^{p} \mathbf{p}_{ik}(\xi)\overline{\mathbf{p}_{k\ell}(\mu)}K_{i\ell}(\xi,\mu)\right)=\mathbf{a}(\xi)^*\mathbf{a}(\mu)-\mathbf{c}(\xi)^*\mathbf{c}(\mu) \quad (2.25)$$

and to the constraint

$$\sum_{i,j=1}^{p} \Lambda_{ij}(\xi)K_{ij}(\xi,\xi)\leq \Psi(\xi) \quad (2.26)$$

for every $\xi,\mu\in\Omega$ *and such that the kernel*

$$K(\xi,\mu)=\begin{bmatrix} K_{11}(\xi,\mu) & \dots & K_{1p}(\xi,\mu)\\ \vdots & & \vdots \\ K_{p1}(\xi,\mu) & \dots & K_{pp}(\xi,\mu)\end{bmatrix} \quad (2.27)$$

is positive on Ω*:*

$$K(\xi,\mu)\succeq 0. \quad (2.28)$$

Taking advantage of operators (2.22) and functions (2.23) and (2.24), one can rewrite relation (2.25) as

$$\sum_{k=1}^{p} E_k^* K(\xi,\mu)E_k-\sum_{k=1}^{q} N_k(\xi)^*K(\xi,\mu)N_k(\mu)=\mathbf{a}(\xi)^*\mathbf{a}(\mu)-\mathbf{c}(\xi)^*\mathbf{c}(\mu). \quad (2.29)$$

Proof. Here we check the necessity of conditions (2.25)–(2.28). The proof of the sufficiency part is postponed until Section 4 where it will be obtained as a consequence of slightly stronger results. Let S be a solution of Problem 1.2, that is, let S be an element of $\mathcal{SA}_{\mathbf{P}}(\mathcal{E},\,\mathcal{E}_*)$ which satisfies the interpolation conditions (1.17) and (1.18). Since S belongs to $\mathcal{SA}_{\mathbf{P}}(\mathcal{E},\,\mathcal{E}_*)$, the identity (1.9) holds for some $\mathcal{L}(\mathbb{C}^p\otimes\mathcal{H},\mathcal{E}_*)$-valued function H which is analytic on $\mathcal{D}_{\mathbf{P}}$. Let H be decomposed into blocks $H_1,\dots,H_p$ as in (1.8) and let $T_j(\xi)$ stand for the following strong limit

$$T_j(\xi):=\lim_{r\to 1} H_j(r\xi)^*\mathbf{a}(\xi) \quad (j=1,\dots,p;\ \xi\in\Omega), \quad (2.30)$$

which exists at every point $\xi\in\Omega$, by Lemma 2.1. Then the block operator-valued function

$$T(\xi):=\begin{bmatrix} T_1(\xi)\\ \vdots \\ T_p(\xi)\end{bmatrix}=\lim_{r\to 1} H(r\xi)^*\mathbf{a}(\xi) \quad (\xi\in\Omega) \quad (2.31)$$

satisfies (again by Lemma 2.1)

$$T(\xi)^*\Lambda_{\mathbf{P}}(\xi)T(\xi)=L(\xi):=\lim_{r\to 1}\mathbf{a}(\xi)^*\,\frac{I_{\mathcal{E}_*}-S(r\xi)S(r\xi)^*}{1-r^2}\,\mathbf{a}(\xi) \quad (2.32)$$

for each $\xi\in\Omega$. Let

$$K_{ij}(\xi,\mu)=T_i(\xi)^*T_j(\mu) \quad (i,j=1,\dots,p;\ \xi,\mu\in\Omega). \quad (2.33)$$

Then the kernel $K(\xi,\mu)$ defined in (2.27) admits a representation

$$K(\xi,\mu)=\begin{bmatrix} T_1(\xi)^* \\ \vdots \\ T_p(\xi)^* \end{bmatrix}\begin{bmatrix} T_1(\mu) & \dots & T_p(\mu) \end{bmatrix} \tag{2.34}$$

and is clearly positive on Ω. Furthermore, condition (2.26) holds by (1.18), (2.32) and (2.34):

$$\begin{aligned}\sum_{i,j=1}^{p}\Lambda_{ij}(\xi)K_{ij}(\xi,\xi) &= T(\xi)^*\Lambda_{\mathbf{P}}(\xi)T(\xi)\\ &= \lim_{r\to 1}\mathbf{a}(\xi)^*\frac{I_{\mathcal{E}_*}-S(r\xi)S(r\xi)^*}{1-r^2}\mathbf{a}(\xi)\leq\Psi(\xi).\end{aligned}$$

Finally, setting $z=r\xi$ and $w=r\mu$ ($\xi\neq\mu$) in (1.9) and multiplying both sides in the resulting identity by $\mathbf{a}(\xi)^*$ on the left and by $\mathbf{a}(\mu)$ on the right, we get

$$\mathbf{a}(\xi)^*\left(I_{\mathcal{E}_*}-S(r\xi)S(r\mu)^*\right)\mathbf{a}(\mu)=\mathbf{a}(\xi)^*H(r\xi)(I-\mathbf{P}(r\xi)\mathbf{P}(r\mu)^*)H(r\mu)^*\mathbf{a}(\mu). \tag{2.35}$$

Making use of interpolation condition (1.17), we get

$$\lim_{r\to 1}\mathbf{a}(\xi)^*\left(I_{\mathcal{E}_*}-S(r\xi)S(r\mu)^*\right)\mathbf{a}(\mu)=\mathbf{a}(\xi)^*\mathbf{a}(\mu)-\mathbf{c}(\xi)^*\mathbf{c}(\mu),$$

whereas relations (2.31), (2.34) together with (1.13) and partitionings (1.7) and (1.8) lead to

$$\begin{aligned}&\lim_{r\to 1}\mathbf{a}(\xi)^*H(r\xi))\left(I_{\mathbb{C}^p\otimes\mathcal{H}}-\mathbf{P}(r\xi)\mathbf{P}(r\mu)^*\right)H(r\mu)^*\mathbf{a}(\mu)\\ &=\lim_{r\to 1}\sum_{j=1}^{p}\mathbf{a}(\xi)^*H_j(r\xi)H_j(r\mu)^*\mathbf{a}(\mu)\\ &\quad-\lim_{r\to 1}\sum_{i,\ell=1}^{p}\mathbf{a}(\xi)^*H_i(r\xi)\left(\sum_{k=1}^{q}\mathbf{p}_{ik}(r\xi)\overline{\mathbf{p}_{k\ell}(r\mu)}\right)H_\ell(r\mu)^*\mathbf{a}(\mu)\\ &=\sum_{j=1}^{p}T_j(\xi)^*T_j(\mu)-\sum_{i,\ell=1}^{p}\left(\sum_{k=1}^{q}\mathbf{p}_{ik}(\xi)\overline{\mathbf{p}_{k\ell}(\mu)}\right)T_i(\xi)^*T_\ell(\mu)\\ &=\sum_{j=1}^{p}K_{jj}(\xi,\mu)-\sum_{k=1}^{q}\left(\sum_{i,\ell=1}^{p}\mathbf{p}_{ik}(\xi)\overline{\mathbf{p}_{k\ell}(\mu)}K_{i\ell}(\xi,\mu)\right).\end{aligned}$$

Taking limits as $r\to 1$ on both sides in (2.35) and making use of the last two equalities, we get (2.25). $\square$

3. Solutions to the interpolation problem and unitary extensions

In this section we analyze the structure of the set of solutions of Problem 1.2 under the assumption that the necessary conditions (2.25)–(2.28) hold. We therefore assume throughout this section: *we are given an interpolation data set* $(\mathbf{a}, \mathbf{c}, \Psi)$ *as in* (1.16) *and there exists an* $\mathcal{L}(\mathcal{E}_L)$*-valued positive kernel* $K(\xi, \mu)$ *on* $\Omega \times \Omega$ *satisfying the necessary conditions* (2.25)–(2.28) *in Theorem* 2.2.

We define a $\mathbf{P}$-colligation as a quadruple

$$\mathcal{C} = \{\mathcal{H},\ \mathcal{E},\ \mathcal{E}_*,\ \mathbf{U}\} \tag{3.1}$$

consisting of three Hilbert spaces $\mathcal{H}$ (*the state space*), $\mathcal{E}$ (*the input space*) and $\mathcal{E}_*$ (*the output space*), together with a connecting operator

$$\mathbf{U} = \begin{bmatrix} A & B \\ C & D \end{bmatrix} : \begin{bmatrix} \mathbb{C}^p \otimes \mathcal{H} \\ \mathcal{E} \end{bmatrix} \to \begin{bmatrix} \mathbb{C}^q \otimes \mathcal{H} \\ \mathcal{E}_* \end{bmatrix}. \tag{3.2}$$

The colligation is said to be *unitary* if the connecting operator $\mathbf{U}$ is unitary. A colligation

$$\widetilde{\mathcal{C}} = \{\widetilde{\mathcal{H}},\ \mathcal{E},\ \mathcal{E}_*,\ \widetilde{\mathbf{U}}\} \tag{3.3}$$

is said to be *unitarily equivalent* to the colligation $\mathcal{C}$ if there is a unitary operator $\alpha :\ \mathcal{H} \to \widetilde{\mathcal{H}}$ such that

$$\begin{bmatrix} \alpha \otimes I_q & 0 \\ 0 & I_{\mathcal{E}_*} \end{bmatrix} \mathbf{U} = \widetilde{\mathbf{U}} \begin{bmatrix} \alpha \otimes I_p & 0 \\ 0 & I_{\mathcal{E}} \end{bmatrix}$$

The *characteristic function* of the colligation $\mathcal{C}$ is defined as

$$S_{\mathcal{C}}(z) = D + C\left(I_{\mathbb{C}^p \otimes \mathcal{H}} - \mathbf{P}(z)A\right)^{-1} \mathbf{P}(z)B. \tag{3.4}$$

Thus, Theorem 1.1 asserts that a $\mathcal{L}(\mathcal{E},\ \mathcal{E}_*)$-valued function S analytic on $\mathcal{D}_{\mathbf{P}}$ belongs to the class $\mathcal{SA}_{\mathbf{P}}(\mathcal{E},\ \mathcal{E}_*)$ if and only it is the characteristic function of some unitary $\mathbf{P}$-colligation $\mathcal{C}$ of the form (3.2).

Remark 3.1. *Unitarily equivalent colligations have the same characteristic function.*

In this section, under the assumption that the necessary conditions (2.25)–(2.28) in Theorem 2.2 hold, we identify a class of unitary colligations, the characteristic functions of which give rise exactly to the set of all solutions of Problem 1.2. These unitary colligations coincide with unitary extensions of certain partial isometry $\mathbf{V}_K$ constructed from some positive kernel meeting the conditions (2.25)–(2.28) for Problem 1.2. We therefore assume that there exist positive kernel solutions K of conditions (2.25)–(2.28) and we let E_j and N_j be defined as in (2.22) and (2.23), respectively.

Let $\mathcal{H}_0$ be the linear space of $\mathcal{E}_L$-valued functions $f(\xi)$ defined on Ω which take nonzero values at at most finitely many points and let $\mathcal{H}_1$ be the linear space of $\mathbb{C}^p \otimes \mathcal{E}_L$-valued functions $\xi \mapsto h(\xi)$ defined on Ω which take nonzero values at at most finitely many points Let $X \in \mathcal{L}(\mathcal{H}_0, \mathcal{E}_*)$ and $Y \in \mathcal{L}(\mathcal{H}_0, \mathcal{E})$ be operators defined by

$$Xf = \sum_{\xi \in \Omega} \mathbf{a}(\xi) f(\xi), \qquad Yf = \sum_{\xi \in \Omega} \mathbf{c}(\xi) f(\xi). \tag{3.5}$$

For a fixed choice of a positive kernel K meeting conditions (2.25)–(2.28), let $D_K(h,g)$ be the quadratic form on $\mathcal{H}_1 \times \mathcal{H}_1$ defined as

$$D_K(h,g) = \sum_{\xi_i, \xi_\ell \in \Omega} \langle K(\xi_i, \xi_\ell) h(\xi_\ell), \, g(\xi_i) \rangle_{\mathbb{C}^p \otimes \mathcal{E}_L}. \tag{3.6}$$

It follows by the definitions of the spaces $\mathcal{H}_0$ and $\mathcal{H}_1$ and by the definitions (2.22) and (2.23) of E_j and N_k, that for every $f \in \mathcal{H}_0$, the functions $E_j f$ and $N_k f$ belong to $\mathcal{H}_1$. Furthermore, it follows from (2.25) that

$$\sum_{j=1}^{p} D_K(E_j f, \, E_j g) - \sum_{k=1}^{q} D_K(N_k f, \, N_k g) = \langle Xf, \, Xg \rangle_{\mathcal{E}_*} - \langle Yf, \, Yg \rangle_{\mathcal{E}}. \tag{3.7}$$

We say that $h_1 \sim h_2$ if and only if $D_K(h_1 - h_2, \, y) = 0$ for all $y \in \mathcal{H}_0$ and denote $[h]$ the equivalence class of h with respect to the above equivalence. The linear space of equivalence classes endowed with the inner product

$$\langle [h], \, [g] \rangle = D_K(h, \, g) \tag{3.8}$$

is a prehilbert space, whose completion we denote by $\widehat{\mathcal{H}}_K$. Rewriting (3.7) as

$$\sum_{j=1}^{p} \langle [E_j f], \, [E_j g] \rangle_{\widehat{\mathcal{H}}} + \langle Yf, \, Yg \rangle_{\mathcal{E}} = \sum_{k=1}^{q} \langle [N_k f], \, [N_k g] \rangle_{\widehat{\mathcal{H}}} + \langle Xf, \, Xg \rangle_{\mathcal{E}_*},$$

we conclude that the linear map

$$\mathbf{V}_K : \begin{bmatrix} [E_1 f] \\ \vdots \\ [E_p f] \\ Yf \end{bmatrix} \to \begin{bmatrix} [N_1 f] \\ \vdots \\ [N_q f] \\ Xf \end{bmatrix} \tag{3.9}$$

is an isometry from

$$\mathcal{D}_{\mathbf{V}_K} = \operatorname{Clos} \left\{ \begin{bmatrix} [E_1 f] \\ \vdots \\ [E_p f] \\ Yf \end{bmatrix}, \; f \in \mathcal{H}_0 \right\} \subset \begin{bmatrix} \mathbb{C}^p \otimes \widehat{\mathcal{H}}_K \\ \mathcal{E} \end{bmatrix} \tag{3.10}$$

onto

$$\mathcal{R}_{\mathbf{V}_K} = \operatorname{Clos} \left\{ \begin{bmatrix} [N_1 f] \\ \vdots \\ [N_q f] \\ Xf \end{bmatrix}, \; f \in \mathcal{H}_0 \right\} \subset \begin{bmatrix} \mathbb{C}^q \otimes \widehat{\mathcal{H}}_K \\ \mathcal{E}_* \end{bmatrix}. \tag{3.11}$$

The next two lemmas establish a correspondence between solutions S to Problem 1.2 and unitary extensions of the partially defined isometry $\mathbf{V}_K$ associated with some positive kernel solution K of (2.25)–(2.28) given in (3.6).

Lemma 3.2. *Let S be a solution of Problem 1.2. Then there exists a kernel K satisfying conditions (2.25)–(2.28) such that S is a characteristic function of a unitary colligation*

$$\widetilde{\mathbf{U}} = \begin{bmatrix} \widetilde{A} & \widetilde{B} \\ \widetilde{C} & \widetilde{D} \end{bmatrix} : \begin{bmatrix} \mathbb{C}^p \otimes (\widehat{\mathcal{H}}_K \oplus \widetilde{\mathcal{H}}) \\ \mathcal{E} \end{bmatrix} \to \begin{bmatrix} \mathbb{C}^q \otimes (\widehat{\mathcal{H}}_K \oplus \widetilde{\mathcal{H}}) \\ \mathcal{E}_* \end{bmatrix}, \tag{3.12}$$

which is an extension of the isometry $\mathbf{V}_K$ given in (3.9).

Proof. Let S be a solution to Problem 1.2. In particular, S belongs to $\mathcal{SA}_{\mathbf{P}}(\mathcal{E}, \mathcal{E}_*)$ and by Theorem 1.1, it is the characteristic function of some unitary colligation $\mathcal{C}$ of the form (3.2). In other words, S admits a unitary realization (1.13) with the state space $\mathcal{H}$ and representation (1.9) holds for the function H defined via (1.15) and decomposed as in (1.8). This function is analytic and $\mathcal{L}(\mathbb{C}^p \otimes \mathcal{H}, \mathcal{E}_*)$-valued on $\mathcal{D}_{\mathbf{P}}$ and leads to the following representation

$$S(z) = D + H(z)\mathbf{P}(z)B \tag{3.13}$$

of S, which is equivalent to (1.13).

The interpolation conditions (1.17) and (1.18) which are assumed to be satisfied by S, force certain restrictions on the connecting operator $\mathbf{U} = \begin{bmatrix} A & B \\ C & D \end{bmatrix}$. By Lemma 2.1, the strong limit (2.31) exists. Substituting (3.13) into (1.17) we get

$$\lim_{r \to 1} \left(D^* + B^*\mathbf{P}(r\xi)^* H(r\xi)^*\right) \mathbf{a}(\xi) = \mathbf{c}(\xi) \quad (\xi \in \Omega),$$

where the limit is understood in the strong sense. It also follows from (1.15) that $C + H(z)Z(z)A = H(z)$ and therefore, that (strongly)

$$C^*\mathbf{a}(\xi) + \lim_{r \to 1} A^*\mathbf{P}(r\xi)^* H(r\xi)^*\mathbf{a}(\xi) = \lim_{r \to 1} H(r\xi)^*\mathbf{a}(\xi).$$

By (2.31), the two last (displayed) equalities are equivalent to

$$D^*\mathbf{a}(\xi) + B^*\mathbf{P}(\xi)^*T(\xi) = \mathbf{c}(\xi) \tag{3.14}$$

and

$$C^*\mathbf{a}(\xi) + A^*\mathbf{P}(\xi)^*T(\xi) = T(\xi), \tag{3.15}$$

respectively, which can be written in matrix form as

$$\begin{bmatrix} A^* & C^* \\ B^* & D^* \end{bmatrix} \begin{bmatrix} \mathbf{P}(\xi)^*T(\xi) \\ \mathbf{a}(\xi) \end{bmatrix} = \begin{bmatrix} T(\xi) \\ \mathbf{c}(\xi) \end{bmatrix} \quad (\xi \in \Omega).$$

Since the operator $\begin{bmatrix} A & B \\ C & D \end{bmatrix}$ is unitary, we conclude from the last equality that

$$\begin{bmatrix} A & B \\ C & D \end{bmatrix} \begin{bmatrix} T(\xi) \\ \mathbf{c}(\xi) \end{bmatrix} = \begin{bmatrix} \mathbf{P}(\xi)^*T(\xi) \\ \mathbf{a}(\xi) \end{bmatrix}. \tag{3.16}$$

Let K_{ij} and K be defined as in (2.33) and (2.34), let $\mathbf{V}_K$ be the isometry given in (3.9) and let

$$R(\xi) = \begin{bmatrix} T_1(\xi) & \dots & T_p(\xi) \end{bmatrix} \tag{3.17}$$

so that

$$K(\xi, \mu) = R(\xi)^* R(\mu). \tag{3.18}$$

Define the operator $\mathbf{T}: \mathcal{H}_1 \to \mathcal{H}$ by the rule

$$\mathbf{T}h = \sum_{\xi} R(\xi) h(\xi). \tag{3.19}$$

Upon making subsequent use of (3.8), (3.6), (3.18) and (3.19), we get

$$\begin{aligned} \langle [h],\, [y] \rangle_{\widehat{\mathcal{H}}_K} = D_K(h,\, y) &= \sum_{\xi_i, \xi_\ell} \langle K(\xi_i, \xi_\ell) h(\xi_\ell),\, y(\xi_i) \rangle_{\mathcal{E}} \\ &= \sum_{\xi_i, \xi_\ell} \langle R(\xi_\ell) h(\xi_\ell),\, R(\xi_i) y(\xi_i) \rangle_{\mathcal{H}} \\ &= \left\langle \sum_{\xi_\ell} R(\xi_\ell) h(\xi_\ell),\, \sum_{\xi_i} R(\xi_i) y(\xi_i) \right\rangle_{\mathcal{H}} \\ &= \langle \mathbf{T}h,\, \mathbf{T}y \rangle_{\mathcal{H}}. \end{aligned}$$

Therefore, the linear transformation U defined by the rule

$$U: \ \mathbf{T}f \to [f] \qquad (f \in \mathcal{H}_0) \tag{3.20}$$

can be extended to the unitary map (which still is denoted by U) from $\overline{\operatorname{Ran} \mathbf{T}}$ onto $\widehat{\mathcal{H}}_K$. Noticing that $\overline{\operatorname{Ran} \mathbf{T}}$ is a subspace of $\mathcal{H}$ and setting

$$\mathcal{N} := \mathcal{H} \ominus \overline{\operatorname{Ran} \mathbf{T}} \quad \text{and} \quad \widetilde{\mathcal{H}} := \widehat{\mathcal{H}}_K \oplus \mathcal{N},$$

we define the unitary map $\widetilde{U}: \ \mathcal{H} \to \widetilde{\mathcal{H}}$ by the rule

$$\widetilde{U} g = \begin{cases} Ug & \text{for} \quad g \in \overline{\operatorname{Ran} \mathbf{T}} \\ g & \text{for} \quad g \in \mathcal{N}. \end{cases} \tag{3.21}$$

Introducing the operators

$$\widetilde{A} = (\widetilde{U} \otimes I_q) A (\widetilde{U} \otimes I_p)^*, \quad \widetilde{B} = (\widetilde{U} \otimes I_q) B, \quad \widetilde{C} = C (\widetilde{U} \otimes I_p)^*, \quad \widetilde{D} = D$$

we construct the colligation $\widetilde{\mathcal{C}}$ via (3.3) and (3.12). By definition, $\widetilde{\mathcal{C}}$ is unitarily equivalent to the initial colligation $\mathcal{C}$ defined in (3.1). By Remark 3.1, $\widetilde{\mathcal{C}}$ has the same characteristic function as $\mathcal{C}$, that is, $S(z)$. It remains to check that the connecting operator of $\widetilde{\mathcal{C}}$ is an extension of $\mathbf{V}_K$, i.e.,

$$\begin{bmatrix} \widetilde{A} & \widetilde{B} \\ \widetilde{C} & \widetilde{D} \end{bmatrix} \begin{bmatrix} [E_1 f] \\ \vdots \\ [E_p f] \\ Yf \end{bmatrix} = \begin{bmatrix} [N_1 f] \\ \vdots \\ [N_q f] \\ Xf \end{bmatrix}, \qquad f \in \mathcal{H}_0. \tag{3.22}$$

To this end, note that by (3.20), (3.21), it holds for every $f \in \mathcal{H}_0$ that

$$\widetilde{U}^* ([E_j f]) = \mathbf{T}(E_j f) = \sum_{\xi} R(\xi) E_j f(\xi) = \sum_{\xi} T_j(\xi) f(\xi) \quad (j = 1, \ldots, p)$$

and therefore,

$$(\widetilde{U} \otimes I_p)^* \left(\begin{bmatrix} [E_1 f] \\ \vdots \\ [E_p f] \end{bmatrix} \right) = \sum_{\xi} \begin{bmatrix} T_1(\xi) \\ \vdots \\ T_p(\xi) \end{bmatrix} f(\xi) = \sum_{\xi} T(\xi) f(\xi). \tag{3.23}$$

Similarly,

$$\widetilde{U} \left(\sum_{\xi} R(\xi) N_k(\xi) f(\xi) \right) = \widetilde{U} \mathbf{T}(N_k f) = [N_k f] \quad (k = 1, \ldots, q)$$

and since $R(\xi)N_k(\xi)$ is equal to the k-th block row of $\mathbf{P}(\xi)T(\xi)$ (which is clear from the definitions (3.17), (2.32) and (2.23) of R, T and N_k), it follows that

$$(\widetilde{U} \otimes I_q) \left(\mathbf{P}(\xi)^* T(\xi) f(\xi) \right) = \begin{bmatrix} [N_1 f] \\ \vdots \\ [N_p f] \end{bmatrix}. \tag{3.24}$$

Thus, by (3.16), (3.23) and (3.24),

$$\begin{aligned} \begin{bmatrix} \widetilde{A} & \widetilde{B} \\ \widetilde{C} & \widetilde{D} \end{bmatrix} \begin{bmatrix} [E_1 f] \\ \vdots \\ [E_p f] \\ Y f \end{bmatrix} &= \begin{bmatrix} \widetilde{U} \otimes I_q & 0 \\ 0 & I \end{bmatrix} \begin{bmatrix} A & B \\ C & D \end{bmatrix} \begin{bmatrix} (\widetilde{U} \otimes I_p)^* & 0 \\ 0 & I \end{bmatrix} \begin{bmatrix} [E_1 f] \\ \vdots \\ [E_p f] \\ Y f \end{bmatrix} \\ &= \begin{bmatrix} \widetilde{U} \otimes I_q & 0 \\ 0 & I \end{bmatrix} \begin{bmatrix} A & B \\ C & D \end{bmatrix} \left(\sum_{\xi} \begin{bmatrix} T(\xi) \\ \mathbf{c}(\xi) \end{bmatrix} f(\xi) \right) \\ &= \begin{bmatrix} \widetilde{U} \otimes I_q & 0 \\ 0 & I \end{bmatrix} \left(\sum_{\xi} \begin{bmatrix} A & B \\ C & D \end{bmatrix} \begin{bmatrix} T(\xi) \\ \mathbf{c}(\xi) \end{bmatrix} f(\xi) \right) \\ &= \begin{bmatrix} \widetilde{U} \otimes I_q & 0 \\ 0 & I \end{bmatrix} \left(\sum_{\xi} \begin{bmatrix} \mathbf{P}(\xi)^* T(\xi) \\ \mathbf{a}(\xi) \end{bmatrix} f(\xi) \right) = \begin{bmatrix} [N_1 f] \\ \vdots \\ [N_q f] \\ X f \end{bmatrix}, \end{aligned} \tag{3.25}$$

which proves (3.22) and completes the proof of the lemma. □

The converse statement will be proved in Theorem 3.4 below. We start with some auxiliary results.

Lemma 3.3. *Let $A: \mathbb{C}^p \otimes \mathcal{H} \to \mathbb{C}^q \otimes \mathcal{H}$ be a contraction, and let β be a point in $\partial\mathcal{D}_{\mathbf{P}}$. Then the following limit*

$$\Upsilon_\beta := \lim_{r\to 1}(1-r)(I - A\mathbf{P}(r\beta))^{-1} \tag{3.26}$$

exists in the strong sense and satisfies

$$\Upsilon_\beta \geq 0, \quad A\mathbf{P}(\beta)\Upsilon_\beta = \Upsilon_\beta \quad \text{and} \quad \Upsilon_\beta A = \Upsilon_\beta \mathbf{P}(\beta)^*. \tag{3.27}$$

Furthermore, the following limits also exist in the strong sense:

$$\lim_{r\to 1}(I - A\mathbf{P}(r\beta))^{-1}(I - A\mathbf{P}(\beta)) = I - \Upsilon_\beta \mathbf{P}(\beta)^* W_\beta, \tag{3.28}$$

$$\lim_{r\to 1}(I - \mathbf{P}(\beta)A)(I - \mathbf{P}_\beta(r)A)^{-1} = I - W_\beta \Upsilon_\beta \mathbf{P}(\beta)^*, \tag{3.29}$$

where

$$W_\beta = \lim_{r\to 1}\frac{\mathbf{P}(\beta) - \mathbf{P}(r\beta)}{1-r}. \tag{3.30}$$

Proof. To show the strong convergence of the limit in (3.26), consider a unitary realization (2.12) with the state space $\mathcal{G}_\beta$ of the Schur function $\mathbf{P}_\beta(\zeta)$. By a calculation similar to (2.19),

$$(I - A\mathbf{P}_\beta(\zeta))^{-1} = (I - AD_1)^{-1}\left[I + \zeta AC_1(I - \zeta A_2)^{-1}B_1(I - AD_1)^{-1}\right], \tag{3.31}$$

where A_2 is given in (2.17). Since A_2 is a contraction on $\mathcal{G}_\beta$, the limit

$$\lim_{r\to 1}(1-r)\left(I_{\mathcal{G}_\beta} - rA_2\right)^{-1} = P_{\mathrm{Ker}(I-A_2)}$$

converges strongly to the orthogonal projection of $\mathcal{G}_\beta$ onto the eigenspace $\{x \in \mathcal{G}_\beta : \ A_2 x = x\}$ of fixed vectors of A_2 (for the proof see [10, Section 2]). Then it follows from the two last relations that the limit in (3.26) converges strongly to the operator

$$\Upsilon_\beta = (I - AD_1)^{-1}AC_1 P_{\mathrm{Ker}(I-A_2)}B_1(I - AD_1)^{-1}.$$

To show that this operator satisfies conditions (3.27), consider the function

$$\Phi_\beta(\zeta) = (I - A\mathbf{P}_\beta(\zeta))^{-1}(I + A\mathbf{P}_\beta(\zeta)). \tag{3.32}$$

Since

$$\Phi_\beta(\zeta) + \Phi_\beta(\zeta)^* = 2(I - A\mathbf{P}_\beta(\zeta))^{-1}\left[I - A\mathbf{P}_\beta(\zeta)\mathbf{P}_\beta(\zeta)^*A^*\right](I - \mathbf{P}_\beta(\zeta)^*A^*)^{-1} \geq 0,$$

the real part of Φ_β is positive semidefinite on $\mathbb{D}$ and therefore, Φ admits a Herglotz representation

$$\Phi_\beta(\zeta) = i\Im\Phi_\beta(0) + \int_{\mathbb{T}}\frac{t+\zeta}{t-\zeta}d\Sigma_\beta(t)$$

with a positive operator-valued measure Σ_β. Then a consequence of the Lebesgue Dominated Convergence Theorem is that

$$\lim_{r\to 1}(1-r)\Phi_\beta(r) = 2\Sigma_\beta(\{1\}) \tag{3.33}$$

where $\Sigma_\beta(\{1\}) \geq 0$ is the measure assigned by Σ_β at the point $t = 1$. Therefore,

$$\lim_{r\to 1}(1-r)\frac{\Phi_\beta(r)+I}{2} = \Sigma_\beta(\{1\}). \tag{3.34}$$

By (3.32),

$$\frac{\Phi_\beta(r)+I}{2} = (I - A\mathbf{P}_\beta(r))^{-1} = (I - A\mathbf{P}(r\beta))^{-1}$$

and upon comparing (3.26) with (3.34) we conclude that the limit in (3.26) exists and Υ_β is given by

$$\Upsilon_\beta = \Sigma_\beta(\{1\}) \geq 0. \tag{3.35}$$

From (3.32), (3.33) and (3.35) we see that

$$\lim_{r\to 1}(1-r)(I - A\mathbf{P}(r\beta))^{-1}(I + A\mathbf{P}(r\beta)) = 2\Upsilon_\beta,$$

which together with (3.26) implies

$$\lim_{r\to 1}(1-r)(I - A\mathbf{P}(r\beta))^{-1}A\mathbf{P}(r\beta) = \lim_{r\to 1}(1-r)A\mathbf{P}(r\beta)(I - A\mathbf{P}(r\beta))^{-1} = \Upsilon_\beta.$$

The limits in the latter relations can be split into products which leads us to

$$A\mathbf{P}(\beta)\Upsilon_\beta = \Upsilon_\beta A\mathbf{P}(\beta) = \Upsilon_\beta.$$

Multiplying the second equality by $\mathbf{P}(\beta)^*$ and taking into account that $\mathbf{P}(\beta)\mathbf{P}(\beta)^* = I$ we come to

$$\Upsilon_\beta A = \Upsilon_\beta \mathbf{P}(\beta)^*,$$

which completes the proof of (3.27).

Furthermore, on account of (3.26) and the third relation in (3.27),

$$\begin{aligned}\lim_{r\to 1}(I - A\mathbf{P}(r\beta))^{-1}(I - A\mathbf{P}(\beta)) &= I - \lim_{r\to 1}(1-r)(I - A\mathbf{P}(r\beta))^{-1}A\frac{\mathbf{P}(\beta)-\mathbf{P}(r\beta)}{1-r}\\ &= I - \Upsilon_\beta A W_\beta\\ &= I - \Upsilon_\beta \mathbf{P}(\beta)^* W_\beta\end{aligned}$$

and quite similarly,

$$\begin{aligned}\lim_{r\to 1}(I - \mathbf{P}(\beta)A)(I - \mathbf{P}(r\beta)A)^{-1} &= I - \lim_{r\to 1}\frac{\mathbf{P}(\beta)-\mathbf{P}(r\beta)}{1-r}(1-r)(I - A\mathbf{P}(r\beta))^{-1}A\\ &= I - W_\beta\Upsilon_\beta A\\ &= I - W_\beta\Upsilon_\beta\mathbf{P}(\beta)^*,\end{aligned}$$

which completes the proof of the lemma. □

Lemma 3.4. *Let K be a kernel on Ω satisfying conditions* (2.25)–(2.28) *and let* $\widetilde{\mathbf{U}}$ *of the form* (3.12) *be a unitary extension of the partially defined isometry* $\mathbf{V}_K$ *given in* (3.9). *Then the characteristic function S of the unitary colligation* $\widetilde{\mathcal{C}} = \{\widehat{\mathcal{H}} \oplus \widetilde{\mathcal{H}},\ \mathcal{E},\ \mathcal{E}_*,\ \widetilde{\mathbf{U}}\}$,

$$S(z) = \widetilde{D} + \widetilde{C}\left(I_{\mathbb{C}^p\otimes(\widehat{\mathcal{H}}\oplus\widetilde{\mathcal{H}})} - \mathbf{P}(z)\widetilde{A}\right)^{-1}\mathbf{P}(z)\widetilde{B},$$

is a solution to Problem 1.2.

Proof. We start with a factorization of the form (3.18) for the kernel K

$$K(\xi,\mu)=R(\xi)^*R(\mu) \text{ with } R(\xi)=\begin{bmatrix}T_1(\xi) & \dots & T_p(\xi)\end{bmatrix}\in\mathcal{L}(\mathbb{C}^p\otimes\mathcal{E}_L,\mathcal{H})$$

and then set

$$T(\xi)=\begin{bmatrix}T_1(\xi)\\ \vdots \\ T_d(\xi)\end{bmatrix}\in\mathcal{L}(\mathcal{E}_L,\mathbb{C}^p\otimes\mathcal{H}).$$

By the assumption that $\widetilde{U}$ is a unitary map of the form (3.12) which extends $\mathbf{V}_K$, by reversing the argument in the proof of Lemma 3.2 we see that the operator $\mathbf{U}$ defined by

$$\mathbf{U}=\begin{bmatrix}A & B\\ C & D\end{bmatrix}=\begin{bmatrix}(\widetilde{U}\otimes I_q)^* & 0\\ 0 & I\end{bmatrix}\begin{bmatrix}\widetilde{A} & \widetilde{B}\\ \widetilde{C} & \widetilde{D}\end{bmatrix}\begin{bmatrix}\widetilde{U}\otimes I_p & 0\\ 0 & I\end{bmatrix}$$

satisfies (3.16) (or equivalently, (3.14) and (3.15)), which can be easily seen from (3.25). By Remark 3.1, the colligations $\mathcal{C}$ and $\widetilde{\mathcal{C}}$ defined in (3.1) and (3.3) have the same characteristic functions and thus, S can be taken in the form (1.13). Let $H(z)$ be defined as in (1.15) and decomposed as in (1.8). We shall use the representation (3.13) of $S(z)$ which is equivalent to (1.13).

Fix $\xi\in\Omega\subseteq\partial\mathcal{D}_{\mathbf{P}}$ and consider the Schur function $\mathbf{P}_\xi(\zeta)=\mathbf{P}(\zeta\xi)$, $(\zeta\in\mathbb{D})$. Then it follows from (3.26) and (3.30) that

$$\lim_{r\to 1}\left(\mathbf{P}(\xi)-\mathbf{P}(r\xi)\right)(I-A\mathbf{P}(r\xi))^{-1}=W_\xi\Upsilon_\xi. \tag{3.36}$$

Since the operator $\mathbf{U}$ is unitary, we have in particular,

$$I-AA^*=BB^*. \tag{3.37}$$

Combining the second and the third relations in (3.27) we conclude that

$$\Upsilon_\xi=\Upsilon_\xi\mathbf{P}(\xi)^*A^*=\Upsilon_\xi AA^*. \tag{3.38}$$

It follows now from (3.38) by (3.37) that $\Upsilon_\xi BB^*=0$ and therefore, that

$$\Upsilon_\xi B=0. \tag{3.39}$$

Multiplying both parts in (3.36) by B on the right we and taking into account (3.39) we get

$$\lim_{r\to 1}\left(\mathbf{P}(\xi)-\mathbf{P}(r\xi)\right)(I-A\mathbf{P}(r\xi))^{-1}B=0. \tag{3.40}$$

Using (1.13) and expressions for $D^*\mathbf{a}(\xi)$ and $C^*\mathbf{a}(\xi)$ derived from (3.14) and (3.15), respectively, we get

$$\begin{aligned}S(r\xi)^*\mathbf{a}(\xi)&=D^*\mathbf{a}(\xi)+B^*\mathbf{P}(r\xi)^*\left(I-A^*\mathbf{P}(r\xi)^*\right)^{-1}C^*\mathbf{a}(\xi)\\&=\mathbf{c}(\xi)-B^*\mathbf{P}(\xi)^*T(\xi)\\&\quad+B^*\mathbf{P}(r\xi)^*\left(I-A^*\mathbf{P}(r\xi)^*\right)^{-1}\left(I-A^*\mathbf{P}(\xi)^*\right)T(\xi)\\&=\mathbf{c}(\xi)-B^*\left(I-\mathbf{P}(r\xi)^*A^*\right)^{-1}(\mathbf{P}(\xi)^*-\mathbf{P}(r\xi)^*)T(\xi).\end{aligned}\tag{3.41}$$

Taking limits in the last identity as r tends to one and taking into account (3.40), we come to (1.17).

It remains to verify (1.18). By (3.41),

$$\mathbf{c}(\xi)^*\mathbf{c}(\xi) - \mathbf{c}(\xi)^*S(r\xi)^*\mathbf{a}(\xi) = \mathbf{c}(\xi)^*B^*\left(I - \mathbf{P}(r\xi)^*A^*\right)^{-1}\left(\mathbf{P}(\xi)^* - \mathbf{P}(r\xi)^*\right)T(\xi). \tag{3.42}$$

It follows from (3.16) that

$$AT(\xi) + B\mathbf{c}(\xi) = \mathbf{P}(\xi)^*T(\xi)$$

and therefore, since $\mathbf{P}(\xi)\mathbf{P}(\xi)^* = I$,

$$\mathbf{c}(\xi)^*B^* = T(\xi)^*(\mathbf{P}(\xi) - A^*) = T(\xi)^*\mathbf{P}(\xi)\left(I - \mathbf{P}(\xi)^*A^*\right).$$

Substituting the latter equality into (3.42) and dividing both parts of the resulting equality by $(1-r)$, we get

$$\begin{aligned}&\frac{\mathbf{c}(\xi)^*\mathbf{c}(\xi) - \mathbf{c}(\xi)^*S(r\xi)^*\mathbf{a}(\xi)}{1-r}\\ &= T(\xi)^*\mathbf{P}(\xi)\left(I - \mathbf{P}(\xi)^*A^*\right)\left(I - \mathbf{P}(r\xi)^*A^*\right)^{-1}\frac{\mathbf{P}(\xi)^* - \mathbf{P}(r\xi)^*}{1-r}T(\xi).\end{aligned}$$

Taking limits in the last identity as $r \to 1$ and using (3.28) we get

$$\lim_{r\to 1}\frac{\mathbf{c}(\xi)^*\mathbf{c}(\xi) - \mathbf{c}(\xi)^*S(r\xi)^*\mathbf{a}(\xi)}{1-r} = T(\xi)^*\mathbf{P}(\xi)\left(I - W_\xi^*\mathbf{P}(\xi)\Upsilon_\xi\right)W_\xi^*T(\xi). \tag{3.43}$$

The operator $\Lambda_{\mathbf{P}}(\xi) \geq 0$ given in (2.7), admits, by (2.10), representations

$$\Lambda_{\mathbf{P}}(\xi) = W_\xi\mathbf{P}(\xi)^* = \mathbf{P}(\xi)W_\xi^*. \tag{3.44}$$

Setting for short

$$U_\xi := \mathbf{P}(\xi)\Upsilon_\xi W_\xi^* \tag{3.45}$$

and making use of (3.44) we rewrite (3.43) as

$$\lim_{r\to 1}\frac{\mathbf{c}(\xi)^*\mathbf{c}(\xi) - \mathbf{c}(\xi)^*S(r\xi)^*\mathbf{a}(\xi)}{1-r} = T(\xi)^*\Lambda_{\mathbf{P}}(\xi)(I - U_\xi)T(\xi). \tag{3.46}$$

On the other hand, it follows from (1.14) that

$$\begin{aligned}&\mathbf{a}(\xi)^*\frac{I_{\mathcal{E}_*} - S(r\xi)S(r\xi)^*}{1-r^2}\mathbf{a}(\xi)\\ &= \mathbf{a}(\xi)^*C\left(I - \mathbf{P}(r\xi))A\right)^{-1}\frac{I - \mathbf{P}(r\xi)\mathbf{P}(r\xi)^*}{1-r^2}\left(I - A^*\mathbf{P}(r\xi)^*\right)^{-1}C^*\mathbf{a}(\xi).\end{aligned} \tag{3.47}$$

It follows from (3.15) that

$$C^*\mathbf{a}(\xi) = (I - A^*\mathbf{P}(\xi)^*)T(\xi),$$

which being substituted into (3.47), leads to

$$\begin{aligned}\mathbf{a}(\xi)^*\frac{I_{\mathcal{E}_*}-S(r\xi)S(r\xi)^*}{1-r^2}\mathbf{a}(\xi) &= T(\xi)^*(I-\mathbf{P}(\xi)A)\left(I-\mathbf{P}(r\xi))A\right)^{-1}\\ &\quad\times\frac{I-\mathbf{P}(r\xi)\mathbf{P}(r\xi)^*}{1-r^2}\\ &\quad\times\left(I-A^*\mathbf{P}(r\xi)^*\right)^{-1}(I-A^*\mathbf{P}(\xi)^*)T(\xi).\end{aligned}$$

Taking limits in the last identity as $r\to 1$ and using (3.29), (2.24) and (3.45) we get

$$\lim_{r\to 1}\mathbf{a}(\xi)^*\frac{I_{\mathcal{E}_*}-S(r\xi)S(r\xi)^*}{1-r^2}\mathbf{a}(\xi)=T(\xi)^*\left(I-U_\xi^*\right)\Lambda_{\mathbf{P}}(\xi)\left(I-U_\xi\right)T(\xi). \quad (3.48)$$

By Lemma 2.1,

$$\lim_{r\to 1}\mathbf{a}(\xi)^*\frac{I_{\mathcal{E}_*}-S(r\xi)S(r\xi)^*}{1-r^2}\mathbf{a}(\xi)=\lim_{r\to 1}\frac{\mathbf{c}(\xi)^*\mathbf{c}(\xi)-\mathbf{c}(\xi)^*S(r\xi)^*\mathbf{a}(\xi)}{1-r},$$

which implies, on account of (3.46) and (3.48), that

$$T(\xi)^*\Lambda_{\mathbf{P}}(\xi)(I-U_\xi)T(\xi)=T(\xi)^*\left(I-U_\xi^*\right)\Lambda_{\mathbf{P}}(\xi)\left(I-U_\xi\right)T(\xi).$$

The last equality implies

$$T(\xi)^*\left(I-U_\xi^*\right)\Lambda_{\mathbf{P}}(\xi)\left(I-U_\xi\right)T(\xi)=T(\xi)^*\left(\Lambda_{\mathbf{P}}(\xi)-U_\xi^*\Lambda_{\mathbf{P}}(\xi)U_\xi\right)T(\xi)$$

and now we conclude from (3.48) and (2.26), that for every $\xi\in\Omega$,

$$\begin{aligned}\lim_{r\to 1}\mathbf{a}(\xi)^*\frac{I_{\mathcal{E}_*}-S(r\sigma(\xi))S(r\sigma(\xi))^*}{1-r^2}\mathbf{a}(\xi) &= T(\xi)^*\left(\Lambda_{\mathbf{P}}(\xi)-U_\xi^*\Lambda_{\mathbf{P}}(\xi)U_\xi\right)T(\xi)\\ &\le T(\xi)^*\Lambda_{\mathbf{P}}(\xi)T(\xi)\le\Psi(\xi),\end{aligned}$$

where we used the assumption (2.26) for the last step. This proves (1.18) and completes the proof of the theorem. □

4. The universal unitary colligation associated with the interpolation problem

A general result of Arov and Grossman (see [7], [8]) describes how to parametrize the set of all unitary extensions of a given partially defined isometry $\mathbf{V}$. Their result has been extended to the multivariable case in [16] (for the case of the polydisk), in [17] (for the case of the unit ball) and in [12] for $\mathbf{P}$-colligations. We recall the result from [12] for the reader's convenience.

We assume that we are given an interpolation data set $(\mathbf{a},\mathbf{c},\Psi)$ as in (1.16) and a kernel $K(\xi,\mu)$ on $\Omega\times\Omega$ satisfying the conditions (2.25)–(2.28) as in the previous section. Let $\mathbf{V}_K\colon\mathcal{D}_{\mathbf{V}_K}\to\mathcal{R}_{\mathbf{V}_K}$ with $\mathcal{D}_{\mathbf{V}_K}$ and $\mathcal{R}_{\mathbf{V}_K}$ given as in (3.10) and (3.11) be the isometry associated with K as in (3.9). Introduce the defect spaces

$$\Delta_K=\begin{bmatrix}\mathbb{C}^p\otimes\widehat{\mathcal{H}}\\ \mathcal{E}\end{bmatrix}\ominus\mathcal{D}_{\mathbf{V}_K}\quad\text{and}\quad\Delta_{K*}=\begin{bmatrix}\mathbb{C}^q\otimes\widehat{\mathcal{H}}\\ \mathcal{E}_*\end{bmatrix}\ominus\mathcal{R}_{\mathbf{V}_K}$$

and let $\widetilde{\Delta}_K$ be another copy of Δ_K and $\widetilde{\Delta}_{K*}$ be another copy of Δ_{K*} with unitary identification maps

$$i_K\colon \Delta_K \to \widetilde{\Delta}_K \quad \text{and} \quad i_{K*}\colon \Delta_{K*} \to \widetilde{\Delta}_{K*}. \tag{4.1}$$

Define a unitary operator $\mathbf{U}_{K,0}$ from $\mathcal{D}_{\mathbf{V}_K} \oplus \Delta_K \oplus \widetilde{\Delta}_{K*}$ onto $\mathcal{R}_{\mathbf{V}_K} \oplus \Delta_{K*} \oplus \widetilde{\Delta}_K$ by the rule

$$\mathbf{U}_{K,0}x = \begin{cases} \mathbf{V}_K x, & \text{if } x \in \mathcal{D}_{\mathbf{V}_K} \\ i_K(x) & \text{if } x \in \Delta_K, \\ i_{K*}^{-1}(x) & \text{if } x \in \widetilde{\Delta}_{K*}. \end{cases} \tag{4.2}$$

Identifying $\begin{bmatrix} \mathcal{D}_{\mathbf{V}_K} \\ \Delta_K \end{bmatrix}$ with $\begin{bmatrix} \mathbb{C}^p \otimes \widehat{\mathcal{H}} \\ \mathcal{E} \end{bmatrix}$ and $\begin{bmatrix} \mathcal{R}_{\mathbf{V}_K} \\ \Delta_{K*} \end{bmatrix}$ with $\begin{bmatrix} \mathbb{C}^q \otimes \widehat{\mathcal{H}} \\ \mathcal{E}_* \end{bmatrix}$, we decompose $\mathbf{U}_{K,0}$ defined by (4.2) according to

$$\mathbf{U}_{K,0} = \begin{bmatrix} U_{K,11} & U_{K,12} & U_{K,13} \\ U_{K,21} & U_{K,22} & U_{K,23} \\ U_{K,31} & U_{K,32} & 0 \end{bmatrix} : \begin{bmatrix} \mathbb{C}^p \otimes \widehat{\mathcal{H}} \\ \mathcal{E} \\ \widetilde{\Delta}_{K*} \end{bmatrix} \to \begin{bmatrix} \mathbb{C}^q \otimes \widehat{\mathcal{H}} \\ \mathcal{E}_* \\ \widetilde{\Delta}_K \end{bmatrix}. \tag{4.3}$$

The $(3,3)$ block in this decomposition is zero, since (by definition (4.2)), for every $x \in \widetilde{\Delta}_{K*}$, the vector $\mathbf{U}_{K,0}x$ belongs to Δ_K, which is a subspace of $\begin{bmatrix} \widehat{\mathcal{H}} \\ \mathcal{E}_* \end{bmatrix}$ and therefore, is orthogonal to $\widetilde{\Delta}_K$ (in other words $\mathbf{P}_{\widetilde{\Delta}_K}\mathbf{U}_{K,0}|_{\widetilde{\Delta}_{K*}} = 0$, where $\mathbf{P}_{\widetilde{\Delta}_K}$ stands for the orthogonal projection of $\mathcal{R}_{\mathbf{V}_K} \oplus \Delta_{K*} \oplus \widetilde{\Delta}_K$ onto $\widetilde{\Delta}_K$).

The unitary operator $\mathbf{U}_{K,0}$ is the connecting operator of the unitary colligation

$$\mathcal{C}_{K,0} = \left\{\widehat{\mathcal{H}}, \begin{bmatrix} \mathcal{E} \\ \widetilde{\Delta}_{K*} \end{bmatrix}, \begin{bmatrix} \mathcal{E}_* \\ \widetilde{\Delta}_K \end{bmatrix}, \mathbf{U}_{K,0}\right\}, \tag{4.4}$$

which is called *the universal unitary colligation* associated with the kernel K satisfying the necessary conditions (2.25)–(2.28) for existence of solutions of the interpolation problem.

According to (3.4), the characteristic function of the $\mathbf{P}$-colligation $\mathcal{C}_{K,0}$ defined in (4.4) is given by

$$\begin{aligned} \Sigma_K(z) &= \begin{bmatrix} \Sigma_{K,11}(z) & \Sigma_{K,12}(z) \\ \Sigma_{K,21}(z) & \Sigma_{K,22}(z) \end{bmatrix} \\ &= \begin{bmatrix} U_{K,22} & U_{K,23} \\ U_{K,32} & 0 \end{bmatrix} + \begin{bmatrix} U_{K,21} \\ U_{K,31} \end{bmatrix} (I - \mathbf{P}(z)U_{K,11})^{-1}\mathbf{P}(z)\begin{bmatrix} U_{K,12} & U_{K,13} \end{bmatrix} \end{aligned} \tag{4.5}$$

and belongs to the class $\mathcal{S}_{\mathbf{P}}(\mathcal{E} \oplus \widetilde{\Delta}_{K*},\ \mathcal{E}_* \oplus \widetilde{\Delta}_K)$, by Theorem 1.1.

Theorem 4.1. *Let $\mathbf{V}_K$ be the isometry defined in (3.9) associated with a positive kernel K meeting the necessary conditions (2.25)–(2.28) for existence of solutions to the interpolation Problem 1.2, let Σ_K be the function constructed as above from $\mathbf{V}_K$, and let S be a $\mathcal{L}(\mathcal{E}, \mathcal{E}_*)$-valued function. Then the following are equivalent:*

1. *S is a solution of Problem 1.2.*
2. *There exists a kernel K satisfying conditions* (2.25)–(2.28) *so that S is the characteristic function of a* $\mathbf{P}$*-colligation* $\mathcal{C} = \{\widehat{\mathcal{H}} \oplus \widetilde{\mathcal{H}},\ \mathcal{E},\ \mathcal{E}_*,\ \mathbf{U}\}$ *with the connecting operator* $\mathbf{U}$ *being a unitary extension of* $\mathbf{V}_K$.
3. *There exists a positive kernel K satisfying conditions* (2.25)–(2.28) *such that S is of the form*

$$S(z) = \Sigma_{K,11}(z) + \Sigma_{K,12}(z)\left(I_{\widetilde{\Delta}_*} - \mathcal{T}(z)\Sigma_{K,22}(z)\right)^{-1}\mathcal{T}(z)\Sigma_{K,21}(z) \tag{4.6}$$

for a function $\mathcal{T}$ from the class $\mathcal{S}A_{\mathbf{P}}(\widetilde{\Delta}_K,\ \widetilde{\Delta}_{K*})$.

Proof. The equivalence **1** $\Leftrightarrow$ **2** follows by Lemmas 3.2 and 3.3. For the proof of **2** $\Leftrightarrow$ **3** (for a given fixed choice of K) see [12, Theorem 6.1]. □

As a corollary we obtain the sufficiency part of Theorem 2.2: under assumptions (2.25)–(2.28) the set of all solutions of Problem 1.2 is parametrized by formula (4.6) as one sweeps through all functions $\mathcal{T}$ from the class $\mathcal{S}A_{\mathbf{P}}(\widetilde{\Delta}_K,\ \widetilde{\Delta}_{K*})$ and through all positive kernels K meeting the conditions (2.25)–(2.28), and hence in particular is nonempty.

References

[1] J. Agler. *On the representation of certain holomorphic functions defined on a polydisk*, in *Topics in Operator Theory: Ernst D. Hellinger memorial Volume* (Ed. L. de Branges, I. Gohberg and J. Rovnyak), volume 48 of *Operator Theory: Advances and Applications*, pages 47–66. Birkhäuser Verlag, Basel, 1990, pp. 47–66.

[2] J. Agler. Some interpolation theorems of Nevanlinna-Pick type. Preprint.

[3] J. Agler and J.E. McCarthy, *Nevanlinna-Pick interpolation on the bidisk*, J. Reine Angew. Math. 506 (1999), 191–204.

[4] J. Agler and J.E. McCarthy, *Complete Nevanlinna-Pick kernels*, J. Funct. Anal., **175** (2000), 111–124.

[5] C.-G. Ambrozie and D. Timotin. *A von Neumann type inequality for certain domains in* $\mathbb{C}^n$, Proc. Amer. Math. Soc., **2003** (2003), 859–869.

[6] C. Ambrozie and J. Eschmeier, A commutant lifting theorem on analytic polyhedra, preprint, 2002.

[7] D.Z. Arov and L.Z. Grossman, *Scattering matrices in the theory of unitary extensions of isometric operators*, Soviet Math. Dokl., **270** (1983), 17–20.

[8] D.Z. Arov and L.Z. Grossman, *Scattering matrices in the theory of unitary extensions of isometric operators*, Math. Nachr., **157** (1992), 105–123.

[9] J.A. Ball, *Interpolation problems of Pick-Nevanlinna and Loewner type for meromorphic matrix functions*, Integral Equations Operator Theory, **6** (1983), 804–840.

[10] J.A. Ball and V. Bolotnikov, *A bitangential interpolation problem on the closed unit ball for multipliers of the Arveson space*, Integral Equations Operator Theory, **46** (2003), no. 2, 125–164.

[11] J.A. Ball and V. Bolotnikov, *A tangential interpolation problem on the distinguished boundary of the polydisk for the Schur-Agler class*, J. Math. Anal. Appl., **273** (2002), no. 2, 328–348.

[12] J. Ball and V. Bolotnikov, *Realization and interpolaton for Schur-Agler-class functions on domains with matrix polynomial defining function in* $\mathbb{C}^n$, J. Funct. Anal., to appear.

[13] J.A. Ball, I. Gohberg and L. Rodman, *Interpolation of Rational Matrix Functions*, Birkhäuser Verlag, Basel, 1990.

[14] J. Ball and J.W. Helton, *Interpolation problems of Pick-Nevanlinna and Loewner types for meromorphic matrix-functions: parametrization of the set of all solutions*, Integral Equations Operator Theory, **9** (1986), 155–203.

[15] J.A. Ball, W.S. Li, D. Timotin and T.T. Trent, *A commutant lifting theorem on the polydisc*, Indianan University Mathematics Journal **48** (1999), 653–675.

[16] J. Ball and T. Trent, *Unitary colligations, reproducing kernel Hilbert spaces and Nevanlinna-Pick interpolation in several variables*, J. Funct. Anal., **157** (1998), no.1, 1–61.

[17] J.A. Ball, T.T. Trent and V. Vinnikov, *Interpolation and commutant lifting for multipliers on reproducing kernels Hilbert spaces*, in: Operator Theory and Analysis: The M.A. Kaashoek Anniversary Volume (Workshop in Amsterdam, Nov. 1997), pages 89–138, **OT 122**, Birkhäuser Verlag, Basel-Boston-Berlin, 2001.

[18] V. Bolotnikov, *Interpolation for multipliers on reproducing kernel Hilbert spaces*, Proc. Amer. Math. Soc., **131** (2003), 1373–1383.

[19] V. Bolotnikov and H. Dym, *On boundary interpolation for matrix Schur functions*, Mem. Amer. Math. Soc., to appear.

[20] L. de Branges and J. Rovnyak, Canonical models in quantum scattering theory, in *Perturbation Theory and its Applications in Quantum Mechanics* (Ed. C.H. Wilcox), Wiley, New York, 1966, pp. 295–392.

[21] M.S. Brodskiĭ, *Triangular and Jordan Representations of Linear Operators*, Volume Thirty-Two, Translations of Mathematical Monographs, American Mathematical Society, Providence, 1971.

[22] C. Cowen and B. MacCluer, *Composition Operators on Spces of Analytic Functions*, CRC Press, Boca Raton-New York-London-Tokyo, 1995.

[23] R.E. Curto, Applications of several complex variables to multiparameter spectral theory, , *Surveys of some recent resutls in operator theory, Volume II* (Ed. J.B. Conway and B.B. Morrel), pp. 25–90, Longman Scientific & Technical, Essex, UK, 1988.

[24] K.R. Davidson and D.R. Pitts, *Nevanlinna-Pick interpolation for non-commutative analytic Toeplitz algebras*, Integral Equations Operator Theory **31** (1998), no. 3, 321–337.

[25] P. Dewilde and H. Dym, *Lossless inverse scattering, digital filters, and estimation theory*, IEEE Trans. Inform. Theory, **30** (1984), no. 4, 644–662.

[26] H. Dym, *J contractive matrix functions, reproducing kernel spaces and interpolation*, CBMS Lecture Notes, vol. 71, Amer. Math. Soc., Rhodes Island, 1989.

[27] D.R. Georgijević, *Solvability condition for a boundary value interpolation problem of Loewner type*, J. Anal. Math. **74** (1998), 213–234.

[28] I. Gohberg (ed.), *I. Schur Metods in Operator Theory and Signal Processing*, OT18 Birkhäuser Verlag, Basel-Boston, 1986.

[29] J.W. Helton, *The characteristic functions of operator theory and electrical network realization*, Indiana Univ. Math. J. **22** (1972/73), 403–414.

[30] A. Kheifets, *The abstract interpolation problem and applications*, in: *Holomorphic spaces* (Ed. D. Sarason, S. Axler, J. McCarthy), pages 351–379, Cambridge Univ. Press, Cambridge, 1998.

[31] I.V. Kovalishina, Loewner problem in the sight of J-theory of analytic matrix functions. in: *Analysis in infinite-dimensional spaces and operator theory*, pp. 87–97 Naukova-Dumka, Kiev, 1983 (Edited by V.A. Marchenko).

[32] S. McCullough and T.T. Trent, *Invariant subspaces and Nevanlinna-Pick kernels*, J. Functional Analysis **178** (2000), 226–249.

[33] B. Sz.-Nagy and C. Foiaş, *Harmonic Analysis of Operators on Hilbert Space*, North Holland/American Elsevier, 1970.

[34] N.K. Nikol'skiĭ, *Treatis on the Shift Operator: Spectral Function Theory*, Springer-Verlag, Berlin, 1986.

[35] G. Popescu, *Interpolation problems in several variables*, J. Math. Anal. Appl., **227** (1998), 227–250.

[36] P. Quiggin, *For which reproducing kernel Hilbert spaces is Pick's theorem true?* Integral Equations Operator Theory **16** (1993), no. 2, 244–266.

[37] D. Sarason, *Nevanlinna-Pick interpolation with boundary data*, Integral Equations Operator Theory, **30** (1998), 231–250.

[38] A.T. Tomerlin, *Products of Nevanlinna-Pick kernels and operator colligations*, Integral Equations Operator Theory **38** (2000), no. 3, 350–356.

Joseph A. Ball
Department of Mathematics
Virginia Polytechnic Institute
Blacksburg, VA 24061-0123, USA
e-mail: ball@math.vt.edu

Vladimir Bolotnikov
Department of Mathematics
The College of William and Mary
Williamsburg, VA 23187-8795, USA
e-mail: vladi@math.wm.edu

Operator Theory:
Advances and Applications, Vol. 149, 133–138

On Realizations of Rational Matrix Functions of Several Variables III

M.F. Bessmertnyĭ*

Abstract. It is shown that any rational scalar positive real (in the sense of [1]) function of several variables, having degree two of its numerator polynomial, admits the representation in the form of Schur complement of a block of some linear homogeneous matrix bundle with real positive semidefinite matrix coefficients.

1. Preliminaries

Recall the necessary notations, definitions and facts from the first part [1] of this work (note that the present paper does not depend on the second part [2] of the author's Ph. D. thesis which has been translated recently by V. Katsnelson). Let $\mathbb{C}^n = \mathbb{C} \times \cdots \times \mathbb{C}$ denotes the direct product of n copies of the complex plane $\mathbb{C}$. If $z = (z_1, \ldots, z_n) \in \mathbb{C}^n$ then $\overline{z} := (\overline{z}_1, \ldots, \overline{z}_n)$. If A is a matrix with complex entries, then A' is its transpose, $\overline{A}$ is the matrix with complex conjugate entries, and $A^* := \overline{A'} = (\overline{A})'$.

Let $A_0, A_1, \ldots, A_n$ be square matrices with complex entries. The linear matrix bundle $A(z) := A_0 + z_1 A_1 + \cdots + z_n A_n$ is said to be *non-singular* if the matrix $A(z^0)$ is invertible for at least one point $z^0 \in \mathbb{C}^n$. If the linear matrix bundle $A(z)$ is non-singular, then $A(z)^{-1}$ is well-defined for all $z \in \mathbb{C}^n$ except some "thin" subset (which is a zero-set of a non-zero polynomial $\det A(z)$). In this case the matrix-valued function $A(z)^{-1}$ is said to be the *long resolvent* of the linear matrix bundle $A(z) = A_0 + z_1 A_1 + \cdots z_n A_n$.

Let k and N be natural numbers, $k \leq N$. We use the $k \times N$ matrix π and its adjoint π^*:

$$\pi = \begin{bmatrix} I_k & 0 \end{bmatrix}, \quad \pi^* = \begin{bmatrix} I_k \\ 0 \end{bmatrix}, \tag{1.1}$$

Received by the editors February 21, 2003. Submitted by L. Rodman.
Mathematics Subject Classification (2000). Primary 47A48; Secondary 26C15, 47A56.
Key words and phrases. Rational, positive real, functions of degree two, long resolvent representations.

* This paper is a translation from the Russian, prepared by D. Kalyuzhnyĭ-Verbovetzkiĭ, of a part of the second chapter from the author's Ph. D. thesis entitled "Functions of Several Variables in the Theory of Finite Linear Structures", Kharkov, 1982. The manuscript, entitled "Realization of Rational Functions of Several Variables by Means of a Long Resolvent" was deposited at VINITI (86 pages, submitted by the Kharkov University, July 8, 1981, No. 3352-81).

where I_k is the identity $k \times k$ matrix. Let $f(z) = f(z_1, \ldots, z_n)$ be a rational $k \times k$ matrix-valued function, which is *non-degenerate* in the sense that $\det f(z) \not\equiv 0$. Then the representation of $f(z)$ in the form

$$f(z) = (\pi A(z)^{-1} \pi^*)^{-1}, \tag{1.2}$$

where π is defined in (1.1), and

$$A(z) = A_0 + z_1 A_1 + \cdots + z_n A_n \tag{1.3}$$

is a linear $N \times N$ matrix bundle, is said to be a *long resolvent representation.* Consider the block decomposition of the matrix-valued function (1.3):

$$A(z) = \begin{bmatrix} A_{11}(z) & A_{12}(z) \\ A_{21}(z) & A_{22}(z) \end{bmatrix}, \tag{1.4}$$

where the block $A_{11}(z)$ is of size $k \times k$. If the linear matrix bundle $A_{22}(z)$ is non-singular, then using the formula for the inverse of a block matrix, we obtain

$$f(z) = A_{11}(z) - A_{12}(z) A_{22}(z)^{-1} A_{21}(z), \tag{1.5}$$

that is equivalent to (1.2) in this case (in other words, in (1.2) and (1.5) $f(z)$ is represented in the form of Schur complement of the block entry $A_{22}(z)$ of the linear matrix bundle $A(z)$).

In the sequel we use the following subsets of $\mathbb{C}^n$:

$$\begin{aligned}
\mathcal{D}_R^+ &= \{z : \ z \in \mathbb{C}^n, \operatorname{Re} z_1 > 0, \ldots, \operatorname{Re} z_n > 0\}, \\
\mathcal{D}_R^- &= \{z : \ z \in \mathbb{C}^n, \operatorname{Re} z_1 < 0, \ldots, \operatorname{Re} z_n < 0\}, \\
\mathcal{D}_J^+ &= \{z : \ z \in \mathbb{C}^n, \operatorname{Im} z_1 > 0, \ldots, \operatorname{Im} z_n > 0\}, \\
\mathcal{D}_J^- &= \{z : \ z \in \mathbb{C}^n, \operatorname{Im} z_1 < 0, \ldots, \operatorname{Im} z_n < 0\}.
\end{aligned}$$

A rational matrix-valued function $f(z) = f(z_1, \ldots, z_n)$ of the complex variables $z_1, \ldots, z_n$ is said to be *positive* if the following *positivity conditions* hold:

$$\begin{aligned}
f(z) + f(z)^* &\geq 0 \quad \text{for} \quad z \in \mathcal{D}_R^+, \\
f(z) + f(z)^* &\leq 0 \quad \text{for} \quad z \in \mathcal{D}_R^-, \\
i\,(f(z)^* - f(z)) &\geq 0 \quad \text{for} \quad z \in \mathcal{D}_J^+, \\
i\,(f(z)^* - f(z)) &\leq 0 \quad \text{for} \quad z \in \mathcal{D}_J^-.
\end{aligned}$$

Let us remark that this notion of positivity is much stronger than the customary one which requires only the first of these conditions on $f(z)$. A matrix-valued function $f(z) = f(z_1, \ldots, z_n)$ is said to be *real* if its entries take real values for $z : \operatorname{Im} z_1 = 0, \ldots, \operatorname{Im} z_n = 0$. The class of all rational positive real matrix-valued functions is denoted by $\mathcal{P}$.

Let a $k \times k$ matrix-valued function $f(z)$ be represented in the form (1.2) where $A_0 = 0$, and A_j $(j = 1, \ldots, n)$ are non-negative $N \times N$ matrices in (1.3). Then f is positive. If, moreover, the matrices A_j $(j = 1, \ldots, n)$ have real entries, then f is real as well, i.e., $f \in \mathcal{P}$.

Now, for the clearness and self-sufficiency of the exposition, we formulate here the statements from [1] which are used in the sequel. We preserve their original numeration, e.g., Lemma I.1.4 means Lemma 1.4 from the first part [1].

Lemma I.1.1. *Assume that the matrix-valued functions*

$$A^0(z) = \begin{bmatrix} a^0(z) & b^0(z) \\ c^0(z) & d^0(z) \end{bmatrix} \quad \text{and} \quad \Psi(z) = \begin{bmatrix} I_k \\ \Phi(z) \end{bmatrix},$$

where $a^0(z)$ is $k \times k$ matrix-valued and $d^0(z) \not\equiv 0$, satisfy the conditions

$$A^0(z)\Psi(z) = \begin{bmatrix} f(z) \\ 0 \end{bmatrix}, \qquad \Psi(z)' A^0(z) = [f(z) \ 0]. \tag{1.6}$$

Then there exists a submatrix

$$A(z) = \begin{bmatrix} a(z) & b(z) \\ c(z) & d(z) \end{bmatrix}$$

of $A^0(z)$, which is symmetric when $A^0(z)$ is symmetric, and such that:

a) $\det d(z) \not\equiv 0$,

b) $f(z) = a(z) - b(z)d(z)^{-1}c(z)$.

Lemma I.1.4. *Let $f(z) = P(z)/Q(z)$ be a rational scalar-valued function in the variables $z_1, \ldots, z_n$, satisfying the condition*

$$f(\lambda z_1, \ldots, \lambda z_n) = \lambda f(z_1, \ldots, z_n), \quad \forall \lambda \in \mathbb{C}.$$

Then there exists a matrix bundle

$$A(z) = z_1 A_1 + \cdots + z_n A_n$$

such that

$$A(z) \begin{bmatrix} Q(z) \\ z^{\alpha_2} \\ \vdots \\ z^{\alpha_M} \end{bmatrix} = \begin{bmatrix} P(z) \\ 0 \\ \vdots \\ 0 \end{bmatrix},$$

where the z^{α_j} $(j = 1, 2, \ldots, M)$ are all monomials in the variables $z_1, \ldots, z_n$, of the same degree as the polynomial $Q(z)$, taken in some (arbitrary) order of multi-indices α_j's. The matrices A_j are symmetric[1] *$(A_j = A_j'$, $j = 1, \ldots, n)$. If, moreover, $\overline{f(\overline{z})} = f(z)$, then the matrices A_j may be chosen real $(\overline{A_j} = A_j$, $j = 1, \ldots, n)$.*

Theorem I.2.3. *A rational matrix-valued function $f(z)$ is positive real if and only if the following three conditions are satisfied:*

(i) $f(z) + f(z)^* \geq 0$ *for* $z \in \mathcal{D}_R^+$;

(ii) $f(\lambda z) = \lambda f(z)$ *for* $\lambda \in \mathbb{C}, z \in \mathbb{C}^n$;

(iii) $f(z) = f(z)' = \overline{f(\overline{z})}$ *for* $z \in \mathbb{C}^n$.

---

[1] A scalar function is always symmetric.

Theorem I.2.5. *If $f(z)$ is a positive real matrix-valued function, then*

$$\frac{\partial f}{\partial z_j}(x_1, \ldots, x_n) \geq 0 \qquad \forall j \in \{1, \ldots, n\},\ \forall x = (x_1, \ldots, x_n) \in \mathbb{R}^n.$$

2. Rational positive real functions of degree two

In [2] a class of so-called *primary* rational positive real matrix-valued functions was treated as a subclass of those functions from $\mathcal{P}$ which admit a long resolvent representation (1.2) with $A_0 = 0$ and matrices $A_j = \overline{A_j} = A_j' \geq 0$ $(j = 1, \ldots, n)$ of rank one. Here we consider one more class of positive real functions which admit a representation in the form of the long resolvent of a linear matrix bundle with positive semidefinite real matrix coefficients. The latter is the class of rational scalar-valued positive real functions of the form

$$f(z) = \frac{P(z)}{Q(z)},$$

where the total degree of the polynomial P(z) in the variables $z_1, \ldots, z_n$ equals two. We shall refer to functions of this kind as *rational functions of degree two.* The following statement takes place.

Theorem 2.1. *Every rational scalar-valued positive real function of degree two is representable in the form*

$$f(z) = [\pi(z_1 A_1 + \cdots + z_n A_n)^{-1}\pi^*]^{-1}, \tag{2.1}$$

where π, π^* are given in (1.1) with $k = 1$, and A_j $(j = 1, \ldots, n)$ are constant positive semidefinite real matrices.

Proof. By Theorem I.2.3, conditions (i)–(iii) are satisfied for $f(z)$. Then, by Lemma I.1.4, there exists a matrix bundle $A^0(z) = z_1 A_1^0 + \cdots + z_n A_n^0$ with $A_j^0 = \overline{A_j^0} = (A_j^0)'$, such that

$$A^0(z) \begin{bmatrix} Q(z) \\ z_2 \\ \vdots \\ z_n \end{bmatrix} = \begin{bmatrix} P(z) \\ 0 \\ \vdots \\ 0 \end{bmatrix}. \tag{2.2}$$

Without loss of generality, we may consider $Q_1 \neq 0$ for the linear homogeneous function $Q(z) = z_1 Q_1 + \cdots + z_n Q_n$, since otherwise we may choose some $j \in \{2, \ldots, n\}$ for which $Q_j \neq 0$, and replace the $(n-1)$-tuple $(z_2, \ldots, z_n)$ in (2.2) by the $(n-1)$-tuple $(z_1, \ldots, z_{j-1}, z_{j+1}, \ldots, z_n)$, that corresponds to another ordering of multi-indices α_j's in Lemma I.1.4. Now, for $j = 1, \ldots, n$ we have:

$$\frac{\partial P}{\partial z_j}(z) Q(z) = Q(z) \frac{\partial P}{\partial z_j}(z) = \begin{bmatrix} Q(z) & z_2 & \ldots & z_n \end{bmatrix} \begin{bmatrix} \frac{\partial P}{\partial z_j}(z) \\ 0 \\ \vdots \\ 0 \end{bmatrix} =$$

$$= \begin{bmatrix} Q(z) & z_2 & \dots & z_n \end{bmatrix} A_j^0 \begin{bmatrix} Q(z) \\ z_2 \\ \vdots \\ z_n \end{bmatrix} +$$

$$+ \begin{bmatrix} Q(z) & z_2 & \dots & z_n \end{bmatrix} A^0(z) \begin{bmatrix} \frac{\partial Q}{\partial z_j}(z) \\ \delta_{2j} \\ \vdots \\ \delta_{nj} \end{bmatrix} =$$

$$= \begin{bmatrix} Q(z) & z_2 & \dots & z_n \end{bmatrix} A_j^0 \begin{bmatrix} Q(z) \\ z_2 \\ \vdots \\ z_n \end{bmatrix} + \begin{bmatrix} P(z) & 0 & \dots & 0 \end{bmatrix} \begin{bmatrix} \frac{\partial Q}{\partial z_j}(z) \\ \delta_{2j} \\ \vdots \\ \delta_{nj} \end{bmatrix} =$$

$$= \begin{bmatrix} Q(z) & z_2 & \dots & z_n \end{bmatrix} A_j^0 \begin{bmatrix} Q(z) \\ z_2 \\ \vdots \\ z_n \end{bmatrix} + P(z)\frac{\partial Q}{\partial z_j}(z)$$

(here we used the symmetry property $A^0(z)' = A^0(z)$, and notation δ_{kj} for the Kronecker symbol). Together with Theorem I.2.5, the latter implies for any $j \in \{1, \dots, n\}$ and $x \in \mathbb{R}^n$:

$$0 \leq Q(x)^2 \tfrac{\partial f}{\partial z_j}(x) = \tfrac{\partial P}{\partial z_j}(x)Q(x) - P(x)\tfrac{\partial Q}{\partial z_j}(x) =$$
$$\begin{bmatrix} Q(x) & x_2 & \dots & x_n \end{bmatrix} A_j^0 \begin{bmatrix} Q(x) \\ x_2 \\ \vdots \\ x_n \end{bmatrix}.$$

Since $Q_1 \neq 0$, we have

$$\operatorname{span} \left\{ \begin{bmatrix} Q(x) \\ x_2 \\ \vdots \\ x_n \end{bmatrix}, \ x = (x_1, \dots, x_n) \in \mathbb{R}^n \right\} = \mathbb{R}^n,$$

and then $A_j^0 \geq 0, \ j = 1, \dots, n$.

Since $A^0(z)' = A^0(z)$, we may rewrite (2.2) as (1.6) with

$$\Psi(z)' = \begin{bmatrix} 1/Q(z) & z_2/Q(z) & \dots & z_n/Q(z) \end{bmatrix}.$$

If the right lower $(n-1) \times (n-1)$ matrix block $d^0(z)$ of $A^0(z)$ is not identically zero, then we apply Lemma I.1.1 and get a linear matrix bundle

$$A(z) = z_1 A_1 + \dots + z_n A_n = \begin{bmatrix} a(z) & b(z) \\ c(z) & d(z) \end{bmatrix}$$

such that $\det d(z) \not\equiv 0$, and $f(z) = a(z) - b(z)d(z)^{-1}c(z)$. Moreover, $A_j = A_j' = \overline{A_j} \geq 0$, $j = 1, \ldots, n$. Since any non-zero scalar-valued function $f(z)$ is non-degenerate, the linear matrix bundle $A(z)$ is non-singular (this was mentioned in Section 1). Thus, (1.2) holds with $A_0 = 0$, i.e., (2.1) holds, and the claim is proved.

If $d^0(z) \equiv 0$, then $d_j^0 = 0$ for all $j = 1, \ldots, n$, and since $A_j^0 \geq 0$ for all $j = 1, \ldots, n$, we get $b_j^0 = 0$, $c_j^0 = 0$ for all $j = 1, \ldots, n$. Then (1.6) implies

$$f(z) = a^0(z) = z_1 a_1^0 + \ldots + z_n a_n^0.$$

Since $a_j^0 = (a_j^0)' = \overline{a_j^0}$ for all $j = 1, \ldots, n$, we get (2.1) with $A(z) = a^0(z)$, $\pi = \pi^* = 1$. The proof is complete. □

References

[1] M.F. Bessmertnyĭ, *On realization of rational matrix functions of several variables.* Oper. Theory: Adv. Appl. **134** (2002), 157–185.

[2] M.F. Bessmertnyĭ, *On realization of rational matrix functions of several variables. II.* Oper. Theory: Adv. Appl. **143** (2003).

M. F. Bessmertnyĭ
Svobody Square 4
Department of Mathematics
Faculty of Physics
Kharkov National University
61077 Kharkov, Ukraine

Operator Theory:
Advances and Applications, Vol. 149, 139–152

Operator-valued Extension of the Theorem of Helson and Szegö

Ramón Bruzual and Marisela Domínguez

Abstract. We consider the Hilbert transform for Hilbert space-valued trigonometric polynomials on the unit circle and we obtain operator-valued versions of theorems of Helson-Szegö and of Helson-Sarason.

1. Introduction

Let μ be a finite Borel measure on the unit circle $\mathbb{T} \approx [0, 2\pi]$ and let dx be the Lebesgue measure in $\mathbb{T}$. We denote with $L^p(\mathbb{T})$ the usual Lebesgue spaces and with $H^p(\mathbb{T})$ the usual Hardy spaces.

The theorem of Helson and Szegö [7] provides a necessary and sufficient condition that the Hilbert transform (see [5]) be bounded on $L^2(\mathbb{T}, \mu)$. The condition is that μ must be absolutely continuous, $d\mu = w\,dx$, and the density w must have the form

$$w = \exp(u + \tilde{v})$$

with $u \in L^\infty(\mathbb{T})$ and $\|v\|_\infty < \pi/2$ ($\tilde{v}$ denotes a harmonic conjugate of v).

The condition of Helson and Szegö also provides a necessary and sufficient condition for the subspace spanned by $\{e^{ikx} : k \in \mathbb{Z}, k \geq 0\}$ and the subspace spanned by $\{e^{ikx} : k \in \mathbb{Z}, k \leq -1\}$ to be at positive angle in $L^2(\mathbb{T}, \mu)$.

Helson and Sarason [6] extended the results of Helson and Szegö, their theorem states that in the Hilbert space $L^2(\mathbb{T}, \mu)$ the subspaces spanned by

$$\{e^{ikx} : k \in \mathbb{Z}, k \geq n\} \qquad \text{and} \qquad \{e^{ikx} : k \in \mathbb{Z}, k \leq -1\}$$

are at positive angle in $L^2(\mathbb{T}, \mu)$ if and only if μ is absolutely continuous, $d\mu = w\,dx$, and the density w has the form

$$w = |p_1|^2 \exp(u + \tilde{v})$$

where p_1 is a polynomial of degree less than n, $u \in L^\infty(\mathbb{T})$ and $\|v\|_\infty < \pi/2$.

Received by the editors February 12, 2003; revised July 18, 2003. Submitted by L. Rodman.
Mathematics Subject Classification (2000). Primary 60G25, 43A32; Secondary 47A56, 42A50.
Key words and phrases. Helson Szegö theorem, Helson Sarason Theorem, Hilbert transform, operator-valued measure.
Both authors were supported in part by the CDCH of the Univ. Central de Venezuela and by CONICIT grant G-97000668. Both authors were visitors at IVIC during the realization of this paper.

In this paper we give operator-valued versions of this result of Helson and Sarason (see Theorems 3.1, 3.5 and 3.7). For the case $n = 0$ they are operator-valued versions of the theorem of Helson and Szegö.

For vectorial cases, results related with the Helson-Szegö theorem can be found in [1], [4], [8] and related with the Helson-Sarason theorem can be found in [4].

2. The basic setting

Throughout this paper:

(a) $(\mathcal{G}, \langle\ ,\ \rangle_{\mathcal{G}})$ is a separable Hilbert space, $\|.\|_{\mathcal{G}}$ is the associated norm and $L(\mathcal{G})$ stand for the space of the continuous linear operators on $\mathcal{G}$.

(b) $\mathbb{T} \approx [0, 2\pi]$ is the one-dimensional torus and $\mathcal{B}(\mathbb{T})$ is the Borel σ-algebra of $\mathbb{T}$.

(c) dx denotes the normalized Lebesgue measure on $\mathbb{T}$.

(d) For $n \in \mathbb{Z}$, the set of the integers,

$$e_n(x) = e^{inx}.$$

2.1. Operator-valued trigonometric polynomials and Hilbert transform

A *$\mathcal{G}$-valued trigonometric polynomial* is a function $\varphi : \mathbb{T} \to \mathcal{G}$, such that

$$\varphi(x) = \sum_{n\in\mathbb{Z}} \xi_n\, e_n(x)$$

where $\{\xi_n\}_{n\in\mathbb{Z}} \subset \mathcal{G}$ has finite support. In this case $\widehat{\varphi}(n)$ denotes ξ_n.

Let $\mathcal{P}(\mathcal{G})$ be the space of all trigonometric polynomials. It is known that $\mathcal{P}(\mathcal{G})$ is a subalgebra of the space of all continuous $\mathcal{G}$-valued functions on $\mathbb{T}$.

The Hilbert transform H can be defined using the Fourier coefficients, for $\varphi \in \mathcal{P}(\mathcal{G})$ set

$$(\mathsf{H}\varphi)\widehat{\ }(n) = -i \ \mathrm{sgn}\, n \ \widehat{\varphi}(n)$$

where $\mathrm{sgn}\, n = 1$ if $n \geq 0$ and $\mathrm{sgn}\, n = -1$ if $n < 0$.

2.2. Hilbert space-valued functions

We denote by $L^\infty_{\mathcal{G}}(\mathbb{T})$ the set of all weakly measurable essentially bounded functions $F : \mathbb{T} \mapsto \mathcal{G}$.

For $1 \leq p < \infty$ let $L^p_{\mathcal{G}}(\mathbb{T})$ denote the set of all weakly measurable functions $f : \mathbb{T} \mapsto \mathcal{G}$ such that

$$\int_{\mathbb{T}} \|f(x)\|^p_{\mathcal{G}}\, dx < \infty.$$

Then $L^2_{\mathcal{G}}(\mathbb{T})$ is a Hilbert space under the inner product

$$\langle f, g\rangle_{L^2_{\mathcal{G}}(\mathbb{T})} = \int_{\mathbb{T}} \langle f(x), g(x)\rangle_{\mathcal{G}} dx.$$

$\mathcal{P}(\mathcal{G})$ is also dense in $L^p_{\mathcal{G}}(\mathbb{T})$, for $1 \leq p \leq \infty$.

Consider a function $f \in L^1_{\mathcal{G}}(\mathbb{T})$ and $\zeta \in \mathcal{G}$, take the complex-valued function $f_\zeta : \mathbb{T} \mapsto \mathbb{C}$ given by

$$f_\zeta(x) = \langle \zeta, f(x) \rangle_{\mathcal{G}}.$$

Then $f_\zeta \in L^1(\mathbb{T})$.

For $f \in L^1_{\mathcal{G}}(\mathbb{T})$ and $\zeta \in \mathcal{G}$ let $\widehat{f_\zeta}$ be the usual Fourier transform of f_ζ. By linearity for every $n \in \mathbb{Z}$ there exists $\widehat{f}(n) \in \mathcal{G}$ such that

$$\widehat{f_\zeta}(n) = \langle \zeta, \widehat{f}(n) \rangle_{\mathcal{G}}.$$

The map $n \mapsto \widehat{f}(n)$ is called the Fourier transform of f and the vectors $\widehat{f}(n)$ are the Fourier coefficients.

More details about these spaces can be found in [10].

2.3. Operator-valued functions

Let $F : \mathbb{T} \to L(\mathcal{G})$ be a function, recall that F is weakly measurable if, for all $\zeta, \xi \in \mathcal{G}$, the scalar-valued function

$$x \mapsto \langle F(x)\zeta, \xi \rangle_{\mathcal{G}}$$

is measurable.

Remark 2.1. Suppose that $F : \mathbb{T} \to L(\mathcal{G})$ is weakly measurable function. If $x \in \mathbb{T}$ then

$$\|F(x)\|_{L(\mathcal{G})} = \sup_{\|\zeta\|_{\mathcal{G}} = \|\xi\|_{\mathcal{G}} = 1} |\langle F(x)\zeta, \xi \rangle_{\mathcal{G}}|.$$

Since $\mathcal{G}$ is supposed to be separable we have that the function

$$x \mapsto \|F(x)\|_{L(\mathcal{G})}$$

is measurable.

We denote by $L^\infty_{L(\mathcal{G})}(\mathbb{T})$ the set of all weakly measurable essentially bounded functions $F : \mathbb{T} \mapsto L(\mathcal{G})$.

For $1 \leq p < \infty$ let $L^p_{L(\mathcal{G})}(\mathbb{T})$ denote the set of all funcions $F : \mathbb{T} \mapsto L(\mathcal{G})$ such that

$$\int_{\mathbb{T}} \|F(x)\|^p_{L(\mathcal{G})}\, dx < \infty.$$

For a function $F \in L^1_{L(\mathcal{G})}(\mathbb{T})$ and $\zeta, \xi \in \mathcal{G}$, $F_{\zeta\xi}$ will denote the complex-valued function $F_{\zeta\xi} : \mathbb{T} \mapsto \mathbb{C}$ given by

$$F_{\zeta\xi}(x) = \langle F(x)\zeta, \xi \rangle_{\mathcal{G}}.$$

Note that $F_{\zeta\xi} \in L^1(\mathbb{T})$.

For $F \in L^1_{\mathcal{G}}(\mathbb{T})$ and $\zeta, \xi \in \mathcal{G}$ let $\widehat{F_{\zeta\xi}}$ be the usual Fourier transform of $F_{\zeta\xi}$. By linearity for every $n \in \mathbb{Z}$ there is $\widehat{F}(n) \in L(\mathcal{G})$ such that

$$\widehat{F_{\zeta\xi}}(n) = \langle \widehat{F}(n)\zeta, \xi \rangle_{\mathcal{G}}.$$

2.4. Operator-valued measures

Definition 2.2. A *$L(\mathcal{G})$-valued measure* is a function $\mu : \mathcal{B}(\mathbb{T}) \to L(\mathcal{G})$ such that for all $\zeta, \xi \in \mathcal{G}$ the function defined by

$$\mu_{\zeta\xi}(\Delta) = \langle \mu(\Delta)\zeta, \xi \rangle_{\mathcal{G}}$$

is a scalar Radon finite measure on $\mathbb{T}$.

If $\mu(\Delta)$ is a positive operator for all $\Delta \in \mathcal{B}(\mathbb{T})$ we say that the measure μ is positive.

Set

$$\mathfrak{M}(\mathcal{G}) = \left\{ \mu : \mu \text{ is a } L(\mathcal{G})\text{-valued measure and } \sup_{\Delta \in \mathcal{B}(\mathbb{T})} \|\mu(\Delta)\|_{L(\mathcal{G})} < +\infty \right\}$$

Observe that each $W \in L^1_{L(\mathcal{G})}(\mathbb{T})$ gives rise to a measure $\mu \in \mathfrak{M}$, taking

$$d\langle \mu(x)\zeta, \xi \rangle_{\mathcal{G}} = \langle W(x)\zeta, \xi \rangle_{\mathcal{G}}\, dx = W_{\zeta\xi}(x)\, dx.$$

As it is natural the measure μ associated to $W \in L^1_{L(\mathcal{G})}(\mathbb{T})$ will be denoted with $d\mu = W\, dx$. In this case the measure μ is positive if and only if the operator $W(x)$ is nonnegative for almost every x.

Let $\mathcal{P}$ denote the linear space of all scalar-valued trigonometric polynomials.

Proposition 2.3. *Let $\mu \in \mathfrak{M}(\mathcal{G})$ and let $c = \sup_{\Delta \in \mathcal{B}(\mathbb{T})} \|\mu(\Delta)\|_{L(\mathcal{G})}$. There exists a unique linear operator $T_\mu : \mathcal{P} \to L(\mathcal{G})$ such that:*

(a) *For all $p \in \mathcal{P}$ and $\zeta, \xi \in \mathcal{G}$ we have that*

$$\langle T_\mu(p)\zeta, \xi \rangle_{\mathcal{G}} = \int_{\mathbb{T}} p(x)\, d\langle \mu(x)\zeta, \xi \rangle_{\mathcal{G}}.$$

(b) $\|T_\mu(p)\|_{\mathcal{G}} \le 4\, c\, \|p\|_\infty$ *for all $p \in \mathcal{P}$.*

Proof. For $\zeta, \xi \in \mathcal{G}$ let

$$B(\zeta, \xi) = \int_{\mathbb{T}} p(x)\, d\langle \mu(x)\zeta, \xi \rangle_{\mathcal{G}}.$$

Then B is a sesquilinear form on $\mathcal{G} \times \mathcal{G}$. Using the Hahn decomposition of the real part of μ and of the imaginary part of μ we obtain that

$$|B(\zeta, \xi)| \le 4\, c\, \|p\|_\infty \|\zeta\|\, \|\xi\|.$$

So the result follows from the Riesz representation theorem. □

Let $\mu \in \mathfrak{M}(\mathcal{G})$. We define $\widehat{\mu} : \mathbb{Z} \to L(\mathcal{G})$ by

$$\widehat{\mu}(n) = T_\mu(e_{-n}) \quad \text{if } n \in \mathbb{Z}.$$

Note that if $\zeta, \xi \in \mathcal{G}$ and $n \in \mathbb{Z}$ we have that

$$\langle \widehat{\mu}(n)\zeta, \xi \rangle_{\mathcal{G}} = \int_{\mathbb{T}} e^{-inx}\, d\langle \mu(x)\zeta, \xi \rangle_{\mathcal{G}} = \widehat{\mu_{\zeta\xi}}(n).$$

For $\mu \in \mathfrak{M}(\mathcal{G})$ and $\varphi, \psi \in \mathcal{P}(\mathcal{G})$ we define

$$\langle \varphi, \psi \rangle_\mu = \sum_{n,m \in \mathbb{Z}} \langle \widehat{\mu}(m-n)\, \widehat{\varphi}(n), \widehat{\psi}(m) \rangle_{\mathcal{G}}.$$

Note that

$$\langle \varphi, \psi \rangle_\mu = \sum_{n,m \in \mathbb{Z}} \int_{\mathbb{T}} e^{-i(m-n)x}\, d\langle \mu(x)\, \widehat{\varphi}(n), \widehat{\psi}(m) \rangle_{\mathcal{G}}.$$

Remark 2.4. In the scalar case (when $\mathcal{G} = \mathbb{C}$) it holds that

$$\langle p, q \rangle_\mu = \int_{\mathbb{T}} p(x) \overline{q(x)}\, d\mu(x)$$

where $p, q \in \mathcal{P}$.

Suppose $d\mu = W\, dx$ where $W \in L^1_{L(\mathcal{G})}(\mathbb{T})$.
If $\varphi, \psi \in \mathcal{P}(\mathcal{G})$ are given by

$$\varphi(x) = \sum_{n \in \mathbb{Z}} \widehat{\varphi}(n) e^{inx} \qquad\qquad \psi(x) = \sum_{m \in \mathbb{Z}} \widehat{\psi}(m) e^{imx}$$

then

$$\begin{aligned}
\langle \varphi, \psi \rangle_\mu &= \sum_{n,m \in \mathbb{Z}} \int_{\mathbb{T}} e^{-i(m-n)x}\, d\langle \mu(x)\, \widehat{\varphi}(n), \widehat{\psi}(m) \rangle_{\mathcal{G}} \\
&= \sum_{n,m \in \mathbb{Z}} \int_{\mathbb{T}} e^{-i(m-n)x}\, \langle W(x)\, \widehat{\varphi}(n), \widehat{\psi}(m) \rangle_{\mathcal{G}} dx \\
&= \int_{\mathbb{T}} \Big\langle W(x) \sum_{n \in \mathbb{Z}} \widehat{\varphi}(n) e^{inx}, \sum_{m \in \mathbb{Z}} \widehat{\psi}(m) e^{imx} \Big\rangle_{\mathcal{G}} dx \\
&= \int_{\mathbb{T}} \langle W(x)\, \varphi(x), \psi(x) \rangle_{\mathcal{G}}\, dx.
\end{aligned}$$

Proposition 2.5. *Let $\mu \in \mathfrak{M}(\mathcal{G})$. Then"*
(a) *$\langle\, ,\, \rangle_\mu$ is a sesquilinear form on $\mathcal{P}(\mathcal{G})$,*
(b) *$\langle\, ,\, \rangle_\mu$ is nonnegative if and only if μ is a positive measure.*

Proof. Part (a) follows from the definition of $\langle\, ,\, \rangle_\mu$ and part (b) follows from Lemma 1 of [1]. □

If $\mu \in \mathfrak{M}(\mathcal{G})$ and $\varphi, \psi \in \mathcal{P}(\mathcal{G})$ we write

$$\int_{\mathbb{T}} \langle d\mu(x)\, \varphi(x), \psi(x) \rangle_{\mathcal{G}} = \langle \varphi, \psi \rangle_\mu.$$

If $\mu \in \mathfrak{M}(\mathcal{G})$ is a positive measure $\|\ \|_\mu$ will denote the norm (possibly degenerated) associated with $\langle\, ,\, \rangle_\mu$.

Note that if $\varphi, \psi \in \mathcal{P}(\mathcal{G})$ and $n \in \mathbb{Z}$ then

$$\langle e_n\, \varphi, e_n\, \psi \rangle_\mu = \langle \varphi, \psi \rangle_\mu.$$

In particular

$$\|e_n\varphi\|_\mu = \|\varphi\|_\mu.$$

Let

$$\mathfrak{M}(\mathcal{G})_+ = \{\mu \in \mathfrak{M}(\mathcal{G}) : \widehat{\mu}(n) = 0 \text{ if } n < 0\}.$$

Remark 2.6. If $\mu \in \mathfrak{M}(\mathcal{G})_+$ and $\zeta, \xi \in \mathcal{G}$, then for the scalar measure $\mu_{\zeta\xi}$, we have that

$$\widehat{\mu_{\zeta\xi}}(n) = 0 \quad \text{for } n < 0,$$

so, by the F. and M. Riesz theorem, there exists $h_{\zeta\xi} \in H^1(\mathbb{T})$ such that

$$d\mu_{\zeta\xi}(x) = h_{\zeta\xi}(x)\, dx.$$

In [1, page 4] we found that it is not possible to prove that there is an $L(\mathcal{G})$-valued function H such that, for almost all x,

$$h_{\zeta\xi}(x) = \langle H(x)\zeta, \xi\rangle_{\mathcal{G}}.$$

However the family $\{h_{\zeta\xi}\}_{\zeta,\xi\in\mathcal{G}}$ satisfies

$$h_{\lambda\zeta_1+\zeta_2,\xi} = \lambda h_{\zeta_1\xi} + h_{\zeta_2\xi}$$
$$h_{\zeta\xi} = \overline{h_{\xi\zeta}}$$

for all $\lambda \in \mathbb{C}$, $\zeta_1, \zeta_2, \zeta, \xi \in \mathcal{G}$.

3. Weighted inequalities

In this section we give a characterization of those $\mu \in \mathfrak{M}(\mathcal{G})$ positive, such that $d\mu = W\,dx$ where $W \in L^1_{L(\mathcal{G})}(\mathbb{T})$ and

$$\|\,\mathsf{H}\varphi\|_\mu \le M\,\|\varphi\|_\mu$$

for every $\varphi \in \mathcal{P}(\mathcal{G})$. That is

$$\int_{\mathbb{T}} \langle W(x)\mathsf{H}\varphi(x), \mathsf{H}\varphi(x)\rangle_{\mathcal{G}}\, dx \le M^2 \int_{\mathbb{T}} \langle W(x)\varphi(x), \varphi(x)\rangle_{\mathcal{G}}\, dx$$

for every $\varphi \in \mathcal{P}(\mathcal{G})$.

The dual group of $\mathbb{T}$ is the group of the integers $\mathbb{Z} = \{0, \pm1, \pm2, \dots\}$. Set $\mathbb{Z}_1 = \{n \in \mathbb{Z} : n \ge 0\}$ and $\mathbb{Z}_2 = \{n \in \mathbb{Z} : n < 0\}$.

Let

$$\mathcal{P}_1(\mathcal{G}) = \{\varphi \in \mathcal{P}(\mathcal{G}) : \widehat{\varphi}(n) = 0 \text{ if } n < 0\},$$
$$\mathcal{P}_2(\mathcal{G}) = \{\varphi \in \mathcal{P}(\mathcal{G}) : \widehat{\varphi}(n) = 0 \text{ if } n \ge 0\}.$$

For $\mu \in \mathfrak{M}(\mathcal{G})$ positive and n a nonnegative integer, let

$$\rho_n(\mu) = \sup\{\, |\langle e_n\varphi_1, \varphi_2\rangle_\mu| \;:\varphi_1 \in \mathcal{P}_1(\mathcal{G}), \varphi_2 \in \mathcal{P}_2(\mathcal{G}), \|\varphi_1\|_\mu = \|\varphi_2\|_\mu = 1\}.$$

Set $\rho_n(W) = \rho_n(\mu)$ when $d\mu = W(x)dx$ for $W \in L^1_{L(\mathcal{G})}(\mathbb{T})$.

It is clear that $0 \le \rho_n(\mu) \le 1$.

Consider the closures of $\overline{e_n\mathcal{P}_1(\mathcal{G})}$ and $\overline{\mathcal{P}_2(\mathcal{G})}$ with respect to $\|\;\|_\mu$. Usually $\overline{\mathcal{P}_1(\mathcal{G})}$ is called the *future* and $\overline{\mathcal{P}_2(\mathcal{G})}$ is called the *past.*

Observe that:

(a) The spaces $\mathcal{P}_1(\mathcal{G})$ and $\mathcal{P}_2(\mathcal{G})$ are orthogonal with respect to $\langle\, ,\, \rangle_\mu$ if and only if $\rho_0(\mu) = 0$.

(b) If there is a nonzero vector in both $\overline{\mathcal{P}_1(\mathcal{G})}$ and $\overline{\mathcal{P}_2(\mathcal{G})}$ then $\rho_0(\mu) = 1$.

Recall that the closed subspaces $\overline{e_n\mathcal{P}_1(\mathcal{G})}$ and $\overline{\mathcal{P}_2(\mathcal{G})}$ are said to be *at positive angle* if $\rho_n(\mu) < 1$. The *angle* between these subspaces is $\arccos(\rho_n(\mu))$.

Theorem 3.1. *Let $\mu \in \mathfrak{M}(\mathcal{G})$ positive, suppose that $d\mu = W\,dx$ where $W \in L^1_{L(\mathcal{G})}(\mathbb{T})$ and let $M \geq 1$. The following conditions are equivalent:*

(a) $0 < \rho_n(W) \leq \dfrac{M^2-1}{M^2+1}$.

(b) *For every* $\varphi \in e_n\mathcal{P}_1(\mathcal{G}) + \mathcal{P}_2(\mathcal{G})$:

$$\|\,\mathsf{H}\varphi\|_\mu \leq M\,\|\varphi\|_\mu .$$

(c) *There exists $\nu \in \mathfrak{M}(\mathcal{G})_+$ such that, for each $\zeta, \xi \in \mathcal{G}$*

$$d\nu_{\zeta\xi}(x) = h_{\zeta\xi}(x)\,dx,$$

where $h_{\zeta\xi} \in H^1(\mathbb{T})$ and

$$(M^2+1)^2\,|\,e_n(x)W_{\zeta\xi}(x) - h_{\zeta\xi}(x)\,|^2 \leq (M^2-1)^2\,W_{\zeta\zeta}(x)\,W_{\xi\xi}(x) \quad a.e.$$

Proof. (a) $\Leftrightarrow$ (b) Let $\varphi = e_n\,\varphi_1 + \varphi_2$ where $\varphi_1 \in \mathcal{P}_1(\mathcal{G})$ and $\varphi_2 \in \mathcal{P}_2(\mathcal{G})$, then

$$\frac{1}{M^2+1}\left(M^2\ \|\varphi\|^2_\mu - \|\,\mathsf{H}\varphi\|^2_\mu\right)$$
$$= \left(\langle e_n\,\varphi_1, \varphi_2\rangle_\mu + \langle\,\varphi_2, e_n\varphi_1\rangle_\mu\right) + \frac{M^2-1}{M^2+1}\left(\|\varphi_1\|^2_\mu + \|\varphi_2\|^2_\mu\right). \quad (3.1)$$

Therefore

$$\rho_n(\mu) \leq \frac{M^2-1}{M^2+1} \qquad \text{if and only if} \qquad \|\,\mathsf{H}\varphi\|^2_\mu \leq M^2\,\|\varphi\|^2_\mu$$

for every $\varphi \in e_n\mathcal{P}_1(\mathcal{G}) + \mathcal{P}_2(\mathcal{G})$.

(b) $\Rightarrow$ (c) Let $\varphi = e_n\,\varphi_1 + \varphi_2$ where $\varphi_1 \in \mathcal{P}_1(\mathcal{G})$ and $\varphi_2 \in \mathcal{P}_2(\mathcal{G})$. From (3.1) it follows that

$$\left(\langle e_n\,\varphi_1, \varphi_2\rangle_\mu + \langle\,\varphi_2, e_n\varphi_1\rangle_\mu\right) + \frac{M^2-1}{M^2+1}\left(\|\varphi_1\|^2_\mu + \|\varphi_2\|^2_\mu\right) \geq 0 \qquad (3.2)$$

Consider the measures given by

$$d\mu_{11} = d\mu_{22} = \frac{M^2-1}{M^2+1}d\mu \qquad \text{and} \qquad d\mu_{12} = d\mu^*_{21} = e_n d\mu,$$

where μ^*_{21} is defined by

$$\langle\mu^*_{21}(\Delta)\zeta, \xi\rangle_{\mathcal{G}} = \langle\zeta, \mu_{21}(\Delta)\xi\rangle_{\mathcal{G}}.$$

An easy calculation shows that

$$\langle\varphi_1, \varphi_2\rangle_{\mu_{12}} = \langle e_n\,\varphi_1, \varphi_2\rangle_\mu,$$

so from equation (3.2) we have that

$$\sum_{\alpha,\beta=1}^{2} \langle \varphi_\alpha, \varphi_\beta \rangle_{\mu_{\alpha\beta}} \geq 0$$

for every $(\varphi_1, \varphi_2) \in \mathcal{P}_1(\mathcal{G}) \times \mathcal{P}_2(\mathcal{G})$. That is, the matrix $\{\mu_{\alpha,\beta}\}_{\alpha,\beta=1,2}$ is weakly positive (for the definition of weakly positive see [1, section II]).

A direct application of Theorem II of [1] provides $\nu \in \mathfrak{M}(\mathcal{G})_+$ such that

$$\sum_{\alpha,\beta=1}^{2} \langle \varphi_\alpha, \varphi_\beta \rangle_{\mu_{\alpha\beta}} + \langle \varphi_1, \varphi_2 \rangle_\nu + \langle \varphi_2, \varphi_1 \rangle_{\nu*} \geq 0$$

for φ_1, $\varphi_2 \in \mathcal{P}(\mathcal{G})$.

Take $\zeta, \xi \in \mathcal{G}$ and let $\Delta \subset \mathbb{T}$ be a Borel set. By Lemma 3 of [1]

$$|\langle \mu_{12}(\Delta)\,\zeta, \xi\rangle_{\mathcal{G}} - \langle \nu(\Delta)\,\zeta, \xi\rangle_{\mathcal{G}}|^2 \leq \langle \mu_{11}(\Delta)\,\zeta, \zeta\rangle_{\mathcal{G}}\ \langle \mu_{22}(\Delta)\,\xi, \xi\rangle_{\mathcal{G}}. \tag{3.3}$$

Since $\nu \in \mathfrak{M}(\mathcal{G})_+$, for each $\zeta, \xi \in \mathcal{G}$ there exists $h_{\zeta\xi} \in H^1(\mathbb{T})$ such that

$$d\nu_{\zeta\xi}(x) = h_{\zeta\xi}(x)dx.$$

It follows from equation (3.3) that

$$\left|\int_\Delta (e_n W_{\zeta\xi} - h_{\zeta\xi})(x)dx\right|^2 \leq \left(\frac{M^2-1}{M^2+1}\right)^2 \left(\int_\Delta W_{\zeta\zeta}(x)dx\right)\left(\int_\Delta W_{\xi\xi}(x)dx\right),$$

so

$$(M^2+1)^2\,|\,e_n(x)W_{\zeta\xi}(x) - h_{\zeta\xi}(x)\,|^2 \leq (M^2-1)^2\,W_{\zeta\zeta}(x)\,W_{\xi\xi}(x)$$

for almost every $x \in \mathbb{T}$.

(c) $\Rightarrow$ (b) We have that

$$(M^2+1)\,|\,e_n(x)W_{\zeta\xi}(x) - h_{\zeta\xi}(x)\,| \leq (M^2-1)\,(W_{\zeta\zeta}(x))^{\frac{1}{2}}\ (W_{\xi\xi}(x))^{\frac{1}{2}}$$

for almost every $x \in \mathbb{T}$, so

$$\begin{aligned}(M^2+1)\int_\Delta |\,e_n(x)W_{\zeta\xi}(x) - h_{\zeta\xi}(x)\,|\,dx &\leq (M^2-1)\int_\Delta (W_{\zeta\zeta}(x))^{\frac{1}{2}}\ (W_{\xi\xi}(x))^{\frac{1}{2}}\,dx\\ &\leq (M^2-1)\left(\int_\Delta W_{\zeta\zeta}(x)\,dx\right)^{\frac{1}{2}}\left(\int_\Delta W_{\xi\xi}(x)\,dx\right)^{\frac{1}{2}},\end{aligned}$$

for every Borel set $\Delta \subset \mathbb{T}$.

To complete the proof, we just notice that at the rest of the steps of the proof of (b) $\Rightarrow$ (c) we have equivalent statements. □

Proposition 3.2. *Let $\mu \in \mathfrak{M}(\mathcal{G})$ positive such that $d\mu = W dx$ where $W \in L^1_{L(\mathcal{G})}(\mathbb{T})$ and let $M \geq 1$. Suppose that*

$$\|\,\mathsf{H}\varphi\|_\mu \leq M\,\|\varphi\|_\mu$$

for every $\varphi \in e_n\mathcal{P}_1(\mathcal{G}) + \mathcal{P}_2(\mathcal{G})$.

Let $\mathrm{H}_1 : \mathcal{P} \to \mathcal{P}$ *be the Hilbert transform and let* $\zeta \in \mathcal{G}$. *Then*

$$\| \mathrm{H}_1 p\|_{L^2(W_{\zeta\zeta})} \leq M \, \|p\|_{L^2(W_{\zeta\zeta})}$$

for every scalar-valued trigonometric polynomial $p \in e_n\mathcal{P}_1 + \mathcal{P}_2$.

Proof. Let $p \in e_n\mathcal{P}_1 + \mathcal{P}_2$, if we take $\varphi(x) = p(x)\zeta$, then $\varphi \in e_n\mathcal{P}_1(\mathcal{G}) + \mathcal{P}_2(\mathcal{G})$ and

$$\mathsf{H}\varphi(x) = \mathrm{H}_1 p(x)\zeta.$$

So

$$\begin{aligned}
&\int_{\mathbb{T}} |\mathrm{H}_1 p(x)|^2 \, W_{\zeta\zeta}(x) \, dx = \int_{\mathbb{T}} |\mathrm{H}_1 p(x)|^2 \, \langle W(x)\zeta, \zeta\rangle_{\mathcal{G}} \, dx \\
&= \int_{\mathbb{T}} \langle W(x)\mathrm{H}_1 p(x)\zeta, \mathrm{H}_1 p(x)\zeta\rangle_{\mathcal{G}} \, dx = \int_{\mathbb{T}} \langle W(x)\mathsf{H}\varphi(x), \mathsf{H}\varphi(x)\rangle_{\mathcal{G}} \, dx \\
&\leq M^2 \int_{\mathbb{T}} \langle W(x)\varphi(x), \varphi(x)\rangle_{\mathcal{G}} \, dx = M^2 \int_{\mathbb{T}} |p(x)|^2 \, W_{\zeta\zeta}(x) \, dx
\end{aligned}$$ □

Remark 3.3. It follows from Proposition 3.2 that if $W \in L^1_{L(\mathcal{G})}(\mathbb{T})$ is nonnegative and the Hilbert transform is continuous with respect to $d\mu = W dx$ then, for each $\zeta \in \mathcal{G}$, we have that either $W_{\zeta\zeta}(x) = 0$ almost everywhere or $W_{\zeta\zeta}(x) > 0$ almost everywhere. In the following results we will suppose that $W_{\zeta\zeta}(x) > 0$ almost everywhere.

With $\tilde{v}$ we will indicate the harmonic conjugate of a function v.

Lemma 3.4. *Let* $\mu \in \mathfrak{M}(\mathcal{G})$ *be positive, suppose that* $d\mu = W dx$ *where* $W \in L^1_{L(\mathcal{G})}(\mathbb{T})$ *and let* $M > 1$. *If*

$$\| \mathsf{H}\varphi\|_\mu \leq M \, \|\varphi\|_\mu$$

for every $\varphi \in e_n\mathcal{P}_1(\mathcal{G}) + \mathcal{P}_2(\mathcal{G})$.

Then for $\zeta \in \mathcal{G}$ *there exists* $h_1 \in H^1(\mathbb{T})$ *such that:*

$$W_{\zeta\zeta} = |h_1| \exp(u) = |p_1|^2 \exp(u + \tilde{v}) \quad a.e.$$

where $v = -arg\,(e_{-n}\,h_1)$, $u \in L^\infty(\mathbb{T})$, p_1 *is a polynomial of degree less than* n,

$$|u| \leq \operatorname{arcosh}\left(\frac{M^2+1}{2M} \cos v\right) \quad a.e. \tag{3.4}$$

and

$$\|v\|_\infty \leq \frac{\pi}{2} - \arccos\left(\frac{M^2-1}{M^2+1}\right) < \frac{\pi}{2} \quad a.e. \tag{3.5}$$

Proof. Using Proposition 3.2 the proof can be reduced to the one-dimensional case ($\mathcal{G} = \mathbb{C}$). Therefore the result follows from the one-dimensional theorem of Helson and Sarason ([6, Theorem 6]) in the form given in [2] (Theorem 3A).

For a self-contained treatment we give a simple proof of this result.

Let $\zeta \in \mathcal{G}$. We have that there exists $h \in H^1(\mathbb{T})$ such that

$$(M^2+1)^2 \, | \, W_{\zeta\zeta}(x) - e_{-n}(x)\, h(x) \,|^2 \, \leq (M^2-1)^2 \, W_{\zeta\zeta}(x) \, W_{\zeta\zeta}(x) \quad \text{a.e.}$$

Take $w = W_{\zeta\zeta}$. Observe that $w \in L^1(\mathbb{T})$ and

$$(M^2+1)^2 \, |e_{-n} h - w|^2 \le (M^2-1)^2 w^2 \quad \text{a.e.} \tag{3.6}$$

Let

$$\phi = \frac{M^2+1}{2M} \, e_{-n} h, \qquad u = \log\left(w/|\phi|\right).$$

Then

$$w = |\phi| \exp(u) \quad \text{a.e.}$$

Let

$$r = \frac{M^2-1}{M^2+1}.$$

From (3.6) it follows that

$$\left|1 - \frac{e_{-n} h}{w}\right| \le r < 1 \quad \text{a.e.}$$

Let

$$v = -\arg(e_{-n} h/(r\,w)),$$

then $|v| < \pi/2$ and

$$\sin|v| \le r \quad \text{a.e.}$$

Then

$$\|v\|_\infty \le \frac{\pi}{2} - \arccos r < \frac{\pi}{2} \quad \text{a.e.}$$

and the inequality (3.5) follows.

Using (3.6) we obtain

$$4M^2 w^2 - 2(M^2+1)^2 \mathrm{Re}\,(e_{-n} h)\, w + (M^2+1)^2 |h|^2 \le 0 \quad \text{a.e.}$$

Since $v = -\arg\,(e_{-n} h)$, if for $k = 1, 2$ we consider

$$\begin{aligned} w_k &= \frac{M^2+1}{2M} |h| \left(\frac{M^2+1}{2M} \cos v + (-1)^k \sqrt{\left(\frac{M^2+1}{2M}\right)^2 \cos^2 v - 1} \right) \\ &= |\phi| \exp\left((-1)^k \mathrm{arcosh} \left(\frac{M^2+1}{2M} \cos v \right) \right), \end{aligned}$$

we obtain

$$w_1 \le w \le w_2 \quad \text{a.e.}$$

And the inequality (3.4) follows easily.

If we set

$$h_1 = \frac{M^2+1}{2M} h$$

obviously $h_1 \in H^1(\mathbb{T})$,

$$v = -arg\,(e_{-n} h_1)$$

and

$$W_{\zeta\zeta} = |\phi| \exp(u) = \frac{M^2+1}{2M} |h| \exp(u) = |h_1| \exp(u) \quad \text{a.e.}$$

Since $v = -arg\,(\phi)$ we have

$$w = \phi\ \exp(u + iv) \quad \text{a.e.}$$

Thus

$$0 < w\ \exp(-u - \tilde{v}) = \phi\ \exp(-\tilde{v} + iv) \quad \text{a.e.}$$

Set

$$S = \phi\ \exp(-\tilde{v} + iv).$$

By the Helson and Sarason analytic continuation principle (see [6]) we have that $S = |p_1|^2$ for a polynomial p_1 of degree less than n.

Thus

$$w\ \exp(-u - \tilde{v}) = |p_1|^2 \quad \text{a.e.}$$

that is

$$W_{\zeta\zeta} = |p_1|^2 \exp(u + \tilde{v}) \quad \text{a.e.}$$

□

Theorem 3.5. *Let $\mu \in \mathfrak{M}(\mathcal{G})$ positive, suppose that $d\mu = W dx$ where $W \in L^1_{L(\mathcal{G})}(\mathbb{T})$ and let $M > 1$.*

If

$$\|\,\mathsf{H}\varphi\|_\mu \leq M\ \|\varphi\|_\mu$$

for every $\varphi \in e_n\mathcal{P}_1(\mathcal{G}) + \mathcal{P}_2(\mathcal{G})$ and if

$$0 < (M^2 - 1)\sqrt{W_{\zeta\zeta}W_{\xi\xi}} < (M^2 + 1)\,|W_{\zeta\xi}| \quad a.e.$$

then

(a) *There exist $u \in L^\infty(\mathbb{T})$ and $h \in H^1(\mathbb{T})$ such that*

$$|W_{\zeta\xi}| = \frac{|q|}{\sqrt{|q|^2 - 1}}\,|h| \exp(u) \quad a.e.$$

where

$$q(x) = \frac{M^2 + 1}{M^2 - 1}\,\frac{e_n(x) W_{\zeta\xi}(x)}{\sqrt{W_{\zeta\zeta}(x)\ W_{\xi\xi}(x)}}$$

and

$$|u| \leq \text{arcosh}\left(\frac{|q|}{\sqrt{|q|^2 - 1}}\ \cos v\right) \quad a.e. \tag{3.7}$$

Also

$$v(x) = -\arg\left(e_{-n}(x)\,h(x)\ \exp(-ix\ \arg W_{\zeta\xi}(x))\right)$$

satisfies

$$\|v\|_\infty \leq \frac{\pi}{2} - \arccos\left(\frac{1}{|q(x)|}\right) < \frac{\pi}{2}. \tag{3.8}$$

(b) *If there exists a nonnegative integer N such that $W_{\zeta\xi}(x) = |W_{\zeta\xi}(x)|\,e_N(x)$ a.e. then there exists a polynomial p of degree less than $n + N$ such that*

$$|h| = |p|^2 \exp(\tilde{v}) \qquad and \qquad |W_{\zeta\xi}| = \frac{|q|}{\sqrt{|q|^2 - 1}}\,|p|^2 \exp(u + \tilde{v}) \quad a.e.$$

Proof. Let $\zeta, \xi \in \mathcal{G}$.

(a) From Theorem 3.1 it follows that there exists $h \in H^1(\mathbb{T})$ such that

$$(M^2+1)^2\,|\,e_n(x)W_{\zeta\xi}(x) - h(x)\,|^2 \;\le (M^2-1)^2\,W_{\zeta\zeta}(x)\,W_{\xi\xi}(x) \quad \text{a.e.}$$

Let

$$\begin{aligned} z_{11}(x) &= (M^2-1)\,W_{\zeta\zeta}(x),\\ z_{22}(x) &= (M^2-1)\,W_{\xi\xi}(x),\\ z_{12}(x) &= \overline{z_{21}(x))} = (M^2+1)\,e_n(x)\,W_{\zeta\xi}(x). \end{aligned}$$

Then $z_{\alpha,\beta} \in L^1(\mathbb{T})$ for $\alpha, \beta = 1, 2$, $z_{11}(x) > 0$ a.e., $z_{22}(x) > 0$ a.e. and the matrix $\{z_{\alpha,\beta}\}_{\alpha,\beta=1,2}$ is weakly positive (see [4, page 291] and [1]).

By hypothesis

$$0 < \sqrt{(z_{11}(x)z_{22}(x)} < |z_{12}(x)| \quad \text{a.e.}$$

so from Proposition 2.2 of [4] we have that

$$\frac{|z_{12}(x)|}{|q(x)|} = \sqrt{z_{11}(x)z_{22}(x)} = \frac{1}{\sqrt{|q(x)|^2-1}}\,|h(x)|\exp(u(x)) \quad \text{a.e.}$$

where

$$q(x) = \frac{z_{12}(x)}{\sqrt{z_{11}(x)z_{22}(x)}} = \frac{M^2+1}{M^2-1}\,\frac{e_n(x)W_{\zeta\xi}(x)}{\sqrt{W_{\zeta\zeta}(x)\,W_{\xi\xi}(x)}},$$

$u \in L^\infty(\mathbb{T})$ satisfies

$$|u(x)| \le \operatorname{arcosh}\left(\frac{1}{\sqrt{1-(1/|q(x)|^2)}}\,cos\,v(x)\right) \quad \text{a.e.}$$

and

$$v(x) = -\arg\left(e_{-n}(x)\,h(x)\,\exp(-ix\arg W_{\zeta\xi}(x))\right)$$

satisfies

$$|v(x)| \le \arcsin\left(\frac{1}{|q(x)|}\right) < \frac{\pi}{2} \quad \text{a.e..}$$

Then

$$|W_{\zeta\xi}| = \frac{|q|}{\sqrt{|q|^2-1}}\,|h|\exp(u) \quad \text{a.e.}$$

where

$$q(x) = \frac{M^2+1}{M^2-1}\,\frac{e_n(x)W_{\zeta\xi}(x)}{\sqrt{W_{\zeta\zeta}(x)\,W_{\xi\xi}(x)}},$$

u satisfies equation (3.7) and v satisfies equation (3.8).

(b) Suppose there exists N such that $W_{\zeta\xi}(x) = |W_{\zeta\xi}(x)|\, e_N(x)$. Then using Lemma 5 of [2] we have that

$$|h| = |p|^2 \exp(\tilde{v})$$

where p is a polynomial of degree less than $n + N$. □

Remark 3.6. Lemma 3.4 can be obtained from Theorem 3.5.

For the finite-dimensional case the following result was given in [4].

Theorem 3.7. *Let $\mu \in \mathfrak{M}(\mathcal{G})$ be positive, suppose that $d\mu = W dx$ where $W \in L^1_{L(\mathcal{G})}(\mathbb{T})$, and let $M > 1$.*

Suppose that for every $\zeta, \xi \in \mathcal{G}$

$$0 < (M^2 - 1)\sqrt{W_{\zeta\zeta} W_{\xi\xi}} < (M^2 + 1)\,|W_{\zeta\xi}| \quad a.e.$$

Then

$$\|\,\mathsf{H}\varphi\|_\mu \leq M\,\|\varphi\|_\mu$$

for every $\varphi \in e_n \mathcal{P}_1(\mathcal{G}) + \mathcal{P}_2(\mathcal{G})$ if and only if there exists $\nu \in \mathfrak{M}(\mathcal{G})_+$ such that, for each $\zeta, \xi \in \mathcal{G}$

$$d\nu_{\zeta\xi}(x) = h_{\zeta\xi}(x)\, dx,$$

where $h_{\zeta\xi} \in H^1(\mathbb{T})$ and for

$$q_{\zeta\xi}(x) = \frac{M^2+1}{M^2-1} \frac{e_n(x) W_{\zeta\xi}(x)}{\sqrt{W_{\zeta\zeta}(x)\, W_{\xi\xi}(x)}}$$

the following conditions hold:

(i) *for every $\zeta \in \mathcal{G}$*

$$W_{\zeta\zeta} = \frac{M^2+1}{2M}\,|h_{\zeta\zeta}| \exp(u_{\zeta\zeta}) = \frac{M^2+1}{2M} |p_{\zeta\zeta}|^2 \exp(u_{\zeta\zeta} + \tilde{v_{\zeta\zeta}}) \quad a.e.$$

where $p_{\zeta\zeta}$ is a polynomial of degree less than n, $u_{\zeta\zeta} \in L^\infty(\mathbb{T})$ satisfies equation (3.4) *and*

$$v_{\zeta\xi} = -arg\,(e_{-n}\, h_{\zeta\xi})$$

satisfies equation (3.5).

(ii) *for every $\zeta, \xi \in \mathcal{G}$*

$$|W_{\zeta\xi}| = \frac{|q_{\zeta\xi}|}{\sqrt{|q_{\zeta\xi}|^2 - 1}}\,|h_{\zeta\xi}| \exp(u_{\zeta\xi}) \quad a.e.$$

where $u_{\zeta\xi} \in L^\infty(\mathbb{T})$ satisfies equation (3.7) *and*

$$v_{\zeta\xi}(x) = -arg\,(h_{\zeta\xi}(x)\, exp(-ix(n + \arg W_{\zeta\xi}(x)))\,)$$

satisfies equation (3.8).

Furthermore, if there exists N such that $W_{\zeta\xi}(x) = |W_{\zeta\xi}(x)|\, e_N(x)$ then

$$|W_{\zeta\xi}| = \frac{|q_{\zeta\xi}|}{\sqrt{|q_{\zeta\xi}|^2 - 1}}\,|p_{\zeta\xi}|^2 \exp(u_{\zeta\xi} + \tilde{v_{\zeta\xi}}) \quad a.e.$$

where $p_{\zeta\xi}$ is a polynomial of degree less than $n + N$.

Proof. Let $\zeta, \xi \in \mathcal{G}$. From Theorem 3.1 it follows that there exist $\nu \in \mathfrak{M}(\mathcal{G})_+$ and $h_{\zeta\xi} \in H^1(\mathbb{T})$ such that

$$d\nu_{\zeta\xi}(x) = h_{\zeta\xi}(x)\, dx$$

and

$$(M^2+1)^2 \,|\, e_n(x) W_{\zeta\xi}(x) - h_{\zeta\xi}(x) \,|^2 \;\le (M^2-1)^2\, W_{\zeta\zeta}(x)\, W_{\xi\xi}(x) \quad \text{a.e.}$$

Parts (i) and (ii) follow as in Lemma 3.4 and Theorem 3.5 (a).

The last part follows from Theorem 3.5 (b). □

References

[1] R. Arocena, M. Cotlar, *Dilation of generalized Toeplitz kernels and some vectorial moment and weighted problems.* Lecture Notes in Math. Springer **908**, (1982), 169–188.

[2] R. Arocena, M. Cotlar, C. Sadosky, *Weighted inequalities in L^2 and lifting properties.* Adv. Math. Suppl. Stud. **7A**, (1981), 95–128.

[3] M. Cotlar, C. Sadosky, *On the Helson-Szegö theorem and a related class of modified Toeplitz kernels.* Proc. Symp. Pure Math. AMS. **35-I**, (1979), 383–407.

[4] M. Domínguez, *A matricial extension of the Helson-Sarason theorem and a characterization of some multivariate linearly completely regular processes.* Journal of Multivariate Analysis, **31-2**, (1989), 289–310.

[5] J. Garnett, *Bounded analytic functions.* Academic Press, 1981.

[6] H. Helson, D. Sarason, *Past and future.* Math. Scand. **21**, (1967), 5–16.

[7] H. Helson, G. Szegö, *A problem in prediction theory.* Ann. Math. Pura Appl. **51**, (1960), 107–138.

[8] M. Pourahmadi, *A matricial extension of the Helson-Szegö theorem and its application in multivariate prediction.* J. Multivariate Anal. **16**, (1985), 265–275.

[9] W. Rudin, *Fourier Analysis on Groups*, Interscience Publishers, 1962.

[10] B. Sz.-Nagy and C. Foias, *Harmonic analysis of operators on Hilbert space*, North Holland Publishing Co. 1970.

Ramón Bruzual and Marisela Domínguez
Escuela de Matemática
Fac. Ciencias
Universidad Central de Venezuela
Current address:
R. Bruzual, Apartado Postal 47686, Caracas 1041-A, Venezuela
e-mail: rbruzual@euler.ciens.ucv.ve, rbruzual@reacciun.ve

Current address:
M. Domínguez, Apartado Postal 47159, Caracas 1041-A, Venezuela
e-mail: mdomin@euler.ciens.ucv.ve

Operator Theory:
Advances and Applications, Vol. 149, 153–165

On Super-wavelets

Q. Gu, X. Dai, and Y. Diao

Abstract. A super wavelet of length m is an m-tuple $(\psi_1, \psi_2, \ldots, \psi_m)$ in the direct sum Hilbert space $D_m = \sum_{j=1}^{m} \oplus L^2(\mathbb{R})$, such that the *coordinated dilates* of all its *coordinated translates* form an orthonormal basis for D_m. The concept of super-wavelet was first introduced by Han and Larson. It is known that there exist super-wavelets with very interesting properties and some of these super-wavelets may have important applications. On the other hand, many questions concerning general properties of super-wavelets remain open. In this paper, we will answer some of these questions.

1. Super-wavelets and some related problems

Let T and D be translation and dilation operators on $L^2(\mathbb{R})$ respectively. Namely, for each $f \in L^2(\mathbb{R})$, $Tf(t) = f(t-1)$, $Df(t) = \sqrt{2}f(2t)$. One may phrase the concept of wavelet ([6]) using unitary operators D, T. A function $\psi \in L^2(\mathbb{R})$ is an orthonormal wavelet for $L^2(\mathbb{R})$ if the set $\{D^n T^\ell \psi : n, \ell \in \mathbb{Z}\}$ is an orthonormal basis for $L^2(\mathbb{R})$. When ψ is an orthonormal wavelet, the following equation holds for any $f \in L^2(\mathbb{R})$,

$$f = \sum_{n,\ell \in \mathbb{Z}} \langle f, D^n T^\ell \psi \rangle D^n T^\ell \psi, \quad \forall f \in L^2(\mathbb{R}). \tag{1}$$

However, the converse of the above statement is not true. That is, even if (1) holds for all $f \in L^2(\mathbb{R})$, $\{D^n T^\ell \psi : n, \ell \in \mathbb{Z}\}$ may not be an orthonormal basis for $L^2(\mathbb{R})$ in general.

If there exist two positive constants $0 < A \le B$ such that for any $f \in L^2(\mathbb{R})$,

$$A\|f\|^2 \le \sum_{n,\ell \in \mathbb{Z}} |\langle f, D^n T^\ell \psi \rangle|^2 \le B\|f\|^2, \tag{2}$$

then $\{D^n T^\ell \psi : n, \ell \in \mathbb{Z}\}$ is called a *frame* of $L^2(\mathbb{R})$ and ψ is called a *frame wavelet* of $L^2(\mathbb{R})$. Although study of frames as generalization of orthonormal bases can be traced back to half a century ago ([5], [12]), frame wavelets, as generalization of orthonormal wavelets, were introduced and studied quite recently. See for example [7] by Daubechies, Grossman and Meyer. If one can choose $A = B = 1$ in (2), then

Received by the editors December 20, 2002; revised May 6, 2003. Submitted by L. Rodman.
Mathematics Subject Classification (2000). Primary 42C40.
Key words and phrases. Wavelet, super-wavelet, frames, normalized tight frames.

ψ is called a *normalized tight frame wavelet.* It can be shown that if $\psi \in L^2(\mathbb{R})$ is a normalized tight frame wavelet, then (1) holds for any $f \in L^2(\mathbb{R})$.

Throughout this article, we will use $\mathcal{F}$ to denote the Fourier-Plancherel transform on $L^2(\mathbb{R})$. This is a unitary operator. If $f \in L^2(\mathbb{R}) \cap L^1(\mathbb{R})$, then

$$(\mathcal{F}f)(s) = \frac{1}{\sqrt{2\pi}} \int_{\mathbb{R}} e^{-ist} f(t)dt. \tag{3}$$

We also write $\widehat{f}$ for $\mathcal{F}f$. For a subset X of $L^2(\mathbb{R})$, $\widehat{X}$ is the set of the Fourier-Plancherel transforms of all elements in X. For a bounded linear operator S on $L^2(\mathbb{R})$, we will denote $\mathcal{F}S\mathcal{F}^{-1}$ by $\widehat{S}$. It is left to the reader to verify that we have $\widehat{D} = D^{-1}$ and $\widehat{T}f = e^{-is}f$. Furthermore, (1) and the following condition

$$f = \sum_{n,\ell \in \mathbb{Z}} \langle f, \widehat{D}^n \widehat{T}^\ell \widehat{\psi} \rangle \widehat{D}^n \widehat{T}^\ell \widehat{\psi}, \quad \forall f \in L^2(\mathbb{R}) \tag{4}$$

are equivalent.

The following theorem characterizes orthonormal wavelets and normalized tight frame wavelets. Its proof can be found in [10] by Hernández and Weiss.

Theorem 1. *$\psi \in L^2(\mathbb{R})$ is an orthonormal wavelet if and only if $\|\psi\|^2 = 1$ and for almost all $s \in \mathbb{R}$, we have*

$$\sum_{j \in \mathbb{Z}} |\widehat{\psi}(2^j s)|^2 = \frac{1}{2\pi}, \tag{5}$$

$$\sum_{j=0}^{\infty} \widehat{\psi}(2^j s) \cdot \overline{\widehat{\psi}(2^j(s + 2m\pi))} = 0, \ m \in 2\mathbb{Z} + 1. \tag{6}$$

On the other hand, $\psi \in L^2(\mathbb{R})$ is a normalized tight frame wavelet if and only if the above two equations hold.

Notice that the only difference in the above two characterizations is the extra condition $\|\psi\|^2 = 1$ satisfied by orthonormal wavelets.

Some orthonormal wavelets and normalized tight frame wavelets are closely related to certain subsets of $\mathbb{R}$. Let E be a measurable subset of $\mathbb{R}$ with finite measure. If the inverse Fourier transform of $\frac{1}{\sqrt{2\pi}}\chi_E$ is an orthonormal wavelet (resp. normalized tight frame wavelet), then E is called a wavelet set (resp. normalized tight frame wavelet set). Readers are referred to [3] for details on wavelet sets.

Let ψ_1, ψ_2, $\ldots$, ψ_m be normalized tight frame wavelets in $L^2(\mathbb{R})$. The m-tuple $(\psi_1, \psi_2, \ldots, \psi_m)$ is called a *super-wavelet* in the direct sum space $D_m = L^2(\mathbb{R}) \oplus \cdots \oplus L^2(\mathbb{R})$ if the set $\{(D^n T^\ell \psi_1, D^n T^\ell \psi_2, \ldots, D^n T^\ell \psi_m) : n, \ell \in \mathbb{Z}\}$ is an orthonormal basis for D_m. We say that a normalized tight frame wavelet ψ_1 in $L^2(\mathbb{R})$ can be *extended* to a super-wavelet of length m if there exist $\psi_2, \ldots, \psi_m \in L^2(\mathbb{R})$ such that $(\psi_1, \psi_2, \ldots, \psi_m)$ is a super-wavelet. The following theorem is a characterization of super-wavelets due to Han and Larson ([9]).

Theorem 2. *Let $\psi_1, \psi_2, \dots, \psi_n$ be functions in $L^2(\mathbb{R})$. Then $(\psi_1, \psi_2, \dots, \psi_n)$ is a super-wavelet if and only if the following four equations hold for each $i = 1, 2, \dots, n$ and almost all $s \in \mathbb{R}$,*

$$\sum_{j\in\mathbb{Z}} |\widehat{\psi}_i(2^j s)|^2 = \frac{1}{2\pi}, \tag{7}$$

$$\sum_{j=0}^{\infty} \widehat{\psi}_i(2^j s) \cdot \overline{\widehat{\psi}_i(2^j(s+2k\pi))} = 0, \; k \in 2\mathbb{Z}+1, \tag{8}$$

$$\sum_{k\in\mathbb{Z}} \sum_{i=1}^{n} |\widehat{\psi}_i(s+2k\pi)|^2 = \frac{1}{2\pi}, \tag{9}$$

$$\sum_{k\in\mathbb{Z}} \sum_{i=1}^{n} \widehat{\psi}_i(2^j(s+2k\pi)) \cdot \overline{\widehat{\psi}_i(s+2k\pi)} = 0, \qquad \forall j \in \mathbb{N}. \tag{10}$$

By Theorems 1 and 2, it is clear that each component of a super-wavelet of length n with $n \geq 2$ must be a function which is a normalized tight frame wavelet but not an orthonormal wavelet. However, it is not clear from these theorems whether every such function can be a component of a super-wavelet of length $n \geq 2$. This turns out to be nontrivial. The following problem is raised in [9] (Problem B).

Problem 1: Let η_1 be a normalized tight frame wavelet which is not an orthonormal wavelet. Is there always a super-wavelet $(\eta_1, \eta_2, \dots, \eta_n)$ for some n? for all $n \geq 2$?

In this paper, we will show that the answer to the above question is negative. This result then leads us to the following question.

Problem 2: Does there exist a normalized tight frame wavelet η_1 that can always be extended to a super-wavelet of length n for any $n \geq 2$?

Problem 2 is partially answered affirmatively in Theorem 3 ([9]) below. The normalized tight frame wavelets used there are the so-called *Frazier-Jawerth type frame wavelets.* A function $\psi \in L^2(\mathbb{R})$ is a Frazier-Jawerth type frame wavelet if $\text{supp}(\widehat{\psi}) \subset \{s : \frac{1}{2} \leq |s| \leq 2\}$ and

$$\sum_{j\in\mathbb{Z}} |\widehat{\psi}(2^j s)|^2 = \frac{1}{2\pi}, \quad \forall s \neq 0.$$

Functions of such form were first studied by M. Frazier and B. Jawerth ([8]). It is easy to see that the Frazier-Jawerth type frame wavelets satisfy the conditions in Theorem 1, hence they are normalized tight frame wavelets.

Theorem 3. *Let ψ be any Frazier-Jawerth type frame wavelet. Then ψ can be extended to a super-wavelet of length n for any $n \geq 3$.*

In this paper, we will improve the above theorem to include the case $n = 2$. In other words, we give an affirmative answer to the following question raised in [9] (Problem B_1):

Problem 3: Is it true that any Frazier-Jawerth type frame wavelet can be extended to a length 2 super-wavelet?

The last problem discussed in this paper concerns *MRA super-wavelets.* Recall that a *multiresolution analysis (MRA)* for $L^2(\mathbb{R})$ consists of a sequence $\{V_j : j \in \mathbb{Z}\}$ of closed subspaces of $L^2(\mathbb{R})$ satisfying

(1) $V_j \subset V_{j+1},\ j \in \mathbb{Z}$,
(2) $\cap_{j\in\mathbb{Z}} V_j = \{0\},\ \overline{\cup_{j\in\mathbb{Z}} V_j} = L^2(\mathbb{R})$,
(3) $f \in V_j$ if and only if $Df \in V_{j+1},\ j \in \mathbb{Z}$,
(4) there exists $\phi \in V_0$ such that $\{T^k\phi : k \in \mathbb{Z}\}$ is an orthonormal basis for V_0 (ϕ is called a *scaling function*).

It is a well known fact that there always exists an orthonormal wavelet ψ in the difference space $W_0 = V_1 \ominus V_0$. Such wavelets are called *MRA wavelets.* It is also known that not all orthonormal wavelets are MRA wavelets. Given that ψ is an orthonormal wavelet (not necessarily an MRA wavelet), if we let W_j be the subspace generated by $\{D^jT^l\psi : l \in \mathbb{Z}\}$ and let $V_j = \bigoplus_{k<j} W_k$, then ψ is an MRA wavelet if conditions (1), (2), (3), and (4) above hold. If in the above discussion, ψ is replaced by a normalized tight frame wavelet and the subspaces $V_j = \bigoplus_{k<j} W_k$ also satisfy conditions (1), (2), (3), and (4) (with the phrase "orthonormal basis" replaced by "normalized tight frame" in condition (4)), then ψ is called an MRA frame wavelet according to [9]. It is shown in [9] that MRA super-wavelets in the "normal sense" do not exist. Instead, a super-wavelet $(\eta_1, \eta_2, \ldots, \eta_n)$ is called an *MRA super-wavelet* ([9]) if each η_i is an MRA frame wavelet. Naturally, if one component in a super-wavelet is an MRA frame wavelet, one may ask whether it implies that every component in the super-wavelet is an MRA frame wavelet. This is phrased formally as the following question (Problem E in [9]):

Problem 4. Suppose that $(\eta_1, \eta_2, \ldots, \eta_n)$ is a super-wavelet, is it true that $(\eta_1, \eta_2, \ldots, \eta_n)$ is an MRA super-wavelet if one of the η_i's is an MRA frame wavelet?

We will show that the answer to this question is negative.

2. Some preliminary lemmas and theorems

In this section, we introduce a few lemmas and theorems that we will need for the next section. Some of these results are new and may be interesting and useful in their own right.

Let E, F be two measurable sets on $\mathbb{R}$. We say that the set F is 2-dilation congruent to E if there is a one-to-one measurable mapping δ from F onto E (modulus a null set) such that for each $x \in F$, $\delta(x) = 2^{l_x}x$ for some $l_x \in \mathbb{Z}$. Likewise, we say that the set F is 2π-translation congruent to E if there is a one-to-one measurable mapping τ from F onto E (modulus a null set) such that for each $x \in F$, $\tau(x) = x + 2k_x\pi$ for some $k_x \in \mathbb{Z}$. A set E is called a 2π-translation generator of $\mathbb{R}$ if $\{E + 2k\pi : k \in \mathbb{Z}\}$ is a partition of $\mathbb{R}$. Similarly, E is called a 2-dilation generator of $\mathbb{R}$ if $\{2^k E : k \in \mathbb{Z}\}$ is a partition of $\mathbb{R}$. We also let $\tau(E) = \cup_{k\in\mathbb{Z}}(E + 2k\pi)$ and $\delta(E) = \cup_{k\in\mathbb{Z}} 2^k E$. The following lemma is a direct consequence of Theorem 1 in [4] and we state it without a proof.

Lemma 1. *Let E and F be bounded measurable sets in $\mathbb{R}$ such that E contains a neighborhood of 0, F has nonempty interior, and F is bounded away from 0. Then there is a measurable set $G \subset \mathbb{R}$, which is 2-dilation congruent to F and 2π-translation congruent to E.*

For any set E in $\mathbb{R}$, define $E^s = \cup_{k\geq 1} 2^{-k}E$. The next lemma concerning E^s is a direct application of Lemma 6 in [1].

Lemma 2. *Let E be a bounded normalized tight frame wavelet set. Then for any $f \in L^2(\mathbb{R})$, we have*

$$\sum_{\ell\in\mathbb{Z}} \langle f(s), \frac{1}{\sqrt{2\pi}} e^{-i\ell s} \cdot \chi_{E^s} \rangle \frac{1}{\sqrt{2\pi}} e^{-i\ell s} \cdot \chi_{E^s} = \sum_{k\in\mathbb{Z}} \left(f(s + 2k\pi) \cdot \chi_{E^s} \right).$$

In particular, if E^s is 2π-translation congruent to a subset of $[-\pi, \pi)$, then

$$\sum_{\ell\in\mathbb{Z}} \langle f(s), \frac{1}{\sqrt{2\pi}} e^{-i\ell s} \cdot \chi_{E^s} \rangle \frac{1}{\sqrt{2\pi}} e^{-i\ell s} \cdot \chi_{E^s} = f(s) \cdot \chi_{E^s}.$$

Let E be a measurable set with finite measure, we will use ψ_E to denote the inverse Fourier transform of $\frac{1}{\sqrt{2\pi}}\chi_E$. The following theorem characterizes bounded MRA frame wavelet sets.

Theorem 4. *Let E be a bounded normalized tight frame wavelet set. Then ψ_E is an MRA frame wavelet if and only if E^s is 2π-translation congruent to a subset of $[-\pi, \pi)$.*

Proof. Clearly the subspace generated by $\{\widehat{T}^l \frac{1}{\sqrt{2\pi}}\chi_E : l \in \mathbb{Z}\}$ is contained in $L^2(E)$. Since E is a normalized tight frame wavelet set, the subspace W_0 generated by $\{\widehat{T}^l \frac{1}{\sqrt{2\pi}}\chi_E : l \in \mathbb{Z}\}$ is exactly $L^2(E)$. So $\mathrm{span}\{\widehat{D}^j \widehat{T}^l \frac{1}{\sqrt{2\pi}}\chi_E : l \in \mathbb{Z}\} = L^2(2^j E)$ for each $j \in \mathbb{Z}$. So $W_j = L^2(2^j E)$ and the W_j's are mutually orthogonal because $E \cap (2^k E) = \emptyset$ for any $k \neq 0$. Hence the subspaces

$$\widehat{V_j} = \oplus_{k<j} \widehat{W_k} = \oplus_{k<j} L^2(2^k E) = L^2(\cup_{k<j} 2^k E)$$

satisfy conditions (1), (2), and (3) in the definition of MRA (frame) wavelets.

Now if E^s is 2π-translation congruent to a subset of $[-\pi, \pi)$, then the set $\{\frac{1}{\sqrt{2\pi}} e^{-i\ell s}\chi_{E^s} : \ell \in \mathbb{Z}\}$ is a normalized tight frame on $L^2(E^s) = \widehat{V_0}$ by Lemma 2. Hence condition (4) (with the phrase "orthonormal basis" replaced by "normalized tight frame") holds. Therefore, ψ_E is an MRA frame wavelet by definition.

On the other hand, if E^s is not 2π-translation congruent to a subset of $[-\pi, \pi)$, we want to show that ψ_E is not an MRA frame wavelet. It suffices to show that there does not exist any single function ϕ on $L^2(E^s)$ such that $\{e^{-i\ell s}\phi\}$ is a normalized tight frame on $L^2(E^s)$, since this implies that condition (4) in the definition of MRA frame wavelets will not hold. Assume that this is not the case, i.e., there exists a function $\phi \in L^2(E^s)$ such that $\{e^{-i\ell s}\phi\}$ is a normalized tight frame on $L^2(E^s)$. Thus, for any $f \in L^2(E^s)$, we have

$$\sum_{\ell\in\mathbb{Z}} \langle f(s), e^{-i\ell s}\phi\rangle e^{-i\ell s}\phi = f(s). \tag{11}$$

Note that $\langle f(s), e^{-i\ell s}\phi(s)\rangle = \langle f(s)\overline{\phi(s)}, e^{-i\ell s}\chi_{E^s}\rangle$, so by Lemma 2, equation (11) can be rewritten as

$$\begin{aligned}
&\sum_{\ell\in\mathbb{Z}} \langle f(s), e^{-i\ell s}\phi\rangle e^{-i\ell s}\phi \\
&= 2\pi\phi(s)\sum_{\ell\in\mathbb{Z}} \langle f(s)\overline{\phi(s)}, \frac{1}{\sqrt{2\pi}}e^{-i\ell s}\chi_{E^s}\rangle \frac{1}{\sqrt{2\pi}}e^{-i\ell s}\chi_{E^s} \\
&= 2\pi\phi(s)\sum_{k\in\mathbb{Z}} f(s+2k\pi)\overline{\phi(s+2k\pi)}\chi_{E^s} \\
&= f(s).
\end{aligned} \tag{12}$$

On the other hand, since E^s is not 2π-translation congruent to a subset of $[-\pi,\pi)$, there exist two subsets F_1, F_2 of E^s and an integer k_0 such that F_1, F_2 are of positive measures, F_1 is 2π-translation congruent to a subset of $[-\pi,\pi)$ and $F_2 = F_1 + 2k\pi$. Apparently, if $\phi = 0$ a.e. on F_1 and F_2, then equation (12) will not hold for $f = \chi_{F_1}$. So we must have $\|\phi\cdot\chi_{F_1\cup F_2}\| > 0$. In this case, we define

$$f(s) = \begin{cases} -\overline{\phi(s+2k_0\pi)}, & s\in F_1, \\ \overline{\phi(s-2k_0\pi)}, & s\in F_2, \\ 0, & \text{otherwise.} \end{cases} \tag{13}$$

Thus, for any $s \in F_1$, the last equality in (12) becomes

$$\begin{aligned}
&2\pi\phi(s)\big(f(s)\overline{\phi(s)} + f(s+2k_0\pi)\overline{\phi(s+2k_0\pi)}\big) \\
&= 2\pi\phi(s)\big(-\overline{\phi(s+2k_0\pi)\phi(s)} + \overline{\phi(s)\phi(s+2k_0\pi)}\big) \\
&= 0 = f(s).
\end{aligned}$$

Similarly, $f(s) = 0$ for any $s \in F_2$ as well. Since $\|f\| = \|\phi\cdot\chi_{F_1\cup F_2}\| > 0$ by its definition, this gives the contradiction we need. □

We close this section by the following lemma. Its proof can be found in [9].

Lemma 3. *Let E, F be two normalized tight frame wavelet sets. Then (ψ_E, ψ_F) is a super-wavelet if and only if for any $k \in \mathbb{Z}$, we have $(E+2k\pi)\cap F = \emptyset$ and $E\cup F$ is 2π-translation congruent to $[-\pi,\pi)$, i.e., $E\cup F$ is a 2π-translation generator of $\mathbb{R}$.*

3. Answers to the problems

Let $E = [-\pi, \frac{1}{2}\pi)$ and $F = (-2,-1] \cup [1,2)$. By Lemma 1, there is a set $E^* \subset \mathbb{R}$ which is 2-dilation congruent to F and 2π-translation congruent to E. By Theorem 1, E^* is a normalized tight frame wavelet set. The following is our first main result. It gives a negative answer to Problem 1 raised in the last section.

Theorem 5. *There exists a normalized tight frame wavelet ψ in $L^2(\mathbb{R})$ which cannot be extended to a super-wavelet of any length $n \geq 2$.*

Proof. We will prove that the function ψ_{E^*} defined by the set E^* is such a function. We will first consider the case $n = 2$. Assume that ψ_{E^*} can be extended to a super-wavelet of length 2, that is, there exists a normalized tight frame wavelet ψ such that (ψ_{E^*}, ψ) is a super-wavelet of length 2. Since E^* is 2π-translation congruent to $[-\pi, \frac{1}{2}\pi)$, we have

$$\sum_{k\in\mathbb{Z}} |\widehat{\psi}_{E^*}(s+2k\pi)|^2 = \frac{1}{2\pi}, \quad \forall s \in [-\pi, \frac{1}{2}\pi), \tag{14}$$

$$\sum_{k\in\mathbb{Z}} |\widehat{\psi}_{E^*}(s+2k\pi)|^2 = 0, \quad \forall s \in [\frac{1}{2}\pi, \pi). \tag{15}$$

By equation (9) in Theorem 2, we have

$$\sum_{k\in\mathbb{Z}} |\widehat{\psi}_{E^*}(s+2k\pi)|^2 + \sum_{k\in\mathbb{Z}} |\widehat{\psi}(s+2k\pi)|^2 = \frac{1}{2\pi}, \qquad \forall s \in \mathbb{R}. \tag{16}$$

Therefore, ψ satisfies the following equations:

$$\sum_{k\in\mathbb{Z}} |\widehat{\psi}(s+2k\pi)|^2 = 0, \quad \forall s \in [-\pi, \frac{1}{2}\pi), \tag{17}$$

$$\sum_{k\in\mathbb{Z}} |\widehat{\psi}(s+2k\pi)|^2 = \frac{1}{2\pi}, \quad \forall s \in [\frac{1}{2}\pi, \pi). \tag{18}$$

Thus, the support of $\widehat{\psi}$ is contained in the set $\cup_{j\in\mathbb{Z}}([\frac{1}{2}\pi, \pi) + 2j\pi)$. Let F be the support of $\widehat{\psi}$ and let $F^+ = F \cap (0, +\infty)$. Notice that the set $(0, +\infty)$ is the disjoint union of $2^j[\frac{1}{2}\pi, \pi), j \in \mathbb{Z}$, we have

$$\begin{aligned}\int_{(0,+\infty)} |\widehat{\psi}(s)|^2 \cdot \frac{1}{s} ds &= \sum_{j\in\mathbb{Z}} \int_{2^j[\frac{1}{2}\pi,\pi)} |\widehat{\psi}(s)|^2 \cdot \frac{1}{s} ds = \sum_{j\in\mathbb{Z}} \int_{[\frac{1}{2}\pi,\pi)} |\widehat{\psi}(2^j s)|^2 \cdot \frac{1}{s} ds \\ &= \int_{[\frac{1}{2}\pi,\pi)\sum_{j\in\mathbb{Z}}} |\widehat{\psi}(2^j s)|^2 \cdot \frac{1}{s} ds = \frac{\ln 2}{2\pi}.\end{aligned}$$

On the other hand, we have

$$\begin{aligned}
&\int_{(0,+\infty)} |\widehat{\psi}(s)|^2 \cdot \frac{1}{s} ds \\
&= \int_{F^+} |\widehat{\psi}(s)|^2 \cdot \frac{1}{s} ds \\
&\leq \sum_{k\in\mathbb{N}} \int_{[\frac{1}{2}\pi,\pi)+2(k-1)\pi} |\widehat{\psi}(s)|^2 \cdot \frac{1}{s} ds \\
&= \sum_{k\in\mathbb{N}} \int_{[\frac{1}{2}\pi,\pi)} |\widehat{\psi}(s+2(k-1)\pi)|^2 \cdot \frac{1}{s+2(k-1)\pi} ds \\
&\leq \sum_{k\in\mathbb{N}} \int_{[\frac{1}{2}\pi,\pi)} |\widehat{\psi}(s+2(k-1)\pi)|^2 \cdot \frac{1}{s} ds \\
&= \int_{[\frac{1}{2}\pi,\pi)} \sum_{k\in\mathbb{N}} |\widehat{\psi}(s+2(k-1)\pi)|^2 \cdot \frac{1}{s} ds \\
&< \int_{[\frac{1}{2}\pi,\pi)} \sum_{k\in\mathbb{Z}} |\widehat{\psi}(s+2(k-1)\pi)|^2 \cdot \frac{1}{s} ds \\
&= \frac{\ln 2}{2\pi}.
\end{aligned}$$

The last inequality is due to the fact that the set $F \cap (-\infty, 0)$ must have positive measure. So we have a contradiction. This proves the case $n = 2$.

Now assume that there are $n-1$ functions $\psi_2, \ldots, \psi_n$ (with $n \geq 3$) such that $(\psi_{E^*}, \psi_2, \ldots, \psi_n)$ is a super-wavelet. By Theorem 2 and the properties of E^* we have

$$\sum_{k\in\mathbb{Z}} \sum_{i=2}^{n} |\widehat{\psi}_i(s+2k\pi)|^2 = \frac{1}{2\pi}, \quad \forall s \in [\frac{1}{2}\pi, \pi),$$

$$\sum_{k\in\mathbb{Z}} \sum_{i=2}^{n} |\widehat{\psi}_i(s+2k\pi)|^2 = 0, \quad \forall s \in [-\pi, \frac{1}{2}\pi).$$

So the support of each $\widehat{\psi}_i$ is contained in $\bigcup_{k\in\mathbb{Z}} [\frac{1}{2}\pi, \pi)$. Since $[\frac{1}{2}\pi, \pi)$ is a 2-dilation generator for $(0, +\infty)$, we have

$$\begin{aligned}
\int_{(0,+\infty)} \sum_{i=2}^{n} |\widehat{\psi}_i(s)|^2 \cdot \frac{1}{s} ds &= \sum_{j\in\mathbb{Z}} \int_{2^j[\frac{1}{2}\pi,\pi)} \sum_{i=2}^{n} |\widehat{\psi}_i(s)|^2 \cdot \frac{1}{s} ds = \cdots \\
&= \int_{[\frac{1}{2}\pi,\pi)} \sum_{i=2}^{n} \sum_{j\in\mathbb{Z}} |\widehat{\psi}_i(2^j s)|^2 \cdot \frac{1}{s} ds = \frac{\ln 2}{2\pi}.
\end{aligned}$$

On the other hand, similarly to the case $n = 2$, we also have

$$\int_{(0,+\infty)} \sum_{i=2}^{n} |\widehat{\psi}_i(s)|^2 \cdot \frac{1}{s} ds = \sum_{k \in \mathbb{N}} \int_{[\frac{1}{2}\pi,\pi)+2(k-1)\pi} \sum_{i=2}^{n} |\widehat{\psi}_i(s)|^2 \cdot \frac{1}{s} ds$$
$$\leq \cdots < \frac{\ln 2}{2\pi},$$

which gives us the expected contradiction. Some detailed calculations are left to the reader. □

The following theorem answers Problem 3 from last section.

Theorem 6. *Every Frazier-Jawerth type frame can be extended to a super-wavelet of length* 2.

Proof. Let η be a Frazier-Jawerth frame wavelet. By the definition of η, we have $\text{supp}(\widehat{\eta}) \subset \{s : \frac{1}{2} \leq |s| \leq 2\}$ and for almost all $s \in \mathbb{R}$,

$$\sum_{j \in \mathbb{Z}} |\widehat{\eta}(2^j s)|^2 = \frac{1}{2\pi}, \tag{19}$$

$$\sum_{j=0}^{\infty} \widehat{\eta}(2^j s)\overline{\widehat{\eta}(2^j(s+2k\pi))} = 0, \quad \forall k \in 2\mathbb{Z}+1. \tag{20}$$

We need to find a function ψ such that (η, ψ) is a super-wavelet, i.e., a function ψ satisfying the following conditions for almost all $s \in \mathbb{R}$ (by Theorem 2):

$$\sum_{j \in \mathbb{Z}} |\widehat{\psi}(2^j s)|^2 = \frac{1}{2\pi}, \tag{21}$$

$$\sum_{j=0}^{\infty} \widehat{\psi}(2^j s)\overline{\widehat{\psi}(2^j(s+2k\pi))} = 0, \quad \forall k \in 2\mathbb{Z}+1, \tag{22}$$

$$\sum_{k \in \mathbb{Z}} |\widehat{\eta}(s+2k\pi)|^2 + |\widehat{\psi}(s+2k\pi)|^2 = \frac{1}{2\pi}, \tag{23}$$

$$\sum_{k \in \mathbb{Z}} \widehat{\eta}(2^j(s+2k\pi)) \cdot \overline{\widehat{\eta}(s+2k\pi)} +$$
$$\sum_{k \in \mathbb{Z}} \widehat{\psi}(2^j(s+2k\pi)) \cdot \overline{\widehat{\psi}(s+2k\pi)} = 0, \quad \forall j \in \mathbb{N}. \tag{24}$$

Notice that (19) implies

$$|\widehat{\eta}(s)|^2 + |\widehat{\eta}(2s)|^2 = \frac{1}{2\pi}, \quad \forall s \in [-1, -\frac{1}{2}) \cup [\frac{1}{2}, 1).$$

So the set $[-1,-\frac{1}{2})\cup[\frac{1}{2},1)$ can be expressed as the disjoint union of the following three measurable sets A, B, and C:

$$\begin{aligned} A &= \{s\in[-1,-\frac{1}{2})\cup[\frac{1}{2},1): 0<|\widehat{\eta}(s)|^2<\frac{1}{2\pi}\}.\\ B &= \{s\in[-1,-\frac{1}{2})\cup[\frac{1}{2},1): |\widehat{\eta}(s)|^2=0\}.\\ C &= \{s\in[-1,-\frac{1}{2})\cup[\frac{1}{2},1): |\widehat{\eta}(s)|^2=\frac{1}{2\pi}\}.\end{aligned}$$

Consequently, $[-2,-\frac{1}{2})\cup[\frac{1}{2},2)$ is the disjoint union of the following six sets A, B, C, $2A$, $2B$ and $2C$. Furthermore, we have $0<|\widehat{\eta}(s)|^2<\frac{1}{2\pi}$ for any $s\in 2A$, $|\widehat{\eta}(s)|^2=\frac{1}{2\pi}$ for any $s\in 2B$, and $|\widehat{\eta}(s)|^2=0$ for any $s\in 2C$.

Note 1. Since the 2π-translates of $[-2,2)$ are mutually disjoint, the 2π-translates of the sets $A,B,C,2A,2B,2C$ are also mutually disjoint. On the other hand, the 2-dilates of the interval $[-2+2\pi,2+2\pi)$ are mutually disjoint since $-2+2\pi>(\frac{1}{2})(2+2\pi)$. It follows that the 2-dilates of the set $A+2\pi$ are also mutually disjoint since $A+2\pi\subset[-2+2\pi,2+2\pi)$.

Let $\Omega=\mathbb{R}\backslash\bigcup_k\big((A\cup C\cup 2A\cup 2B)+2k\pi\big)$ and $\Lambda=\mathbb{R}\backslash\bigcup_j 2^j(A+2\pi)$. Observe that the set $\Omega_0=[-\pi,\pi)\cap\Omega$ is a 2π-translation generator for Ω with 0 in its interior and the set $\Lambda_0=\big([-\pi,-\frac{1}{2}\pi)\cup[\frac{1}{2}\pi,\pi)\big)\cap\Lambda$ is a 2-dilation generator of Λ with at least one interior point. By Lemma 1, there is a set $F\subset\mathbb{R}$ such that F is 2π-translation congruent to Ω_0 and 2-dilation congruent to Λ_0. Clearly, F is disjoint from any 2π-translate of $C\cup A\cup 2A\cup 2B$ and any 2-dilate of $A+2\pi$. Define

$$\widehat{\psi}(s)=\begin{cases}\overline{\widehat{\eta}(2(s-2\pi))}, & s\in A+2\pi,\\ -\overline{\widehat{\eta}(\frac{1}{2}(s-4\pi))}, & s\in 2A+4\pi,\\ \frac{1}{\sqrt{2\pi}}, & s\in F,\\ 0, & \text{otherwise.}\end{cases}\tag{25}$$

We will show that equations (21), (22), (23) and (24) hold.

First, we verify that equation (22) holds. For any integer $j\ge 0$ and odd integer k, $2^j(s+2k\pi)$ is a 2π-translate of 2^js. By the definition of F and Note 1, the 2π-translates of $A+2\pi$, $2A+4\pi$ and F are mutually disjoint, so we must have either $\widehat{\psi}(2^js)=0$ or $\widehat{\psi}(2^j(s+2k\pi))=0$ by the definition of $\widehat{\psi}$. Hence equation (22) holds.

To see that equations (21), (23) and (24) hold, we need to use the fact that $F\cup(A+2\pi)$ is a 2-dilation generator of $\mathbb{R}$, $F\cup C\cup A\cup 2A\cup 2B$ is a 2π-translation generator of $\mathbb{R}$. This fact is guaranteed by the way F is obtained.

Due to the nature of equation (21), we only need to verify it for all real numbers in the 2-dilation generator $F\cup(A+2\pi)$ of $\mathbb{R}$. For $s\in F$, we have

$$\sum_{j\in\mathbb{Z}}|\widehat{\psi}(2^js)|^2=|\widehat{\psi}(s)|^2=\frac{1}{2\pi}.$$

For $s \in (A+2\pi)$, we have

$$\sum_{j\in\mathbb{Z}} |\widehat{\psi}(2^j s)|^2 = |\widehat{\psi}(s)|^2 + |\widehat{\psi}(2s)|^2 = |\widehat{\eta}(2(s-2\pi))|^2 + |\widehat{\eta}(s-2\pi)|^2 = \frac{1}{2\pi}.$$

Thus equation (21) holds as well.

Similarly, to verify that equations (23), (24) hold for all $s \in \mathbb{R}$, we only need to verify them for all real numbers in the 2π-translation generator $F \cup C \cup A \cup 2A \cup 2B$ of $\mathbb{R}$. For equation (23), if $s \in F$, since $\widehat{\eta}(s+2k\pi) = 0$ for all integers k and $\widehat{\psi}(s+2k\pi) = 0$ for all k except for $k = 0$, we have

$$\sum_{k\in\mathbb{Z}} (|\widehat{\eta}(s+2k\pi)|^2 + |\widehat{\psi}(s+2k\pi)|^2) = |\widehat{\psi}(s)|^2 = \frac{1}{2\pi}.$$

If $s \in C \cup 2B$, since $\widehat{\psi}(s+2k\pi) = 0$ for all integer k and $\widehat{\eta}(s+2k\pi) = 0$ for all k except for $k = 0$, we have

$$\sum_{k\in\mathbb{Z}} (|\widehat{\eta}(s+2k\pi)|^2 + |\widehat{\psi}(s+2k\pi)|^2) = |\widehat{\eta}(s)|^2 = \frac{1}{2\pi}.$$

For $s \in A$, we have

$$\begin{aligned} & \sum_{k\in\mathbb{Z}} (|\widehat{\eta}(s+2k\pi)|^2 + |\widehat{\psi}(s+2k\pi)|^2) \\ = \; & |\widehat{\eta}(s)|^2 + |\widehat{\psi}(s+2\pi)|^2 \\ = \; & |\widehat{\eta}(s)|^2 + |\widehat{\eta}(2s)|^2 = \frac{1}{2\pi}. \end{aligned}$$

Finally, for $s \in 2A$, we have

$$\begin{aligned} & \sum_{k\in\mathbb{Z}} (|\widehat{\eta}(s+2k\pi)|^2 + |\widehat{\psi}(s+2k\pi)|^2) \\ = \; & |\widehat{\eta}(s)|^2 + |\widehat{\psi}(s+4\pi)|^2 \\ = \; & |\widehat{\eta}(s)|^2 + |\widehat{\eta}(\frac{1}{2}s)|^2 = \frac{1}{2\pi}. \end{aligned}$$

Lastly, we will show that equation (24) holds for all $s \in F \cup C \cup A \cup 2A \cup 2B$. For $s \in F \cup C \cup 2B$, this follows easily using the same observation we made when checking equation (23) and the fact that (19) and (21) hold. For $s \in A$, the equation holds trivially if $j \neq 1$. For $j = 1$, by the definition of ψ, we have

$$\begin{aligned} & \sum_{k\in\mathbb{Z}} \widehat{\eta}(2^j(s+2k\pi)) \cdot \overline{\widehat{\eta}(s+2k\pi)} \\ + \; & \sum_{k\in\mathbb{Z}} \widehat{\psi}(2^j(s+2k\pi)) \cdot \overline{\widehat{\psi}(s+2k\pi)} \\ = \; & \widehat{\eta}(2s) \cdot \overline{\widehat{\eta}(s)} + \widehat{\psi}(2s+4\pi) \cdot \overline{\widehat{\psi}(s+2\pi)} \\ = \; & \widehat{\eta}(2s) \cdot \overline{\widehat{\eta}(s)} - \overline{\widehat{\eta}(s)} \cdot \widehat{\eta}(2s) = 0. \end{aligned}$$

And for $s \in 2A$, the equation holds trivially if $j \neq -1$. For $j = -1$,we have

$$\begin{aligned} & \sum_{k\in\mathbb{Z}} \widehat{\eta}(2^j(s+2k\pi)) \cdot \overline{\widehat{\eta}(s+2k\pi)} \\ + & \sum_{k\in\mathbb{Z}} \widehat{\psi}(2^j(s+2k\pi)) \cdot \overline{\widehat{\psi}(s+2k\pi)} \\ = & \widehat{\eta}(\frac{1}{2}s) \cdot \overline{\widehat{\eta}(s)} + \widehat{\psi}(\frac{1}{2}(s+4\pi)) \cdot \overline{\widehat{\psi}(s+4\pi)} \\ = & \widehat{\eta}(\frac{1}{2}s) \cdot \overline{\widehat{\eta}(s)} - \overline{\widehat{\eta}(s)} \cdot \widehat{\eta}(\frac{1}{2}s) = 0. \end{aligned}$$

□

We end our paper with an answer to Problem 4 raised in the first section. In the next theorem, we will see that a "hybrid" super frame wavelet is possible. We will actually construct two sets E, F such that ψ_E is an MRA frame wavelet, ψ_F is a normalized tight frame wavelet but not an MRA frame wavelet, and (ψ_E, ψ_F) is a super-wavelet.

Theorem 7. *There exists a super-wavelet (ψ_1, ψ_2) such that ψ_1 is an MRA frame wavelet and ψ_2 is not an MRA frame wavelet.*

Proof. Let $E = [-\pi, -\frac{1}{2}\pi) \cup [\frac{1}{2}\pi, \pi)$ and $F_0 = [2\pi + \frac{1}{16}\pi, 2\pi + \frac{1}{8}\pi) \cup [36\pi + \frac{1}{6}\pi, 36\pi + \frac{1}{2}\pi)$. We leave the following to the reader to verify.

(a) $\tau(E) \cap F_0 = \emptyset$;
(b) $(F_0 + 2k\pi) \cap F_0 = \emptyset$ for any $k \neq 0$;
(c) $(2^k F_0) \cap F_0 = \emptyset$ for any $k \neq 0$;
(d) $\tau(F_0^s) \cap F_0^s = [2\pi + \frac{1}{12}\pi, 2\pi + \frac{1}{8}\pi)$;
(e) $\tau(F_0 \cup E)$ is a proper subset of $\mathbb{R}$;
(f) $\delta(F_0)$ is a proper subset of $\mathbb{R}$.

Let $F_1 = E \backslash \delta(F_0)$ and $F_2 = [-\frac{1}{2}\pi, \frac{1}{2}\pi) \backslash \tau(F_0)$. Observe that $F_0 \cup F_1$ is a 2-dilation generator of $\mathbb{R}$ and $F_0 \cup F_2 \cup E$ is a 2π-translation generator of $\mathbb{R}$. Since 0 is an interior point of F_2 and F_1 has nonempty interior, by Lemma 1, there exists a measurable set ΔF such that ΔF is 2π-translation congruent to F_2 and 2-dilation congruent to F_1. Now we define $F = F_0 \cup \Delta F$. It is clear that F is a normalized tight frame wavelet set and (ψ_E, ψ_F) is a super-wavelet by Lemma 3. By condition (d) above, ψ_F is not an MRA frame wavelet by Theorem 4. □

References

[1] X. Dai, Y. Diao and Q. Gu, *Frame wavelet sets in* $\mathbb{R}$, Proc. AMS, Vol. **129**, No. 7 (2000), 2045–2055.

[2] X. Dai, Y. Diao, Q. Gu and D. Han, *Frame Wavelets in Subspaces of $L^2(\mathbb{R}^d)$*, Proc. Amer. Math. Soc., **130** (2002), No. 6, 1661–1667.

[3] X. Dai and D. Larson, *Wandering vectors for unitary systems and orthogonal wavelets,* Memoirs Amer. Math. Soc., **134** (1998), No. 640.

[4] X. Dai, D. Larson and D. Speegle, *Wavelet sets in* $\mathbb{R}^n$, J. Fourier Analysis and Applications, **3** (1997), 451–456.

[5] R. J. Duffin and A. C. Shaffer, *A class of nonharmonic Fourier Series,* Tran. Amer. Math. Soc., **72** (1952), 341–366.

[6] I. Daubechies, *Ten Lectures on Wavelets,* CBMS-NSF Reg. Conf. Series in Appl. Math. **61**, SIAM, Philadelphia 1992.

[7] I. Daubechies, A. Grossman and Y. Meyer, *Painless nonorthogonal expansions,* J. Math. Phys, **27** (1986), 1271–1283.

[8] M. Frazier and B. Jawerth, *A discrete transform and decomposition of distribution spaces,* J. Funct. Anal., **93** (1990), 34–170.

[9] D. Han and D. Larson, *Frames, bases and group representations,* Memoirs Amer. Math. Soc., **147** (2000), No. 697.

[10] E. Hernández, G. Weiss, *A first course on wavelets,* CRC Press, Boca Raton, 1996.

[11] S. Mallat, *Multiresolution approximations and wavelet orthonormal basis of* $L^2(\mathbb{R})$, Trans. Amer. Math. Soc. **315** (1989) 69–87.

[12] B. Sz-Nagy, *Expansion theorems of Paley-Wiener type,* Duke Math. J., **14** (1947), 975–978.

Q. Gu
Department of Mathematics
East China Normal University
Shanghai, China
e-mail: guqing2218@vip.sina.com

X. Dai
Department of Mathematics
University of North Carolina at Charlotte
Charlotte, NC 28223, USA
e-mail: xdai@uncc.edu

Y. Diao
Department of Mathematics
University of North Carolina at Charlotte
Charlotte, NC 28223, USA
e-mail: ydiao@uncc.edu

Operator Theory:
Advances and Applications, Vol. 149, 167–197

Fast Algorithms for Toeplitz Least Squares Problems

Georg Heinig

Abstract. New direct algorithms of Levinson and Schur type for the solution of $m \times n$ Toeplitz least squares problems with computational complexity $O(n^2)+O(m \log m)$ are presented. The approach is analogous to the two-stage approach for Toeplitz matrix inversion and is based, firstly, on a formula for the pseudo inverse involving a fundamental system and, secondly, a recursive computation of a fundamental system. The paper is focused on algebraic properties and complexity rather than problems of accuracy and stability.

1. Introduction

In this paper we consider the least squares (LS) problem for Toeplitz matrices which reads as follows. Let $T = [\, a_{i-j} \,]_{i=1}^{m} \, _{j=1}^{n}$ be a $m \times n$ Toeplitz matrix, $m \geq n$, with complex entries and full column rank n. For a given $\mathbf{b} \in \mathbb{C}^m$, find an $\mathbf{x} \in \mathbb{C}^n$ such that $\|T\mathbf{x} - \mathbf{b}\|$ is minimal. Here $\| \cdot \|$ denotes the (Euclidean) 2-norm.

The solution of this problem is given by $\mathbf{x} = T^{\dagger}\, \mathbf{b}$, where $T^{\dagger} = (T^*T)^{-1}T^*$ is the (Moore-Penrose) pseudo inverse of T and T^* denotes the complex conjugate transpose of T. That means that solving LS problems is closely related to pseudo inversion of matrices.

Toeplitz LS problems appear in many applications and are, therefore, subject of many research activities and publications. The main attention is dedicated to the problem how the structure of the matrix can be exploited. The solution of a general unstructured LS problem, for example via QR factorization, requires $O(mn^2)$ arithmetic operations. It is desirable to reduce this complexity to $O(mn)$ or less taking the Toeplitz structure into account.

An algorithm for the solution of Toeplitz LS problems that requires only $O(mn)$ flops was first designed by D. Sweet in his PhD thesis [23]. This algorithm was further improved in [24], [4] and [6]. Further references concerning algorithms with complexity $O(mn)$ include the papers [21], [7], [10], [20], [27] and many others. In [28] a "superfast" algorithm with complexity $O(m \log^2 m)$ was proposed. One drawback of the superfast algorithm and other algorithms based on matrix

---

Received by the editors February 3, 2003; revised June 9, 2003. Submitted by L. Rodman.
Mathematics Subject Classification (2000). 15A09, 65F20, 47B35 (15A23, 15A57).
Key words and phrases. Toeplitz matrix, Moore-Penrose pseudoinverse, fast algorithms, least squares problems, QR-factorization.
This research was supported by Kuwait University Research Grant SM07/00.

transformations, like those in [10] and [27], is that no simple update of the solution is possible if the matrix is slightly modified, like one column or one row of is added to T or deleted from T. In contrast to this, the algorithms in the present paper are based on updates of this kind (see Section 5). The overall computational complexity of most of them is equal to $O(n^2) + O(m \log m)$.

In most of the papers dedicated to Toeplitz LS problem a major part is dedicated to numerical issues like stability and accuracy. One reason for this is that many Toeplitz LS problems in applications are ill-conditioned, another one is that the fast algorithms, in general, do not enjoy the stability properties of the standard algorithms, like Householder or modified Gram-Schmidt QR factorization, for unstructured problems. In the present paper we do not discuss the problems of stability and accuracy in detail. The main focus will be on algebraic issues and complexity.

The motivation for the present work arose from a comparison of the algebraic theory of Toeplitz matrix inversion and the literature on Toeplitz LS problems. In fact, for $m = n$ the Toeplitz LS problem turns into the problem to solve a linear system with a Toeplitz coefficient matrix. Concerning the solution of Toeplitz systems there exists a rich algebraic structure theory. This theory can be exploited to design efficient fast algorithms. Comparing the literature of the algebraic theory of square Toeplitz matrices on one hand and of Toeplitz LS problem on the other hand one can observe that in the latter there are several gaps. The main goal of the present paper is to fill this gaps. We will present an approach to pseudo inversion of Toeplitz matrices which is analogous to that one of inversion.

A standard approach for solving Toeplitz systems of equations consists in two stages. The first stage consists in computing a "fundamental system" for the matrix, which is a systems of two vectors containing the full information about matrix. The second stage consists in fast matrix-vector multiplication using formulas that represent the inverse matrix with the help of the fundamental system and FFT.

Let us first discuss the second stage. Formulas for the inverses of a Toeplitz matrix are based on the fact that inverses of Toeplitz matrices have a displacement structure or, more precisely, they are Bezoutians of two polynomials. The coefficient vectors of these polynomials form a fundamental system. Bezoutians have various matrix representations. The Gohberg-Semencul formula [9] is the classical and most famous representative of them. This formula represents Bezoutians with the help of triangular Toeplitz matrices. More efficient formulas are those that are built from circulants and skew-circulants and were first designed by Ammar-Gader [1, 2] (see also [8], [12]). These formulas can be written in a form that contains only diagonal matrices and Discrete Fourier Transformations, which implies that the solution can be found with complexity $O(n \log n)$, provided that the $O(n)$ parameters in the formula, i.e., the components of a fundamental system, are known. The formulas are particularly important if problems involving several different right-hand sides $\mathbf{b}$ and one and the same coefficient matrix have to be solved. They are also important for the design of efficient iteration methods.

In [13] and, using a different approach, in [14] (see also [11]) it was shown that pseudo inverses of Toeplitz matrices (including also the rank deficient case) are generalized Bezoutians, more precisely 4-Bezoutians. A third approach to obtain this result for the full rank case via the augmented matrix approach was presented in [18]. This approach will be also the basis of the considerations in the present paper. The result claims that the pseudo inverse of T can be written in the form

$$T^{\dagger} = [\, I_n \; 0\,] \sum_{l=1}^{4} C_l^{+} C_l^{-},$$

where C_l^+ are $m \times m$ circulant and C_l^- $m \times m$ skewcirculant matrices. Matrix-vector multiplication by circulants and skewcirculants can be carried out with complexity $O(m \log m)$ if FFT is applied.

Note that the formula for the pseudo inverse can also be used to solve minimal norm problems. A minimal norm problem is to find, for a given vector $\mathbf{c}$, a vector $\mathbf{r}$ with minimal norm for which $T^*\mathbf{r} = \mathbf{c}$.

Let us discuss now the first stage of our approach. The fundamental system for a nonsingular Toeplitz matrix T that is involved in the inversion formula can be characterized in different ways. One of them is the following. A fundamental system consists of the first and last columns of the inverse of an $(n+1) \times (n+1)$ nonsingular extension of T. The Levinson algorithm [1] computes recursively the first and last columns of the inverses of the principal submatrices $[\, a_{i-j}\,]_{i,j=1}^{k}$ for $k = 1, \ldots, n+1$ and so a fundamental system. From computational view point it is more efficient to compute the normalized vectors, i.e., the solutions of the Yule-Walker equations. In the algorithm some "residuals" are computed via inner products. To avoid inner product calculations, which could be the bottleneck in parallel processing, the residuals can be precomputed by the Schur algorithm[2].

The Schur algorithm also computes the LU-factorization of T. The LU-factorization can be used to solve a Toeplitz system via back substitution without using the inversion formula. This way is in particular preferable for ill-conditioned Toeplitz matrices.

The situation is similar for pseudo inversion, which is closely related to the inversion of the augmented matrix. The variety of possible recursions is much larger than for inversion. In Section 5 we will discuss 9 versions of updating the fundamental systems after the matrix T is slightly changed.

From the updating formulas in particular 4 types of algorithms emerge: 1. algorithms based on row-and-column extensions, which is the most straightforward generalization of the classical recursions, 2. recursions based on pure column extensions, which are related to algorithms for QR-factorization, 3. recursions based on pure row extensions and 4. recursion based on order enlargement. The latter

[1]This algorithm is also often attributed to Durbin.

[2]The original Schur algorithm deals with functions rather than matrices. For Toeplitz matrices this algorithm is often attributed to Bareiss.

is in particular important in systems identification and model reduction. The 4 types of recursions can be visualized for a 5×3 Toeplitz matrix as follows:

$$
1.\ \begin{array}{ccc} \bullet & \circ & * \\ \bullet & \circ & * \\ \bullet & \circ & * \\ \circ & \circ & * \\ * & * & * \end{array}
\qquad
2.\ \begin{array}{ccc} \bullet & \circ & * \\ \bullet & \circ & * \\ \bullet & \circ & * \\ \bullet & \circ & * \\ \bullet & \circ & * \end{array}
\qquad
3.\ \begin{array}{ccc} \bullet & \bullet & \bullet \\ \bullet & \bullet & \bullet \\ \bullet & \bullet & \bullet \\ \circ & \circ & \circ \\ * & * & * \end{array}
\qquad
4.\ \begin{array}{ccc} \bullet & & \\ \bullet & \circ & \\ \bullet & \circ & * \\ \bullet & \circ & * \\ \bullet & \circ & * \\ \bullet & \circ & * \\ \bullet & \circ & * \end{array}
$$

Here the black circles mark the entries of the initial matrix, and black and white circles mark the matrix after the first recursion step.

In Section 6 we present the Levinson-type algorithms for these recursions, which include inner product calculations and in Section 7 the corresponding Schur-type recursions for the residuals. The Schur-type algorithm for column extensions computes also the QR-factorization $T = QR$ of the Toeplitz matrix T, in which $Q^*Q = I_n$ and R is upper triangular, whereas the Levinson-type algorithm computes Q and R^{-1}. These factorizations can be used to solve Toeplitz LS problems without using the formula for the pseudo inverse.

We are aware that the algorithms in this paper are, in general, unstable. One source of instability is the fact that in the fundamental systems vectors occur with magnitude $(T^*T)^{-1}$ that means with magnitude $\kappa(T)^2$, where $\kappa(T)$ denotes the condition number of the matrix. However, it is shown in [18] that there exists representations of the pseudo inverse in which only quantities with magnitude $\kappa(T)$ occur. Furthermore, it is known how to stabilize the general Schur algorithm (see for example Chapters 2 and 3 in [19]). For this reason we are confident that stable versions of the algorithms in this paper exist. This will be subject of forthcoming research.

Notations. The following notations will be used throughout the paper.

$\mathbf{e}_k$ will denote the kth vector, $\mathbf{e}_-$ the first, and $\mathbf{e}_+$ the last vector in the standard basis of $\mathbb{C}^n$. We do not distinguish between the column $\mathbf{e}_\pm$ and the row $\mathbf{e}_\pm^T$.

$\mathbf{0}$ stands for a zero row or column, O for any zero matrix, I_n is the $n \times n$ identity.

We use quasi-Matlab notation. If $\mathbf{a} = (a_i)$, then we denote $\mathbf{a}(k:l) = (\,a_i\,)_{i=k}^{l}$. $\mathbf{a}(k : l)$ can be a row or a column, depending on the context. If a_j is not a component of $\mathbf{a}$, then we set $a_j = 0$. We include cases in which k and l are negative or $l < k$. In the latter case the step size for i is -1.

Occasionally we will use polynomial language. For a matrix $A = [\,a_{ij}\,]$, $A(t,s)$ will denote the bivariate polynomial $A(t,s) = \sum_{i,j} a_{ij} t^{i-1} s^{j-1}$, and for $\mathbf{u} = (u_i)_{i=1}^n$ we set $\mathbf{u}(t) = \sum_{i=1}^n u_i t^{i-1}$. $A(t,s)$, $\mathbf{u}(t)$ is called the generating polynomial of A, $\mathbf{u}$, respectively.

2. Toeplitz Matrix Inversion – Revised

In this section we present some well-known facts concerning Toeplitz matrix inversion. This should serve as a guide for our approach to Toeplitz matrix pseudo inversion.

2.1. Formula for the inverse

The inverse of a Toeplitz matrix can be described in terms of the "Toeplitz displacement" operator ∇ which is defined as follows. For an $n \times m$ matrix $B = [b_{ij}]_{i=1}^{n}\,{}_{j=1}^{m}$, $\nabla(B)$ is, by definition the $(n+1) \times (m+1)$ matrix

$$\nabla(B) = [\, b_{ij} - b_{i-1,j-1} \,]_{i=1}^{n+1}\,{}_{j=1}^{m+1}.$$

Here we set $b_{ij} = 0$ if b_{ij} is not an entry of the matrix B. The transformation ∇ is left invertible and B can be recursively reconstructed from $\nabla(B)$.

In terms of generating polynomials we have

$$\nabla(B)(t,s) = (1 - ts)\, B(t,s)\ .$$

Thus, if $\nabla(B)$ has rank r and $\nabla(B) = \sum_{j=1}^{r} \mathbf{g}_j \mathbf{h_j}^T$, then

$$B(t,s) = \frac{1}{1-ts} \sum_{j=1}^{r} \mathbf{g}_j(t)\mathbf{h}_j(s)\ .$$

For an $m \times n$ Toeplitz matrix $T = [\, a_{i-j} \,]_{i=1}^{m}\,{}_{j=1}^{n}$, let ∂T denote the $(m-1) \times (n+1)$ Toeplitz matrix

$$\partial T = [\, a_{i-j} \,]_{i=1}^{m-1}\,{}_{j=0}^{n}\ .$$

If $m = n$ and T is nonsingular, then ∂T has full rank. Hence this matrix has a 2-dimensional kernel. The importance of the transformations ∇ and ∂ becomes clear in the following proposition, which is well known (see [15]).

Proposition 2.1. *The matrix B is the inverse of a Toeplitz matrix T if and only if the matrix $\nabla(B)$ has rank 2. Furthermore, for a nonsingular Toeplitz matrix T,*

$$\text{range}\, \nabla(T^{-1}) = \ker \partial T.$$

Any basis of $\ker \partial T$ will be called *fundamental system* for T and the matrix the columns of which form a fundamental system a *fundamental matrix.*

In view of the persymmetry of Toeplitz matrices, i.e., the fact that $J_n T J_n = T^T$, we have the following. If U is a fundamental matrix for T, then $J_{n+1}U$ is a fundamental matrix for T^T. Hence, for a fundamental matrix U for T we have a representation

$$\nabla(T^{-1}) = UC(J_{n+1}U)^T$$

for some nonsingular 2×2 matrix C.

For different choices of the fundamental system we have different representation. A "canonical" choice is the following. The fundamental matrix U for T is called *canonical* (with respect to a_{-n}) if $FU = I_2$, where

$$F = \begin{bmatrix} \mathbf{e}_+ \\ \mathbf{a}(-1 : -n) \end{bmatrix}.$$

If U is a canonical fundamental matrix, then

$$\nabla(T^{-1}) = U \begin{bmatrix} 0 & 1 \\ -1 & 0 \end{bmatrix} (J_{n+1}U)^T .$$

If $\mathbf{u}$ and $\mathbf{v}$ denote the columns of U, then we have

$$T^{-1}(t,s) = \frac{1}{1-ts}\left(\mathbf{u}(t)\mathbf{v}^{\#}(s) - \mathbf{v}(t)\mathbf{u}^{\#}(s)\right),$$

where $\mathbf{u}^{\#} = J_{n+1}\mathbf{u}$. That means that the inverse of a Toeplitz matrix is the Toeplitz-Bezoutian of the vectors of the canonical fundamental system.

Another characterization of a fundamental system uses a $(n+1)\times(n+1)$ extension of T. Let $\widetilde{T} = [\, a_{i-j}\,]_{i,j=1}^{n+1}$ be such an extension which is nonsingular, and let $\mathbf{x}_-$ be the first and $\mathbf{x}_+$ the last column of $\widetilde{T}^{-1}$. Then we have

$$\nabla(T^{-1}) = \frac{1}{\gamma}\left(\mathbf{x}_-(\mathbf{x}_+^{\#})^T - \mathbf{x}_+(\mathbf{x}_-^{\#})^T\right), \tag{2.1}$$

where $\gamma = \mathbf{e}_-\mathbf{x}_- = \mathbf{e}_+\mathbf{x}_+ \neq 0$. Below we will prefer to work with the normalized vectors $\mathbf{u}_\pm = \frac{1}{\gamma}\mathbf{x}_\pm$. These vectors are the solutions of equations

$$\widetilde{T}\mathbf{u}_\pm = \rho\,\mathbf{e}_\pm \,, \tag{2.2}$$

where ρ is some real number, with the additional condition that the first component of $\mathbf{u}_-$ and the last component of $\mathbf{u}_+$ are equal to 1, in other words, $\mathbf{u}_+(t)$ is monic and $\mathbf{u}_-(t)$ is comonic. We obtain the following formula for the inverse matrix.

Theorem 2.1. *The inverse of a nonsingular Toeplitz matrix T is given by*

$$T^{-1}(t,s) = \frac{1}{\rho}\,\frac{\mathbf{u}_-(t)\mathbf{u}_+^{\#}(s) - \mathbf{u}_+(t)\mathbf{u}_-^{\#}(s)}{1-ts} .$$

The formulas for the inverse matrix can be applied in different ways. The can be used to construct the inverse matrix recursively (see [25]). More important are matrix representations like the Gohberg-Semencul formula, which is built from triangular Toeplitz matrices. For more matrix representations of Bezoutians we refer to [8], [12], [26], [16].

Here we show how this formula can be brought into a form that only contains Fourier and diagonal matrices (see also [16] and [26] for slightly different representations). Let $N \geq n$, preferably a power of 2, and let $\omega_j^+ = \exp(\frac{(2j-2)\pi\sqrt{-1}}{N})$, $\omega_j^- = \exp(\frac{(2j-1)\pi\sqrt{-1}}{N})$ $(j = 1,\ldots,N)$ be the N th roots of 1 and -1, respectively. We introduce the Fourier matrices $\mathcal{F}_\pm = [\,(\omega_j^\pm)^k]_{j,k=1}^N$.

Let B be the $N\times N$ matrix obtained from T^{-1} by adding $N-n$ zero rows at the bottom and $N-n$ zero columns on the right. Since $\mathcal{F}_\pm\mathbf{x} = (\mathbf{x}(\omega_j^\pm))_{j=1}^N$, we have

$$\mathcal{F}_+ B\mathcal{F}_-^* = \frac{1}{\rho}\left[\frac{(\mathbf{u}_-(\omega_j^+)\mathbf{u}_+(\omega_k^-) - \mathbf{u}_+(\omega_j^+)\mathbf{u}(\omega_k^-))(\omega_k^-)^{-n}}{1-\omega_j^+(\omega_k^-)^*}\right]_{j,k=1}^N .$$

In view of and $\mathcal{F}_{\pm}^{-1} = \frac{1}{N}\mathcal{F}_{\pm}^{*}$ this can be written as

$$B = \frac{1}{\rho N^2}\mathcal{F}_{+}^{*}(D_{-+}\Omega D_{+-} - D_{++}\Omega D_{--})D\mathcal{F}_{-}\ , \tag{2.3}$$

where

$$D_{\pm+} = \operatorname{diag}(\mathbf{u}_{\pm}(\omega_j^{+}))_{j=1}^{N}, \quad D_{\pm-} = \operatorname{diag}(\mathbf{u}_{\pm}(\omega_j^{-}))_{j=1}^{N}, \quad D = \operatorname{diag}((\omega_j^{-})^{-n})_{j=1}^{N}$$

and $\Omega = [\,(1-\omega_j^{+}(\omega_k^{-})^{*})^{-1}\,]_{j,k=1}^{N}$. It remains to observe that $\Omega = \frac{1}{2}\mathcal{F}_{+}\mathcal{F}_{-}^{*}$ to derive the desired representation from (2.3).

2.2. Levinson algorithm

Let $T = T_n = [\,a_{i-j}\,]_{i,j=1}^{n}$ be a strongly nonsingular Toeplitz matrix, which means that all leading principal submatrices $T_k = [\,a_{i-j}\,]_{i,j=1}^{k}$ are nonsingular ($k = 1,\ldots,n$). Furthermore, let $T_{n+1} = [\,a_{i-j}\,]_{i,j=1}^{n+1}$ be a nonsingular Toeplitz extension of T_n. Let $\mathbf{x}_k^{\pm}$ denote the first and last columns of T_k^{-1}, respectively. According to (2.1) the inverse of T_n can be built from $\mathbf{x}_{n+1}^{\pm}$.

The Levinson algorithm computes $\mathbf{x}_k^{\pm}$ recursively via

$$[\,\mathbf{x}_{k+1}^{-}(t)\ \ \mathbf{x}_{k+1}^{+}(t)\,] = [\,\mathbf{x}_k^{-}(t)\ \ t\mathbf{x}_k^{+}(t)\,]\,\Phi_k\ ,$$

where

$$\Phi_k = \frac{1}{1-\gamma_k^{-}\gamma_k^{+}}\begin{bmatrix} 1 & -\gamma_k^{+} \\ -\gamma_k^{-} & 1 \end{bmatrix}$$

and

$$\gamma_k^{-} = \mathbf{a}(k:1)\mathbf{x}_k^{-}, \quad \gamma_k^{+} = \mathbf{a}(1:-k)\mathbf{x}_k^{+}\ .$$

the recursion can be started with $\mathbf{x}_1^{\pm} = \frac{1}{a_0}$.

The number of multiplications can be reduced if instead of the $\mathbf{x}_k^{\pm}$ the normalized vectors $\mathbf{u}_k^{\pm}$ are computed, which are the solutions of the equations

$$T_k\mathbf{u}_k^{\pm} = \rho_k^{\pm}\mathbf{e}_{\pm}$$

for which the first component of $\mathbf{u}_k^{-}$ and the last component of $\mathbf{u}_k^{+}$ are equal to 1. In other words, $\mathbf{u}_k^{-}(t)$ is comonic and $\mathbf{u}_k^{+}(t)$ is monic. We have

$$\rho_k^{-} = \mathbf{a}(0:1-k)\mathbf{u}_k^{-} = \rho_k^{+} = \mathbf{a}(k-1:0)\mathbf{u}_k^{+}.$$

Theorem 2.2. *For the vectors $\mathbf{u}_k^{\pm}$ and the numbers $\rho_k = \rho_k^{\pm}$ the recursions*

$$[\,\mathbf{u}_{k+1}^{-}(t)\ \ \mathbf{u}_{k+1}^{+}(t)\,] = [\,\mathbf{u}_k^{-}(t)\ \ t\mathbf{u}_k^{+}(t)\,]\,\Theta_k$$

and

$$\rho_{k+1} = \rho_k - \frac{\alpha_k^{-}\alpha_k^{+}}{\rho_k}$$

hold, where

$$\alpha_k^{-} = \mathbf{a}(k:1)\mathbf{u}_k^{-}, \quad \alpha_k^{+} = \mathbf{a}(1:-k)\mathbf{u}_k^{+} \qquad \text{and} \qquad \Theta_k = \begin{bmatrix} 1 & -\frac{\alpha_k^{+}}{\rho_k} \\ -\frac{\alpha_k^{-}}{\rho_k} & 1 \end{bmatrix}.$$

To start the recursion we observe that $\mathbf{u}_1^{\pm} = 1$ and $\rho_1 = a_0$.

In each step of the Levinson algorithm one has to compute 2 inner products of length k and multiply 2 vectors with length k by a scalar. This results in about $2\,n^2$ (complex) multiplications and the same number of additions.

Note that the upper triangular matrix formed by the vectors $\mathbf{u}_k^+$ represent the upper triangular factor of a UL-factorization of T^{-1}. The lower triangular factor is obtained from the $\mathbf{u}_k^-$ after some permutation and scaling. Using the persymmetry of Toeplitz matrices one can obtain also the LU-factorization of T^{-1} from it.

2.3. Schur Algorithm

In order to avoid inner product calculations, the parameters in the recursions for the vectors $\mathbf{u}_k$ (or $\mathbf{x}_k$) can be precomputed by the Schur algorithm.

We introduce numbers

$$\rho_{k,j}^- = \mathbf{a}(-j : -j-k+1)\,\mathbf{u}_k^-, \quad \alpha_{k,j}^+ = \mathbf{a}(-j : -j-k+1)\,\mathbf{u}_k^+,$$

$$\alpha_{k,j}^- = \mathbf{a}(k+j-1 : j)\,\mathbf{u}_k^-, \quad \rho_{k,j}^+ = \mathbf{a}(k+j-1 : j)\,\mathbf{u}_k^+,$$

for $j = 0, \ldots, n-k+1$. Then the numbers $\rho_{k,0}^{\pm}$ and $\alpha_{k,1}^{\pm}$ are the numbers $\rho_k^{\pm}$ and $\alpha_k^{\pm}$, respectively, that occur in the Levinson recursion, and $\alpha_{k,0}^{\pm} = 0$. In this subsection we denote by $\rho_k^{\pm}$ and $\alpha_k^{\pm}$ the vectors $\rho_k^{\pm} = (\rho_{k,j}^{\pm})_{j=0}^{n-k+1}$ and $\alpha_k^{\pm} = (\alpha_{k,j}^{\pm})_{j=0}^{n-k+1}$.

Theorem 2.3. *The polynomials $\rho_k^{\pm}(t)$ and $\alpha_k^{\pm}(t)$ satisfy the recursion*

$$\begin{bmatrix} \rho_{k+1}^-(t) & \alpha_{k+1}^+(t) \\ \alpha_{k+1}^-(t) & \rho_{k+1}^+(t) \end{bmatrix} = \begin{bmatrix} \rho_k^-(t) & t^{-1}\alpha_k^+(t) \\ t^{-1}\alpha_k^-(t) & \rho_k^+(t) \end{bmatrix} \Theta_k \pmod{t^{n-k+1}},$$

where

$$\Theta_k = \begin{bmatrix} 1 & -\frac{\alpha_{1,k}^+}{\rho_{0,k}} \\ -\frac{\alpha_{1,k}^-}{\rho_{0,k}} & 1 \end{bmatrix}.$$

Here "mod t^{n-k+1}" means that the terms with t^{n-k+1} has to be deleted, so that the length of the vectors $\rho_k^{\pm}$ and $\alpha_k^{\pm}$ shrinks by 1 in each step.

For initialization we mention that $\rho_{1,j}^- = \alpha_{1,j}^+ = a_{-j}$ and $\rho_{1,j}^+ = \alpha_{1,j}^- = a_j$ for $j = 0, \ldots, n$.

To compute a fundamental system using the Schur algorithm instead of inner product calculations requires about $3\,n^2$ multiplications and about the same number of additions.

Note that the matrix built by the vectors ρ_k^+ is the upper triangular factor of an LU-factorization of T. The lower triangular factor can be obtained from ρ_k^- after some permutations and scaling. That means that the Schur algorithm computes the LU-factorization of T with $2\,n^2$ multiplications and about the same number of additions. The factorization can be used to solve a Toeplitz system via back substitution.

3. Augmented Matrix Approach

In this section we describe the familiar augmented matrix approach for general matrices (see [22] and [3], Section 2.5.3). Let A be a complex $m \times n$ matrix with $m \geq n$. Let, for any $m \times n$ matrix A, aug(A) denote "augmented" $(m+n) \times (m+n)$ matrix

$$\operatorname{aug}(A) = \begin{bmatrix} I_m & A \\ A^* & O \end{bmatrix}.$$

The relation between the matrices A and aug(A) is described by the following lemma, which is well-known and easily verified.

Proposition 3.1. *The matrix* aug(A) *is nonsingular if and only if the matrix* A *has full rank. If the latter is the case, then*

$$\operatorname{aug}(A)^{-1} = \begin{bmatrix} I_m - AA^\dagger & (A^\dagger)^* \\ A^\dagger & -(A^*A)^{-1} \end{bmatrix},$$

where $A^\dagger := (A^*A)^{-1}A^*$ *is the pseudo inverse of* A. *In particular,*

$$A^\dagger = [O \;\; I_n] \operatorname{aug}(A)^{-1} \begin{bmatrix} I_m \\ O \end{bmatrix}.$$

Note that $I_m - AA^\dagger$ is the orthoprojector onto the kernel of A^*.

Corollary 3.1. *The solution of the least squares problem is equal to the second block component* $\mathbf{x}$ *of the solution of the system*

$$\operatorname{aug}(A) \begin{bmatrix} \mathbf{r} \\ \mathbf{x} \end{bmatrix} = \begin{bmatrix} \mathbf{b} \\ \mathbf{0} \end{bmatrix}.$$

Note that first block component of the solution of

$$\operatorname{aug}(A) \begin{bmatrix} \mathbf{r} \\ \mathbf{x} \end{bmatrix} = \begin{bmatrix} \mathbf{0} \\ \mathbf{c} \end{bmatrix}.$$

provides the solution of the minimal norm problem, which is to find $\mathbf{r}$ with minimal norm satisfying $T^*\mathbf{r} = \mathbf{c}$.

Let us clarify what means QR-factorization of A in the augmented matrix framework. Suppose that $A = QR$, where $Q^*Q = I_n$ and R is upper triangular. Then we have

$$\operatorname{aug}(A) \begin{bmatrix} -Q \\ R^{-1} \end{bmatrix} = \begin{bmatrix} O \\ -R^* \end{bmatrix}.$$

Vice versa, from a representation

$$\operatorname{aug}(A) \begin{bmatrix} W \\ U \end{bmatrix} = \begin{bmatrix} O \\ L \end{bmatrix}$$

in which U is $n \times n$ upper and L is $n \times n$ lower triangular one can get the QR-factorization of A by scaling the columns of W, U and L.

4. Formula for the pseudo inverse

In this section we present a formula for the pseudo inverse of a full rank Toeplitz matrix that can be considered as generalization of Theorem 2.1. The material is based on some result in [18] in which a formula based on a "canonical" fundamental systems is given. We adopt this to the purpose of the present paper by representing the pseudo inverse by columns of the inverse of the augmented matrix of a $(m+1)\times(n+1)$ extension of T.

To begin with we mention a result from [17] (see also [11]) claiming that Proposition 2.1 can be generalized to Toeplitz block matrices. Let for a block matrix $B = [\,B_{ij}\,]_{i=1}^{q}\,{}_{j=1}^{p}$ and for a block matrix $A = [\,T_{ij}\,]_{i=1}^{p}\,{}_{j=1}^{q}$ with Toeplitz blocks T_{ij}, $\nabla(B)$ and ∂A be defined block-wise:

$$\nabla(B) = [\,\nabla(B_{ij})\,]_{i=1}^{q}\,{}_{j=1}^{p}, \quad \partial A = [\,\partial T_{ij}\,]_{i=1}^{p}\,{}_{j=1}^{q}.$$

Proposition 4.1. [17] *The $q\times p$ block matrix B is the inverse of a $p\times q$ block matrix with Toeplitz blocks A if and only if $\nabla(B)$ has rank $p+q$. Furthermore, for a nonsingular matrix A,*

$$\operatorname{range}\nabla(A^{-1}) = \ker\partial A.$$

From now on, let T be an $m\times n$ Toeplitz matrix with rank n. Applying Proposition 4.1 to the augmented matrix $\operatorname{aug}(T)$ we obtain the following.

Corollary 4.1. *For the Toeplitz matrix T,*

$$\operatorname{rank}\nabla(T^{\dagger}) \le \operatorname{rank}\nabla(\operatorname{aug}(T)^{-1}) = 4$$

and

$$\operatorname{range}\nabla(\operatorname{aug}(T)^{-1}) = \ker\partial\operatorname{aug}(T),$$

$$\operatorname{range}\nabla(T^{\dagger}) = [\,O\;\; I_{n+1}\,]\ker\partial\operatorname{aug}(T)\,.$$

Any $(m+n+2)\times 4$ matrix the columns of which form a basis of the kernel of $\partial\operatorname{aug}(T)$ will be called a *(complete) fundamental matrix for* $\operatorname{aug}(T)$ and the columns of U will be called *fundamental system.* The $(n+1)\times 4$ lower part of a complete fundamental matrix will be called *reduced fundamental matrix.*

As consequence of Corollary 4.1 and the fact that $\operatorname{aug}(T)$ and $\nabla(\operatorname{aug}(T))$ are hermitian, we conclude that for any fundamental matrix U there is an hermitian 4×4 matrix M such that

$$\nabla(\operatorname{aug}(T)^{-1}) = UMU^{*}\,. \tag{4.1}$$

There are different possibilities to characterize fundamental matrices. A "canonical" one is the following. Let F_{right} and F_{left} be defined by

$$F_{\text{right}} = \begin{bmatrix} \mathbf{e}_- & \mathbf{a}(0:-n) \\ \mathbf{a}(0:m)^* & \mathbf{0} \\ \mathbf{e}_+ & \mathbf{0} \\ \mathbf{0} & \mathbf{e}_+ \end{bmatrix}, \; F_{\text{left}} = \begin{bmatrix} \mathbf{e}_- & \mathbf{0} \\ \mathbf{0} & \mathbf{e}_- \\ \mathbf{e}_+ & -\mathbf{a}(m:m-n) \\ -\mathbf{a}(-n:m-n) & \mathbf{0} \end{bmatrix}.$$

The fundamental matrix U will be called right-canonical if $F_{\text{right}}U = I_4$ and left-canonical if $F_{\text{left}}U = I_4$.

In [18] the following is shown. For convenience of the reader we sketch the proof.

Proposition 4.2. *If U_{right} is the right and U_{left} the left canonical matrix for* aug (T), *then*

$$\nabla(\text{aug}\,(T)^{-1}) = U_{\text{right}}U^*_{\text{left}}\,. \tag{4.2}$$

Proof. Let S_n denote the matrix of the forward shift with ones on the diagonal below the main diagonal and zeros elsewhere, and let S_{mn} denote the direct sum of S_m and S_n. Then

$$\begin{aligned} TS_n - S_mT &= \mathbf{e}_-\mathbf{a}(-1: \ n)^* - \mathbf{a}(-n, m-n-1)\,\mathbf{e}_+^*\,, \\ T^*S_m - S_nT^* &= \mathbf{e}_-\overline{\mathbf{a}(1:m)}^* - \overline{\mathbf{a}(m, m-n-1)}\,\mathbf{e}_+^*\,. \end{aligned}$$

This implies

$$\text{aug}\,(T)S_{mn} - S_{mn}\,\text{aug}\,(T) = G_+G_-^*\,,$$

where

$$\begin{aligned} G_+ &= \begin{bmatrix} \mathbf{e}_- & -\mathbf{a}(-n:m-n-1) & \mathbf{0} & \mathbf{0} \\ \mathbf{0} & \mathbf{0} & \mathbf{e}_- & -\overline{\mathbf{a}(m:m-n-1)} \end{bmatrix}, \\ G_- &= \begin{bmatrix} \mathbf{0} & \mathbf{0} & \mathbf{a}(1:m) & \mathbf{e}_+ \\ \overline{\mathbf{a}(-1:-n)} & \mathbf{e}_+ & \mathbf{0} & \mathbf{0} \end{bmatrix}. \end{aligned}$$

Hence

$$\Delta = \text{aug}\,(T)^{-1}S_{mn} - S_{mn}\,\text{aug}\,(T)^{-1} = -W_+W_-^*\,,$$

where $W_\pm = \text{aug}\,(T)^{-1}G_\pm$. To complete the proof we observe that for any $m \times n$ matrix B the matrix $BS_n - S_mB$ is the $m \times n$ upper right corner of the $(m+1) \times (n+1)$ matrix $\nabla(B)$. Comparing the entries we find that the block matrix consisting of upper right corners of the blocks in $U_{\text{right}}U^*_{\text{left}}$ is equal to Δ. It remains to check (4.2) for the remaining entries. □

If U is any other fundamental matrix, then (4.1) holds, where

$$M^{-1} = (F_{\text{left}}U)^*(F_{\text{right}}U)\,. \tag{4.3}$$

As in Section 2.1 for the inverse, we are going to design a formula for the pseudo inverse that includes data from an extension of T. Let $\widetilde{T} = [\,a_{i-j}\,]_{i=1}^{m+1}\,{}_{j=1}^{n+1}$ be an $(m+1) \times (n+1)$ extension of T with full rank $n+1$. Clearly, $\widetilde{T}$ depends on the choice of a_m and a_{-n}. Let X_- be the $(m+n+2) \times 2$ matrix consisting of the first and the $(m+2)$th columns of aug $(\widetilde{T})^{-1}$, and X_+ the $(m+n+2) \times 2$ matrix consisting of the $(m+1)$th and the last columns of this matrix. Then it is easily checked that $X = [\,X_-\ X_+\,]$ is a fundamental matrix.

The columns of $X_\pm$ can be interpreted in terms of $\widetilde{T}$ as follows. Suppose that $X_\pm = \begin{bmatrix} \mathbf{s}_1^\pm & \mathbf{s}_2^\pm \\ \mathbf{x}_1^\pm & \mathbf{x}_2^\pm \end{bmatrix}$. Then $\mathbf{x}_1^\pm$ is the first and the last column of $\widetilde{T}^\dagger$, and $\mathbf{x}_2^\pm$ is the last and first column of $(\widetilde{T}^*\widetilde{T})^{-1}$, respectively. Furthermore, $\widetilde{T}\mathbf{x}_1^{\pm} = \mathbf{s}_1^{\pm} - \mathbf{e}_{\pm}$ and $\widetilde{T}\mathbf{x}_2^{\pm} = \mathbf{s}_2^{\pm}$. The vector $(\mathbf{s}_2^{\pm})^*$ is also the last and first row of $T^{\dagger}$, respectively.

We introduce the 2×2 matrices

$$\gamma_{\pm} = \begin{bmatrix} \mathbf{e}_{\pm} & \mathbf{0} \\ \mathbf{0} & \mathbf{e}_{\pm} \end{bmatrix} X_{\pm}, \quad \delta_{\pm} = \begin{bmatrix} \mathbf{e}_{\pm} & \mathbf{0} \\ \mathbf{0} & \mathbf{e}_{\pm} \end{bmatrix} X_{\mp}. \tag{4.4}$$

Since $\operatorname{aug}(\widetilde{T})$ is hermitian, $\gamma_{\pm}$ are hermitian and $\delta_- = \delta_+^*$.

Then we have

$$F_{\text{right}} X = \begin{bmatrix} I_2 & O \\ \delta_+ & \gamma_+ \end{bmatrix}, \quad F_{\text{left}} X = \begin{bmatrix} \gamma_- & \delta_- \\ O & -I_2 \end{bmatrix}.$$

Hence the matrix in (4.3) is given by

$$M^{-1} = \begin{bmatrix} \gamma_- & O \\ O & -\gamma_+ \end{bmatrix}.$$

From (4.1) we conclude now the following.

Proposition 4.3. *The matrix* $\operatorname{aug}(T)^{-1}$ *is given by*

$$\nabla(\operatorname{aug}(T)^{-1}) = X_-\gamma_-^{-1}X_-^* - X_+\gamma_+^{-1}X_+^* .$$

The matrix $\nabla(T^{\dagger})$ is the left lower corner of $\nabla(\operatorname{aug}(T)^{-1})$. Let $X_{\pm} = \begin{bmatrix} \mathbf{q}_{\pm} \\ \mathbf{x}_{\pm} \end{bmatrix}$. Then matrix $T^{\dagger}$ is given by the first and last columns of $\widetilde{T}^{\dagger}$ and $(\widetilde{T}^*\widetilde{T})^{-1}$ via

$$\nabla(T^{\dagger}) = \mathbf{x}_+\gamma_+^{-1}\mathbf{q}_+^* - \mathbf{x}_-\gamma_-^{-1}\mathbf{q}_-^* , \tag{4.5}$$

where

$$\mathbf{q}_{\pm} = \mathbf{e}_{\pm}\,[1 \;\; 0] - \widetilde{T}\mathbf{x}_{\pm} .$$

As for inversion, we will prefer to consider the normalized block vectors $U_{\pm} = X_{\pm}\gamma_{\pm}^{-1}$. Then

$$\operatorname{aug}(\widetilde{T})\, U_{\pm} = \begin{bmatrix} \mathbf{e}_{\pm}\rho_1^{\pm} \\ \mathbf{e}_{\pm}\rho_2^{\pm} \end{bmatrix}, \tag{4.6}$$

where $\rho_{\pm} = \begin{bmatrix} \rho_1^{\pm} \\ \rho_2^{\pm} \end{bmatrix} = \gamma_{\pm}^{-1}$. Observe that the matrix polynomial $U_+(t)$ is monic and $U_-(t)$ is comonic.

Rewriting (4.5) in terms of $U_{\pm}$ we arrive at the following.

Theorem 4.1. *The pseudo inverse $T^{\dagger}$ of the Toeplitz matrix T is given by the normalized block vectors* $U_{\pm} = \begin{bmatrix} \mathbf{r}_{\pm} \\ \mathbf{u}_{\pm} \end{bmatrix}$ *via*

$$T^{\dagger}(t,s) = \frac{\mathbf{u}_+(t)\rho_+^{-1}\mathbf{r}_+^*(s) - \mathbf{u}_-(t)\rho_-^{-1}\mathbf{r}_-^*(s)}{1 - ts},$$

where,

$$\mathbf{r}_{\pm} = \mathbf{e}_{\pm}\rho_1^{\pm} - \widetilde{T}\mathbf{u}_{\pm} .$$

It can be checked that in case that $m = n$, i.e., $\widetilde{T}$ is nonsingular, the formula in Theorem 4.1 goes over into the formula of Theorem 2.1.

Like the formula for the inverse of a Toeplitz matrix, this formula can be transformed into a form that contains only diagonal and Fourier matrices. To show this we use the notations introduced in 2.1.

We choose $N \geq m$, preferable a power of 2. Let B denote the matrix obtained from $T^\dagger$ after adding $N - n$ zero rows at the bottom and $N - m$ zero columns at the right. Let

$$\mathbf{u}_\pm(\omega_j^+)\rho_\pm^{-1} = [\,\xi_{1j}^\pm \;\; \xi_{2j}^\pm\,], \quad \mathbf{r}_\pm(\omega_j^-) = [\,\eta_{1j}^\pm \;\; \eta_{2j}^\pm\,]\,.$$

Then we obtain

$$B = \frac{1}{N^2}\,\mathcal{F}_+^* \sum_{l=1}^{2} \left(D(\xi_l^+)\Omega D(\eta_l^+)^* - D(\xi_l^-)\Omega D(\eta_l^-)^*\right)\mathcal{F}_-\,,$$

where $\xi_l^\pm = (\xi_{lj})_{j=1}^N$ and, for any vector $\xi = (\xi_j)_{j=1}^N$, $D(\xi)$ denotes the diagonal matrix $\operatorname{diag}(\xi_j)_{j=1}^N$. Recall that $\Omega = \frac{1}{2}\mathcal{F}_+\mathcal{F}_-^*$.

5. Updating and Downdating the Fundamental Matrix

In this section we study how the fundamental system of a Toeplitz matrix changes after the matrix is modified. In all cases T will denote the original $m \times n$ matrix and $\widetilde{T}$ the modification. U will be a $(m + n + 2) \times 4$ fundamental matrix for $\operatorname{aug}(T)$ and $\widetilde{U}$ a fundamental matrix for $\operatorname{aug}(\widetilde{T})$. $U(t)$ and $\widetilde{U}(t)$ will stand for the corresponding generating 2×4 matrix polynomials. Furthermore, we denote by $\mathcal{C}$ the kernel of $\partial \operatorname{aug}(T)$ and by $\widetilde{\mathcal{C}}$ the kernel of $\partial \operatorname{aug}(\widetilde{T})$. $\mathcal{C}(t)$ and $\widetilde{\mathcal{C}}(t)$ will stand for the spaces of corresponding generating polynomials.

We utilize the fact that with U the matrix $U\Theta$ is again a fundamental matrix, if Θ is a nonsingular 4×4 matrix. The idea is first to adapt the fundamental matrix to the desired modification by multiplying U from the right by an appropriate nonsingular Θ. Then the modification can easily be carried out.

We will consider 9 cases of modifications. In all cases we will introduce test matrices F_- and F_+, and in all cases F will denote the matrix $F = \begin{bmatrix} F_- \\ F_+ \end{bmatrix}$.

Without explicitly mentioning it we frequently will make use of the following fact.

Lemma 5.1. *If a nonsingular matrix A and its inverse has the block representation*

$$A = \begin{bmatrix} A_{11} & A_{12} \\ A_{21} & A_{22} \end{bmatrix}, \quad A^{-1} = \begin{bmatrix} C_{11} & C_{12} \\ C_{21} & C_{22} \end{bmatrix},$$

where A_{11} and C_{11} are square and of the same size, then A_{11} is nonsingular if and only if C_{22} is nonsingular.

5.1. Row-and-Column Deletion

First we consider the change of U after the deletion of a row at the bottom and a column at the right of the matrix, i.e., $\widetilde{T} = [\,a_{i-j}\,]_{i=1}^{m-1}\,{}_{j=1}^{n-1}$.

We form an $2 \times (m+n+2)$ matrix of test 2 functionals

$$F_- = \begin{bmatrix} \mathbf{e}_- & \mathbf{0} \\ \mathbf{0} & \mathbf{e}_- \end{bmatrix}, \quad F_+ = \begin{bmatrix} \mathbf{e}_+ & \mathbf{0} \\ \mathbf{0} & \mathbf{e}_+ \end{bmatrix}$$

Then the following is easily checked.

Lemma 5.2. *Let $U_0(t) \in \mathcal{C}(t)$. If $F_+U_0 = \mathbf{0}$, then $U_0(t) \in \widetilde{\mathcal{C}}(t)$. If $F_-U_0 = \mathbf{0}$, then $t^{-1}U_0(t) \in \widetilde{\mathcal{C}}(t)$.*

We consider the 4×4 matrix $M = FU$. M is formed by entries of the matrix. M is nonsingular if $\operatorname{aug}(\widetilde{T})$ is nonsingular, i.e., if and only if $\widetilde{T}$ has full rank.

To obtain a fundamental matrix for $\operatorname{aug}(\widetilde{T})$ we need that M has the form

$$\begin{bmatrix} \times & \times & 0 & 0 \\ \times & \times & 0 & 0 \\ 0 & 0 & \times & \times \\ 0 & 0 & \times & \times \end{bmatrix}. \tag{5.1}$$

Theorem 5.1. *If Θ is a matrix such that $M\Theta$ has the form* (5.1), *then a fundamental matrix for* $\operatorname{aug}(\widetilde{T})$ *is given by*

$$\widetilde{U}(t) = U(t)\,\Theta \begin{bmatrix} I_2 & O \\ O & t^{-1}I_2 \end{bmatrix}.$$

5.2. Row-and-Column Extension

Next we consider the change of U after T is extended by a row at the bottom and a column at the right, i.e., $\widetilde{T} = [\,a_{i-j}\,]_{i=1}^{m+1}\,{}_{j=1}^{n+1}$, where a_m and a_{-n} are new parameters. We form $2 \times (m+n+2)$ matrices of test functionals (depending on a_m and a_{-n})

$$F_- = \begin{bmatrix} \mathbf{e}_- & \mathbf{a}(0:-n) \\ \mathbf{a}(0:m)^* & \mathbf{0} \end{bmatrix}, \quad F_+ = \begin{bmatrix} \mathbf{e}_+ & \mathbf{a}(m:m-n) \\ \mathbf{a}(-n:m-n)^* & \mathbf{0} \end{bmatrix}.$$

Note that the rows of $F_\pm$ are rows of $\operatorname{aug}(\widetilde{T})$. The following is easily checked.

Lemma 5.3. *Let $U_0(t) \in \mathcal{C}(t)$. If $F_+U_0 = \mathbf{0}$, then $U_0(t) \in \widetilde{\mathcal{C}}(t)$. If $F_-U_0 = \mathbf{0}$, then $tU_0(t) \in \widetilde{\mathcal{C}}(t)$.*

We consider the matrix $M = FU$. This matrix is nonsingular if $\operatorname{aug}(\widetilde{T})$ is nonsingular, i.e., if $\widetilde{T}$ has full rank.

Theorem 5.2. *If Θ is a matrix such that $M\Theta$ has the form* (5.1), *then a fundamental matrix for* $\operatorname{aug}(\widetilde{T})$ *is given by*

$$\widetilde{U}(t) = U(t)\,\Theta \begin{bmatrix} I_2 & O \\ O & tI_2 \end{bmatrix}.$$

5.3. Row Deletion

Now we consider the deletion of a row, i.e., $\widetilde{T} = [\,a_{i-j}\,]_{i=1}^{m-1}\,{}_{j=1}^{n}$. We introduce

$$F_- = \begin{bmatrix} \mathbf{e}_- & \mathbf{0} \\ \mathbf{0} & \mathbf{e}_+ \\ \mathbf{a}(-n:m-n)^* & \mathbf{0} \end{bmatrix}, \quad F_- = \begin{bmatrix} \mathbf{e}_+ & \mathbf{0} \end{bmatrix}.$$

With these test matrices Lemma 5.2 holds.

With the same arguments as above, one can show the following.

Theorem 5.3. *The matrix $M = FU$ is nonsingular if $\widetilde{T}$ has full rank. Furthermore, if Θ is a matrix such that $M\Theta$ has the form*

$$\begin{bmatrix} \times & \times & \times & 0 \\ \times & \times & \times & 0 \\ \times & \times & \times & 0 \\ 0 & 0 & 0 & \times \end{bmatrix}, \tag{5.2}$$

then a fundamental matrix for aug $(\widetilde{T})$ *is given by*

$$\widetilde{U}(t) = U(t)\,\Theta \begin{bmatrix} I_3 & \mathbf{0} \\ \mathbf{0} & t^{-1} \end{bmatrix}.$$

5.4. Row Extension

Next we consider the extension of T by a row at the bottom, i.e., $\widetilde{T} = [a_{i-j}]_{i=1}^{m+1}\,{}_{j=1}^{n}$.

$$F_- = \begin{bmatrix} \mathbf{e}_- & \mathbf{a}(0:-n) \\ \mathbf{a}(0:m)^* & \mathbf{0} \\ \mathbf{0} & \mathbf{e}_+ \end{bmatrix}, \quad F_+ = \begin{bmatrix} \mathbf{0} & \mathbf{a}(m:m-n) \end{bmatrix}.$$

The Lemma 5.2 holds for this setting, which implies the following.

Theorem 5.4. *The matrix $M = FU$ is nonsingular, and if Θ is a matrix such that $M\Theta$ has the form* (5.2), *then a fundamental matrix for* aug $(\widetilde{T})$ *is given by*

$$\widetilde{U}(t) = U(t)\Theta \begin{bmatrix} I_3 & 0 \\ 0 & t \end{bmatrix}.$$

5.5. Column Deletion

Now let $\widetilde{T}$ be obtained by deleting one column at the right, i.e., $\widetilde{T} = [\,a_{i-j}\,]_{i=1}^{m}\,{}_{j=1}^{n-1}$ and

$$F_- = \begin{bmatrix} \mathbf{e}_- & \mathbf{0} \\ \mathbf{0} & \mathbf{e}_- \\ \mathbf{0} & \mathbf{a}(m:m-n) \end{bmatrix}, \quad F_+ = \begin{bmatrix} \mathbf{0} & \mathbf{e}_+ \end{bmatrix}.$$

Then Lemma 5.2 holds with this setting and the following.

Theorem 5.5. *The matrix $M = FU$ is nonsingular, and if Θ is a matrix such that $M\Theta$ has the form* (5.2), *then a fundamental matrix for* aug $(\widetilde{T})$ *is given by*

$$\widetilde{U}(t) = U(t)\,\Theta \begin{bmatrix} I_3 & \mathbf{0} \\ \mathbf{0} & t^{-1} \end{bmatrix}.$$

5.6. Column extension

Next we consider column extensions at the right, i.e., $\widetilde{T} = [\, a_{i-j} \,]_{i=1}^{m} {}_{j=1}^{n+1}$. We define

$$F_- = \begin{bmatrix} \mathbf{e}_- & \mathbf{a}(0:-n) \\ \mathbf{a}(0:m)^* & \mathbf{0} \\ \mathbf{e}_+ & \mathbf{0} \end{bmatrix}, \quad F_+ = \begin{bmatrix} \mathbf{a}(-n:m-n)^* & 0 \end{bmatrix}.$$

Then Lemma 5.3 holds and the following.

Theorem 5.6. *The matrix $M = FU$ is nonsingular if $\widetilde{T}$ has full rank, and if Θ is a matrix such that $M\Theta$ has the form* (5.2), *then a fundamental matrix for* aug $(\widetilde{T})$ *is given by*

$$\widetilde{U}(t) = U(t)\,\Theta \begin{bmatrix} I_3 & \mathbf{0} \\ \mathbf{0} & t \end{bmatrix}.$$

5.7. Sliding Window

Now we consider the so called "sliding window", which means that $\widetilde{T}$ is obtained from T after a row is added at the bottom and the first row is deleted, i.e., $\widetilde{T} = [\, a_{i-j} \,]_{i=2}^{m+1} {}_{j=1}^{n}$. In this case we define

$$F_- = \begin{bmatrix} 0 & \mathbf{e}_+ \\ \mathbf{a}(0:m)^* & \mathbf{0} \end{bmatrix}, \quad F_+ = \begin{bmatrix} \mathbf{e}_- & 0 \\ \mathbf{e}_+ & \mathbf{a}(m:m-n) \end{bmatrix}$$

and consider $M = FU$.

Lemma 5.4. *Let $U_0(t) \in \mathcal{C}(t)$. If $F_+U_0 = \mathbf{0}$, then* $\operatorname{diag}(t^{-1}, 1)\, U_0(t) \in \widetilde{\mathcal{C}}(t)$. *If $F_-U_0 = \mathbf{0}$, then* $\operatorname{diag}(1, t)\, U_0(t) \in \widetilde{\mathcal{C}}(t)$.

Theorem 5.7. *If Θ is a matrix such that $M\Theta$ has the form* (5.1), *then a fundamental matrix for* aug $(\widetilde{T})$ *is given by*

$$\widetilde{U}(t) = \begin{bmatrix} 1 & 0 \\ 0 & t \end{bmatrix} U(t)\,\Theta \begin{bmatrix} I_2 & O \\ O & t^{-1}I_2 \end{bmatrix}.$$

5.8. Order Enlargement

Now we consider a modification of T which is important for systems identification. Let $\widetilde{T} = [\, a_{i-j} \,]_{i=1}^{m-1} {}_{j=0}^{n}$. Note that the $(m-1) \times (n+1)$ matrix $\widetilde{T}$ is built from the same data $(a_j)_{j=1-n}^{m-1}$ as T.

We define

$$F_- = \begin{bmatrix} \mathbf{e}_+ & \mathbf{0} \\ \mathbf{a}(0:m)^* & \mathbf{0} \end{bmatrix}, \quad F_+ = \begin{bmatrix} \mathbf{e}_- & \mathbf{0} \\ \mathbf{a}(-n:m-n)^* & \mathbf{0} \end{bmatrix}.$$

Then Lemma 5.4 holds with this setting.

Theorem 5.8. *If Θ is a matrix such that $M\Theta$ has the form* (5.1), *then a fundamental matrix for* aug $(\widetilde{T})$ *is given by*

$$\widetilde{U}(t) = \begin{bmatrix} t^{-1} & 0 \\ 0 & 1 \end{bmatrix} U(t)\,\Theta \begin{bmatrix} I_2 & O \\ O & tI_2 \end{bmatrix}.$$

5.9. Order Reduction

Finally we consider the $(m+1) \times (n-1)$ matrix $\widetilde{T} = [\, a_{i-j} \,]_{i=0}^{m} {}_{j=1}^{n-1}$. We define

$$F_- = \begin{bmatrix} \mathbf{0} & \mathbf{e}_- \\ \mathbf{0} & \mathbf{a}(m:m-n) \end{bmatrix}, \quad F_+ = \begin{bmatrix} \mathbf{0} & \mathbf{e}_+ \\ \mathbf{0} & \mathbf{a}(-1:-n) \end{bmatrix}.$$

Then the following is true.

Lemma 5.5. *Let $U_0(t) \in \mathcal{C}(t)$. If $F_+U_0 = \mathbf{0}$, then* $\operatorname{diag}(t, 1)\, U_0(t) \in \widetilde{\mathcal{C}}(t)$. *If $F_-U_0 = \mathbf{0}$, then* $\operatorname{diag}(1, t^{-1})\, U_0(t) \in \widetilde{\mathcal{C}}(t)$.

Theorem 5.9. *Let U be a fundamental matrix for* $\operatorname{aug}(T)$ *and $M = FU$. If Θ is a matrix such that $M\Theta$ has the form* (5.1), *then a fundamental matrix for* $\operatorname{aug}(\widetilde{T})$ *is given by*

$$\widetilde{U}(t) = \begin{bmatrix} t & 0 \\ 0 & 1 \end{bmatrix} U(t)\, \Theta \begin{bmatrix} I_2 & O \\ O & t^{-1} I_2 \end{bmatrix}.$$

6. Levinson-type Algorithms

Throughout this section, let $T = [\, a_{i-j} \,]_{i=1}^{m} {}_{j=1}^{n}$ be an $m \times n$ Toeplitz matrix with full rank n, and let $l = m - n$. In order to solve a LS problem for T using a formula for $T^\dagger$ we have to compute a fundamental matrix for $\operatorname{aug}(T)$. Recall that a fundamental systems can be characterized, for example, by columns of $\widetilde{T}^\dagger$ and $(\widetilde{T}^*\widetilde{T})^{-1}$, where $\widetilde{T}$ is a full rank $(m+1) \times (n+1)$ extension of T. Like for the classical Levinson algorithm, the number of multiplications in the algorithms can be reduced if these vectors are normalized. For this reason we present only this version of the recursions.

6.1. Algorithm Based on Row-and-Column Extensions

First we present a recursion that seems to be the most straightforward generalization of the classical Levinson algorithms to LS problems. In this subsection we denote by T_k the $(l+k) \times k$ submatrices $T_k = [\, a_{i-j} \,]_{i=1}^{l+k} {}_{j=1}^{k}$ for $k = 1, \dots, n$, where $l = m - n$. For $m = n$ the matrices T_k are just the leading principal submatrices of T occurring in the classical Levinson algorithm. We assume that all matrices T_k have full rank k. The $(m+1) \times (n+1)$ Toeplitz extension of T will, naturally, be denoted by T_{n+1}.

In the algorithm we recursively compute fundamental matrices for $\operatorname{aug}(T_k)$. We denote by U_k a fundamental matrix for $\operatorname{aug}(T_{k-1})$ and by $U_k(t)$ the corresponding generating 2×4 matrix polynomial. The fundamental matrices are normalized in the following way. We split U_k into two part $U_k = [\, U_k^- \; U_k^+ \,]$, where $U_k^\pm(t)$ are 2×2 matrix polynomials and assume that

$$\begin{bmatrix} \mathbf{e}_\pm & \mathbf{0} \\ \mathbf{0} & \mathbf{e}_\pm \end{bmatrix} U_k^\pm = I_2\,. \tag{6.1}$$

The condition (6.1) means that $U_k^-(t)$ is comonic and $U_k^+(t)$ is monic. Furthermore, we introduce the matrices

$$H_k^- = \begin{bmatrix} \mathbf{e}_- & \mathbf{a}(0:1-k) \\ \mathbf{a}(0:k+l-1)^* & \mathbf{0} \end{bmatrix}, \quad H_k^+ = \begin{bmatrix} \mathbf{e}_+ & \mathbf{a}(l+k-1:l) \\ \mathbf{a}(1-k:l)^* & \mathbf{0} \end{bmatrix}$$

and $\rho_k^\pm = H_k^\pm U_k^\pm$, and assume that

$$H_k^\pm U_k^\mp = O\,.$$

With this condition and the normalization (6.1) $U_k^\pm$ are uniquely defined.

In other words, the matrices $U_k^\pm$ satisfy

$$\mathrm{aug}\,(T_k)\,[\,U_k^- \;\; U_k^+\,] = \begin{bmatrix} \rho_{k,11}^- & \rho_{k,12}^- & 0 & 0 \\ \mathbf{0} & \mathbf{0} & \mathbf{0} & \mathbf{0} \\ 0 & 0 & \rho_{k,11}^+ & \rho_{k,12}^+ \\ \rho_{k,21}^- & \rho_{k,22}^- & 0 & 0 \\ \mathbf{0} & \mathbf{0} & \mathbf{0} & \mathbf{0} \\ 0 & 0 & \rho_{k,21}^+ & \rho_{k,22}^+ \end{bmatrix},$$

where $\rho_k^\pm = [\,\rho_{k,ij}^\pm\,]_{i,j=1}^2 = H_k^\pm U_k^\pm$.

Now we discuss the recursion $k \to k+1$. We introduce the test matrices

$$F_k^- = \begin{bmatrix} \mathbf{0} & \mathbf{a}(k+l:l+1) \\ \mathbf{a}(-k:l-1)^* & \mathbf{0} \end{bmatrix} \quad F_k^+ = \begin{bmatrix} \mathbf{0} & \mathbf{a}(-1:-k) \\ \mathbf{a}(1:k+l)^* & \mathbf{0} \end{bmatrix}$$

that depend on the new parameters a_{-k} and a_{l+k}.

We have

$$\mathrm{aug}\,(T_{k+1})\,[\,U_k^- \;\; tU_k^+\,] = \begin{bmatrix} \rho_{k,11}^- & \rho_{k,12}^- & \alpha_{k,11}^+ & \alpha_{k,12}^+ \\ \mathbf{0} & \mathbf{0} & \mathbf{0} & \mathbf{0} \\ \alpha_{k,11}^- & \alpha_{k,12}^- & \rho_{k,11}^+ & \rho_{k,12}^+ \\ \rho_{k,21}^- & \rho_{k,22}^- & \alpha_{k,21}^+ & \alpha_{k,22}^+ \\ \mathbf{0} & \mathbf{0} & \mathbf{0} & \mathbf{0} \\ \alpha_{k,21}^- & \alpha_{k,22}^- & \rho_{k,21}^+ & \rho_{k,22}^+ \end{bmatrix}, \tag{6.2}$$

where $\alpha_k^\pm = [\,\alpha_{k,ij}^\pm\,]_{i,j=1}^2 = F_k^\pm U_k^\pm$. Here tU_k^+ denotes the block vector with generating polynomial $tU_k^+(t)$.

In analogy to the classical Levinson recursion, this leads to the following.

Theorem 6.1. *The matrix polynomials $U_k^\pm(t)$ and the matrices $\rho_k^\pm$ satisfy the recursion*

$$[\,U_{k+1}^-(t) \;\; U_{k+1}^+(t)\,] = [\,U_k^-(t) \;\; tU_k^+(t)\,]\,\Theta_k,$$

where

$$\Theta_k = \begin{bmatrix} I_2 & -(\rho_k^-)^{-1}\alpha_k^+ \\ -(\rho_k^+)^{-1}\alpha_k^- & I_2 \end{bmatrix}.$$

and

$$\rho_{k+1}^\pm = \rho_k^\pm - \alpha_k^\mp(\rho_k^\mp)^{-1}\alpha_k^\pm\,.$$

Proof. From (6.2) we obtain that $[U_{k+1}^-(t)\ U_{k+1}^+(t)]$ given in Theorem 6.1 is a fundamental matrix for aug (T_k). It remains to check that $U_{k+1}^{\pm}$ satisfy the normalization conditions (6.1) and the condition (6.8) for k being replaced by $k+1$. The former is obvious and the latter follows from the definition of Θ_k. □

With the help of this theorem a fundamental matrix for aug (T_{n+1}) and so the displacement $\nabla(T^\dagger)$ of the pseudo inverse matrix can be computed. For initialization we observe that

$$U_1^- = \begin{bmatrix} \mathbf{e}_- & a_0\mathbf{e}_- - \mathbf{a}(0:l) \\ 0 & 1 \end{bmatrix}, \quad U_1^+ = \begin{bmatrix} \mathbf{e}_+ & a_l\mathbf{e}_+ - \mathbf{a}(0:l) \\ 0 & 1 \end{bmatrix},$$

and

$$\rho_1^- = \begin{bmatrix} 1 & a_0 \\ \bar{a}_0 & |a_0|^2 - \gamma \end{bmatrix}, \quad \rho_1^+ = \begin{bmatrix} 1 & a_l \\ \bar{a}_l & |a_l|^2 - \gamma \end{bmatrix},$$

where $\gamma = \|\mathbf{a}(0:l)\|^2$.

We discuss now how the number of operations in the recursion of Theorem 6.1 can be reduced. For this we observe that it is sufficient to compute the lower $k \times 2$ parts of $U_k^{\pm}$, since the upper part, i.e., the residuals, can be obtained from the lower part. However, the residuals are needed to compute the matrices $\alpha_k^{\pm}$. In fact, let $U_k^{\pm} = \begin{bmatrix} \mathbf{r}_k^{\pm} \\ \mathbf{u}_k^{\pm} \end{bmatrix}$. Then the second rows $\alpha_{k,2}^{\pm}$ of $\alpha_k^{\pm}$ are given by

$$\alpha_{k,2}^- = \mathbf{a}(-k:l-1)^*\mathbf{r}_k^-, \quad \alpha_{k,2}^+ = \mathbf{a}(1:k+l)^*\mathbf{r}_k^+ .$$

The $\mathbf{r}_k^{\pm}$ can be expressed in terms of $\mathbf{u}_k^{\pm}$ as

$$\mathbf{r}_k^{\pm} = -T_k\mathbf{u}_k^{\pm} + \mathbf{e}_{\pm}\rho_{k,1}^{\pm} , \tag{6.3}$$

where $\rho_{k,1}^{\pm}$ denote the first rows of $\rho_k^{\pm}$. Thus we have

$$\alpha_{k,2}^- = -\mathbf{g}_k^-\mathbf{u}_k^- + \bar{a}_{-k}\rho_{k,1}^-, \quad \alpha_{k,2}^+ = -\mathbf{g}_k^+\mathbf{u}_k^+ + \bar{a}_{k+l}\rho_{k,1}^+ ,$$

where

$$\mathbf{g}_k^- = \mathbf{a}(-k:l-1)^*T_k , \quad \mathbf{g}_k^+ = \mathbf{a}(1:k+l)^*T_k . \tag{6.4}$$

That means, knowing the $\mathbf{g}_k^{\pm}$, we have to compute only inner products of length k.

The vectors $\mathbf{g}_k^{\pm}$ can be computed recursively. In fact, the following can easily be checked.

Lemma 6.1. *The vectors $\mathbf{g}_k^{\pm}$ given by (6.4) satisfy the recursion*

$$\mathbf{g}_k^- = \bar{a}_{-k}\mathbf{a}(0:1-k) + [\lambda_k\ \ \mathbf{g}_{k-1}^-], \quad \mathbf{g}_k^+ = \bar{a}_{k+l}\mathbf{a}(k+l-1:l) + [\mathbf{g}_{k-1}^+\ \ \bar{\lambda}_k],$$

where

$$\lambda_k = \mathbf{a}(1-k:l-1)^*\mathbf{a}(1:k+l-1) .$$

In case that the complete matrix T_{n+1} is given from the beginning we can find the vector $(\lambda_k)_{k=1}^{n+1}$ by matrix-vector multiplication

$$(\lambda_k)_{k=1}^{n+1} = \begin{bmatrix} a_0 & \dots & a_{l-1} & & & 0 \\ a_{-1} & \dots & a_{l-2} & a_{l-1} & & \\ \vdots & & \vdots & \vdots & \ddots & \\ a_{-n} & \dots & a_{l-n-1} & a_{l-n} & \dots & a_{l-1} \end{bmatrix} \mathbf{a}(1:m)\,.$$

Let us estimate the complexity of the resulting algorithm. For simplicity we count only (complex) multiplications and neglect lower-order term. The number of additions is in the same range. First we compute the parameters $\gamma_k^{\pm}$. If FFT is used, then this can be done with complexity $m \log m$. If the data a_k of the matrix come in successively, then in each step the numbers $\gamma_k^{\pm}$ have to be computed by 2 inner products of length $k+l$.

Furthermore, in one step of the recursion $k \to k+1$ we need for the recursion of the $\mathbf{g}_k^{\pm}$ 2 multiplications of vectors of length k by a scalar. Furthermore, we have to compute 8 inner products of length k and, in order to apply the recursion formula in Theorem 6.1 on $\mathbf{u}_k^{\pm}$, 8 multiplications of a vector by a scalar are needed. Thus the amount for one recursion step is $18\,k$ multiplications. This leads to a complexity of $9\,n^2 + O(m \log m)$ multiplications to compute a reduced fundamental matrix for $\operatorname{aug}(T)$. In order to get the full fundamental matrix we have to compute $\mathbf{r}_{n+1}^{\pm}$ with the help of (6.3) for $k = n+1$. If FFT is applied this requires $O(m \log m)$ operations. That means the overall complexity will be, provided that the complete matrix is given from the very beginning of the calculations, $9\,n^2 + O(m \log m)$ multiplications and a few more additions.

6.2. Algorithm Based on Column Extensions

In this subsection we present an algorithm for computing a fundamental matrix for $\operatorname{aug}(T)$ by column extensions. This version is in particular important because it computes also the Q-factor and inverse R-factor of the QR-factorization of T. In contrast to the row-and-column extension in 6.1 and the row extension in 6.3, we do not have an extra condition on the matrix T.

Let T and T_{n+1} be as above. In this subsection we denote by T_k the submatrices $T_k = [\,a_{i-j}\,]_{i=1}^{m+1}{}_{j=1}^{k}$. These matrices automatically have full rank if T_{n+1} has full rank. Let U_k denote a fundamental matrix for $\operatorname{aug}(T_{k-1})$.

We split $U_k(t)$ into two parts, $U_k(t) = [\,U_k^-(t)\ U_k^+(t)\,]$, but this time $U_k^-(t)$ is a 2×3 matrix polynomial and $U_k(t)^+$ a 2×1 matrix polynomial.

We introduce the matrices

$$H_k^- = \begin{bmatrix} \mathbf{e}_- & \mathbf{a}(0:1-k) \\ \mathbf{a}(0:m)^* & \mathbf{0} \\ \mathbf{e}_+ & \mathbf{0} \end{bmatrix}, \quad H_k^+ = [\ \mathbf{a}(1-k:m-k+1)^* \quad \mathbf{0}\ ],$$

and $\rho_k^{\pm} = H_k^{\pm} U_k^{\pm}$. Note that $\rho_k^- = [\,\rho_{k,ij}^-\,]_{i,j=1}^3$ is a 3×3 matrix and ρ_k^+ is a number.

We normalize $U_k^{\pm}$ by demanding that

$$\begin{bmatrix} \mathbf{e}_- & \mathbf{0} \\ \mathbf{0} & \mathbf{e}_- \\ \mathbf{e}_+ & \mathbf{a}(m:m-k+1) \end{bmatrix} U_k^- = I_3 , \quad [\, \mathbf{0} \;\; \mathbf{e}_+ \,] U_k^+ = 1 \tag{6.5}$$

and

$$H_k^- U_k^+ = \mathbf{0}, \quad H_k^+ U_k^- = \mathbf{0} .$$

That means that

$$\operatorname{aug}(T_k)\,[\, U_k^- \;\; U_k^+ \,] = \begin{bmatrix} \rho_{k,11}^- & \rho_{k,12}^- & \rho_{k,13}^- & 0 \\ \mathbf{0} & \mathbf{0} & \mathbf{0} & \mathbf{0} \\ 0 & 0 & 1 & * \\ \rho_{k,21}^- & \rho_{k,22}^- & \rho_{k,23}^- & 0 \\ \mathbf{0} & \mathbf{0} & \mathbf{0} & \mathbf{0} \\ 0 & 0 & 0 & \rho_k^+ \end{bmatrix}$$

and

$$[\, \mathbf{e}_+ \;\; \mathbf{0} \,]\,[\, U_k^- \;\; U_k^+ \,] = [\, \rho_{k,31}^- \;\; \rho_{k,32}^- \;\; \rho_{33}^- \;\; 0 \,] ,$$

where the asterisk denotes a number that is not important for the calculations. With these conditions the $U_k^{\pm}$ are unique.

The test matrices for the recursion $k \to k+1$ are

$$F_k^- = [\; \mathbf{a}(-k:m-k)^* \;\;\; \mathbf{0} \;] , \quad F_k^+ = \begin{bmatrix} \mathbf{0} & \mathbf{a}(-1:-k) \\ \mathbf{a}(1:m+1)^* & \mathbf{0} \\ \mathbf{e}'_+ & \mathbf{0} \end{bmatrix} ,$$

where $\mathbf{e}'_+$ denotes the last but one vector in the standard basis.

We have now

$$\operatorname{aug}(T_{k+1})\,[\, U_k^- \;\; tU_k^+ \,] = \begin{bmatrix} \rho_{k,11}^- & \rho_{k,12}^- & \rho_{k,13}^- & \alpha_{k,1}^+ \\ \mathbf{0} & \mathbf{0} & \mathbf{0} & \mathbf{0} \\ 0 & 0 & 1 & 0 \\ \rho_{k,21}^- & \rho_{k,22}^- & \rho_{k,23}^- & \alpha_{k,2}^+ \\ \mathbf{0} & \mathbf{0} & \mathbf{0} & \mathbf{0} \\ \alpha_{k,1}^- & \alpha_{k,2}^- & \alpha_{k,3}^- & \rho_k^+ \end{bmatrix}$$

and

$$[\, \mathbf{e}_+ \;\; \mathbf{0} \,]\,[\, U_k^- \;\; tU_k^+ \,] = [\, \rho_{k,31}^- \;\; \rho_{k,32}^- \;\; \rho_{k,33}^- \;\; \alpha_{k,3}^+ \,] ,$$

where $F_k^{\pm} U_k^{\pm} = \alpha_k^{\pm} = (\alpha_{k,j})_{j=1}^3$. Here tU_k^+ denote the block vector with generating polynomial $tU_k^+(t)$.

From these relations we conclude the following.

Theorem 6.2. *The matrix polynomials $U_k^{\pm}$ and the matrices $\rho_k^{\pm}$ satisfy the recursion*

$$[\, U_{k+1}^-(t) \; U_{k+1}^+(t) \,] = [\, U_k^-(t) \; tU_k^+(t) \,]\, \Theta_k ,$$

where

$$\Theta_k = \begin{bmatrix} I_3 & -(\rho_k^-)^{-1} \alpha_k^+ \\ -(\rho_k^+)^{-1} \alpha_k^- & 1 \end{bmatrix} .$$

and

$$\rho_{k+1}^{\pm} = \rho_k^{\pm} - \alpha_k^{\mp}(\rho_k^{\mp})^{-1}\alpha_k^{\pm} \, .$$

To start the recursion we find $U_1^{\pm}$ as

$$U_1^- = \begin{bmatrix} \mathbf{e}_- & a_0\mathbf{e}_- - \mathbf{a}(0:m) & \mathbf{e}_+ \\ 0 & 1 & 0 \end{bmatrix}, \quad U_1^+ = \begin{bmatrix} a_m\mathbf{e}_+ - \mathbf{a}(0:m) \\ 1 \end{bmatrix}.$$

Furthermore,

$$\rho_1^- = \begin{bmatrix} 1 & a_0 & 0 \\ \overline{a}_0 & |a_0|^2 - \gamma & \overline{a}_m \\ 0 & -a_m & 1 \end{bmatrix}, \quad \rho_1^+ = |a_m|^2 - \gamma \, ,$$

where $\gamma = \|\mathbf{a}(0:m)\|^2$.

In order to avoid long inner products we can take advantage of the fact that the vectors $\mathbf{g}_k = \mathbf{a}(-k:m-k)^* \, T_k = [\, g_{kj} \,]_{j=0}^n$ satisfy the relation

$$g_{k+1,j+1} = g_{kj} + \overline{a}_{-k}a_{-j} - \overline{a}_{m+1-k}a_{m+1-j} \, .$$

Thus the $\mathbf{g}_k$ can be computed recursively using the facts that $[\, g_{0j} \,] = \mathbf{a}(0:m)^* \, T_{n+1}$ and $g_{jk} = \overline{g_{kj}}$.

We show that the algorithm presented in this subsection is related to the QR-factorization of T_n. Suppose that $U_k^+ = \begin{bmatrix} \mathbf{r}_k^+ \\ \mathbf{u}_k^+ \end{bmatrix}$, $\mathbf{r}_k^+ \in \mathbb{C}^{m+1}$, $\mathbf{u}_k^+ \in \mathbb{C}^k$ and $\nu_k = \|\mathbf{r}_k^+\|$. By construction, the last component of $\mathbf{r}_k^+$ vanishes. Let $\mathbf{q}_k \in \mathbb{C}^m$ be the vector obtained by deleting this last zero component of $\mathbf{r}_k^+$ and dividing by ν_k. Furthermore, let $\mathbf{z}_k \in \mathbb{C}^n$ denote the vector obtained from $\mathbf{u}_k^+$ by adding $n-k+1$ zeros at the bottom and dividing by $-\nu_k$. We form the $m \times n$ matrix Q with the columns $\mathbf{q}_k$ and the upper triangular $n \times n$ matrix Z with the columns $\mathbf{z}_k$ $(k = 1, \ldots, n)$.

Then we have

$$\operatorname{aug}(T_n) \begin{bmatrix} Q \\ -Z \end{bmatrix} = \begin{bmatrix} O \\ L \end{bmatrix}, \tag{6.6}$$

where L is $n \times n$ lower triangular. Hence $Q = T_nZ$ and $T_n^*Q = L$. We conclude that $T_n = QZ^{-1}$ is the QR-factorization of the matrix T_n. Furthermore, $\nu_k = (\rho_k^+)^{-\frac{1}{2}}$. That means that the algorithm emerging from Theorem 6.2 computes the Q-factor and the inverse R-factor of the QR-factorization of T_n.

Let us estimate the number of multiplications and compare it with the algorithm in 6.1. In each step we have to compute 6 inner products instead of 8 in 6.1. But the length of the vectors $\mathbf{g}_k$ is $n+1$ instead of k. This gives a total number of multiplications of $8n^2 + O(m \log m)$ compared with $9n^2 + O(m \log m)$ in 6.1.

6.3. Algorithm Based on Row Extensions

An algorithm based on row extension is recommended if $l = m - n$ is small. Let T be as above. The $(m+1) \times (n+1)$ extension of T be denoted this time by $\widetilde{T}$. In this subsection we consider the submatrices $T_k = [\, a_{i-j} \,]_{i=1}^{n+k}\,{}_{j=1}^{n+1}$ $(k = 1, \ldots, l+1$ of $\widetilde{T}$. In particular, $\widetilde{T} = T_{l+1}$. We have to assume that the square matrix T_1 is nonsingular.

As in Section 6.2 we split $U_k(t)$ into two parts, $U_k(t) = [\, U_k^-(t)\; U_k^+(t) \,]$, where $U_k^-(t)$ is a 2×3 and $U_k^+(t)$ a 2×1 matrix polynomial.

We introduce the matrices

$$H_k^- = \begin{bmatrix} \mathbf{e}_- & \mathbf{a}(0:-n) \\ \mathbf{a}(0:k+n-1)^* & \mathbf{0} \\ \mathbf{0} & \mathbf{e}_+ \end{bmatrix}, \quad H_k^+ = [\; \mathbf{0} \quad \mathbf{a}(k+n-1:k-1) \;]$$

and $\rho_k^\pm = H_k^\pm U_k^\pm$.

We normalize $U_k^\pm$ by demanding that

$$\begin{bmatrix} \mathbf{e}_- & \mathbf{0} \\ \mathbf{0} & \mathbf{e}_- \\ \mathbf{a}(-n:k-1)^* & \mathbf{0} \end{bmatrix} U_k^- = I_3\,, \quad [\mathbf{0}\;\; \mathbf{e}_+\,]\, U_k^+ = 1$$

and

$$H_k^- U_k^+ = \mathbf{0}, \quad H_k^+ U_k^- = \mathbf{0}\,.$$

That means that

$$\operatorname{aug}(T_k)\,[\, U_k^- \;\; U_k^+ \,] = \begin{bmatrix} \rho_{k,11}^- & \rho_{k,12}^- & \rho_{k,13}^- & 0 \\ \mathbf{0} & \mathbf{0} & \mathbf{0} & \mathbf{0} \\ 0 & 0 & 0 & \rho_k^+ \\ \rho_{k,21}^- & \rho_{k,22}^- & \rho_{k,23}^- & 0 \\ \mathbf{0} & \mathbf{0} & \mathbf{0} & \mathbf{0} \\ 0 & 0 & 1 & * \end{bmatrix}$$

and

$$[\mathbf{0}\;\; \mathbf{e}_+\,]\,[\, U_k^- \;\; U_k^+ \,] = [\, \rho_{31}^- \;\; \rho_{32}^- \;\; \rho_{33}^- \;\; 0 \,]\,.$$

With this normalization $U_k^\pm$ are unique.

The test matrices for the recursion $k \to k+1$ are

$$F_k^- = [\; \mathbf{0} \quad \mathbf{a}(k+n:k) \;]\,, \quad F_k^+ = \begin{bmatrix} \mathbf{0} & \mathbf{a}(-1:-n-1) \\ \mathbf{a}(1:k+n)^* & \mathbf{0} \\ \mathbf{0} & \mathbf{e}_+' \end{bmatrix},$$

where $\mathbf{e}'_+$ denotes the last but one vector in the standard basis. We set $\alpha_k^\pm = F_k^\pm U_k^\pm$. Then

$$\operatorname{aug}(T_{k+1})\,[\,U_k^- \quad tU_k^+\,] = \begin{bmatrix} \rho_{k,11}^- & \rho_{k,12}^- & \rho_{k,13}^- & \alpha_{k,1}^+ \\ \mathbf{0} & \mathbf{0} & \mathbf{0} & \mathbf{0} \\ \alpha_{k,1}^- & \alpha_{k,2}^- & \alpha_{k,3}^- & \rho_k^+ \\ \rho_{k,21}^- & \rho_{k,22}^- & \rho_{k,23}^- & \alpha_{k,2}^+ \\ \mathbf{0} & \mathbf{0} & \mathbf{0} & \mathbf{0} \\ 0 & 0 & 1 & 0 \end{bmatrix}$$

and

$$[\,\mathbf{0} \ \ \mathbf{e}_+\,]\,[\,U_k^- \ \ U_k^+\,] = [\,\rho_{k,31}^- \ \ \rho_{k,32}^- \ \ \rho_{k,33}^- \ \ \alpha_{k,3}^+\,]\,,$$

where $\alpha_k^\pm = (\alpha_{k,j})_{j=1}^3$.

As in Section 6.2 we conclude the following.

Theorem 6.3. *The matrix polynomials $U_k^\pm$ and the matrices $\rho_k^\pm$ satisfy the recursion in Theorem 6.2.*

For initialization we have to find a fundamental matrix U_1 for $\operatorname{aug}(T_1)$. For this we find first the first and last columns $\mathbf{w}_\pm$ of T_1^{-1} by the classical Levinson algorithm. Then we compute $\mathbf{z}_\pm = -(T_1^* T_1)^{-1}\mathbf{e}_\pm = -T_1^{-1} J_{n+1}\overline{\mathbf{w}}_\mp$ using the inversion formula. Now

$$U_1 = \begin{bmatrix} \mathbf{0} & \mathbf{0} & J_{n+1}\overline{\mathbf{w}}_+ & J_{n+1}\overline{\mathbf{w}}_- \\ \mathbf{w}_- & \mathbf{w}_+ & \mathbf{z}_- & \mathbf{z}_+ \end{bmatrix}.$$

is a fundamental matrix for $\operatorname{aug}(T_1)$, which still has to be normalized to start the recursion.

In the algorithm we recursively compute only the lower part $\mathbf{u}_k^\pm$ of $U_k^\pm$, since the upper part $\mathbf{r}_k^\pm$ is given by

$$\mathbf{r}_k^\pm = -T_k \mathbf{u}_k^\pm + \mathbf{e}_- \rho_{k,1}^- + \mathbf{e}_+ \rho_k^+ \,.$$

To apply $F_k^\pm$ we only have to replace the inner product $\alpha_{k,2}^+ = \mathbf{a}(1:k+n)^* \mathbf{r}_k^+$ by an inner product in terms of $\mathbf{u}_k^+$. Here $\alpha_{k,2}^+$ denotes the second component of α_k^+.

We have $\mathbf{r}_k^+ = -T_k \mathbf{u}_k^+ + \rho_k^+ \mathbf{e}_+$. Hence

$$\alpha_{k,2}^+ = -\mathbf{g}_k \mathbf{u}_k^+ \,,$$

where $\mathbf{g}_k = \mathbf{a}(1:k+n)^* T_k$. Obviously, the $\mathbf{g}_k$ satisfy the recursion

$$\mathbf{g}_k = \mathbf{g}_{k-1} + \overline{a}_{k+n}\mathbf{a}(k+n-1:k-1)\,.$$

Let us estimate the number of multiplications. In one step we have to compute 6 inner products of vectors of length $n+1$ and 6 multiplications of vectors of this length by scalars for the recursion of the $U_k^\pm$. Furthermore, we have one multiplication of a vector by a scalar for the recursion of $\mathbf{g}_k$. Since we have l steps, this results in about $13\,n(m-n)$ multiplications. For the initialization we need another $2\,n^2$ multiplications. Thus the overall complexity will be $13\,n(m-n) + 2\,n^2 + O(m \log m)$ multiplications and about the same number of additions.

6.4. Algorithm Based on Order Enlargement

In this section it is convenient to number the elements of T in a different way. Let $T = [\,a_{i-j}\,]_{i=n}^{N}{}_{j=1}^{n}$, where $N = m+n-1$. We consider the family of $(N+1-k)\times k$ matrices $T_k = [\,a_{i-j}\,]_{i=k}^{N}{}_{j=1}^{k}$ for $k = 1,\dots,n$. Note that the T_k run over all Toeplitz matrices that are generated by one and the same vector $\mathbf{a} = (a_j)_{j=0}^{N-1}$. Clearly, the matrices T_k have full rank if T has full rank.

We split U_k into two part $U_k = [\,U_k^- \; U_k^+\,]$, where $U_k^{\pm}(t)$ are 2×2 matrix polynomials and assume, for normalization, that

$$\begin{bmatrix} \mathbf{0} & \mathbf{e}_- \\ \mathbf{e}_+ & \mathbf{a}(N-1:N-k) \end{bmatrix} U_k^- = I_2\,, \quad \begin{bmatrix} \mathbf{0} & \mathbf{e}_+ \\ \mathbf{e}_- & \mathbf{a}(k-1:1) \end{bmatrix} U_k^+ = I_2\,. \tag{6.7}$$

Furthermore, we introduce thc matrices

$$H_k^- = \begin{bmatrix} \mathbf{e}_+ & \mathbf{0} \\ \mathbf{a}(k-1:N-1)^* & \mathbf{0} \end{bmatrix}, \quad H_k^+ = \begin{bmatrix} \mathbf{e}_- & \mathbf{0} \\ \mathbf{a}(0:N-k)^* & \mathbf{0} \end{bmatrix}.$$

We set $\rho_k^{\pm} = H_k^{\pm} U_k^{\pm}$ and assume that

$$H_k^{\pm} U_k^{\mp} = O\,. \tag{6.8}$$

With this condition and the normalization (6.7) the $U_k^{\pm}$ are uniquely defined.

We have

$$\operatorname{aug}(T_k)\,[\,U_k^- \; U_k^+\,] = \begin{bmatrix} * & * & 0 & 1 \\ \mathbf{0} & \mathbf{0} & \mathbf{0} & \mathbf{0} \\ 0 & 1 & * & * \\ \rho_{k,21}^- & \rho_{k,22}^- & 0 & 0 \\ \mathbf{0} & \mathbf{0} & \mathbf{0} & \mathbf{0} \\ 0 & 0 & \rho_{k,21}^+ & \rho_{k,22}^+ \end{bmatrix}$$

and

$$\begin{bmatrix} \mathbf{e}_+ & \mathbf{0} \\ \mathbf{e}_- & \mathbf{0} \end{bmatrix} [\,U_k^- \; U_k^+\,] = \begin{bmatrix} \rho_{k,11}^- & \rho_{k,12}^- & 0 & 0 \\ 0 & 0 & \rho_{k,11}^+ & \rho_{k,12}^+ \end{bmatrix},$$

where $\rho_k^{\pm} = [\,\rho_{k,ij}^{\pm}\,]_{i,j=1}^{2} = H_k^{\pm} U_k^{\pm}$.

Wc introduce the test matrices

$$F_k^- = \begin{bmatrix} \mathbf{e}'_- & \mathbf{0} \\ \mathbf{a}(-1:N-k-1)^* & \mathbf{0} \end{bmatrix} \quad F_k^+ = \begin{bmatrix} \mathbf{e}'_+ & \mathbf{0} \\ \mathbf{a}(k:N)^* & \mathbf{0} \end{bmatrix},$$

where $\mathbf{e}'_-$ stands for the second and $\mathbf{e}'_+$ for the last but one vector in the standard basis, and introduce $\alpha_k^{\pm} = F_k^{\pm} U_k^{\pm}$.

Let $\widetilde{U}_k^{\pm}$ be defined by

$$\widetilde{U}_k^-(t) = \begin{bmatrix} t^{-1} & 0 \\ 0 & 1 \end{bmatrix} U_k^-(t), \quad \widetilde{U}_k^+(t) = \begin{bmatrix} 1 & 0 \\ 0 & t \end{bmatrix} U_k^+(t)\,.$$

Then we have

$$\operatorname{aug}(T_k)\,[\,\widetilde{U}_k^- \;\; \widetilde{U}_k^+\,] = \begin{bmatrix} 0 & 0 & 0 & 1 \\ \mathbf{0} & \mathbf{0} & \mathbf{0} & \mathbf{0} \\ 0 & 1 & 0 & 0 \\ \rho_{k,21}^- & \rho_{k,22}^- & \alpha_{k,21}^+ & \alpha_{k,22}^+ \\ \mathbf{0} & \mathbf{0} & \mathbf{0} & \mathbf{0} \\ \alpha_{k,21}^- & \alpha_{k,22}^- & \rho_{k,21}^+ & \rho_{k,22}^+ \end{bmatrix}$$

and

$$\begin{bmatrix} \mathbf{e}_+ & \mathbf{0} \\ \mathbf{e}_- & \mathbf{0} \end{bmatrix} [\,\widetilde{U}_k^- \;\; \widetilde{U}_k^+\,] = \begin{bmatrix} \rho_{k,11}^- & \rho_{k,12}^- & \alpha_{k,11}^+ & \alpha_{k,12}^+ \\ \alpha_{k,11}^- & \alpha_{k,12}^- & \rho_{k,12}^+ & \rho_{k,12}^+ \end{bmatrix}.$$

From these relations we conclude the following.

Theorem 6.4. *The matrix polynomials $U_k^\pm$ and the matrices $\rho_k^\pm$ satisfy the recursion*

$$[\,U_{k+1}^-(t)\; U_{k+1}^+(t)\,] = \begin{bmatrix} t^{-1} & 0 \\ 0 & 1 \end{bmatrix} [\,U_k^-(t)\; U_k^+(t)\,] \begin{bmatrix} I_2 & 0 \\ O & tI_2 \end{bmatrix} \Theta_k,$$

where

$$\Theta_k = \begin{bmatrix} I_2 & -(\rho_k^-)^{-1}\alpha_k^+ \\ -(\rho_k^+)^{-1}\alpha_k^- & I_2 \end{bmatrix} \quad \textit{and} \quad \rho_{k+1}^\pm = \rho_k^\pm - \alpha_k^\mp(\rho_k^\mp)^{-1}\alpha_k^\pm\,.$$

For initialization we observe that

$$U_1^- = \begin{bmatrix} \mathbf{e}_+ & a_0\mathbf{e}_- - \mathbf{a} \\ 0 & 1 \end{bmatrix}, \quad U_1^+ = \begin{bmatrix} \mathbf{e}_- & a_{N-1}\mathbf{e}_+ - \mathbf{a} \\ 0 & 1 \end{bmatrix}$$

and

$$\rho_k^- = \begin{bmatrix} 1 & -a_{N-1} \\ \overline{a}_{N-1} & -\|\mathbf{a}(1:N-1)\|^2 \end{bmatrix}, \quad \rho_k^+ = \begin{bmatrix} 1 & -a_0 \\ \overline{a}_0 & -\|\mathbf{a}(0:N-2)\|^2 \end{bmatrix}.$$

7. Schur-type Algorithms

Instead of computing the parameters in the Levinson-type recursions via inner products they can be computed recursively via a Schur-type algorithm. In the column extension version the Schur-type algorithm also provides the R-factor of the QR-factorization of T. For the other versions we could not find a reasonable factorization interpretation of the Schur-type algorithm.

7.1. Algorithm Based on Row-and-Column Extensions

We introduce the 2×2 matrices $\rho_{k,j}^\pm$ and $\alpha_{k,j}^\pm$ for $j = 1,\ldots,n-k+1$ by

$$\begin{bmatrix} \rho_{k,j}^- & \alpha_{k,j}^+ \end{bmatrix} = \begin{bmatrix} \mathbf{0} & \mathbf{a}(-j:-j-k+1) \\ \mathbf{a}(j:l+k+j-1)^* & \mathbf{0} \end{bmatrix} [\,U_k^- \;\; U_k^+\,]$$

and

$$\begin{bmatrix} \alpha_{k,j}^- & \rho_{k,j}^+ \end{bmatrix} = \begin{bmatrix} \mathbf{0} & \mathbf{a}(l+j-1:l+j-k) \\ \mathbf{a}(j-k+1:l+j)^* & \mathbf{0} \end{bmatrix} [\,U_k^- \;\; U_k^+\,].$$

Furthermore, we set $\rho^{\pm}_{k,0} = H^{\pm}_k U^{\pm}_k$, where $H^{\pm}_k$ is defined as in Section 6.1, and $\alpha^{\pm}_{k,0} = O$. Note that $\alpha^{\pm}_{k,1}$ are the $\alpha^{\pm}_k$ that occur in Theorem 6.1.

Differently to Section 6.1, in this section we denote by $\rho^{\pm}_k$ and $\alpha^{\pm}_k$ the block vectors $\rho^{\pm}_k = (\rho^{\pm}_{k,j})_{j=0}^{n-k+1}$ and $\alpha^{\pm}_k = (\alpha^{\pm}_{k,j})_{j=0}^{n-k+1}$.

In analogy to the classical Schur recursion the following can easily be checked.

Theorem 7.1. *The* 2×2 *matrix polynomials* $\rho^{\pm}_k(t)$ *and* $\alpha^{\pm}_k(t)$ *satisfy the recursion*

$$\begin{bmatrix} \rho^{-}_{k+1}(t) & \alpha^{+}_{k+1}(t) \\ \alpha^{-}_{k+1}(t) & \rho^{+}_{k+1}(t) \end{bmatrix} = \begin{bmatrix} \rho^{-}_{k}(t) & t^{-1}\alpha^{+}_{k}(t) \\ t^{-1}\alpha^{-}_{k}(t) & \rho^{+}_{k}(t) \end{bmatrix} \Theta_k \pmod{t^{n-k+1}},$$

where

$$\Theta_k = \begin{bmatrix} I_2 & -(\rho^{+}_{0,k})^{-1}\alpha^{+}_{1,k} \\ -(\rho^{-}_{0,k})^{-1}\alpha^{-}_{1,k} & I_2 \end{bmatrix}.$$

Theorem 7.1 can be used to compute the recursion parameters $\alpha^{\pm}_k$ and $\rho^{\pm}_k$ occurring in Theorem 6.1. As for the Levinson-type algorithm, it is sufficient to compute the lower parts of the matrices $U^{\pm}_k$.

For initialization we introduce the $(l+1)\times(2n+1)$ Toeplitz matrix T^0 built from the parameters of T_{n+1} as

$$T^0 = [\, a_{j-k+n} \,]_{j=0}^{l}\,{}_{k=0}^{2n} .$$

We have to compute the vector

$$(T^0)^* \mathbf{a}(0:l) = (\gamma_j)_{j=-n}^{n}.$$

Then we have

$$\rho^{-}_0 = \begin{bmatrix} 1 & a_0 \\ \bar{a}_0 & |a_0|^2 - \gamma_0 \end{bmatrix}, \quad \rho^{+}_0 = \begin{bmatrix} 1 & a_l \\ \bar{a}_l & |a_l|^2 - \gamma_0 \end{bmatrix},$$

and, for $j = 1, \ldots, n$,

$$\rho^{-}_j = \begin{bmatrix} 0 & a_{-j} \\ \bar{a}_j & a_0\bar{a}_{-j} - \gamma_{-j} \end{bmatrix}, \quad \alpha^{+}_j = \begin{bmatrix} 0 & a_{l-j} \\ \bar{a}_{l+j} & a_l\bar{a}_{l+j} - \gamma_{-j} \end{bmatrix},$$

$$\alpha^{-}_j = \begin{bmatrix} 0 & a_{j} \\ \bar{a}_{-j} & a_0\bar{a}_{j} - \gamma_{j} \end{bmatrix}, \quad \rho^{+}_j = \begin{bmatrix} 0 & a_{l+j} \\ \bar{a}_{l-j} & a_l\bar{a}_{l-j} - \gamma_{j} \end{bmatrix}.$$

Let us estimate the complexity. Again we count only multiplications and neglect lower-order terms. In one recursion step we have 2 times 8 multiplications of vectors of length $n-k+1$ by scalars in order to find the vectors $\alpha^{\pm}_{k+1}$ and $\rho^{\pm}_{k+1}$. Furthermore we have 8 multiplications of vectors of length k by scalars in order to find $\mathbf{u}^{\pm}_{k+1}$. Together this will be $16\,n - 8\,k$ multiplications per step. Thus the overall complexity will be $12\,n^2$ plus $O(m\log m)$ for the computation of the numbers γ_j in the initialization.

7.2. Algorithm Based on Column Extension

For the Schur-type recursion based on column extension we introduce 3×3 matrices $\rho_{k,j}^-$, 3×1 matrices $\alpha_{k,j}^+$, 1×3 matrices $\alpha_{k,j}^-$ and the numbers $\rho_{k,j}^+$ $(j=1,\dots,n-k)$ by

$$\begin{bmatrix} \rho_{k,j}^- & \alpha_{k,j}^+ \end{bmatrix} = \begin{bmatrix} \mathbf{0} & \mathbf{a}(-j:-j-k+1) \\ \mathbf{a}(j:m+j)^* & \mathbf{0} \\ \mathbf{e}_{m-j} & \mathbf{0} \end{bmatrix} [U_k^- \;\; U_k^+]$$

and

$$\begin{bmatrix} \alpha_{k,j}^- & \rho_{k,j}^+ \end{bmatrix} = [\; \mathbf{a}(j-k+1:m+j-k+1)^* \;\; \mathbf{0} \;] \, [U_k^- \;\; U_k^+]\,.$$

Furthermore, we set $\rho_{k,0}^\pm = H_k^\pm U_k^\pm$, where $H_k^\pm$ is defined as in Section 6.2, and $\alpha_{k,0}^\pm = O$ and introduce block vectors $\rho_k^\pm = (\rho_{k,j}^\pm)_{j=0}^{n-k+1}$ and $\alpha_k^\pm = (\alpha_{k,j}^\pm)_{j=0}^{n-k+1}$.

Theorem 7.2. *The matrix polynomials $\rho_k^\pm(t)$ and $\alpha_k^\pm(t)$ satisfy the recursion*

$$\begin{bmatrix} \rho_{k+1}^-(t) & \alpha_{k+1}^+(t) \\ \alpha_{k+1}^-(t) & \rho_{k+1}^+(t) \end{bmatrix} = \begin{bmatrix} \rho_k^-(t) & t^{-1}\alpha_k^+(t) \\ t^{-1}\alpha_k^-(t) & \rho_k^+(t) \end{bmatrix} \Theta_k \pmod{t^{n-k+1}},$$

where

$$\Theta_k = \begin{bmatrix} I_3 & -(\rho_{0,k}^+)^{-1}\alpha_{1,k}^+ \\ -(\rho_{0,k}^-)^{-1}\alpha_{1,k}^- & 1 \end{bmatrix}.$$

The R-factor of the QR-factorization is obtained as follows. We form n column vectors of length n with the numbers $\rho_{k,j}^+$ $(j=0,\dots,n-k)$ and $k-1$ zeros at the top, and we divide these vectors by ν_k (see 6.2). The lower triangular matrix formed by these vectors $(k=1,\dots,n)$ is just the matrix L appearing in (6.6). Now the R-factor of the QR-factorization of T is the upper triangular matrix L^*.

Estimating the complexity we observe that in one recursion step we have 12 multiplications of vectors of length $n-k+1$ by scalars to find $\alpha_{k+1}^\pm$ and $\rho_{k+1}^\pm$ and 6 multiplications of vectors of length k to find $\mathbf{u}_{k+1}^\pm$. This results in complexity $9\,n^2 + O(m\log m)$. That means this algorithm requires about 25% less operations than that in 7.1.

7.3. Algorithm Based on Row Extension

For the Schur-type recursion based on row extensions we introduce 3×3 matrices $\rho_{k,j}^-$, 3×1 matrices $\alpha_{k,j}^+$, 1×3 matrices $\alpha_{k,j}^-$ and the numbers $\rho_{k,j}^+$ $(j=1,\dots,l-k+1)$ by

$$\begin{bmatrix} \rho_{k,j}^- & \alpha_{k,j}^+ \end{bmatrix} = \begin{bmatrix} \mathbf{0} & \mathbf{a}(-j:-j-n) \\ \mathbf{a}(j:j+k+n-1)^* & \mathbf{0} \\ \mathbf{0} & \mathbf{e}_{n-j} \end{bmatrix} [U_k^- \;\; U_k^+]$$

and

$$\begin{bmatrix} \alpha_{k,j}^- & \rho_k^+ \end{bmatrix} = [\; \mathbf{0} \;\; \mathbf{a}(j+k+n-1:j+k-1) \;] \, [U_k^- \;\; U_k^+]\,.$$

Furthermore, we set $\rho_{k,0}^\pm = H_k^\pm U_k^\pm$, where $H_k^\pm$ is defined as in Section 6.3, and $\alpha_{k,0}^\pm = O$, and introduce block vectors $\rho_k^\pm = (\rho_{k,j}^\pm)_{j=0}^{n-k+1}$ and $\alpha_k^\pm = (\alpha_{k,j}^\pm)_{j=0}^{n-k+1}$.

Theorem 7.3. *The matrix polynomials $\rho_k^{\pm}(t)$ and $\alpha_k^{\pm}(t)$ satisfy the recursion in Theorem 7.2, where "mod t^{n-k+1}" has to be replaced by "mod t^{l-k+1}".*

For initialization one has to compute the matrix-vector products

$$\begin{bmatrix} a_{-n} & & 0 \\ \vdots & \ddots & \\ a_0 & \dots & a_{-n} \end{bmatrix} [\mathbf{w}_{\pm} \ \ \mathbf{z}_{\pm}], \quad \begin{bmatrix} a_n & \dots & a_0 \\ & \ddots & \vdots \\ 0 & & a_n \end{bmatrix} \mathbf{w}_{\pm}$$

and

$$\begin{bmatrix} a_{n+1} & \dots & a_1 \\ \vdots & & \vdots \\ a_m & \dots & a_l \end{bmatrix} [\mathbf{w}_{\pm} \ \ \mathbf{z}_{\pm}]$$

and normalize accordingly.

Let us estimate the complexity. We need 12 multiplications of vectors of length $l-k+1$ by scalars to find $\alpha_{k+1}^{\pm}$ and $\rho_{k+1}^{\pm}$, and 6 multiplications of vectors of length n to find $\mathbf{u}_{k+1}^{\pm}$. Since we have $l = m - n$ step, this results in $6\,(m-n)m$ multiplications. In addition we need $2\,n^2$ multiplications for initialization using the classical Levinson algorithm plus $O(m \log m)$ operations.

7.4. Algorithm Based on Order Enlargement

We introduce the 2×2 matrices $\rho_{k,j}^{\pm}$ and $\alpha_{k,j}^{\pm}$ for $j = 1, \dots, n-k+1$ by

$$\begin{bmatrix} \rho_{k,j}^{-} & \alpha_{k,j}^{+} \end{bmatrix} = \begin{bmatrix} \mathbf{e}_{N-k-j} & \mathbf{0} \\ \mathbf{a}(k-1+j : N-1+j)^* & \mathbf{0} \end{bmatrix} [U_k^- \ \ U_k^+]$$

and

$$\begin{bmatrix} \alpha_{k,j}^{-} & \rho_{k,j}^{+} \end{bmatrix} = \begin{bmatrix} \mathbf{e}_{j+1} & \mathbf{0} \\ \mathbf{a}(-j : N-k-j)^* & \mathbf{0} \end{bmatrix} [U_k^- \ \ U_k^+] \, .$$

Furthermore, we set $\rho_{k,0}^{\pm} = H_k^{\pm} U_k^{\pm}$, where $H_k^{\pm}$ is defined as in Section 6.4, and $\alpha_{k,0}^{\pm} = O$. Note that $\alpha_{k,1}^{\pm}$ are the $\alpha_k^{\pm}$ that occur in Theorem 6.4.

We denote by $\rho_k^{\pm}$ and $\alpha_k^{\pm}$ the block vectors $\rho_k^{\pm} = (\rho_{k,j}^{\pm})_{j=0}^{n-k+1}$ and $\alpha_k^{\pm} = (\alpha_{k,j}^{\pm})_{j=0}^{n-k+1}$.

Theorem 7.4. *The polynomials $\rho_k^{\pm}(t)$ and $\alpha_k^{\pm}$ satisfy the recursions in Theorem* 7.1.

This theorem can be combined with Theorem 6.4. In the recursions of Theorem 6.4 it is sufficient to compute the lower parts of the matrices $U_k^{\pm}$.

To discuss the initialization of the algorithm we introduce the polynomial $\gamma(t)$ in t and t^{-1} defined by

$$\gamma(t) = \sum_{j=1-N}^{N-1} \gamma_j t^j = \overline{\mathbf{a}}(t^{-1})\mathbf{a}(t)$$

which can be computed with complexity $O(N \log N)$ $(= O(m \log m))$. We have

$$\begin{bmatrix} \overline{a}_{N-1} & & \\ \vdots & \ddots & \\ \overline{a}_0 & \dots & \overline{a}_{N-1} \\ & \ddots & \vdots \\ & & \overline{a}_0 \end{bmatrix} \begin{bmatrix} a_0 \\ \vdots \\ a_{N-1} \end{bmatrix} = \begin{bmatrix} \gamma_{1-N} \\ \vdots \\ \gamma_0 \\ \vdots \\ \gamma_{N-1} \end{bmatrix}.$$

Using this relation we obtain

$$\rho_{1,j}^- = \begin{bmatrix} 0 & -a_{N-j-1} \\ 0 & a_0\overline{a}_j - \gamma_{-j} \end{bmatrix}, \quad \alpha_{1,j}^+ = \begin{bmatrix} 0 & -a_{N-j-1} \\ \overline{a}_j & -\gamma_{-j} \end{bmatrix}$$

and

$$\alpha_{1,j}^- = \begin{bmatrix} 0 & -a_j \\ a_{N-1-j} & -\gamma_j \end{bmatrix}, \quad \rho_{1,j}^+ = \begin{bmatrix} 0 & -a_j \\ 0 & a_{N-1}\overline{a}_{N-1-j} - \gamma_j \end{bmatrix}.$$

The complexity of the algorithm is about the same as that in Section 7.1. However, its advantage is that it works without any additional condition on the matrix T.

References

[1] G. Ammar and P. Gader, *A variant of the Gohberg-Semencul formula involving circulant matrices,* SIAM J. Matrix Anal.Appl., 12 (1991), 534–540.

[2] G. Ammar and P. Gader, *New decompositions of the inverse of a Toeplitz matrix,* In: *Signal Processing, Scattering and Operator Theory,* Proceedings of the Internat. Sympos. MTNS-89, vol.3, Birkhäuser Boston 1990, pp.421–428.

[3] A. Björck, *Numerical methods for least squares problems,* SIAM Publ., Philadelphia 1996.

[4] A. Bojanczyk, R. Brent, F. de Hoog, *QR factorization of Toeplitz matrices,* Numer. Math., 49 (1986), 81–94.

[5] S. Chadrasekaran, A. Sayed, *Stabilized Schur algorithms,* In: T. Kailath, A. Sayed (Ed.), *Fast reliable algorithms for matrices with structure,* SIAM Publ. 1999, pp. 57–81.

[6] J. Chun, T. Kailath, H. Lev-ari, *Fast parallel algorithms for QR and triangular factorization,* SIAM J. Sci. Statist. Comput., 8 (1987), 899–913.

[7] G. Cybenko, *Fast Toeplitz orthogonalization using inner products,* SIAM J. Sci. Stat. Comput., 8 (1987), 734–740.

[8] I. Gohberg, V. Olshevsky, *Circulants, displacements and decomposition of matrices,* Integral Equations and Operator Theory, 15 (1992), 730–743.

[9] I. Gohberg and A. Semencul, *On the inversion of finite-section Toeplitz matrices and their continuous analogues,* Matemat. Issledovaniya, 7 (1972), 201–224. (in Russian)

[10] M. Gu, *New fast algorithms for structured linear least squares problems,* SIAM J. Matrix Anal. Appl., 20(1) (1998), 244–269.

[11] G. Heinig, *Generalized inverses of Hankel and Toeplitz mosaic matrices*, Linear Algebra Appl., 216 (1995), 43–60.

[12] G. Heinig, *Matrix representation of Bezoutians*, Linear Algebra Appl., 223/224 (1995), 337–354.

[13] G. Heinig, F. Hellinger, *The Bezoutian structure of the Moore-Penrose inverse of Hankel matrices*, SIAM J. Matrix Anal.Appl., 14 (1993), 629–645.

[14] G. Heinig, F. Hellinger, *Displacement structure of pseudo inverses*, Linear Algebra Appl., 197/198 (1994), 623–650.

[15] G. Heinig, K. Rost, *Algebraic Methods for Toeplitz-like Matrices and Operators*, Birkhäuser Verlag, Basel, Boston, Stuttgart 1984.

[16] G. Heinig, K. Rost, *DFT representations of Toeplitz-plus-Hankel Bezoutians with application to fast matrix-vector multiplication*, Linear Algebra Appl., 284 (1998), 157–175.

[17] G. Heinig, A. Tewodros, *On the inverses of Hankel and Toeplitz mosaic matrices*, Operator Equ. and Numer.Analysis, K.-Weierstraß-Inst. für Math., Berlin 1988, 53–65.

[18] G. Heinig, M. Van Barel, P. Kravanja, *A formula for the pseudo inverse of a Toeplitz matrix and its applications*, J. of Comp. and Appl. Math. (submitted).

[19] T. Kailath, A. Sayed (eds.), *Fast reliable algorithms for matrices with structure*, SIAM, Philadelphia, 1999.

[20] H. Park, L. Elden, *Schur-type methods for solving least squares problems with Toeplitz structure*, SIAM J. Sci. Comput., 22(2) (2000), 406–430.

[21] Qiao, *Hybrid algorithm for fast Toeplitz orthogonalization*, Numer. Mathem., 53 (1988), 351–366.

[22] I. H. Siegel, *Deferment of computation in the methods of least squares*, Math. Comp., 19 (1965), 329–331.

[23] D. Sweet, *Numerical methods for Toeplitz matrices*, PhD thesis, University of Adelaide, Australia, 1982.

[24] D. Sweet, *Fast Toeplitz orthogonalization*, Numer. Mathem., 43(1) (1984), 1–21.

[25] W. Trench, *An algorithm for the inversion of finite Toeplitz matrices*, J. of SIAM, 13, 3 (1964), 515–522.

[26] M. Van Barel, G. Heinig, P. Kravanja, *A stabilized superfast solver for non-symmetric Toeplitz systems*, SIAM J. Matrix Analysis Appl., 23, 2 (2001), 494–510.

[27] M. Van Barel, G. Heinig, P. Kravanja, *An algorithm based on orthogonal vector polynomials for Toeplitz least squares problems*, In: Vulkov, Lubin et al (eds.), *Numerical Analysis and Its Appl.*, Springer Lecture Notes in Comp. Sci. 1988 (2000), 27–34.

[28] M. Van Barel, G. Heinig, P. Kravanja, *A superfast method for solving Toeplitz least squares problems*, Linear Algebra Appl., 366 (2003), 257–281.

Georg Heinig
Department of Mathematics and Computer Science,
P.O.Box 5969, Safat 1306, Kuwait
e-mail: georg@mcs.sci.kuniv.edu.kw

Operator Theory:
Advances and Applications, Vol. 149, 199–221

Admissibility of Control and Observation Operators for Semigroups: A Survey

Birgit Jacob and Jonathan R. Partington

Abstract. This paper reviews the literature on admissibility of control and observation operators for semigroups, presenting many recent results in this approach to infinite-dimensional systems theory. The themes discussed include duality between control and observation, conditions for admissibility expressed in terms of the resolvent of the infinitesimal generator, results for normal semigroups and their links with Carleson measures, properties of shift semigroups and Hankel operators, contraction semigroups and functional models, Hille-Yosida conditions on the resolvent, and weak admissibility.

1. Introduction

There are two different approaches to linear infinite-dimensional systems: one approach is to develop theories and results which are applicable to particular classes of such systems. For example, authors such as Lasiecka, Lions and Triggiani [46], Lions [52], Butkovsky [4], and Lasiecka and Triggiani [47, 48] have very successfully studied systems described by partial differential equations, whereas other authors including Mitter and Delfour [13], Marshall [53], Dym et al. [16] and Gluesing-Luerssen [25] have analysed systems described by delay equations. The second approach abstracts the linear model to an abstract ordinary differential equation on an infinite-dimensional space. This unifying approach enables us to consider all the classes of systems in the same formulation, and a linear system is then described by

$$\dot{x}(t) = Ax(t) + Bu(t), \quad x(0) = x_0, \qquad t \geq 0, \tag{1}$$

$$y(t) = Cx(t), \qquad t \geq 0, \tag{2}$$

where $x(t)$ is the state at time t, $u(t)$ is the input at time t and $y(t)$ denotes the output at time t. We assume that state space, input space and output space, denoted by H, U and Y, respectively, are separable Hilbert spaces, that A generates a strongly continuous semigroup (C_0-semigroup in short) $(T(t))_{t\geq 0}$ on H, and that x_0 is an arbitrary element of the state space. For the theory on C_0-semigroups we refer the reader to [12], [20], [26], [32], [54], [60].

Received by the editors November 29, 2002; revised April 7, 2003. Submitted by L. Rodman.
Mathematics Subject Classification (2000). Primary 93C25; Secondary 93B28, 47D06, 47B35.
Key words and phrases. Admissible control operators, admissible observation operators, resolvent condition, Carleson measures, Hankel operators, Lyapunov equation.

Systems with bounded input and output operators, that is, $B \in \mathcal{L}(U,H)$ and $C \in \mathcal{L}(H,Y)$, have been studied and used for example in Curtain and Pritchard [8], Curtain and Zwart [10] and Fuhrmann [21]. However, in most applications the control and observation operators are unbounded. This can happen for example if our system (1)–(2) is a model for a linear partial differential equation with point or boundary observation. There are mainly three different classes of systems which allow different degrees of unboundedness of the operators B and C: the Pritchard-Salamon class, see, e.g., Pritchard and Salamon [61], Salamon [64], the Weiss class of regular systems, see, e.g., Curtain and Weiss [9] and Weiss [76], and the Salamon class of well-posed linear systems, see, e.g., Salamon [65, 66]. The Pritchard-Salamon class is the smallest class and unfortunately does not contain most of the interesting examples. On the other hand, many PDEs can be written as Weiss systems. How this can be done in general is explained, e.g., in [17], [64], [65], and for particular examples see, e.g., [1], [5], [81]. More information on these classes can be found in Staffans [69] and in the survey article by Curtain [6]. In this survey we will deal mainly with the case $B = 0$, and the results we obtain are needed for the study of a general system in the Salamon or Weiss class. Thus the systems we study have the special form

$$\dot{x}(t) = Ax(t), \qquad x(0) = x_0, \qquad t \geq 0, \tag{3}$$

$$y(t) = Cx(t), \qquad t \geq 0. \tag{4}$$

By a solution of (3) we mean the so-called mild solution

$$x(t) = T(t)x_0, \quad t \geq 0,$$

which is actually the classical solution if and only if $x_0 \in D(A)$, where $D(A)$ denotes the domain of A. If the observation operator C is a linear bounded operator from H to Y, then the output is clearly given by

$$y(t) = CT(t)x_0, \quad t \geq 0.$$

However, in the case of unbounded observation operators a condition has to be imposed to guarantee that the output function is locally square integrable with values in the output space. A well-known sufficient condition is (see for example [15], [45], [62], [64], [65], [66]) that the operator C is an infinite-time admissible observation operator for $(T(t))_{t\geq 0}$, defined as follows.

Definition 1.1. *An operator* $C \in \mathcal{L}(D(A), Y)$ *is called an* (infinite-time) admissible observation operator for $(T(t))_{t\geq 0}$ *if there exists a constant* $M > 0$ *such that*

$$\|CT(\cdot)x_0\|_{L^2(0,\infty;Y)} \leq M\|x_0\|_H, \quad x_0 \in D(A). \tag{5}$$

Note that $D(A)$ equipped with the graph norm is a Hilbert space which is densely embedded in H. Infinite-time admissibility guarantees that the mapping $\mathbb{L} : D(A) \to L^2(0,\infty;Y)$, given by

$$(\mathbb{L}x_0)(t) := CT(t)x_0, \qquad t \geq 0, \quad x_0 \in D(A),$$

has a unique extension (again denoted by $\mathbb{L}$) to a linear bounded operator from H to $L^2(0,\infty;Y)$. In this case equation (4) can be interpreted as

$$y(t) = (\mathbb{L}x_0)(t), \quad t \geq 0,$$

which makes sense for every $x_0 \in H$. There are slightly different definitions of (infinite-time) admissibility available in the literature (see for example [6], [29], [75]) and we provide a discussion of the definition in Section 2.1.

The aim of this article is to review the literature on admissibility of observation operators for semigroups. We proceed as follows. In Section 2 we summarize some useful properties of admissibility such as duality, the Lebesgue extension and invariance properties. Section 3 and 4 are devoted to equivalent conditions for admissibility. In Section 3 we study the special case that the dimension of Y is finite and in Section 4 we summarize general conditions for admissibility. Finally in Section 5 we discuss some extensions and applications.

2. Admissible observation operators

2.1. Discussion of the definition

We start this section with a discussion of the definition of infinite-time admissibility. First of all it seems to be quite restrictive to require L^2-outputs rather than L^2_{loc}-outputs, and one might suggest the following definition.

Definition 2.1. *An operator $C \in \mathcal{L}(D(A), Y)$ is called a* finite-time admissible observation operator *for $(T(t))_{t\geq 0}$ if for some $t_0 > 0$ (and hence for all $t_0 > 0$) there exists a constant $M(t_0) > 0$ such that $\|CT(\cdot)x_0\|_{L^2(0,t_0;Y)} \leq M(t_0)\|x_0\|_H$, for all $x_0 \in D(A)$.*

If the observation operator is finite-time admissible then equation (4) can be interpreted in a similar way as for infinite-time admissible observation operators. The relation of finite- and infinite-time admissibility is as follows: Clearly, infinite-time admissibility implies finite-time admissibility, but more importantly the notions are equivalent if the semigroup $(T(t))_{t\geq 0}$ is exponentially stable, see [28]. It is easy to see that C is a finite-time admissible observation operator for $(T(t))_{t\geq 0}$ if and only if C is a finite-time admissible observation operator for $(e^{-\alpha t}T(t))_{t\geq 0}$. This leads to the following two observations. On the one hand if we only want to check whether a given operator $C \in \mathcal{L}(D(A), Y)$ is finite-time admissible, we first multiply our semigroup by $e^{-\alpha t}$, where $\alpha \geq 0$ is chosen such that we obtain an exponentially stable semigroup, and then we can apply a test for infinite-time admissibility. On the other hand if the semigroup is not exponentially stable, but C is infinite-time admissible then the system satisfies some stability requirements as well, and thus the notion of infinite-time admissibility is more powerful.

Secondly, it seems to be restrictive to require that C is a linear bounded operator from $D(A)$ to Y, and one would suggest the following definition.

Definition 2.2. *Let $\tilde{H}$ be a dense $(T(t))_{t\geq 0}$-invariant subspace of H. Then a linear (not necessarily bounded) operator $C : \tilde{H} \to Y$ is called an* admissible observation operator for $(T(t))_{t\geq 0}$ *if for some $t_0 > 0$ (and hence for all $t_0 > 0$) there exists a constant $M(t_0) > 0$ such that $\|CT(\cdot)x_0\|_{L^2(0,t_0;Y)} \leq M(t_0)\|x_0\|_H$, $x_0 \in \tilde{H}$.*

If C is admissible (in the sense of Definition 2.2) then again we can define operators $\mathbb{L}_t : \tilde{H} \to L^2(0,t;Y)$, $t > 0$, which have a unique extension to a bounded operator on H and which describe the output of the system. However, it may happen that different admissible observation operators (in the sense of Definition 2.2) lead to the same operators $(\mathbb{L}_t)_{t\geq 0}$, and we call those observation operators equivalent. Weiss [75] (see also Salamon [66]) showed that every admissible observation operator (in the sense of Definition 2.2) is equivalent to a finite-time admissible observation operator in $\mathcal{L}(D(A), Y)$. Thus without loss of generality we may assume that $C \in \mathcal{L}(D(A), Y)$.

In particular, if C is the observation operator of a well-posed linear system with semigroup $(T(t))_{t\geq 0}$, then C is infinite-time admissible for $(e^{-\alpha t}T(t))_{t\geq 0}$ for some $\alpha > 0$.

2.2. Admissible control operators and duality

In this paper we deal with observation operators rather than control operators, but most of the results mentioned translate directly into results concerning control operators as explained in the following. We consider the system

$$\dot{x}(t) = Ax(t) + Bu(t), \quad x(0) = x_0, \qquad t \geq 0.$$

Here $u(t) \in U$ is the input at time t, the input space U is a separable Hilbert space and B is an unbounded operator from U to H. However, we then need to impose a condition guaranteeing that the state is still in H. A well-known sufficient condition (see, e.g., [34], [45], [62], [65], [66], [74]) is that B is an *infinite-time admissible control operator for* $(T(t))_{t\geq 0}$, that is, $B \in \mathcal{L}(U, D(A^*)')$ and there exists a constant $M > 0$ such that

$$\left\| \int_0^\infty T(t)Bu(t)\, dt \right\|_H \leq M\|u\|_{L^2(0,\infty;U)}, \quad u \in L^2(0,\infty;U).$$

Note that, in the integral on the left-hand side, the operator $T(t)$ has been extended to a bounded operator acting on $D(A^*)'$. A simple calculation shows that B is an infinite-time admissible control operator for $(T(t))_{t\geq 0}$ if and only if B^* is an infinite-time admissible observation operator for the dual semigroup $(T(t)^*)_{t\geq 0}$. A proof of this result can be found in [75]. However, the duality of admissible control and observation operators has already been discussed earlier in Curtain and Pritchard [8], Dolecki and Russell [15], Pritchard and Wirth [62], Salamon [65] and others. This duality result allows us to translate most of the results given in this paper for observation operators into results concerning control operators. In particular, if B is the control operator of a well-posed linear system with semigroup $(T(t))_{t\geq 0}$, then B is infinite-time admissible for $(e^{-\alpha t}T(t))_{t\geq 0}$ for some $\alpha > 0$.

2.3. Lebesgue extension

Working with infinite-time admissible observation operators we only know that the output is given by $y := \mathbb{L}x_0$ and we do not have a priori a pointwise interpretation of $y(t) = CT(t)x_0$ for every $x_0 \in H$ and a.e. $t \geq 0$. However, such a pointwise interpretation is required in some applications, see for example [76]. This problem can be solved by replacing the operator C by its Lebesgue extension C_L. For an infinite-time admissible observation operator $C \in \mathcal{L}(D(A), Y)$, Weiss [75, 78] defined the Lebesgue extension C_L as follows:

$$\begin{aligned} C_L x &:= \lim_{t\to 0} \frac{1}{t} C \int_0^t T(s)x\, ds, \\ D(C_L) &:= \left\{ x \in H \mid \lim_{t\to 0} \frac{1}{t} C \int_0^t T(s)x\, ds \text{ exists} \right\}. \end{aligned}$$

Equipping $D(C_L)$ with the norm

$$\|x\|_{D(C_L)} := \|x\|_H + \sup_{t\in(0,1]} \left\| \frac{1}{t} C \int_0^t T(s)x\, ds \right\|,$$

$D(C_L)$ is a Banach space, $D(A) \subset D(C_L) \subset H$ with continuous embeddings and $C_L \in \mathcal{L}(D(C_L), Y)$. In [75] it is further noted that in general C_L is not closed or closable. In particular, if $D(C_L) \neq H$ and $\dim Y < \infty$ then C_L cannot be closable. Moreover, $D(A)$ is dense in $D(C_L)$, see [78]. Now the operator C_L possesses the following useful property: let $x_0 \in H$, then $T(t)x_0 \in D(C_L)$ for a.e. $t \geq 0$ and $y(t) = C_L T(t)x_0$ for a.e. $t \geq 0$. Thus after replacing C by C_L we have a pointwise interpretation of (4).

2.4. Λ-extension

Beside the Lebesgue extension, Weiss [79, 78] introduced another extension of C, the *Λ-extension:*

$$\begin{aligned} C_\Lambda x &:= \lim_{\lambda\to+\infty,\lambda>0} C\lambda(\lambda I - A)^{-1}x, \quad x \in D(C_\Lambda), \\ D(C_\Lambda) &:= \{x \in H \mid \text{ the above limit exists}\}. \end{aligned}$$

It is shown that C_Λ extends the extension C_L, but in general the inclusion $D(C_L) \subset D(C_\Lambda)$ is strict, see [43]. Equipping $D(C_L)$ with the norm

$$\|x\|_{D(C_\Lambda)} := \|x\|_H + \sup_{\lambda>0} \|C\lambda(\lambda I - A)^{-1}x\|,$$

$D(C_\Lambda)$ is a Banach space, $D(A) \subset D(C_L) \subset D(C_\Lambda) \subset H$ with continuous embeddings and $C_\Lambda \in \mathcal{L}(D(C_\Lambda), Y)$. However, it an open problem whether or not $D(A)$ is dense in $D(C_\Lambda)$. In most cases it makes no difference if one works with C_L or C_Λ. However, the advantage of the Λ-extension is that it is invariant under admissible feedback, as discussed in the following section.

2.5. Invariance properties of admissibility

Admissibility is stable under stable perturbations as follows. Let $C \in \mathcal{L}(D(A), Y)$ and $P \in \mathcal{L}(H)$ with $\sigma(P) \subset \mathbb{C}_-$, where $\mathbb{C}_-$ denotes the open left half-plane. Then it is easy to see, and mentioned in [75], that if C is infinite-time admissible for $(T(t))_{t\geq 0}$ then C is infinite-time admissible for the C_0-semigroup generated by $A+P$ if the latter is exponentially stable. In Weiss [78] (see also Hansen and Weiss [31]) this result is generalized to admissible feedback operators. In particular, it is shown that under admissible feedback the Λ-extension of C is invariant. Further, we note that admissibility is invariant under restrictions to $(T(t))_{t\geq 0}$-invariant subspaces.

Using a simple trick the question whether a C is infinite-time admissible can be reduced to a question which involves bounded operators only. Namely, if zero is an element of the resolvent set of A then C is infinite-time admissible for $(T(t))_{t\geq 0}$ if and only if CA^{-1} is infinite-time admissible for $(e^{tA^{-1}})_{t\geq 0}$ [7]. Note that CA^{-1} and A^{-1} are bounded operators. The system (A^{-1}, CA^{-1}), called *reciprocal system*, was introduced and studied by Curtain [7].

If we fix a semigroup and an output space Y, then the set of all infinite-time admissible observation operators C, denoted by $\mathcal{C}$, is a Banach space with the norm $\|C\|$ given by the smallest $M \geq 0$ for which (5) holds and we have $\mathcal{L}(H, Y) \subset \mathcal{C} \subset \mathcal{L}(D(A), Y)$ with continuous embeddings, see [75].

3. Criteria for admissibility: Finite-rank operators

In this section and in the following one we summarize criteria for an operator C to be admissible. All criteria which do not require a finite-dimensional output space can be found in Section 4, whereas in this section we assume that the dimension of Y is finite. It is easy to see that without loss of generality we may assume that the dimension of Y is one, that is, we assume $Y = \mathbb{C}$ within this section.

3.1. Carleson measures and normal operators

We first consider diagonal semigroups. Thus let $\{\lambda_n\}_n$ be a sequence in the open left half-plane, $H := \ell^2(\mathbb{N})$, and

$$\begin{aligned} Au &:= \{\lambda_n u_n\}_n = \begin{pmatrix} \lambda_1 & & \\ & \lambda_2 & \\ & & \ddots \end{pmatrix} \begin{pmatrix} u_1 \\ u_2 \\ \vdots \end{pmatrix} \\ D(A) &:= \{u \in H \mid \{\lambda_n u_n\}_n \in H\}. \end{aligned}$$

A generates an asymptotically stable C_0-semigroup of contractions $(T(t))_{t\geq 0}$. We write C as $\{c_n\}_n$, $c_n \in \mathbb{C}$. In this subsection we do not assume a priori that C is a bounded linear operator from $D(A)$ to $\mathbb{C}$. However, we first assume that the semigroup is exponentially stable.

The first result was given in [34], where Ho and Russell showed that if $\nu = \sum_{n=1}^{\infty} |c_n|^2 \delta_{-\lambda_n}$ is a Carleson measure then C is infinite-time admissible. In [72] it was then shown by Weiss that this condition is also necessary, that is, if C is infinite-time admissible then ν is a Carleson measure.

It can be easily seen that ν is a Carleson measure if and only if C satisfies the *Carleson measure criterion*, that is, there is a constant $M > 0$ such that

$$\sum_{-\lambda_k \in R(h,\omega)} |c_k|^2 \leq Mh \tag{6}$$

holds for any $h > 0$ and $\omega \in \mathbb{R}$. Here $R(h, \omega)$ is the rectangle given by

$$R(h,\omega) = \{z \in \mathbb{C} : 0 \leq \operatorname{Re} z \leq h, \, |\operatorname{Im} z - \omega| \leq h\}. \tag{7}$$

In particular, the Carleson measure criterion implies $C \in \mathcal{L}(D(A), \mathbb{C})$, that is,

$$\sum_{k=1}^{\infty} \left| \frac{c_k}{\lambda_k} \right|^2 < \infty.$$

Thus a $\mathbb{C}$-valued sequence $\{c_n\}_n$ defines an infinite-time admissible observation operator if and only if it satisfies the Carleson measure criterion.

Using the spectral theorem for normal semigroups this result was generalized in [77] (see also [30, Remark 5.2 on page 226]) to exponentially stable normal semigroups.

Without assuming exponential stability Grabowski [28] proved later that an operator $C = \{c_n\} \in \mathcal{L}(D(A), \mathbb{C})$ is infinite-time admissible for a normal semigroup if and only if $\{c_n\}$ satisfies the Carleson measure criterion.

We now assume $C \in \mathcal{L}(D(A), \mathbb{C})$ in the following. In [77], [80] it is shown for normal semigroups that C is infinite-time admissible if and only if there exists a constant $M > 0$ such that

$$\|C(\lambda I - A)^{-1}\| \leq \frac{M}{\sqrt{\operatorname{Re} \lambda}}, \qquad \lambda \in \mathbb{C}_+. \tag{8}$$

We shall refer to (8) as the *resolvent condition*. Moreover, if additionally the semigroup is analytic then (8) is equivalent to the existence of $M > 0$ such that

$$\|C(\lambda I - A)^{-1}\| \leq \frac{M}{\sqrt{\lambda}}, \qquad \lambda > 0.$$

Later Grabowski [28] showed the equivalence of the resolvent condition and infinite-time admissibility for diagonal semigroups without assuming exponentially stability, and he proved further for these semigroups that it is enough to test the resolvent condition for the sequence $\{-\overline{\lambda_n}\}$, that is, C is infinite-time admissible if and only if there exists a constant $M > 0$ such that

$$\|C(-\overline{\lambda_n} I - A)^{-1}\| \leq \frac{M}{\sqrt{\operatorname{Re} \lambda_n}}, \qquad n \in \mathbb{N}.$$

The resolvent condition will be further studied in Section 3.2.

3.2. The Weiss conjecture

It is easy to verify (see Weiss [77] for a proof) that the resolvent condition (8) is necessary for infinite-time admissibility for an arbitrary C_0-semigroup. The equivalence of the two conditions for normal semigroups (and exponentially stable right-invertible semigroups, as we explain in Section 4.5) led Weiss [77] to formulate a general conjecture (with a \$100 reward offered for a proof or disproof), namely the equivalence of the two conditions for all strongly continuous semigroups. This we shall refer to as the *Weiss conjecture.* This conjecture was reformulated more recently in [80].

Probably the simplest semigroup for which admissibility had not been analysed by 1998 was the right-shift semigroup on $L^2(0,\infty)$, given for $f \in L^2(0,\infty)$ and t, $\tau > 0$ by

$$(\widetilde{T}(t)f)(\tau) = \begin{cases} 0 & \text{for } \tau < t, \\ f(\tau - t) & \text{for } \tau > t. \end{cases}$$

This was considered by Partington and Weiss [59], by looking at the unitarily equivalent semigroup $(T(t))_{t\geq 0}$ on $H^2(\mathbb{C}_+)$ defined for $x \in H^2(\mathbb{C}_+)$ by

$$(T(t)x)(s) = e^{-st}x(s), \qquad (\operatorname{Re} s > 0).$$

Here $(Ax)(s) = -sx(s)$ for $x \in D(A)$, and the A-bounded functionals $C : D(A) \to \mathbb{C}$ are represented by

$$Cx = \int_{-\infty}^{\infty} \overline{c(i\omega)}x(i\omega)\, d\omega, \qquad (x \in D(A)),$$

where the function $s \mapsto c(s)/(1+s)$ lies in $H^2(\mathbb{C}_+)$. The infinite-time admissibility condition turns out to be equivalent to the boundedness of the Hankel operator $\Gamma_c : H^2(\mathbb{C}_-) \to H^2(\mathbb{C}_+)$ defined by

$$\Gamma_c u = P_{H^2(\mathbb{C}_+)}(c \cdot u), \qquad (u \in H^2(\mathbb{C}_-)),$$

where $P_{H^2(\mathbb{C}_+)}$ is the orthogonal projection from $L^2(i\mathbb{R})$ onto its closed subspace $H^2(\mathbb{C}_+)$. This is equivalent to the condition that c lies in $BMOA(\mathbb{C}_+)$, the space of analytic functions of bounded mean oscillation (see, for example, the books by Garnett [23] and Koosis [44]).

The resolvent condition (8) may also be interpreted in terms of Γ_c. Note first that for $\lambda \in \mathbb{C}_+$, one has

$$\|C(\lambda I - A)^{-1}\| = \|\phi_\lambda\|_{H^2(\mathbb{C}_+)},$$

where

$$\phi_\lambda : s \mapsto \frac{c(s) - c(\overline{\lambda})}{s - \overline{\lambda}}, \qquad (s \in \mathbb{C}_+).$$

From this the resolvent condition translates into the condition that Γ_c be bounded on the set of normalized reproducing kernels of $H^2(\mathbb{C}_-)$, namely functions of the form $s \mapsto \frac{1}{\sqrt{\pi}}(\operatorname{Re}\lambda)/(s - \overline{\lambda})$, with $\lambda \in \mathbb{C}_+$.

The result that boundedness of Hankel operators can be tested on normalized reproducing kernels was given in the case of Hardy spaces on the disc by Bonsall [2] (see also some related results due to Holland and Walsh [35]). It is a deep result, apparently requiring Fefferman's duality theorem to prove. It is now sometimes referred to as the *reproducing kernel thesis*, as it can be formulated in the context of other reproducing kernel Hilbert spaces (see [57]).

Thus the Weiss conjecture holds for the right-shift semigroup, and soon afterwards the methods above were extended by Jacob and Partington [38] to give a generalization to the class of all contraction semigroups. The basic idea was to decompose the contraction semigroup into two parts, as follows:

1. a unitary part, which, being normal, could be handled by the Carleson measure approach outlined above; and
2. a completely non-unitary (c.n.u.) part, to which the Sz.-Nagy-Foiaş model theory [71] could be applied.

Explicitly, a completely non-unitary contraction semigroup $(T(t))_{t\geq 0}$ is necessarily equivalent to a semigroup constructed as follows. Let E and F be Hilbert spaces and $\Theta : \mathbb{C}_+ \to \mathcal{L}(E,F)$ be holomorphic and satisfying $\|\Theta(s)\| \leq 1$ for $s \in \mathbb{C}_+$. Then $\Delta : i\mathbb{R} \to \mathcal{L}(E)$ is the defect operator-valued function

$$\Delta(i\omega) = [I_E - \Theta(i\omega)^*\Theta(i\omega)]^{1/2}, \qquad (\omega \in \mathbb{R}).$$

Write

$$\begin{aligned} X &= H^2(\mathbb{C}_+, F) \oplus \overline{\Delta L^2(i\mathbb{R}, E)}, \qquad \text{and} \\ U &= X \ominus \{(\Theta f, \Delta f) : f \in H^2(\mathbb{C}_+, E)\}. \end{aligned}$$

Then any c.n.u. contraction semigroup is equivalent to some semigroup on a space U, defined by $(T(t))u = P_U[e^{i\omega t}u(i\omega)]$ for $t \geq 0$ and $u \in U$. Thus it can be regarded as the compression of a shift semigroup. This gives a way of adapting the methods of [59] to show that the Weiss conjecture holds for all contraction semigroups (a result that simultaneously generalizes the Carleson embedding theorem and the Bonsall theorem on Hankel operators).

An alternative proof of this result was later given by Staffans [69].

It is known, however, that there are bounded semigroups that are not unitarily equivalent to contraction semigroups, with examples given by Packel [58], Simard [67] and others. The Weiss conjecture as stated above was finally resolved in the negative with a construction due to Jacob and Zwart [41], which we now describe briefly.

Let H be a Hilbert space with a conditional basis $(e_n)_{n\in\mathbb{N}}$ such that

1. $\inf_{n\in\mathbb{N}} \|e_n\| > 0$; and
2. $(e_n)_{n\in\mathbb{N}}$ is not Besselian.

A standard reference for such properties is the book of Singer [68].

Define the semigroup $(T(t))_{t\geq 0}$ on H by

$$T(t)e_n = e^{-4^n t}e_n.$$

Thus its infinitesimal generator A satisfies $Ae_n = -4^n e_n$ for each n. Indeed, $(T(t))_{t\geq 0}$ is both analytic and exponentially stable. Finally one defines $C : D(A) \to \mathbb{C}$ by

$$Cx = \sum_{n=1}^{\infty} 2^n x_n, \qquad \text{for} \quad x = \sum_{n=1}^{\infty} x_n e_n \in D(A).$$

Detailed calculations show that C is an A-bounded operator satisfying the resolvent condition, but the fact that $(e_n)_{n\in\mathbb{N}}$ is not Besselian can be used to show that C is not infinite-time admissible.

Further work by Le Merdy [49] has focused on the case of bounded analytic semigroups. It is shown that the equivalence of admissibility and the resolvent condition (referred to as the *Weiss property* of the semigroup) holds also in the case of bounded analytic semigroups, whenever $(-A)^{1/2}$ is an admissible observation operator. Since this properly falls into the subject of infinite-rank observation operators, we defer further details to Section 4, remarking here only that it enabled Le Merdy to provide an example of a bounded analytic semigroup satisfying the Weiss property, which is not equivalent to a contraction semigroup.

4. General criteria for admissibility

In this section we further summarize equivalent conditions for admissibility for operators $C \in \mathcal{L}(D(A), Y)$. In particular, we do not impose conditions on Y besides being a separable Hilbert space.

4.1. Semigroup criterion

Grabowski and Callier [29] characterized infinite-time admissibility via a semigroup on an extended Hilbert space. More precisely, under the assumption that A generates an exponentially stable C_0-semigroup they showed that an observation operator $C \in \mathcal{L}(D(A), Y)$ is infinite-time admissible if and only if the operator matrix

$$A_e := \begin{pmatrix} A & 0 \\ 0 & -\frac{d}{ds} \end{pmatrix}, \tag{9}$$

$$D(A_e) := \left\{ \begin{pmatrix} x \\ g \end{pmatrix} \in D(A) \times W^{1,2}(0,\infty; Y) : g(0) = Cx \right\} \tag{10}$$

generates a C_0- semigroup on $H \times L^2(0,\infty; Y)$. The proof is lengthy, but the methods used are elementary. Later Engel [18] presented a shorter proof for this semigroup criterion using a method he developed in [19].

Gao and Hou [22] generalized this result to arbitrary C_0-semigroups. They prove that an observation operator $C \in \mathcal{L}(D(A), Y)$ is infinite-time admissible if and only if the operator matrix given by (9)–(10) generates a C_0-semigroup

$$T_e(t) = \begin{pmatrix} T_{11}(t) & T_{12}(t) \\ T_{21}(t) & T_{22}(t) \end{pmatrix}$$

on $H \times L^2(0,\infty;Y)$ satisfying $\sup\{\|T_{21}(t)\| \mid t \geq 0\} < \infty$.

4.2. A Hille-Yosida type condition

Under the assumption that A generates an exponentially stable C_0-semigroup, Grabowski and Callier [29] applied the Hille-Yosida theorem to the extended semigroup (9)-(10) and showed that C is infinite-time admissible if and only if there exists a constant $M > 0$ such that

$$\left\| \sum_{k=0}^{n} \frac{\cdot^k e^{-\lambda \cdot}}{k!} C(\lambda I - A)^{-(n-k+1)} \right\|_{\mathcal{L}(H, L^2(0,\infty;Y))} \leq \frac{M}{(\operatorname{Re}\lambda)^{n+1}}, \quad n \in \mathbb{N}_0, \quad \lambda \in \mathbb{C}_+, \tag{11}$$

where $\mathbb{C}_+$ denotes the open right half-plane. For $n = 1$, (11) is exactly the resolvent condition (8). It is sufficient to test the Hille-Yosida condition for all $\lambda > 0$ only, since the result was obtained via the Hille-Yosida theorem.

Later, Gao and Hou [22] were able to show that the Hille-Yosida condition is also equivalent to infinite-time admissibility under the weaker condition that the open right half-plane belongs to the resolvent set of A. Further comments on the extended semigroup and the Hille-Yosida condition can be found in Staffans [70, 69].

4.3. Lyapunov equation type criterion

The connection between infinite-time admissibility and the Lyapunov equation has been investigated for $C \in \mathcal{L}(H, Y)$ by Datko [11] and Levan [50]. Grabowski [27] was the first to study this connection for unbounded observation operators. In particular, Grabowski showed that C is infinite-time admissible if and only if the operator Lyapunov equation

$$\langle Ax, Xx \rangle + \langle Xx, Ax \rangle = -\langle Cx, Cx \rangle, \quad x \in D(A)$$

has a nonnegative solution $X \in \mathcal{L}(H)$, and that strong stability of the semigroup $(T(t))_{t\geq 0}$ guarantees a unique selfadjoint solution. The problem was further investigated by Hansen and Weiss [31]. They proved further that C is infinite-time admissible if and only if there exists an operator $P \in \mathcal{L}(H)$ such that, for any $x \in D(A)$,

$$Px = \lim_{\tau \to \infty} \int_0^{\tau} T^*(t) C^* C T(t) x \, dt, \quad \text{in } H,$$

and if C is infinite-time admissible, then P is the smallest nonnegative solution of the operator Lyapunov equation. Thus if C is infinite-time admissible and the semigroup $(T(t))_{t\geq 0}$ is strongly stable then P is the unique self-adjoint solution of the operator Lyapunov equation. Moreover, Hansen and Weiss showed that

$(T(t))_{t\geq 0}$ is strongly stable if P is invertible. The operator P is called the *observability Gramian.* Note that in general P being invertible does not imply that $(T^*(t))_{t\geq 0}$ is strongly stable [31], and $(T^*(t))_{t\geq 0}$ being strongly stable does not in general imply the existence of a unique solution [31]. Further results on the Lyapunov equation are given in [22], where all nonnegative solutions (all injective nonnegative solutions, all invertible nonnegative solutions) of the operator Lyapunov equation are characterized in terms of the invariant subspaces of A^*.

4.4. Diagonal semigroups

Admissibility of infinite-rank control and observation operators for diagonal semigroups was considered in [30], where Hansen and Weiss formulated the *operator Carleson measure* criterion (OCM). An infinite matrix C with columns c_k satisfies (OCM) for a diagonal semigroup $(T(t))_{t\geq 0}$ given by $(T(t)x)_k = e^{\lambda_k t}x_k$, for $k \in \mathbb{N}$, if all the c_k lie in ℓ^2 and there is a constant $M > 0$ such that

$$\left\| \sum_{-\lambda_k \in R(h,\omega)} c_k c_k^* \right\|_{\mathcal{L}(\ell^2)} \leq Mh \tag{12}$$

for every $h > 0$ and $\omega \in \mathbb{R}$. Here $R(h,\omega)$ is defined in (7). It is clear that this provides an operatorial generalization of the scalar condition given in (6). The fact that each operator $c_k c_k^*$ has rank 1 means that it is possible to check the criterion in some specific cases.

The first positive results involving (OCM) were the following, given in [30]:

1. If C represents an admissible observation operator for the exponentially stable semigroup $(T(t))_{t\geq 0}$, then C satisfies (OCM).
2. Conversely, if the semigroup is analytic or invertible as well as exponentially stable, then every C satisfying (OCM) is admissible.

Dropping the assumption of exponential stability, further results on the link between admissibility and (OCM) were proved by the same authors in [31], of which we mention two that can be stated simply:

1. C is an admissible observation operator for $(T(t))_{t\geq 0}$ if and only if the infinite matrix $P = (p_{jk})_{j,k=1}^{\infty}$ with entries given by
$$p_{jk} = -\frac{\langle c_j, c_k\rangle}{\lambda_j + \overline{\lambda}_k}$$
represents a bounded operator on ℓ^2 (indeed, it is the observability Gramian). Here $\langle \cdot\,, \cdot\rangle$ denotes the usual inner product in ℓ^2.
2. C satisfies (OCM) if and only if there is a constant $M \geq 0$ for which
$$\sup_{\|v\|_{\ell^2}\leq 1} \sum_{k=1}^{\infty} \left| \frac{\langle v, c_k\rangle}{s - \lambda_k} \right|^2 \leq \frac{M}{\operatorname{Re} s} \tag{13}$$
for all $s \in \mathbb{C}_+$.

If C satisfies (OCM), then the left-hand side of (13) is equal to $\|C(sI-A)^{-1}\|^2$, and thus a link is established with the Weiss conjecture, which is described further in Section 4.5.

An interesting special case is when the diagonal entries of $T(t)$ are e^{-kt}, $k \in \mathbb{N}$ (so that the semigroup is analytic), and C has one nonzero row, consisting entirely of ones. The entries of P are now $p_{jk} = 1/(j+k)$, that is, the admissibility of C is equivalent to the boundedness of Hilbert's Hankel matrix as an operator on ℓ^2.

4.5. The Weiss conjecture

In [77, Thm. 4.1] Weiss gave the equivalence of infinite-time admissibility and the resolvent condition (8) for the case of exponentially stable right-invertible semigroups, even in the case $\dim Y = \infty$. Indeed, a formally weaker condition is sufficient for admissibility, namely that for some constants $M > 0$ and $\alpha > 0$ the inequality

$$\|C(sI-A)^{-1}\| \le M$$

holds for all s with $\operatorname{Re} s > \alpha$. The methods used are elementary, and do not require more than the Paley-Wiener theorem, unlike the analysis of normal semigroups in the same paper. An even shorter proof was given recently by Zwart [83].

Further, the equivalence of infinite-time admissibility and the resolvent condition (8) holds for normal analytic exponentially stable C_0-semigroups, see [30], [31], and [77], even in the case $\dim Y = \infty$.

A more ambitious conjecture than the scalar Weiss conjecture described in Section 3.2 was the so-called *combined conjecture*, suggesting that admissibility would be equivalent to the resolvent condition (8), even in the case of infinite-rank C. This was formulated in [77]. The label "combined conjecture" arises because it would have been a consequence of the scalar Weiss conjecture and a further conjecture on weak admissibility, which we describe in Section 4.7.

In Weiss [73], see also Staffans [69, Lemma 3.11.7], it was shown that the combined conjecture holds for $C = I$ and arbitrary C_0-semigroups.

As in the scalar case, it became of interest to consider non-normal semigroups, and, as in the case of finite-rank C, the right-shift semigroup on $L^2(0,\infty)$ (or its unitarily equivalent form on $H^2(\mathbb{C}_+)$) was a natural choice.

The Bonsall result on the boundedness of Hankel operators [2] does not extend to the vector-valued case, as the following result given by Jacob, Partington and Pott [40] explains. As usual H denotes an infinite-dimensional separable Hilbert space.

There exists an analytic operator-valued function $\phi : i\mathbb{R} \to \mathcal{L}(H)$ such that the Hankel operator $\Gamma_\phi : H^2(\mathbb{C}_-, H) \to H^2(\mathbb{C}_+, H)$ densely defined by

$$\Gamma_\phi x = P_{H^2(\mathbb{C}_+,H)}(\phi \cdot x), \qquad (x \in D(\Gamma_\phi) \subset H^2(\mathbb{C}_-, H))$$

cannot be extended to a bounded operator on $H^2(\mathbb{C}_-, H)$, although there is a constant $m > 0$ such that

$$\|\Gamma_\phi^*(k_\lambda h)\| \le m\|h\| \qquad \text{for every} \quad \lambda \in \mathbb{C}_+,\ h \in H,$$

where $k_\lambda \in H^2(\mathbb{C}_+)$ is given by

$$k_\lambda(s) = \frac{\sqrt{\operatorname{Re}\lambda}}{s+\lambda}.$$

A still more general result is given in [24]: even the simultaneous boundedness of Γ_ϕ and Γ_ϕ^* on sets of normalized reproducing kernels (which are inequivalent conditions) is insufficient to guarantee the boundedness of Γ_ϕ. It is also possible to derive another counterexample from the results of Nazarov, Treil and Volberg [56], who show that an infinite-dimensional analogue of the Carleson embedding theorem fails.

A further counterexample was given by Zwart, Jacob and Staffans [84], as follows. (In fact the example they give was expressed in terms of control operators, so we give now the dual version.)

Let $(v_n)_{n\in\mathbb{N}}$ be a non-Besselian basis of a Hilbert space H and define the semigroup $(T(t))_{t\geq 0}$ by $T(t)v_n = e^{-2^n t}v_n$, extending continuously and linearly to H. Then $C : D(C) \subset H \to \ell^2$ is defined by $Cv_n = 2^{n/2}e_n$, where $(e_n)_{n\in\mathbb{N}}$ is the standard orthonormal basis of ℓ^2. It can be verified that C is not admissible, but nonetheless it satisfies the resolvent condition (8). In this case the semigroup is analytic.

Let us now present some positive results on the link between admissibility and the resolvent condition. In [39] an adjoint version of the resolvent condition was introduced, motivated by the fact that a Hankel operator is bounded if and only if its adjoint is bounded. Thus, if K is finite-dimensional, one can study the shift semigroup on $L^2(0,\infty;K)$ with an observation operator C mapping into an infinite-dimensional space Y by dualizing. This then reduces to the case of a finite-rank observation operator on the bigger semigroup $L^2(0,\infty;Y)$. The upshot of all this is that C is admissible for the right shift semigroup on $L^2(0,\infty;K)$ for K finite-dimensional *if and only if* there is a constant $M > 0$ such that

$$\|C(sI-A)^{-1}\|_{HS} \leq \frac{M}{\sqrt{\operatorname{Re} s}}, \qquad (s \in \mathbb{C}_+), \tag{14}$$

where $\|\cdot\|_{HS}$ is the Hilbert-Schmidt norm (and, in the case of finite-rank C, conditions (8) and (14) are equivalent).

In general, this new condition (14) is too strong to be necessary, but it is sufficient for admissibility in the case of contraction semigroups $(T(t))_{t\geq 0}$ on separable Hilbert spaces. It is also necessary and sufficient if the semigroup is completely non-unitary and has finite multiplicity: that is, if we let $F = (A+I)(A-I)^{-1}$, the co-generator of the semigroup in the sense of [71, Ch. III.8], then the defect indices $\dim(I-F^*F)^{1/2}$ and $\dim(I-FF^*)^{1/2}$ are both finite. The standard example here is the right shift semigroup on $L^2(0,\infty;K)$ with K finite-dimensional.

Le Merdy [49] takes a rather different approach. His main result is that for bounded analytic semigroups, defined on general Banach spaces, the Weiss property that (8) implies admissibility holds if and only if the fractional power

$C = (-A)^{1/2}$ is admissible for the semigroup. Note that $(-A)^{1/2}$ is a closed operator whose domain contains $D(A)$. In particular, he showed that the combined conjecture holds for analytic contraction semigroups.

Thus Le Merdy's result extends the results of Hansen and Weiss [30] for bounded, analytic and normal semigroups. The main tool used is the McIntosh H^∞ functional calculus, introduced in [55] as an extension of the Riesz-Dunford functional calculus. One further consequence of this work is the construction of a semigroup on a Hilbert space, with infinitesimal generator A, such that $(-A)^{1/2}$ is admissible for the semigroup but A does not have a bounded $H^\infty(\Sigma_\alpha)$ functional calculus on any open sector $\Sigma_\alpha = \{z \in \mathbb{C} : \pi - \alpha < \arg z < \pi + \alpha\}$ including its spectrum. In particular, A is not similar to a contraction semigroup, even though in this case the Weiss property holds.

4.6. Alternative forms of the resolvent condition

We recall that, for the right-shift semigroup, the fact that the Weiss conjecture holds for scalar observation functionals is equivalent to the Bonsall theorem characterizing bounded Hankel operators in terms of their action on reproducing kernels. There are other conditions available for the boundedness of Hankel operators, using alternative sets of test functions, and these may be found in [2, 35, 3]. Similarly one may formulate alternative conditions for admissibility, as in [39].

Consider the following conditions:

1. There is a constant $K > 0$ such that

$$\left\| \tau^{-1/2} \int_0^\tau e^{i\omega t} CT(t)x \, dt \right\| \le K\|x\|, \qquad (x \in D(A),\ \tau > 0,\ \omega \in \mathbb{R}). \tag{15}$$

2. There is a constant $K > 0$ such that

$$\left\| \tau^{-1/2} \int_\tau^{2\tau} e^{i\omega t} CT(t)x \, dt \right\| \le K\|x\|, \qquad (x \in D(A),\ \tau > 0,\ \omega \in \mathbb{R}). \tag{16}$$

The first two conditions are always equivalent to each other, and for bounded semigroups they are also equivalent to the resolvent condition (8) – the more difficult part of this is to show that the resolvent condition implies (15). The result for the right-shift semigroup asserts that one can check the boundedness of a Hankel integral operator on $L^2(0,\infty)$ by means of test functions that are undamped sinusoids of finite duration, rather than damped oscillations of infinite duration.

A frequency-domain test is often more useful, and if $0 \notin \sigma_p(A)$, where $\sigma_p(A)$ denotes the point spectrum of A, we may reformulate (15) in the following equivalent form.

There is a constant $K > 0$ such that

$$\left\| \tau^{-1/2} C(i\omega I - A)^{-1}(I - e^{-i\omega\tau}T(\tau)) \right\|_{\mathcal{L}(X,Y)} \le K, \qquad (\tau > 0,\ \omega \in \mathbb{R}).$$

The resolvent condition (8) was obtained by integrating against the test function $t \mapsto e^{-\lambda t}$. By integrating against $t \mapsto t^{n-1}e^{-\lambda t}$ for $n \in \mathbb{N}$ it is easy to derive a more general necessary condition, namely, that there exists a constant $M > 0$ such that

$$\|C(sI-A)^{-n}\| \leq \frac{M}{n^{1/4}(\operatorname{Re} s)^{n-1/2}}, \qquad (n \in \mathbb{N},\ s \in \mathbb{C},\ \operatorname{Re} s > 0).$$

Details can be found in [69, Prop. 4.2.9, p. 177].

In [84] it is shown that the slightly stronger condition

$$\|C(sI-A)^{-n}\| \leq \frac{M}{n^{1/2}(\operatorname{Re} s)^{n-1/2}}, \qquad (n \in \mathbb{N},\ s \in \mathbb{C},\ \operatorname{Re} s > 0).$$

is still not sufficient.

However, Zwart [83] proved that the condition

$$\|C(sI-A)^{-1}\| \leq \frac{M}{\log(\operatorname{Re} s)(\operatorname{Re} s)^{1/2}}, \qquad (s \in \mathbb{C},\ \operatorname{Re} s > \omega),$$

is sufficient for infinite-time admissibility under the extra assumption that the semigroup is exponentially stable. In Zwart's condition the function log can be replaced by a monotonically increasing function g satisfying

$$\sum_{n=1}^{\infty} g(\alpha^n)^{-2} < \infty,$$

where $\alpha > 1$ is fixed.

4.7. Weak admissibility

As related in Section 4.5, Weiss formulated a combined conjecture in [77]. One half of this concerned weak admissibility, which is the passage from the vector case to the scalar case as follows:

An observation operator $C : D(A) \to Y$ *is said to be* weakly admissible *for* $(T(t))_{t\geq 0}$ *if and only if* $\phi C : D(A) \to \mathbb{C}$ *is an admissible functional for every bounded linear functional* $\phi : Y \to \mathbb{C}$.

Dually, a control operator B is weakly admissible if Bu is an admissible input element for every $u \in U$.

The property of weak admissibility is clearly an immediate consequence of admissibility. In some special cases one has equivalence of the two properties, and it was conjectured in [77] that they were always equivalent.

Hansen and Weiss [30] proved that weak admissibility implies full admissibility whenever $(T(t))_{t\geq 0}$ is a normal analytic semigroup, and also if it is invertible; this latter was strengthened to "left invertible" in [77].

However, during the IWOTA 2000 conference in Bordeaux, two examples were found that show that weak admissibility is not equivalent to admissibility.

1. In the example of Zwart, Jacob and Staffans [84], described in Section 4.5, of an analytic semigroup failing the Weiss property, the observation operator C is weakly admissible, but not admissible.

2. In the example of Jacob, Partington and Pott [40] the observation operator C constructed for the right shift semigroup is weakly admissible (this follows from [59] since each ϕC satisfies the resolvent condition and is finite-rank) but C is not admissible.

Le Merdy [49] gives a further example, with a bounded analytic semigroup and infinitesimal generator A such that $(-A)^{1/2}$ is weakly admissible but not admissible for A.

5. Extensions, applications and conclusions

To generalize the above results, one can generalize the definition of admissibility, or one can look at semigroups on more general Banach spaces. We now discuss both of these briefly, using the same notation as before. At present the body of results available is much less complete than for the situation we have described above, and the question of characterizing admissibility in more general situations is still wide open.

For $1 \leq p \leq \infty$, an operator $C \in \mathcal{L}(D(A), Y)$ is called *(infinite-time) p-admissible for* $(T(t))_{t\geq 0}$ if there exists a constant $M > 0$ such that

$$\|CT(\cdot)x_0\|_{L^p(0,\infty;Y)} \leq M\|x_0\|_H, \quad x_0 \in D(A). \tag{17}$$

Weiss [75] showed that the set of all p-admissible observation operators, denoted by $\mathcal{C}_p$, is a Banach space with the norm $\|C\|$ given by the smallest $M \geq 0$ for which (17) holds, that $\mathcal{C}_{p_1} \subset \mathcal{C}_{p_2}$ for $p_1 \geq p_2$ with continuous embedding and that $\mathcal{C}_\infty = \mathcal{L}(H, Y)$.

In a similar manner infinite-time p-admissible control operators can be introduced [74], and a simple calculation shows that B is an infinite-time p-admissible control operator for $(T(t))_{t\geq 0}$ if and only if B^* is an infinite-time q-admissible observation operator for $(T^*(t))_{t\geq 0}$, where $1/p + 1/q = 1$.

Zwart [83] showed that the resolvent condition (8) for some exponentially stable C_0-semigroup and some $C \in \mathcal{L}(D(A), Y)$ implies p-admissibility for $p \in [1, 2)$, but in general not 2-admissibility as we have seen earlier.

Further, p-admissibility is discussed briefly in the draft book of Staffans [69].

In the work of Desch et al. [14], a condition closely related to 1-admissibility is found, namely, that for some continuous nondecreasing function $\gamma : [0, \infty) \to [0, \infty)$ one has the condition

$$\int_0^t \|CT(s)Ax_0\|\, ds \leq \gamma(t)\|x_0\|.$$

This arises in a study of the abstract Cauchy problem $\dot{x}(t) = A(I + C)x(t)$.

Infinite-time p-admissibility, and in particular infinite-time admissibility, has also been studied for arbitrary Banach spaces rather than Hilbert spaces, but much less is known for the general situation and some of the result mentioned above do not hold. For example the resolvent condition (8) is no longer sufficient for infinite-time admissibility in the case of contraction semigroups and scalar outputs.

We mention also some very recent work in non-Hilbert space frameworks. Le Merdy [49], as part of his analysis of the Weiss conjecture, considered Banach spaces of cotype 2: these are spaces X for which there is a constant $c > 0$ such that for any finite sequence $x_1, \dots, x_n \in X$ one has the inequality

$$\mathbb{E}\left\|\sum_{k=1}^{n} \varepsilon_k x_k\right\|^2 \geq c \sum_{k=1}^{n} \|x_k\|^2,$$

where the symbol $\mathbb{E}$ on the left-hand side denotes taking the mean value over all 2^n possible sign choices $\varepsilon_k = \pm 1$. Such spaces include the L^p spaces for $1 \leq p \leq 2$ (only); for more information on cotype, we refer to the book [51]. In [49] it is shown that if the semigroup is analytic and the underlying space has cotype 2, then the existence of a bounded $H^\infty(\Sigma_\alpha)$ functional calculus (cf. Subsection 4.5) implies the Weiss property that admissibility is equivalent to the resolvent condition (8).

Another recent paper [82] discusses admissibility of unbounded control operators on a non-reflexive Banach space. It is currently available only in Chinese.

Apart from the standard semigroup systems described in this paper, it is possible to consider questions of admissibility for systems with different input-output relations. Of these we mention first bilinear systems, given by the equation

$$\begin{aligned} x'(t) &= Ax(t) + u(t)Bx(t), \qquad x(0) = x_0, \\ y(t) &= C(t)x(t). \end{aligned}$$

These have been considered recently by Idrissi [36], who establishes sufficient conditions for the admissibility of a family of time dependent observations, which are given in terms of the generator of the system. The results are illustrated with an application to the heat equation with unbounded control.

Jung [42] considers systems arising from Volterra integral equations, namely systems of the form

$$x'(t) = (da * Ax)(t) + bu(t), \qquad x(0) = x_0, \tag{18}$$

where a is a scalar-valued function of bounded variation. Such equations are described in detail in the book of Prüss [63]. Jung shows that under certain hypotheses on a (involving completely positivity) the functional b is admissible for the system (18) if and only if it is admissible for the standard linear system $x'(t) = Ax(t) + bu(t)$.

Finally, we mention that the concept of admissibility has been successfully generalized to time-varying infinite-dimensional systems of the form

$$x'(t) = A(t)x(t), \quad y(t) = C(t)x(t), \qquad x(0) = x_0,$$

see Hinrichsen and Pritchard [33], and Jacob, Dragan, Pritchard [37].

References

[1] G. Avalos, I. Lasiecka, and R. Rebarber. Well-posedness of a structural acoustics control model with point observation of the pressure. *J. Differential Equations*, 173(1):40–78, 2001.

[2] F.F. Bonsall. Boundedness of Hankel matrices. *J. London Math. Soc. (2)*, 29(2):289–300, 1984.

[3] F.F. Bonsall. Conditions for boundedness of Hankel matrices. *Bull. London Math. Soc.*, 26(2):171–176, 1994.

[4] A.G. Butkovskiy. *Distributed control systems.* Translated from the Russian by Scripta Technica, Inc. Translation Editor: George M. Kranc. Modern Analytic and Computational Methods in Science and Mathematics, No. 11. American Elsevier Publishing Co., Inc., New York, 1969.

[5] C.I. Byrnes, D.S. Gilliam, V.I. Shubov, and G. Weiss. Regular linear systems governed by a boundary controlled heat equation. *J. Dynam. Control Systems*, 8(3):341–370, 2002.

[6] R.F. Curtain. The Salamon-Weiss class of well-posed infinite-dimensional linear systems: a survey. *IMA J. Math. Control Inform.*, 14(2):207–223, 1997.

[7] R.F. Curtain. Reciprocals of regular systems: a survey. 2002. Submitted.

[8] R.F. Curtain and A.J. Pritchard. *Infinite-dimensional linear systems theory*, volume 8 of *Lecture Notes in Control and Information Sciences.* Springer-Verlag, Berlin, 1978.

[9] R.F. Curtain and G. Weiss. Well posedness of triples of operators (in the sense of linear systems theory). In *Control and estimation of distributed parameter systems (Vorau, 1988)*, volume 91 of *Internat. Ser. Numer. Math.*, pages 41–59. Birkhäuser, Basel, 1989.

[10] R.F. Curtain and H. Zwart. *An introduction to infinite-dimensional linear systems theory*, volume 21 of *Texts in Applied Mathematics.* Springer-Verlag, New York, 1995.

[11] R. Datko. Extending a theorem of Liapunov to Hilbert spaces. *J. Math. Anal. Appl.*, 32:610–616, 1970.

[12] E.B. Davies. *One-parameter semigroups*, volume 15 of *London Mathematical Society Monographs.* Academic Press Inc. [Harcourt Brace Jovanovich Publishers], London, 1980.

[13] M.C. Delfour and S.K. Mitter. Controllability and observability for infinite dimensional systems. *SIAM J. Control*, 10:329–333, 1972.

[14] W. Desch, I. Lasiecka, and W. Schappacher. Feedback boundary control problems for linear semigroups. *Israel J. Math.*, 51(3):177–207, 1985.

[15] S. Dolecki and D.L. Russell. A general theory of observation and control. *SIAM J. Control Optimization*, 15(2):185–220, 1977.

[16] H. Dym, T.T. Georgiou, and M.C. Smith. Explicit formulas for optimally robust controllers for delay systems. *IEEE Trans. Automat. Control*, 40(4):656–669, 1995.

[17] Z. Emirasjlow and S. Townley. From PDEs with boundary control to the abstract state equation with an unbounded input operator: a tutorial. *Eur. J. Control*, 6(1):27–53, 2000.

[18] K.-J. Engel. On the characterization of admissible control- and observation operators. *Systems Control Lett.*, 34(4):225–227, 1998.

[19] K.-J. Engel. Spectral theory and generator property for one-sided coupled operator matrices. *Semigroup Forum*, 58(2):267–295, 1999.

[20] K.-J. Engel and R. Nagel. *One-parameter semigroups for linear evolution equations*, volume 194 of *Graduate Texts in Mathematics*. Springer-Verlag, New York, 2000.

[21] P.A. Fuhrmann. *Linear systems and operators in Hilbert space*. McGraw-Hill International Book Co., New York, 1981.

[22] M.-C. Gao and J.-C. Hou. The infinite-time admissibility of observation operators and operator Lyapunov equations. *Integral Equations Operator Theory*, 35(1):53–64, 1999.

[23] J.B. Garnett. *Bounded analytic functions*, volume 96 of *Pure and Applied Mathematics*. Academic Press Inc. [Harcourt Brace Jovanovich Publishers], New York, 1981.

[24] T.A. Gillespie, S. Pott, S. Treil, and A. Volberg. Logarithmic growth for weighted Hilbert transforms and vector Hankel operators. 2001.

[25] H. Gluesing-Luerssen. *Linear delay-differential systems with commensurate delays: an algebraic approach*, volume 1770 of *Lecture Notes in Mathematics*. Springer-Verlag, Berlin, 2002.

[26] J.A. Goldstein. *Semigroups of linear operators and applications*. Oxford Mathematical Monographs. The Clarendon Press Oxford University Press, New York, 1985.

[27] P. Grabowski. On the spectral-Lyapunov approach to parametric optimization of distributed-parameter systems. *IMA J. of Math. Control & Information*, 7:317–338, 1991.

[28] P. Grabowski. Admissibility of observation functionals. *Int. J. Control*, 62(5):1161–1173, 1995.

[29] P. Grabowski and F.M. Callier. Admissible observation operators, semigroup criteria of admissibility. *Integral Equations Operator Theory*, 25(2):182–198, 1996.

[30] S. Hansen and G. Weiss. The operator Carleson measure criterion for admissibility of control operators for diagonal semigroups on ℓ_2. *Systems Control Lett.*, 16(3):219–227, 1991.

[31] S. Hansen and G. Weiss. New results on the operator Carleson measure criterion. *IMA J. Math. Control Inform.*, 14:3–32, 1997.

[32] E. Hille and R.S. Phillips. *Functional analysis and semi-groups*. American Mathematical Society, Providence, R. I., 1974.

[33] D. Hinrichsen and A.J. Pritchard. Robust stability of linear evolution operators on Banach spaces. *SIAM J. Control Optim.*, 32(6):1503–1541, 1994.

[34] L.F. Ho and D.L. Russell. Admissible input elements for systems in Hilbert space and a Carleson measure criterion. *SIAM J. Control Optim.*, 21(4):614–640, 1983. Erratum, ibid. 985–986.

[35] F. Holland and D. Walsh. Boundedness criteria for Hankel operators. *Proc. Roy. Irish Acad. Sect. A*, 84(2):141–154, 1984.

[36] A. Idrissi. Admissibilité de l'observation pour des systèmes bilinéaires contrôlés par des opérateurs non bornés. *Ann. Math. Blaise Pascal*, 8(1):73–92, 2001.

[37] B. Jacob, V. Drăgan, and A.J. Pritchard. Infinite-dimensional time-varying systems with nonlinear output feedback. *Integral Equations Operator Theory*, 22(4):440–462, 1995.

[38] B. Jacob and J.R. Partington. The Weiss conjecture on admissibility of observation operators for contraction semigroups. *Integral Equations Operator Theory*, 40(2):231–243, 2001.

[39] B. Jacob, J.R. Partington, and S. Pott. Conditions for admissibility of observation operators and boundedness of Hankel operators. *Integral Equations Operator Theory*, 47(3):315–338, 2003.

[40] B. Jacob, J.R. Partington, and S. Pott. Admissible and weakly admissible observation operators for the right shift semigroup. *Proc. Edinburgh Math. Soc.*, 45(2):353–362, 2002.

[41] B. Jacob and H. Zwart. Counterexamples concerning observation operators for C_0-semigroups. SIAM Journal of Optimization and Control, to appear.

[42] M. Jung. Admissibility of control operators for solution families to Volterra integral equations. *SIAM J. Control Optim.*, 38(5):1323–1333, 2000.

[43] V. Katsnelson and G. Weiss. A counterexample in Hardy spaces with an application to systems theory. *Z. Anal. Anwendungen*, 14(4):705–730, 1995.

[44] P. Koosis. *Introduction to H_p spaces*, volume 115 of *Cambridge Tracts in Mathematics.* Cambridge University Press, Cambridge, 1998.

[45] I. Lasiecka. Unified theory for abstract parabolic boundary problems – a semigroup approach. *Appl. Math. Optim.*, 6(4):287–333, 1980.

[46] I. Lasiecka, J.-L. Lions, and R. Triggiani. Nonhomogeneous boundary value problems for second order hyperbolic operators. *J. Math. Pures Appl. (9)*, 65(2):149–192, 1986.

[47] I. Lasiecka and R. Triggiani. *Control theory for partial differential equations: continuous and approximation theories. I*, volume 74 of *Encyclopedia of Mathematics and its Applications.* Cambridge University Press, Cambridge, 2000.

[48] I. Lasiecka and R. Triggiani. *Control theory for partial differential equations: continuous and approximation theories. II*, volume 75 of *Encyclopedia of Mathematics and its Applications.* Cambridge University Press, Cambridge, 2000.

[49] C. Le Merdy. The Weiss conjecture for bounded analytic semigroups. *J. London Math. Soc.*, 67(3):715–738, 2003.

[50] N. Levan. Stability, stabilizability, and the equation $[PAx, x]+[x, PAx] = -\|B^*x\|^2$. In R. E. Kalman, G. I. Marchuk, A. E. Ruberti, and A. J. Viterbi, editors, *Recent advances in Communication and Control Theory*, pages 214–226, New York. Optimization Software, Inc.

[51] J. Lindenstrauss and L. Tzafriri. *Classical Banach spaces. II*, volume 97 of *Ergebnisse der Mathematik und ihrer Grenzgebiete [Results in Mathematics and Related Areas].* Springer-Verlag, Berlin, 1979.

[52] J.-L. Lions. *Optimal control of systems governed by partial differential equations.* Translated from the French by S. K. Mitter. Die Grundlehren der mathematischen Wissenschaften, Band 170. Springer-Verlag, New York, 1971.

[53] J. E. Marshall. *Control of time-delay systems*, volume 10 of *IEE Control Engineering Series.* Peter Peregrinus Ltd., Stevenage, 1979.

[54] A.C. McBride. *Semigroups of linear operators: an introduction*, volume 156 of *Pitman Research Notes in Mathematics Series*. Longman Scientific & Technical, Harlow, 1987.

[55] A. McIntosh. Operators which have an H_∞ functional calculus. In *Miniconference on operator theory and partial differential equations (North Ryde, 1986)*, volume 14 of *Proc. Centre Math. Anal. Austral. Nat. Univ.*, pages 210–231. Austral. Nat. Univ., Canberra, 1986.

[56] F. Nazarov, S. Treil, and A. Volberg. Counterexample to the infinite-dimensional Carleson embedding theorem. *C. R. Acad. Sci. Paris Sér. I Math.*, 325(4):383–388, 1997.

[57] N.K. Nikolski. *Operators, functions, and systems: an easy reading. Vol. 1*, volume 92 of *Mathematical Surveys and Monographs*. American Mathematical Society, Providence, RI, 2002.

[58] E.W. Packel. A semigroup analogue of Foguel's counterexample. *Proc. Amer. Math. Soc.*, 21:240–244, 1969.

[59] J.R. Partington and G. Weiss. Admissible observation operators for the right shift semigroup. *Math. Control Signals Syst.*, 13(3):179–192, 2000.

[60] A. Pazy. *Semigroups of linear operators and applications to partial differential equations*, volume 44 of *Applied Mathematical Sciences*. Springer-Verlag, New York, 1983.

[61] A.J. Pritchard and D. Salamon. The linear quadratic control problem for infinite-dimensional systems with unbounded input and output operators. *SIAM J. Control Optim.*, 25(1):121–144, 1987.

[62] A.J. Pritchard and A. Wirth. Unbounded control and observation systems and their duality. *SIAM J. Control Optim.*, 16(4):535–545, 1978.

[63] J. Prüss. *Evolutionary integral equations and applications*, volume 87 of *Monographs in Mathematics*. Birkhäuser Verlag, Basel, 1993.

[64] D. Salamon. *Control and observation of neutral systems*, volume 91 of *Research Notes in Mathematics*. Pitman (Advanced Publishing Program), Boston, MA, 1984.

[65] D. Salamon. Infinite-dimensional linear systems with unbounded control and observation: a functional analytic approach. *Trans. Amer. Math. Soc.*, 300(2):383–431, 1987.

[66] D. Salamon. Realization theory in Hilbert space. *Math. Systems Theory*, 21:147–164, 1989.

[67] A. Simard. Counterexamples concerning powers of sectorial operators on a Hilbert space. *Bull. Austral. Math. Soc.*, 60(3):459–468, 1999.

[68] I. Singer. *Bases in Banach spaces. I*. Springer-Verlag, New York, 1970.

[69] O.J. Staffans. Well-posed linear systems I: General theory. Manuscript, Draft available at `http://www.abo.fi/ staffans/publ.htm`. Edition dated July 2002.

[70] O.J. Staffans. Well-posed linear systems, Lax-Phillips scattering and L_p-multipliers. In A. A. Borichev and N. K. Nikolski, editors, *Systems, Approximation, Singular Integral Operators, and Related Topics*, Operator Theory: Advances and Applications Vol 129, pages 445–464, Basel, 2001. Birkhäuser Verlag.

[71] B. Sz.-Nagy and C. Foiaş. *Harmonic analysis of operators on Hilbert space*. Translated from the French and revised. North-Holland Publishing Co., Amsterdam, 1970.

[72] G. Weiss. Admissibility of input elements for diagonal semigroups on l^2. *Systems Control Lett.*, 10:79–82, 1988.

[73] G. Weiss. Weak L^p-stability of a linear semigroup on a Hilbert space implies exponential stability. *J. Differential Equations*, 76(2):269–285, 1988.

[74] G. Weiss. Admissibility of unbounded control operators. *SIAM J. Control Optim.*, 27(3):527–545, 1989.

[75] G. Weiss. Admissible observation operators for linear semigroups. *Israel Journal of Mathematics*, 65(1):17–43, 1989.

[76] G. Weiss. The representation of regular linear systems on Hilbert spaces. In *Control and estimation of distributed parameter systems (Vorau, 1988)*, volume 91 of *Internat. Ser. Numer. Math.*, pages 401–416. Birkhäuser, Basel, 1989.

[77] G. Weiss. Two conjectures on the admissibility of control operators. In F. Kappel W. Desch, editor, *Estimation and Control of Distributed Parameter Systems*, pages 367–378, Basel, 1991. Birkhäuser Verlag.

[78] G. Weiss. Regular linear systems with feedback. *Math. Control Signals Systems*, 7(1):23–57, 1994.

[79] G. Weiss. Transfer functions of regular linear systems. I. Characterizations of regularity. *Trans. Amer. Math. Soc.*, 342(2):827–854, 1994.

[80] G. Weiss. A powerful generalization of the Carleson measure theorem? In V. Blondel, E. Sontag, M. Vidyasagar, and J. Willems, editors, *Open Problems in Mathematical Systems Theory and Control.* Springer Verlag, 1998.

[81] G. Weiss and M. Tucsnak. How to get a conservative well-posed linear system out of thin air. Part I: Well-posedness and energy balance. *ESAIM-COCV*, 9:247–274, 2003.

[82] H. Zhang, Q. Hu, and Y. Wang. Admissibility of unbounded control operators on non-reflexive Banach space. *Acta Anal. Funct. Appl.*, 1(1):92–96, 1999.

[83] H. Zwart. Sufficient conditions for admissibility. Submitted.

[84] H. Zwart, B. Jacob, and O. Staffans. Weak admissibility does not imply admissibility for analytic semigroups. *Systems Control Lett.*, 48(3–4):341–350, 2003.

Birgit Jacob
Fachbereich Mathematik
Universität Dortmund
D-44221 Dortmund, Germany
e-mail: `birgit.jacob@math.uni-dortmund.de`

Jonathan R. Partington
School of Mathematics
University of Leeds
Leeds LS2 9JT, UK
e-mail: `J.R.Partington@leeds.ac.uk`

Operator Theory:
Advances and Applications, Vol. 149, 223–253

Closed Subspaces which are Attractors for Representations of the Cuntz Algebras

Palle E.T. Jorgensen

Abstract. We analyze the structure of co-invariant subspaces for representations of the Cuntz algebras $\mathcal{O}_N$ for $N = 2, 3, \ldots, N < \infty$, with special attention to the representations which are associated to orthonormal and tight-frame wavelets in $L^2(\mathbb{R})$ corresponding to scale number N.

1. Introduction: Wavelets

A particular construction of wavelets on the real line $\mathbb{R}$ is based on what is called subband filters. The idea is that a wavelet decomposition of $L^2(\mathbb{R})$ can be organized in frequency bands with adjustment to a system of subspaces (identifying a cascade of resolutions) in $L^2(\mathbb{R})$, and each frequency band having its scaling resolution. The subband filters may be realized as functions on the torus $\mathbb{T} = \{z \in \mathbb{C} : |z| = 1\}$. If the scaling is $N > 1$ then the $L^2(\mathbb{R})$ functions needed in the wavelet decomposition may be obtained (under favorable conditions) as solutions to a system of cocycle conditions, see (1.1) below, involving Fourier transform $\psi \mapsto \hat{\psi}$ on $\mathbb{R}$, or rather $L^2(\mathbb{R})$. The $L^2(\mathbb{R})$-system consists of a scaling function φ and the wavelet generators $\psi_1, \ldots, \psi_{N-1}$. The cocycle conditions (see [BrJo02] for details) are

$$\begin{aligned} \sqrt{N}\hat{\varphi}(N\xi) &= m_0\left(e^{i\xi}\right)\hat{\varphi}(\xi), && \xi \in \mathbb{R}, \text{ and} \\ \sqrt{N}\hat{\psi}_j(N\xi) &= m_j\left(e^{i\xi}\right)\hat{\varphi}(\xi), && j = 1, \ldots, N-1. \end{aligned} \tag{1.1}$$

For more details on this, we refer the reader to [Dau92], [Mal99], [Jor99], [BrJo99], [Jor01], and [BrJo02]. After the problem is discretized, and a Fourier series is introduced, we then arrive at a certain system of operators on $L^2(\mathbb{T})$, where the one-torus $\mathbb{T}$ is equipped with the usual normalized Haar measure. The operators are defined from a fixed system of functions $m_0, \ldots, m_{N-1}$ on $\mathbb{T}$ as follows:

$$(S_j f)(z) = m_j(z) f\left(z^N\right), \qquad f \in L^2(\mathbb{T}),\ z \in \mathbb{T},\ j = 0, 1, \ldots, N-1. \tag{1.2}$$

Received by the editors December 27, 2002; revised February 19, 2003. Submitted by J. A. Ball.
Mathematics Subject Classification (2000). Primary 46L60, 47L30, 42A16, 43A65; Secondary 33C45, 42C10, 94A12, 46L45, 42A65, 41A15.
Key words and phrases. Wavelet, Cuntz algebra, representation, orthogonal expansion, quadrature mirror filter, isometry in Hilbert space.
Work supported in part by the U.S. National Science Foundation under grants DMS-9987777, INT-9724781, and DMS-0139473.

The orthogonality conditions which are usually imposed are known to imply the following relations on the operators $(S_j)_{j=0}^{N-1}$ in (1.2):

$$S_j^* S_i = \delta_{j,i} I_{L^2(\mathbb{T})} \quad \text{and} \quad \sum_{j=0}^{N-1} S_j S_j^* = I_{L^2(\mathbb{T})}. \tag{1.3}$$

The reader is referred to [Jor01], [Jor00], and [BrJo02] for additional details on this point. Specifically, the properties which must be imposed on the functions $m_0, \dots, m_{N-1}$ from (1.2) are known in signal processing as the subband quadrature (if $N = 2$) conditions. They can be checked to be equivalent to the operator relations (1.3), also known as the Cuntz relations. They are satisfied if and only if the $N \times N$ matrix $(i, j = 0, \dots, N-1)$

$$A_{i,j}(z) := \frac{1}{N} \sum_{\substack{w \in \mathbb{T} \\ w^N = z}} m_i(w) w^{-j}, \qquad z \in \mathbb{T}, \tag{1.4}$$

is unitary for all (or almost all, with respect to Haar measure on $\mathbb{T}$) $z \in \mathbb{T}$. The relations (1.3) are called the Cuntz relations, and they are special cases of relations which are defined axiomatically in the theory of representation of C^*-algebras. But they have an independent life in the science of signal processing; see [BrJo02], [Mal99], and the references given there.

2. The Cuntz relations

It is known [Cun77] that there is a simple C^*-algebra $\mathcal{O}_N$ such that the representations ρ of $\mathcal{O}_N$ are in a one-to-one correspondence with systems of operators on Hilbert space satisfying the Cuntz relations. If the generators of $\mathcal{O}_N$ are denoted s_i, then

$$s_i^* s_j = \delta_{i,j} 1 \quad \text{and} \quad \sum_{j=1}^{N} s_j s_j^* = 1, \tag{2.1}$$

where 1 denotes the unit-element in the C^*-algebra $\mathcal{O}_N$. The system of operators $S_j := \rho(s_j)$ will then satisfy the Cuntz relations on the Hilbert space $\mathcal{H}$ which carries the representation ρ. Conversely, every system of operators S_j on a Hilbert space $\mathcal{H}$ which satisfies the Cuntz relations comes from a representation ρ of $\mathcal{O}_N$ via the formula $\rho(s_j) = S_j$, $j = 1, \dots, N$. It is known that not every representation of $\mathcal{O}_N$ is of the form (1.2). $\mathcal{O}_N$ has many type III representations, and (1.2) implies type I (see [BJO99]). But it was shown in [Jor01] that the analysis of the wavelet representations (1.2) predicts a number of global and geometric properties of the variety of all wavelets subject to a fixed scaling.

In addition the representations of $\mathcal{O}_N$ and their functional models are used in multivariable scattering theory: see, for example, [DKS01], [BaVi1], [BaVi2], and [Kribs].

While our present results apply to general representations of $\mathcal{O}_N$, they are motivated by (1.2). In particular, we will be interested in smaller subspaces of $L^2(\mathbb{T})$ which determine the representation. Further, when a wavelet representation is given, a result in Section 10 makes precise a sense in which these subspaces are attractors, i.e., they arise as limits of a dynamical iteration. It is known from [Jor01] that, if the wavelet system consists of compactly supported functions $\varphi, \psi_1, \dots, \psi_{N-1}$ on $\mathbb{R}$, then there is a finite-dimensional subspace $\mathcal{L}$ which determines the representation, and therefore the wavelet analysis. The construction of wavelets from multiresolutions (i.e., scales of closed subspaces in $L^2(\mathbb{R})$) is explained in several books on wavelets, starting with [Dau92]. A geometric approach which is close to the present one, based on wandering subspaces, was first outlined in [DaLa98]; see also [BaMe99]. In this paper, we study such determining minimal closed subspaces in the context of the most general representations of the Cuntz algebras $\mathcal{O}_N$.

3. Subspaces of the Hilbert space

The Hilbert space which carries a representation of $\mathcal{O}_N$ must be infinite-dimensional, but we show that it contains "small" distinguished closed subspaces.

First some definitions: there will be eleven in all.

Definitions 3.1.

(i) If $\{S_i\}_{i=1}^N$ is a representation of $\mathcal{O}_N$ on a Hilbert space, set $S_I = S_{i_1} \cdots S_{i_k}$ for all *multi-indices* $I = (i_1, \dots, i_k)$, and define the *multi-index length* as $|I| := k$. The set of all such multi-indices I with $i_\nu \in \{1, \dots, N\}$ is denoted $\mathcal{I}(N)$. The set of multi-indices of length k is denoted $\mathcal{I}_k(N)$.

(ii) If $\mathcal{F}$ is a set of vectors in a Hilbert space $\mathcal{H}$, the notation $\bigvee \mathcal{F}$ stands for the closed linear span of $\mathcal{F}$.

(iii) If $\mathcal{M}$ is a family of subspaces in $\mathcal{H}$, then $\bigvee \mathcal{M}$ stands for the closed linear span of these subspaces.

(iv) If $\{S_i\}$ are the isometries which define a representation of $\mathcal{O}_N$, acting on a fixed Hilbert space $\mathcal{H}$, and if $\mathcal{L}$ is a closed subspace in $\mathcal{H}$, then we set

$$S\mathcal{L} := \bigvee_{i=1}^{N} S_i\mathcal{L}, \quad \text{and} \quad S^*\mathcal{L} := \bigvee_{i=1}^{N} S_i^*\mathcal{L}.$$

(Note that, since the operators S_i are isometries, it follows that each space $S_i\mathcal{L}$ is closed. From the identity

$$\sum_{i=1}^{N} S_i S_i^* = I_{\mathcal{H}},$$

it follows that there are natural conditions which imply that each of the spaces $S_i^*\mathcal{L}$ is also closed. We spell out this point in Lemma 3.2 below.)

(v) A closed subspace $\mathcal{L}$ is said to be *co-invariant* for a fixed representation (S_i) of $\mathcal{O}_N$ acting in a Hilbert space $\mathcal{H}$ if $S^*\mathcal{L} \subset \mathcal{L}$.

(vi) A closed subspace $\mathcal{L}$ is said to be *saturated* if

$$\bigvee_{k=1}^{\infty} S^k\mathcal{L} = \mathcal{H},$$

where, for every k,

$$S^k\mathcal{L} := \bigvee_{(i_1,\dots,i_k)\in\mathcal{I}_k(N)} S_{i_1}S_{i_2}\cdots S_{i_k}\mathcal{L}.$$

(vii) A co-invariant closed subspace $\mathcal{L}$ is said to be *minimal* if the corresponding complementing space $\mathcal{W} := S\mathcal{L} \ominus \mathcal{L}$ generates a maximal subspace, i.e., the family

$$\mathcal{F}(\mathcal{W}) := \{\mathcal{W},\ S_I\mathcal{W} : I = (i_1,\dots,i_k),\ k \geq 1,\ I \in \mathcal{I}(N)\}$$

is such that

$$\sum\nolimits^{\oplus} \{\mathcal{K} : \mathcal{K} \in \mathcal{F}(\mathcal{W})\}$$

is maximal in $\mathcal{H}$.

(viii) We say that $\mathcal{L} \subset \mathcal{H}$ is a *core* for the representation if it is co-invariant, saturated, and minimal.

(ix) A closed subspace $\mathcal{W} \subset \mathcal{H}$ is called *wandering* if all the spaces in the family $\mathcal{F}(\mathcal{W})$ are mutually orthogonal.

(x) A subspace (not necessarily closed) $\mathcal{V}$ in $\mathcal{H}$ is said to *reduce to a closed subspace $\mathcal{L}$ in a representation* (S_i) if, for every $v \in \mathcal{V}$, there is a $k_0 \in \mathbb{N}$ such that $S_{i_1}^*S_{i_2}^*\cdots S_{i_k}^*v \in \mathcal{L}$ whenever $k \geq k_0$.

(xi) A subspace $\mathcal{L}$ is said to be *stable* with respect to a representation (S_i) of $\mathcal{O}_N$ if it is invariant under each of the projections $E_i := S_iS_i^*$, $i = 1,\dots,N$.

Our main result is Theorem 10.1. It states that the standard wavelet representation (1.2) corresponding to filter functions m_j which are Lipschitz always has a natural finite-dimensional co-invariant subspace $\mathcal{L}_{\text{fin}}$. The results leading up to Section 10 throw light on the properties of the subspace $\mathcal{L}_{\text{fin}}$, and on co-invariant subspaces more generally.

Lemma 3.2. *Let a representation (S_i) of $\mathcal{O}_N$ on a Hilbert space $\mathcal{H}$ be given. Let $\mathcal{L} \subset \mathcal{H}$ be a closed subspace which is stable. Then each of the linear spaces $S_i^*\mathcal{L} = \{S_i^*x : x \in \mathcal{L}\}$ is closed in $\mathcal{H}$.*

Proof. Set $i = 1$ for specificity. Let x_n be a sequence in $\mathcal{L}$ such that $S_1^*x_n \underset{n\to\infty}{\longrightarrow} y$. We introduce the projection $E_1 = S_1S_1^*$. Since $S_j^*E_1 = \delta_{j,1}S_1^*$, we have $S_1^*E_1x_n = S_1^*x_n$, $S_j^*E_1x_n = 0$ if $j \neq 1$, and $\|S_1^*E_1(x_n - x_m)\| = \|E_1(x_n - x_m)\| \underset{n,m\to\infty}{\longrightarrow} 0$. The sequence (E_1x_n) is convergent, and its limit is in $\mathcal{L}$ since $\mathcal{L}$ is closed and stable. If $E_1x_n \underset{n\to\infty}{\longrightarrow} z$, then $S_1^*z = y$, and we conclude that $S_1^*\mathcal{L}$ is closed. $\square$

4. Co-invariant closed subspaces

A representation of $\mathcal{O}_N$ on a Hilbert space $\mathcal{H}$ is specified by operators $S_1, \ldots, S_N$ on $\mathcal{H}$ subject to the Cuntz relations

$$S_i^* S_j = \delta_{i,j} I_{\mathcal{H}}, \qquad \sum_{j=1}^{N} S_j S_j^* = I_{\mathcal{H}}, \tag{4.1}$$

where $I_{\mathcal{H}}$ denotes the identity operator in the Hilbert space $\mathcal{H}$. Intrinsic to this is the set of N commuting projections $E_j := S_j S_j^*$. They enter into the statement of the next lemmas.

In the next lemma we record some general properties about co-invariant subspaces. They are stated in terms of projections. Recall there is a one-to-one correspondence between closed subspaces $\mathcal{L} \subset \mathcal{H}$ and *projections* P in $\mathcal{H}$, i.e., $P = P^* = P^2$. If $\mathcal{L}$ is given, there is a unique P such that $\mathcal{L} = P\mathcal{H} = \{x \in \mathcal{H} : Px = x\}$, and conversely.

Lemma 4.1. *Let $\{S_i\}_{i=1}^N$ be a representation of $\mathcal{O}_N$ on a Hilbert space $\mathcal{H}$, and let $\mathcal{L}$ be a closed subspace in $\mathcal{H}$. Set $\alpha(A) = \sum_{i=1}^N S_i A S_i^*$, $A \in \mathcal{B}(\mathcal{H})$.*

(a) *Then $\alpha\colon \mathcal{B}(\mathcal{H}) \to \mathcal{B}(\mathcal{H})$ is an endomorphism satisfying $\alpha(I_{\mathcal{H}}) = I_{\mathcal{H}}$.*
(b) *If P denotes the projection onto $\mathcal{L}$, then $\mathcal{L}$ is co-invariant if and only if $P \leq \alpha(P)$; and $\alpha(P)$ is then the projection onto $S\mathcal{L}$.*
(c) *If $\mathcal{L}$ is co-invariant, then $Q = \alpha(P) - P$ is a projection. Its range is the wandering subspace $\mathcal{W} := (S\mathcal{L}) \ominus \mathcal{L}$.*

Proof. The details are left to the reader. They are based on standard geometric facts about projections in Hilbert space. □

Lemma 4.2. *Let $\{S_i\}_{i=1}^N$ be a representation of $\mathcal{O}_N$, and let $\mathcal{L}$ be a closed co-invariant subspace. Set $\mathcal{W} = (S\mathcal{L}) \ominus \mathcal{L}$, and $\mathcal{F}(\mathcal{W}) = \mathcal{W} \oplus S\mathcal{W} \oplus S^2\mathcal{W} \oplus \cdots$. Let $P_{\mathcal{L}}$ be the projection onto $\mathcal{L}$, and let α be the endomorphism of Lemma 4.1. Then the limit $P_\alpha = \lim_{n\to\infty} \alpha^n(P_{\mathcal{L}})$ exists, and the projection $P_{\mathcal{F}}$ onto $\mathcal{F}(\mathcal{W})$ is given by the formula*

$$P_{\mathcal{F}} = P_\infty - P_{\mathcal{L}}. \tag{4.2}$$

Moreover, the operators $T_i = P_{\mathcal{F}} S_i P_{\mathcal{F}}$, $i = 1, \ldots, N$, satisfy the following (Cuntz-Toeplitz-Fock) relations:

(a) *$T_i^* T_j = \delta_{i,j} P_{\mathcal{F}}$, $i, j = 1, \ldots, N$,*
(b) *$\sum_{i=1}^N T_i T_i^* \leq P_{\mathcal{F}}$,*
(c) *$T_i^* w = 0$ for all $i = 1, \ldots, N$, and all $w \in \mathcal{W}$.*

Proof. The details amount to direct verifications and are left for the reader. □

Lemma 4.3. *Let $\{S_i\}_{i=1}^N$ be a representation of $\mathcal{O}_N$ on a Hilbert space $\mathcal{H}$. Let $\mathcal{W} \subset \mathcal{H}$ be a closed subspace such that $\langle x \mid S_J y \rangle = 0$ for all $x, y \in \mathcal{W}$ and all multi-indices $J \in \mathcal{I}(N)$. Then the operators $T_i := P_{\mathcal{F}(\mathcal{W})} S_i P_{\mathcal{F}(\mathcal{W})}$ satisfy the conditions (a)–(c) in Lemma 4.2, i.e., the subspace $\mathcal{F}(\mathcal{W}) = \mathcal{W} \oplus S\mathcal{W} \oplus S^2\mathcal{W} \oplus \cdots$ induces a Fock-space representation of the Cuntz-Toeplitz relations.*

Proof. It suffices to show that each operator S_i^* maps $\mathcal{W}$ into $\mathcal{L} = \mathcal{H} \ominus \mathcal{F}(\mathcal{W})$. Suppose $x \in \mathcal{W}$ and $y \in \mathcal{F}(\mathcal{W})$: then we claim that $\langle S_i^* x \mid y \rangle = 0$. The assertion follows from this. To establish the claim, suppose first that $y \in \mathcal{W}$. Then $\langle S_i^* x \mid y \rangle = \langle x, S_i y \rangle = 0$ holds on account of the definition of $\mathcal{W}$. Similarly $\langle S_i^* x \mid S_J y \rangle = \langle x \mid S_i S_J y \rangle = 0$ for all $x, y \in \mathcal{W}$, and all multi-indices J. It follows, in particular, that the operators $T_i = P_{\mathcal{F}(\mathcal{W})} S_i P_{\mathcal{F}(\mathcal{W})}$ satisfy $T_i^* x = 0$ for all $x \in \mathcal{W}$, in addition to the Cuntz-Toeplitz relations (a)–(c). □

Our next result is a partial converse to Lemma 4.2.

Proposition 4.4. *Let $\{S_i\}_{i=1}^N$ be a representation of $\mathcal{O}_N$ on a Hilbert space $\mathcal{H}$, and let $\mathcal{W}_m$ be chosen as in Lemma 4.7 below to be maximal (in the sense of Zorn's lemma) with respect to the property*

$$\langle x \mid S_J y \rangle = 0, \qquad x, y \in \mathcal{W}_m, \text{ and } J \in \mathcal{I}(N). \tag{4.3}$$

Then every co-invariant closed subspace $\mathcal{L}$ such that $S^* \mathcal{W}_m \subset \mathcal{L} \subset \mathcal{W}_m^\perp$ satisfies

$$S\mathcal{L} \ominus \mathcal{L} = \mathcal{W}_m. \tag{4.4}$$

Conversely, if equality holds in (4.4) for some wandering subspace $\mathcal{W}$ and all $\mathcal{L}$ with $S^* \mathcal{W} \subset \mathcal{L} \subset \mathcal{W}^\perp$, then $\mathcal{W}$ is maximal.

Proof. First note that, if $\mathcal{W}$ is any wandering subspace, then the inclusions

$$S^* \mathcal{W} \subset \mathcal{F}(\mathcal{W})^\perp \subset \mathcal{W}^\perp$$

are automatic; and $\mathcal{L}_\mathcal{W} := \mathcal{F}(\mathcal{W})^\perp$ is co-invariant. Also $\mathcal{L}_\mathcal{W}$ is maximal among the co-invariant subspaces $\mathcal{L}$ satisfying

$$S^* \mathcal{W} \subset \mathcal{L} \subset \mathcal{W}^\perp.$$

Recall that, if $P_\mathcal{L}$ is the projection onto $\mathcal{L}$, then by Lemma 4.1, $\alpha(P_\mathcal{L}) = \sum_{i=1}^N S_i P_\mathcal{L} S_i^*$ is the projection onto $S\mathcal{L}$. If we show that $\alpha(P_\mathcal{L}) x = x$ for all $x \in \mathcal{W}_m$, it follows that $\mathcal{W}_m \subseteq (S\mathcal{L}) \ominus \mathcal{L}$. But the vectors in $(S\mathcal{L}) \ominus \mathcal{L}$ satisfy the orthogonality relations (4.3), and we may then conclude that (4.4) holds by virtue of the maximality of $\mathcal{W}_m$; see Lemma 4.7 below. The proof is now completed since $S_i^* x \in \mathcal{L}$ for all $x \in \mathcal{W}_m$, and we get $\alpha(P_\mathcal{L}) x = \sum_{i=1}^N S_i P_\mathcal{L} S_i^* x = \sum_{i=1}^N S_i S_i^* x = x$. We leave the verification of the converse implication to the reader, i.e., that equality in (4.4) for some $\mathcal{W}$ and all $\mathcal{L}$ as specified implies that $\mathcal{W}$ is maximal with respect to (4.3). □

Remark 4.5. The argument shows in particular that if $\mathcal{W}$ is any closed wandering subspace, then $S^* \mathcal{W} \subset \mathcal{W}^\perp$, and further that there will always be co-invariant subspaces $\mathcal{L}$ satisfying

$$S^* \mathcal{W} \subset \mathcal{L} \subset \mathcal{W}^\perp. \tag{4.5}$$

In fact, if $\mathcal{W}$ is given, then $\mathcal{L} = \mathcal{F}(\mathcal{W})^\perp$ is such an intermediate co-invariant subspace. If $\mathcal{W}$ is also assumed maximal in the sense of Proposition 4.4, then it follows that $\mathcal{L} = \mathcal{F}(\mathcal{W})^\perp$ is the only closed and co-invariant saturated subspace which is intermediate as stated in (4.5).

Remark 4.6. While, on the face of it, the relations (4.1) appear to depend on a choice of coordinates, this is not so: the Cuntz C^*-algebra is in fact a functor from Hilbert space to C^*-algebras, as noted in [Cun77]. Specifically, consider the mapping

$$z = (z_1, \dots, z_N) \longmapsto s(z) = \sum_{j=1}^{N} z_j s_j \tag{4.6}$$

from the Hilbert space $\mathbb{C}^N$ into the C^*-algebra $\mathcal{O}_N$. If 1 denotes the unit element in $\mathcal{O}_N$, we get $s(z)^* s(w) = \langle z \mid w \rangle 1$ where $\langle z \mid w \rangle = \sum_{j=1}^{N} \bar{z}_j w_j$ is the usual inner product of $\mathbb{C}^N$. If $\|\cdot\|$ denotes the C^*-norm on $\mathcal{O}_N$, i.e., satisfying $\|1\| = 1$ and $\|A^*A\| = \|A\|^2$, $A \in \mathcal{O}_N$, then $\|s(z)\| = \left(\sum_{j=1}^{N} |z_j|^2\right)^{\frac{1}{2}} = \|z\|_{\mathbb{C}^N}$.

States on $\mathcal{O}_N$ are linear functionals $\omega\colon \mathcal{O}_N \to \mathbb{C}$ such that $\omega(1) = 1$, $\omega(A^*A) \geq 0$, $A \in \mathcal{O}_N$. A special state ω on $\mathcal{O}_N$ is determined by

$$\omega\left(s_{i_1} s_{i_2} \cdots s_{i_k} s_{j_l}^* \cdots s_{j_2}^* s_{j_1}^*\right) = \prod_s \delta_{i_s, j_s} \cdot \delta_{k,l} \cdot N^{-k}. \tag{4.7}$$

As is well known, states induce cyclic representations. A representation ρ of $\mathcal{O}_N$ in a Hilbert space $\mathcal{H}$ is *cyclic* if there is a vector Ω in $\mathcal{H}$ such that $\{\rho(A)\Omega : A \in \mathcal{O}_N\}$ is dense in $\mathcal{H}$. If the state is ω, the corresponding representation ρ is determined by $\omega(A) = \langle \Omega \mid \rho(A)\Omega \rangle$. When ρ is given, we set $S_i := \rho(s_i)$, $i = 1, \dots, N$. We shall often identify a representation ρ of $\mathcal{O}_N$ with the corresponding operators $S_i = \rho(s_i)$.

The representation $\{S_i\}_{i=1}^N$ of the state (4.7) does not have a closed co-invariant and saturated subspace which is minimal with respect to the usual ordering of closed subspaces. Nonetheless the following general result holds.

Lemma 4.7. *Let $\{S_i\}_{i=1}^N$ be a representation of $\mathcal{O}_N$ on a Hilbert space $\mathcal{H}$.*

(i) *Then $\mathcal{H}$ contains a wandering subspace $\mathcal{W}_m$ which is maximal with respect to inclusion. Specifically, $\mathcal{W}_m$ satisfies the implication: Whenever $\mathcal{W}$ is a wandering subspace such that $\mathcal{W}_m \subseteq \mathcal{W}$, then $\mathcal{W}_m = \mathcal{W}$.*
(ii) *Maximal wandering subspaces are non-unique.*

Proof. Recall that a closed subspace $\mathcal{W}$ is wandering if and only if the family of spaces $\{\mathcal{W}, S_I\mathcal{W}\}$ is orthogonal as I varies over all the multi-indices, i.e., $I \in \mathcal{I}(N)$. Equivalently,

$$P_{\mathcal{W}} S_I P_{\mathcal{W}} = 0 \qquad \text{for all } I \in \mathcal{I}(N) \tag{4.8}$$

where $P_{\mathcal{W}}$ denotes the projection onto $\mathcal{W}$. If $\{\mathcal{W}_\alpha\}$ is a linearly ordered family of wandering subspaces (i.e., for all α, β, one of the inclusions $\mathcal{W}_\alpha \subseteq \mathcal{W}_\beta$ or $\mathcal{W}_\beta \subseteq \mathcal{W}_\alpha$ holds), then it can be checked that $\mathcal{W} = \bigvee_\alpha \mathcal{W}_\alpha$ satisfies the condition (4.8). The existence of $\mathcal{W}_m$ now follows from Zorn's lemma. The example of (4.7) shows that $\mathcal{W}_m$ will typically be non-unique. □

Lemma 4.8. *Let $(S_i)_{i=1}^N$ be a representation of $\mathcal{O}_N$ on a Hilbert space $\mathcal{H}$, and let $\mathcal{L}$ be a closed subspace of $\mathcal{H}$. Then conditions* (i) *and* (ii) *are equivalent:*

(i) $S^*\mathcal{L} \subseteq \mathcal{L}$;
(ii) $\mathcal{L} \subseteq S\mathcal{L}$.

Proof. (i) $\Rightarrow$ (ii). Assume (i). Let $x \in \mathcal{L}$. Then $x = \sum_i S_i S_i^* x \in S\mathcal{L}$ since $S_i^* x \in \mathcal{L}$ for $i = 1, \dots, N$.

(ii) $\Rightarrow$ (i). Assume (ii). If $x \in \mathcal{L}$, then there are $y_i \in \mathcal{L}$ such that $x = \sum_i S_i y_i$. We get $S_j^* x = \sum_i S_j^* S_i y_i = \sum_i \delta_{j,i} y_i = y_j$, which proves (i). □

Remark 4.9. Example 5.3 below shows that there are representations $(S_i)_{i=1}^N$ of $\mathcal{O}_N$ for every $N = 2, 3, \dots$, such that the second inclusion in the lemma is sharp, but not the first; i.e., it may happen that $S^*\mathcal{L} = \mathcal{L}$ while $\mathcal{L} \subsetneqq S\mathcal{L}$.

Example 4.10. The cyclic representation $(S_i)_{i=1}^N$ defined from the state ω in (4.7) has a co-invariant infinite-dimensional subspace $\mathcal{L}$ spanned by the vectors $S_I^*\Omega$ where $I = (i_1, \dots, i_k)$ runs over all the multi-indices $k \geq 1$, and the relative orthocomplement $S\mathcal{L} \ominus \mathcal{L}$ is spanned by the following N^2 orthogonal vectors:

$$\left\{ S_i S_j^* \Omega : i, j = 1, \dots, N \right\}.$$

Proof. The details are left to the reader. □

Lemma 4.11. *Let $(S_i)_{i=1}^N$ be a representation of $\mathcal{O}_N$ on a Hilbert space $\mathcal{H}$, and let $\mathcal{L}$ be a closed subspace in $\mathcal{H}$ which is co-invariant. Then the subspace*

$$\mathcal{M} := \bigvee \{E_i \mathcal{L} : i = 1, \dots, N\} \tag{4.9}$$

is co-invariant and stable. *We use the notation $E_i = S_i S_i^*$ for the projection onto $S_i\mathcal{H}$.*

Proof. Since $E_i^2 = E_i$, it is clear that the space $\mathcal{M}$ is invariant under each E_i. Since $S_j^* E_i = \delta_{j,i} S_i^*$, it is clear that $\mathcal{M}$ is also invariant under each S_j^*. Hence $\mathcal{M}$ is co-invariant. For every $x \in \mathcal{L}$, we have $x = \sum_i E_i x$ by (4.1), and it follows then from (4.9) that $\mathcal{L} \subset \mathcal{M}$, which was used in the argument above. □

5. Resolution subspaces of $\mathcal{H}$

We begin this section with an explicit isomorphism between the family Co-inv of all closed co-invariant subspaces, and the family Wan of all closed wandering subspaces. The definitions refer to a specified representation $\{S_i\}_{i=1}^N$ of $\mathcal{O}_N$ on a Hilbert space $\mathcal{H}$, and we shall work with the corresponding endomorphism $\alpha(A) = \sum_{i=1}^N S_i A S_i^*$, $A \in \mathcal{B}(\mathcal{H})$. Note that if P is a projection of $\mathcal{H}$ onto a closed subspace $\mathcal{K} \subset \mathcal{H}$, then $\alpha(P)$ is the projection onto $S\mathcal{K}$.

Lemma 5.1. *For $\mathcal{L} \in$ Co-inv define $\mu(\mathcal{L}) = (S\mathcal{L}) \ominus \mathcal{L}$. Then $\mu(\mathcal{L}) \in$ Wan. For $\mathcal{W} \in$ Wan, set $\lambda(\mathcal{W}) = \mathcal{H} \ominus \mathcal{F}(\mathcal{W})$; then $\lambda(\mathcal{W}) \in$ Co-inv. Moreover*

$$\mathcal{W} = \mu(\lambda(\mathcal{W})). \tag{5.1}$$

Proof. If $\mathcal{K} \subset \mathcal{H}$ is a closed subspace, then we denote by $P_{\mathcal{K}}$ the projection onto $\mathcal{K}$, i.e., $\mathcal{K} = P_{\mathcal{K}}\mathcal{H}$, and $P_{\mathcal{K}} = P_{\mathcal{K}}^* = P_{\mathcal{K}}^2$. Thus, identifying closed subspaces with projections, we arrive at the formulas

$$\mu(P_{\mathcal{L}}) = \alpha(P_{\mathcal{L}}) - P_{\mathcal{L}}$$

and

$$\lambda(P_{\mathcal{W}}) = I_{\mathcal{H}} - P_{\mathcal{F}(\mathcal{W})} = I_{\mathcal{H}} - \sum_{n=0}^{\infty} \alpha^n(P_{\mathcal{W}}).$$

If $\mathcal{W}$ is wandering, then the projections $\alpha^n(P_{\mathcal{W}})$ in the sum are mutually orthogonal, and it follows that $\lim_{n\to\infty} \alpha^n(P_{\mathcal{W}}) = 0$. Hence it follows that the two terms in the difference below are convergent, and that

$$\mu(\lambda(P_{\mathcal{W}})) = \sum_{n=0}^{\infty} \alpha^n(P_{\mathcal{W}}) - \sum_{n=1}^{\infty} \alpha^n(P_{\mathcal{W}}) = P_{\mathcal{W}}. \tag{5.2}$$

□

Lemma 5.2. *Let a representation (S_i) of $\mathcal{O}_N$ be given, and let $\mathcal{L}$ be a closed subspace which is co-invariant. Let $\mathcal{V}$ denote the linear span of the spaces $S^k\mathcal{L}$. Then $\mathcal{V}$ reduces to $\mathcal{L}$ in the representation.*

Proof. We have $S^*\mathcal{L} \subset \mathcal{L}$. Using $\sum_i S_i S_i^* = I_{\mathcal{H}}$ we conclude that $\mathcal{L} \subset S\mathcal{L}$, and by induction $S^k\mathcal{L} \subset S^{k+1}\mathcal{L}$ for $k \in \mathbb{N}$. Since, clearly, $S_i^*\left(S^{k+1}\mathcal{L}\right) \subset S^k\mathcal{L}$, we conclude that $\mathcal{V}$ reduces to $\mathcal{L}$ in the sense of Definition 3.1(x). □

The next two examples are the representation of $\mathcal{O}_2$ on the Hilbert space $L^2(\mathbb{T})$ which are defined from the two best-known examples of wavelets in $L^2(\mathbb{R})$.

The first is the Haar wavelet. It has $\varphi = \chi_I$, where I is the unit interval $I = [0, 1)$, and

$$\psi = \chi_{[0,\frac{1}{2})} - \chi_{[\frac{1}{2},1)} \tag{5.3}$$

The corresponding functions m_j, $j = 0, 1$, which define the wavelet filters are $m_0(z) = \frac{1}{\sqrt{2}}(1+z)$ and $m_1(z) = \frac{1}{\sqrt{2}}(1-z)$; see (1.1) and (1.2). The other wavelet is not localized in x-space, but rather in the dual Fourier variable ξ of the Fourier transform $\hat{\psi}(\xi) = \int_{-\infty}^{\infty} e^{-i2\pi\xi x}\psi(x)\,dx$. These wavelets are called frequency localized; see [BrJo99]. It is known that there is a wavelet (named after Shannon) for which the wavelet generator ψ_S is characterized by

$$\hat{\psi}_S(\xi) = \chi_{[-\frac{1}{2},-\frac{1}{4})\cup[\frac{1}{4},\frac{1}{2})}(\xi), \qquad \xi \in \mathbb{R}. \tag{5.4}$$

We now compare the $\mathcal{O}_2$-representation of ψ_S with that of the Haar wavelet from (5.3).

More generally, we note that the examples of representations of $\mathcal{O}_N$ which primarily motivate our results are those which arise from discretizing wavelet problems, in the sense of [BrJo02] and [Jor99]. They may be realized on the Hilbert space $\mathcal{H} = L^2(I)$ where I is a compact interval with Lebesgue measure. Using the Fourier series on functions on I, it will be convenient to view $\mathcal{H}$ alternately as $\ell^2(\mathbb{Z})$, or as $L^2(\mathbb{T})$, where $\mathbb{T}$ is the one-torus, equipped with the usual normalized Haar measure. When working with $L^2(\mathbb{T})$, it will be convenient to have the orthonormal basis $e_n(z) = z^n$, $n \in \mathbb{Z}$, for use in computations.

The following two examples of representations of $\mathcal{O}_N$ on $\mathcal{H}$ illustrate the concepts in Section 1. We will give them just for $N = 2$, but the reader can easily write out the general case.

Example 5.3. Let $S_0 f(z) = f(z^2)$, and $S_1 f(z) = z f(z^2)$, $f \in \mathcal{H} = L^2(\mathbb{T})$.

In terms of the Fourier basis, this representation may be identified as follows:

$$S_0 e_n = e_{2n}, \qquad S_1 e_n = e_{2n+1}, \qquad n \in \mathbb{Z},$$

and with the adjoint operators

$$S_0^* e_n = \begin{cases} e_{n/2} & \text{if } n \text{ is even}, \\ 0 & \text{if } n \text{ is odd}, \end{cases} \qquad \text{and} \qquad \begin{cases} S_1^* e_{2n+1} = e_n, \\ S_1^* e_m = 0 \text{ if } m \text{ is even}. \end{cases}$$

It is immediate that the system of operators $(S_i)_{i=0}^1$ defines a representation of $\mathcal{O}_2$, and that $\mathcal{L} = \operatorname{span}\{e_{-1}, e_0\}$ is a *core* for the representation, in the sense of Definition 3.1(viii). Since

$$\begin{aligned} S_0^* e_{-1} &= 0, & S_1^* e_{-1} &= e_{-1}, \\ S_0^* e_0 &= e_0, & S_1^* e_0 &= 0, \end{aligned} \tag{5.5}$$

it is clear that $\mathcal{L}$ is *stable* in the sense of Definition 3.1(xi).

Remark 5.4. For Haar's representation $(S_i)_{i=0}^1$ in Example 5.3 we have two natural finite-dimensional co-invariant subspaces,

$$\mathcal{L}_{\min} = \bigvee \{e_{-1}, e_0\} \qquad \text{and} \qquad \mathcal{L} = \bigvee \{e_{-2}, e_{-1}, e_0, e_1\};$$

and $\mathcal{L}_{\min}$ is minimal in the sense of Definition 3.1(viii), while $\mathcal{L}$ is not. This follows from the observations

$$\begin{aligned} (S\mathcal{L}_{\min}) \ominus \mathcal{L}_{\min} &= \bigvee \{e_{-2}, e_1\}, \\ S\mathcal{L} \ominus \mathcal{L} &= \bigvee \{e_{-4}, e_{-3}, e_2, e_3\}, \end{aligned}$$

and

$$S^* \mathcal{L} = \mathcal{L}_{\min}.$$

Example 5.5. Working with $I = \left[-\frac{1}{2}, \frac{1}{2}\right]$, we may view $L^2(I)$ as a space of $\mathbb{Z}$-periodic functions on $\mathbb{R}$, and we will work with the operations $x \mapsto 2x$, $x \mapsto \frac{x}{2}$, $x \mapsto \frac{x+1}{2}$ modulo $\mathbb{Z}$. The last two are branches of inverses of the doubling $x \mapsto 2x \bmod 1$, or equivalently of $z \mapsto z^2$ on $\mathbb{T}$. In the second incarnation the two branches of the inverse may be thought of as $\pm\sqrt{z}$ restricted to $\mathbb{T} \subset \mathbb{C}$, i.e., the two branches of the complex square root. Consider the subsets A and B of I:

$A = \{x \in I : |x| \leq \frac{1}{4}\}$, and $B = \{x : -\frac{1}{2} \leq x < -\frac{1}{4}\} \cup \{x : \frac{1}{4} < x \leq \frac{1}{2}\}$, forming a dyadic partition of I. Then set

$$\begin{aligned} S_0 f(x) &:= \sqrt{2}\chi_A(x) f(2x \bmod \mathbb{Z}), \qquad \text{and} \\ S_1 f(x) &:= \sqrt{2}\chi_B(x) f(2x \bmod \mathbb{Z}). \end{aligned} \tag{5.6}$$

It will be convenient for us to omit the mod $\mathbb{Z}$ notation when it is otherwise implicit in the formulas. Then it is immediate that the isometries $(S_i)_{i=0}^{1}$ define a representation of $\mathcal{O}_2$, and that $\mathcal{L} := \mathbb{C}\mathbb{1}$ is a *core* for the representation, acting on $\mathcal{H} = L^2(I)$, and viewing I as $\mathbb{R}/\mathbb{Z}$. Also we use the notation $\mathbb{1}$ for the function which is constant $\equiv 1$ on I. The formulas

$$S_0^*\mathbb{1} = \frac{1}{\sqrt{2}}\mathbb{1}, \qquad S_1^*\mathbb{1} = \frac{1}{\sqrt{2}}\mathbb{1}, \tag{5.7}$$

$$S_0\mathbb{1} = \sqrt{2}\chi_A, \qquad S_1\mathbb{1} = \sqrt{2}\chi_B \tag{5.8}$$

now make it clear that this one-dimensional core subspace $\mathcal{L}$ is not *stable* in the sense of Definition 3.1(xi). Specifically, we have:

$$E_0\mathbb{1} = \chi_A, \qquad \text{and} \qquad E_1\mathbb{1} = \chi_B. \tag{5.9}$$

As in the discussion below Lemma 5.2, it follows that

$$\begin{aligned} S_0^*\chi_A &= \frac{1}{\sqrt{2}}\mathbb{1}, & S_0^*\chi_B &= 0, \\ S_1^*\chi_A &= 0, & S_1^*\chi_B &= \frac{1}{\sqrt{2}}\mathbb{1}. \end{aligned} \tag{5.10}$$

Therefore the two-dimensional space $\mathcal{M} := \operatorname{span}\{\chi_A, \chi_B\}$ is co-invariant, and it is *saturated* in the sense of Definition 3.1(vi). The space $\mathcal{M}$ is also stable.

6. Nontrivial co-invariant subspaces

Let $(S_i)_{i=1}^{N}$ be a representation of $\mathcal{O}_N$ acting on a Hilbert space $\mathcal{H}$. Suppose a closed subspace $\mathcal{L}$ in $\mathcal{H}$ is co-invariant. We then get the resolution

$$\mathcal{L} \subset S\mathcal{L} \subset S^2\mathcal{L} \subset \cdots \subset S^n\mathcal{L} \subset S^{n+1}\mathcal{L} \subset \cdots \subset \mathcal{H}. \tag{6.1}$$

If $\mathcal{L}$ is saturated, then $\bigvee_n S^n\mathcal{L} = \mathcal{H}$.

There are two trivial cases where both conditions are satisfied for a given $\mathcal{L}$. First, if $\mathcal{L} = \mathcal{H}$, there isn't much to say. Secondly, if $\mathcal{L} = S\mathcal{L}$, then the given representation on $\mathcal{H}$ simply restricts to a representation on $\mathcal{L}$. Recall, if $\mathcal{L}$ is co-invariant, and if $\mathcal{L} = S\mathcal{L}$, then $\mathcal{L}$ is an invariant subspace for all the $2N$ operators

$$\underbrace{S_1, \dots, S_N}_{\text{isometries}}, \qquad \underbrace{S_1^*, \dots, S_N^*}_{\text{co-isometries}}.$$

We say that $\mathcal{L}$ is a reducing subspace for the given representation. We then get two orthogonal representations of $\mathcal{O}_N$ by restriction to each of the two spaces, $\mathcal{L}$ and its orthocomplement $\mathcal{H} \ominus \mathcal{L} = \{x \in \mathcal{H} : \langle x \mid y \rangle = 0,\ y \in \mathcal{L}\}$. We say that a co-invariant subspace $\mathcal{L}$ for the given representation on $\mathcal{H}$ is *nontrivial* if $\mathcal{L} \neq \mathcal{H}$, and $\mathcal{L} \subsetneqq S\mathcal{L}$.

Theorem 6.1. *There is a two-way correspondence between the nontrivial, closed, co-invariant, and saturated subspaces $\mathcal{L}$ for a representation $(S_i)_{i=1}^N$ of $\mathcal{O}_N$ on $\mathcal{H}$, and closed subspaces $\mathcal{W} \neq 0$ in $\mathcal{H} \ominus \mathcal{L}$ which provide an orthogonal decomposition*

$$\mathcal{H} = \mathcal{L} \oplus \mathcal{W} \oplus S\mathcal{W} \oplus S^2\mathcal{W} \oplus \cdots \tag{6.2}$$

Note the subspace $\mathcal{W}$ in (6.2) *is called* wandering, *and the conditions imply pairwise orthogonality of all the closed subspaces:*

$$\mathcal{L},\ \mathcal{W},\ S_i\mathcal{W},\ S_{i_1}S_{i_2}\mathcal{W},\ S_{i_1}S_{i_2}S_{i_3}\mathcal{W},\ \dots,\ S_{i_1}\cdots S_{i_k}\mathcal{W},\ \dots, \tag{6.3}$$

where all multi-indices $(i_1, i_2, \dots, i_k)$ are considered for $k = 1, 2, \dots$.

Proof. Suppose first that $0 \neq \mathcal{W}$ satisfies the conditions from (6.2). Then set $\mathcal{L} := \mathcal{H} \ominus \sum_{k=0}^{\infty}{}^{\oplus} S^k\mathcal{W}$, with the understanding that $S^0\mathcal{W} := \mathcal{W}$ and

$$S^k\mathcal{W} = \bigvee_{(i_1,i_2,\dots,i_k)} S_{i_1}\cdots S_{i_k}\mathcal{W}. \tag{6.4}$$

Since $\sum_{k=0}^{\infty}{}^{\oplus} S^k\mathcal{W}$ is invariant under all the isometries S_i, $i = 1, \dots, N$, it is clear that $\mathcal{L}$ is invariant under the adjoints S_i^*, and so $\mathcal{L}$ is co-invariant. Since $\mathcal{W} \neq 0$, clearly $\mathcal{L} \neq \mathcal{H}$. It remains to verify that the strict inclusion $\mathcal{L} \subsetneqq S\mathcal{L}$ holds. Suppose, indirectly, that $\mathcal{L} = S\mathcal{L}$. Then $\mathcal{L}$ is reducing, and so is $\sum_{k=0}^{\infty}{}^{\oplus} S^k\mathcal{W}$. To see that this is impossible, let $\mathcal{I}(k)$ denote the set of all multi-indices $(i_1, \dots, i_k)$, where $i_1, i_2, \dots \in \{1, 2, \dots, N\}$. If $\sum_{k=0}^{\infty}{}^{\oplus} S^k\mathcal{W}$ were S_i^*-invariant for all i, then

$$\lim_{k\to\infty} \sum_{I\in\mathcal{I}(k)} \|S_I^*x\|^2 = 0 \qquad \text{for all } x \in \mathcal{W}, \tag{6.5}$$

where $S_I^* = S_{i_k}^* \cdots S_{i_1}^*$ for $I = (i_1, \dots, i_k)$. But this is impossible by the Cuntz relations. Recall that

$$\sum_{I\in\mathcal{I}(k)} \|S_I^*x\|^2 = \|x\|^2 \qquad \text{holds for all } x. \tag{6.6}$$

This contradicts (6.5) and the condition $\mathcal{W} \neq 0$.

Suppose now that $\mathcal{L}$ is a given nontrivial, closed, co-invariant and saturated subspace for the given representation in $\mathcal{H}$. As noted, then $\mathcal{L} \subsetneqq S\mathcal{L}$; so the relative orthocomplement $\mathcal{W} := (S\mathcal{L}) \ominus \mathcal{L}$ is nonzero. We first prove that all the multi-indexed subspaces listed in (6.3) are mutually orthogonal. The argument is by induction, starting with $\mathcal{W} \perp S\mathcal{W}$. We must check the inner products

$$\left\langle \sum_i S_i x_i \,\middle|\, \sum_{j_1,j_2} S_{j_1}S_{j_2}y_{j_1,j_2} \right\rangle \tag{6.7}$$

where $x_i, y_{j_1,j_2} \in \mathcal{L}$, under the conditions

$$\sum_i S_i x_i \perp \mathcal{W} \qquad \text{and} \qquad \sum_{j_2} S_{j_2} y_{j_1,j_2} \perp \mathcal{W} \quad \text{for all } j_1 = 1, \dots, N. \tag{6.8}$$

But the term in (6.7) simplifies to

$$\sum_{i,j} \langle x_i \mid S_j y_{i,j} \rangle = 0,$$

where the vanishing results from the conditions (6.8). We leave the remaining recursive argument to the reader.

It remains to check that the sequence of closed subspaces

$$\mathcal{L},\ \mathcal{W},\ S\mathcal{W},\ S^2\mathcal{W},\ \dots$$

is total in $\mathcal{H}$, i.e., that (6.2) holds.

Since $\mathcal{L}$ is given to be saturated, the conclusion will follow from

$$S^{k+1}\mathcal{L} \subset \mathcal{L} \oplus \mathcal{W} \oplus S\mathcal{W} \oplus \cdots \oplus S^k\mathcal{W}. \tag{6.9}$$

If $k = 0$, this holds from the ansatz which defines $\mathcal{W}$ in terms of the given closed subspace $\mathcal{L}$, i.e., $\mathcal{W} := (S\mathcal{L}) \ominus \mathcal{L}$. Suppose (6.9) has been verified up to $k-1$. Then

$$\begin{aligned} S^{k+1}\mathcal{L} &\subset S\left(\mathcal{L} \oplus \mathcal{W} \oplus \cdots \oplus S^{k-1}\mathcal{W}\right) \\ &\subset S\mathcal{L} \oplus S\mathcal{W} \oplus \cdots \oplus S^k\mathcal{W} \\ &\subset \overbrace{\mathcal{L} \oplus \mathcal{W}} \oplus S\mathcal{W} \oplus \cdots \oplus S^k\mathcal{W}, \end{aligned}$$

where we used the induction hypothesis and the ansatz. □

7. Existence

While the co-invariant subspaces for the wavelet representations are relatively well understood for the special representations of $\mathcal{O}_N$ which derive from wavelet analysis, the situation is somewhat mysterious in the case of the most general $\mathcal{O}_N$-representations. Even the existence of nontrivial co-invariant subspaces which form cores for a given $\mathcal{O}_N$-representation appears to be open in general. However:

Corollary 7.1. *A representation $(S_i)_{i=1}^N$ of $\mathcal{O}_N$ on a Hilbert space $\mathcal{H}$ has a nontrivial* core subspace *$\mathcal{L}$, i.e., a closed co-invariant subspace $\mathcal{L} \subset \mathcal{H}$ which is nontrivial and saturated, if and only if it has a nonzero wandering vector. Further, $\mathcal{L}$ may be chosen to be minimal with respect to being co-invariant and saturated.*

Proof. We showed in Theorem 6.1 that the general case may be reduced to the consideration of nontrivial co-invariant subspaces. In Theorem 6.1, we also showed that the nontrivial co-invariant subspaces $\mathcal{L}$ may be understood from the corresponding wandering subspaces $\mathcal{W}$. If $\mathcal{L}$ is a nontrivial co-invariant subspace, then $\mathcal{W} := (S\mathcal{L}) \ominus \mathcal{L}$ is wandering for the representation, and conversely every wandering subspace $\mathcal{W}$ comes from a co-invariant subspace $\mathcal{L}$. The idea is to prove the existence claim in the corollary by establishing the existence of a maximal wandering subspace $\mathcal{W}_{\max}$. We will do this by Zorn's lemma. We say that a wandering subspace $\mathcal{W}_{\max}$ is maximal if for every wandering subspace $\mathcal{W}$ such that $\mathcal{W} \supseteq \mathcal{W}_{\max}$, we may conclude that $\mathcal{W} = \mathcal{W}_{\max}$. The existence of $\mathcal{W}_{\max}$ follows by an application of Zorn's lemma to the family of all wandering subspaces $\mathcal{W} \neq 0$ ordered by inclusion. To be able to do this, we need to know that every representation of $\mathcal{O}_N$ has at least one wandering subspace $\mathcal{W} \neq 0$. Some details are given below for the benefit of the reader.

To start the transfinite induction we must assume that there are solutions

$$x \in \mathcal{H}, \qquad \|x\| = 1,$$

to the system

$$\langle x \mid S_I x \rangle = 0, \qquad \text{for all } I = (i_1, \ldots, i_k)\,,\ k \geq 1. \tag{7.1}$$

This is the stated condition in the formulation of Corollary 7.1, and it is needed for the start of the transfinite induction. It is conceivable that such a nonzero vector x automatically exists, but we have not been abe to show this. We thank Ken Davidson for calling this point to our attention. □

Remark 7.2. It follows from Section 4 above that $\|1 - s_I\| \geq \sqrt{2}$ for all I where $s_I \in \mathcal{O}_N$ satisfies $\rho(s_I) = S_I = S_{i_1} \cdots S_{i_k}$. Hence it follows that there is a state ω on $\mathcal{O}_N$ such that $\omega(S_I) = 0$ for all $I \in \mathcal{I}(N) \setminus \{\varnothing\}$, but it is not known if such a normal state ω may be found. If so we would have

$$\operatorname{trace}(|x\rangle\langle x| S_I) = \langle x \mid S_I x \rangle = 0,$$

where $|x\rangle\langle x|$ is the rank-1 projection of $\mathcal{H}$ onto $\mathbb{C}x$, and where we used Dirac's bra-ket notation.

Remark 7.3. The existence problem for wandering subspaces $\mathcal{W}$ may be reformulated as a fixed-point problem. This is made clear by the equivalent identities (5.1) and (5.2) of Lemma 5.1. Introducing the transformation $\mathcal{W} \mapsto \mu(\lambda(\mathcal{W}))$ of the right-hand side in (5.1), or $P_{\mathcal{W}} \mapsto \mu(\lambda(P_{\mathcal{W}}))$ in (5.2), it is clear that the existence of a wandering subspace $\mathcal{W}$ is equivalent to the existence of a fixed point $\neq 0$ in the identity (5.1).

8. Pure co-invariant subspaces

Let $(S_i)_{i=1}^N$ be a representation of $\mathcal{O}_N$ on a Hilbert space $\mathcal{H}$, and let $\mathcal{L}$ be a co-invariant subspace. For $x \in \mathcal{H}$, set

$$S^n x := \bigvee \left[S_{i_1} \cdots S_{i_n} x\,;\, I = (i_1, \ldots, i_k) \in \mathcal{I}_n(N)\right], \qquad S^0 x := x,$$

$$\text{and} \qquad S^\infty x := \bigvee_{n \geq 0} S^n x.$$

The special case $N = 1$ covers a single isometry.

Definition 8.1. We say that a co-invariant subspace $\mathcal{L}$ is *pure* if the following implication holds:

$$x \in \mathcal{L},\ S^{\infty}x \subset \mathcal{L} \Longrightarrow x = 0. \tag{8.1}$$

To motivate the next result, we recall first the case $N = 1$, i.e., a single isometry S in a Hilbert space $\mathcal{H}$. The Wold decomposition states that $\mathcal{H}$ decomposes as $\mathcal{H} = \mathcal{H}_0 \oplus \mathcal{H}_\infty$, $\mathcal{H}_\infty = \bigwedge_{n\geq 1} S^n\mathcal{H}$, $\mathcal{H}_0 = \mathcal{H} \ominus \mathcal{H}_\infty$, both subspaces invariant, with $S|_{\mathcal{H}_\infty}$ unitary, and $S_0 := S|_{\mathcal{H}_0}$ a shift. To say that S_0 is a *shift* means ${S_0^*}^n \underset{n\to\infty}{\longrightarrow} 0$. A shift is determined up to unitary equivalence by its multiplicity, i.e., $\dim\{x \in \mathcal{H}_0 \,;\, S_0^* x = 0\}$. The space $\mathcal{W}_0 := \ker(S_0^*) = \mathcal{H}_0 \ominus S_0\mathcal{H}_0$ is wandering, and $(\mathcal{H}_0, S_0)$ is unitarily equivalent to the obvious shift on vectors, i.e., to $\sum_{n\geq 0}^{\oplus} \mathcal{W}_0 := \mathcal{W}_0 \oplus \mathcal{W}_0 \oplus \cdots$ and the operator given by $(x_0, x_1, \dots) \mapsto (0, x_0, x_1, \dots)$. So for $N = 1$, our understanding of the invariant and the co-invariant subspaces amounts to the corresponding issues for the shift. Since a subspace $\mathcal{L}$ is co-invariant if and only if its orthocomplement is invariant, the problem is solved by Beurling's theorem. Identify $\sum_{n\geq 0}^{\oplus} \mathcal{W}_0$ by the Hardy space $H_+(\mathcal{W}_0)$ of analytic $\mathcal{W}_0$-valued functions, i.e.,

$$f\colon \mathbb{T} \longrightarrow \mathcal{W}_0,\ f(z) = \sum_{n=0}^{\infty} \xi_n z^n, \qquad \xi_n \in \mathcal{W}_0,\ \|f\|^2 = \sum_{n=0}^{\infty} \|\xi_n\|^2. \tag{8.2}$$

An inner function u is a function on $\mathbb{T}$, taking values in the unitary operators on $\mathcal{W}_0$, i.e., $u(z)^* u(z) = I_{\mathcal{W}_0}$, a.e. $z \in \mathbb{T}$, with u having an analytic operator-valued continuation to $\{z \in \mathbb{C} \,;\, |z| < 1\}$. Since, by (6.2), the shift is represented by

$$M_z f(z) = z f(z), \qquad f \in H_+(\mathcal{W}_0),$$

it follows that the space

$$H_+(u) := \{uf \,;\, f \in H_+(\mathcal{W}_0)\}$$

is invariant when u is an inner function. Beurling's theorem [Hel64] states that every invariant subspace for the shift has this form.

Proposition 8.2. *If an inner function u is given, then the co-invariant subspace*

$$\mathcal{L}(u) := H_+(u)^{\perp} = H_+(\mathcal{W}_0) \ominus H_+(u) \tag{8.3}$$

is pure.

Proof. If u is constant, $u \equiv I_{\mathcal{W}_0}$, then $\mathcal{L}(u) = 0$ satisfies the condition, so we assume u to be non-constant. Suppose $f \in \mathcal{L}(u)$ and $S^\infty f \subset \mathcal{L}(u)$. Then

$$f(z),\ zf(z),\ z^2 f(z),\ \dots \qquad \text{are all in } \mathcal{L}(u). \tag{8.4}$$

Let $f = \sum_{n=0}^{\infty} \xi_n z^n$, $\xi_n \in \mathcal{W}_0$, and $u(z) = \sum_{n=0}^{\infty} A_n z^n$ where $A_n \colon \mathcal{W}_0 \to \mathcal{W}_0$ is a system of operators in $\mathcal{W}_0$ such that

$$\sum_{n=0}^{\infty} A_n A_n^* = I_{\mathcal{W}_0} \qquad \text{and} \tag{8.5}$$

$$\sum_{n=0}^{\infty} A_n A_{n+k}^* = 0 \qquad \text{for } k = 1, 2, \dots . \tag{8.6}$$

Using (8.3) and (8.4), we get the system:

$$\begin{aligned} A_n^* \xi_0 + A_{n+1}^* \xi_1 + \dots &= 0, \\ A_n^* \xi_1 + A_{n+1}^* \xi_2 + \dots &= 0, \\ A_n^* \xi_2 + A_{n+1}^* \xi_3 + \dots &= 0, \\ &\vdots \end{aligned}$$

Multiply each of these equations by A_n and sum over $n = 0, 1, 2, \dots$. We then get

$$\sum_{n=0}^{\infty} A_n A_n^* \xi_k + \sum_{n=0}^{\infty} A_n A_{n+1}^* \xi_{k+1} + \cdots = 0$$

for $k = 0, 1, \dots,$ An application of (8.5)–(8.6) now yields $\xi_k = 0$ for all $k = 0, 1, \dots$. This proves the conclusion. □

If $N > 1$, we are considering representations of $\mathcal{O}_N$, and the examples in Section 3 show that not all co-invariant subspaces are pure. But we do have the following:

Theorem 8.3. *Let $(S_i)_{i=1}^{N}$, $N > 1$, be a representation of $\mathcal{O}_N$ on a Hilbert space $\mathcal{H}$, and let $\mathcal{L}$ be a co-invariant subspace. Then $\mathcal{L}$ contains a pure co-invariant subspace.*

Proof. If $\mathcal{L} = 0$, we are done. If not, consider vectors (if any) $x \in \mathcal{L}$, $\|x\| = 1$, such that $S^\infty x \subset \mathcal{L}$. If there are no such vectors, $\mathcal{L}$ is pure. Otherwise pick a family $\mathcal{F} = \{x\}$ such that $S^\infty x \subset \mathcal{L}$ and $S^\infty x \perp S^\infty y$ for all $x, y \in \mathcal{F}$, $x \neq y$. By Zorn's lemma, we may pick $\mathcal{F}$ to be maximal with respect to these properties. Now set $\mathcal{L}_p := \mathcal{L} \ominus [S^\infty x \,;\, x \in \mathcal{F}]$. This space is clearly co-invariant. If some vector $x_p \in \mathcal{L}_p$ satisfies $S^\infty x_p \subset \mathcal{L}_p$, then $\mathcal{F} \cup \{x_p\}$ satisfies the orthogonality property. Since $\mathcal{F}$ was chosen maximal, we conclude that $x_p = 0$. This proves that $\mathcal{L}_p$ is pure. □

Examples 8.4.

(a) Consider the representation $(S_i)_{i=0}^{1}$ of $\mathcal{O}_2$ acting on $\ell^2(\mathbb{Z}) \simeq L^2(\mathbb{T})$ outlined in Example 5.3. From (5.5) we note that the closed subspace spanned by $\{e_0, e_1, e_2, \dots\}$ is invariant for the representation, i.e., invariant under all four of the operators S_i, S_j^*, $i, j = 0, 1$; and moreover this restricted representation, ρ_+

say, is irreducible on this space. The space is $H_+ \simeq \ell^2(\{0,1,2,\dots\}) \simeq$ the Hardy space. Recall the Hardy space $H_+ \subset L^2(\mathbb{T})$ has the representation

$$H_+ = \left\{ f \in L^2(\mathbb{T}) : f(z) = \sum_{n=0}^{\infty} \xi_n z^n,\ z \in \mathbb{T},\ \sum_{n=0}^{\infty} |\xi_n|^2 \ (= \|f\|^2) < \infty \right\}. \quad (8.7)$$

Since, in general, *every finite-dimensional* co-invariant subspace is pure, we note that each one of the following subspaces $\mathcal{L}_1$, $\mathcal{L}_2$, and $\mathcal{L}_3$ in H_+ is pure: $\mathcal{L}_1 = [e_0]$, $\mathcal{L}_2 = [e_0, e_1]$, $\mathcal{L}_3 = [e_0, e_1, e_2]$. The corresponding three wandering subspaces, i.e.,

$$\mathcal{W}_i := (S\mathcal{L}_i) \ominus \mathcal{L}_i, \qquad i = 1,2,3, \quad (8.8)$$

are:

$$\mathcal{W}_1 = [e_1], \qquad \mathcal{W}_2 = [e_2, e_3], \qquad \mathcal{W}_3 = [e_3, e_4, e_5]. \quad (8.9)$$

An example of a proper (i.e., nontrivial) co-invariant subspace ($\subset H_+$) which is not pure is

$$\mathcal{L} := [e_0, e_1, e_3, e_6, e_7, e_{12}, e_{13}, e_{14}, e_{15}, e_{24}, e_{25}, \dots]. \quad (8.10)$$

The stated properties for this last subspace $\mathcal{L}$ follow from its representation as

$$\mathcal{L} = \mathcal{L}_2 \oplus S^{\infty} e_3. \quad (8.11)$$

(b) Based on (a), one might think that a pure co-invariant subspace cannot be infinite-dimensional. This is not so, as we now illustrate with the representation ρ from Remark 4.6. This is the representation ρ of $\mathcal{O}_N$ with cyclic vector Ω which acts on the Hilbert space $\mathcal{H}$, i.e., $\mathcal{H}$ is the Hilbert space spanned by the following vectors: $S_I S_J^* \Omega$, where I and J run over the set of all multi-indices $\mathcal{I}(N)$, i.e., $(i_1, i_2, \dots, i_n)$, $n = 1, 2, \dots$, $i_\nu \in \{1, 2, \dots, N\}$, with the convention $n = 0$ corresponding to $I = \varnothing$, and $S_\varnothing = I$. The two indices I and J might have different length. It is immediate from (4.7) that the following subspace,

$$\mathcal{L} := \bigvee \left[S_I^* \Omega : I \in \mathcal{I}(N) \setminus \{\varnothing\} \right], \quad (8.12)$$

is a nontrivial co-invariant subspace for the representation ρ, and that the corresponding wandering subspace $\mathcal{W}$ is N^2-dimensional; in fact, $\mathcal{W} = \bigvee_{i,j} \left[S_i S_j^* \Omega \right]$; see Example 4.10.

Observation 8.5. The subspace $\mathcal{L}$ in (8.12) is a *pure* co-invariant subspace for the representation ρ of Remark 4.6.

Proof. Let $P = P_{\mathcal{L}}$ denote the (orthogonal) projection onto $\mathcal{L}$. We show that

$$\|P S_i P\| \leq N^{-\frac{1}{2}} \qquad \text{for } i = 1, \dots, N. \quad (8.13)$$

Since $N > 1$, the result follows; in fact, if x is any vector $x \in \mathcal{L}$ for which $S_i x \in \mathcal{L}$ for some i, then $x = 0$. Indeed, $\|x\| = \|S_i x\| = \|P S_i P x\| \leq N^{-\frac{1}{2}} \|x\|$.

To prove (8.13), we note that the following normalized vectors,

$$\left\{ N^{\frac{k}{2}} S_I^* \Omega : |I| = k' \right\}, \quad (8.14)$$

are mutually orthogonal when I varies over $\mathcal{I}_{k'}(N)$, and also when the respective length of indices I, I' are different, i.e., $|I| = k \neq k' = |I'|$. Hence, by (8.12), we have an orthonormal basis for $\mathcal{L}$. Now let P_k denote the projection onto the closed subspace spanned by the vectors in (8.14) for all values k' such that $k' \leq k$. Then we get

$$\lim_{k\to\infty} P_k = P_{\mathcal{L}} \ (= P). \tag{8.15}$$

If $x \in \mathcal{L}$, then

$$\begin{aligned} \|P_k S_i x\|^2 &= N^k \sum_{|I|=k} |\langle S_I^* \Omega \mid S_i x\rangle|^2 \\ &= N^{-1} N^{k+1} \sum_{|I|=k} |\langle \underbrace{S_i^* S_I}_{(k+1)\text{-index}} \Omega \mid x\rangle|^2 \\ &\leq N^{-1} \|P_{k+1} x\|^2 . \end{aligned}$$

Letting $k \to \infty$, and using (8.15), we now arrive at the conclusions: $x = P_{\mathcal{L}} x$ $(= \lim_{k\to\infty} P_k x)$, $P_{\mathcal{L}} S_i x = \lim_{k\to\infty} P_k S_i x$, and $\|P_{\mathcal{L}} S_i x\| \leq N^{-\frac{1}{2}} \|x\|$, the last estimate being equivalent to the desired one (8.13). □

9. Tight frames of wavelets

We now turn to the representations (1.2)–(1.4) which define tight frames of multiresolution wavelets, and we give a representation of pure co-invariant subspaces in the Hilbert space $L^2(\mathbb{T})$. It is shown in [BrJo02] that to get solutions φ, ψ_i as in (1.1) which are in $L^2(\mathbb{R})$, the following condition must be satisfied by the matrix function $\mathbb{T} \ni z \mapsto A(z) \in \mathrm{U}_N(\mathbb{C})$ from formula (1.4). Specifically, if $\rho = \rho_N = e^{i2\pi/N}$, $i = \sqrt{-1}$, and if

$$A_{j,k}(1) = \frac{1}{\sqrt{N}} \rho^{j\cdot k} = \frac{1}{\sqrt{N}} e^{i\frac{j\cdot k 2\pi}{N}}, \tag{9.1}$$

then the solutions $\hat{\varphi}$, $\hat{\psi}_i$ in (1.4) are in $L^2(\mathbb{R})$, and their inverse Fourier transforms φ, ψ_i are in $L^2(\mathbb{R})$ as well. Introducing the triple-indexed functions

$$\psi_{i,j,k}(x) = N^{\frac{j}{2}} \psi_i\left(N^j x - k\right), \qquad i = 1, \dots, N-1,\ j \in \mathbb{Z},\ k \in \mathbb{Z}, \tag{9.2}$$

we say that $\{\psi_{i,j,k}\}$ is a *tight frame* if and only if, for all $F \in L^2(\mathbb{R})$, the following (Bessel) identity holds:

$$\int_{\mathbb{R}} |F(x)|^2 \, dx = \sum_{i,j,k} \left| \langle \psi_{i,j,k} \mid F \rangle_{L^2(\mathbb{R})} \right|^2 \tag{9.3}$$

The next result is also proved in [BrJo02], but is included here for the convenience of the reader:

Theorem 9.1. *If* $\mathbb{T} \ni z \mapsto A(z) \in \mathrm{U}_N(\mathbb{C})$ *is a unitary matrix function satisfying* (9.1) *then the wavelet functions* $\psi_{i,j,k}$ *defined in* (1.1) *and* (9.2) *form a tight frame in* $L^2(\mathbb{R})$.

Remark 9.2. (The stretched Haar wavelet.) The following example for $N = 2$ shows that this system (9.2) might not in fact be an orthonormal basis: Take $A(z) = \frac{1}{\sqrt{2}}\begin{pmatrix} 1 & z \\ 1 & -z \end{pmatrix}$. Then $A(1) = \frac{1}{\sqrt{2}}\begin{pmatrix} 1 & 1 \\ 1 & -1 \end{pmatrix}$,

$$\varphi(x) = \frac{1}{3}\chi_{[0,3)}(x), \qquad \text{and} \tag{9.4}$$

$$\psi(x) = \frac{1}{3}\left(\chi_{[0,\frac{3}{2})}(x) - \chi_{[\frac{3}{2},3)}(x)\right). \tag{9.5}$$

It follows from a direct verification, or from the theorem, that this function ψ in (9.5) makes

$$\psi_{j,k}(x) := 2^{\frac{j}{2}}\psi\left(2^j x - k\right), \qquad j,k \in \mathbb{Z}, \tag{9.6}$$

into a tight frame. But since $\|\psi\| = \frac{1}{\sqrt{3}}$, and since the different functions in (9.6) are *not* orthogonal, we see that this is a wavelet tight frame which is not an orthonormal basis in $L^2(\mathbb{R})$.

Not all representations as in (1.2) and (1.4) satisfy condition (9.1); for example, the representation $(T_i)_{i=0}^{N-1}$ defined from the constant matrix function $\mathbb{T} \ni z \mapsto I_N = \begin{pmatrix} 1 & & 0 \\ & \ddots & \\ 0 & & 1 \end{pmatrix} \in \mathrm{U}_N(\mathbb{C})$ clearly does not satisfy (9.1). Yet as we show, we may use this simple representation as a base-point for a comparison with all other representations. Specifically, we have the following lemma.

Lemma 9.3. *If $S_i = S_i^{(A)}$ is any representation defined from some matrix function $\mathbb{T} \ni z \mapsto A(z) \in \mathrm{U}_N(\mathbb{C})$, then*

$$\left(T_j^* S_i^{(A)}\right) f(z) = A_{i,j}(z) f(z), \qquad i,j = 0,\dots,N-1,\ f \in L^2(\mathbb{T}); \tag{9.7}$$

i.e., when i, j are given, then the operator $T_j^* S_i^{(A)}$ is a multiplication operator on $L^2(\mathbb{T})$, in fact multiplication by the matrix entry $A_{i,j}(z)$ of the unitary matrix $A(z)$.

Proof. Apply formula (1.4) and the fact that the operators T_j and T_j^* are given on $L^2(\mathbb{T})$ as

$$T_j f(z) = z^j f\left(z^N\right), \qquad \text{and} \tag{9.8}$$

$$\left(T_j^* f\right)(z) = \frac{1}{N} \sum_{\substack{w \in \mathbb{T} \\ w^N = z}} w^{-j} f(w), \qquad f \in L^2(\mathbb{T}). \tag{9.9}$$

□

As a corollary we get the following formula for the adjoint $S_i^{(A)*}$ in general:

$$S_i^{(A)*} f(z) = \sum_{j=0}^{N-1} \overline{A_{i,j}(z)}\, T_j^* f(z). \tag{9.10}$$

When A is given we introduce the subspace

$$\mathcal{L} = \mathcal{L}^{(A)} = \bigvee_{I \in \mathcal{I}(N)} \left[S_I^{(A)*} \mathbb{1} \right], \tag{9.11}$$

where $\mathbb{1} = e_0$ is the constant function on $\mathbb{T}$, consistent with the terminology $e_n(z) = z^n$, $n \in \mathbb{Z}$. If the matrix entries in (9.7) are Fourier polynomials, it is clear that $\mathcal{L}^{(A)}$ is a finite-dimensional co-invariant subspace, and therefore also pure. In the next result, we give a necessary and sufficient condition for $\mathcal{L}^{(A)}$ to be a pure co-invariant subspace for the wavelet representation $\left(S_j^{(A)}\right)_{j=0}^{N-1}$ on $L^2(\mathbb{T})$.

Lemma 9.4. *Let $\mathbb{T} \ni z \mapsto A(z) \in \mathrm{U}_N(\mathbb{C})$ be a measurable unitary matrix-valued function, and let $(S_j) = \left(S_j^{(A)}\right)$ be the corresponding representation of $\mathcal{O}_N$ on the Hilbert space $L^2(\mathbb{T})$. Then the following three conditions are equivalent. (We consider $0 \leq j < N$ and functions in $L^2(\mathbb{T})$.)*

(i)
$$S_j f = g. \tag{9.12}$$

(ii)
$$S_j^* g = f \qquad \text{and} \qquad \|g\| = \|f\|.$$

(iii)
$$T_i^* g = f \cdot A_{j,i} \qquad \text{for all } i. \tag{9.13}$$

Here each of the identities in (i)–(iii) *is taken in the pointwise sense, i.e., identity for the functions on $\mathbb{T}$ pointwise* a.e. *with respect to Haar measure on $\mathbb{T}$. The product on the right-hand side in* (9.13) *is $f(z) A_{j,i}(z)$* a.e. $z \in \mathbb{T}$.

Proof. (i) ⇒ (ii): This is clear since S_j is an isometry. Hence $\|f\| = \|g\|$, and $S_j^* g = S_j^* S_j f = f$, which is the combined assertion in (ii).

(ii) ⇒ (iii): Assuming (ii), and using (9.10), we get

$$\sum_{i=0}^{N-1} \overline{A_{j,i}(z)}\, T_i^* g(z) = f(z), \qquad z \in \mathbb{T}. \tag{9.14}$$

Using unitarity of the matrix function A, and the Schwarz inequality for the Hilbert space $\mathbb{C}^N$, we get the pointwise estimate

$$|f(z)|^2 \leq \sum_{i=0}^{N-1} |T_i^* g(z)|^2, \qquad \text{a.e. } z \in \mathbb{T}. \tag{9.15}$$

Integration of this over $\mathbb{T}$ with respect to Haar measure yields

$$\|f\|^2 \leq \sum_{i=0}^{N-1} \|T_i^* g\|^2 = \|g\|^2. \tag{9.16}$$

But the second condition in (ii) then states that we have equality in Schwarz's inequality. First we have it in the vector form (9.16). But this means that

$$\int_{\mathbb{T}} \left(\sum_{i=0}^{N-1} \left|T_i^* g(z)\right|^2 - \left|f(z)\right|^2 \right) d\mu(z) = 0.$$

In view of (9.15), this means that in fact, (9.15) is an a.e. identity, i.e., that

$$|f|^2 = \sum_{i=0}^{N-1} \left|T_i^* g\right|^2 \qquad \text{a.e. on } \mathbb{T},$$

and that therefore

$$\left| \sum_{i=0}^{N-1} \overline{A_{j,i}(z)}\, T_i^* g(z) \right|^2 = \sum_{i=0}^{N-1} \left|T_i^* g(z)\right|^2 \qquad \text{for a.e. } z \in \mathbb{T}.$$

Hence there is a function h_j on $\mathbb{T}$ such that $T_i^* g = h_j A_{j,i}$. But an application of (ii) and $\sum_{i=0}^{N-1} T_i T_i^* = I_{L^2(\mathbb{T})}$ shows that $h_j = f$, which is the desired conclusion (iii), i.e., the formula (9.13).

(iii) ⇒ (i): If (iii) holds, we get

$$\begin{aligned} S_j f(z) &= \sum_{i=0}^{N-1} A_{j,i}\left(z^N\right) T_i f(z) = \sum_{i=0}^{N-1} z^i T_i^* g\left(z^N\right) \\ &= \sum_{i=0}^{N-1} T_i T_i^* g(z) - g(z). \end{aligned}$$

□

We note two consequences deriving from the condition (9.1). It is a condition on the given measurable matrix function $\mathbb{T} \ni z \mapsto A(z) \in \mathrm{U}_N(\mathbb{C})$, and therefore on the corresponding representation $\left(S_j^{(A)}\right)_{j=0}^{N-1}$ of $\mathcal{O}_N$. This representation acts on the Hilbert space $L^2(\mathbb{T})$. But the wavelet system φ, ψ_j from (1.1) relates to the line $\mathbb{R}$, and not directly to $\mathbb{T}$. Indeed, condition (9.1) ensures that the wavelet functions of the system φ, ψ_j, derived from A, are in $L^2(\mathbb{R})$, and (9.1) is called the frequency-subband condition. The functions

$$m_j^{(A)}(z) = \sum_{k=0}^{N-1} A_{j,k}\left(z^N\right) z^k \tag{9.17}$$

are called subband filters: $m_0^{(A)}$ is the low-pass filter, and the others $m_j^{(A)}$, $j \geq 1$, are the higher-pass filter bands.

Notations 9.5. Condition (9.1) gives the distribution of the N cases with probabilities $\frac{1}{N}\left|m_j^{(A)}(\cdot)\right|^2$ on the frequencies $\frac{j}{N}$, $j = 0, 1, \ldots, N-1$, which represent the bands. Recall if $\rho = e^{i2\pi/N}$, then $\left\{\rho^j : 0 \leq j < N\right\}$ are the N'th roots of unity, i.e., $\left(\rho^j\right)^N = 1$. The frequency passes for the bands $0, \frac{1}{N}, \dots, \frac{N-1}{N}$, referring to the low-pass filter, are $(\underbrace{1,}_{\text{pass}} \underbrace{0,0,\dots,0}_{\text{halt}})$, and similarly

$$\frac{1}{N}\left|m_j^{(A)}\left(\rho^k\right)\right|^2 = \delta_{j,k}. \tag{9.18}$$

In fact, these conditions (9.18) are equivalent to the single matrix condition (9.1) for A.

We also note that (9.1) implies that each one of the N isometries $S_j^{(A)}$ on $L^2(\mathbb{T})$ is a shift, i.e., that

$$\lim_{n\to\infty} S_j^{(A)*n} = 0. \tag{9.19}$$

This conclusion, while nontrivial, is contained in the result Theorem 3.1 in [BrJo97]. Note that each of the N shift operators has infinite multiplicity in the sense of Proposition 8.2 above. Recall if $S\colon \mathcal{H} \to \mathcal{H}$ is a shift in a Hilbert space $\mathcal{H}$, then the multiplicity space is $\mathcal{W}_S := (S\mathcal{H})^\perp = \ker(S^*)$: specifically, $\mathcal{H} \simeq \sum_{n=0}^{\infty}{}^{\oplus} \mathcal{W}_S$ with S represented as

$$(x_0, x_1, \dots) \xrightarrow{\hat{S}} (0, x_0, x_1, \dots) \tag{9.20}$$

where $x_0, x_1, \dots \in \mathcal{W}_S$. For the particular application to $S_j^{(A)}$, we have

$$\mathcal{W}_{S_j^{(A)}} = \sum_{k\neq j}{}^{\oplus} S_k^{(A)}\left(L^2(\mathbb{T})\right). \tag{9.21}$$

The next result is a corollary of Proposition 8.2 and the discussion above.

Corollary 9.6. (A Dichotomy.) *Let the matrix function $\mathbb{T} \ni z \mapsto A(z) \in \mathrm{U}_N(\mathbb{C})$, $N > 1$, satisfy (9.1), and let $\left(S_j^{(A)}\right)_{j=0}^{N-1}$ be the corresponding representation of $\mathcal{O}_N$ on $L^2(\mathbb{T})$. If $\mathcal{L} \subset L^2(\mathbb{T})$ is a co-invariant subspace, then the following two conditions are equivalent.*

(i) *$\mathcal{L}$ is pure.*
(ii) *$\mathcal{L} \neq L^2(\mathbb{T})$.*

Proof. Clearly (i) $\Rightarrow$ (ii), and it is immediate from Definition 8.1 that every finite-dimensional co-invariant subspace $\mathcal{L}$ is not pure. That follows since, if $f \in \mathcal{L}$ satisfies $S^\infty f \subset \mathcal{L}$, then for each $k \in \mathbb{Z}_+$, the family $\{S_I f : I \in \mathcal{I}(N),\ |I| = k\} \subset S^\infty f \subset \mathcal{L}$ consists of orthogonal vectors. If $f \neq 0$, then there are N^k such vectors. Hence the result holds whenever the entries in A are Fourier polynomials; see also [JoKr02]. We now turn to (ii) $\Rightarrow$ (i) in the general case: Let A satisfy (9.1), and let $\mathcal{L}$ be a co-invariant subspace, referring to the representation of $\mathcal{O}_N$ given by $\left(S_j^{(A)}\right)_{j=0}^{N-1}$. Then $\mathcal{L}^\perp = L^2(\mathbb{T}) \ominus \mathcal{L}$ is invariant for each of the N shift operators $S_j^{(A)}$, $j = 0, \dots, N-1$. Now suppose $\mathcal{L} \neq L^2(\mathbb{T})$, or equivalently that $\mathcal{L}^\perp \neq 0$.

Setting $j = 0$, we see that there is a unitary operator-valued function $\mathbb{T} \ni z \mapsto u_0(z) \in \mathrm{U}\left(\mathcal{W}_{S_0^{(A)}}\right)$ such that

$$\mathcal{L}^{\perp} = \left\{u_0 h_+ : h_+ \in H_+\left(\mathcal{W}_{S_0^{(A)}}\right)\right\} \tag{9.22}$$

where $\mathcal{W}_{S_0^{(A)}}$ is given by (9.21).

If $f \in \mathcal{L}$ satisfies $S^{(A)\,\infty} f \subset \mathcal{L}$, then $f, S_0^{(A)} f, S_0^{(A)\,2} f, \dots, S_0^{(A)\,n} f, \dots \in \mathcal{L}$. Relative to the representation (9.20), applied to $S_0^{(A)}$, we get

$$\hat{f}, z\hat{f}, z^2\hat{f}, \dots, z^n\hat{f}, \dots \in \mathcal{L}.$$

But then we get $f = 0$ by an application of Proposition 8.2 to $S_0^{(A)}$. Hence, $\mathcal{L}$ is pure; see Definition 8.1. Note that [BrJo97, Theorem 3.1] was used as well. This result implies that $S_0^{(A)}$ is a shift, and so it has the representation (9.20). □

We now turn to the space of Lipschitz functions on $\mathbb{T}$. Via the coordinate $z = e^{-i\xi}$, $\xi \in \mathbb{R}$, we identify functions on $\mathbb{T}$ with 2π-periodic functions on $\mathbb{R}$ and we define the *Lipschitz space* Lip_1 by

$$\|f\|_{\mathrm{Lip}_1} := |f(0)| + \sup_{-\pi \leq \xi < \eta < \pi} \frac{|f(\xi) - f(\eta)|}{|\xi - \eta|} < \infty. \tag{9.23}$$

A matrix function is said to be Lipschitz if its matrix entries are in Lip_1.

Theorem 9.7. *A matrix function, $\mathbb{T} \ni z \mapsto A(z) \in \mathrm{U}_N(\mathbb{C})$ is given. We assume it is in the Lipschitz class, and that it satisfies (9.1). Consider the co-invariant subspace $\mathcal{L}^{(A)} \subset L^2(\mathbb{T})$ defined from the corresponding representation $\left(S_j^{(A)}\right)_{j=0}^{N-1}$ of $\mathcal{O}_N$ as follows:*

$$\mathcal{L}^{(A)} := \bigvee \left[S_I^{(A)\,*} e_0 : I \in \mathcal{I}(N)\right] \tag{9.24}$$

where $e_0 = \mathbb{1}$ is the constant function 1 on $\mathbb{T}$. Then $\mathcal{L}^{(A)}$ is pure.

Proof. In view of Corollary 9.6, it is enough to show that $\mathcal{L}^{(A)} \neq L^2(\mathbb{T})$, or equivalently that $\left(\mathcal{L}^{(A)}\right)^{\perp} \neq 0$. The argument is indirect. If $\mathcal{L}^{(A)} = L^2(\mathbb{T})$, we get a contradiction as follows (we have suppressed the superscript A in the notation): The set of vectors $\{S_I^* e_0 : I \in \mathcal{I}(N)\}$ is relatively compact in $C(\mathbb{T})$ by Arzelà-Ascoli. To see this, we use the Lipschitz property, and the formula

$$S_I^* e_0(z) = \frac{1}{N^k} \sum_{w^{N^k} = z} \overline{m_I^{(k)}(w)}$$

where $m_I^{(k)}(z) = m_{i_1}(z)\, m_{i_2}(z^N) \cdots m_{i_k}\left(z^{N^{k-1}}\right)$, $I = (i_1, \dots, i_k)$. Further note that $\left(\sqrt{N} S_0^* f\right)(1) = f(1)$ for all Lipschitz functions f, and that the analogous conditions hold for $S_1^*, \dots, S_{N-1}^*$. Since $m_j\left(\rho^k\right) = \delta_{j,k}\sqrt{N}$, there is a Lipschitz solution f to the following system of equations:

$$f(1) = 1, \qquad S_0^* f = \frac{1}{\sqrt{N}} f, \qquad S_j^* f = 0,\ j \geq 1. \tag{9.25}$$

Using $f = \sum_j S_j S_j^* f = \frac{1}{\sqrt{N}} S_0 f$, it follows that S_0 has $\sqrt{N}$ as eigenvalue, contradicting that S_0 is isometric. The contradiction proves that $\mathcal{L}^{(A)\perp} \neq 0$. □

In the next section, we give additional details on the existence question for the eigenvalue problems related to the operators $S_j^{(A)*}$ in the case when the matrix function A is assumed to be Lipschitz.

10. Finite dimensions

In this section we offer a construction of a finite-dimensional co-invariant (nonzero, and nontrivial) subspace for the representation of $\mathcal{O}_N$ on $L^2(\mathbb{T})$ which is associated with a multiresolution wavelet of scale N. It is both natural and optimal with respect to the conditions of Sections 6–9. The setting is as in the previous section: Recall, a Lipschitz mapping $\mathbb{T} \ni z \mapsto A(z) \in \mathrm{U}_N(\mathbb{C})$ is given, and it is assumed that the subbands are ordered according to (9.1), i.e., that $A(1)$ is the Hadamard matrix, or equivalently that

$$m_j\left(\rho_N^k\right) = \delta_{j,k}\sqrt{N}, \qquad j,k = 0, \dots, N-1, \tag{10.1}$$

where $\rho_N := \exp\left(i\frac{2\pi}{N}\right)$, and

$$m_j(z) = \sum_{k=0}^{N-1} A_{j,k}\left(z^N\right) z^k. \tag{10.2}$$

Then the operators $S_j = S_j^{(A)}$ on $L^2(\mathbb{T})$ are

$$\left(S_j f\right)(z) = m_j(z)\, f\left(z^N\right), \qquad j = 0, \dots, N-1,\ f \in L^2(\mathbb{T}),\ z \in \mathbb{T}. \tag{10.3}$$

Theorem 10.1. *Let $\left(S_j\right)_{j=0}^{N-1}$ be a representation as in (10.3) determined by a Lipschitz system and subject to conditions (10.1) and (10.2). Then the following two conditions are equivalent.*

(i) *There is a finite-dimensional co-invariant subspace $\mathcal{L} \subset L^2(\mathbb{T})$ which contains the solutions f to the following affine conditions:*

$$f \in \mathrm{Lip}_1, \qquad f(1) = 1, \qquad \sqrt{N} S_0^* f = f. \tag{10.4}$$

(ii) *There is a finite constant K such that, for all $f \in \mathrm{Lip}_1$,*

$$\sup_{J \in \mathcal{I}(N)} \left| N^{\frac{|J|}{2}} \left(S_J^* f\right)(1) \right| \leq K \left\| f \right\|.$$

The affine dimension of the convex set (10.4) of Lipschitz functions is at least one.

Proof. We begin with the conditions (10.4). We will show that there is a well defined linear operator T on Lip_1 with finite-dimensional range (dimension at least one) such that

$$(Tf)(1) = f(1) \qquad \text{for all } f \in \mathrm{Lip}_1 \tag{10.5}$$

and

$$(Tf)(z) = \lim_{k\to\infty} \left(\sqrt{N} S_0^*\right)^k f(z), \qquad f \in \mathrm{Lip}_1, \tag{10.6}$$

where the limit in (10.6) is uniform for $z \in \mathbb{T}$.

Lemma 10.2. *Let the operators S_j, $j = 0, \dots, N-1$, be as specified above in* (10.3), *i.e., the Lipschitz property is assumed, as is* (10.1). *For bounded functions f on $\mathbb{T}$, set $\|f\| := \sup_{z\in\mathbb{T}} |f(z)|$; and if f is differentiable, set*

$$\tilde{f}(x) := f\left(e^{-i2\pi x}\right), \tag{10.7}$$

and $f' := \tilde{f}'$. Finally, let

$$M_1 := N^{-\frac{1}{2}} \max_{0\leq j<N} \|m_j'\|. \tag{10.8}$$

Then we have the following estimate:

$$N^{\frac{k}{2}} \left\| \left(S_{j_k}^* \cdots S_{j_2}^* S_{j_1}^* f\right)' \right\| \leq N^{-\frac{k}{2}} \|f'\| + M_1 \|f\|$$
$$\text{for all } k \in \mathbb{Z}_+, \text{ all } f \in \mathrm{Lip}_1,$$
$$\text{and all multi-indices } J = (j_1, j_2, \dots, j_k) \in \mathcal{I}_k(N). \tag{10.9}$$

Proof. Using (10.7), we shall pass freely between any of the four equivalent formulations, functions on $\mathbb{T}$, functions on $[0,1)$, functions on $\mathbb{R}/\mathbb{Z}$, or one-periodic functions on $\mathbb{R}$, omitting the distinction between f and $\tilde{f}$ in (10.7). With the multi-index notation in (10.9), we set

$$m_J(z) := m_{j_1}(z)\, m_{j_2}\left(z^N\right) \cdots m_{j_k}\left(z^{N^{k-1}}\right), \qquad z \in \mathbb{T}. \tag{10.10}$$

Setting $S_J := S_{j_1} \cdots S_{j_k}$, and $S_J^* := S_{j_k}^* \cdots S_{j_2}^* S_{j_1}^*$, we find

$$(S_J f)(z) = m_J(z)\, f\left(z^{N^k}\right), \tag{10.11}$$

and

$$(S_J^* f)(z) = \frac{1}{N^k} \sum_{\substack{w\in\mathbb{T} \\ w^{N^k}=z}} \overline{m_J(w)}\, f(w). \tag{10.12}$$

The sum in (10.12) contains N^k terms, which is evident from the rewrite in the form below, using instead one-periodic functions on $\mathbb{R}$, $x \in \mathbb{R}$:

$$(S_J^* f)(x) = \frac{1}{N^k} \sum_{\substack{y\in\mathbb{R} \\ N^k y \equiv x \bmod 1}} \overline{m_J(y)}\, f(y). \tag{10.13}$$

In this form $m_J(y) = m_{j_1}(y)\, m_{j_2}(Ny) \cdots m_{j_k}(N^{k-1}y)$ (compare with (10.10)), and the points y may be represented as

$$y = \frac{x + l_0 + l_1 N + l_2 N^2 + \cdots + l_{k-1} N^{k-1}}{N^k}, \tag{10.14}$$

with the integers $l_0, l_1, \ldots, l_{k-1}$ taking values over the mod N residue classes $0, 1, \ldots, N-1$. Given this, it is clear how the general form of (10.9) follows from the case $k = 1$. We will do $k = 1$, and leave the induction and the multi-index gymnastics to the reader. Using (10.13)–(10.14) we have

$$N^{\frac{1}{2}} \left(S_j^* f\right)'(x) = \frac{1}{N\sqrt{N}} \sum_{l=0}^{N-1} \bar{m}_j\left(\frac{x+l}{N}\right) f'\left(\frac{x+l}{N}\right) + \frac{1}{N\sqrt{N}} \sum_{l=0}^{N-1} \bar{m}_j'\left(\frac{x+l}{N}\right) f\left(\frac{x+l}{N}\right). \tag{10.15}$$

For the individual terms on the right-hand side, we use Schwarz's inequality, as follows: First,

$$\begin{aligned} \frac{1}{N}\left|\sum_{l=0}^{N-1} \bar{m}_j\left(\frac{x+l}{N}\right) f'\left(\frac{x+l}{N}\right)\right| &\le \underbrace{\left(\frac{1}{N}\sum_{l=0}^{N-1}\left|m_j\left(\frac{x+l}{N}\right)\right|^2\right)^{\frac{1}{2}}}_{=1} \left(\frac{1}{N}\sum_{l=0}^{N-1}\left|f'\left(\frac{x+l}{N}\right)\right|^2\right)^{\frac{1}{2}} \\ &\le \|f'\| ; \end{aligned}$$

and second,

$$\begin{aligned} \frac{1}{N}\left|\sum_{l=0}^{N-1} \bar{m}_j'\left(\frac{x+l}{N}\right) f\left(\frac{x+l}{N}\right)\right| &\le \left(\frac{1}{N}\sum_{l=0}^{N-1}\left|m_j'\left(\frac{x+l}{N}\right)\right|^2\right)^{\frac{1}{2}} \left(\frac{1}{N}\sum_{l=0}^{N-1}\left|f\left(\frac{x+l}{N}\right)\right|^2\right)^{\frac{1}{2}} \\ &\le \sqrt{N} M_1 \|f\|, \end{aligned}$$

where the terms on the right are given in (10.8) and the discussion in that paragraph.

Introducing the last two estimates back into (10.15), we get

$$N^{\frac{1}{2}} \left|\left(S_j^* f\right)'(x)\right| \le N^{-\frac{1}{2}} \|f'\| + M_1 \|f\|,$$

which is the desired estimate. □

Note that if the functions f in (10.9) are restricted by $\|f\| \le 1$, then all the functions $N^{\frac{k}{2}} S_J^* f$ are contained in a compact subset in the Banach space $C(\mathbb{T})$.

This follows from [DuSc, vol. I, p. 245, Theorem IV.3.5]; see also [IoMa50, Section 4], or [Bal00]. The conclusion is that there is a fixed finite-dimensional subspace $\mathcal{L}$ which contains all the functions obtained as limits of the terms $N^{\frac{k}{2}} S_J^* f$ as the multi-indices $J = (j_1, \dots, j_k)$ vary. Introducing $\mathcal{I}_\infty (N) = \{0, 1, \dots, N-1\}^{\mathbb{Z}_+}$, for each $\xi = (j_1, j_2, \dots) \in \mathcal{I}_\infty (N)$ we may pass to infinity via an ultrafilter $u(\xi)$, i.e.,

$$T_\xi f := \lim_{\substack{u(\xi) \\ \xi = (J_\to * \cdots * \cdots)}} N^{\frac{|J|}{2}} S_J^* f, \tag{10.16}$$

and we note that all the operators T_ξ, $\xi \in \mathcal{I}_\infty (N)$, have their range contained in $\mathcal{L}$. If $j \in \{0, 1, \dots, N-1\}$, then the extension $(j\xi)$ is a point in $\mathcal{I}_\infty (N)$ for every $\xi \in \mathcal{I}_\infty (N)$, and

$$N^{\frac{1}{2}} S_j^* T_\xi f = T_{(j\xi)} f \in \mathcal{L}. \tag{10.17}$$

This proves that $\mathcal{L}$ is co-invariant.

For a reference to ultrafilters and compactifications, we suggest [Ency] and [Bou89].

The limit (10.6) is covered by this discussion, since the point $\xi = (0, 0, \dots) \in \mathcal{I}_\infty (N)$, so the corresponding operator T in (10.6) is just $T_{(0,0,\dots)}$. Hence (10.6) is a special case of (10.16). Finally, (10.5) follows from

$$\sqrt{N} (S_0^* f)(0) = \frac{1}{\sqrt{N}} \sum_{l=0}^{N-1} \overline{m_0 \left(\frac{l}{N}\right)} f\left(\frac{l}{N}\right) = f(0), \tag{10.18}$$

which is based on (10.1). In the additive form, (10.1) implies $m_0(0) = \sqrt{N}$, and $m_0\left(\frac{1}{N}\right) = \cdots = m_0\left(\frac{N-1}{N}\right) = 0$. Note $z = \exp(-i2\pi(x=0)) = 1$ is used in (10.5).

Let $f \in \mathrm{Lip}_1$, and set $f_J = N^{\frac{|J|}{2}} S_J^* f$. We noted that each f_J is in Lip_1, and we gave a uniform estimate on the derivatives, i.e., on $f_J'(x)$. But the sequence is also bounded in $C(\mathbb{T})$, relative to the usual norm $\|\cdot\|$ on $C(\mathbb{T})$. Moreover,

$$\|f_J\| \leq N^{-\frac{|J|}{2}} \|f'\| + (M_1 + K) \|f\|. \tag{10.19}$$

To see this, first note the estimate from (ii), i.e.,

$$\sup_{J \in \mathcal{I}(N)} |f_J(1)| \leq K \|f\|. \tag{10.20}$$

The details are understood best in the additive formulation, i.e., with $\mathbb{T} \cong [0, 1)$ and the identification $z \cong x$ via $z = \exp(-i2\pi x)$. The first iteration step, starting with f, yields $(x = 0)$

$$N^{\frac{1}{2}} (S_0^* f)(0) = N^{-\frac{1}{2}} \sum_{l=0}^{N-1} \bar{m}_0 \left(\frac{l}{N}\right) f\left(\frac{l}{N}\right) = f(0),$$

and subsequent steps yield

$$N^{\frac{1}{2}} (S_j^* f)(0) = N^{-\frac{1}{2}} \sum_{l=0}^{N-1} \bar{m}_j \left(\frac{l}{N}\right) f\left(\frac{l}{N}\right) = f\left(\frac{j}{N}\right),$$

where conditions (10.1) are used. So if $|J| = 1$, (10.20) holds, and $\leq$ is equality. Note that in general, if $|J| = k > 1$, then

$$F_J(0) = N^{-\frac{k}{2}} \sum_{N^k x \equiv 0 \bmod 1} \overline{m_J(x)}\, f(x), \tag{10.21}$$

where x is an N-adic fraction $0 \leq x < 1$. Note that x varies over the solutions $N^k x \equiv 0 \bmod 1$, and J is fixed. For real values of x, $0 \leq x < 1$, we have

$$|f_J(x) - f_J(0)| \leqq \int_0^x |f'_J(y)|\, dy \leq x\left(N^{-\frac{k}{2}} \|f'\| + M_1 \|f\|\right);$$

and, using (10.20), we get

$$|f_J(x)| \leq N^{-\frac{k}{2}} \|f'\| + (M_1 + K)\|f\|, \tag{10.22}$$

which is the desired estimate (10.19).

If $\xi \in \mathcal{I}_\infty(N)$, the operator T_ξ is defined on f as in (10.16) by a limit over an ultrafilter,

$$T_\xi f = \lim_{\substack{u(\xi) \\ \underbrace{(J_\rightarrow * \cdots * \cdots)}_{\xi}}} f_J, \tag{10.23}$$

as noted in (10.16). Using (10.19), and passing to the limit, we get $\|T_\xi f\| \leq (M_1 + K)\|f\|$. Similarly, an application of (10.20) yields $|(T_\xi f)(1)| \leq K\|f\|$. The fact that the range of T_ξ is finite-dimensional results as noted from Arzelà-Ascoli in view of the uniform estimate on the derivatives f'_J. □

Remark 10.3. Even for the case of the Haar wavelet, both the standard one where the polyphase matrix $A(z)$ is $A(z) = \begin{pmatrix} 1 & 1 \\ 1 & -1 \end{pmatrix}$, or the stretched Haar wavelet where $A(z) = \begin{pmatrix} 1 & z \\ 1 & -z \end{pmatrix}$, the space $\left\{f \in \mathrm{Lip}_1 : \sqrt{2} S_0^* f = f\right\}$ is of dimension more than 1. In the first case, the dimension can be checked to be 2, and in the second case, it is 3. The respective subspaces $\mathcal{L}_1$ and $\mathcal{L}_3$ are $\mathcal{L}_1 = [e_0, e_{-1}]$ and $\mathcal{L}_3 = [e_0, e_{-3}, e_{-1} + e_{-2}]$, where $e_n(z) = z^n$, $n \in \mathbb{Z}$, $z \in \mathbb{T}$. Two independent functions f in $\mathcal{L}_1$ satisfying $f(1) = 1$ are e_0 and $\frac{1}{2}(e_0 + e_{-1})$. For $\mathcal{L}_3$, three such independent solutions to the system $\sqrt{2} S_0^* f = f$, $f(1) = 1$ may be taken to be e_0, $\frac{1}{2}(e_0 + e_3)$, and $\frac{1}{3}(e_0 + e_{-1} + e_{-2})$.

Despite the fact that $\left\{f \in \mathrm{Lip}_1 : S_0^* f = \left(1/\sqrt{N}\right) f\right\}$ is automatically finite-dimensional for the representations of Theorem 10.1, we show that for the same class of representations, and for every $\lambda \in \mathbb{C}$, $|\lambda| < 1$, the space

$$\left\{f \in L^2(\mathbb{T}) : S_0^* f = \bar{\lambda} f\right\} \tag{10.24}$$

is infinite-dimensional. If $N = 2$, then in fact it is isomorphic to $\mathcal{W} = S_1 L^2(\mathbb{T})$. Introducing the projection $P_{\mathcal{W}} := S_1 S_1^*$, it can easily be checked that the operator

$(I - \bar{\lambda} S_0)^{-1} P_{\mathcal{W}}$ is well defined and maps into $\ker(\bar{\lambda} I - S_0^*)$. To see that inner products are preserved, note that

$$\left\langle (I - \bar{\lambda} S_0)^{-1} w_1 \,\middle|\, (I - \bar{\lambda} S_0)^{-1} w_2 \right\rangle = \frac{1}{1 - |\lambda|^2} \langle w_1 \mid w_2 \rangle \tag{10.25}$$

holds for all $w_1, w_2 \in \mathcal{W}$. The easiest way to see this is via the following known theorem.

Theorem 10.4. *For every representation $(S_j)_{j=0}^1$ of $\mathcal{O}_2$ on a Hilbert space $\mathcal{H}$ there is a unique unitary isomorphism $T \colon H_+(\mathcal{W}) \to \mathcal{H}$ such that $Tw = w$ for $w \in \mathcal{W}$, and*

$$S_0 T = T M_z \tag{10.26}$$

if and only if $S_0^{*\,n} x \underset{n \to \infty}{\longrightarrow} 0$, $x \in \mathcal{H}$.

Proof. This result is essentially contained in [BrJo97, Theorem 9.1]. We will just give the formula for T and its adjoint T^*, and then leave the remaining verifications to the reader. $T \colon H_+(\mathcal{W}) \to \mathcal{H}$ is

$$T\left(\sum_{n=0}^{\infty} z^n w_n\right) = \sum_{n=0}^{\infty} S_0^n w_n \tag{10.27}$$

where $w_n \in \mathcal{W}$ and

$$\left\| \sum_{n=0}^{\infty} z^n w_n \right\|_{H_+(\mathcal{W})}^2 = \sum_{n=0}^{\infty} \|w_n\|_{\mathcal{W}}^2, \tag{10.28}$$

and if $x \in \mathcal{H}$,

$$T^* x = \sum_{n=0}^{\infty} z^n P_{\mathcal{W}} S_0^{*\,n} x \in H_+(\mathcal{W}). \tag{10.29}$$

The fact that $\|T^* x\| = \|x\|$ is based on the assumption that $S_0^{*\,n} x \to 0$, i.e., that S_0 is a shift. □

Introducing the Szegö kernel

$$C_\lambda(z) := \frac{1}{1 - \bar{\lambda} z} \tag{10.30}$$

and the inner function

$$u_\lambda(z) = \frac{\lambda - z}{1 - \bar{\lambda} z}, \tag{10.31}$$

we arrive at

$$T(C_\lambda \otimes w) = (I - \bar{\lambda} S_0)^{-1} w, \tag{10.32}$$

which makes (10.25) immediate. For the orthocomplement $\{x \in \mathcal{H} : S_0^* x = \bar{\lambda} x\}^\perp$, the $H_+(\mathcal{W})$ representation is

$$\{u_\lambda(z) F_+(z) : F_+ \in H_+(\mathcal{W})\} = \{F \in H_+(\mathcal{W}) : F(\lambda) = 0\}. \tag{10.33}$$

Acknowledgements

We gratefully acknowledge discussions with David Kribs, Joe Ball and Victor Vinnikov about representations of the Cuntz relations, at the 2002 IWOTA conference at Virginia Tech in Blacksburg, VA, and an e-mail with corrections from Ken Davidson. We thank Brian Treadway for beautiful typesetting, for corrections, and for helpful suggestions.

References

[BaMe99] L.W. Baggett and K.D. Merrill, *Abstract harmonic analysis and wavelets in* $\mathbf{R}^n$, The Functional and Harmonic Analysis of Wavelets and Frames (San Antonio, 1999) (L.W. Baggett and D.R. Larson, eds.), Contemp. Math., vol. 247, American Mathematical Society, Providence, 1999, pp. 17–27.

[Bal00] V. Baladi, *Positive Transfer Operators and Decay of Correlations*, World Scientific, River Edge, NJ, Singapore, 2000.

[BaVi1] J.A. Ball and V. Vinnikov, *Functional models for representations of the Cuntz algebra*, preprint, 2002, Virginia Polytechnic Institute and State University.

[BaVi2] J.A. Ball and V. Vinnikov, *Lax-Phillips scattering and conservative linear systems: A Cuntz-algebra multi-dimensional setting*, preprint, 2002, Virginia Polytechnic Institute and State University.

[Bou89] N. Bourbaki, *General Topology*, Elements of Mathematics, Springer, 1989, Chapters 1–2 (translated from the French).

[BrJo97] O. Bratteli and P.E.T. Jorgensen, *Isometries, shifts, Cuntz algebras and multiresolution wavelet analysis of scale* N, Integral Equations Operator Theory **28** (1997), 382–443.

[BrJo99] O. Bratteli and P.E.T. Jorgensen, *Convergence of the cascade algorithm at irregular scaling functions*, The Functional and Harmonic Analysis of Wavelets and Frames (San Antonio, 1999) (L.W. Baggett and D.R. Larson, eds.), Contemp. Math., vol. 247, American Mathematical Society, Providence, 1999, pp. 93–130.

[BrJo02] O. Bratteli and P.E.T. Jorgensen, *Wavelets through a Looking Glass: The World of the Spectrum*, Applied and Numerical Harmonic Analysis, Birkhäuser, Boston, 2002.

[BJO99] O. Bratteli, P.E.T. Jorgensen, and V. Ostrovs'kyĭ, *Representation theory and numerical AF-invariants: The representations and centralizers of certain states on* $\mathcal{O}_d$, Mem. Amer. Math. Soc. **168**, No. 797.

[Cun77] J. Cuntz, *Simple* C^**-algebras generated by isometries*, Comm. Math. Phys. **57** (1977), 173–185.

[DaLa98] X. Dai and D.R. Larson, *Wandering vectors for unitary systems and orthogonal wavelets*, Mem. Amer. Math. Soc. **134** (1998), no. 640.

[Dau92] I. Daubechies, *Ten Lectures on Wavelets*, CBMS-NSF Regional Conf. Ser. in Appl. Math., vol. 61, SIAM, Philadelphia, 1992.

[DKS01] K.R. Davidson, D.W. Kribs, and M.E. Shpigel, *Isometric dilations of non-commuting finite rank* n*-tuples*, Canad. J. Math. **53** (2001), 506–545.

[DuSc] N. Dunford and J.T. Schwartz, *Linear Operators*, Interscience Publishers, Inc., New York, 1958–1971.

[Ency] *Ultrafilter*, Encyclopaedia of Mathematics, vol. 9 (M. Hazewinkel, ed.), Kluwer Academic Publishers, Dordrecht, Boston, London, 1993, pp. 301–302.

[Hel64] H. Helson, *Lectures on Invariant Subspaces*, Academic Press, New York, 1964.

[IoMa50] C.T. Ionescu Tulcea and G. Marinescu, *Théorie ergodique pour des classes d'opérations non complètement continues*, Ann. of Math. (2) **52** (1950), 140–147.

[Jor99] P.E.T. Jorgensen, *A geometric approach to the cascade approximation operator for wavelets*, Integral Equations Operator Theory **35** (1999), 125–171.

[Jor00] P.E.T. Jorgensen, *Compactly supported wavelets and representations of the Cuntz relations, II*, Wavelet Applications in Signal and Image Processing VIII (San Diego, 2000) (A. Aldroubi, A.F. Laine, and M.A. Unser, eds.), Proceedings of SPIE, vol. 4119, SPIE, Bellingham, WA, 2000, pp. 346–355.

[Jor01] P.E.T. Jorgensen, *Minimality of the data in wavelet filters*, Adv. Math. **159** (2001), 143–228.

[JoKr02] P.E.T. Jorgensen and D. Kribs, *Wavelet representations and Fock space on positive matrices*, J. Funct. Anal. **197** (2003), 526–559.

[JSW95] P.E.T. Jorgensen, L.M. Schmitt, and R.F. Werner, *Positive representations of general commutation relations allowing Wick ordering*, J. Funct. Anal. **134** (1995), 33–99.

[Kribs] D. Kribs, *Quantum channels, wavelets, dilations, and representations of $\mathcal{O}_N$*, Proc. Edinburgh Math. Soc., to appear.

[Mal99] S.G. Mallat, *A Wavelet Tour of Signal Processing*, 2nd ed., Academic Press, Orlando – San Diego, 1999.

Palle E.T. Jorgensen
Department of Mathematics
The University of Iowa
14 MacLean Hall
Iowa City, IA 52242-1419, USA
e-mail: jorgen@math.uiowa.edu

Operator Theory:
Advances and Applications, Vol. 149, 255–289

On the Bessmertnyĭ Class of Homogeneous Positive Holomorphic Functions of Several Variables

Dmitry S. Kalyuzhnyĭ-Verbovetzkiĭ

Abstract. The class of operator-valued functions which are homogeneous of degree one, holomorphic in the open right polyhalfplane, have positive semidefinite real parts there and take selfadjoint operator values at real points, and its subclass consisting of functions representable in the form of Schur complement of a block of a linear pencil of operators with positive semidefinite operator coefficients, are investigated. The latter subclass is a generalization of the class of characteristic matrix functions of passive 2n-poles considered as functions of impedances of its elements, which was introduced by M. F. Bessmertnyĭ. Several equivalent characterizations of the generalized Bessmertnyĭ class are given, and its intimate connection with the Agler-Schur class of holomorphic contractive operator-valued functions on the unit polydisk is established.

1. Introduction

In the Ph. D. Thesis of M. F. Bessmertnyĭ [9], which appeared in Russian about twenty years ago and until very recent time was unknown to Western readers (the translations of some its parts into English are appearing now: see [10, 7, 8]), rational $n \times n$ matrix-valued functions representable in the form

$$f(z) = a(z) - b(z)d(z)^{-1}c(z), \quad z \in \mathbb{C}^N, \tag{1.1}$$

with a linear $(n+p) \times (n+p)$ matrix-valued function

$$A(z) = A_0 + z_1 A_1 + \cdots + z_N A_N = \begin{bmatrix} a(z) & b(z) \\ c(z) & d(z) \end{bmatrix} \tag{1.2}$$

were considered. Another form of such a representation is

$$f(z) = \left(\begin{bmatrix} I_{n\times n} & 0_{n\times p} \end{bmatrix} A(z)^{-1} \begin{bmatrix} I_{n\times n} \\ 0_{p\times n} \end{bmatrix} \right)^{-1}, \tag{1.3}$$

Received by the editors January 5, 2003; revised May 7, 2003. Submitted by J. A. Ball.
Mathematics Subject Classification (2000). Primary 47A48; Secondary 32A10, 47A56, 47A60.
Key words and phrases. Operator-valued functions, several complex variables, homogeneous, positive, holomorphic, long resolvent representations, positive semidefinite kernels, reproducing kernel Hilbert spaces.

and both (1.1) and (1.3) were called by Bessmertnyĭ a *long resolvent representation.* It is easy to see that this is nothing but the *Schur complement* of the block $d(z)$ in the linear matrix pencil $A(z)$. In [9] (see also [10]) Bessmertnyĭ constructed a long resolvent representation for an arbitrary rational matrix function, and in the homogeneous case, i.e., when

$$f(\lambda z_1, \dots, \lambda z_N) = \lambda f(z_1, \dots, z_N), \quad \lambda \in \mathbb{C} \backslash \{0\}, \ z = (z_1, \dots, z_N) \in \mathbb{C}^N, \quad (1.4)$$

one has $A_0 = 0$.

A particular role in his thesis is played by functions of the form (1.1) or (1.3) with $A_0 = 0$ and $A_k = A_k^T \geq 0, \ k = 1, \dots, N$ (matrices A_k are assumed to have real entries), due to their relation to electrical circuits. He proved that such functions constitute the class (let us denote it by $\mathbb{R}\mathcal{B}_N^{n \times n}$) of characteristic matrix functions of passive 2n-poles, where impedances of elements (resistances, capacitances, inductances and ideal transformers are allowed) are considered as independent variables (let us note, that in the analytic theory of electrical circuits it is customary to consider characteristic matrices as functions of frequency, e.g., see [17, 20, 12, 18]). It is easy to verify that every $f \in \mathbb{R}\mathcal{B}_N^{n \times n}$ satisfies (1.4) and the following two properties:

$$f(z) + f(z)^* \geq 0, \quad z \in \Pi^N, \quad (1.5)$$

where $\Pi^N := \{z \in \mathbb{C}^N : \ \mathrm{Re} z_k > 0, \ k = 1, \dots, N\}$ is the *open right polyhalfplane*, i.e., the Cartesian product of N copies of the open right half-plane $\Pi \subset \mathbb{C}$, and

$$f(\bar{z}) = f(z)^* = f(\bar{z})^T, \quad z \in \mathbb{C}^N, \quad (1.6)$$

where $\bar{z} := (\overline{z_1}, \dots, \overline{z_N})$. Denote by $\mathbb{R}\mathcal{P}_N^{n \times n}$ the class of rational *homogeneous positive real* $n \times n$ matrix functions, i.e., rational functions taking $n \times n$ matrix values and satisfying conditions (1.4)–(1.6). Then $\mathbb{R}\mathcal{B}_N^{n \times n} \subset \mathbb{R}\mathcal{P}_N^{n \times n}$. Let us remark that replacement of the requirement $A_k = A_k^T \geq 0$ by $A_k = A_k^* \geq 0, \ k = 1, \dots, N$ (i.e., removing the assumption that positive semidefinite matrices A_k have only real entries), in the definition of $\mathbb{R}\mathcal{B}_N^{n \times n}$, and removing the second equality in condition (1.6) in the definition of $\mathbb{R}\mathcal{P}_N^{n \times n}$ define the classes $\mathcal{B}_N^{n \times n} = \mathbb{C}\mathcal{B}_N^{n \times n}$ and $\mathcal{P}_N^{n \times n} = \mathbb{C}\mathcal{P}_N^{n \times n}$, and $\mathcal{B}_N^{n \times n} \subset \mathcal{P}_N^{n \times n}$. It is clear that in the case $N = 1$ one has $\mathbb{R}\mathcal{B}_N^{n \times n} = \mathbb{R}\mathcal{P}_N^{n \times n} = \{f(z) = zA : \ A = A^T \geq 0\}$ and $\mathcal{B}_N^{n \times n} = \mathcal{P}_N^{n \times n} = \{f(z) = zA : \ A = A^* \geq 0\}$, where $z \in \mathbb{C}$ and A is an $n \times n$ matrix with real (resp., complex) entries, thus this case is trivial. It was shown in [9] that in the case $N = 2$ one has $\mathbb{R}\mathcal{B}_N^{n \times n} = \mathbb{R}\mathcal{P}_N^{n \times n}$, too (and we shall prove in the present paper that $\mathcal{B}_N^{n \times n} = \mathcal{P}_N^{n \times n}$). For $N \geq 3$ the question whether the inclusion $\mathbb{R}\mathcal{B}_N^{n \times n} \subset \mathbb{R}\mathcal{P}_N^{n \times n}$ (as well as $\mathcal{B}_N^{n \times n} \subset \mathcal{P}_N^{n \times n}$) is proper, is still open.

Bessmertnyĭ constructed the long resolvent representations for the following special cases of functions from $\mathbb{R}\mathcal{P}_N^{n \times n}$:

- every $f \in \mathbb{R}\mathcal{P}_2^{n \times n}$;
- every rational scalar function $f = \frac{P}{Q} \in \mathbb{R}\mathcal{P}_N$ with co-prime polynomials P and Q, where $\deg P = 2$ (see [9], and also [8]);

- every rational matrix function $f = \frac{P}{Q} \in \mathbb{R}\mathcal{P}_N^{n\times n}$ such that the matrix-valued polynomial P and the scalar polynomial Q are co-prime and of degree at most one with respect to each variable, and for each $j,k = 1,\dots,N$ there exist scalar polynomials $\varphi_{jk}, \psi_1^{(j)}(z),\dots,\psi_n^{(j)}(z)$ such that

$$\left(\frac{\partial Q}{\partial z_j}\right)\frac{\partial}{\partial z_k}\left(\frac{Q}{\frac{\partial Q}{\partial z_j}}\right)(z) = \varphi_{jk}^2(z);$$

$$Q^2(z)\frac{\partial f}{\partial z_j}(z) = \begin{bmatrix}\psi_1^{(j)}(z)\\ \vdots \\ \psi_n^{(j)}(z)\end{bmatrix}\begin{bmatrix}\psi_1^{(j)}(z) & \dots & \psi_n^{(j)}(z)\end{bmatrix};$$

  such functions are called *primary* in Bessmertnyĭ's work, and it was shown that f is primary if and only if there exists its long resolvent representation where coefficients A_k, $k = 1,\dots,N$, of $A(z)$ in (1.2) are positive semidefinite (PSD) $(n+p)\times(n+p)$ matrices of rank one with real entries, and $A_0 = 0$ (see [9], and also [7]).

However, no intrinsic characterization of the class $\mathbb{R}\mathcal{B}_N^{n\times n}$ appears in [9], i.e., Bessmertnyi's thesis doesn't give an idea how to distinguish functions from $\mathbb{R}\mathcal{P}_N^{n\times n}$ which admit a long resolvent representation, except the cases mentioned above and those functions which arise as characteristic functions of certain concrete passive electrical 2n-poles.

The purpose of our paper is to give such a characterization. However, our framework is more general. Namely, we extend Bessmertnyi's considerations from rational matrix-valued functions to holomorphic operator-valued functions, and from finite-dimensional long resolvent representations to infinite-dimensional ones.

We start with the "complex case", i.e., generalize the classes $\mathcal{B}_N^{n\times n} = \mathbb{C}\mathcal{B}_N^{n\times n}$ and $\mathcal{P}_N^{n\times n} = \mathbb{C}\mathcal{P}_N^{n\times n}$. In Section 2 we introduce the corresponding classes $\mathcal{B}_N(\mathcal{U})$ and $\mathcal{P}_N(\mathcal{U})$ of homogeneous positive holomorphic $L(\mathcal{U})$-valued functions (throughout this paper $L(\mathcal{U},\mathcal{V})$ denotes the Banach space of bounded linear operators mapping a Hilbert space $\mathcal{U}$ into a Hilbert space $\mathcal{V}$, and $L(\mathcal{U}) := L(\mathcal{U},\mathcal{U})$; all Hilbert spaces are supposed to be complex). We obtain the characterization of functions from the class $\mathcal{B}_N(\mathcal{U})$ (which we call the *Bessmertnyĭ class*) via a couple of identities which involve certain PSD kernels. In Section 3 we show that one of these identities turns under the Cayley transform over the variables into the Agler identity for holomorphic functions on the *unit polydisk* $\mathbb{D}^N := \{z \in \mathbb{C}^N : |z_k| < 1,\ k = 1,\dots,N\}$ taking operator values with PSD real parts. The latter means that the image of the $\mathcal{B}_N(\mathcal{U})$ under the Cayley transform over the variables is a subclass in the Agler-Herglotz class $\mathcal{AH}_N(\mathcal{U})$, introduced in [1]. Using the characterization of $\mathcal{AH}_N(\mathcal{U})$ in terms of functional calculus of N-tuples of commuting strictly contractive linear operators on a Hilbert space, we obtain the characterization of $\mathcal{B}_N(\mathcal{U})$ in terms of functional calculus of N-tuples of commuting bounded strictly accretive operators on a Hilbert space. In Section 4 we characterize the image of the Bessmertnyĭ class $\mathcal{B}_N(\mathcal{U})$ under the double Cayley transform ("double" means that this linear-fractional transform is applied to the variables and to the operator values, simultaneously), as a subclass in the Agler-Schur class $\mathcal{AS}_N(\mathcal{U})$, also introduced in [1]. This characterization turns out to be pretty surprising: a function belongs to this subclass if and only if it is representable as a transfer function of an Agler unitary colligation for which the colligation operator is not only unitary, but also selfadjoint. In Section 5 we establish a natural one-to-one correspondence between $\mathcal{B}_N(\mathcal{U})$ (which consists of homogeneous functions of N variables) and certain class of (generically) non-homogeneous functions of $N-1$ variables. However, a special complicated structure of the latter class does rather convince us that $\mathcal{B}_N(\mathcal{U})$ is more likeable to deal with. In Section 6 we turn to the "real" case. We introduce the notions of ι-real operator and ι-real operator-valued function for an anti-unitary involution $\iota = \iota_{\mathcal{U}}$ on a Hilbert space $\mathcal{U}$ which plays a role analogous to the complex conjugation in $\mathbb{C}$, and then introduce the subclasses $\iota\mathbb{R}\mathcal{B}_N(\mathcal{U})$ and $\iota\mathbb{R}\mathcal{P}_N(\mathcal{U})$ in the classes $\mathcal{B}_N(\mathcal{U})$ and $\mathcal{P}_N(\mathcal{U})$, respectively, consisting of ι-real operator-valued functions. These subclasses generalize the classes $\mathbb{R}\mathcal{B}_N^{n\times n}$ and $\mathbb{R}\mathcal{P}_N^{n\times n}$, respectively. We adapt the results for the "complex case" of the preceding sections to this "real case", i.e., give the characterizations of $\iota\mathbb{R}\mathcal{B}_N(\mathcal{U})$ in terms of long resolvent representations, in terms of identities involving PSD kernels, in terms of Agler's unitary colligations and their transfer function representations for images of its elements under the double Cayley transform. In Section 7 we summarize the results obtained in this paper, and also formulate and briefly discuss the most important open problems arising in connection with our investigation.

2. The classes of homogeneous positive holomorphic functions

2.1. The class $\mathcal{P}_N(\mathcal{U})$

Let $\mathcal{U}$ be a Hilbert space. Consider the class $\mathcal{P}_N(\mathcal{U})$ consisting of all $L(\mathcal{U})$-valued functions f holomorphic in the domain $\Omega_N := \bigcup_{\lambda\in\mathbb{T}}(\lambda\Pi)^N \subset \mathbb{C}^N$ (here for a fixed $\lambda \in \mathbb{T}$ a polyhalfplane $(\lambda\Pi)^N$ is the product of N copies of the half-plane $\lambda\Pi := \{\lambda z = (\lambda z_1, \ldots, \lambda z_N) : z \in \Pi^N\}$), such that the following conditions are satisfied:

$$f(\lambda z_1, \ldots, z_N) = \lambda f(z_1, \ldots, z_N), \quad \lambda \in \mathbb{C}\backslash\{0\},\ z \in \Omega_N; \tag{2.1}$$

$$f(z) + f(z)^* \geq 0, \qquad z \in \Pi^N; \tag{2.2}$$

$$f(\bar{z}) = f(z)^*, \qquad z \in \Omega_N. \tag{2.3}$$

Let us formulate and prove some geometrical properties of the domain Ω_N which appears naturally in this definition of the class $\mathcal{P}_N(\mathcal{U})$ and the definition of the class $\mathcal{B}_N(\mathcal{U})$ given subsequently (see Remark 2.5 below for the motivation), even though we will not use these properties explicitly in this paper.

Proposition 2.1. *The domain Ω_N has the following properties:*

(i) *Ω_N is an (open)* **cone** *in $\mathbb{C}^N$, i.e., for each $z = (z_1, \dots, z_N) \in \Omega_N$ and $t > 0$ one has $tz = (tz_1, \dots, tz_N) \in \Omega_N$;*

(ii) *Ω_N is a* **circular domain**, *i.e., for each $z = (z_1, \dots, z_N) \in \Omega_N$ and $\lambda \in \mathbb{T}$ one has $\lambda z = (\lambda z_1, \dots, \lambda z_N) \in \Omega_N$;*

(iii) *for $N = 1$ and $N = 2$ one has $clos(\Omega_N) = \mathbb{C}^N$, and for $N \geq 3$ one has $clos(\Omega_N) \neq \mathbb{C}^N$;*

(iv) *$clos(\Omega_N)$, as well as Ω_N, is not convex;*

(v) *Ω_N is* **pseudo-convex** *(i.e., there exists a plurisubharmonic function on Ω_N, which tends to $+\infty$ as its variable point approaches to the boundary $\partial\Omega_N$); or equivalently,*

(vi) *Ω_N is a* **holomorphy domain** *(i.e., there exists a holomorphic function on Ω_N, which is not holomorphically extendable to any larger domain); or equivalently,*

(vii) *Ω_N is* **not holomorphically extendable to any boundary point** *(i.e., for every point $a \in \partial\Omega_N$ there exist a neighborhood Γ of a, and a function which is holomorphic in $\Omega_N \cap \Gamma$ and not holomorphically extendable to a).*

For the proof of equivalence of properties (v)–(vii), *see* [22].

Proof. The properties (i) and (ii) are evident.

(iii). $N = 1$: $\mathrm{clos}(\Omega_1) = \mathrm{clos}(\mathbb{C}\backslash\{0\}) = \mathbb{C}$.

$N = 2$: for every $z = (z_1, z_2) \in \mathbb{C}^2$ there is a $\lambda \in \mathbb{T}$ such that $z_1 \in \mathrm{clos}(\lambda\Pi), z_2 \in \mathrm{clos}(\lambda\Pi)$, thus $z = (z_1, z_2) \in (\mathrm{clos}(\lambda\Pi))^2 \subset \mathrm{clos}(\Omega_2)$, i.e., $\mathrm{clos}(\Omega_2) = \mathbb{C}^2$.

$N \geq 3$: a point $z = (1, \exp(2\pi i/N), \dots, \exp(2\pi i(N-1)/N))$, together with some neighborhood, is disjoint with any open polyhalfplane, and therefore with Ω_N, thus $z \notin \mathrm{clos}(\Omega_N)$.

(iv). Clearly, Ω_N is not convex because for any $z \in \Omega_N$ one has $-z \in \Omega_N$, and $\frac{z+(-z)}{2} = 0 \notin \Omega_N$.

Let us show that $\mathrm{clos}(\Omega_N) = \bigcup_{\lambda\in\mathbb{T}}(\lambda\mathrm{clos}(\Pi))^N$.

Indeed, if $z \in (\lambda\mathrm{clos}(\Pi))^N$ for some $\lambda \in \mathbb{T}$ then, clearly, $z \in \mathrm{clos}(\Omega_N)$. Conversely, if $z \notin \bigcup_{\lambda\in\mathbb{T}}(\lambda\mathrm{clos}(\Pi))^N$ then for every $\lambda \in \mathbb{T}$ one has $\frac{z}{\lambda} = (\frac{z_1}{\lambda}, \dots, \frac{z_N}{\lambda}) \notin (\mathrm{clos}(\Pi))^N$, moreover, there exist a neighborhood $\Gamma_\lambda \subset \mathbb{C}^N$ of z, and a neighborhood $\Delta_\lambda \subset \mathbb{C}$ of λ such that for every $z' \in \Gamma_\lambda$ and $\lambda' \in \Delta_\lambda$ one has $\frac{z'}{\lambda'} \notin (\mathrm{clos}(\Pi))^N$. Since the collection of open sets $\{\Delta_\lambda\}_{\lambda\in\mathbb{T}}$ is a covering of a compact set $\mathbb{T}$, one may choose its finite subcovering $\{\Delta_{\lambda_j}\}_{j=1}^m$. Set $\Gamma := \bigcap_{j=1}^m \Gamma_{\lambda_j}$. Then for every $z' \in \Gamma$ and $\lambda \in \mathbb{T}$ one has $\frac{z'}{\lambda} \notin (\mathrm{clos}(\Pi))^N$, i.e., $\Gamma \cap (\lambda\mathrm{clos}(\Pi))^N = \emptyset$. Thus, $\Gamma \cap \Omega_N = \emptyset$, i.e., $z \notin \mathrm{clos}(\Omega_N)$.

A point $z = (1, \exp(2\pi i/N), \dots, \exp(2\pi i(N-1)/N)) \notin \mathrm{clos}(\Omega_N)$ can be expressed as

$$z = \frac{1}{N}\sum_{k=1}^{N} z^{(k)},$$

where

$$z^{(k)} = (0,\dots,0,\underbrace{N\exp(2\pi i(k-1)/N)}_{k-\text{th place}},0,\dots,0) \in \bigcup_{\lambda\in\mathbb{T}} (\lambda \operatorname{clos}(\Pi))^N = \operatorname{clos}(\Omega_N),$$

thus the set $\operatorname{clos}(\Omega_N)$ is not convex.

(vii). First of all, let us show that the boundary points of Ω_N can be of two different types:

1. $z^\circ = (z_1^\circ,\dots,z_{j-1}^\circ,0,z_{j+1}^\circ,\dots,z_N^\circ)$ for some $j\in\{1,\dots,N\}$, where $z_j^{\circ\prime} := (z_1^\circ,\dots,z_{j-1}^\circ,z_{j+1}^\circ,\dots,z_N^\circ)\in \operatorname{clos}(\Omega_{N-1})$;
2. $z^\circ = (z_1^\circ,\dots,z_N^\circ)\in\mathbb{C}^N$ such that $z_k^\circ\neq 0$ for all $k\in\{1,\dots,N\}$, and there exist $l,j\in\{1,\dots,N\}$, $l<j$, and $t>0$ for which $z_l^\circ + tz_j^\circ = 0$ and $z_{lj}^{\circ\prime\prime} := (z_1^\circ,\dots,z_{l-1}^\circ,z_{l+1}^\circ,\dots,z_{j-1}^\circ,z_{j+1}^\circ,\dots,z_N^\circ)\in (e^{i\arg z_l^\circ}\operatorname{clos}(\Pi))^{N-2}$ (or $\in (e^{-i\arg z_l^\circ}\operatorname{clos}(\Pi))^{N-2}$).

It is clear that points of these two types are boundary for Ω_N. Assume $z^\circ\in\partial\Omega_N$ is neither of these types. Then $z^\circ\in(\lambda\operatorname{clos}(\Pi))^N$ for some $\lambda\in\mathbb{T}$, moreover $z_{j_\mu}\in\partial(\lambda\Pi)$, $\mu=1,\dots,s$, for some $j_1,\dots,j_s\in\{1,\dots,N\}$,

$$\arg z_{j_1} = \ldots = \arg z_{j_s} = \arg\lambda + \frac{\pi}{2}\ \left(\text{or } = \arg\lambda - \frac{\pi}{2}\right),$$

and $z_k\in(\lambda\Pi)$ for $k\in\{1,\dots,N\}\backslash\{j_1,\dots,j_s\}$. In this case, there exists $\varepsilon>0$ such that $z^\circ\in(\lambda e^{i\varepsilon}\Pi)^N$ (resp., $z^\circ\in(\lambda e^{-i\varepsilon}\Pi)^N$), i.e., z° is an interior point of Ω_N, that contradicts to our assumption.

Let Γ be a neighborhood of a point z° of type 1. Set $f(z) := \frac{1}{z_j}$. Since for all points of Ω_N one has $z_j\neq 0$, f is holomorphic in $\Omega_N\cap\Gamma$, and is not holomorphically extendable to z°.

Let Γ be a neighborhood of a point z° of type 2. Set $f(z) := \frac{1}{z_l+tz_j}$. Since for all points of Ω_N one has $z_l+tz_j\neq 0$, f is holomorphic in $\Omega_N\cap\Gamma$, and not holomorphically extendable to z°.

The proof is complete. □

An equivalent definition of the class $\mathcal{P}_N(\mathcal{U})$ will be given below. To this end let us prove, first of all, the following lemma.

Lemma 2.2. *Let a scalar-valued function f be holomorphic on $\mathbb{C}\backslash\{0\}$ and satisfy $f(i^n\Pi)\subset i^n\Pi$ for $n=0,1,2,3$. Then $f(z)=\alpha z$, $z\in\mathbb{C}$, with some number $\alpha>0$.*

Proof. Set $h(z) := \frac{f(z)}{z}$. From the assumptions of the lemma it follows that h doesn't take values in the negative real semi-axis. Hence, the function $g(z) := \sqrt{h(z)}$ (with the principal branch of the square root) is well defined and holomorphic on $\mathbb{C}\backslash\{0\}$, and $g(\mathbb{C}\backslash\{0\})\subset\Pi$. Since for any number w in the open left half-plane there is no sequence z_j such that $\lim_{j\to\infty} z_j = 0$ and $\lim_{j\to\infty} g(z_j) = w$, the point $z=0$ can not be essentially singular for g (by the Sokhotsky theorem, see, e.g., [21]). Hence, this point is not essentially singular for h, too. But $z=0$ can not be a pole of h, since in this case $\arg h(z)$ would take all values from $[-\pi,\pi)$, including the value $-\pi$, which is banned. Thus, h has a removable singularity at $z=0$.

Now consider the function $\widetilde{h}(z) := h(\frac{1}{z}) = zf(\frac{1}{z})$. This function is holomorphic on $\mathbb{C}\backslash\{0\}$, and also doesn't take values in the negative real semi-axis. Applying the same reasoning, we obtain that $\widetilde{h}$ has a removable singularity at $z = 0$, and therefore h has a removable singularity at infinity. Finally, we have got the entire function h which has no singularity at infinity. Thus, by the Liouville theorem, h is a constant. Therefore, $f(z) = \alpha z$, and since $f(\Pi) = \alpha\Pi \subset \Pi$, we get $\alpha > 0$. □

Corollary 2.3. *For an $L(\mathcal{U})$-valued function f holomorphic on the domain Ω_N, conditions* (2.1) *and* (2.2) *together are equivalent to the following set of conditions:*

$$\begin{array}{ll} f(z) + f(z)^* \geq 0, & z \in \Pi^N, \\ f(z) + f(z)^* \leq 0, & z \in (-\Pi)^N, \\ i(f(z)^* - f(z)) \geq 0, & z \in (i\Pi)^N, \\ i(f(z)^* - f(z)) \leq 0, & z \in (-i\Pi)^N. \end{array} \tag{2.4}$$

Proof. If conditions (2.1) and (2.2) are fulfilled, then (2.4) follows immediately. Conversely, let conditions (2.4) are fulfilled. Consider for each fixed $x \in \Pi^N \cap \mathbb{R}^N$ the slice-function $\varphi_x(\lambda) := f(\lambda x_1, \dots, \lambda x_N)$, $\lambda \in \mathbb{C}\backslash\{0\}$, and for each fixed $x \in \Pi^N \cap \mathbb{R}^N$, $u \in \mathcal{U}$ the function $\psi_{x,u}(\lambda) = \langle\varphi_x(\lambda)u, u\rangle$, $\lambda \in \mathbb{C}\backslash\{0\}$. If $\psi_{x,u}(\lambda) \equiv 0$ set $\alpha_{x,u} := 0$. If $\psi_{x,u}(\lambda) \not\equiv 0$ then $\psi_{x,u}$ satisfies the conditions of Lemma 2.2. Thus, $\psi_{x,u}(\lambda) = \alpha_{x,u}\lambda$, $\lambda \in \mathbb{C}\backslash\{0\}$ for some number $\alpha_{x,u} > 0$. Since $\psi_{x,u}(1) = \alpha_{x,u} = \langle f(x)u, u\rangle$, one has $\psi_{x,u}(\lambda) = \langle\varphi_x(\lambda)u, u\rangle = \langle f(x)u, u\rangle\lambda$ for any fixed $x \in \Pi^N \cap \mathbb{R}^N$, $u \in \mathcal{U}$. Therefore, $\varphi_x(\lambda) = \lambda f(x)$, i.e., $f(\lambda x_1, \dots, \lambda x_N) = \lambda f(x_1, \dots, x_N)$, $\lambda \in \mathbb{C}\backslash\{0\}$ for every $x \in \Pi^N \cap \mathbb{R}^N$. By the uniqueness theorem for holomorphic functions of several variables (see, e.g., [22]), we obtain (2.1). Since (2.2) is contained in the set of conditions (2.4), the proof is complete. □

Finally, we get the following theorem.

Theorem 2.4. *An $L(\mathcal{U})$-valued holomorphic function f on Ω_N belongs to the class $\mathcal{P}_N(\mathcal{U})$ if and only if conditions* (2.4) *and* (2.3) *are satisfied.*

2.2. The class $\mathcal{B}_N(\mathcal{U})$

Let $\mathcal{U}$ be a Hilbert space. Define the *Bessmertnyĭ class* $\mathcal{B}_N(\mathcal{U})$ as a class of $L(\mathcal{U})$-valued functions f holomorphic on the domain Ω_N and representable in the form

$$f(z) = a(z) - b(z)d(z)^{-1}c(z), \quad z \in \Omega_N, \tag{2.5}$$

where

$$A(z) = z_1A_1 + \cdots + z_NA_N = \begin{bmatrix} a(z) & b(z) \\ c(z) & d(z) \end{bmatrix} \in L(\mathcal{U} \oplus \mathcal{H}), \quad z \in \Omega_N, \tag{2.6}$$

for some Hilbert space $\mathcal{H}$, and bounded linear operators $A_k = A_k^*$, $k = 1, \dots, N$, on $\mathcal{U} \oplus \mathcal{H}$ are PSD. It is clear that $c(z) = \widetilde{b}(z) := b(\overline{z})^*$, i.e., $z_1c_1 + \cdots + z_Nc_N = z_1b_1^* + \cdots + z_Nb_N^*$, $z \in \Omega_N$, and linear pencils $a(z) = z_1a_1 + \cdots + z_Na_N = \widetilde{a}(z)$, $d(z) = z_1d_1 + \cdots + z_Nd_N = \widetilde{d}(z)$ have PSD coefficients $a_k = a_k^*$, $d_k = d_k^*$, $k = 1, \dots, N$, from $L(\mathcal{U})$ and $L(\mathcal{H})$, respectively. For a function $f \in \mathcal{B}_N(\mathcal{U})$ it is easy to check properties (2.1)–(2.3), thus $\mathcal{B}_N(\mathcal{U}) \subset \mathcal{P}_N(\mathcal{U})$.

Remark 2.5. If a function f is holomorphic on Π^N and has the representation (2.5) there, then f can be extended to Ω_N by homogeneity of degree one, and this extension is, clearly, holomorphic and has a representation (2.5) in Ω_N. That is why we define the class $\mathcal{B}_N(\mathcal{U})$ straight away as a class of functions on Ω_N. Keeping in mind the possibility and uniqueness of such extension, we will write sometimes $f \in \mathcal{B}_N(\mathcal{U})$ for functions defined originally on Π^N.

Remark 2.6. If $f \in \mathcal{B}_N(\mathcal{U})$ and $f(z)$ is boundedly invertible in Π^N (and hence in Ω_N) then (2.5) can be rewritten in the form

$$f(z) = \left(P_{\mathcal{U}} A(z)^{-1}|\mathcal{U}\right)^{-1}, \quad z \in \Omega_N, \tag{2.7}$$

where $P_{\mathcal{U}}$ is the orthogonal projector onto $\mathcal{U}$ in $\mathcal{U} \oplus \mathcal{H}$, and $A(z)$ is given by (2.6). This follows from the equality

$$\begin{bmatrix} a(z) - b(z)d(z)^{-1}c(z) & 0 \\ 0 & d(z) \end{bmatrix} = \begin{bmatrix} I_{\mathcal{U}} & -b(z)d(z)^{-1} \\ 0 & I_{\mathcal{H}} \end{bmatrix} \begin{bmatrix} a(z) & b(z) \\ c(z) & d(z) \end{bmatrix} \begin{bmatrix} I_{\mathcal{U}} & 0 \\ -d(z)^{-1}c(z) & I_{\mathcal{H}} \end{bmatrix}. \tag{2.8}$$

Let Ω be a set. Recall that an $L(\mathcal{U})$-valued function Φ on $\Omega \times \Omega$ is called a *positive semidefinite kernel* (or, for the shortness, a *PSD kernel*) if for every $m \in \mathbb{N}$, $z^{(1)}, \ldots, z^{(m)} \in \Omega$, $u^{(1)}, \ldots, u^{(m)} \in \mathcal{U}$ one has

$$\sum_{\mu=1}^{m}\sum_{\nu=1}^{m} \langle \Phi(z^{(\mu)}, z^{(\nu)})u^{(\mu)}, u^{(\nu)}\rangle \geq 0. \tag{2.9}$$

In particular, in this case

$$\Phi(z,\zeta) = \Phi(\zeta, z)^*, \quad (z,\zeta) \in \Omega \times \Omega. \tag{2.10}$$

If Ω is a domain in $\mathbb{C}^N$ and a PSD kernel $\Phi(z,\zeta)$ on $\Omega \times \Omega$ is holomorphic in z and anti-holomorphic in ζ, then $\Phi(z,\zeta)$ is said to be a *holomorphic PSD kernel on* $\Omega \times \Omega$ (not to be confused with a holomorphic function on $\Omega \times \Omega$).

Theorem 2.7. *Let f be an $L(\mathcal{U})$-valued function holomorphic on Π^N. Then $f \in \mathcal{B}_N(\mathcal{U})$ if and only if there exist holomorphic PSD kernels $\Phi_k(z,\zeta)$, $k = 1, \ldots, N$, on $\Pi^N \times \Pi^N$ such that*

$$f(z) = \sum_{k=1}^{N} z_k \Phi_k(z,\zeta), \quad (z,\zeta) \in \Pi^N \times \Pi^N, \tag{2.11}$$

holds. In this case the kernels $\Phi_k(z,\zeta)$, $k = 1, \ldots, N$, can be uniquely extended to the holomorphic PSD kernels on $\Omega_N \times \Omega_N$ (we use the same notation for the extended kernels) which are homogeneous of degree zero, i.e., for every $\lambda \in \mathbb{C}\backslash\{0\}$

$$\Phi_k(\lambda z, \lambda\zeta) = \Phi_k(z,\zeta), \quad (z,\zeta) \in \Omega_N \times \Omega_N, \tag{2.12}$$

and identity (2.11) is extended to all of $(z,\zeta) \in \Omega_N \times \Omega_N$.

Proof. **Necessity.** This part of the theorem was proved by Bessmertnyĭ in [9] (see also [10]) for rational matrix-valued functions f by playing with a long resolvent representation of f in the form (1.3). We follow the same idea, however we use in our proof the representation (2.5), which does exist for our case without an additional assumption on the invertibility of $f(z)$ (see Remark 2.6).

Let $f \in \mathcal{B}_N(\mathcal{U})$. Then (2.5) holds for some Hilbert space $\mathcal{H}$ and a linear pencil of operators $A(z)$ of the form (2.6). Define

$$\psi(z) := \begin{bmatrix} I_{\mathcal{U}} \\ -d(z)^{-1}c(z) \end{bmatrix} \in L(\mathcal{U}, \mathcal{U} \oplus \mathcal{H}), \quad z \in \Omega_N.$$

Then for all $(z, \zeta) \in \Omega_N \times \Omega_N$ one has

$$\begin{aligned} f(z) &= a(z) - b(z)d(z)^{-1}c(z) \\ &= \begin{bmatrix} I_{\mathcal{U}} & -c(\zeta)^* d(\zeta)^{*-1} \end{bmatrix} \begin{bmatrix} a(z) - b(z)d(z)^{-1}c(z) \\ 0 \end{bmatrix} \\ &= \begin{bmatrix} I_{\mathcal{U}} & -c(\zeta)^* d(\zeta)^{*-1} \end{bmatrix} \begin{bmatrix} a(z) & b(z) \\ c(z) & d(z) \end{bmatrix} \begin{bmatrix} I_{\mathcal{U}} \\ -d(z)^{-1}c(z) \end{bmatrix} \\ &= \psi(\zeta)^* A(z) \psi(z). \end{aligned}$$

Set $\Phi_k(z, \zeta) := \psi(\zeta)^* A_k \psi(z)$, $k = 1, \ldots, N$. Clearly, functions $\Phi_k(z, \zeta)$, $k = 1, \ldots, N$, satisfy (2.9) for $\Omega = \Omega_N$. Since ψ is holomorphic on Ω_N we obtain that $\Phi_k(z, \zeta)$, $k = 1, \ldots, N$, are holomorphic PSD kernels on $\Omega_N \times \Omega_N$. Rewriting the equality

$$f(z) = \psi(\zeta)^* A(z) \psi(z), \quad (z, \zeta) \in \Omega_N \times \Omega_N,$$

in the form

$$f(z) = \sum_{k=1}^{N} z_k \Phi_k(z, \zeta), \quad (z, \zeta) \in \Omega_N \times \Omega_N, \tag{2.13}$$

we obtain, in particular, (2.11).

Sufficiency. Let us note that the assumption that a Hilbert space $\mathcal{H}$ involved implicitly in the representation (2.5) can be infinite-dimensional, is essential in our proof of this part of the theorem (Bessmertnyĭ proved only the necessity part, under his assumptions[1]).

Let f be an $L(\mathcal{U})$-valued function holomorphic on Π^N and representable there in the form (2.11) with some holomorphic PSD kernels $\Phi_k(z, \zeta)$, $k = 1, \ldots, N$. For these kernels there exist auxiliary Hilbert spaces $\mathcal{M}_k$ and holomorphic $L(\mathcal{U}, \mathcal{M}_k)$-valued functions φ_k on Π^N such that $\Phi_k(z, \zeta) = \varphi_k(\zeta)^* \varphi_k(z)$, $(z, \zeta) \in \Pi^N \times \Pi^N$ (see [4]). Set $\mathcal{M} := \bigoplus_{k=1}^N \mathcal{M}_k$, $P_k := P_{\mathcal{M}_k}$, $\varphi(z) := \operatorname{col} \begin{bmatrix} \varphi_1(z) & \ldots & \varphi_N(z) \end{bmatrix} \in L(\mathcal{U}, \mathcal{M})$, $e := (1, \ldots, 1) \in \Pi^N$. From (2.11) we get

$$f(e) = \sum_{k=1}^{N} \varphi_k(\zeta)^* \varphi_k(e), \quad \zeta \in \Pi^N. \tag{2.14}$$

[1]Recall that his class $\mathcal{B}_N^{n \times n}$ consists of rational $n \times n$ matrix-valued functions which admit a long resolvent representation with a *finite-dimensional* space $\mathcal{H} = \mathbb{C}^p$, see (1.1)–(1.3).

In particular,

$$f(e)=\sum_{k=1}^{N}\varphi_k(e)^*\varphi_k(e). \tag{2.15}$$

By subtracting (2.15) from (2.14) we get

$$\sum_{k=1}^{N}[\varphi_k(\zeta)-\varphi_k(e)]^*\varphi_k(e)=0,\quad \zeta\in\Pi^N,$$

i.e., the following orthogonality relation holds:

$$\mathcal{H}:=\operatorname{clos\,span}_{\zeta\in\Pi^N}\{[\varphi(\zeta)-\varphi(e)]\mathcal{U}\}\perp\operatorname{clos}\{\varphi(e)\mathcal{U}\}=:\mathcal{X}.$$

For any $\zeta\in\Pi^N$ and $u\in\mathcal{U}$ one can represent now $\varphi(\zeta)u$ as

$$\operatorname{col}\begin{bmatrix}\varphi(e) & \varphi(\zeta)-\varphi(e)\end{bmatrix}u\in\mathcal{X}\oplus\mathcal{H}.$$

On the other hand, for any $u\in\mathcal{U}$, $\zeta\in\Pi^N$ one has

$$\varphi(e)u\in\operatorname{clos\,span}_{\zeta\in\Pi^N}\{\varphi(\zeta)\mathcal{U}\},\quad [\varphi(\zeta)-\varphi(e)]u\in\operatorname{clos\,span}_{\zeta\in\Pi^N}\{\varphi(\zeta)\mathcal{U}\}.$$

Thus, $\operatorname{clos\,span}_{\zeta\in\Pi^N}\{\varphi(\zeta)\mathcal{U}\}=\mathcal{X}\oplus\mathcal{H}$. Let $\kappa:\mathcal{X}\oplus\mathcal{H}\longrightarrow\mathcal{M}$ be the natural embedding defined by

$$\kappa:\begin{bmatrix}\varphi(e)u\\(\varphi(\zeta)-\varphi(e))u\end{bmatrix}\longmapsto\varphi(\zeta)u=\begin{bmatrix}\varphi_1(\zeta)u\\\vdots\\\varphi_N(\zeta)u\end{bmatrix},\quad \zeta\in\Pi^N,\ u\in\mathcal{U}, \tag{2.16}$$

and extended to the whole $\mathcal{X}\oplus\mathcal{H}$ by linearity and continuity. Set

$$A_k:=\begin{bmatrix}\varphi(e)^* & 0\\0 & I_{\mathcal{H}}\end{bmatrix}\kappa^*P_k\kappa\begin{bmatrix}\varphi(e) & 0\\0 & I_{\mathcal{H}}\end{bmatrix}\in L(\mathcal{U}\oplus\mathcal{H}),\quad k=1,\dots,N.$$

Clearly, $A_k=A_k^*$ are PSD operators. Set

$$\psi(\zeta):=\begin{bmatrix}I_{\mathcal{U}}\\\varphi(\zeta)-\varphi(e)\end{bmatrix}\in L(\mathcal{U},\mathcal{U}\oplus\mathcal{H}),\quad \zeta\in\Pi^N.$$

Then $f(z)=\psi(\zeta)^*A(z)\psi(z)$, $(z,\zeta)\in\Pi^N\times\Pi^N$. Indeed,

$\psi(\zeta)^*A(z)\psi(z)=$

$$\begin{bmatrix}I_{\mathcal{U}}\\\varphi(\zeta)-\varphi(e)\end{bmatrix}^*\begin{bmatrix}\varphi(e) & 0\\0 & I_{\mathcal{H}}\end{bmatrix}^*\kappa^*\left(\sum_{k=1}^{N}z_kP_k\right)\kappa\begin{bmatrix}\varphi(e) & 0\\0 & I_{\mathcal{H}}\end{bmatrix}\begin{bmatrix}I_{\mathcal{U}}\\\varphi(z)-\varphi(e)\end{bmatrix}$$

$$=\sum_{k=1}^{N}z_k\varphi_k(\zeta)^*\varphi_k(z)=f(z).$$

Now, let A_k, $k=1,\dots,N$, have the block partitioning:

$$A_k=\begin{bmatrix}a_k & b_k\\c_k & d_k\end{bmatrix}\in L(\mathcal{U}\oplus\mathcal{H}).$$

Then

$$\begin{aligned} A(z)\psi(z) &= \begin{bmatrix} a(z) & b(z) \\ c(z) & d(z) \end{bmatrix} \begin{bmatrix} I_{\mathcal{U}} \\ \varphi(z)-\varphi(e) \end{bmatrix} \\ &= \begin{bmatrix} a(z)+b(z)(\varphi(z)-\varphi(e)) \\ c(z)+d(z)(\varphi(z)-\varphi(e)) \end{bmatrix} =: \begin{bmatrix} f_1(z) \\ f_2(z) \end{bmatrix}. \end{aligned}$$

Since for every $(z,\zeta) \in \Pi^N \times \Pi^N$ one has

$$\psi(\zeta)^*A(z)\psi(z) = \begin{bmatrix} I_{\mathcal{U}} & \varphi(\zeta)^*-\varphi(e)^* \end{bmatrix} \begin{bmatrix} f_1(z) \\ f_2(z) \end{bmatrix} = f(z),$$

by setting $\zeta := e$ in this equality we get $f_1(z) = f(z),\ z \in \Pi^N$. Therefore, for every $(z,\zeta) \in \Pi^N \times \Pi^N$ one has $[\varphi(\zeta)-\varphi(e)]^*f_2(z) = 0$. This implies that for every $z \in \Pi^N$ and $u \in \mathcal{U}$ one has $f_2(z)u \perp \mathcal{H}$. But $f_2(z)u \in \mathcal{H}$. Therefore, $f_2(z)u = 0$, and $f_2(z) \equiv 0$, i.e.,

$$c(z)+d(z)[\varphi(z)-\varphi(e)] \equiv 0. \tag{2.17}$$

Since for every $z \in \Pi^N$ the operator $P(z) := \sum_{k=1}^N z_kP_k$ has positive definite real part, the operator $d(z) = P_{\mathcal{H}}\kappa^*P(z)\kappa|\mathcal{H}$ has positive definite real part, too. Therefore, $d(z)$ is boundedly invertible for all $z \in \Pi^N$. From (2.17) we get $\varphi(z)-\varphi(e) = -d(z)^{-1}c(z),\ z \in \Pi^N$, and

$$f(z) = f_1(z) = a(z)-b(z)d(z)^{-1}c(z), \quad z \in \Pi^N.$$

Taking into account Remark 2.5, we get $f \in \mathcal{B}_N(\mathcal{U})$.

Since $\varphi(z)-\varphi(e) = -d(z)^{-1}c(z)$ and, hence, $\psi(z)$ are well-defined, holomorphic and homogeneous of degree zero functions on Ω_N, the kernels $\Phi_k(z,\zeta) = \psi(\zeta)^*A_k\psi(z)$, $k = 1,\dots,N$, are extended to $\Omega_N \times \Omega_N$, and (2.12) holds. One can easily verify that these extended functions are holomorphic PSD kernels on $\Omega_N \times \Omega_N$, and (2.13) holds. The proof is complete. □

3. The class $\mathcal{B}_N(\mathcal{U})$ and functional calculus

In this section we will give a characterization of the class $\mathcal{B}_N(\mathcal{U})$ via the functional calculus of N-tuples of commuting bounded strictly accretive operators. First of all, let us observe that the identity (2.11) is equivalent to the pair of the following identities:

$$f(z)+f(\zeta)^* = \sum_{k=1}^N (z_k+\overline{\zeta_k})\Phi_k(z,\zeta), \quad (z,\zeta) \in \Pi^N \times \Pi^N, \tag{3.1}$$

$$f(z)-f(\zeta)^* = \sum_{k=1}^N (z_k-\overline{\zeta_k})\Phi_k(z,\zeta), \quad (z,\zeta) \in \Pi^N \times \Pi^N. \tag{3.2}$$

We will show that the Cayley transform over all the variables turns the first of these identities into Agler's identity which characterizes the Agler-Herglotz class of holomorphic functions on the unit polydisk $\mathbb{D}^N$. The latter admits also, due to

[1], another characterization, via the functional calculus of N-tuples of commuting strict contractions, that will give us the desired result for the class $\mathcal{B}_N(\mathcal{U})$.

Let us recall some needed definitions. Denote by $\mathcal{C}^N$ the set of all N-tuples $\mathbf{T} = (T_1, \ldots, T_N)$ of commuting linear operators on some common Hilbert space $\mathcal{H}$ subject to the condition $\|T_k\| < 1,\ k = 1, \ldots, N$ (*strict contractions*). Then for any holomorphic $L(\mathcal{U})$-valued function

$$F(w) = \sum_{t \in \mathbb{Z}_+^N} \widehat{F}_t w^t, \quad w \in \mathbb{D}^N$$

(here $\mathbb{Z}_+^N := \{t \in \mathbb{Z}^N : \ t_k \geq 0, k = 1, \ldots, N\}$ and $w^t := \prod_{k=1}^N w_k^{t_k}$), and every $\mathbf{T} \in \mathcal{C}^N$ the operator

$$F(\mathbf{T}) := \sum_{t \in \mathbb{Z}_+^N} \widehat{F}_t \otimes \mathbf{T}^t \in L(\mathcal{U} \otimes \mathcal{H}) \tag{3.3}$$

is well-defined as a sum of a series convergent in the operator norm. The *Agler-Herglotz* class $\mathcal{AH}_N(\mathcal{U})$ consists of all $L(\mathcal{U})$-valued functions F which are holomorphic on $\mathbb{D}^N$ and satisfy the condition

$$F(\mathbf{T}) + F(\mathbf{T})^* \geq 0, \quad \mathbf{T} \in \mathcal{C}^N, \tag{3.4}$$

where the inequality "$\geq$" is considered in the sense of positive semi-definiteness of a selfadjoint operator. It was proved in [1] that $F \in \mathcal{AH}_N(\mathcal{U})$ if and only if there exist holomorphic PSD kernels $\Xi_k(w, \omega),\ k = 1, \ldots, N$, on $\mathbb{D}^N \times \mathbb{D}^N$ such that

$$F(w) + F(\omega)^* = \sum_{k=1}^N (1 - \overline{\omega_k} w_k) \Xi_k(w, \omega), \quad (w, \omega) \in \mathbb{D}^N \times \mathbb{D}^N. \tag{3.5}$$

Denote by $\mathcal{A}^N$ the set of all N-tuples $\mathbf{R} = (R_1, \ldots, R_N)$ of commuting bounded linear operators on some common Hilbert space $\mathcal{H}$ for which there exists a real constant $s > 0$ such that $R_k + R_k^* \geq sI_{\mathcal{H}},\ k = 1, \ldots, N$ (*strictly accretive operators*). The *operator Cayley transform*, defined by

$$R_k := (I_{\mathcal{H}} + T_k)(I_{\mathcal{H}} - T_k)^{-1}, \quad k = 1, \ldots, N, \tag{3.6}$$

maps the set $\mathcal{C}^N$ onto $\mathcal{A}^N$, and its inverse

$$T_k := (R_k - I_{\mathcal{H}})(R_k + I_{\mathcal{H}})^{-1}, \quad k = 1, \ldots, N, \tag{3.7}$$

maps $\mathcal{A}^N$ onto $\mathcal{C}^N$.

Let f be an $L(\mathcal{U})$-valued function holomorphic on Π^N. Then the *Cayley transform over variables* applied to f gives

$$F(w) := f\left(\frac{1 + w_1}{1 - w_1}, \ldots, \frac{1 + w_N}{1 - w_N}\right), \quad w \in \mathbb{D}^N, \tag{3.8}$$

which is a holomorphic $L(\mathcal{U})$-valued function on $\mathbb{D}^N$.

For any $\mathbf{R} \in \mathcal{A}^N$ let us define $f(\mathbf{R}) := F(\mathbf{T})$ where $\mathbf{T} = (T_1, \ldots, T_N)$ is defined by (3.7).

Theorem 3.1. *Let f be an $L(\mathcal{U})$-valued function holomorphic on Π^N. Then $f \in \mathcal{B}_N(\mathcal{U})$ if and only if the following conditions are satisfied:*

(i) $f(tz_1,\dots,tz_N) = tf(z_1,\dots,z_N), \quad t>0,\ z=(z_1,\dots,z_N)\in\Pi^N;$

(ii) $f(\mathbf{R}) + f(\mathbf{R})^* \geq 0, \quad \mathbf{R}\in\mathcal{A}^N;$

(iii) $f(\bar{z}) = f(z)^*, \quad z\in\Pi^N.$

Proof. **Necessity.** Let $f\in\mathcal{B}_N(\mathcal{U})$. Then (i) and (iii) easily follow from the representation (2.5) of f. Condition (ii) on f is equivalent to condition (3.4) on F which is defined by (3.8), i.e., to $F\in\mathcal{AH}_N(\mathcal{U})$. Let us show the latter. Since f satisfies (3.1), one can set $z_k := \frac{1+w_k}{1-w_k}$, $\zeta_k = \frac{1+\omega_k}{1-\omega_k}$, $k=1,\dots,N$, in (3.1) and get

$$F(w)+F(\omega)^* =$$
$$\sum_{k=1}^N \left(\frac{1+w_k}{1-w_k}+\frac{1+\overline{\omega_k}}{1-\overline{\omega_k}}\right)\Phi_k\left(\frac{1+w_1}{1-w_1},\dots,\frac{1+w_N}{1-w_N};\frac{1+\omega_1}{1-\omega_1},\dots,\frac{1+\omega_N}{1-\omega_N}\right)$$
$$=\sum_{k=1}^N (1-\overline{\omega_k}w_k)\Xi_k(w,\omega), \quad (w,\omega)\in\mathbb{D}^N\times\mathbb{D}^N, \tag{3.9}$$

where for $k=1,\dots,N$:

$$\Xi_k(w,\omega) = \frac{2}{(1-w_k)(1-\overline{\omega_k})}\Phi_k\left(\frac{1+w_1}{1-w_1},\dots,\frac{1+w_N}{1-w_N};\frac{1+\omega_1}{1-\omega_1},\dots,\frac{1+\omega_N}{1-\omega_N}\right). \tag{3.10}$$

Since for each $k=1,\dots,N$ one has $\Phi_k(z,\zeta) = \varphi_k(\zeta)^*\varphi_k(z)$, $(z,\zeta)\in\Pi^N\times\Pi^N$, where φ_k is a holomorphic function on Π^N with values in $L(\mathcal{U},\mathcal{M}_k)$ for an auxiliary Hilbert space $\mathcal{M}_k$ (again, see [4]), we get $\Xi_k(w,\omega) = \xi_k(\omega)^*\xi_k(w)$, $(w,\omega)\in\mathbb{D}^N\times\mathbb{D}^N$, where

$$\xi_k(w) = \frac{\sqrt{2}}{1-w_k}\varphi_k\left(\frac{1+w_1}{1-w_1},\dots,\frac{1+w_N}{1-w_N}\right), \tag{3.11}$$

are holomorphic $L(\mathcal{U},\mathcal{M}_k)$-valued functions on $\mathbb{D}^N$. Thus, $\Xi_k(w,\omega)$, $k=1,\dots,N$, are holomorphic PSD kernels on $\mathbb{D}^N\times\mathbb{D}^N$, and (3.9) means that $F\in\mathcal{AH}_N(\mathcal{U})$.

Sufficiency. Let f satisfy conditions (i)–(iii). Since (ii) is equivalent to $F\in\mathcal{AH}_N(\mathcal{U})$, where F is defined by (3.8), the identity (3.5) holds with holomorphic PSD kernels $\Xi_k(w,\omega)$ on $\mathbb{D}^N\times\mathbb{D}^N$. Let $\Xi_k(w,\omega)=\xi_k(\omega)^*\xi_k(w)$, $(w,\omega)\in\mathbb{D}^N\times\mathbb{D}^N$, where ξ_k are holomorphic functions on $\mathbb{D}^N$ taking values in $L(\mathcal{U},\mathcal{M}_k)$ for some auxiliary Hilbert spaces $\mathcal{M}_k$, $k=1,\dots,N$. Set $w_k := \frac{z_k-1}{z_k+1}$, $\omega_k = \frac{\xi_k-1}{\xi_k+1}$, $k=1,\dots,N$, in (3.5), and by virtue of (3.8) get:

$$f(z)+f(\zeta)^*$$
$$=\sum_{k=1}^N\left(1-\frac{\overline{\zeta_k}-1}{\overline{\zeta_k}+1}\cdot\frac{z_k-1}{z_k+1}\right)\xi_k\left(\frac{\zeta_1-1}{\zeta_1+1},\dots,\frac{\zeta_N-1}{\zeta_N+1}\right)^*\xi_k\left(\frac{z_1-1}{z_1+1},\dots,\frac{z_N-1}{z_N+1}\right)$$
$$=\sum_{k=1}^N(z_k+\overline{\zeta_k})\varphi_k(\zeta)^*\varphi_k(z), \quad (z,\zeta)\in\Pi^N\times\Pi^N, \tag{3.12}$$

where for $k = 1, \dots, N$

$$\varphi_k(z) := \frac{\sqrt{2}}{z_k + 1}\xi_k\left(\frac{z_1 - 1}{z_1 + 1}, \dots, \frac{z_N - 1}{z_N + 1}\right) \tag{3.13}$$

are holomorphic $L(\mathcal{U}, \mathcal{M}_k)$-valued functions on Π^N. It follows that $\Phi_k(z, \zeta) := \varphi_k(\zeta)^*\varphi_k(z)$, $k = 1, \dots, N$, are holomorphic PSD kernels on $\Pi^N \times \Pi^N$, and (3.1) holds. The property (iii) implies $f(x) = f(x)^*$ for every $x \in \mathbb{R}^N \cap \Pi^N$, and for every such x, and $t > 0$ by (3.12) one has:

$$\begin{aligned} f(x) + f(tx) &= (1+t)\sum_{k=1}^{N} x_k\varphi_k(tx)^*\varphi_k(x), \\ f(tx) + f(x) &= (1+t)\sum_{k=1}^{N} x_k\varphi_k(x)^*\varphi_k(tx), \\ \frac{1+t}{2}[f(x) + f(x)] &= \frac{1+t}{2}\sum_{k=1}^{N} 2x_k\varphi_k(x)^*\varphi_k(x), \\ \frac{1+t}{2t}[f(tx) + f(tx)] &= \frac{1+t}{2t}\sum_{k=1}^{N} 2tx_k\varphi_k(tx)^*\varphi_k(tx). \end{aligned}$$

By (i), the left-hand sides of these equalities coincide and equal $(1+t)f(x)$, hence

$$\begin{aligned} f(x) &= \sum_{k=1}^{N} x_k\varphi_k(tx)^*\varphi_k(x) = \sum_{k=1}^{N} x_k\varphi_k(x)^*\varphi_k(tx) \\ &= \sum_{k=1}^{N} x_k\varphi_k(x)^*\varphi_k(x) = \sum_{k=1}^{N} x_k\varphi_k(tx)^*\varphi_k(tx). \end{aligned}$$

It follows from the latter equalities that

$$\begin{aligned} 0 &\leq \sum_{k=1}^{N} x_k[\varphi_k(tx) - \varphi_k(x)]^*[\varphi_k(tx) - \varphi_k(x)] \\ &= \sum_{k=1}^{N} x_k\varphi_k(tx)^*\varphi_k(tx) - \sum_{k=1}^{N} x_k\varphi_k(tx)^*\varphi_k(x) \\ &- \sum_{k=1}^{N} x_k\varphi_k(x)^*\varphi_k(tx) + \sum_{k=1}^{N} x_k\varphi_k(x)^*\varphi_k(x) = 0. \end{aligned}$$

Thus $\varphi_k(tx) - \varphi_k(x) = 0$ for every $x \in \mathbb{R}^N \cap \Pi^N$, $t > 0$ and $k = 1, \dots, N$. For fixed $k \in \{1, \dots, N\}$ and $t > 0$ the function $h_{k,t}(z) := \varphi_k(tz) - \varphi_k(z)$ is holomorphic on Π^N and takes values in $L(\mathcal{U}, \mathcal{M}_k)$. Then for any fixed $k \in \{1, \dots, N\}$, $t > 0$, $u \in \mathcal{U}$ and $m \in \mathcal{M}_k$ the scalar function $h_{k,t,u,m}(z) := \langle h_{k,t}(z)u, m\rangle_{\mathcal{M}_k}$ is holomorphic on Π^N and vanishes on $\mathbb{R}^N \cap \Pi^N$. By the uniqueness theorem for holomorphic functions of several variables (see, e.g., [22]), $h_{k,t,u,m}(z) \equiv 0$, hence $h_{k,t}(z) \equiv 0$, that means:

$$\varphi_k(tz) = \varphi_k(z), \quad t > 0, \ z \in \Pi^N.$$

It follows from the latter equality that for every $(z, \zeta) \in \Pi^N \times \Pi^N$ and $t > 0$ one has

$$\begin{aligned} f(z) + tf(\zeta)^* &= f(z) + f(t\zeta)^* = \sum_{k=1}^N (z_k + t\overline{\zeta_k})\varphi_k(t\zeta)^*\varphi_k(z) \\ &= \sum_{k=1}^N (z_k + t\overline{\zeta_k})\varphi_k(\zeta)^*\varphi_k(z) \\ &= \sum_{k=1}^N z_k\varphi_k(\zeta)^*\varphi_k(z) + t\sum_{k=1}^N \overline{\zeta_k}\varphi_k(\zeta)^*\varphi_k(z), \end{aligned}$$

and the comparison of the coefficients of the two linear functions in t, at the beginning and at the end of this chain of equalities, gives:

$$f(z) = \sum_{k=1}^N z_k\varphi_k(\zeta)^*\varphi_k(z), \quad (z, \zeta) \in \Pi^N \times \Pi^N,$$

i.e., (2.11). By Theorem 2.7, $f \in \mathcal{B}_N(\mathcal{U})$. The proof is complete. □

Corollary 3.2. *Let f be an $L(\mathcal{U})$-valued function holomorphic on Ω_N. Then $f \in \mathcal{B}_N(\mathcal{U})$ if and only if the following conditions are satisfied:*

(i) $f(\lambda z_1, \ldots, \lambda z_N) = \lambda f(z_1, \ldots, z_N), \quad \lambda \in \mathbb{C}\backslash\{0\}, \ z = (z_1, \ldots, z_N) \in \Omega_N$;
(ii) $f(\mathbf{R}) + f(\mathbf{R})^* \geq 0, \quad \mathbf{R} \in \mathcal{A}^N$;
(iii) $f(\bar{z}) = f(z)^*, \quad z \in \Omega_N$.

Proof. If $f \in \mathcal{B}_N(\mathcal{U})$ then conditions (i) and (iii) follow from the representation (2.5) of f, and condition (ii) follows from Theorem 3.1. Conversely, conditions (i)–(iii) of the Corollary imply conditions (i)–(iii) of Theorem 3.1, which in turn imply $f \in \mathcal{B}_N(\mathcal{U})$. □

Remark 3.3. By Corollary 3.2, its conditions (i)–(iii) on holomorphic $L(\mathcal{U})$-valued functions on Ω_N give an equivalent definition of the class $\mathcal{B}_N(\mathcal{U})$, which seems to be more natural than the original definition given in Section 2 in "existence" terms. The definition of the class $\mathcal{P}_N(\mathcal{U})$ is obtained upon replacing condition (ii) by a weaker condition (2.2).

4. The image of the Bessmertnyĭ class under the double Cayley transform

In Section 3 it was established that the Cayley transform over the variables maps the Bessmertnyĭ class $\mathcal{B}_N(\mathcal{U})$ into the Agler-Herglotz class $\mathcal{AH}_N(\mathcal{U})$. Since the Cayley transform over the values of functions maps the class $\mathcal{AH}_N(\mathcal{U})$ into the

Agler-Schur class $\mathcal{AS}_N(\mathcal{U})$ (see [1]), the composition of these two transforms (the *double Cayley transform*) maps $\mathcal{B}_N(\mathcal{U})$ into $\mathcal{AS}_N(\mathcal{U})$. The class $\mathcal{AS}_N(\mathcal{U})$ is important in the interpolation theory and systems theory in several variables (see, e.g., [1, 2, 6, 15, 16, 5]), that is why it is of interest to describe the image of $\mathcal{B}_N(\mathcal{U})$ in $\mathcal{AS}_N(\mathcal{U})$ under the double Cayley transform.

Given $f \in \mathcal{B}_N(\mathcal{U})$, define for $w \in \mathbb{D}^N$:

$$\mathcal{F}(w) = (F(w) - I_\mathcal{U})(F(w) + I_\mathcal{U})^{-1} =$$
$$\left(f\left(\frac{1+w_1}{1-w_1}, \dots, \frac{1+w_N}{1-w_N}\right) - I_\mathcal{U}\right)\left(f\left(\frac{1+w_1}{1-w_1}, \dots, \frac{1+w_N}{1-w_N}\right) + I_\mathcal{U}\right)^{-1} \quad (4.1)$$

We shall write down $\mathcal{F} = \mathcal{C}(f)$, and call $\mathcal{C}(\cdot)$ the *double Cayley transform.*

Let us recall the definition, and resume the main results of [1] on the Agler-Schur class $\mathcal{AS}_N(\mathcal{U})$.

Theorem 4.1. *Let $\mathcal{F}$ be a holomorphic $L(\mathcal{U})$-valued function on $\mathbb{D}^N$. The following statements are equivalent:*

(i) $\|\mathcal{F}(\mathbf{T})\| \le 1$ *for every* $\mathbf{T} \in \mathcal{C}^N$;

(ii) *there exist holomorphic PSD kernels $\Theta_k(w,\omega)$ on $\mathbb{D}^N \times \mathbb{D}^N$, $k = 1, \dots, N$, such that for every $(w,\omega) \in \mathbb{D}^N \times \mathbb{D}^N$ one has*

$$I_\mathcal{U} - \mathcal{F}(\omega)^*\mathcal{F}(w) = \sum_{k=1}^N (1 - \overline{\omega_k} w_k)\Theta_k(w,\omega); \quad (4.2)$$

(iii) *there exist Hilbert spaces $\mathcal{X}, \mathcal{X}_1, \dots, \mathcal{X}_N$ such that $\mathcal{X} = \bigoplus_{k=1}^N \mathcal{X}_k$, and a unitary operator $U = \begin{bmatrix} A & B \\ C & D \end{bmatrix} \in L(\mathcal{X} \oplus \mathcal{U})$ such that*

$$\mathcal{F}(w) = D + CP(w)(I_\mathcal{X} - AP(w))^{-1}B, \quad (4.3)$$

where $P(w) := \sum_{k=1}^N w_k P_{\mathcal{X}_k}$, i.e., $\mathcal{F}$ is the **transfer function** *of an* **Agler unitary colligation** $\alpha = (N; U; \mathcal{X} = \bigoplus_{k=1}^N \mathcal{X}_k, \mathcal{U}, \mathcal{U})$ *(we will write $\mathcal{F} = \mathcal{F}_\alpha$ in this case) with the* **state space** *$\mathcal{X}$, and the same* **input and output spaces** *equal to $\mathcal{U}$.*

The *Agler-Schur class $\mathcal{AS}_N(\mathcal{U})$* consists of all functions satisfying any (and, hence, all) of conditions (i)–(iii) of Theorem 4.1.

Theorem 4.2. *A holomorphic $L(\mathcal{U})$-valued function $\mathcal{F}$ on $\mathbb{D}^N$ can be represented as $\mathcal{F} = \mathcal{C}(f)$ for some $f \in \mathcal{B}_N(\mathcal{U})$ if and only if the following conditions are fulfilled:*

(i) *$\mathcal{F} = \mathcal{F}_\alpha$ for an Agler unitary colligation $\alpha = (N; U; \mathcal{X} = \bigoplus_{k=1}^N \mathcal{X}_k, \mathcal{U}, \mathcal{U})$ with the additional condition $U = U^*$;*

(ii) $1 \notin \sigma(\mathcal{F}(0))$.

Proof. **Necessity.** Let $f \in \mathcal{B}_N(\mathcal{U})$. Then (3.1) and (3.2) hold. As we have shown in Theorem 3.1, the identity (3.1) implies the identity (3.9) for the holomorphic $L(\mathcal{U})$-valued function F on $\mathbb{D}^N$ which is defined by (3.8), with holomorphic PSD kernels $\Xi_k(w,\omega)$, $k=1,\dots,N$, on $\mathbb{D}^N\times\mathbb{D}^N$ defined by (3.10). Analogously, the identity (3.2) implies

$$F(w)-F(\omega)^* = \sum_{k=1}^N\left(\frac{1+w_k}{1-w_k}-\frac{1+\overline{\omega_k}}{1-\overline{\omega_k}}\right)\Phi_k\left(\frac{1+w_1}{1-w_1},\dots,\frac{1+w_N}{1-w_N};\frac{1+\omega_1}{1-\omega_1},\dots,\frac{1+\omega_N}{1-\omega_N}\right)$$
$$=\sum_{k=1}^N(w_k-\overline{\omega_k})\Xi_k(w,\omega),\quad (w,\omega)\in\mathbb{D}^N\times\mathbb{D}^N, \tag{4.4}$$

with the same kernels $\Xi_k(w,\omega)$, $k=1,\dots,N$. Let $\mathcal{F}=\mathcal{C}(f)$, i.e., $\mathcal{F}$ is determined by f or F in accordance with (4.1). Then

$$I_{\mathcal{U}}-\mathcal{F}(\omega)^*\mathcal{F}(w)=I_{\mathcal{U}}-(F(\omega)^*+I_{\mathcal{U}})^{-1}(F(\omega)^*-I_{\mathcal{U}})(F(w)-I_{\mathcal{U}})(F(w)+I_{\mathcal{U}})^{-1}$$
$$=(F(\omega)^*+I_{\mathcal{U}})^{-1}[(F(\omega)^*+I_{\mathcal{U}})(F(w)+I_{\mathcal{U}})-(F(\omega)^*-I_{\mathcal{U}})(F(w)-I_{\mathcal{U}})]$$
$$\times(F(w)+I_{\mathcal{U}})^{-1}=2(F(\omega)^*+I_{\mathcal{U}})^{-1}(F(w)+F(\omega)^*)(F(w)+I_{\mathcal{U}})^{-1}.$$

According to (3.9), we get

$$I_{\mathcal{U}}-\mathcal{F}(\omega)^*\mathcal{F}(w)=\sum_{k=1}^N(1-\overline{\omega_k}w_k)\Theta_k(w,\omega),\quad (w,\omega)\in\mathbb{D}^N\times\mathbb{D}^N, \tag{4.5}$$

where

$$\Theta_k(w,\omega)=2\,(F(\omega)^*+I_{\mathcal{U}})^{-1}\,\Xi_k(w,\omega)\,(F(w)+I_{\mathcal{U}})^{-1}. \tag{4.6}$$

Since $\Xi_k(w,\omega)=\xi_k(\omega)^*\xi_k(w)$, $(w,\omega)\in\mathbb{D}^N\times\mathbb{D}^N$, where ξ_k are holomorphic $L(\mathcal{U},\mathcal{M}_k)$-valued functions on $\mathbb{D}^N$, one has

$$\Theta_k(w,\omega)=\theta_k(\omega)^*\theta_k(w),\quad (w,\omega)\in\mathbb{D}^N\times\mathbb{D}^N, \tag{4.7}$$

where

$$\theta_k(w)=\sqrt{2}\xi_k(w)\,(F(w)+I_{\mathcal{U}})^{-1},\quad (w,\omega)\in\mathbb{D}^N\times\mathbb{D}^N, \tag{4.8}$$

are also holomorphic $L(\mathcal{U},\mathcal{M}_k)$-valued functions on $\mathbb{D}^N$. Thus, $\Theta_k(w,\omega)$, $k=1,\dots,N$, are holomorphic PSD kernels on $\mathbb{D}^N\times\mathbb{D}^N$, and due to (4.5), $\mathcal{F}\in\mathcal{AS}_N(\mathcal{U})$. Analogously,

$$\mathcal{F}(w)-\mathcal{F}(\omega)^*=2\,(F(\omega)^*+I_{\mathcal{U}})^{-1}\,(F(w)-F(\omega)^*)\,(F(w)+I_{\mathcal{U}})^{-1},$$

and according to (4.4) we get

$$\mathcal{F}(w)-\mathcal{F}(\omega)^*=\sum_{k=1}^N(w_k-\overline{\omega_k})\Theta_k(w,\omega),\quad (w,\omega)\in\mathbb{D}^N\times\mathbb{D}^N, \tag{4.9}$$

with the same set of kernels $\Theta_k(w,\omega)$, $k=1,\dots,N$, defined by (4.6). Let us rewrite (4.5) and (4.9) in a somewhat different way. Since $f\in\mathcal{B}_N(\mathcal{U})$ satisfies $f(\overline{z})=f(z)^*$, $z\in\Pi^N$, one has also

$$F(\overline{w})=F(w)^*,\quad \mathcal{F}(\overline{w})=\mathcal{F}(w)^*,\quad w\in\mathbb{D}^N.$$

Therefore, (4.5) and (4.9) are equivalent to the following two identities, respectively:

$$I_{\mathcal{U}} - \mathcal{F}(w)\mathcal{F}(\omega)^* = \sum_{k=1}^{N}(1 - w_k\overline{\omega_k})\widetilde{\theta_k}(w)\widetilde{\theta_k}(\omega)^*, \quad (w,\omega) \in \mathbb{D}^N \times \mathbb{D}^N, \tag{4.10}$$

$$\mathcal{F}(w) - \mathcal{F}(\omega)^* = \sum_{k=1}^{N}(w_k - \overline{\omega_k})\widetilde{\theta_k}(w)\widetilde{\theta_k}(\omega)^*, \quad (w,\omega) \in \mathbb{D}^N \times \mathbb{D}^N, \tag{4.11}$$

where $\widetilde{\theta_k}(w) = \theta_k(\bar{w})^*$ are holomorphic $L(\mathcal{M}_k, \mathcal{U})$-valued functions on $\mathbb{D}^N$. We will show that the identities (4.10) and (4.11) allow us to construct an Agler unitary colligation satisfying condition (i) of this theorem. For this purpose, we will use the functional model of an Agler colligation by J. A. Ball and T. T. Trent [6].

Let us recall their construction. Let $W \in \mathcal{AS}_N(\mathcal{U},\mathcal{Y})$ (the definition of $\mathcal{AS}_N(\mathcal{U},\mathcal{Y})$ is the same as of $\mathcal{AS}_N(\mathcal{U})$, with only difference that values of functions from this class are in $L(\mathcal{U},\mathcal{Y})$). Then there exist Hilbert spaces $\mathcal{L}_k$ and holomorphic functions H_k, H_{*k} on $\mathbb{D}^N$ taking values in $L(\mathcal{L}_k,\mathcal{Y})$ and $L(\mathcal{L}_k,\mathcal{U})$, $k = 1,\dots,N$, respectively, such that

$$\begin{aligned} &\begin{bmatrix} I_{\mathcal{U}} - W(\bar{w})^*W(\bar{\omega}) & W(\bar{w})^* - W(\omega)^* \\ W(w) - W(\bar{\omega}) & I_{\mathcal{Y}} - W(w)W(\omega)^* \end{bmatrix} \\ &= \sum_{k=1}^{N} \begin{bmatrix} 1 - w_k\overline{\omega_k} & w_k - \overline{\omega_k} \\ w_k - \overline{\omega_k} & 1 - w_k\overline{\omega_k} \end{bmatrix} \circ \begin{bmatrix} H_{*k}(w)H_{*k}(\omega)^* & H_{*k}(w)H_k(\omega)^* \\ H_k(w)H_{*k}(\omega)^* & H_k(w)H_k(\omega)^* \end{bmatrix} \end{aligned} \tag{4.12}$$

for all $(w,\omega) \in \mathbb{D}^N \times \mathbb{D}^N$, where "$\circ$" is a *Schur (entry-wise) matrix multiplication.* For every $k = 1,\dots,N$ and $(w,\omega) \in \mathbb{D}^N \times \mathbb{D}^N$, set

$$\begin{aligned} K_{*k}(w,\omega) &:= H_{*k}(w)H_{*k}(\omega)^*, & K_k(w,\omega) &:= H_k(w)H_k(\omega)^*, \\ L_k(w,\omega) &:= H_{*k}(w)H_k(\omega)^*, & L_{*k}(w,\omega) &:= L_k(\omega,w)^* = H_k(w)H_{*k}(\omega)^*, \end{aligned}$$

and

$$\widehat{K_k}(w,\omega) := \begin{bmatrix} K_{*k}(w,\omega) & L_k(w,\omega) \\ L_{*k}(w,\omega) & K_k(w,\omega) \end{bmatrix}.$$

The latter function is a holomorphic $L(\mathcal{U} \oplus \mathcal{Y})$-valued PSD kernel on $\mathbb{D}^N \times \mathbb{D}^N$, which serves as the reproducing kernel for the Hilbert space $\mathcal{H}(\widehat{K_k})$ of holomorphic $(\mathcal{U} \oplus \mathcal{Y})$-valued functions on $\mathbb{D}^N$; this space is determined uniquely by $\widehat{K_k}$ (for the theory of reproducing kernel Hilbert spaces see, e.g., [4]). Set $\mathcal{D}(\{\widehat{K_k}\}_{k=1}^N) := \bigoplus_{k=1}^N \mathcal{H}(\widehat{K_k})$. The latter is a Hilbert space of holomorphic $(\mathcal{U} \oplus \mathcal{Y})^N$-valued functions on $\mathbb{D}^N$, with the reproducing kernel $\widehat{K}(w,\omega) := \bigoplus_{k=1}^N \widehat{K_k}(w,\omega)$. Define the lineals $\mathcal{D}_0 \subset \mathcal{D}(\{\widehat{K_k}\}_{k=1}^N) \oplus \mathcal{U}$ and $\mathcal{R}_0 \subset \mathcal{D}(\{\widehat{K_k}\}_{k=1}^N) \oplus \mathcal{Y}$ as

$$\mathcal{D}_0 := \operatorname{span}\left\{ \begin{bmatrix} \overline{w_1}K_{*1}(\cdot,w) \\ \overline{w_1}L_{*1}(\cdot,w) \\ \vdots \\ \overline{w_N}K_{*N}(\cdot,w) \\ \overline{w_N}L_{*N}(\cdot,w) \\ I_{\mathcal{U}} \end{bmatrix} u, \begin{bmatrix} L_1(\cdot,w) \\ K_1(\cdot,w) \\ \vdots \\ L_N(\cdot,w) \\ K_N(\cdot,w) \\ W(w)^* \end{bmatrix} y : \; u \in \mathcal{U}, y \in \mathcal{Y}, w \in \mathbb{D}^N \right\},$$

$$\mathcal{R}_0 := \operatorname{span}\left\{ \begin{bmatrix} K_{*1}(\cdot,w) \\ L_{*1}(\cdot,w) \\ \vdots \\ K_{*N}(\cdot,w) \\ L_{*N}(\cdot,w) \\ W(\overline{w}) \end{bmatrix} u, \begin{bmatrix} \overline{w_1}L_1(\cdot,w) \\ \overline{w_1}K_1(\cdot,w) \\ \vdots \\ \overline{w_N}L_N(\cdot,w) \\ \overline{w_N}K_N(\cdot,w) \\ I_{\mathcal{Y}} \end{bmatrix} y : \; u \in \mathcal{U}, y \in \mathcal{Y}, w \in \mathbb{D}^N \right\}.$$

The operator $U_0 : \mathcal{D}_0 \to \mathcal{R}_0$, correctly defined by

$$U_0 : \begin{bmatrix} \overline{w_1}K_{*1}(\cdot,w) \\ \overline{w_1}L_{*1}(\cdot,w) \\ \vdots \\ \overline{w_N}K_{*N}(\cdot,w) \\ \overline{w_N}L_{*N}(\cdot,w) \\ I_{\mathcal{U}} \end{bmatrix} u \longmapsto \begin{bmatrix} K_{*1}(\cdot,w) \\ L_{*1}(\cdot,w) \\ \vdots \\ K_{*N}(\cdot,w) \\ L_{*N}(\cdot,w) \\ W(w) \end{bmatrix} u, \quad u \in \mathcal{U}, w \in \mathbb{D}^N,$$

$$U_0 : \begin{bmatrix} L_1(\cdot,w) \\ K_1(\cdot,w) \\ \vdots \\ L_N(\cdot,w) \\ K_N(\cdot,w) \\ W(w)^* \end{bmatrix} y \longmapsto \begin{bmatrix} \overline{w_1}L_1(\cdot,w) \\ \overline{w_1}K_1(\cdot,w) \\ \vdots \\ \overline{w_N}L_N(\cdot,w) \\ \overline{w_N}K_N(\cdot,w) \\ I_{\mathcal{Y}} \end{bmatrix} y, \quad y \in \mathcal{Y}, w \in \mathbb{D}^N,$$

is uniquely extended to the (correctly defined) unitary operator $\widetilde{U_0} : \operatorname{clos}(\mathcal{D}_0) \to \operatorname{clos}(\mathcal{R}_0)$.

In the case when

$$\dim\{(\mathcal{D}(\{\widehat{K_k}\}_{k=1}^N) \oplus \mathcal{U}) \ominus \operatorname{clos}(\mathcal{D}_0)\} = \dim\{(\mathcal{D}(\{\widehat{K_k}\}_{k=1}^N) \oplus \mathcal{Y}) \ominus \operatorname{clos}(\mathcal{R}_0)\} \quad (4.13)$$

there exists a (non-unique!) unitary operator $U : \mathcal{D}(\{\widehat{K_k}\}_{k=1}^N) \oplus \mathcal{U} \to \mathcal{D}(\{\widehat{K_k}\}_{k=1}^N) \oplus \mathcal{Y}$ such that $U|\operatorname{clos}(\mathcal{D}_0) = \widetilde{U_0}$. The corresponding Agler unitary colligation is $\alpha = (N; U; \mathcal{D}(\{\widehat{K_k}\}_{k=1}^N) = \bigoplus_{k=1}^N \mathcal{H}(\widehat{K_k}), U, Y)$, and $W = W_\alpha$, i.e., W is the transfer function of α.

Let us apply this construction to a function $\mathcal{F} = \mathcal{C}(f)$ where $f \in \mathcal{B}_N(U)$. In this case $\mathcal{Y} = \mathcal{U}$, $W = \mathcal{F}$. Since $\mathcal{F}(\overline{w}) = \mathcal{F}(w)^*$, $w \in \mathbb{D}^N$, it follows from (4.10) and (4.11) that (4.12) holds with $H_k = H_{*k} = \widetilde{\theta_k}$, $k = 1, \dots, N$. Therefore, for $k = 1, \dots, N$ one has

$$K_k(w, \omega) = K_{*k}(w, \omega) = L_k(w, \omega) = L_{*k}(w, \omega), \ (w, \omega) \in \mathbb{D}^N \times \mathbb{D}^N. \tag{4.14}$$

In turn, this means $\mathcal{D}_0 = \mathcal{R}_0$, and

$$(\mathcal{D}(\{\widehat{K_k}\}_{k=1}^N) \oplus \mathcal{U}) \ominus \operatorname{clos}(\mathcal{D}_0) = (\mathcal{D}(\{\widehat{K_k}\}_{k=1}^N) \oplus \mathcal{Y}) \ominus \operatorname{clos}(\mathcal{R}_0).$$

In particular, (4.13) holds. Define the operator

$$U := \widetilde{U_0} \oplus I_{(\mathcal{D}(\{\widehat{K_k}\}_{k=1}^N) \oplus \mathcal{U}) \ominus \operatorname{clos}(\mathcal{D}_0)} \in L(\mathcal{D}(\{\widehat{K_k}\}_{k=1}^N) \oplus \mathcal{U}).$$

Clearly, U is unitary. Let us show that $U = U^*$. From the definition of U_0 and (4.14) it follows that $U_0 = U_0^{-1}$. Therefore $\widetilde{U_0} = \widetilde{U_0}^{-1}$ and $U = U^{-1}$. Since $U^* = U^{-1}$, we obtain $U = U^*$. Thus, condition (i) is fulfilled.

Since $f(e) = f(e)^* \geq 0$ where $e = (1, \dots, 1)$, the operator $f(e) \in L(\mathcal{U})$ has a spectral decomposition (see, e.g., [3])

$$f(e) = \int_0^{\|f(e)\|} t \, dE_t,$$

therefore,

$$\mathcal{F}(0) = (f(e) - I_{\mathcal{U}})(f(e) + I_{\mathcal{U}})^{-1} = \int_0^{\|f(e)\|} \frac{t-1}{t+1} \, dE_t.$$

Since the function $s(t) = \frac{t-1}{t+1}$ increases on the segment $[0, \|f(e)\|]$, one has $\mathcal{F}(0) \leq \frac{\|f(e)\|-1}{\|f(e)\|+1} I_{\mathcal{U}}$. Since $\frac{\|f(e)\|-1}{\|f(e)\|+1} < 1$, we conclude that $1 \notin \sigma(\mathcal{F}(0))$, i.e., condition (ii) is also fulfilled.

Sufficiency. Let conditions (i) and (ii) for the function $\mathcal{F}$ be satisfied. Since $U = U^* = U^{-1}$, one has $D = D^*$ and D is a contraction in $\mathcal{U}$. Since $\mathcal{F}(0) = D$ and $1 \notin \sigma(\mathcal{F}(0))$, we get $\sigma(\mathcal{F}(0)) \subset [-1, a]$ with some $a: \ -1 \leq a < 1$. Hence, $\sigma(\frac{I_{\mathcal{U}} + \mathcal{F}(0)}{2}) \subset [0, \frac{1+a}{2}]$, thus $\left\| \frac{I_{\mathcal{U}} + \mathcal{F}(0)}{2} \right\| \leq \frac{1+a}{2} < 1$. By the maximum principle for holomorphic operator-valued functions of several variables (e.g., see [19]), $\left\| \frac{I_{\mathcal{U}} + \mathcal{F}(w)}{2} \right\| < 1$ for all $w \in \mathbb{D}^N$. Indeed, (i) implies $\mathcal{F} \in \mathcal{AS}_N(\mathcal{U})$, and therefore $\frac{I_{\mathcal{U}} + \mathcal{F}}{2} \in \mathcal{AS}_N(\mathcal{U})$. Thus $\left\| \frac{I_{\mathcal{U}} + \mathcal{F}(w)}{2} \right\| \leq 1$ for all $w \in \mathbb{D}^N$. If for some $w_0 \in \mathbb{D}^N$ one had $\left\| \frac{I_{\mathcal{U}} + \mathcal{F}(w_0)}{2} \right\| = 1$, then the maximum principle would imply $\left\| \frac{I_{\mathcal{U}} + \mathcal{F}(w)}{2} \right\| = 1$ everywhere in $\mathbb{D}^N$. In particular, $\left\| \frac{I_{\mathcal{U}} + \mathcal{F}(0)}{2} \right\| = 1$, that is not true. Finally, we get $1 \notin \sigma(\frac{I_{\mathcal{U}} + \mathcal{F}(w)}{2})$, and therefore $1 \notin \sigma(\mathcal{F}(w))$ for all $w \in \mathbb{D}^N$. Thus, the function $F(w) = (I_{\mathcal{U}} + \mathcal{F}(w))(I_{\mathcal{U}} - \mathcal{F}(w))^{-1}$ is correctly defined and holomorphic on $\mathbb{D}^N$. It is easy to see that

$$F(w) + F(\omega)^* = 2(I_{\mathcal{U}} - \mathcal{F}(\omega)^*)^{-1}(I_{\mathcal{U}} - \mathcal{F}(\omega)^* \mathcal{F}(w))(I_{\mathcal{U}} - \mathcal{F}(w))^{-1}, \tag{4.15}$$

$$F(w) - F(\omega)^* = 2(I_{\mathcal{U}} - \mathcal{F}(\omega)^*)^{-1}(\mathcal{F}(w) - \mathcal{F}(\omega)^*)(I_{\mathcal{U}} - \mathcal{F}(w))^{-1} \tag{4.16}$$

for all $(w,\omega) \in \mathbb{D}^N \times \mathbb{D}^N$. Since $U = U^{-1} = U^*$, due to (4.3) one has

$$
\begin{aligned}
I_{\mathcal{U}} - \mathcal{F}(\omega)^*\mathcal{F}(w) &= I_{\mathcal{U}} - [D + CP(\omega)(I_{\mathcal{X}} - AP(\omega))^{-1}B]^* \\
&\times[D + CP(w)(I_{\mathcal{X}} - AP(w))^{-1}B] = I_{\mathcal{U}} - D^*D \\
&-D^*CP(w)(I_{\mathcal{X}} - AP(w))^{-1}B - B^*(I_{\mathcal{X}} - P(\bar{\omega})A^*)^{-1}P(\bar{\omega})C^*D \\
&-B^*(I_{\mathcal{X}} - P(\bar{\omega})A^*)^{-1}P(\bar{\omega})C^*CP(w)(I_{\mathcal{X}} - AP(w))^{-1}B \\
&= B^*B + B^*AP(w)(I_{\mathcal{X}} - AP(w))^{-1}B + B^*(I_{\mathcal{X}} - P(\bar{\omega})A^*)^{-1}P(\bar{\omega})A^*B \\
&-B^*(I_{\mathcal{X}} - P(\bar{\omega})A^*)^{-1}P(\bar{\omega})(I_{\mathcal{X}} - A^*A)P(w)(I_{\mathcal{X}} - AP(w))^{-1}B \\
&= B^*(I_{\mathcal{X}} - P(\bar{\omega})A)^{-1}[(I_{\mathcal{X}} - P(\bar{\omega})A)(I_{\mathcal{X}} - AP(w)) + (I_{\mathcal{X}} - P(\bar{\omega})A)AP(w) \\
&+P(\bar{\omega})A(I_{\mathcal{X}} - AP(w)) - P(\bar{\omega})(I_{\mathcal{X}} - A^2)P(w)](I_{\mathcal{X}} - AP(w))^{-1}B \\
&= B^*(I_{\mathcal{X}} - P(\omega)A)^{-1}(I_{\mathcal{X}} - P(\bar{\omega})P(w))(I_{\mathcal{X}} - AP(w))^{-1}B \\
&= \sum_{k=1}^{N}(1 - \overline{\omega_k}w_k)B^*(I_{\mathcal{X}} - P(\bar{\omega})A)^{-1}P_{\mathcal{X}_k}(I_{\mathcal{X}} - AP(w))^{-1}B.
\end{aligned}
$$

Analogously,

$$
\mathcal{F}(w) - \mathcal{F}(\omega)^* = \sum_{k=1}^{N}(w_k - \overline{\omega_k})B^*(I_{\mathcal{X}} - P(\bar{\omega})A)^{-1}P_{\mathcal{X}_k}(I_{\mathcal{X}} - AP(w))^{-1}B.
$$

Thus, from (4.15) and (4.16) we obtain that F satisfies (3.9) and (4.4) with $\Xi_k(w,\omega) = \xi_k(\omega)^*\xi_k(w)$, where

$$
\xi_k(w) = \sqrt{2}P_{\mathcal{X}_k}(I_{\mathcal{X}} - AP(w))^{-1}B(I_{\mathcal{U}} - \mathcal{F}(w))^{-1}, \quad w \in \mathbb{D}^N,\ k = 1,\dots,N.
$$

Since for $z_k \in \Pi, \zeta_k \in \Pi$ one has

$$
1 - \frac{\overline{\zeta_k} - 1}{\overline{\zeta_k} + 1}\cdot\frac{z_k - 1}{z_k + 1} = \frac{2(z_k + \overline{\zeta_k})}{(\overline{\zeta_k} + 1)(z_k + 1)}, \qquad \frac{z_k - 1}{z_k + 1} - \frac{\overline{\zeta_k} - 1}{\overline{\zeta_k} + 1} = \frac{2(z_k - \overline{\zeta_k})}{(\overline{\zeta_k} + 1)(z_k + 1)},
$$

by setting $w_k = \frac{z_k - 1}{z_k + 1}$ and $\omega_k = \frac{\zeta_k - 1}{\zeta_k + 1}$, $k = 1,\dots,N$, in (3.9) and (4.4), we get for $f(z) = F(\frac{z_1 - 1}{z_1 + 1},\dots,\frac{z_N - 1}{z_N + 1})$ the identities (3.1) and (3.2) with $\Phi_k(z,\zeta) = \varphi_k(\zeta)^*\varphi_k(z)$, $(z,\zeta) \in \Pi^N \times \Pi^N$, and

$$
\begin{aligned}
\varphi_k(z) &= \frac{\sqrt{2}}{z_k + 1}\xi_k\left(\frac{z_1 - 1}{z_1 + 1},\dots,\frac{z_N - 1}{z_N + 1}\right) \\
&= \frac{2}{z_k + 1}P_{\mathcal{X}_k}\left(I_{\mathcal{X}} - A\sum_{k=1}^{N}\frac{z_k - 1}{z_k + 1}P_{\mathcal{X}_k}\right)^{-1} B\left(I_{\mathcal{U}} - \mathcal{F}\left(\frac{z_1 - 1}{z_1 + 1},\dots,\frac{z_N - 1}{z_N + 1}\right)\right)^{-1},
\end{aligned}
$$

$k = 1,\dots,N$. Thus, finally we get $\mathcal{F} = \mathcal{C}(f)$ where $f \in \mathcal{B}_N(\mathcal{U})$. The proof is complete. □

5. De-homogenization

In this section we establish a one-to-one correspondence between the classes $\mathcal{P}_N(\mathcal{U})$, $\mathcal{B}_N(\mathcal{U})$ and certain classes of non-homogeneous holomorphic functions of $N-1$ variables.

Theorem 5.1. *Let $f \in \mathcal{P}_N(\mathcal{U})$. Then the function*

$$g(z_1,\dots,z_{N-1}) := f(z_1,\dots,z_{N-1},1) \tag{5.1}$$

is holomorphic in the domain $\Omega_{N-1}^+ := \bigcup_{\lambda\in\mathbb{T}\cap\Pi}(\lambda\Pi)^{N-1} \subset \mathbb{C}^{N-1}$ and has the properties:

(i) $g(\overline{z'}) = g(z')^*, \quad z' \in \Omega_{N-1}^+$;

(ii) $Re\left[z_N g\left(\frac{z_1}{z_N},\dots,\frac{z_{N-1}}{z_N}\right)\right] \geq 0, \quad z=(z_1,\dots,z_{N-1},z_N)\in\Pi^N$.

Conversely, let an $L(\mathcal{U})$-valued function $g(z') = g(z_1,\dots,z_{N-1})$ be holomorphic in the domain Ω_{N-1}^+ and satisfy conditions (i) and (ii). Then the function

$$f(z_1,\dots,z_{N-1},z_N) := z_N g\left(\frac{z_1}{z_N},\dots,\frac{z_{N-1}}{z_N}\right) \tag{5.2}$$

is correctly defined on Ω_N and belongs to the class $\mathcal{P}_N(\mathcal{U})$.

Proof. The function g defined by (5.1) is holomorphic in the domain

$$\begin{aligned} & \{z'=(z_1,\dots,z_{N-1})\in\mathbb{C}^{N-1}:\ (z_1,\dots,z_{N-1},1)\in\Omega_N\} \\ = & \{z'\in\mathbb{C}^{N-1}:\ \exists\lambda\in\mathbb{T}:\ (z_1,\dots,z_{N-1},1)\in(\lambda\Pi)^N\} \\ = & \{z'\in\mathbb{C}^{N-1}:\ \exists\lambda\in\mathbb{T}\cap\Pi:\ z'\in(\lambda\Pi)^{N-1}\} = \Omega_{N-1}^+, \end{aligned}$$

since $1\in\lambda\Pi$ means $\lambda\in\Pi$. Clearly, (i) is valid for g since (2.3) is valid for f. For every $z=(z_1,\dots,z_{N-1},z_N)\in\Pi^N$, due to (2.1) and (2.2) for f, one has

$$\begin{aligned} \mathrm{Re}\left[z_N g\left(\frac{z_1}{z_N},\dots,\frac{z_{N-1}}{z_N}\right)\right] &= \mathrm{Re}\left[z_N f\left(\frac{z_1}{z_N},\dots,\frac{z_{N-1}}{z_N},1\right)\right] \\ &= \mathrm{Re} f(z_1,\dots,z_{N-1},z_N) \geq 0. \end{aligned}$$

Thus, (ii) is valid, too.

Conversely, let an $L(\mathcal{U})$-valued function $g(z') = g(z_1,\dots,z_{N-1})$ be holomorphic in Ω_{N-1}^+ and satisfy (i) and (ii). Then the function f is correctly defined by (5.2). Indeed, for any $z\in\Omega_N$ there is a $\lambda\in\mathbb{T}$ such that $z\in(\lambda\Pi)^N$. Hence, $\left(\frac{z_1}{z_N},\dots,\frac{z_{N-1}}{z_N}\right)\in\left(e^{i(\arg\lambda-\arg z_N)}\Pi\right)^{N-1}$. Since $\arg z_N\in(\arg\lambda-\frac{\pi}{2},\arg\lambda+\frac{\pi}{2})$, one has $\arg\lambda-\arg z_N\in(-\frac{\pi}{2},\frac{\pi}{2})$. Therefore, $\left(\frac{z_1}{z_N},\dots,\frac{z_{N-1}}{z_N}\right)\in\Omega_{N-1}^+$. Moreover, f is holomorphic in Ω_N. Properties (2.1)–(2.3) of f are easily verified. Thus, $f\in\mathcal{P}_N(\mathcal{U})$. The proof is complete. □

Theorem 5.2. *Let $f\in\mathcal{B}_N(\mathcal{U})$. Then the function g defined by (5.1) is holomorphic in the domain Ω_{N-1}^+ and has the properties:*

(i) $g(\overline{z'}) = g(z')^*, \quad z' \in \Omega_{N-1}^+$;

(ii) *for every* $\mathbf{R} \in \mathcal{A}^N$ *such that* $R_k \in L(\mathcal{H})$, $k = 1, \dots, N$, *where* $\mathcal{H}$ *is some Hilbert space, the operator* $g(R_N^{-1}R_1, \dots, R_N^{-1}R_{N-1})$ *is correctly defined and* $Re[(I_{\mathcal{U}} \otimes R_N)g(R_N^{-1}R_1, \dots, R_N^{-1}R_{N-1})] \geq 0$.

Conversely, let an $L(\mathcal{U})$*-valued function* $g(z') = g(z_1, \dots, z_{N-1})$ *be holomorphic in the domain* Ω_{N-1}^+ *and satisfy conditions* (i) *and* (ii). *Then the function* f *is correctly defined by* (5.2) *and belongs to the class* $\mathcal{B}_N(\mathcal{U})$.

Proof. Since $f \in \mathcal{B}_N(\mathcal{U}) \subset \mathcal{P}_N(\mathcal{U})$, by Theorem 5.1 the function g is holomorphic in Ω_{N-1}^+ and satisfies (i). Since the function $g\left(\frac{z_1}{z_N}, \dots, \frac{z_{N-1}}{z_N}\right) = f\left(\frac{z_1}{z_N}, \dots, \frac{z_{N-1}}{z_N}, 1\right)$ is holomorphic in $z = (z_1, \dots, z_N) \in \Omega_N$, the operator $g(R_N^{-1}R_1, \dots, R_N^{-1}R_{N-1})$ is correctly defined for every $\mathbf{R} = (R_1, \dots, R_N) \in \mathcal{A}^N$ (see Section 3). Moreover,

$$\begin{aligned} \mathrm{Re}[(I_{\mathcal{U}} \otimes R_N)g(R_N^{-1}R_1, \dots, R_N^{-1}R_{N-1})] &= \\ \mathrm{Re}[(I_{\mathcal{U}} \otimes R_N)f(R_N^{-1}R_1, \dots, R_N^{-1}R_{N-1}, I_{\mathcal{H}})] &= \mathrm{Re} f(R_1, \dots, R_{N-1}, R_N) \geq 0, \end{aligned}$$

due to properties (i) and (ii) of f in Corollary 3.2. Thus g has property (ii) of the present theorem.

Conversely, let an $L(\mathcal{U})$-valued function $g(z')$ be holomorphic in Ω_{N-1}^+ and satisfy conditions (i) and (ii) of this theorem. In particular, g satisfies conditions (i) and (ii) of Theorem 5.1. Then the function f is correctly defined by (5.2) and belongs to the class $\mathcal{P}_N(\mathcal{U})$. According to Corollary 3.2, the only thing we have to check is that $\mathrm{Re} f(\mathbf{R}) \geq 0$ for every $\mathbf{R} \in \mathcal{A}^N$. The latter follows easily:

$$\mathrm{Re} f(R_1, \dots, R_{N-1}, R_N) = \mathrm{Re}[(I_{\mathcal{U}} \otimes R_N)g(R_N^{-1}R_1, \dots, R_N^{-1}R_{N-1})] \geq 0.$$

The proof is complete. □

6. The "real" case

Let us introduce the operator analogues of real matrices, and operator-valued analogues of real matrix-valued functions. To this end, we define an anti-linear operator which generalizes the complex conjugation in $\mathbb{C}^n$.

The operator ι on a Hilbert space $\mathcal{U}$ is called an *anti-unitary involution* (AUI) if the following conditions are satisfied:

$$\iota^2 = I_{\mathcal{U}}; \tag{6.1}$$

$$\langle \iota u_1, \iota u_2 \rangle = \langle u_2, u_1 \rangle, \quad u_1 \in \mathcal{U}, \ u_2 \in \mathcal{U}. \tag{6.2}$$

Proposition 6.1. *An AUI* ι *has the following properties:*

(i) ι *is* **additive**, *i.e., for every* $u_1 \in \mathcal{U}$, $u_2 \in \mathcal{U}$ *one has* $\iota(u_1 + u_2) = \iota u_1 + \iota u_2$;

(ii) ι *is* **anti-homogeneous**, *i.e., for every* $u \in \mathcal{U}$, $\alpha \in \mathbb{C}$ *one has* $\iota(\alpha u) = \bar{\alpha}\iota u$;

(iii) *the operators* $\pi_\pm := \frac{I_{\mathcal{U}} \pm \iota}{2}$ *are idempotents, their ranges* $\pi_\pm \mathcal{U}$ *are closed in* $\mathcal{U}$, *and* $\pi_+\mathcal{U} \dotplus \pi_-\mathcal{U} = \mathcal{U}$.

Proof. Let $u_1 \in \mathcal{U},\ u_2 \in \mathcal{U}$. Then

$$\begin{aligned} \|\iota(u_1+u_2)-\iota u_1-\iota u_2\|^2 &= \|\iota(u_1+u_2)\|^2+\|\iota u_1\|^2+\|\iota u_2\|^2 \\ - &\ 2\mathrm{Re}\langle\iota(u_1+u_2),\iota u_1+\iota u_2\rangle+2\mathrm{Re}\langle\iota u_1,\iota u_2\rangle \\ = &\ \|u_1+u_2\|^2+\|u_1\|^2+\|u_2\|^2-2\mathrm{Re}\langle u_1,u_1+u_2\rangle-2\mathrm{Re}\langle u_2,u_1+u_2\rangle \\ + &\ 2\mathrm{Re}\langle u_2,u_1\rangle=0. \end{aligned}$$

Therefore, $\iota(u_1+u_2)-\iota u_1-\iota u_2=0$, that proves (i). Next, for every $u\in\mathcal{U},\ \alpha\in\mathbb{C}$ one has

$$\begin{aligned} \|\iota(\alpha u)-\bar{\alpha}\iota u\|^2 &= \|\iota(\alpha u)\|^2-2\mathrm{Re}\langle\iota(\alpha u),\bar{\alpha}\iota u\rangle+|\alpha|^2\|\iota u\|^2 \\ &= \|\alpha u\|^2-2\mathrm{Re}(\alpha\langle\iota(\alpha u),\iota u\rangle)+|\alpha|^2\|u\|^2 \\ &= 2|\alpha|^2\|u\|^2-2\mathrm{Re}(\alpha\langle u,\alpha u\rangle)=2|\alpha|^2\|u\|^2-2|\alpha|^2\|u\|^2=0, \end{aligned}$$

that proves (ii). Since

$$\pi_\pm^2=\frac{(I_\mathcal{U}\pm\iota)^2}{4}=\frac{(I_\mathcal{U}\pm2\iota+\iota^2)}{4}=\frac{I_\mathcal{U}\pm\iota}{2}=\pi_\pm,$$

the operators $\pi_\pm$ are idempotents. Since ι preserves the norm, ι is continuous, thus $\pi_\pm$ are also continuous operators. The latter means that if $\pi_\pm u_j\to y$ as $j\to\infty$ then $\pi_\pm u_j=\pi_\pm(\pi_\pm u_j)\to\pi_\pm y$, i.e., $y=\pi_\pm y\in\pi_\pm\mathcal{U}$. Therefore, $\pi_\pm\mathcal{U}$ are closed lineals in $\mathcal{U}$.

If $\pi_+u_1=\pi_-u_2$ then $\pi_+^2u_1=\pi_+\pi_-u_2$. Since $\pi_+^2=\pi_+$ and $\pi_+\pi_-=\frac{(I_\mathcal{U}+\iota)(I_\mathcal{U}-\iota)}{4}=\frac{I_\mathcal{U}-\iota^2}{4}=0$, one has $\pi_+u_1=\pi_-u_2=0$, i.e., $\pi_+\mathcal{U}\cap\pi_-\mathcal{U}=\{0\}$. Since for every $u\in\mathcal{U}$ one has $u=\pi_+u+\pi_-u$, and $\pi_\pm u\in\pi_\pm\mathcal{U}$, we get $\pi_+\mathcal{U}\dot{+}\pi_-\mathcal{U}=\mathcal{U}$. The proof is complete. □

Example 1. Let $\mathcal{U}=\mathbb{C}^n$. For $u=\mathrm{col}(u_1,\dots,u_n)\in\mathbb{C}^n$ denote $\bar{u}:=\mathrm{col}(\overline{u_1},\dots,\overline{u_n})$. Set $\iota u:=\bar{u}$, i.e, ι is the complex conjugation in $\mathbb{C}^n$. Then, clearly, ι is an AUI and

$$\pi_+u=\mathrm{Re}\,u=\mathrm{col}(\mathrm{Re}\,u_1,\dots,\mathrm{Re}\,u_n),\ \ \pi_-u=i\mathrm{Im}\,u=i\mathrm{col}(\mathrm{Im}\,u_1,\dots,\mathrm{Im}\,u_n).$$

The operator $A\in L(\mathcal{U})$ is called *ι-real* (resp., *ι-symmetric*) if $\iota A=A\iota$ (resp., $\iota A=A^*\iota$).

Example 2. Let $\mathcal{U}=\mathbb{C}^n$ and $\iota u=\bar{u}$, as in Example 1. Then the operator $A\in L(\mathcal{U})$ is ι-real (resp., ι-symmetric) if and only if its matrix in the standard basis $\{e_k\}_{k=1}^n$ has real entries (resp., is symmetric, i.e., $A^T=A$). Indeed, in the first case

$$\begin{aligned} a_{kj} &= \langle Ae_j,e_k\rangle=\langle\iota e_k,\iota Ae_j\rangle=\langle\iota e_k,A\iota e_j\rangle \\ &= \langle e_k,Ae_j\rangle=\overline{\langle Ae_j,e_k\rangle}=\overline{a_{kj}},\quad k,j=1,\dots,n; \end{aligned}$$

in the second case

$$\begin{aligned} a_{kj} &= \langle Ae_j,e_k\rangle=\langle\iota e_k,\iota Ae_j\rangle=\langle\iota e_k,A^*\iota e_j\rangle \\ &= \langle e_k,A^*e_j\rangle=\langle Ae_k,e_j\rangle=a_{jk},\quad k,j=1,\dots,n. \end{aligned}$$

Lemma 6.2. *Let $\iota=\iota_\mathcal{U}$ be an AUI on a Hilbert space $\mathcal{U}$, and $A\in L(\mathcal{U})$. For the following three conditions, any two of them imply the third one:*

(i) $\iota A = A\iota$;
(ii) $\iota A = A^*\iota$;
(iii) $A = A^*$.

Proof. (i) & (ii) $\Rightarrow$ (iii). If $\iota A = A\iota$ and $\iota A = A^*\iota$ then $A\iota = A^*\iota$. Therefore, $A\iota^2 = A^*\iota^2$, i.e., $A = A^*$.

Implications (i) & (iii) $\Rightarrow$ (ii) and (ii) & (iii) $\Rightarrow$ (i) are obvious. □

Let ι be an AUI on a Hilbert space $\mathcal{U}$, and $\Omega \subset \mathbb{C}^N$ be a domain invariant under the complex conjugation: $\overline{\Omega} = \Omega$. For a function $f : \Omega \to L(\mathcal{U})$ set $f^\sharp(z) = \iota f(\bar{z})\iota$, $z \in \Omega$. A function $f : \Omega \to L(\mathcal{U})$ is called *ι-real* if for every $z \in \Omega$ one has $f^\sharp(z) = f(z)$.

Example 3. Let $\mathcal{U} = \mathbb{C}^n$ and $\iota u = \bar{u}$, as in Examples 1 and 2, and $\Omega \subset \mathbb{C}^N$ be a domain satisfying $\overline{\Omega} = \Omega$. Then ι-real $L(\mathcal{U})$-valued functions are those matrix functions whose values in the standard basis $\{e_k\}_{k=1}^n$ satisfy the condition $f(\bar{z}) = \overline{f(z)}$. Indeed,

$$\begin{aligned}[f(\bar{z})]_{kj} &= \langle f(\bar{z})e_j, e_k\rangle = \langle f(\bar{z})\iota e_j, \iota e_k\rangle = \langle \iota^2 e_k, \iota f(\bar{z})\iota e_j\rangle \\ &= \langle e_k, f(z)e_j\rangle = \overline{\langle f(z)e_j, e_k\rangle} = \overline{[f(z)]_{kj}}, \quad k, j = 1, \ldots, n.\end{aligned}$$

Let $\mathcal{U}$ be a Hilbert space and $\iota = \iota_{\mathcal{U}}$ be an AUI on $\mathcal{U}$. Denote by $\iota\mathbb{R}\mathcal{P}_N(\mathcal{U})$ and $\iota\mathbb{R}\mathcal{B}_N(\mathcal{U})$ the subclasses of $\mathcal{P}_N(\mathcal{U})$ and $\mathcal{B}_N(\mathcal{U})$, respectively, consisting of ι-real functions. Clearly, $\iota\mathbb{R}\mathcal{B}_N(\mathcal{U}) \subset \iota\mathbb{R}\mathcal{P}_N(\mathcal{U})$.

Proposition 6.3. *The Taylor coefficients of functions from $\iota\mathbb{R}\mathcal{P}_N(\mathcal{U})$, and therefore, functions from $\iota\mathbb{R}\mathcal{B}_N(\mathcal{U})$ satisfy conditions* (i)–(iii) *of Lemma* 6.2.

Proof. According to Lemma 6.2, it is sufficient to verify any two of its conditions, e.g., (i) and (iii), for the Taylor coefficients of an arbitrary function $f \in \iota\mathbb{R}\mathcal{P}_N(\mathcal{U})$. It follows from Proposition 6.1 that

$$f^\sharp(z) = \iota f(\bar{z})\iota = \sum_{t\in\mathbb{Z}_+^N} \iota\widehat{f}_t(\bar{z} - \overline{z^\circ})^t\iota = \sum_{t\in\mathbb{Z}_+^N} \iota\widehat{f}_t\iota(z - z^\circ)^t$$

in some neighborhood Γ of $z^\circ \in \Omega_N$. Since f is ι-real,

$$f^\sharp(z) = f(z) = \sum_{t\in\mathbb{Z}_+^N} \widehat{f}_t(z - z^\circ)^t$$

in some neighborhood Γ' of $z^\circ \in \Omega_N$. By the uniqueness of Taylor's expansion in the neighborhood $\Gamma \cap \Gamma'$ of z°, we get $\iota\widehat{f}_t\iota = \widehat{f}_t$ and $\iota\widehat{f}_t = \widehat{f}_t\iota$, for every $t \in \mathbb{Z}_+^N$, i.e., condition (i) is satisfied for all of the Taylor coefficients of f. Since

$$f(\bar{z})^* = \left(\sum_{t\in\mathbb{Z}_+^N} \widehat{f}_t(\bar{z} - \overline{z^\circ})^t\right)^* = \sum_{t\in\mathbb{Z}_+^N} \widehat{f}_t^{\,*}(z - z^\circ)^t$$

in Γ, and

$$f(\bar{z})^* = f(z) = \sum_{t \in \mathbb{Z}_+^N} \widehat{f}_t (z - z^\circ)^t$$

in Γ', by the uniqueness of Taylor's expansion in the neighborhood $\Gamma \cap \Gamma'$ of z°, we get $\widehat{f_t}^* = \widehat{f_t}$ for every $t \in \mathbb{Z}_+^N$, i.e, condition (iii) is also satisfied for all of the Taylor coefficients of f. □

Theorem 6.4. *Let f be a holomorphic $L(\mathcal{U})$-valued function on Ω_N, and $\iota = \iota_{\mathcal{U}}$ be an AUI on a Hilbert space $\mathcal{U}$. The following statements are equivalent:*

- (i) $f \in \iota\mathbb{R}\mathcal{B}_N(\mathcal{U})$;
- (ii) *there exist a representation* (2.5) *of f and an AUI $\iota_{\mathcal{H}}$ on $\mathcal{H}$ such that the function $A(z)$ given by* (2.6) *is $(\iota_{\mathcal{U}} \oplus \iota_{\mathcal{H}})$-real;*
- (iii) *there exists a representation* (2.11) *of f for which the holomorphic PSD kernels $\Phi_k(z,\zeta)$, $k = 1,\dots,N$, on $\Pi^N \times \Pi^N$ are $\iota_{\mathcal{U}}$-real;*
- (iv) *there exist an Agler unitary colligation $\alpha = (N; U; \mathcal{X} = \bigoplus_{k=1}^N \mathcal{X}_k, \mathcal{U}, \mathcal{U})$ with $U = U^*$, the corresponding representation* (4.3) *of $\mathcal{F} = \mathcal{C}(f)$ (the latter is given by* (4.1)*) and an AUI $\iota_{\mathcal{X}}$ on $\mathcal{X}$ which commutes with the orthogonal projectors $P_k = P_{\mathcal{X}_k}$, $k = 1,\dots,N$, such that the operator U is $(\iota_{\mathcal{X}} \oplus \iota_{\mathcal{U}})$-real.*

Proof. (i)⇒(iii). By Theorem 2.7 there exists a representation (2.11) of f with holomorphic PSD kernels $\Phi_k(z,\zeta)$, $k = 1,\dots,N$, on $\Pi^N \times \Pi^N$. Define

$$\Phi_k^\circ(z,\zeta) := \frac{\Phi_k(z,\zeta) + \Phi_k^\sharp(z,\zeta)}{2}, \quad (z,\zeta) \in \Pi^N \times \Pi^N, \ k = 1,\dots,N.$$

For every $u_1 \in \mathcal{U}$, $u_2 \in \mathcal{U}$, $k = 1,\dots,N$, one has

$$\begin{aligned} \langle \Phi_k^\sharp(z,\overline{\zeta})u_1, u_2\rangle &= \langle \iota\Phi_k(\bar{z},\zeta)\iota u_1, u_2\rangle = \langle \iota u_2, \Phi_k(\bar{z},\zeta)\iota u_1\rangle \\ &= \langle \Phi_k(\bar{z},\zeta)^*\iota u_2, \iota u_1\rangle = \langle \Phi_k(\zeta,\bar{z})\iota u_2, \iota u_1\rangle, \end{aligned}$$

which is, clearly, a holomorphic function in $(z,\zeta) \in \Pi^N \times \Pi^N$ (we used the property (2.10) of PSD kernels). Thus, $\Phi_k^\sharp(z,\overline{\zeta})$ and, therefore, $\Phi_k^\circ(z,\overline{\zeta})$ are holomorphic operator-valued functions on $\Pi^N \times \Pi^N$.

Let $m \in \mathbb{N}$, $z^{(1)},\dots,z^{(m)} \in \Pi^N$, $u^{(1)},\dots,u^{(m)} \in \mathcal{U}$. Then for $k = 1,\dots,N$ one has

$$\begin{aligned} \sum_{\mu=1}^m\sum_{\nu=1}^m \langle \Phi_k^\sharp(z^{(\mu)}, z^{(\nu)})u^{(\mu)}, u^{(\nu)}\rangle &= \sum_{\mu=1}^m\sum_{\nu=1}^m \langle \iota\Phi_k(\overline{z^{(\mu)}}, \overline{z^{(\nu)}})\iota u^{(\mu)}, u^{(\nu)}\rangle \\ \sum_{\mu=1}^m\sum_{\nu=1}^m \langle \iota u^{(\nu)}, \Phi_k(\overline{z^{(\mu)}}, \overline{z^{(\nu)}})\iota u^{(\mu)}\rangle &= \sum_{\mu=1}^m\sum_{\nu=1}^m \langle \iota u^{(\nu)}, \Phi_k(\overline{z^{(\nu)}}, \overline{z^{(\mu)}})^*\iota u^{(\mu)}\rangle \\ &= \sum_{\mu=1}^m\sum_{\nu=1}^m \langle \Phi_k(\overline{z^{(\nu)}}, \overline{z^{(\mu)}})\iota u^{(\nu)}, \iota u^{(\mu)}\rangle \geq 0, \end{aligned}$$

by virtue of (2.9) for $\Phi_k(z,\zeta)$. Thus, $\Phi_k^\sharp(z,\zeta)$ and, therefore, $\Phi_k^\circ(z,\zeta)$ satisfy condition (2.9). Finally, we have proved that $\Phi_k^\sharp(z,\zeta)$ and, therefore, $\Phi_k^\circ(z,\zeta)$ are holomorphic PSD kernels on $\Pi^N \times \Pi^N$.

Since $f^\sharp(z) = \iota f(\bar{z})\iota = f(z)$, the identity (2.11) together with properties (i) and (ii) in Proposition 6.1 of ι imply

$$\begin{aligned} f(z) &= \iota\left(\sum_{k=1}^N \overline{z_k}\Phi_k(\bar{z},\bar{\zeta})\right)\iota = \sum_{k=1}^N z_k \iota\Phi_k(\bar{z},\bar{\zeta})\iota \\ &= \sum_{k=1}^N z_k\Phi_k^\sharp(z,\zeta), \quad (z,\zeta)\in\Pi^N\times\Pi^N. \end{aligned}$$

Therefore, $f(z) = \sum_{k=1}^N z_k\Phi_k^\circ(z,\zeta), \quad (z,\zeta)\in\Pi^N\times\Pi^N$. Moreover,

$$\Phi_k^{\circ\sharp}(z,\zeta) = \frac{\Phi_k^\sharp(z,\zeta)+\Phi_k^{\sharp\sharp}(z,\zeta)}{2} = \frac{\Phi_k^\sharp(z,\zeta)+\Phi_k(z,\zeta)}{2} = \Phi_k^\circ(z,\zeta).$$

Thus, (2.11) holds with $\iota_{\mathcal{U}}$-real holomorphic PSD kernels $\Phi_k^\circ(z,\zeta),\ k=1,\dots,N$, and (iii) follows from (i).

(iii)$\Rightarrow$(ii). Recall that holomorphic PSD kernels $\Phi_k(z,\zeta)$ admit factorizations $\Phi_k(z,\zeta) = \varphi_k(\zeta)^*\varphi_k(z)$, where φ_k are holomorphic $L(\mathcal{U},\mathcal{M}_k)$-valued functions on Π^N, and $\mathcal{M}_k$ are auxiliary Hilbert spaces, $k=1,\dots,N$. These $\mathcal{M}_k$'s and φ_k's can be determined in the following way. Set $\widetilde{\Phi_k}(z,\zeta) := \Phi_k(z,\bar{\zeta})^*$. It is easy to convince oneself that $\widetilde{\Phi_k}(z,\zeta)$ are holomorphic PSD kernels on $\Pi^N\times\Pi^N$. Then set $\mathcal{M}_k := \mathcal{H}_{\widetilde{\Phi_k}}$, where $\mathcal{H}_{\widetilde{\Phi_k}}$ denotes the Hilbert space with the reproducing kernel $\widetilde{\Phi_k}(z,\zeta)$, which is obtained by completion of the linear span of functions of the form $\widetilde{\Phi_k}(\cdot,\zeta)u,\ \zeta\in\Pi^N,\ u\in\mathcal{U}$, with respect to the inner product

$$\langle\widetilde{\Phi_k}(\cdot,\zeta)u,\widetilde{\Phi_k}(\cdot,\zeta')u'\rangle := \langle\widetilde{\Phi_k}(\zeta',\zeta)u,u'\rangle.$$

Clearly, $\widetilde{\Phi_k}(z,\zeta) = H_k(z)H_k(\zeta)^*,\ (z,\zeta)\in\Pi^N\times\Pi^N$, where H_k is a holomorphic $L(\mathcal{M}_k,\mathcal{U})$-valued function on Π^N defined by $H_k(z)^*u := \widetilde{\Phi_k}(\cdot,z)u,\ u\in\mathcal{U}$. Set $\varphi_k(z) := \breve{H}_k(z) = H_k(\bar{z})^*$. Then for $u\in\mathcal{U}$ one has $\varphi_k(z)u = \widetilde{\Phi_k}(\cdot,\bar{z})u$, and

$$\Phi_k(z,\zeta) = \widetilde{\Phi_k}(\bar{z},\bar{\zeta})^* = H_k(\bar{\zeta})H_k(\bar{z})^* = \varphi_k(\zeta)^*\varphi_k(z).$$

Define the AUI $\iota_{\mathcal{M}_k}:\mathcal{M}_k\to\mathcal{M}_k$ on generating vectors by

$$\begin{aligned} (\iota_{\mathcal{M}_k}[\widetilde{\Phi_k}(\cdot,\zeta)u])(z) &:= \iota_{\mathcal{U}}\widetilde{\Phi_k}(\bar{z},\zeta)u \qquad (6.3)\\ &(= \iota_{\mathcal{U}}\Phi_k(z,\bar{\zeta})^*u = \iota_{\mathcal{U}}\Phi_k(\bar{\zeta},z)u = \Phi_k(\zeta,\bar{z})\iota_{\mathcal{U}}u \\ &= \widetilde{\Phi_k}(\bar{\zeta},z)^*\iota_{\mathcal{U}}u = \widetilde{\Phi_k}(z,\bar{\zeta})\iota_{\mathcal{U}}u),\ (z,\zeta)\in\Pi^N\times\Pi^N, u\in\mathcal{U}. \end{aligned}$$

This definition is correct. Indeed,

$$\begin{aligned}\langle\iota_{\mathcal{M}_k}\widetilde{\Phi_k}(\cdot,\zeta)u,\iota_{\mathcal{M}_k}\widetilde{\Phi_k}(\cdot,\zeta')u'\rangle &= \langle\widetilde{\Phi_k}(\cdot,\bar{\zeta})\iota_{\mathcal{U}}u,\widetilde{\Phi_k}(\cdot,\overline{\zeta'})\iota_{\mathcal{U}}u'\rangle = \langle\widetilde{\Phi_k}(\overline{\zeta'},\bar{\zeta})\iota_{\mathcal{U}}u,\iota_{\mathcal{U}}u'\rangle \\ = \langle u',\iota_{\mathcal{U}}\widetilde{\Phi_k}(\overline{\zeta'},\bar{\zeta})\iota_{\mathcal{U}}u\rangle &= \langle u',\widetilde{\Phi_k}(\zeta',\zeta)u\rangle = \langle u',\widetilde{\Phi_k}(\zeta,\zeta')^*u\rangle = \langle\widetilde{\Phi_k}(\zeta,\zeta')u',u\rangle \\ &= \langle\widetilde{\Phi_k}(\cdot,\zeta')u',\widetilde{\Phi_k}(\cdot,\zeta)u\rangle.\end{aligned}$$

Therefore, $\iota_{\mathcal{M}_k}$ preserves the norm of any vector of the form $\sum_{\mu=1}^m \widetilde{\Phi_k}(\cdot,\zeta^{(\mu)})u^{(\mu)}$. The density of such vectors in $\mathcal{M}_k$ implies

$$\langle\iota_{\mathcal{M}_k}m,\iota_{\mathcal{M}_k}m'\rangle = \langle m',m\rangle \quad m\in\mathcal{M}_k, m'\in\mathcal{M}_k,$$

that is an analogue of (6.2). Next,

$$\begin{aligned}(\iota^2_{\mathcal{M}_k}[\widetilde{\Phi_k}(\cdot,\zeta)u])(z) &= (\iota_{\mathcal{M}_k}[\widetilde{\Phi_k}(\cdot,\bar{\zeta})\iota_{\mathcal{U}}u])(z) = \widetilde{\Phi_k}(z,\zeta)\iota^2_{\mathcal{U}}u \\ &= \widetilde{\Phi_k}(z,\zeta)u, \quad (z,\zeta)\in\Pi^N\times\Pi^N, u\in\mathcal{U}.\end{aligned}$$

Therefore, by the additivity of $\iota_{\mathcal{M}_k}$ (and $\iota^2_{\mathcal{M}_k}$) and the linearity and continuity ($\iota_{\mathcal{M}_k}$ is norm-preserving!) argument, we get $\iota^2_{\mathcal{M}_k} = I_{\mathcal{M}_k}$, i.e., an analogue of (6.1).

The identities in (6.3) imply

$$\iota_{\mathcal{M}_k}\varphi_k(z) = \varphi_k(\bar{z})\iota_{\mathcal{U}}, \quad z\in\Pi^N. \tag{6.4}$$

Following the sufficiency part of the proof of Theorem 2.7, we obtain $\varphi(z) = \mathrm{col}[\ \varphi_1(z)\ \ \dots\ \ \varphi_N(z)\ ]$,

$$\mathcal{H} = \mathrm{clos\ span}_{z\in\Pi^N}\{[\varphi(z)-\varphi(e)]\mathcal{U}\}\subset\mathcal{M} = \bigoplus_{k=1}^N\mathcal{M}_k,$$

$$A_k = \begin{bmatrix}\varphi(e)^* & 0\\ 0 & I_{\mathcal{H}}\end{bmatrix}\kappa^*P_k\kappa\begin{bmatrix}\varphi(e) & 0\\ 0 & I_{\mathcal{H}}\end{bmatrix}\in L(\mathcal{U}\oplus\mathcal{H})$$

(here κ is defined by (2.16)) such that (2.5) holds for f, where

$$A_k = \begin{bmatrix}a_k & b_k\\ c_k & d_k\end{bmatrix}\in L(\mathcal{U}\oplus\mathcal{H}), \quad k=1,\dots,N.$$

For $\psi(z) = \begin{bmatrix}I_{\mathcal{U}}\\ \varphi(z)-\varphi(e)\end{bmatrix}$ one has $\psi(e)\mathcal{U} = \mathcal{U}\oplus\{0\}$, therefore the linear span of vectors of the form $\psi(z)u$, $z\in\Pi^N$, $u\in\mathcal{U}$, is dense in $\mathcal{U}\oplus\mathcal{H}$. Set $\iota_{\mathcal{M}} := \bigoplus_{k=1}^N\iota_{\mathcal{M}_k}$. By virtue of (6.4), one has $\iota_{\mathcal{M}}(\varphi(z)-\varphi(e))u = (\varphi(\bar{z})-\varphi(e))\iota_{\mathcal{U}}u\in\mathcal{H}$, hence $\iota_{\mathcal{M}}\mathcal{H}\subset\mathcal{H}$. Set $\iota_{\mathcal{H}} := \iota_{\mathcal{M}}|\mathcal{H}$. Clearly, $\iota_{\mathcal{H}}$ is an AUI on $\mathcal{H}$, and $(\iota_{\mathcal{U}}\oplus\iota_{\mathcal{H}})\psi(z) = \psi(\bar{z})\iota_{\mathcal{U}}, z\in\Pi^N$.

Let us verify $(\iota_{\mathcal{U}} \oplus \iota_{\mathcal{H}})A_k(\iota_{\mathcal{U}} \oplus \iota_{\mathcal{H}}) = A_k$, i.e., the $(\iota_{\mathcal{U}} \oplus \iota_{\mathcal{H}})$-realness of A_k, $k = 1, \dots, N$. For every $z \in \Pi^N$, $\zeta \in \Pi^N$, $u \in \mathcal{U}$, $u' \in \mathcal{U}$ one has

$$\left\langle A_k \begin{bmatrix} I_{\mathcal{U}} \\ \varphi(z) - \varphi(e) \end{bmatrix} u, \begin{bmatrix} I_{\mathcal{U}} \\ \varphi(\zeta) - \varphi(e) \end{bmatrix} u' \right\rangle$$
$$= \left\langle \begin{bmatrix} \varphi(e)^* & 0 \\ 0 & I_{\mathcal{H}} \end{bmatrix} \kappa^* P_k \kappa \begin{bmatrix} \varphi(e) & 0 \\ 0 & I_{\mathcal{H}} \end{bmatrix} \begin{bmatrix} I_{\mathcal{U}} \\ \varphi(z) - \varphi(e) \end{bmatrix} u, \begin{bmatrix} I_{\mathcal{U}} \\ \varphi(\zeta) - \varphi(e) \end{bmatrix} u' \right\rangle$$
$$= \left\langle P_k \kappa \begin{bmatrix} \varphi(e) & 0 \\ 0 & I_{\mathcal{H}} \end{bmatrix} \begin{bmatrix} I_{\mathcal{U}} \\ \varphi(z) - \varphi(e) \end{bmatrix} u, \kappa \begin{bmatrix} \varphi(e) & 0 \\ 0 & I_{\mathcal{H}} \end{bmatrix} \begin{bmatrix} I_{\mathcal{U}} \\ \varphi(\zeta) - \varphi(e) \end{bmatrix} u' \right\rangle$$
$$= \left\langle P_k \kappa \begin{bmatrix} \varphi(e) \\ \varphi(z) - \varphi(e) \end{bmatrix} u, \kappa \begin{bmatrix} \varphi(e) \\ \varphi(\zeta) - \varphi(e) \end{bmatrix} u' \right\rangle = \langle P_k \varphi(z) u, \varphi(\zeta) u' \rangle$$
$$= \langle P_k \varphi(z) u, P_k \varphi(\zeta) u' \rangle = \langle \varphi_k(z) u, \varphi_k(\zeta) u' \rangle;$$

$$\left\langle (\iota_{\mathcal{U}} \oplus \iota_{\mathcal{H}}) A_k (\iota_{\mathcal{U}} \oplus \iota_{\mathcal{H}}) \begin{bmatrix} I_{\mathcal{U}} \\ \varphi(z) - \varphi(e) \end{bmatrix} u, \begin{bmatrix} I_{\mathcal{U}} \\ \varphi(\zeta) - \varphi(e) \end{bmatrix} u' \right\rangle$$
$$= \left\langle (\iota_{\mathcal{U}} \oplus \iota_{\mathcal{H}}) \begin{bmatrix} I_{\mathcal{U}} \\ \varphi(\zeta) - \varphi(e) \end{bmatrix} u', A_k (\iota_{\mathcal{U}} \oplus \iota_{\mathcal{H}}) \begin{bmatrix} I_{\mathcal{U}} \\ \varphi(z) - \varphi(e) \end{bmatrix} u \right\rangle$$
$$= \left\langle \begin{bmatrix} \iota_{\mathcal{U}} \\ \iota_{\mathcal{H}}(\varphi(\zeta) - \varphi(e)) \end{bmatrix} u', A_k \begin{bmatrix} \iota_{\mathcal{U}} \\ \iota_{\mathcal{H}}(\varphi(z) - \varphi(e)) \end{bmatrix} u \right\rangle$$
$$= \left\langle \begin{bmatrix} I_{\mathcal{U}} \\ \varphi(\bar{\zeta}) - \varphi(e) \end{bmatrix} \iota_{\mathcal{U}} u', A_k \begin{bmatrix} I_{\mathcal{U}} \\ \varphi(\bar{z}) - \varphi(e) \end{bmatrix} \iota_{\mathcal{U}} u \right\rangle$$
$$= \left\langle A_k \begin{bmatrix} I_{\mathcal{U}} \\ \varphi(\bar{\zeta}) - \varphi(e) \end{bmatrix} \iota_{\mathcal{U}} u', \begin{bmatrix} I_{\mathcal{U}} \\ \varphi(\bar{z}) - \varphi(e) \end{bmatrix} \iota_{\mathcal{U}} u \right\rangle = \langle \varphi_k(\bar{\zeta}) \iota_{\mathcal{U}} u', \varphi_k(\bar{z}) \iota_{\mathcal{U}} u \rangle$$
$$= \langle \iota_{\mathcal{M}_k} \varphi_k(\zeta) u', \iota_{\mathcal{M}_k} \varphi_k(z) u \rangle = \langle \varphi_k(z) u, \varphi_k(\zeta) u' \rangle$$

(in the second chain of calculations we used the result of the first one and (6.4)). As mentioned above, the linear span of vectors of the form $\begin{bmatrix} I_{\mathcal{U}} \\ \varphi(z) - \varphi(e) \end{bmatrix} u$, $z \in \Pi^N$, $u \in \mathcal{U}$, is dense in $\mathcal{U} \oplus \mathcal{H}$. Operators A_k and $(\iota_{\mathcal{U}} \oplus \iota_{\mathcal{H}})A_k(\iota_{\mathcal{U}} \oplus \iota_{\mathcal{H}})$ are continuous and linear (the second operator is additive because A_k and $\iota_{\mathcal{U}} \oplus \iota_{\mathcal{H}}$ are additive, and homogeneous because A_k is homogeneous, and $\iota_{\mathcal{U}} \oplus \iota_{\mathcal{H}}$ is anti-homogeneous and appears twice). Therefore, by comparison the results of the two chains of calculations above, and linearity and continuity argument we obtain $(\iota_{\mathcal{U}} \oplus \iota_{\mathcal{H}})A_k(\iota_{\mathcal{U}} \oplus \iota_{\mathcal{H}}) = A_k$, $k = 1, \dots, N$. Thus, (ii) follows from (iii).

(ii)⇒(i). Let f satisfies (ii). Then $f \in \mathcal{B}_N(\mathcal{U})$, and the operator-valued linear function $A(z)$ is $\iota_{\mathcal{U}} \oplus \iota_{\mathcal{H}}$-real, i.e.,

$$\begin{bmatrix} \iota_{\mathcal{U}} & 0 \\ 0 & \iota_{\mathcal{H}} \end{bmatrix} \begin{bmatrix} a(\bar{z}) & b(\bar{z}) \\ c(\bar{z}) & d(\bar{z}) \end{bmatrix} \begin{bmatrix} \iota_{\mathcal{U}} & 0 \\ 0 & \iota_{\mathcal{H}} \end{bmatrix} = \begin{bmatrix} a(z) & b(z) \\ c(z) & d(z) \end{bmatrix}, \quad z \in \Omega_N.$$

The latter is equivalent to the identities

$$\begin{matrix} \iota_{\mathcal{U}} a(\bar{z}) \iota_{\mathcal{U}} = a(z), & \iota_{\mathcal{U}} b(\bar{z}) \iota_{\mathcal{H}} = b(z), \\ \iota_{\mathcal{H}} c(\bar{z}) \iota_{\mathcal{U}} = c(z) & \iota_{\mathcal{H}} d(\bar{z}) \iota_{\mathcal{H}} = d(z), \end{matrix} \quad z \in \Omega_N.$$

Since $\iota_{\mathcal{H}}^2 = I_{\mathcal{H}}$, and

$$(\iota_{\mathcal{H}} d(\bar{z})^{-1}\iota_{\mathcal{H}}) \cdot (\iota_{\mathcal{H}} d(\bar{z})\iota_{\mathcal{H}}) = (\iota_{\mathcal{H}} d(\bar{z})\iota_{\mathcal{H}}) \cdot (\iota_{\mathcal{H}} d(\bar{z})^{-1}\iota_{\mathcal{H}}) = I_{\mathcal{H}},$$

one has $\iota_{\mathcal{H}} d(\bar{z})^{-1}\iota_{\mathcal{H}} = (\iota_{\mathcal{H}} d(\bar{z})\iota_{\mathcal{H}})^{-1} = d(z)^{-1}$. Therefore,

$$\begin{aligned} f^{\sharp}(z) &= \iota_{\mathcal{U}} f(\bar{z})\iota_{\mathcal{U}} = \iota_{\mathcal{U}}(a(\bar{z}) - b(\bar{z})d(\bar{z})^{-1}c(\bar{z}))\iota_{\mathcal{U}} \\ &= \iota_{\mathcal{U}} a(\bar{z})\iota_{\mathcal{U}} - (\iota_{\mathcal{U}} b(\bar{z})\iota_{\mathcal{H}}) \cdot (\iota_{\mathcal{H}} d(\bar{z})^{-1}\iota_{\mathcal{H}}) \cdot (\iota_{\mathcal{H}} c(\bar{z})\iota_{\mathcal{U}}) \\ &= a(z) - b(z)d(z)^{-1}c(z) = f(z), \end{aligned}$$

i.e., f is $\iota_{\mathcal{U}}$-real. Thus, (i) follows from (ii).

(iv)⇒(i). Let (iv) hold. Then the operator $U = U^* = U^{-1}$ is $(\iota_{\mathcal{X}} \oplus \iota_{\mathcal{U}})$-real (and, by the way, $(\iota_{\mathcal{X}} \oplus \iota_{\mathcal{U}})$-symmetric due to Lemma 6.2), i.e.,

$$\begin{bmatrix} \iota_{\mathcal{X}} & 0 \\ 0 & \iota_{\mathcal{U}} \end{bmatrix} \begin{bmatrix} A & B \\ C & D \end{bmatrix} \begin{bmatrix} \iota_{\mathcal{X}} & 0 \\ 0 & \iota_{\mathcal{U}} \end{bmatrix} = \begin{bmatrix} A & B \\ C & D \end{bmatrix}.$$

This is equivalent to the following identities:

$$\iota_{\mathcal{X}} A \iota_{\mathcal{X}} = A, \quad \iota_{\mathcal{X}} B \iota_{\mathcal{U}} = B, \quad \iota_{\mathcal{U}} C \iota_{\mathcal{X}} = C, \quad \iota_{\mathcal{U}} D \iota_{\mathcal{U}} = D.$$

Moreover, since $\iota_{\mathcal{X}}$ commutes with $P_k,\ k = 1, \ldots, N$, one has $\iota_{\mathcal{X}}(I_{\mathcal{X}} - AP(w))\iota_{\mathcal{X}} = I_{\mathcal{X}} - AP(\bar{w})$, and $\iota_{\mathcal{X}}(I_{\mathcal{X}} - AP(w))^{-1}\iota_{\mathcal{X}} = (I_{\mathcal{X}} - AP(\bar{w}))^{-1}$ (we already used an analogous argument above). Therefore,

$$\begin{aligned} \mathcal{F}^{\sharp}(w) &= \iota_{\mathcal{U}} \mathcal{F}(\bar{w})\iota_{\mathcal{U}} = \iota_{\mathcal{U}}[D + CP(\bar{w})(I_{\mathcal{X}} - AP(\bar{w}))^{-1}B]\iota_{\mathcal{U}} \\ &= D + CP(w)(I_{\mathcal{X}} - AP(w))^{-1}B = \mathcal{F}(w), \quad w \in \mathbb{D}^N, \end{aligned}$$

i.e., $\mathcal{F}$ is $\iota_{\mathcal{U}}$-real. Applying the inverse double Cayley transform to $\mathcal{F}$, one can see that f is also $\iota_{\mathcal{U}}$-real on Π^N, and hence, on Ω_N. Thus, (i) follows from (iv).

(iii)⇒(iv). Let f satisfy the identity (2.11) with holomorphic $\iota_{\mathcal{U}}$-real PSD kernels $\Phi_k(z,\zeta),\ k = 1, \ldots, N$, on $\Pi^N \times \Pi^N$. Arguing like in the proof of (iii)⇒(ii) above, we get Hilbert spaces $\mathcal{M}_k$, holomorphic $L(\mathcal{U}, \mathcal{M}_k)$-valued functions φ_k on Π^N, such that $\Phi_k(z,\zeta) = \varphi_k(\zeta)^*\varphi_k(z),\ (z,\zeta) \in \Pi^N \times \Pi^N$, AUIs $\iota_{\mathcal{M}_k}$ on $\mathcal{M}_k,\ k = 1, \ldots, N$, and $\iota_{\mathcal{M}} = \bigoplus_{k=1}^N \iota_{\mathcal{M}_k}$ on $\mathcal{M} = \bigoplus_{k=1}^N \mathcal{M}_k$, for which (6.4) holds, and hence $\iota_{\mathcal{M}}\varphi(z) = \varphi(\bar{z})\iota_{\mathcal{U}},\ z \in \Pi^N$. Following the proof of Theorem 4.2, we get consecutively: identities (3.9) and (4.4) with holomorphic PSD kernels $\Xi_k(w,\omega)$ such that $\Xi_k(w,\omega) = \xi_k(\omega)^*\xi_k(w),\ (w,\omega) \in \mathbb{D}^N \times \mathbb{D}^N$, and $\xi_k,\ k = 1, \ldots, N$, are given in (3.11), moreover $\iota_{\mathcal{M}}\xi(w) = \xi(\bar{w})\iota_{\mathcal{U}},\ w \in \mathbb{D}^N$; then identities (4.5) and (4.9) with holomorphic PSD kernels $\Theta_k(w,\omega)$ such that $\Theta_k(w,\omega) = \theta_k(\omega)^*\theta_k(w),\ (w,\omega) \in \mathbb{D}^N \times \mathbb{D}^N$, and $\theta_k,\ k = 1, \ldots, N$, are given in (4.8), moreover $\iota_{\mathcal{M}}\theta(w) = \theta(\bar{w})\iota_{\mathcal{U}},\ w \in \mathbb{D}^N$; then identities (4.10) and (4.11) with holomorphic functions $\widetilde{\theta_k}(w) = \theta_k(\bar{w})^*,\ w \in \mathbb{D}^N$, taking values in $L(\mathcal{M}_k, \mathcal{U})$, moreover, $\iota_{\mathcal{U}}\widetilde{\theta_k}(w) = \widetilde{\theta_k}(\bar{w})\iota_{\mathcal{M}_k},\ w \in \mathbb{D}^N$. The latter equality is valid since for every $m \in \mathcal{M}_k,\ u \in \mathcal{U}$ one has

$$\begin{aligned} \langle \iota_{\mathcal{U}}\widetilde{\theta_k}(w)m, u\rangle &= \langle \iota_{\mathcal{U}}\theta_k(\bar{w})^*m, u\rangle = \langle \iota_{\mathcal{U}}u, \theta_k(\bar{w})^*m\rangle = \langle \theta_k(\bar{w})\iota_{\mathcal{U}}u, m\rangle \\ = \langle \iota_{\mathcal{M}_k}\theta_k(w)u, m\rangle &= \langle \iota_{\mathcal{M}_k}m, \theta_k(w)u\rangle = \langle \theta_k(w)^*\iota_{\mathcal{M}_k}m, u\rangle = \langle \widetilde{\theta_k}(\bar{w})\iota_{\mathcal{M}_k}m, u\rangle. \end{aligned}$$

Denote $\widetilde{\theta}(w) := [\ \widetilde{\theta}_1(w) \ \ \dots \ \ \widetilde{\theta_N}(w)\ ]$. Then $\iota_{\mathcal{U}}\widetilde{\theta}(w) = \widetilde{\theta}(\bar{w})\iota_{\mathcal{M}},\ w \in \mathbb{D}^N$. For the reproducing kernels $K_k(w,\omega) = \widetilde{\theta}(w)\widetilde{\theta}(\omega)^*$ of the spaces $\mathcal{M}_k$ we get the identities $\iota_{\mathcal{U}}K_k(\bar{w},\bar{\omega})\iota_{\mathcal{U}} = K_k(w,\omega),\ (w,\omega) \in \mathbb{D}^N \times \mathbb{D}^N,\ k = 1,\dots,N$. Define in the space $\mathcal{X} = \bigoplus_{k=1}^N \mathcal{D}(\widehat{K_k})$, where $\mathcal{D}(\widehat{K_k}),\ k = 1,\dots,N$, are Hilbert spaces with the reproducing kernels

$$\widehat{K_k}(w,\omega) = \begin{bmatrix} K_k(w,\omega) & K_k(w,\omega) \\ K_k(w,\omega) & K_k(w,\omega) \end{bmatrix},$$

the operator $\iota_{\mathcal{X}} := \bigoplus_{k=1}^N(\iota_{\mathcal{M}_k} \oplus \iota_{\mathcal{M}_k})$. Then $\iota_{\mathcal{X}}$ is an AUI on $\mathcal{X}$, and $\iota_{\mathcal{X}}P_k = P_k\iota_{\mathcal{X}},\ k = 1,\dots,N$. Moreover, it is easy to see that $(\iota_{\mathcal{X}}\oplus\iota_{\mathcal{U}})U_0 = U_0(\iota_{\mathcal{X}}\oplus\iota_{\mathcal{U}})$, and therefore $(\iota_{\mathcal{X}}\oplus\iota_{\mathcal{U}})\widetilde{U_0} = \widetilde{U_0}(\iota_{\mathcal{X}}\oplus\iota_{\mathcal{U}})$. It is clear that $\mathcal{D}_0$ is invariant under $\iota_{\mathcal{X}}\oplus\iota_{\mathcal{U}}$, as well as $\mathcal{D}_0^\perp = (\mathcal{X}\oplus\mathcal{U})\ominus\overline{\mathcal{D}_0}$ (in fact, since $\mathcal{D}_0 \supset \mathcal{U}$, one has $(\mathcal{X}\oplus\mathcal{U})\ominus\overline{\mathcal{D}_0} \subset \mathcal{X}$, and $\mathcal{D}_0^\perp$ is invariant under $\iota_{\mathcal{X}}$). Indeed, for every $h_1 \in \mathcal{D}_0,\ h_2 \in \mathcal{D}_0^\perp$ one has

$$\langle(\iota_{\mathcal{X}}\oplus\iota_{\mathcal{U}})h_2, h_1\rangle = \langle(\iota_{\mathcal{X}}\oplus\iota_{\mathcal{U}})h_1, h_2\rangle = 0,$$

since $(\iota_{\mathcal{X}}\oplus\iota_{\mathcal{U}})h_1 \in \mathcal{D}_0$, thus $(\iota_{\mathcal{X}}\oplus\iota_{\mathcal{U}})h_2 \in \mathcal{D}_0^\perp$. As $U = \widetilde{U}_0 \oplus I_{\mathcal{D}_0^\perp}$, we get $(\iota_{\mathcal{X}}\oplus\iota_{\mathcal{U}})U = U(\iota_{\mathcal{X}}\oplus\iota_{\mathcal{U}})$. So, we see that (iv) follows from (iii).

The proof is complete. □

7. Conclusion and open problems

In this paper the class $\mathbb{R}\mathcal{B}_N^{n\times n}$, which was defined by M. F. Bessmertnyĭ in [9] (see also [10]) as a class of rational $n\times n$ matrix-valued functions having a long resolvent representation (1.1) with matrix coefficients $A_0 = 0,\ A_k = A_k^T \geq 0,\ k = 1,\dots,N$, in (1.2) (note, that matrices A_k have real entries), was generalized in several directions simultaneously. First, one can consider the class $\mathcal{B}_N^{n\times n} = \mathbb{C}\mathcal{B}_N^{n\times n}$ of rational $n\times n$ matrix-valued functions having a long resolvent representation (1.1) with matrix coefficients $A_0 = 0,\ A_k = A_k^* \geq 0,\ k = 1,\dots,N$, in (1.2) (i.e., entries of matrices A_k are complex, not necessarily real). Second, we have introduced the class $\mathcal{B}_N(\mathcal{U})$ (which is a generalization of $\mathcal{B}_N^{n\times n}$) consisting of holomorphic (not necessarily rational) functions on the open right polyhalfplane Π^N (and naturally extendable to the domain $\Omega_N = \bigcup_{\lambda\in\mathbb{T}}(\lambda\Pi)^N)$ which take values in $L(\mathcal{U})$ for a (not necessarily finite dimensional) Hilbert space $\mathcal{U}$ and having representations of the form (1.1) with coefficients $A_0 = 0,\ A_k = A_k^* \geq 0,\ k = 1,\dots,N$, in (1.2), which are linear bounded operators on $\mathcal{U}\oplus\mathcal{H}$, where a Hilbert space $\mathcal{H}$ is, generally speaking, not supposed to be finite-dimensional. We have obtained several equivalent characterizations, scattered in different parts of this paper, of the class $\mathcal{B}_N(\mathcal{U})$, which we call the Bessmertnyĭ class, and for convenience of a reader we collect them now in the following theorem.

Theorem 7.1. *Let f be a holomorphic function on the domain $\Omega_N = \bigcup_{\lambda\in\mathbb{T}}(\lambda\Pi)^N \subset \mathbb{C}^N$ which takes values in $L(\mathcal{U})$ for a Hilbert space $\mathcal{U}$. Then the following statements are equivalent:*

(i) *There exist a Hilbert space $\mathcal{H}$ and a representation*

$$f(z) = a(z) - b(z)d(z)^{-1}c(z), \quad z \in \Omega_N,$$

of f, where

$$A(z) = z_1A_1 + \cdots + z_NA_N = \begin{bmatrix} a(z) & b(z) \\ c(z) & d(z) \end{bmatrix} \in L(\mathcal{U} \oplus \mathcal{H}),$$

with $A_k = A_k^* \geq 0,\ k = 1, \ldots, N$;

(ii) *f admits a representation*

$$f(z) = \sum_{k=1}^{N} z_k \Phi_k(z, \zeta), \quad (z, \zeta) \in \Omega_N \times \Omega_N,$$

where $\Phi_k(z,\zeta),\ k = 1, \ldots, N$, are holomorphic PSD kernels on $\Omega_N \times \Omega_N$;

(iii) *f satisfies the conditions:*

1. $f(\lambda z_1, \ldots, \lambda z_N) = \lambda f(z_1, \ldots, z_N),\ \lambda \in \mathbb{C}\backslash\{0\},\ z \in \Omega_N$;
2. $f(\mathbf{R}) + f(\mathbf{R})^* \geq 0, \quad \mathbf{R} \in \mathcal{A}^N$
   (the set $\mathcal{A}^N$ of N-tuples of commuting bounded strictly accretive operators on a common Hilbert space, and the functional calculus for $f(\mathbf{R})$ are defined in Section 3*);*
3. $f(\bar{z}) = f(z)^*, \quad z \in \Omega_N$;

(iv) *there exist Hilbert spaces $\mathcal{X}, \mathcal{X}_1, \ldots, \mathcal{X}_N$ such that $\mathcal{X} = \bigoplus_{k=1}^N \mathcal{X}_k$, and an Agler representation*

$$\mathcal{F}(w) = D + CP(w)(I_{\mathcal{X}} - AP(w))^{-1}B$$

of a double Cayley transform $\mathcal{F} = \mathcal{C}(f)$ of f (which is defined by (4.1)*), where $P(w) = \sum_{k=1}^N w_k P_{\mathcal{X}_k}$, and*

$$\begin{bmatrix} A & B \\ C & D \end{bmatrix} =: U = U^{-1} = U^* \in L(\mathcal{X} \oplus \mathcal{U}).$$

Thus, $\mathcal{B}_N(\mathcal{U})$ can be defined as a class of functions f satisfying any (and hence, all) of conditions (i)–(iv) of Theorem 7.1.

We have introduced also the class $\iota\mathbb{R}\mathcal{B}_N(\mathcal{U})$ consisting of all ι-real functions from $\mathcal{B}_N(\mathcal{U})$, for an anti-unitary involution $\iota = \iota_{\mathcal{U}}$ on $\mathcal{U}$. The operator ι plays a role analogous to the complex conjugation on $\mathbb{C}^N$, and the class $\iota\mathbb{R}\mathcal{B}_N(\mathcal{U})$ is a generalization of the class $\mathbb{R}\mathcal{B}_N^{n\times n}$. We have obtained several characterizations of the class $\iota\mathbb{R}\mathcal{B}_N(\mathcal{U})$, which are collected in Theorem 6.4.

Let us note that though several descriptions of the classes $\mathcal{B}_N(\mathcal{U})$ and $\iota\mathbb{R}\mathcal{B}_N(\mathcal{U})$ were obtained in this paper, the investigation of these classes is still far from its final point. We formulate and discuss below the most important questions subject to further investigation.

Recall that the class $\mathcal{B}_N(\mathcal{U})$ (resp., $\iota\mathbb{R}\mathcal{B}_N(\mathcal{U})$) is a subclass of $\mathcal{P}_N(\mathcal{U})$ (resp., $\iota\mathbb{R}\mathcal{P}_N(\mathcal{U})$), where the latter is a class of all $L(\mathcal{U})$-valued functions holomorphic in Ω_N and satisfying the conditions:

1. $f(\lambda z_1,\dots,\lambda z_N)=\lambda f(z_1,\dots,z_N),\ \lambda\in\mathbb{C}\backslash\{0\},\ z\in\Omega_N$;
2. $f(z)+f(z)^*\geq 0,\quad z\in\Pi^N$;
3. $f(\bar z)=f(z)^*\quad (=f^\sharp(z):=\iota f(\bar z)\iota),\quad z\in\Omega_N$.

Analogously, the class $\mathcal{B}_N^{n\times n}$ (resp., $\mathbb{R}\mathcal{B}_N^{n\times n}$) is a subclass of $\mathcal{P}_N^{n\times n}$ (resp., $\mathbb{R}\mathcal{P}_N^{n\times n}$), where the latter is a class of all rational $n\times n$ matrix-valued functions satisfying the conditions:

1. $f(\lambda z_1,\dots,\lambda z_N)=\lambda f(z_1,\dots,z_N),\ \lambda\in\mathbb{C}\backslash\{0\},\ z\in\mathbb{C}^N$;
2. $f(z)+f(z)^*\geq 0,\quad z\in\Pi^N$;
3. $f(\bar z)=f(z)^*\quad (=f(\bar z)^T),\quad z\in\mathbb{C}^N$.

It is known that in the cases $N=1$ and $N=2$ condition 2 for all of these classes is equivalent to condition (iii2) of Theorem 7.1, thus for these cases the class $\mathcal{B}_N(\mathcal{U})$ (resp., $\iota\mathbb{R}\mathcal{B}_N(\mathcal{U})$, $\mathcal{B}_N^{n\times n}$, $\mathbb{R}\mathcal{B}_N^{n\times n}$) coincides with $\mathcal{P}_N(\mathcal{U})$ (resp., $\iota\mathbb{R}\mathcal{P}_N(\mathcal{U})$, $\mathcal{P}_N^{n\times n}$, $\mathbb{R}\mathcal{P}_N^{n\times n}$).

Problem 1. For which $N\geq 3$, $\mathcal{U}$, $\iota_{\mathcal{U}}$ and $n\in\mathbb{N}$ the class $\mathcal{B}_N(\mathcal{U})$ (resp., $\iota\mathbb{R}\mathcal{B}_N(\mathcal{U})$, $\mathcal{B}_N^{n\times n}$, $\mathbb{R}\mathcal{B}_N^{n\times n}$) is a proper subclass of $\mathcal{P}_N(\mathcal{U})$ (resp., $\iota\mathbb{R}\mathcal{P}_N(\mathcal{U})$, $\mathcal{P}_N^{n\times n}$, $\mathbb{R}\mathcal{P}_N^{n\times n}$)?

For the classes $\mathcal{B}_N(\mathcal{U})$ and $\iota\mathbb{R}\mathcal{B}_N(\mathcal{U})$, $N\geq 3$ this problem can be reformulated as follows: find $f\in\mathcal{P}_N(\mathcal{U})$ and $\mathbf{R}\in\mathcal{A}^N$ such that the selfadjoint operator $f(\mathbf{R})+f(\mathbf{R})^*$ is not PSD. The latter is equivalent to the inequality $\|\mathcal{F}(\mathbf{T})\|>1$, where $\mathcal{F}=\mathcal{C}(f)$ is a holomorphic contractive $L(\mathcal{U})$-valued function on $\mathbb{D}^N$, and the N-tuple of operators $\mathbf{T}=(T_1,\dots,T_N)\in\mathcal{C}^N$ is defined by $T_k:=(R_k-I)(R_k+I)^{-1}$, $k=1,\dots,N$, i.e., for $\mathcal{F}$ and $\mathbf{T}$ the *generalized von Neumann inequality* fails. There are examples of holomorphic contractive operator valued functions $\mathcal{F}$ and N-tuples $\mathbf{T}$ of commuting strict contractions on a Hilbert space, for which the generalized von Neumann inequality fails (see [23, 11, 13, 14]), however in these examples the requirement $\mathcal{F}=\mathcal{C}(f)$ for some $f\in\mathcal{P}_N(\mathcal{U})$ is not fulfilled. A function $\mathcal{F}$ satisfying this requirement must have a certain complicated structure induced by the homogeneity structure of f. Thus, more sophisticated examples should be found to meet this condition.

Another open problem concerns to characterization of classes $\mathcal{B}_N^{n\times n}$ and $\mathbb{R}\mathcal{B}_N^{n\times n}$ (the formulation below is given for the first of them).

Problem 2. Does the representation

$$f(z)=\sum_{k=1}^N z_k\Phi_k(z,\zeta),\quad (z,\zeta)\in\Omega_N\times\Omega_N,$$

of an arbitrary $f\in\mathcal{P}_N^{n\times n}$, where $\Phi_k(z,\zeta)$, $k=1,\dots,N$, are rational $n\times n$ matrix-valued functions which are holomorphic PSD kernels on $\Omega_N\times\Omega_N$, $N\geq 3$, imply $f\in\mathcal{B}_N^{n\times n}$?

Let us formulate this more accurately.

Problem 3. For which $N\geq 3$, $n\in\mathbb{N}$, and $f\in\mathcal{P}_N^{n\times n}$ the question in Problem 2 has a positive answer?

References

[1] J. Agler. On the representation of certain holomorphic functions defined on a polydisc. In *Topics in Operator Theory: Ernst D. Hellinger Memorial Volume*, volume 48 of *Oper. Theory Adv. Appl.*, pages 47–66. Birkhäuser Verlag, Basel, 1990.

[2] J. Agler and J. McCarthy. Nevanlinna-Pick interpolation on the bidisk. *J. Reine Angew. Math.*, 506:191–204, 1999.

[3] N.I. Akhiezer and I.M. Glazman. *Theory of Linear Operators in Hilbert Space.* Nauka, Moscow, 1966. (Russian); English Transl.: Dover Publications, Inc., New York, 1993, xiv + 147 + iv + 218 pp.

[4] N. Aronszajn. Theory of reproducing kernels. *Trans. Amer. Math. Soc.*, 68(3):337–404, 1950.

[5] J.A. Ball, W.S. Li, D. Timotin, and T. Trent. A commutant lifting theorem on the polydisc. *Indiana Univ. Math. J.*, 48(2):653–675, 1999.

[6] J.A. Ball and T. Trent. Unitary colligations, reproducing kernel Hilbert spaces, and Nevanlinna-Pick interpolation in several variables. *J. Funct. Anal.*, 157:1–61, 1998.

[7] M.F. Bessmertnyĭ. On realization of rational matrix functions of several variables. II. In Volume 143 of Oper. Theory Adv. Appl., Birkhäuser Verlag, Basel, 2003.

[8] M.F. Bessmertnyĭ. On realization of rational matrix functions of several variables. III. In Volume 149 of Oper. Theory Adv. Appl., pages 133–138. Birkhäuser Verlag, Basel, 2004.

[9] M.F. Bessmertnyĭ. *Functions of Several Variables in the Theory of Finite Linear Structures.* Ph. D. Thesis, Kharkov University, Kharkov, 1982. (Russian).

[10] M.F. Bessmertnyĭ. On realizations of rational matrix functions of several complex variables. In Volume 134 of *Oper. Theory Adv. Appl.*, pages 157–185. Birkhäuser Verlag, Basel, 2002.

[11] M. Crabb and A. Davie. Von Neumann's inequality for Hilbert space operators. *Bull. London Math. Soc.*, 7:49–50, 1975.

[12] A.V. Efimov and V.P. Potapov. J-expanding matrix functions and their role in the analytical theory of electrical circuits. *Uspekhi Mat. Nauk*, 28(1):65–130, 1973. (Russian). English Transl.: Russ. Math. Surveys, 28:1 (1973), pp. 69–140.

[13] J.A. Holbrook. Schur norms and the multivariate von Neumann inequality. In *Recent advances in operator theory and related topics (Szeged, 1999)*, volume 127 of *Oper. Theory: Adv. Appl.*, pages 375–386. Birkhäuser Verlag, Basel, 2001.

[14] D.S. Kalyuzhniy. On the von Neumann inequality for linear matrix functions of several variables. *Mat. Zametki*, 64(2):218–223, August 1998. (Russian); English translation: Math. Notes **64** (1998), no. 2, 186–189.

[15] D.S. Kalyuzhniy. Multiparametric dissipative linear stationary dynamical scattering systems: Discrete case. *J. Operator Theory*, 43(2):427–460, 2000.

[16] D.S. Kalyuzhniy. Multiparametric dissipative linear stationary dynamical scattering systems: Discrete case, II: Existence of conservative dilations. *Integral Equations Operator Theory*, 36(1):107–120, 2000.

[17] Sh. Karni. *Network Theory: Analysis and Synthesis.* Allyn and Bacon Series in Electrical Engineering. Allyn and Bacon, Boston, 1966.

[18] M.S. Livšic and M.Š. Flekser. The synthesis of a transmission line from given frequency characteristics. *Zap. Meh.-Mat. Fak. i Har'kov. Mat. Obšč.*, 28(4):149–162, 1961. (Russian).

[19] L. Schwartz. *Analyse Mathématique*, volume II. Hermann, Paris, 1967. xxiv + 475 + 21 + 75 pp.

[20] S. Seshu and M.B. Reed. *Linear Graphs and Electrical Networks*. Addison-Wesley Series in the Engineering Sciences. Electrical and Control Systems. Addison-Wesley, Reading, MA, 1961. x + 315 p.

[21] B.V. Shabat. *Introduction to Complex Analysis. Part I. Functions of a Single Variable.* Nauka, Moscow, 1985. (Russian); French Transl.: Chabat, B. Introduction à l'analyse complexe. Tome 1. Fonctions d'une variable. Traduit du Russe: Mathématiques. Mir, Moscow, 1990. 309 pp.

[22] B.V. Shabat. *Introduction to Complex Analysis. Part II. Functions of Several Variables.* Nauka, Moscow, 1985. (Russian); French Transl.: Chabat, B. Introduction à l'analyse complexe. Tome 2. Fonctions de plusieurs variables. Traduit du Russe: Mathématiques. Mir, Moscow, 1990. 420 pp.; English Transl.: Translations of Mathematical Monographs, 110. American Mathematical Society, Providence, RI, 1992. x + 371 pp.

[23] N. Varopoulos. On an inequality of von Neumann and an application of the metric theory of tensor products to operator theory. *J. Funct. Anal.*, 16:83–100, 1974.

Dmitry S. Kalyuzhnyĭ-Verbovetzkiĭ
Department of Mathematics
The Weizmann Institute
Rehovot 76100, Israel
e-mail: dmitryk@wisdom.weizmann.ac.il

Operator Theory:
Advances and Applications, Vol. 149, 291–348

Rational Solutions of the Schlesinger System and Isoprincipal Deformations of Rational Matrix Functions I

V. Katsnelson* and D. Volok

Abstract. In the present paper we discuss the general facts concerning the Schlesinger system: the τ-function, the local factorization of solutions of Fuchsian equations and holomorphic deformations. We introduce the terminology "isoprincipal" for the deformations of Fuchsian equations with general (not necessarily non-resonant) matrix coefficients corresponding to solutions of the Schlesinger system. Every isoprincipal deformation is isomonodromic. The converse is also true in the non-resonant case, but not in general.

In the forthcoming sequel we shall give an explicit description of a class of rational solutions of the Schlesinger system, based on the techniques developed here, and the realization theory for rational matrix functions.

Notations

- $\mathbb{C}$ stands for the complex plane;
- $\overline{\mathbb{C}}$ stands for the extended complex plane (= the Riemann sphere): $\overline{\mathbb{C}} = \mathbb{C} \cup \infty$;
- $\mathbb{C}^n$ stands for the n-dimensional complex space;
- in the coordinate notation, a point $\boldsymbol{t} \in \mathbb{C}^n$ is written as $\boldsymbol{t} = (t_1, \dots, t_n)$;
- $\mathbb{C}^n_*$ is the set of those points $\boldsymbol{t} = (t_1, \dots, t_n) \in \mathbb{C}^n$, whose coordinates $t_1, \dots, t_n$ are pairwise different: $\mathbb{C}^n_* = \mathbb{C}^n \setminus \left(\bigcup_{1 \leq i,j \leq n, i \neq j} \{\boldsymbol{t} : t_i = t_j\} \right)$;
- $\mathfrak{M}_k$ stands for the set of all $k \times k$ matrices with complex entries;
- $[.,.]$ denotes the commutator: for $A, B \in \mathfrak{M}_k$, $[A, B] = AB - BA$;
- I stands for the unity matrix of the appropriate dimension;
- $\mathcal{R}(\mathfrak{M}_k)$ stands for the set of all rational $\mathfrak{M}_k$-valued functions R with $\det R(z) \not\equiv 0$;
- $\mathcal{P}(R)$ stands for the set of all poles of the function R, $\mathcal{N}(R)$ stands for the set of all poles of the function R^{-1}; $\mathcal{P}(R)$ is said to be **the pole set of the function** R, $\mathcal{N}(R)$ is said to be **the zero set of the function** R.

Received by the editors April 1, 2003; revised October 28, 2003. Submitted by L. Rodman.
Mathematics Subject Classification (2000). 34Mxx, 34M55, 93B15, 47A56.
Key words and phrases. Differential equations in the complex domain, monodromy, Schlesinger system.
* The first author was supported by the Minerva foundation.

1. The Schlesinger system

The Schlesinger system is a Pfaffian system of differential equations of the form:

$$\begin{cases} \dfrac{\partial Q_i}{\partial t_j} = \dfrac{[Q_i, Q_j]}{t_i - t_j}, & 1 \leq i, j \leq n,\ i \neq j, \quad (1.1.a) \\ \dfrac{\partial Q_i}{\partial t_i} = - \displaystyle\sum_{\substack{1 \leq j \leq n \\ j \neq i}} \dfrac{[Q_i, Q_j]}{t_i - t_j}, & 1 \leq i \leq n, \quad (1.1.b) \end{cases} \tag{1.1}$$

where $Q_1, \ldots, Q_n$ are square $k \times k$ matrix functions of $\boldsymbol{t} = (t_1, \ldots, t_n)$, $\boldsymbol{t} \in \mathbb{C}^n_*$. The system of equations (1.1) can be rewritten in a compact way as

$$\mathfrak{d} Q_\nu = \sum_{\substack{\mu=1 \\ \mu \neq \nu}}^{n} [Q_\mu, Q_\nu] \mathfrak{d} \log(t_\mu - t_\nu), \quad \nu = 1, \ldots, n, \tag{1.2}$$

where $\mathfrak{d}$ is the exterior differential. The Schlesinger system (1.1) is over-determined. The compatibility (the integrability) condition for this system is of the form

$$\mathfrak{d}\Bigg(\sum_{\substack{\mu=1 \\ \mu \neq \nu}}^{n} [Q_\mu, Q_\nu] \mathfrak{d} \log(t_\mu - t_\nu) \Bigg) = 0, \quad \nu = 1, \ldots, n. \tag{1.3}$$

The condition (1.3) must be fulfilled if $Q_1(\boldsymbol{t}), \ldots, Q_n(\boldsymbol{t})$ satisfy the equations (1.2). This condition expresses the equality $\mathfrak{d}(\mathfrak{d} Q_v) = 0$, which must be satisfied since $\mathfrak{d} \cdot \mathfrak{d} = 0$.

Let us present the integrability condition (1.3) in a more explicit form. Since $\mathfrak{d}\mathfrak{d} \log(t_\mu - t_\nu) = 0$,

$$\mathfrak{d}\Bigg(\sum_{\substack{\mu=1 \\ \mu \neq \nu}}^{n} [Q_\mu, Q_\nu] \mathfrak{d} \log(t_\mu - t_\nu) \Bigg) =$$

$$= \sum_{\substack{\mu=1 \\ \mu \neq \nu}}^{n} [\mathfrak{d} Q_\mu, Q_\nu] \wedge \mathfrak{d} \log(t_\mu - t_\nu) + \sum_{\substack{\mu=1 \\ \mu \neq \nu}}^{n} [Q_\mu, \mathfrak{d} Q_\nu] \wedge \mathfrak{d} \log(t_\mu - t_\nu), \tag{1.4}$$

where we have to substitute the expression of the form (1.2) for $\mathfrak{d} Q_\mu$ and $\mathfrak{d} Q_\nu$ in the right-hand side of (1.4). Performing the substitutions, we obtain the expressions

$$\sum_{\substack{\mu=1 \\ \mu \neq \nu}}^{n} [\sum_{\substack{\lambda=1 \\ \lambda \neq \mu}}^{n} [Q_\lambda, Q_\mu], Q_\nu] \mathfrak{d} \log(t_\lambda - t_\mu) \wedge \mathfrak{d} \log(t_\mu - t_\nu) +$$

$$+ \sum_{\substack{\mu=1 \\ \mu \neq \nu}}^{n} [Q_\mu, \sum_{\substack{\lambda=1 \\ \lambda \neq \nu}}^{n} [Q_\lambda, Q_\nu]] \mathfrak{d} \log(t_\lambda - t_\nu) \wedge \mathfrak{d} \log(t_\mu - t_\nu). \tag{1.5}$$

Since $\omega \wedge \omega = 0$ for every differential form ω, $\partial \log(t_\mu - t_\nu) \wedge \partial \log(t_\mu - t_\nu) = 0$. Therefore, the integrability condition for the Schlesinger system (1.2) can be presented in the form

$$\sum_{\substack{\lambda=1,\mu=1;\\ \lambda\neq\mu,\lambda\neq\nu,\mu\neq\nu}}^{n} \big[[Q_\lambda, Q_\mu], Q_\nu\big] \partial \log(t_\lambda - t_\mu) \wedge \partial \log(t_\mu - t_\nu) +$$
$$+ \sum_{\substack{\lambda=1,\mu=1;\\ \lambda\neq\mu,\lambda\neq\nu,\mu\neq\nu}}^{n} \big[Q_\mu, [Q_\lambda, Q_\nu]\big] \partial \log(t_\lambda - t_\nu) \wedge \partial \log(t_\mu - t_\nu) = 0, \quad \nu = 1, \dots, n. \tag{1.6}$$

Lemma 1.1. *For every holomorphic square matrix functions $Q_1, \dots, Q_n$, the integrability condition* (1.6) *holds.*

The proof of this lemma is based on two identities. The first of them is the Jakobi identity

$$[[A, B], C] + [[B, C], A] + [[C, A], B] = 0, \tag{1.7}$$

which holds for arbitrary square matrices A, B, C (of the same dimension). The second identity is an identity related to the differential holomorphic 1-form $\omega_{\lambda,\mu}$:

$$\omega_{\lambda,\mu} = \frac{dt_\lambda - dt_\mu}{t_\lambda - t_\mu}, \quad 1 \leq \lambda, \mu \leq n, \lambda \neq \mu. \tag{1.8}$$

Since

$$\omega_{\lambda,\mu} = \omega_{\mu,\lambda}, \quad 1 \leq \lambda, \mu \leq n, \lambda \neq \mu, \tag{1.9}$$

there are $\dfrac{n(n-1)}{2}$ different forms $\omega_{\lambda,\mu}$. The second identity is:

$$\omega_{\lambda,\mu} \wedge \omega_{\mu,\nu} + \omega_{\mu,\nu} \wedge \omega_{\nu,\lambda} + \omega_{\nu,\lambda} \wedge \omega_{\lambda,\mu} = 0, \quad 1 \leq \lambda, \mu, \nu \leq n, \quad \lambda \neq \mu \neq \nu. \tag{1.10}$$

To prove this identity we observe that the left-hand side of (1.10) is the sum of the expressions that are obtained from $\omega_{\lambda,\mu} \wedge \omega_{\mu,\nu}$ by all cyclic permutations of the indices λ, μ, ν. Computing the exterior product $\omega_{\lambda,\mu} \wedge \omega_{\mu,\nu}$:

$$\omega_{\lambda,\mu} \wedge \omega_{\mu,\nu} = \frac{dt_\lambda \wedge dt_\mu + dt_\mu \wedge dt_\nu + dt_\nu \wedge dt_\lambda}{(t_\lambda - t_\mu)(t_\mu - t_\nu)}, \tag{1.11}$$

we see that it is represented as a fraction whose numerator is invariant with respect to cyclic permutations of the indices λ, μ, ν. Therefore, the right-hand side of (1.10) is equal to

$$\left\{ \frac{1}{(t_\lambda - t_\mu)(t_\mu - t_\nu)} + \frac{1}{(t_\mu - t_\nu)(t_\nu - t_\lambda)} + \frac{1}{(t_\nu - t_\lambda)(t_\lambda - t_\mu)} \right\}$$
$$\times \left(dt_\lambda \wedge dt_\mu + dt_\mu \wedge dt_\nu + dt_\nu \wedge dt_\lambda\right)$$

Evidently, the expression in the curly brackets vanishes. The identity (1.10) is proved.

Remark 1.2. The forms $\frac{1}{2\pi i}\omega_{\lambda,\mu}$, $1 \le \lambda, \mu \le n$, $\lambda \ne \mu$, generate the cohomology ring $H^*(\mathbb{C}^n_*, \mathbb{Z})$ of the space $\mathbb{C}^n_*$. This result was established by V. Arnold, [Arn].

Proof of Lemma 1.1. Let us fix $\nu \in 1, , \dots, n$. First, we transform the first summand in (1.6). Applying the Jakobi identity to the factor $\big[[Q_\lambda, Q_\mu], Q_\nu\big]$ in the first sum, we rewrite (1.6) as

$$\sum_{\substack{\lambda=1,\mu=1;\\ \lambda\ne\mu,\lambda\ne\nu,\mu\ne\nu}}^{n} \big[[Q_\mu, Q_\nu], Q_\lambda\big]\omega_{\lambda,\mu} \wedge \omega_{\mu,\nu} + \sum_{\substack{\lambda=1,\mu=1;\\ \lambda\ne\mu,\lambda\ne\nu,\mu\ne\nu}}^{n} \big[[Q_\nu, Q_\lambda], Q_\mu\big]\omega_{\lambda,\mu} \wedge \omega_{\mu,\nu} + \sum_{\substack{\lambda=1,\mu=1;\\ \lambda\ne\mu,\lambda\ne\nu,\mu\ne\nu}}^{n} \big[[Q_\lambda, Q_\nu], Q_\mu\big]\omega_{\lambda,\nu} \wedge \omega_{\mu,\nu} = 0 \quad (1.12)$$

Since the summation runs over all pairs λ, μ, we can interchange λ and μ in the second and third sums of (1.12). Using also (1.9) and the anti-commutativity of the exterior product ($\omega \wedge \sigma = -\sigma \wedge \omega$ for every differential 1-forms ω and σ), we rewrite (1.12) as

$$\sum_{\substack{\lambda=1,\mu=1;\\ \lambda\ne\mu,\lambda\ne\nu,\mu\ne\nu}}^{n} \big[[Q_\mu, Q_\nu], Q_\lambda\big]\omega_{\lambda,\mu} \wedge \omega_{\mu,\nu} + \sum_{\substack{\lambda=1,\mu=1;\\ \lambda\ne\mu,\lambda\ne\nu,\mu\ne\nu}}^{n} \big[[Q_\mu, Q_\nu], Q_\lambda\big]\omega_{\nu,\lambda} \wedge \omega_{\lambda,\mu} + \sum_{\substack{\lambda=1,\mu=1;\\ \lambda\ne\mu,\lambda\ne\nu,\mu\ne\nu}}^{n} \big[[Q_\mu, Q_\nu], Q_\lambda\big]\omega_{\mu,\nu} \wedge \omega_{\nu,\lambda} = 0 \quad (1.13)$$

The last equality holds because its left-hand side is of the form

$$\sum_{\substack{\lambda=1,\mu=1;\\ \lambda\ne\mu,\lambda\ne\nu,\mu\ne\nu}}^{n} \big[[Q_\mu, Q_\nu], Q_\lambda\big] \left\{\omega_{\lambda,\mu} \wedge \omega_{\mu,\nu} + \omega_{\nu,\lambda} \wedge \omega_{\lambda,\mu} + \omega_{\mu,\nu} \wedge \omega_{\nu,\lambda}\right\},$$

where the sums in the curly brackets vanish according to the identity (1.10). □

From Lemma 1.1 and from Frobenius theorem it follows that the Cauchy problem for the Schlesinger system is locally solvable for every initial data. More precisely, the following result holds:

Theorem 1.3. *Let $t^0 \in \mathbb{C}^n_*$, $Q^0_1, \dots, Q^0_n \in \mathfrak{M}_k$. Then there exist a neighborhood $\mathcal{V}(t^0, \delta)$ of the point t^0,*

$$\mathcal{V}(t^0,\delta) = \{t \in \mathbb{C}_n : |t_1 - t_1^0| < \delta, \dots, |t_n - t_n^0| < \delta\}, \mathcal{V}(t^0,\delta) \subset \mathbb{C}^n_*,$$

and matrix functions $Q_1(t), \dots, Q_n(t)$,

$$Q_\nu(t) : \mathcal{V}(t^0,\delta) \mapsto \mathfrak{M}_k, \nu = 1, \dots, n,$$

which are holomorphic in $\mathcal{V}(t^0,\delta)$, satisfy the Schlesinger system (1.1) *for $t \in \mathcal{V}(t^0,\delta)$, and meet the initial condition*

$$Q_\nu(t^0) = Q^0_\nu, \nu = 1, \dots, n \tag{1.14}$$

at the point t^0. Here δ is a positive number which depends on t^0 and $Q^0_\nu, \nu = 1, \dots, n$.

Thus Theorem 1.3 is a local existence result. Actually, a much more stronger global existence result holds for the Schlesinger system. It turns out that the solution $Q_1(t), \dots, Q_n(t)$ of the above introduced Cauchy problem for this system exists *globally* as meromorphic (multi-valued) functions in $\mathbb{C}^n_*$. We do not discuss this more general result in detail – this will lead us too far. However, it is worth mentioning that every solution of the Schlesinger system can be represented as a fraction of holomorphic functions in $\mathbb{C}^n_*$, where some complex-valued holomorphic function $\tau(.)$ on $\mathbb{C}^n_*$ appears in the denominator. This is the so-called τ-function for the considered solution. The singularities of this solution are precisely those $t \in \mathbb{C}^n_*$ where $\tau(t) = 0$. The "zeros" of the function $\tau(.)$ are responsible for the singularities of the solution $Q_1(.), \dots, Q_n(.)$.

2. The tau-function of the Schlesinger system

Lemma 2.1. *Let matrix-functions $Q_1(t), \dots, Q_n(t)$ satisfy the Schlesinger system* (1.1). *Then the differential 1-form ω:*

$$\omega \overset{\text{def}}{=} \frac{1}{2} \sum_{\substack{\lambda,\mu=1 \\ \lambda\neq\mu}}^{n} \operatorname{tr}(Q_\lambda(t) Q_\mu(t)) \partial \log(t_\lambda - t_\mu), \tag{2.1}$$

is closed:

$$\partial\omega = 0. \tag{2.2}$$

Proof. It is clear that

$$\begin{aligned} \partial\omega \;&=\; \frac{1}{2} \sum_{\substack{\lambda,\mu=1 \\ \lambda\neq\mu}}^{n} \operatorname{tr}\Big(\big(\partial Q_\lambda(t)\big) Q_\mu(t)\Big) \partial \log(t_\lambda - t_\mu) \\ &\quad + \frac{1}{2} \sum_{\substack{\lambda,\mu=1 \\ \lambda\neq\mu}}^{n} \operatorname{tr}\Big(Q_\lambda(t) \big(\partial Q_\mu(t)\big)\Big) \partial \log(t_\lambda - t_\mu). \end{aligned}$$

Interchanging the summation indices λ and μ in the first sum (and using the rule $\mathrm{tr}AB = \mathrm{tr}BA$), we obtain (after renaming $\lambda \to \nu, \nu \to \lambda$) that

$$\mathfrak{d}\omega = \sum_{\substack{\lambda,\nu=1\\ \lambda\neq\nu}}^{n} \mathrm{tr}\Big(Q_\lambda(t)\big(\mathfrak{d}Q_\nu(t)\big)\Big)\mathfrak{d}\log(t_\nu - t_\lambda). \tag{2.3}$$

Substituting the expression (1.2) for $\mathfrak{d}Q_\nu$ into (2.3), we obtain the following expression for $\mathfrak{d}\omega$:

$$\mathfrak{d}\omega = \sum_{\substack{\lambda=1,\mu=1,\nu=1\\ \lambda\neq\mu,\lambda\neq\nu,\mu\neq\nu}}^{n} \mathrm{tr}\Big(Q_\lambda\big[Q_\mu, Q_\nu\big]\Big)\omega_{\mu,\nu}\wedge\omega_{\nu,\lambda}, \tag{2.4}$$

where $\omega_{\lambda,\mu}$ is defined by (1.8), or

$$\mathfrak{d}\omega = \sum_{\substack{\lambda=1,\mu=1,\nu=1\\ \lambda\neq\mu,\lambda\neq\nu,\mu\neq\nu}}^{n} \mathrm{tr}\Big(Q_\lambda Q_\mu Q_\nu - Q_\lambda Q_\nu Q_\mu\Big)\omega_{\mu,\nu}\wedge\omega_{\nu,\lambda}. \tag{2.5}$$

The traces in the last formula are invariant under a circular permutation of the three indices λ, μ, ν. So we can replace $\omega_{\mu,\nu}\wedge\omega_{\nu,\lambda}$ by $\frac{1}{3}\big(\omega_{\mu,\nu}\wedge\omega_{\nu,\lambda} + \omega_{\nu,\lambda}\wedge\omega_{\lambda,\mu} + \omega_{\lambda,\mu}\wedge\omega_{\mu,\nu}\big)$. But this quantity vanishes identically according to the identity (1.10). Thus, $\mathfrak{d}\omega = 0$. □

Since a closed form $\mathfrak{d}\omega$ is locally exact, the following theorem is an immediate consequence of Lemma 2.1:

Theorem 2.2. *Let holomorphic matrix-functions $Q_1(\boldsymbol{t}), \dots, Q_n(\boldsymbol{t})$ satisfy the Schlesinger system* (1.1) *in some simply connected domain $\mathcal{D}, \mathcal{D} \subset \mathbb{C}^n_*$. Then there exists a holomorphic non-vanishing function $\tau(\boldsymbol{t})$ in $\mathcal{D}$ such that*

$$\mathfrak{d}\log\tau(\boldsymbol{t}) = \frac{1}{2}\sum_{\substack{\lambda,\mu=1\\ \lambda\neq\mu}}^{n} \mathrm{tr}(Q_\lambda(\boldsymbol{t})Q_\mu(\boldsymbol{t}))\mathfrak{d}\log(t_\lambda - t_\mu) \tag{2.6}$$

Such a function τ is unique up to a constant factor: if $\tau_1(\boldsymbol{t})$ and $\tau_2(\boldsymbol{t})$ are two functions satisfying the relation (2.6)*, then $\tau_2(\boldsymbol{t}) = c\tau_1(\boldsymbol{t})$, where $c \neq 0$ is a constant.*

Definition 2.3. A function τ related to a solution $Q_1, \dots, Q_n$ of the Schlesinger system (1.1) by the equality (2.6) is said to be the τ-function for this solution.

The exactness of the differential form (2.1), stated in our Lemma 2.1, was established by M. Sato, T. Miwa and M. Jimbo in [SMJ2] (see §2.4 there). In [SMJ2], the potential (2.6) was introduced as well as the term "τ-function" for this potential.

Remark 2.4. From Lemma 2.1 it follows that the τ-function for a solution of the Schlesinger equation is a holomorphic non-vanishing (multi-valued) function for those $\boldsymbol{t}$ where the solution is holomorphic. Actually, the τ-function is holomorphic on the whole $\mathbb{C}^n_*$. It vanishes for those $\boldsymbol{t}$ where the solution is singular. (See the last paragraph of the section 1). In a very general setting, the holomorphy of τ-functions was proved in [Miw].

Remark 2.5. The τ-function is a very general concept related to non-linear differential equations. This function was first discovered by R. Hirota as a tool for generating many-soliton solutions. (References to the works of H. Hirota can be found in bibliography of the book [New]). Let us give examples of τ-functions for the Korteveg-deVries equation

$$u_t - 6uu_x + u_{xxx} = 0. \tag{2.7}$$

N-soliton solution

$$u(x,t) = -2\frac{\partial^2}{\partial x^2}\log\Delta(x,t), \tag{2.8}$$

$$\Delta(x,t) = \det\left(\delta_{jk} + \frac{c_j c_k}{\eta_j + \eta_k}\exp[-(\eta_j^3 + \eta_k^3)t - (\eta_j + \eta_k)x]\right)_{j,k=1,\dots,N}; \tag{2.9}$$

N-phase solution

$$u(x,t) = -2\frac{\partial^2}{\partial x^2}\log\vartheta(\boldsymbol{a}x + \boldsymbol{b}t + \boldsymbol{c}), \tag{2.10}$$

where ϑ is the Riemann theta function;
rapidly decreasing solution

$$u(x,t) = -2\frac{\partial^2}{\partial x^2}\log\Delta_{[x,\infty)}(t), \tag{2.11}$$

with $\Delta_{[x,\infty)}(t)$ the Fredholm determinant of the Gelfand-Levitan equation. In these examples the τ-function is $\Delta(x,t), \vartheta(\boldsymbol{a}x + \boldsymbol{b}t + \boldsymbol{c}), \Delta_{[x,\infty)}(t)$, respectively.

In the paper [WNTB], τ-function appeared as the spin-spin correlation function for the two-dimensional Ising model. In [JMMS], the Fredholm determinant $\det(\boldsymbol{I} - \lambda\boldsymbol{K})$ of the integral operator $\boldsymbol{K}$ with the kernel $K(x,x') = \frac{\sin(x-x')}{x-x'}$ on the union of intervals $\bigcup_{1\le j\le n}[a_{2j-1}, a_{2j}]$, $a_1 < \cdots < a_{2n}$, is expressed as the τ-function $\tau(a_1,\dots,a_{2n})$ of the Schlesinger system constructed from the operator. (See in particular section 7 of [JMMS], where the deformation theory of the Fredholm integral operator with the kernel $\frac{\sin(x-x')}{x-x'}$ was related to the deformation theory of linear differential equations). The interrelation between the deformation theory of linear differential operators, the Schlesinger system and its τ-function, correlation function in statistical physics theory, quantum fields and Fredholm determinants are covered in the expository survey [SMJ1]. The τ-function for general matrix linear differential equation with rational coefficients was considered in [JMU], and was studied in great detail in [JM1], [JM2].

However, the importance of the τ-function and its central role in the soliton theory was not understood before the works of the research group from Kioto (M. Sato, T. Miwa, M. Jimbo, E. Date, M. Kashiwara, Y. Sato). The work of this group, surrounding Mikio Sato, reached its peak in the years around 1981. In the paper [SaSa], the time evolution of a solution of a non-linear equation is interpreted as the dynamical motion of a point of the infinite-dimensional Grassman manifold.

In the papers of E. Date, M. Jimbo, M. Kashiwara and T. Miwa, Lie algebras and vertex operators were incorporated into the soliton theory. (See the series of seven notes [DJKM1], and the paper [DJKM2]). In this setting, the τ-function appears in a very general way. The paper [Sa] explains some ideas from [SaSa] in more detail. The further development of the approach related to the consideration of soliton equations as dynamic systems on an infinite-dimensional Grassman manifold is done in [SeWi]. In particular, in section 3 of [SeWi] the geometric meaning of the τ-function is explained: this function appears as an infinite determinant related to the determinant bundle over the Grassman manifold. In [MSW], the τ-function is defined as the generalized cross-ratio of four points of the Grassman manifold. Tau function is discussed in Chapters 7 and 8 of the book [Dick]. The book [MJD] is a presentation of the algebraic analysis of the nonlinear differential equations of the KdV type written for students. In particular, it contains a very clear presentation of the circle of problem related to τ-functions. Chapter 4 of the book [New] is dedicated to the τ-function and its relation to the Painlevé property and to the Bäcklund transformations. This non-formally written book covers a very widespread circle of problems.

In the present work we construct a class of rational solutions of the Schlesinger system (1.1). For such a solution, we calculate its τ-function explicitly. This τ-function is equal to the determinant of some (finite) matrix which is intimately related to this solution. In the considered special case, τ-function is a rational function of the variables $t_1, \dots, t_n$.

3. The potential function for a solution of the Schlesinger system

From (1.1.b) it follows that if the matrix functions $Q_1(\boldsymbol{t}), \dots, Q_n(\boldsymbol{t})$ satisfy the Schlesinger system in a domain $\mathcal{G} \in \mathbb{C}^n_*$, then

$$\frac{\partial Q_i(\boldsymbol{t})}{\partial t_j} \equiv \frac{\partial Q_j(\boldsymbol{t})}{\partial t_i}. \tag{3.1}$$

Hence, there exists a matrix function $V(\boldsymbol{t})$ in $\mathcal{G}$ such that

$$Q_j(\boldsymbol{t}) = \frac{\partial V(\boldsymbol{t})}{\partial t_j}, \quad \boldsymbol{t} \in \mathcal{G}, \quad j = 1, \dots, n. \tag{3.2}$$

Remark 3.1. If the domain $\mathcal{G}$ is multiconnected, the function $V(\boldsymbol{t})$ can be multi-valued. In this case we can consider $V(\boldsymbol{t})$ as single-valued function on the universal covering $\operatorname{cov}(\mathcal{G}, \boldsymbol{t}_0)$ of the domain $\mathcal{G}$, where $\boldsymbol{t}_0$ is a chosen distinguished point of $\mathcal{G}$.

Definition 3.2. The matrix function $V(\boldsymbol{t})$ which satisfies the condition (3.2) is said to be the potential matrix function for the solution $Q_1(\boldsymbol{t}), \dots, Q_n(\boldsymbol{t})$ of the Schlesinger system.

We have proved that the potential matrix function exists for any solution of the Schlesinger system. It is clear that the potential function is unique up to an arbitrary constant summand.

Lemma 3.3. *Let matrices* $Q_1(\boldsymbol{t}),\dots,Q_n(\boldsymbol{t})$ *satisfy the Schlesinger system* (1.1), *and moreover the condition*

$$\sum_{1\le i\le n} Q_i(\boldsymbol{t}) = 0 \tag{3.3}$$

for every $\boldsymbol{t}\in\mathcal{G}$.

Then the function

$$V(\boldsymbol{t}) \stackrel{\text{def}}{=} \sum_{1\le i\le n} t_i\cdot Q_i(\boldsymbol{t}),\quad \boldsymbol{t}\in\mathcal{G}, \tag{3.4}$$

is a potential matrix function for the solution $Q_1(\boldsymbol{t}),\dots,Q_n(\boldsymbol{t})$.

Proof. Differentiating the expression on the left-hand side of (3.4), we obtain

$$\frac{\partial V(\boldsymbol{t})}{\partial t_j} = Q_j(\boldsymbol{t}) + \sum_{1\le i\le n} t_i\cdot\frac{\partial Q_i(\boldsymbol{t})}{\partial t_j}. \tag{3.5}$$

Thus, we have to prove that

$$\sum_{1\le i\le n} t_i\cdot\frac{\partial Q_i(\boldsymbol{t})}{\partial t_j} = 0. \tag{3.6}$$

Substituting the expression (1.1) for $\frac{\partial Q_i(\boldsymbol{t})}{\partial t_j}$ into the left-hand side of (3.6), we obtain

$$\begin{aligned}\sum_{1\le i\le n} t_i\cdot\frac{\partial Q_i(\boldsymbol{t})}{\partial t_j} &= \sum_{\substack{1\le i\le n,\\ i\ne j}} t_i\cdot\frac{[Q_i(\boldsymbol{t}),Q_j(\boldsymbol{t})]}{t_i-t_j} - t_j\cdot\sum_{\substack{1\le i\le n,\\ i\ne j}}\frac{[Q_i(\boldsymbol{t}),Q_j(\boldsymbol{t})]}{t_i-t_j}\\ &= \sum_{\substack{1\le i\le n,\\ i\ne j}}[Q_i(\boldsymbol{t}),Q_j(\boldsymbol{t})] = \sum_{1\le i\le n}[Q_i(\boldsymbol{t}),Q_j(\boldsymbol{t}] = 0. \end{aligned}\tag{3.7}$$

Thus, (3.6) holds. □

Lemma 3.4. *Let* $\mathcal{G}$ *be a domain in* $\mathbb{C}^n_*$, *and let matrix functions* $Q_1(\boldsymbol{t}),\dots,$ $Q_n(\boldsymbol{t})$ *satisfy the Schlesinger system* (1.1) *for* $\boldsymbol{t}\in\mathcal{G}$. *Assume that the equality* (3.3) *holds at least at one point* $\boldsymbol{t}=\boldsymbol{t}_0\in\mathcal{G}$. *Then this equality holds at every point* $\boldsymbol{t}\in\mathcal{G}$.

Proof. The equality (3.3) will be proved if we prove that

$$\frac{\partial}{\partial t_j}\Big(\sum_{1\le i\le n} Q_i(\boldsymbol{t})\Big) = 0 \ \text{ for }\ \boldsymbol{t}\in\mathcal{G}. \tag{3.8}$$

From (1.1) it follows that

$$\sum_{1\leq i\leq n}\frac{\partial Q_i(\boldsymbol{t})}{\partial t_j}=\frac{\partial Q_j(\boldsymbol{t})}{\partial t_j}+\sum_{\substack{1\leq i\leq n\\ i\neq j}}\frac{\partial Q_i(\boldsymbol{t})}{\partial t_j}$$

$$=-\sum_{\substack{1\leq i\leq n\\ i\neq j}}\frac{[Q_i(\boldsymbol{t}),Q_j(\boldsymbol{t})]}{t_i-t_j}+\sum_{\substack{1\leq i\leq n\\ i\neq j}}\frac{[Q_i(\boldsymbol{t}),Q_j(\boldsymbol{t})]}{t_i-t_j}=0\ \text{ for }\ \boldsymbol{t}\in\mathcal{G}.$$

Thus, (3.8) holds. □

The following lemma is an immediate consequence of two previous ones.

Lemma 3.5. *Let the matrix functions $Q_1(\boldsymbol{t}),\dots,Q_n(\boldsymbol{t})$ satisfy the Schlesinger system* (1.1) *for every $\boldsymbol{t}\in\mathcal{G}$, where $\mathcal{G}$ is a domain in $\mathbb{C}^n_*$. Assume that the condition* (3.3) *holds at least for one point $\boldsymbol{t}=\boldsymbol{t}_0\in\mathcal{G}$.*

Then the function $V(\boldsymbol{t})$ which is defined by (3.4) *is a potential matrix function for the solution $Q_1(\boldsymbol{t}),\dots,Q_n(\boldsymbol{t})$.*

The expression $\sum t_iQ_i(\boldsymbol{t})$, which appears on the right-hand side of the equality (3.4), has a useful interpretation in terms of an auxiliary ordinary differential equation. Let $\boldsymbol{t}=(t_1,\dots,t_n)\in\mathbb{C}^n_*$, and $Q_1,\dots,Q_n$ be $k\times k$ matrices. (For the moment, we consider the point $\boldsymbol{t}$ and the matrices $Q_1,\dots,Q_n$ as fixed rather than variable.) Let us introduce an auxiliary complex variable x belonging to the domain [1] $\overline{\mathbb{C}}\setminus\{t_1,\dots,t_n\}$. Consider the linear differential equation

$$\frac{dY(x)}{dx}=\Big(\sum_{1\leq i\leq n}\frac{Q_i}{x-t_i}\Big)\cdot Y(x) \tag{3.9}$$

with respect to a $k\times k$ matrix function $Y(x)$ of the variable x. The coefficient matrix function $\sum\frac{Q_i}{x-t_i}$ of the equation (3.9) is holomorphic in the domain $\overline{\mathbb{C}}\setminus\{t_1,\dots,t_n\}$, and if the condition

$$\sum_{1\leq i\leq n}Q_i=0 \tag{3.10}$$

holds, then the residue of the coefficient matrix function at the point ∞ vanishes. Thus, under the condition (3.10), the Cauchy problem with the initial condition

$$Y(\infty)=I \tag{3.11}$$

for the differential equation (3.9) in the domain $\overline{\mathbb{C}}\setminus\{t_1,\dots,t_n\}$ is solvable. The solution $Y(x)$ of this Cauchy problem is a function holomorphic in the disk $R<|x|\leq\infty$, where $R=\max_j|t_j|$. The function $Y(x)$ admits a Laurent expansion in this disk:

$$Y(x)=I+\sum_{1\leq l\leq\infty}\frac{Y_{-l}}{x^l},\quad R<|x|\leq\infty. \tag{3.12}$$

[1] We recall that $\overline{\mathbb{C}}=\mathbb{C}\cup\infty$ is the extended complex plane.

Lemma 3.6. *Under the condition* (3.10), *the Laurent coefficient* Y_{-1} *of the expansion* (3.12) *for the solution* $Y(x)$ *of the Cauchy problem* (3.9)–(3.11) *can be expressed as*

$$Y_{-1} = \sum_{1\le i\le n} t_i Q_i. \tag{3.13}$$

Proof. Under the condition (3.10), the Laurent expansion at ∞ of the coefficient matrix function $\sum \frac{Q_i}{x-t_i}$ is of the form

$$\sum_{1\le i\le n} \frac{Q_i}{x-t_i} = \Big(\sum_{1\le i\le n} t_i Q_i\Big)\cdot x^{-2} + o(|x|^{-2}) \quad \text{as} \quad x \to \infty. \tag{3.14}$$

Substituting the expansions (3.12) and (3.14) into the differential equation (3.9) and comparing the Laurent coefficients, we obtain (3.13). □

Theorem 3.7. *Let* $k\times k$ *matrix functions* $Q_1(\boldsymbol{t}),\dots,Q_n(\boldsymbol{t})$ *satisfy the Schlesinger system for* $\boldsymbol{t}\in\mathcal{G}$, *where* $\mathcal{G}$ *is a domain in* $\mathbb{C}^n_*$. *Let the equality* (3.3) *hold at least at one point* $\boldsymbol{t}=\boldsymbol{t}_0\in\mathcal{G}$, *or, what is the same* [2], *at every point* $\boldsymbol{t}\in\mathcal{G}$. *Assume that for each fixed* $\boldsymbol{t}\in\mathcal{G}$, $k\times k$ *matrix function* $Y(x,\boldsymbol{t})$ *satisfies the linear differential equation*

$$\frac{dY(x,\boldsymbol{t})}{dx} = \Big(\sum_{1\le i\le n}\frac{Q_i(\boldsymbol{t})}{x-t_i}\Big)\cdot Y(x,\boldsymbol{t}), \quad x\in\mathbb{C}\setminus\{t_1,\dots,t_n\}, \tag{3.15}$$

with respect to x *and the initial condition* [3]

$$Y(x,\boldsymbol{t})_{|x=\infty} = I. \tag{3.16}$$

Then the coefficient [4] $V(\boldsymbol{t})$ *of the Laurent expansion*

$$Y(x,\boldsymbol{t}) = I + \frac{V(\boldsymbol{t})}{x} + \frac{Y_{-2}(\boldsymbol{t})}{x^2} + \frac{Y_{-3}(\boldsymbol{t})}{x^2} + \cdots, \quad (\max_i |t_i|) < |x| \le \infty, \tag{3.17}$$

is the potential matrix function for the solution $Q_1(\boldsymbol{t}),\dots,Q_n(\boldsymbol{t})$, *i.e., the equalities* (3.2) *hold.*

Proof. This theorem is an immediate consequence of Lemmas 3.3 and 3.6. □

The differential equation (3.15) plays an important role in what follows.

4. The Schlesinger system as the compatibility condition for some over-determined system of linear differential equations

The result stated in this section is an important step in the study of the nonlinear Schlesinger system (1.1) by linear methods.

---

[2]Lemma 3.4.

[3]Due to (3.3), the value $Y(x,\boldsymbol{t})_{|x=\infty}$ exists.

[4]Here we redenote the Laurent coefficient $Y_{-1}(\boldsymbol{t})$ by $V(\boldsymbol{t})$.

Theorem 4.1. *Let $Q_1(\boldsymbol{t}),\dots,Q_n(\boldsymbol{t})$ be $k\times k$ matrix functions, holomorphic with respect to $\boldsymbol{t}=(t_1,\dots,t_n)$ in some open domain $\mathcal{G}$, $\mathcal{G}\subseteq\mathbb{C}^n_*$.*

Assume that the following system of linear differential equations

$$\frac{\partial Y}{\partial x}=\Bigg(\sum_{1\le j\le n}\frac{Q_j(\boldsymbol{t})}{x-t_j}\Bigg)Y, \tag{4.1a}$$

$$\frac{\partial Y}{\partial t_k}=-\frac{Q_k(\boldsymbol{t})}{x-t_k}Y \quad (k=1,\dots,n) \tag{4.1b}$$

for a square matrix function Y is compatible, that is, there exists a $k\times k$ matrix function $Y=Y(x;\boldsymbol{t})$, which is holomorphic, satisfies the equations (4.1) for $x\in\mathbb{C}$, $\boldsymbol{t}\in\mathcal{G}$, $x\neq t_1,\dots,t_n$, and, moreover, is not degenerate:

$$\det Y(x;\boldsymbol{t})\neq 0,\quad \textit{for}\quad x\in\mathbb{C},\ \boldsymbol{t}\in\mathcal{G},\ x\neq t_1,\dots,x\neq t_n. \tag{4.2}$$

Then the matrix functions $Q_1(\boldsymbol{t}),\dots,Q_n(\boldsymbol{t})$ satisfy the Schlesinger system (1.1) for $\boldsymbol{t}\in\mathcal{G}$.

Proof. Let us exploit the equalities

$$\frac{\partial}{\partial t_k}\frac{\partial Y}{\partial x}=\frac{\partial}{\partial x}\frac{\partial Y}{\partial t_k},\quad 1\le k\le n. \tag{4.3}$$

Differentiating equation(4.1a) with respect to t_k, we obtain

$$\frac{\partial}{\partial t_k}\frac{\partial Y}{\partial x}=\Bigg(\sum_{1\le j\le n}\frac{Q_j}{x-t_j}\Bigg)\frac{\partial Y}{\partial t_k}+\Bigg(\sum_{1\le j\le n}\frac{\frac{\partial Q_j}{\partial t_k}}{x-t_j}\Bigg)Y+\frac{Q_k}{(x-t_k)^2}Y.$$

Substituting the expression for $\frac{\partial Y}{\partial t_k}$ from (4.1b) into the last equality, we get

$$\frac{\partial}{\partial t_k}\frac{\partial Y}{\partial x}=-\sum_{1\le j\le n}\frac{Q_jQ_k}{(x-t_j)(x-t_k)}Y+\frac{Q_k}{(x-t_k)^2}Y+\sum_{1\le j\le n}\frac{\dfrac{\partial Q_j}{\partial t_k}}{x-t_j}Y. \tag{4.4}$$

Differentiating equation(4.1b) with respect to x, we obtain

$$\frac{\partial}{\partial x}\frac{\partial Y}{\partial t_k}=\frac{Q_k}{(x-t_k)^2}Y-\frac{Q_k}{x-t_k}\frac{\partial Y}{\partial x}.$$

Substituting the expression for $\frac{\partial Y}{\partial x}$ from (4.1a) into the last equality, we get

$$\frac{\partial}{\partial x}\frac{\partial Y}{\partial t_k}=\frac{Q_k}{(x-t_k)^2}Y-\sum_{1\le j\le n}\frac{Q_kQ_j}{(x-t_j)(x-t_k)}Y \tag{4.5}$$

Thus, equation (4.3) takes the form

$$\sum_{1\le l\le n}\frac{\dfrac{\partial Q_l}{\partial t_k}}{x-t_l}=\sum_{\substack{1\le l\le n\\ l\neq k}}\frac{[Q_l,Q_k]}{(x-t_l)(x-t_k)} \tag{4.6}$$

(Since matrix Y is non-degenerate, we can cancel out the factor Y.)

The numerators in (4.6) do not depend on x. Thus, (4.6) expresses the equality of two rational functions. Comparing the residues of both sides of (4.6) at $x = t_j, j \neq k$, we obtain (1.1.a). Comparing the residues at $x = t_k$, we obtain (1.1.b). □

Remark 4.2. In the proof of Theorem 4.1 we have not exploited the equalities

$$\frac{\partial}{\partial t_k}\frac{\partial Y}{\partial t_j} = \frac{\partial}{\partial t_j}\frac{\partial Y}{\partial t_k}, \quad 1 \leq k, j \leq n.$$

They lead to (1.1.a) and provide no additional information on $Q_1, \ldots, Q_n$. Thus, reversing the reasoning in the proof, we can demonstrate that the Schlesinger system (1.1) is not only a necessary, but also a sufficient condition for the solvability of the linear system (4.1a)–(4.1b) for a non-degenerate matrix function Y.

Remark 4.3. The result, which we formulated as Theorem 4.1, is contained in the paper [Sch1] of L. Schlesinger. (See Section II of this paper, especially p. 294. See also p. 324 of the book [Sch2]). This was the paper [Sch1] where the Schlesinger equations (1.1) originally appeared. In [Sch1], the equations (1.1) were derived as the compatibility condition of some over-determined system of linear differential equations. In [Gar], R. Garnier has derived the Painlevé equation P_{VI} from the compatibility condition of the appropriate over-determined system of linear differential equations. (This over-determined system itself appeared in the paper [FuR] of R. Fuchs, which, in its turn, was based on works of his father L.Fuchs published in Sitzungsberichte der Berliner Akademie der Wissenschaften in 1888–1898) Since the publication of the works [ZaSh] and [AKNS], representing a solution of a nonlinear differential equation as the compatibility condition for some over-determined system of linear differential equations has become a common wisdom. In [AKNS], under the strong influence of the paper [ZaSh], M. Ablovitz, D. Kaup, A.Newell and H. Segur, have represented the Korteveg de Vries equation, the nonlinear Schrödinger equation, the Sine-Gordon equation and a number of other non-linear equations of physical significance as the compatibility conditions for the appropriate systems of linear differential equations.

5. Fuchsian equations and behavior of their solutions – global and local

A Fuchsian equation is a linear differential equation of the form

$$\frac{dY}{dx} = \Big(\sum_{1\leq j\leq n} \frac{Q_j}{x - t_j}\Big)Y, \tag{5.1}$$

where $t_1, \ldots, t_n$ are pairwise distinct points of the complex plane ($t_1 \neq t_2 \neq \cdots \neq t_n$), $Q_j, j = 1, \ldots, n$ are square matrices of the same dimension (say, $k \times k$), x is the complex variable which belongs to the punctured Riemann sphere $\overline{\mathbb{C}} \setminus \{t_1, \ldots, t_n\}$. The points $t_1, \ldots, t_n$ are regular singular points for the equation (5.1). The residue of the coefficient matrix $\sum\limits_{1\le k\le n} \frac{Q_j}{x-t_j}$ of the equation (5.1) at the point ∞ is equal to $-\sum\limits_{1\le k\le n} Q_k$. So, if the condition

$$\sum_{1\le k\le n} Q_k = 0 \tag{5.2}$$

is satisfied, the point ∞ is a regular point for the equation (5.1). In what follows we assume that the condition (5.2) is satisfied.

We consider the differential equation (5.1) as a *matricial* one. This means that we consider solutions $Y(x)$ of this equation which are $k\times k$ matrix functions. If x_0 is a regular point for the equation (5.1) (that is $x_0 \neq t_1,\dots,t_n$), then for every $k\times k$ matrix Y_0 there exists the solution $Y(x;x_0,Y_0)$ which satisfies the initial condition

$$Y(x;x_0,Y_0)_{|x=x_0} = Y_0. \tag{5.3}$$

This solution is a holomorphic function in a disk neighborhood of the initial point x_0. It can be continued analytically along any path which does not pass through the singular points $t_1,\dots,t_n$. Thus the solution of the *linear* differential equation (5.1) exists *globally* and is a (multi-valued) holomorphic function on the punctured Riemann sphere $\overline{\mathbb{C}}\setminus\{t_1,\dots,t_n\}$. For a point $x\in\overline{\mathbb{C}}\setminus\{t_1,\dots,t_n\}$, the value of the analytic continuation of the solution $Y(.;x_0,Y_0)$ along a path $\gamma(x_0,x)\subset\overline{\mathbb{C}}\setminus\{t_1,\dots,t_n\}$, which starts at the point x_0 and ends at the point x, can depend on this path γ. If $\gamma'(x_0,x)$ and $\gamma''(x_0,x)$ are two such paths, homotopic in $\overline{\mathbb{C}}\setminus\{t_1,\dots,t_n\}$, then the values of analytic continuations of the solution $Y(.;x_0,Y_0)$ along γ' and γ'' coincide. Thus, the value of the analytic continuation of the solution $Y(.;x_0,Y_0)$ to the point x depends on *the homotopy class of the path* leading from x_0 to x, along which the continuation is made, but *not on the path itself.* In other words, the solution $Y(.,x_0,Y_0)$ of the differential equation (5.1) with the initial condition (5.3) is holomorphic function on the universal covering space $\operatorname{cov}(\overline{\mathbb{C}}\setminus\{t_1,\dots,t_n\};x_0)$ of the space $\overline{\mathbb{C}}\setminus\{t_1,\dots,t_n\}$ with the distinguished point x_0.

Definition 5.1. A solution $Y(x)$ of a linear differential equation in a connected domain is said to be fundamental if $\det Y(x)\not\equiv 0$ (or, what is equivalent, if $\det Y(x_0)\neq 0$ for some point x_0 from this domain).

If $Y(x)$ is a fundamental solution of a linear differential equation and $Z(x)$ is another solution of this equation, then $Z(x)=Y(x)C$, where C is a constant (with respect to x) matrix. In what follows we distinguish the solution $Y(x)$ of the differential equation (5.1), which satisfies the normalizing (initial) condition

$$Y(x_0) = I \tag{5.4}$$

at a distinguished point x_0 which is regular for the equation (5.1).

Let $\gamma(x_0,x_0)$ be a *closed* path, which starts at the point x_0 and ends *at the same point* x_0. We call such a path *a loop with the distinguished point* x_0. Let $x_0^*(\gamma)$ be the point of the universal covering space $\operatorname{cov}(\overline{\mathbb{C}}\setminus\{t_1,\dots,t_n\};x_0)$, which corresponds to the homotopic class γ of the loop $\gamma(x_0, x_0)$. We denote by M_γ the value $Y(x_0^*(\gamma); x_0, I)$ of the solution of the equation (5.1), normalized by the initial condition (5.4), at the point $x_0^*(\gamma)$. The correspondence $\gamma \mapsto M_\gamma$ is a representation [5]of the fundamental group $\pi(\overline{\mathbb{C}} \setminus \{t_1, \dots, t_n\}; x_0)$ into the general linear group $GL(k; \mathbb{C})$:

$$M_{\gamma'\gamma''} = M_{\gamma'} M_{\gamma''} \quad \text{for every} \quad \gamma', \gamma'' \in \pi(\overline{\mathbb{C}} \setminus \{t_1, \dots, t_n\}; x_0)$$

and

$$M_\varepsilon = I, \quad \text{where} \quad \varepsilon \quad \text{is the unity of the group} \quad \pi(\overline{\mathbb{C}} \setminus \{t_1, \dots, t_n\}; x_0)$$

Definition 5.2. The matrix M_γ is said to be **the monodromy matrix** for the equation (5.1), corresponding to the homotopic class γ. The correspondence $\gamma \mapsto M_\gamma$ is said to be **the monodromy representation** of the equation (5.1).

The group $\pi(\overline{\mathbb{C}} \setminus \{t_1, \dots, t_n\}; x_0)$ is finitely generated. Let us choose a special system of generators of this group. Every simple loop (i.e., the loop which is homeomorphic to a circle) divides the Riemann sphere $\overline{\mathbb{C}}$ into two components. Let γ_j be a simple loop in the space $\overline{\mathbb{C}} \setminus \{t_1, \dots, t_n\}$, with the distinguished point x_0, such that the point t_j and the set of all other points $\{t_p\}_{p \neq j}$ belong to the different components and, moreover, the loop γ_j goes around the point t_j in the counter-clockwise direction. (See Figure 1). We denote the homotopic class of this loop by

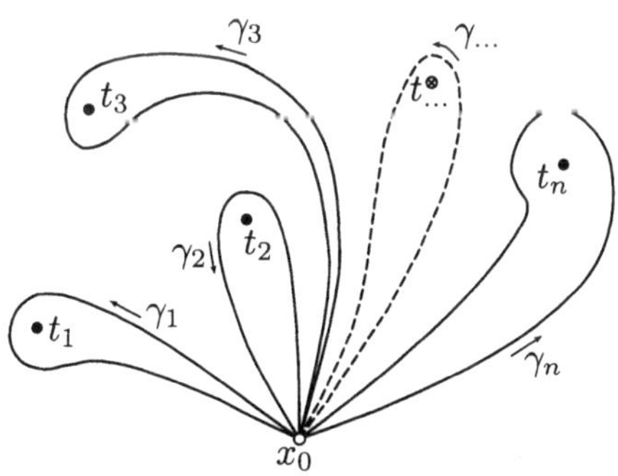

FIGURE 1

γ_j as well. It is known that the elements $\gamma_1, \dots, \gamma_n$ generate the fundamental group $\pi(\overline{\mathbb{C}} \setminus \{t_1, \dots, t_n\}; x_0)$ and that the equality $\gamma_1 \cdot \dots \cdot \gamma_n = \varepsilon$ is the only generating relation [6].

For what follows, it is crucial to know how a solution of a Fuchsian differential equation behaves in a neighborhood of its singular point.

---

[5]As usual, the product $\gamma'\gamma''$ is defined by the composed loop, obtained by traversing first the loop, generating γ'', and then the loop, generating γ'.

[6]The generating relation in this form holds for the loops, arranged as shown in Figure 1. In the general case, this relation is of the form $\gamma_{p_1} \cdot \dots \cdot \gamma_{p_n} = \varepsilon$ where $p_1, \dots, p_n$ is a permutation of the indices $1, \dots, n$ which is determined by the arrangement of the loops.

More generally, let us discuss the behavior of a solution of the differential equation of the form

$$\frac{dY}{dx} = \left(\frac{Q}{x-t} + \Phi(x)\right)Y(x), \tag{5.5}$$

where Q is a constant $k \times k$ matrix, and Φ is a $k \times k$ matrix function, which is holomorphic in the disk $D_{t,\rho}$, centered at the point $x = t$, of radius ρ. Solutions of the equation (5.5) are holomorphic (in general, multi-valued) functions in the punctured disk $D_{t,\rho} \setminus \{t\}$.

The equation (5.5) can be considered as a perturbation of the equation

$$\frac{dY}{dx} = \frac{Q}{x-t}Y(x). \tag{5.6}$$

The "unperturbed" equation (5.6) can be solved explicitly. It has a fundamental solution of the form

$$Y(x) = (x-t)^Q. \tag{5.7}$$

(By definition, $(x-t)^Q \overset{\text{def}}{=} e^{Q\ln(x-t)}$). In general, the function $(x-t)^Q$ considered as a function in the complex plane is multi-valued. It can be considered as a univalued holomorphic matrix function on the universal covering $\operatorname{cov}(\mathbb{C} \setminus \{t\})$ of the complex plane punctured at the point $\{t\}$ (or, in other terminilogy, on the Riemann surface of $\log(x-t)$). The differentiation rule

$$\frac{d}{dx}(x-t)^Q = \frac{Q}{x-t}(x-t)^Q, \quad x \in \operatorname{cov}\big(\mathbb{C} \setminus \{t\}\big), \tag{5.8}$$

holds. (The values of $(x-t)^Q$ in both sides of (5.8) are the same).

The structure of the solution of the original "perturbed" equation (5.5) depends on the nature of the spectrum of the matrix Q.

Definition 5.3. A square matrix Q is said to be non-resonant if no two eigenvalues of Q differ by a nonzero integer, or in other words, if the spectra of the matrices $Q + nI$ and Q are disjoint for every $n \in \mathbb{Z} \setminus 0$.

Remark 5.4. It may happen that some eigenvalues of a non-resonant matrix Q coincide.

Proposition 5.5. *If Q is a non-resonant (constant with respect to x) matrix, and $\Phi(x)$ is a matrix-function holomorphic in the disk $D_{t,\rho}$, then the differential equation (5.5) has a fundamental solution $Y(x)$ of the form*

$$Y(x) = H(x)(x-t)^Q, \quad x \in \operatorname{cov}(D_{t,\rho} \setminus \{t\}), \tag{5.9}$$

defined on the universal covering $\operatorname{cov}(D_{t,\rho} \setminus \{t\})$ of the punctured disc $D_{t,\rho} \setminus \{t\}$, where $H(x)$ is a matrix function, holomorphic and invertible [7] *in the disk $D_{t,\rho}$.*

[7]The inverse matrix function $H^{-1}(x)$ exists and is holomorphic in $D_{t,\rho}$.

The proof of Proposition 5.5, and even of a more general statement, is given in Appendix.

In the general case, when the eigenvalues of the matrix Q can differ by non-zero integers, the situation is more complicated.

Proposition 5.6. *The differential equation* (5.5), *with a constant matrix Q and a matrix function $\Phi(x)$, holomorphic in the disk $D_{t,\rho}$, has a fundamental solution $Y(x)$ of the form*

$$Y(x) = H(x)(x-t)^{L}(x-t)^{\hat{Q}}, \tag{5.10}$$

where $H(x)$ is a matrix function holomorphic and invertible in the disk $D_{t,\rho}$, L is a diagonalizable matrix with integer eigenvalues $l_1, \dots, l_k$, and $\hat{Q}$ is a non-resonant matrix whose eigenvalues $\hat{\lambda}_1, \dots, \hat{\lambda}_k$ are related to the eigenvalues $\lambda_1, \dots, \lambda_k$ of the matrix Q according to

$$\hat{\lambda}_p = \lambda_p - l_p, \qquad 1 \leq p \leq k. \tag{5.11}$$

For a certain choice of the integers $l_1 \geq \cdots \geq l_n$, the matrices L and Q can be simultaneously reduced to the forms

$$L = T^{-1}\mathrm{Diag}(l_1, \dots, l_n)T, \ \hat{Q} = T^{-1}\Big(\mathrm{Diag}(\hat{\lambda}_1, \dots, \hat{\lambda}_n) + U\Big)T,$$

where U is an upper-triangular matrix with zero diagonal.

Proposition 5.6 is applicable to equation (5.5) with the general matrix Q. In particular, it is applicable to the equation with the non-resonant matrix Q. However, if Q is non-resonant, one can choose $L = 0$ and $\hat{Q} = Q$.

Remark 5.7. It should be mentioned that the matrix function $H(x)$ in (5.12) (or in (5.14)) is not quite unique. The matrix $H(t) = H(x)_{x=t}$ can be an arbitrary invertible matrix commuting with the matrix Q. Of course, the solution (5.9) (or (5.10)) depends on this choice. As soon as this choice is made, the solution is determined uniquely (up to the choice of $\arg(x-t)$ which leads to the non-uniqueness of $(x-t)^Q = e^{Q\ln(x-t)}$). In what follows, we choose $H(x)_{|x=t} = I$. The latter equality can be considered as a *normalizing condition* for the solution (5.9) (or (5.10)).

Remark 5.8. There is an essential difference between the non-resonant and the "resonant" cases. If the matrix Q is not non-resonant, then the matrix $\hat{Q}$, which appears in (5.10), depends on the value $\Phi(t) = \Phi(x)_{x=t}$ of the "perturbing" (see (5.5)) function Φ at the point t. In particular, the Jordan structure of the matrix $\hat{Q}$ may depend on the matrix $\Phi(t)$. This Jordan structure, together with the eigenvalues of $\hat{Q}$, determine the multi-valued behavior of the matrix function $(x-t)^{\hat{Q}}$, and finally, the multi-valued behavior of $Y(x)$ in $D_{t,\rho}$. Thus, in the "resonant" case the multi-valued behavior of the solution $Y(x)$ depends not only on the matrix Q (the coefficient of the "leading term") but also on the "perturbing" matrix Φ. As can be seen from (5.9), in the non-resonant case the multi-valued behavior of the solution $Y(x)$ depends only on Q, but not on Φ.

Remark 5.9. Of course, for every matrices A and B, the matrix $e^A \cdot e^B$ can be presented as e^C:

$$e^A \cdot e^B = e^C, \qquad \text{where} \qquad C = \log(e^A \cdot e^B),$$

where an arbitrary value for the matrix $\log(e^A \cdot e^B)$ can be chosen. However, the case of exponential functions is different. In general, a matrix function $e^{A\zeta} \cdot e^{B\zeta}$ of ζ (with constant matrices A and B) is not representable as $e^{C\zeta}$ with a constant matrix C. In particular, in the resonant case the solution (5.10) of the differential equation (5.5) is not, in general, representable in the form $Y(x) = H(x) \cdot (x-t)^R$ with a holomorphic invertible matrix function H and a constant matrix R.

Propositions 5.5 and 5.6 go back to L. Fuchs and G. Frobenius. The systematic study of linear differential equations with analytic coefficients in the neighborhood of a singular point was initiated by Lazarus Fuchs in 1866 and followed by G. Frobenius in 1873. They discussed the "scalar" differential equation of the nth order. Matrix differential equations in the complex domain were considered later. The statement which we formulated as Proposition 5.5 can be found in a lot of excellent textbooks now. See, for example, [CoLe], Theorem 4.1; [Gant], Chapter XV, §10; [Hart], Chapter IV, Corollary 11.2; [IKSY], Chapter I, Theorem 2.3. However, it is difficult to point out the source, where this statement first appeared precisely in the same form as in Proposition 5.5. The earliest reference we know is [Ra] (see Section 3 there). The statement which we formulated as Proposition 5.6 can be found in [Gant], Chapter XV, §10, in [HsSi], Chapter V, Section 5; [Sib], Chapter III, Theorem 3.9.4,page 89. The earliest of this references is [Gant] (the first edition of Gantmacher's book appeared in 1954). However, Y. Sibuya, [Sib], page 89, refers to Hukuhara's book [Huk] (Theorem 41.1, page 143) published in 1950. Unfortunately, the book [Huk] is unavailable to us.

Let us return to the Fuchsian equation (5.1). We consider this equation in a disk neighborhood $D_{t_j,\rho}$ of the singular point t_j, which does not contain other singular points $t_p, p \neq j$. Let us choose and fix a point x_j of the punctured disk $D_{t_j,\rho} \setminus \{t_j\}$. We also choose and fix a path α_j in $\overline{\mathbb{C}} \setminus \{t_1, \dots, t_n\}$, which starts at the distinguished point x_0 and ends at the point x_j. (See Figure 2.a.) We need these x_j and α_j to distinguish a certain analytic continuation Y_{α_j} of the solution of (5.1) normalized by (5.4) into the punctured disk $D_{t_j,\rho} \setminus \{t_j\}$. Let x be an arbitrary point of the punctured disk $D_{t_j,\rho} \setminus \{t_j\}$, and μ be an arbitrary path in $D_{t_j,\rho} \setminus \{t_j\}$, starting at x_j and ending at x. (See Figure 2.b.). So, the pair (x, μ) can be considered as a point of the universal covering $\operatorname{cov}(D_{t_j,\rho} \setminus \{t_j\}; x_j)$. We construct the composed path $\alpha = \mu \cdot \alpha_j$. The path α starts at the point x_0 and ends at the point x. (See Figure 2.c.) We continue analytically the solution Y along the path α from the point x_0 to the point x. In this manner we distinguish the continuation $Y_{\alpha_j}(x)$ of the solution $Y(.)$, normalized by the condition (5.4), from a neighborhood of x_0 to the punctured disk $D_{t_j,\rho} \setminus \{t_j\}$. The function $Y_{\alpha_j}(.)$ can be considered as a multi-valued function in $D_{t_j,\rho} \setminus \{t_j\}$, or, better to say, as a single-valued function in the universal covering $\operatorname{cov}(D_{t_j,\rho} \setminus \{t_j\}; x_j)$ with the

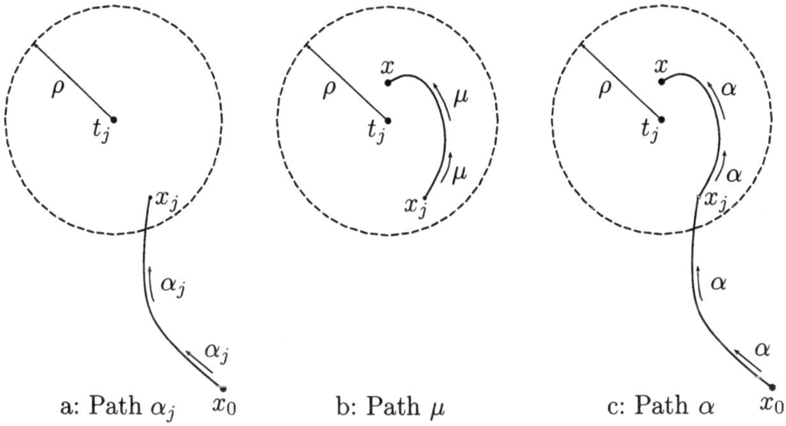

FIGURE 2

distinguished point x_j. The notation Y_{α_j} reflects how the path α_j is involved in the process of the analytic continuation. In fact, this process involves two steps. On the first step, the solution is continued along the path α_j from a neighborhood of the initial point x_0 to a neighborhood of the point x_j, where the point x_j is considered as a distinguished point of the punctured disk $D_{t_j,\rho} \setminus \{t_j\}$. On the second step, the solution is continued along paths which are contained in $D_{t_j,\rho} \setminus \{t_j\}$ from the neighborhood of the point x_j to points of $\mathrm{cov}(D_{t_j,\rho} \setminus \{t_j\}; x_j)$.
The matrix function $Y_{\alpha_j}(.)$ is a fundamental solution of (5.1). Propositions 5.5–5.6 provide another fundamental solution of (5.1). As a preliminary we assume that the matrix Q_j is non-resonant. According to Proposition 5.5, there exists the fundamental solution $Y_j(x)$ of (5.1), which is representable as

$$Y_j(x) = H_j(x)(x - t_j)^{Q_j}, \qquad x \in \mathrm{cov}\big(D_{t_j,\rho} \setminus \{t_j\}, x_j\big), \tag{5.12}$$

where $H_j(x)$ is a matrix function, holomorphic and invertible in $D_{t_j,\rho}$. The matrix function $(x-t_j)^{Q_j}$ is defined up to a right factor of the form $\exp[2\pi i Q_j]$. To avoid this non-uniqueness, we choose a certain value

$$\theta_j = \arg(x_j - t_j), \tag{5.13}$$

and fix this choice. Under the agreement (5.13), the matrix function $(x - t_j)^{Q_j}$ is defined uniquely on $\mathrm{cov}\big(D_{t_j,\rho} \setminus \{t_j\}, x_j\big)$.

If the matrix Q_j is resonant, then, according to Proposition 5.6, there exists the fundamental solution $Y_j(x)$ of (5.1), which is representable as

$$Y_j(x) = H_j(x)(x - t_j)^{L_j}(x - t_j)^{\hat{Q}_j}, \qquad x \in \mathrm{cov}\big(D_{t_j,\rho} \setminus \{t_j\}\big), \tag{5.14}$$

where the matrix $H_j(x)$ has the same properties as before, L_j is a diagonalizable matrix with integer eigenvalues (so, the matrix function $(x-t_j)^{L_j}$ is single-valued in $D_{t_j,\rho} \setminus \{t_j\}$), and $\hat{Q}_j$ is a non-resonant matrix, whose eigenvalues differ from eigenvalues of the matrix Q_j by some integer numbers.

The matrix function $H_j(x)$ in (5.12) (or in (5.14)) is not quite unique. (See Remark 5.7). However, under the normalizing condition

$$H(x)_{|x=t_j} = I \tag{5.15}$$

the matrix function $H_j(x)$, as well as the solution Y_j, are unique in the non-resonant case.

Definition 5.10. A solution Y_j of the Fuchsian equation (5.1) representable in the form (5.12) (in the non-resonant case), or in the form (5.14) (in the resonant case) near $\{t_j\}$, with H_j satisfying the normalizing condition (5.15), is said to be the local solution of (5.1) normalized at the singular point $\{t_j\}$.

Since both solutions, Y_{x_j,α_j} and Y_j, are fundamental, the equality

$$Y_{x_j,\alpha_j}(x) = Y_j(x)C, \tag{5.16}$$

holds, where C is a constant (with respect to x) non-degenerate matrix.

Definition 5.11. The equality (5.16) is said to be the connection relation. The matrix C from (5.16) is said to be the connection coefficient.

According to Definition 5.11, the connection coefficient C depends on the choice of the point $x_j \in D_{t_j,\rho} \setminus \{t_j\}$ and on the choice of the path α_j, connecting the points x_0 and x_j: $C = C_{x_j,\alpha_j}$. Now we will discuss this dependence.

Let α'_j and α''_j be two paths, both of which start at the point x_0 and end at the point x_j, $\alpha'_j, \alpha''_j \subset \overline{\mathbb{C}} \setminus \{t_1, \dots, t_n\}$. It is clear that $\alpha''_j = \alpha'_j \cdot \gamma$, where γ is the loop with the distinguished point x_0, which is composed of the path α''_j and of the path $(\alpha'_j)^{-1}$ (the latter is the path α'_j which is traversed in the opposite direction: from the point x_j to the point x_0): $\gamma = (\alpha'_j)^{-1} \cdot \alpha''_j$. The paths α' and α'', connecting the point x_0 with the point x, are defined as the compositions $\alpha' = \mu \cdot \alpha'_j$ and $\alpha'' = \mu \cdot \alpha''_j$, where, as before, μ , $\mu \subset D_{t_j,\rho} \setminus \{t_j\}$, is a path connecting the points x_j and x. Thus, $\alpha'' = \alpha' \cdot \gamma$, and

$$Y_{\alpha''_j}(x) = Y_{\alpha'_j}(x) M_\gamma,$$

where M_γ is the monodromy matrix corresponding to the loop γ. On the other hand,

$$Y_{\alpha'_j}(x) = Y_j(x) C_{x_j,\alpha'_j}, \qquad Y_{\alpha''_j}(x) = Y_j(x) C_{x_j,\alpha''_j}.$$

Therefore,

$$C_{x_j,\alpha''_j} = C_{x_j,\alpha'_j} M_\gamma, \quad \text{where } \gamma = (\alpha'_j)^{-1} \cdot \alpha''_j. \tag{5.17}$$

Let us discuss, how the connection coefficient C depends on the point x_j. Given two points $x'_j, x''_j \in D_{t_j,\rho} \setminus \{t_j\}$ and a path $\alpha'_j \subset \overline{\mathbb{C}} \setminus \{t_1, \dots, t_n\}$, connecting x_0 and x'_j, there exists a path $\alpha''_j \subset \overline{\mathbb{C}} \setminus \{t_1, \dots, t_n\}$, connecting x_0 and x''_j, such that

$$C_{x'_j,\alpha'_j} = C_{x''_j,\alpha''_j}, \tag{5.18}$$

Indeed, let ν be a path, connecting x'_j and x''_j, $\nu \subset D_{t_j,\rho} \setminus \{t_j\}$. We define the path α''_j connecting x_0 with x''_j as the composition $\alpha''_j = \nu \cdot \alpha'_j$. The pairs (x'_j, α'_j) and

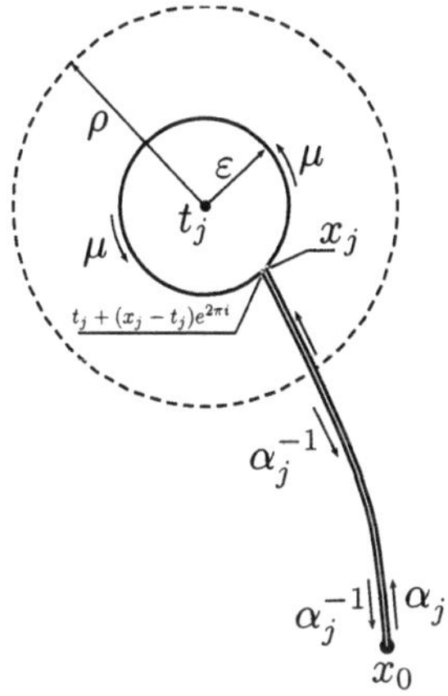

FIGURE 3. Loop γ_{α_j}.

(x''_j, α''_j) allow us to distinguish the solutions $Y_{\alpha'_j}$ and $Y_{\alpha''_j}$ of the equation (5.1) in the punctured neighborhood $D_{t_j,\rho} \setminus \{t_j\}$. To obtain values of these solutions at a point $x \in D_{t_j,\rho} \setminus \{t_j\}$, we have to choose paths μ'_j and μ''_j, connecting the points x'_j and x''_j, respectively, with the point x. If we make this choice coherently, we obtain the equality $Y_{\alpha'_j}(x) = Y_{\alpha''_j}(x)$ which implies (5.18). Such a coherent choice can be done in the following way. If μ' is a path in $D_{t_j,\rho} \setminus \{t_j\}$, connecting x'_j with x, we define the path μ'', connecting x''_j with x, as the composition $\mu'' = \mu' \cdot \nu^{-1}$. To obtain the values of the solutions $Y_{x'_j,\alpha'_j}$ and $Y_{x''_j,\alpha''_j}$ at the point x, we have to continue analytically the solution Y, which is normalized by (5.4), from the point x_0 to the point x along the paths $\alpha' = \mu' \cdot \alpha'_j$ and $\alpha'' = \mu'' \cdot \alpha''_j$, respectively. However, these two paths are homotopic. To see this, we remark that $\alpha'' = (\mu' \cdot \nu^{-1}) \cdot (\nu \cdot \alpha') = \mu' \cdot (\nu^{-1} \cdot \nu) \cdot \alpha'$ and that the path $\nu^{-1} \cdot \nu$ is contractible to the point x'_j. Hence, $Y_{\alpha'_j}(x) = Y_{\alpha''_j}(x)$. Thus, in some sense the connection coefficient does not depend on the choice of the point $x_j \in D_{t_j,\rho} \setminus \{t_j\}$.

Let us calculate the monodromy matrix corresponding to the loop "going around" the point t_j (See Figure 1). We relate such a loop γ_{α_j} to every path α_j, $\alpha_j \subset \overline{\mathbb{C}} \setminus \{t_1, \dots, t_n\}$, which starts at the point x_0 and ends at a point x_j belonging to a punctured disk $D_{t_j,\rho} \setminus \{t_j\}$. The radius ρ must be chosen so small that the disk does not contain any other point $t_p, p \neq j$. Let μ be the circular arc, centered at the point t_j, of the radius $\varepsilon = |x_j - t_j|$, which starts at the point x_j, ends at the point $t_j + (x_j - t_j)e^{2\pi i}$ and is oriented counterclockwise, making precisely one turn around its center t_j. We construct the loop γ_{α_j} as the composition of the path α_j, of the arc μ and of the path α_j^{-1}: $\gamma = (\alpha_j)^{-1} \cdot \mu \cdot \alpha_j$.

Definition 5.12. The loop γ_{α_j} is said to be the loop generated by the path α_j.

Lemma 5.13. *The monodromy matrix $M_{\gamma_{\alpha_j}}$ along the loop γ_{α_j} generated by a path α_j can be represented as*

$$M_{\gamma_{\alpha_j}} = (Y_{\alpha_j}(x_j))^{-1} Y_{\alpha_j}(t_j + (x_j - t_j)e^{2\pi i}), \tag{5.19}$$

where x_j is the endpoint of the path α_j.

Proof. For linear differential equations, the following *semigroup property* holds: Let γ be a path starting at a point a and ending at the point c, and let b is an "intermediate" point of the path α, that is the point which is located "between" the points a and c. Then $Y(c,a) = Y(c,b)Y(b,a)$, where for $\xi, \eta \in \alpha$, $Y(\eta,\xi))$ is the value $Y(x,\xi)_{|x=\eta}$ of the solution of the equation satisfying the initial condition $Y(x,\xi)_{|x=\xi} = I$. Let us take the loop γ_{α_j} as such a path γ. The point a is the point x_0 considered as the starting point of the loop, the point c is the same point x_0, but considered as the end point of the loop. The point b is the point $t_j + (x_j - t_j)e^{2\pi i}$. From the semigroup property it follows that $Y(b,c)Y(c,b) = I$. It is clear that $Y(b,a) = Y_{\alpha_j}(t_j + (x_j - t_j)e^{2\pi i})$, $Y(b,c) = Y_{\alpha_j}(x_j)$, and $Y(c,a) = M_{\gamma_{\alpha_j}}$. □

From (5.16) it follows that in the non-resonant case,

$$Y_{x_j,\alpha_j}(x_j) = H_j(x_j)(x_j - t_j)^{Q_j} C_{\alpha_j}$$

and

$$Y_{x_j,\alpha_j}(t_j + (x_j - t_j)e^{2\pi i}) = H_j(x_j)e^{2\pi i Q_j}(x_j - t_j)^{Q_j} C_{\alpha_j}.$$

(The function $H_j(x)$ is single-valued in $D_{t_j,\rho}$, and the function $(x - t_j)^{Q_j}$ receives the factor $e^{2\pi i Q_j}$, when x traverses the arc μ once, in the counterclockwise direction.) Therefore, according to Lemma 5.13,

$$M_{\gamma_j} = C_{\alpha_j}^{-1}(x_j - t_j)^{-Q_j} e^{2\pi i Q_j}(x_j - t_j)^{Q_j} C_{\alpha_j}. \tag{5.20}$$

Finally, we obtain

Lemma 5.14. *The monodromy matrix $M_{\gamma_{\alpha_j}}$, corresponding to the loop γ_{α_j} generated by a path α_j can be represented as*

$$M_{\gamma_j} = C_{\alpha_j}^{-1} \exp[2\pi i Q_j] C_{\alpha_j} \tag{5.21}$$

in the non-resonant case, and as

$$M_{\gamma_j} = C_{\alpha_j}^{-1} \exp[2\pi i \hat{Q}_j] C_{\alpha_j} \tag{5.22}$$

in the resonant case, where C_{α_j} is the connection coefficient, corresponding to the path α_j.

(The reasoning in the resonant case is analogous. One needs to use the equality (5.14), taking into account that the matrix-function $(x - t_j)^{L_j}$ is single-valued in the disk $D_{t_j,\rho}$).

For what follows, it is very fruitful to rewrite the representations (5.12)–(5.16), (5.14)–(5.16) for a "local" solution [8]of (5.1), as well as the expressions (5.21) and (5.22) for the monodromy matrices, in a different form. The respective

[8]I.e., for a solution in a neighborhood of the singular point t_j.

transformations are trivial, but nevertheless very useful. The form in which we present the local solutions hints at how to introduce the notion of the isoprincipal deformation of the Fuchsian differential equation. This notion will be of crucial importance to us.

Let $D_{t_j,\rho}$ be a disk neighborhood of the singular point t_j of the differential equation (5.1), which is small enough, so that the point t_j is the only singular point which is contained in this neighborhood. In this neighborhood lives a "local" solution Y_j of the equation (5.1), which in the non-resonant case is of the form (5.12), and in the resonant case is of the form (5.14). However, this "local" solution Y_j does not, in general, satisfy the initial condition (5.4). To be precise, we should speak about the analytic continuation of this solution from the neighborhood $D_{t_j,\rho}$ to the point x_0. To discuss such an analytic continuation, we have to choose a point x_j of the punctured neighborhood $D_{t_j,\rho}\setminus\{t_j\}$, and a path α_j, $\alpha_j \subset \overline{\mathbb{C}}\setminus\{t_1,\dots,t_n\}$, leading from the point x_0 to the point x_j. The solution Y_{α_j} which satisfies the initial condition (5.4) at the distinguished point x_0 and which is continued analytically to the neighborhood $D_{t_j,\rho}$ along the path α_j is of the form

$$Y_{\alpha_j}(x) = H_j(x)(x-t_j)^{Q_j}C_{\alpha_j}, \qquad x \in \operatorname{cov}\big(D_{t_j,\rho}\setminus\{t_j\}, x_j\big), \tag{5.23}$$

in the non-resonant case, and of the form

$$Y_{\alpha_j}(x) = H_j(x)(x-t_j)^{L_j}(x-t_j)^{\hat{Q}_j}C_{\alpha_j}, \qquad x \in \operatorname{cov}\big(D_{t_j,\rho}\setminus\{t_j\}, x_j\big), \tag{5.24}$$

in the resonant case, where C_{α_j} is the connection coefficient corresponding to the choice of the homotopic class of the path α_j and the choice of the arguments $\arg(x_j - t_j)$, see (5.13), (5.12) and (5.14).

We introduce the matrices

$$A_{\alpha_j} = C_{\alpha_j}^{-1}Q_jC_{\alpha_j}, \quad H_{\alpha_j} = H_jC_{\alpha_j} \tag{5.25}$$

in the non-resonant case, and the matrices

$$A_{\alpha_j} = C_{\alpha_j}^{-1}\hat{Q}_jC_{\alpha_j}, \quad B_{\alpha_j} = C_{\alpha_j}^{-1}L_jC_{\alpha_j}, \quad H_{\alpha_j} = H_jC_{\alpha_j} \tag{5.26}$$

in the resonant case. Propositions 5.5 and 5.6 can be "developed" in the following way.

Theorem 5.15. *If the matrix Q_j is non-resonant, then the solution Y_{α_j} of the differential equation* (5.1) *that satisfies the initial condition* (5.4) *at the distinguished point x_0 and is continued analytically along a path α_j into a small punctured neighborhood $D_{t_j,\rho}\setminus\{t_j\}$ of the singular point t_j, is representable in the form*

$$Y_{\alpha_j}(x) = H_{\alpha_j}(x)(x-t_j)^{A_{\alpha_j}}, \qquad x \in \operatorname{cov}\big(D_{t_j,\rho}\setminus\{t_j\}, x_j\big), \tag{5.27}$$

where A_{α_j} is a matrix, similar to the matrix Q_j, and H_{α_j} is a matrix function, holomorphic and invertible in the entire neighborhood $D_{t_j,\rho}$ (including the point t_j).

Theorem 5.16. *If the matrix Q_j is resonant, then the solution Y_{α_j} of the differential equation* (5.1), *that satisfies the initial condition* (5.4) *at the distinguished point x_0*

and is continued analytically along a path α_j into a small punctured neighborhood $D_{t_j,\rho} \setminus \{t_j\}$ of the singular point t_j, is representable in the form

$$Y_{\alpha_j}(x) = H_{\alpha_j}(x)(x-t_j)^{Z_{\alpha_j}}(x-t_j)^{A_{\alpha_j}}, \qquad x \in \operatorname{cov}(D_{t_j,\rho} \setminus \{t_j\}, x_j), \tag{5.28}$$

where Z_{α_j} is a diagonalizable matrix with integer eigenvalues $l_1, \dots, l_k$, A_{α_j} is a non-resonant matrix, whose eigenvalues $\hat{\lambda}_p$ are related to the eigenvalues λ_p of the matrix Q_j by the relation

$$\hat{\lambda}_p = \lambda_p - l_j, \quad 1 \le p \le k, \tag{5.29}$$

and $H_{\alpha_j}(x)$ is a matrix function, holomorphic and invertible in the entire neighborhood (including the point t_j). For a certain choice of the integers $l_1 \ge \dots \cdots \ge l_n$, the matrices Z_{α_j} and A_{α_j} can be simultaneously reduced to the forms

$$Z_{\alpha_j} = T_{\alpha_j}^{-1} \operatorname{Diag}(l_1, \dots, l_n) T_{\alpha_j}, \; A_{\alpha_j} = T_{\alpha_j}^{-1}\Big(\operatorname{Diag}(\hat{\lambda}_1, \dots, \hat{\lambda}_n) + U\Big) T_{\alpha_j},$$

where U is an upper-triangular matrix with zero diagonal.

Remark 5.17. If α_j' is another path with the same starting point x_0 and endpoint x_j, then, in the non-resonant case, the matrices A_{α_j} and H_{α_j} from (5.27) are transformed according to the rule

$$A_{\alpha_j'} = M_\gamma^{-1} A_{\alpha_j} M_\gamma, \qquad H_{\alpha_j'}(x) = H_{\alpha_j}(x) M_\gamma, \tag{5.30}$$

and in the resonant case, the matrices Z_{α_j}, A_{α_j} and H_{α_j} from (5.28) are transformed according to the rule

$$Z_{\alpha_j'} = M_\gamma^{-1} Z_{\alpha_j} M_\gamma, \quad A_{\alpha_j'} = M_\gamma^{-1} A_{\alpha_j} M_\gamma, \quad H_{\alpha_j'}(x) = H_{\alpha_j}(x) M_\gamma, \tag{5.31}$$

where M_γ is the monodromy matrix corresponding to the loop $\gamma = (\alpha_j)^{-1}\alpha_j'$. In particular, if the paths α_j and α_j' are homotopic then

$$Z_{\alpha_j'} = Z_{\alpha_j}, \quad A_{\alpha_j'} = A_{\alpha_j}, \quad H_{\alpha_j'}(x) = H_{\alpha_j}(x).$$

Theorem 5.18. *Let $M_{\gamma_{\alpha_j}}$ be the monodromy matrix for the differential equation* (5.1) *corresponding to the loop γ_{α_j}, which goes around the singular point t_j and is constructed from the path α_j, connecting the initial point x_0 to a small neighborhood of t_j: $\gamma_j = (\alpha_j)^{-1} \cdot \mu \cdot \alpha_j$ (see* Figure 3). *Then this monodromy matrix can be expressed as*

$$M_{\gamma_{\alpha_j}} = \exp(2\pi i A_{\alpha_j}), \tag{5.32}$$

where the matrix A_{α_j} is the same as in the exponent in (5.27), if Q_j is non-resonant, and the exponent in (5.28), if Q_j is resonant.

Theorem 5.19. *Let M_{γ_j} be the monodromy matrix for the differential equation* (5.1) *corresponding to a loop γ_j with the distinguished point x_0 which makes one turn counterclockwise around the singular point t_j and makes no turns around the other singular points $t_p, p = 1, \dots, t_n, p \ne j$. Then, whatever the matrix Q_j is, non-resonant or not, the spectra* $\operatorname{spec} M_{\gamma_j}$ *and* $\operatorname{spec} Q_j$ *are related by*

$$\operatorname{spec} M_{\gamma_j} = \exp\{2\pi i \operatorname{spec} Q_j\}, \tag{5.33}$$

where the latter equality holds "with multiplicities", that is the multiplicity κ_μ *of the eigenvalue* μ *of the matrix* M_{γ_j} *and multiplicities* κ_λ *of the eigenvalues* λ *of the matrix* Q_j *are related by* $\kappa_\mu = \sum \kappa_\lambda$*, were the sum is taken over all* λ *such that* $\exp\{\lambda\} = \mu$.

Proof. This theorem is the immediate consequence of Theorem 5.18, of the spectral mapping theorem for matrices (from (5.32) it follows that $\operatorname{spec} M_{\gamma_j} = \exp\{2\pi i \operatorname{spec} A_{\alpha_j}\}$), and of Theorems 5.15 or 5.16, which relate the spectra of the matrices A_{α_j} and Q_j. However, here we can do without Theorem 5.16, restricting ourselves to non-resonant matrices Q_j only. Every matrix can be approximated by a non-resonant one. The spectrum $\operatorname{spec} Q_j$ depends on Q_j continuously. The monodromy matrix M_{γ_j} depends on Q_j continuously as well. □

6. Deformations of Fuchsian differential equations

A *deformation* of a Fuchsian equation is a *family* of Fuchsian differential equations depending on parameters:

$$\frac{dY}{dx} = \Big(\sum_{1 \le j \le n} \frac{Q_p(\boldsymbol{\alpha})}{x - t_p} \Big) Y.$$

If the coefficients $Q_p(\boldsymbol{\alpha})$ depend on the parameters $\boldsymbol{\alpha} = (\alpha_1, \dots, \alpha_m)$ in some open set of the space $\mathbb{C}^m$ holomorphically, such a deformation is said to be a *holomorphic deformation of a Fuchsian differential equation*. The study of linear differential equations that depend on parameters holomorphically was started by Lazarus Fuchs, see [FuL1]–[FuL4]. This study was continued by L. Schlesinger, see [Sch1]–[Sch3], and by Richard Fuchs, see [FuR]. In particular, L. Schlesinger has considered linear equation, where loci $t_1, \dots, t_n$ of singular points serve as the parameters $\boldsymbol{\alpha}$. In other words, L. Schlesinger has considered deformations of Fuchsian differential equations of the form

$$\frac{dY}{dx} = \Big(\sum_{1 \le p \le n} \frac{Q_p(\boldsymbol{t})}{x - t_p} \Big) Y, \tag{6.1}$$

where the coefficients $Q_p(t_1, \dots, t_n), p - 1, \dots, n$, are $k \times k$ matrix functions which are defined and holomorphic for $\boldsymbol{t}$ from a domain $\mathcal{D}$ of the space $\mathbb{C}^n_*$.

In what follows we study the normalized by (5.4) solution $Y(x, \boldsymbol{t}))$ of the family (6.1) of Fuchsian equations as a function of x and $\boldsymbol{t}$. In particular, we consider the monodromy matrices M_{γ_j} for the equation (6.1), as well as the matrices A_{α_j} and functions $H_{\alpha_j}(x)$ from the representations (5.27) and (5.28) for various $t_1, \dots, t_n$: $M_{\gamma_j} = M_{\gamma_j}(\boldsymbol{t})$, $A_{\alpha_j} = A_{\alpha_j}(\boldsymbol{t})$ and $H_{\alpha_j} = H_{\alpha_j}(\boldsymbol{t})$, intending to study how these matrices depend on $\boldsymbol{t} = (t_1, \dots, t_n)$. Here a problem arises. The matrices M_{γ_j}, A_{α_j}, H_{α_j} depend on the homotopy classes of γ_j and α_j, which are considered on the punctured Riemann sphere $\overline{\mathbb{C}} \setminus \{t_1, \dots, t_n\}$. However, for different $\boldsymbol{t} = (t_1, \dots, t_n)$, the sets $\overline{\mathbb{C}} \setminus \{t_1, \dots, t_n\}$ are different. Thus we need to explain what does it mean that, for the deformation (6.1) of the Fuchsian equation, the monodromy matrices M_{γ_j}, or the exponents A_{α_j} from (5.27) (or (5.28)) are holomorphic with respect to $\boldsymbol{t}$, or what does it mean that these matrices do not depend on $\boldsymbol{t}$. The latter problems are of the local nature.

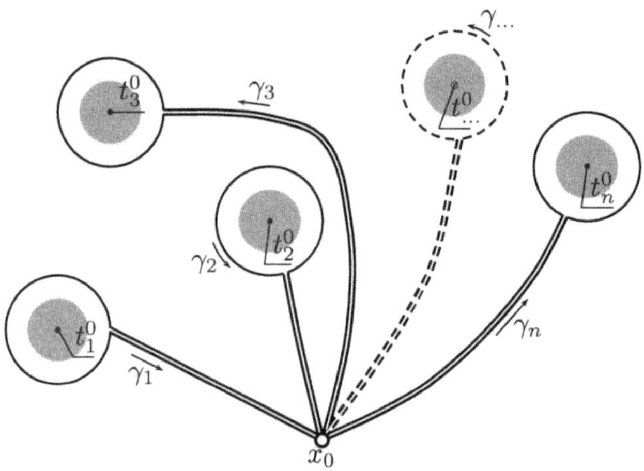

FIGURE 4

Let a point $\boldsymbol{t}^0 = (t_1^0, \dots, t_n^0) \in \mathbb{C}_*^n$ be fixed, $t_j^0 \neq x_0$, $j = 1, \dots, n$, where $x_0 \in \overline{\mathbb{C}}$ is the distinguished point (the same point which appears in the initial condition(5.4), and which serves as the distinguished point for the fundamental group $\pi(\overline{\mathbb{C}} \setminus \{t_1^0, \dots, t_n^0\}; x_0))$. Choose and fix simple loops $\gamma_j, j = 1, \dots, \gamma_n$, each of which starts and ends at the point x_0, the loop γ_j contains the point t_j^0 "inside" and goes around this point counterclockwise, and all other points $\{t_p^0\}_{p \neq j}$ lie "outside" the loop γ_j, $j = 1, \dots, t_n$. Let $\delta > 0$ be a small number such that the disk $D_{t_j^0,\delta} = \{z \in \mathbb{C} : |z - t_j^0| < \delta\}$ does not intersect the loop $\gamma_j, j = 1, \dots, n$ (and hence, is contained inside the loop γ_j). Then for every $\boldsymbol{t} = (t_1, \dots, t_n)$ such that $t_j \in D_{t_j^0,\delta}$, the loops $\gamma_j, j = 1, \dots, \gamma_n$, generate the fundamental group $\pi(\overline{\mathbb{C}} \setminus \{t_1, \dots, t_n\}; x_0))$. See Figure 4, where the disks $D_{t_j^0,\delta}$ are plotted in the light gray color.

Assume that the point $\boldsymbol{t}^0$ belongs to the domain $\boldsymbol{\mathcal{D}}$ of the space $\mathbb{C}_*^n$, where the matrix functions $Q_j(\boldsymbol{t})$ are holomorphic, and that δ is so small that the polydisk

$$\mathsf{D}(\boldsymbol{t}^0, \delta) = \prod_{p=1}^{n} D_{t_p^0,\delta}, \quad \text{where} \quad D_{t_p^0,\delta} = \{t_p : |t_p - t_p^0| < \delta\}, \tag{6.2}$$

is contained in $\boldsymbol{\mathcal{D}}$: $\mathsf{D}(\boldsymbol{t}^0, \delta) \subset \boldsymbol{\mathcal{D}}$. Then, for $x \in \gamma_j$, the coefficient matrix $\sum\limits_{1 \leq k \leq n} \frac{Q_j(\boldsymbol{t})}{x - t_j}$ of the linear equation (6.1) is holomorphic with respect to $\boldsymbol{t}$ in $\mathsf{D}(\boldsymbol{t}^0, \delta)$. From the standard results on dependence of solution of a linear differential equation on parameters follows

Lemma 6.1. *Under above stated assumptions and notation, the monodromy matrices $M_{\gamma_j}(\boldsymbol{t})$, $j = 1, \dots, n$, of the holomorphic family* (6.1) *of Fuchsian equations are holomorphic with respect to $\boldsymbol{t}$ from the polydisk $\mathsf{D}(\boldsymbol{t}^0, \delta)$.*

Now we investigate in more detail the local solution $Y_j(x, \boldsymbol{t})$ of the equation (6.1) which corresponds to a neighborhood of the point t_j. We have already considered such a local solution (see Propositions 5.5–5.6 and Theorems 5.15–5.16). However, until now we considered $\boldsymbol{t}$ as *fixed* and focused our attention on the multiplicative representation (5.12)–(5.14) of the solution *considered as a function of x*. Now we emphasize the dependence of the local solution on $\boldsymbol{t}$, while concentrating on the *non-resonant case*. Given $\boldsymbol{t} \in \mathbb{C}_*^n$ (such that the coefficients $Q_p(\boldsymbol{t}), p = 1, \dots, n$, are defined for this $\boldsymbol{t}$), and assuming that the matrix $Q_j(\boldsymbol{t})$ is non-resonant for certain j, the local solution $Y_j(x, \boldsymbol{t})$ of the equation (6.1) corresponding to the singular point t_j is sought in the form

$$Y_j(x, \boldsymbol{t}) = H_j(x, \boldsymbol{t})(x - t_j)^{Q_j(\boldsymbol{t})}, \quad x \in D_{t_j,\rho} \setminus \{t_j\}. \tag{6.3}$$

The differentiation rule (5.8) for $(x - t_j)^Q$ leads to the differential equation for $H_j(x, \boldsymbol{t})$:

$$\frac{dH_j(x, \boldsymbol{t}))}{dx} = \left(\frac{Q_j(\boldsymbol{t})H_j(x, \boldsymbol{t}) - H_j(x, \boldsymbol{t})Q_j(\boldsymbol{t})}{x - t_j} + \Phi_j(x, \boldsymbol{t}) \right) H_j(x, \boldsymbol{t}), \tag{6.4}$$

where

$$\Phi_j(x, \boldsymbol{t}) = \sum_{\substack{1 \le p \le n, \\ p \ne j}} \frac{Q_p(\boldsymbol{t})}{x - t_p}, \tag{6.5}$$

the function $\Phi_j(x, \boldsymbol{t})$ is holomorphic with respect to x at $x = t_j$. The equation (6.4) does not depend on the choice of branch of the (multivalued, in general) function $(x - t_j)^{Q_j(\boldsymbol{t})}$. The right-hand side of this equation, considered as a function of x, has the singularity at the point $x = t_j$. Nevertheless, it can be shown that, if the matrix $Q_j(\boldsymbol{t})$ is non-resonant, then, under the normalization condition

$$H_j(x, \boldsymbol{t})_{|x=t_j} = I, \tag{6.6}$$

the holomorphic solution $H_j(x, \boldsymbol{t})$ of the equation (6.4) exists for x from a disk centered at t_j, and such a solution is unique. More precise formulation of this result will be done below.

In fact, here we repeat what we have already formulated in the Proposition 5.5 and expressed in the formula (5.12). The only difference between the formulae (6.3) and (5.12) is that the notation in (6.3) reflects the dependence on $\boldsymbol{t}$ explicitly.

Definition 6.2. Let $\boldsymbol{t} = (t_1, \dots, t_n) \in \mathbb{C}_*^n$ be given, and assume that the matrix $Q_j(\boldsymbol{t})$ is non-resonant.

The $k \times k$ matrix function $Y_j(x, \boldsymbol{t})$, possessing the properties

i. $Y_j(x, \boldsymbol{t})$ satisfies the differential equation (6.1) with respect to x in a punctured neighborhood $D_{t_j,\rho} \setminus \{t_j\}$ of the point t_j, $(D_{t_j,\rho} = \{z \in \mathbb{C} : |z - t_j| < \rho\}$, ρ is a positive number);

ii. $Y_j(x,\boldsymbol{t})$ admits the factorization of the form (6.3), where the factor $H_j(x,\boldsymbol{t})$ is a matrix function holomorphic with respect to x in the whole (non-punctured) disk $D_{t_j,\rho}$ and satisfying the normalizing condition (6.6),

is said to be the normalized local solution of the differential equation (6.1), corresponding to the singular point t_j.

Proposition 6.3. *Let $Q_p(\boldsymbol{t}), p = 1,\dots,n$, be $k \times k$ matrix functions, which are holomorphic with respect to $\boldsymbol{t} = (t_1,\dots,t_n)$ for $\boldsymbol{t}$ from an open set $\mathcal{D}, \mathcal{D} \in \mathbb{C}^n_*$. Let $\boldsymbol{t}^0 = (t^0_1,\dots,t^0_n)$ be a point from $\mathcal{D}$, and let*

$$\rho_q(\boldsymbol{t}^0) = \frac{1}{2} \min_{\substack{1\le p\le n\\ p\neq q}} \left|t^0_q - t^0_p\right|, \quad 1 \le q \le n. \tag{6.7}$$

Assume that, for certain j, the matrix $Q_j(\boldsymbol{t}^0)$ is non-resonant. Then there exists $\delta, \delta > 0$, such that for every $\boldsymbol{t} = (t_1,\dots,t_n)$ from the polydisk [9] *$\mathsf{D}(\boldsymbol{t}^0,\delta)$, the normalized local solution $Y_j(x,\boldsymbol{t})$ of the equation (6.1), which corresponds to the singular point t_j, exists for x from the punctured disk* [10] *$D_{t_j,\rho_j(\boldsymbol{t}^0)} \setminus \{t_j\}$, and the factor $H_j(x,\boldsymbol{t})$ in (6.3) is a matrix function holomorphic with respect to $x,\boldsymbol{t}$ for $x \in D_{t_j,\rho_j(\boldsymbol{t}^0)}$, $\boldsymbol{t} \in \mathsf{D}(\boldsymbol{t}^0,\delta)$. Moreover, the matrix $H_j(x,\boldsymbol{t})$ is invertible for these x and $\boldsymbol{t}$.*

We put the proof of this proposition to Appendix, which is located in the end of this paper.

Remark 6.4. Proposition 6.3 can be considered as a version of Proposition 5.5, with the emphasis being placed on dependence on $\boldsymbol{t}$.

Remark 6.5. Proposition 6.3 belongs to the class of statements of the following nature: if coefficients of a differential equations and a normalizing condition depend on some parameters analytically, then the solution of this equation depends on these parameters analytically as well. However, in Proposition 6.3 the normalizing condition is posed at the point t_j which is a singular point for the equation. (This normalizing condition, (6.6), can be presented in the form $\lim_{x\to t_j} Y_j(x,\boldsymbol{t})(x-t_j)^{-Q_j(\boldsymbol{t})} = I$). Therefore, Proposition 6.3 is not a consequence of standard general results about analytic dependence of a solution of a differential equation on parameters. The proof makes explicit use of the assumption of Proposition 6.3 that the matrix $Q_j(\boldsymbol{t}^0)$ is non-resonant. However, the result fails if the matrix $Q_j(\boldsymbol{t}^0)$ is resonant. This shows that the result is rather delicate.

In section 5 the notion of connection coefficients was introduced. The connection coefficients relate the solution of the Fuchsian equation, which is normalized at the distinguished point x_0, to its local solutions, which are normalized at the singular points of the equation. However, in section 5 we did not care about the dependence of connection coefficients on $\boldsymbol{t}$. Now we focus our attention on this

[9] The polydisk $\mathsf{D}(\boldsymbol{t}^0,\delta)$ was defined in (6.2) .

[10] $D_{t_j,\rho_j(\boldsymbol{t}^0)} = \{z \in \mathbb{C} : |z - t_j| < \rho_j(\boldsymbol{t}^0)\}$.

dependence, concentrating on the non-resonant case. Given [11] $t^0, t^0 \in \mathcal{D}$, and given $j, 1 \le j \le n$, let us choose $\rho_j(t^0)$ according to (6.7). Then choose $\delta > 0$, satisfying the conditions

$$2\delta < \rho_j(t^0), \tag{6.8}$$

and choose and *fix* the point $x_j \in \mathbb{C}$, satisfying the condition

$$\delta < |x_j - t_j^0| < \rho_j(t^0) - \delta. \tag{6.9}$$

With such a choice [12],

$$x_j \in D_{t_j,\rho_j(t^0)} \quad \text{for every} \quad t_j \in D_{t_j^0,\delta}. \tag{6.10}$$

(See Figure 5.a, where the disk $D_{t_j^0,\delta}$ is plotted in the light gray color, and the dashed circle is the boundary of the disk $D_{t_j^0,\rho_j(t^0)}$). Choose a simple path α_j, leading from the distinguished point x_0 to the point x_j and separated from the points $t_p^0, p = 1, \dots, t_n$:

$$|z - t_p^0| > \delta, \quad \forall z \in \alpha_j, \quad p = 1, \dots, n. \tag{6.11}$$

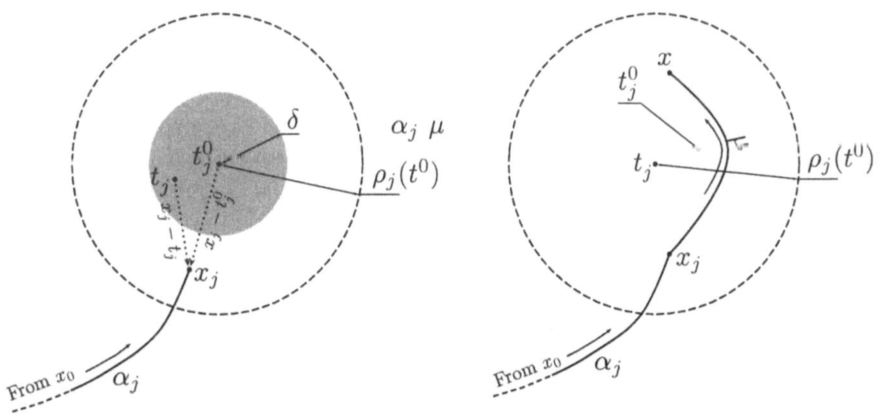

FIGURE 5. a: disk $D_{t_j^0,\rho_j(t^0)}$. b: disk $D_{t_j,\rho_j(t^0)}$.

Then *fix* the choice of the path α_j. (*In particular, let the path α_j be independent of* $\boldsymbol{t}$.) From (6.11) it follows that for [13] $\boldsymbol{t} = (t_1, \dots, t_n) \in \mathsf{D}(t^0, \delta)$, the condition $|x - t_p| > 0 \ \forall x \in \alpha_j, p = 1, \dots, n$, holds. Thus, the function $\sum_{1 \le p \le n} \frac{Q_p(\boldsymbol{t})}{x - t_p}$ is holomorphic with respect to $x, \boldsymbol{t}$ for x from a neighborhood of the path α_j and for

[11] $\mathcal{D}$ is the open set in $\mathbb{C}_*^n$ where the coefficients $Q_p, 1 \le p \le n$, are defined and holomorphic.
[12] For $z \in \mathbb{C}$ and $\varepsilon > 0$, $D_{z,\varepsilon}$ stands for the open disk of the radius ε centered at z.
[13] $\mathsf{D}(t^0, \delta) = D_{t_1^0,\delta} \times \cdots \times D_{t_n^0,\delta}$ is a polydisk.

$\boldsymbol{t} \in \mathsf{D}(\boldsymbol{t}^0, \delta)$. Therefore, the solution $Y(x, \boldsymbol{t})$ of the differential equation (6.1) that satisfies the initial condition

$$Y(x, \boldsymbol{t})_{x=x_0} = I \tag{6.12}$$

can be continued analytically along the path α_j from a neighborhood $\mathcal{V}_{x_0}$ of the initial point x_0 to a neighborhood $\mathcal{V}_{x_j}$ of the point x_j. ($\mathcal{V}_{x_j}$ does not depend on $\boldsymbol{t} \in \mathsf{D}(\boldsymbol{t}^0, \delta)$). We denote by $Y_{\alpha_j}(x, \boldsymbol{t})$ the function in $\mathcal{V}_{x_j}$ which is defined by means of such an analytic continuation. Since the initial condition in the (regular) point x_0 and the path α_j do not depend on $\boldsymbol{t}$, then for every $x \in \alpha_j$, the solution $Y_{\alpha_j}(x, \boldsymbol{t})$ is holomorphic with respect to $\boldsymbol{t}$ for $x \in \mathcal{V}_{x_j}$, $\boldsymbol{t} \in \mathsf{D}(\boldsymbol{t}^0, \delta)$. In particular, the matrix $Y_{\alpha_j}(x_j, \boldsymbol{t})$, as well as the inverse matrix $Y_{\alpha_j}^{-1}(x_j, \boldsymbol{t})$, is holomorphic with respect to $\boldsymbol{t}$ for $\boldsymbol{t} \in \mathsf{D}(\boldsymbol{t}^0, \delta)$. It is clear that the solutions $Y_{\alpha'_j}(x, \boldsymbol{t})$ and $Y_{\alpha''_j}(x, \boldsymbol{t})$ corresponding to the analytic continuations into $\mathcal{V}_{x_j}$ along two different paths α'_j and α''_j (with the same endpoints x_0 and x_j), are related by the equality

$$Y_{\alpha''_j}(x, \boldsymbol{t}) = Y_{\alpha'_j}(x, \boldsymbol{t}) \cdot M_\gamma(\boldsymbol{t}), \tag{6.13}$$

where $M_\gamma(\boldsymbol{t})$ is the monodromy matrix for the equation (6.1) corresponding to the loop γ, $\gamma = \left(\alpha'_j\right)^{-1} \alpha''_j$, composed from the paths α'_j, α''_j. (The loop γ does not depend on $\boldsymbol{t}$, but the equation depends. So, the monodromy matrix $M_\gamma(\boldsymbol{t})$ can depend on $\boldsymbol{t}$).

Now we turn our attention to the local solution $Y_j(x, \boldsymbol{t})$, corresponding to the singular point t_j. Assume that the matrix $Q_j(\boldsymbol{t}^0)$ is non-resonant. Choosing a smaller δ, if necessary, we can apply Proposition 6.3. According to this proposition, the local solution $Y_j(x, \boldsymbol{t})$ exists in the punctured disk $D_{t_j, \rho_j(\boldsymbol{t}^0)} \setminus \{t_j\}$. In particular, it is defined at the point x_j, and even in the neighborhood $\mathcal{V}_{x_j}$, if this neighborhood is small enough. The local solution $Y_j(x, \boldsymbol{t})$ is the product of two factors: $H_j(x, \boldsymbol{t})$ and $(x - t_j)^{Q_j(\boldsymbol{t})}$. Under the normalizing condition (6.6), the first factor is determined uniquely. The second factor is not quite unique. In general, the function $(x - t_j)^{Q_j(\boldsymbol{t})}$ is a multi-valued function of x. This function, and hence the solution $Y_j(x, \boldsymbol{t})$, are determined up to right factor $\exp\{2\pi i Q_j(\boldsymbol{t})\}$. To avoid this non-uniqueness, we must come to an agreement on the value of $\arg(x - t_j)$ for some fixed x, for example, for $x = x_j$. However, now $\boldsymbol{t} = (t_1, \dots, t_n)$ is not a constant but a variable with values in the polydisk $\mathsf{D}(\boldsymbol{t}^0, \delta)$. In particular, the j-th coordinate t_j takes the values in the disk $D_{t_j^0, \rho}$. In order for the function $(x_j - t_j)^{Q_j(\boldsymbol{t})}$ to be holomorphic with respect to $\boldsymbol{t}$, we need to choose $\arg(x_j - t_j)$ *coherently* with respect to t_j.

Let us choose somehow and *fix* the value $\arg(x_j - t_j^0)$, say

$$\arg(x_j - t_j^0) = \vartheta_j^0. \tag{6.14}$$

For $t_j \in D_{t_j^0, \delta}$, choose the value of $\arg(x_j - t_j)$ according to the rule:

$$\vartheta_0 - \pi/2 < \arg(x_j - t_j) < \vartheta_0 + \pi/2. \tag{6.15}$$

The choice (6.15) is possible and unique (see Figure 5.a). With such a choice, the matrix $(x_j - t_j)^{Q_j(\boldsymbol{t})}$ is *determined uniquely* and is holomorphic with respect to t for $t \in \mathsf{D}(t^0, \delta)$. Therefore, the matrix $Y_j(x_j, t) = H_j(x, t)(x_j - t_j)^{Q_j(t)}$ is *determined uniquely* and is holomorphic with respect to t for $t \in \mathsf{D}(t^0, \delta)$.

Moreover, the function $Y_j(x, t)$, the local solution corresponding to the singular point t_j, is determined uniquely on the universal covering $\mathrm{cov}(D_{t_j,\rho_j(t^0)} \setminus \{t_j\}, x_j)$ with the distinguished point x_j. The conditions (6.14)–(6.15), together with the condition (6.6), can be considered as normalizing condition for the local solution $Y_j(x, t)$: the condition (6.6) normalized the factor $H_j(x, t)$), the condition (6.15) normalized the factor $(x - t_j)^{Q_j(t)}$ coherently with respect to t_j.

Let us define the connection coefficient $C_{\alpha_j}(t)$ as

$$C_{\alpha_j}(t) = Y_j^{-1}(x_j, t) \cdot Y_{x_j,\alpha_j}(x_j, t) \tag{6.16}$$

Since both factors in the right-hand side of (6.16) are holomorphic with respect to $t \in \mathsf{D}(t^0, \delta)$ and invertible, *the matrix $C_{\alpha_j}(t)$ is holomorphic with respect to $t \in \mathsf{D}(t^0, \delta)$ and invertible as well.* According to (6.13), the connection coefficients $C'_{\alpha_j}(t)$ and $C''_{\alpha_j}(t)$, corresponding to two different paths α'_j and α''_j, are related by the equality

$$C_{\alpha''_j}(x, t) = C_{\alpha'_j}(x, t) \cdot M_\gamma(t), \tag{6.17}$$

where $M_\gamma(t)$ is the monodromy matrix for the equation (6.1) corresponding to the loop γ, $\gamma = \left(\alpha'_j\right)^{-1} \alpha''_j$.

Since the matrix $C_{\alpha_j}(t)$ does not depend on x, the product $Y_j(x, t) \cdot C_{\alpha_j}$ is a solution of the differential equation (6.1) for $x \in D_{t_j,\rho_j(t^0)}$, in particular, for $x \in \mathcal{V}_{x_j}$. The function $Y_{x_j,\alpha_j}(x, t)$ is also a solution of the differential equation (6.1) for $x \in \mathcal{V}_{x_j}$. Both these solutions coincide at the point x_j. Hence they coincide for all $x \in \mathcal{V}_{x_j}$. Therefore, the product $Y_j(x, t) \cdot C_{\alpha_j}$ can be considered as the *analytic continuation* of the solution $Y_{x_j,\alpha_j}(x, t)$ from $\mathcal{V}_{x_j}$ to the punctured disk $D_{t_j,\rho_j(t^0)} \setminus \{t_j\}$ (or, better to say, to the universal covering $\mathrm{cov}(D_{t_j,\rho_j(t^0)} \setminus \{t_j\}; x_j)$ of this punctured disk with the distinguished point x_j). In its turn, the solution $Y_{x_j,\alpha_j}(x, t)$, which is defined in $\mathcal{V}_{x_j}$, was obtained by the described above analytic continuation of the solution,"living" in $\mathcal{V}_{x_0}$ and satisfying the initial condition (6.12). In other words, the solution $Y_j(x, t) \cdot C_{\alpha_j}$ can be considered as a solution which is obtained as the result of the "two step" analytic continuation from $\mathcal{V}_{x_0}$ to $\mathrm{cov}(D_{t_j,\rho_j(t^0)} \setminus \{t_j\}; x_j)$ of the solution, satisfying the initial condition (6.12). However, it is better to talk not about the *two-step* analytic continuation but about the analytic continuation *along the composed path.* The representative of a point of the universal covering $\mathrm{cov}(D_{t_j,\rho_j(t^0)} \setminus \{t_j\}; x_j)$ is a pair (x, μ), where x is a point of the punctured disk $D_{t_j,\rho_j(t^0)} \setminus \{t_j\}$, and $\mu \subset (D_{t_j,\rho_j(t^0)} \setminus \{t_j\})$ is a path starting at the point x_j and ending at x. The path corresponding to the above-mentioned two-step analytic continuation is the composition [14]. $\alpha = \mu \cdot \alpha_j$. Thus, to describe the dependence on t of the solution $Y_{\alpha_j}(x, t)$, which is normalized at

[14]The path α starts at the point x_0 and ends at the point x. Traversing the path α, we traverse first the path α_j, and then the path μ. See Figure 5.b

the initial point x_0 by the condition

$$Y_{\alpha_j}(x,\boldsymbol{t})_{|x=x_0} = I, \tag{6.18}$$

in the neighborhood of the singular point t_j, we continue analytically this solution from the neighborhood $\mathcal{V}_{x_0}$ of the initial point to the punctured neighborhood $D_{t_j,\rho_j(\boldsymbol{t}^0)}$ along those paths α which reach first the point x_j (x_j *does not depend on* t_j and is common for all these paths), and then go, remaining within the punctured neighborhood $D_{t_j,\rho_j(\boldsymbol{t}^0)} \setminus \{t_j\}$, to the destination points .

Definition 6.6. Such a family [15] $\{Y_{\alpha_j}(x,\boldsymbol{t})\}_{\boldsymbol{t}\in\mathsf{D}(\boldsymbol{t}^0,\delta)}$ of analytic continuations (with respect to x) of the normalized by (6.18) solutions of the equation (6.1) from a neighborhood $\mathcal{V}_{x_0}$ of the initial point x_0 to the family $\{D_{t_j,\rho(\boldsymbol{t}^0)}\}_{t_j\in D_{t_j^0,\delta}}$ of the punctured disk neighborhoods of the points $t_j, t_j \in D_{t_j^0,\delta}$, is said to be **coherent** with respect to $\boldsymbol{t}$. The path α_j along which such continuations are carried out is said to be **the path determining the coherent family of analytic continuations.**

The solution

$$Y_j(x,\boldsymbol{t})\cdot C_{\alpha_j}(\boldsymbol{t}) = H_j(x,\boldsymbol{t})(x-t_j)^{Q_j(\boldsymbol{t})}C_{\alpha_j}(\boldsymbol{t}) \tag{6.19}$$

can be presented in another form – more suitable for our goal:

$$Y_{\alpha_j}(x,\boldsymbol{t}) = H_{\alpha_j}(x,\boldsymbol{t})(x-t_j)^{A_{\alpha_j}(\boldsymbol{t})}, \tag{6.20}$$

where

$$\text{a). } A_{\alpha_j}(\boldsymbol{t}) = C_{\alpha_j}^{-1}(\boldsymbol{t})Q_j(\boldsymbol{t})C_{\alpha_j}(\boldsymbol{t}); \qquad \text{b). } H_{\alpha_j}(x,\boldsymbol{t}) = H_j(x,\boldsymbol{t})C_{\alpha_j}(\boldsymbol{t}). \tag{6.21}$$

Of course, the relation (6.17) leads immediately to the relations

$$\text{a). } A_{\alpha_j''}(\boldsymbol{t}) = (M_\gamma(\boldsymbol{t}))^{-1}A_{\alpha_j'}(\boldsymbol{t})M_\gamma(\boldsymbol{t}), \qquad \text{b). } H_{\alpha_j''}(x,\boldsymbol{t}) = H_{\alpha_j'}(x,\boldsymbol{t})M_\gamma(\boldsymbol{t}), \tag{6.22}$$

where α_j' and α_j'' are two paths with the same endpoints x_0 and x_j, and M_γ is the monodromy matrix for the equation (6.1) corresponding to the loop γ, $\gamma = (\alpha_j')^{-1}\alpha_j''$.

Definition 6.7. The matrix functions $H_{\alpha_j}(x,\boldsymbol{t})$ and $(x-t_j)^{A_{\alpha_j}(\boldsymbol{t})}$ are said to be the *regular factor* and the **principal factor** of the representation (6.20) of the solution $Y_{\alpha_j}(x,\boldsymbol{t})$ near the point t_j.

Let us summarize the results, obtained above, as

Theorem 6.8. *Let the $k\times k$ matrix functions $Q_p(\boldsymbol{t}), p=1,\dots,n$, which appear in the Fuchsian system (6.1) be holomorphic with respect to $\boldsymbol{t}=(t_1,\dots,t_n)$ for $\boldsymbol{t}$ from an open set $\boldsymbol{\mathcal{D}}, \boldsymbol{\mathcal{D}} \in \mathbb{C}^n_*$. Let $\boldsymbol{t}^0=(t_1^0,\dots,t_n^0)$ be a point from $\boldsymbol{\mathcal{D}}$. Assume that, for certain j, the matrix $Q_j(\boldsymbol{t}^0)$ is non-resonant, and let $\rho_j(\boldsymbol{t}^0)$ be defined according to (6.7)) (with $q=j$).*

Then there exists a number δ, $0<\delta<\frac{1}{2}\rho_j(\boldsymbol{t}^0)$, such that, for $\boldsymbol{t}\in\mathsf{D}(\boldsymbol{t}^0,\delta)$, the solutions $\{Y_{\alpha_j}(x,\boldsymbol{t})\}_{\boldsymbol{t}\in\mathsf{D}(\boldsymbol{t}^0,\delta)}$ of the holomorphic family of equation (6.1) which are distinguished by the properties:

[15] Indexed by the parameter $\boldsymbol{t}\in\mathsf{D}(\boldsymbol{t}^0,\delta)$.

i. *Each of these solutions is normalized by* (6.18) *at the initial point* x_0 ,
ii *Each of these solution is continued analytically from a neighborhood of* x_0 *to the family* $\{D_{t_j,\rho(\boldsymbol{t}^0)}\}_{t_j \in D_{t_j^0,\delta}}$ *of the punctured disk neighborhoods of the points* $t_j, t_j \in D_{t_j^0,\delta}$, *and such analytic continuations are performed coherently with respect to* $\boldsymbol{t} \in \boldsymbol{t} \in \mathsf{D}(\boldsymbol{t}^0, \delta)$,

are representable in $D_{t_j,\rho(\boldsymbol{t}^0)} \setminus \{t_j\}$ *in the form* (6.20), *where:*

1. *The matrix function* $H_{\alpha_j}(x, \boldsymbol{t})$ *– the regular factor of the representation* (6.20) *– is holomorphic together with its inverse* $H_{\alpha_j}^{-1}(x, \boldsymbol{t})$ *with respect to* $x, \boldsymbol{t}$ *for* $x \in D_{t_j,\rho(\boldsymbol{t}^0)},$, $\boldsymbol{t} \in \mathsf{D}(\boldsymbol{t}^0, \delta)$;
2. *The exponent* $A_{\alpha_j}(\boldsymbol{t})$ *of the principal factor* $(x-t_j)^{A_{\alpha_j}(\boldsymbol{t})}$ *is a matrix function holomorphic with respect to* $\boldsymbol{t}$ *from the polydisk* $\boldsymbol{t} \in \mathsf{D}(\boldsymbol{t}^0, \delta)$;
3. *The exponent* $A_{\alpha_j}(\boldsymbol{t})$ *is similar to the matrix* $Q_j(\boldsymbol{t})$, *(6.21.a), and the similarity matrix* $C_{\alpha_j}(\boldsymbol{t})$ *can be chosen* [16] *to be holomorphic for* $\boldsymbol{t} \in \mathsf{D}(\boldsymbol{t}^0, \delta)$.
4. *By a different choice of the paths* α_j' *and* α_j'', *along which the coherent analytic continuations are carried out, the matrices* $A_{\alpha_j}(\boldsymbol{t})$ *and* $H_{\alpha_j}(x, \boldsymbol{t})$ *are transformed according to the rule* (6.22), *where* $M_\gamma(\boldsymbol{t})$ *is the monodromy matrix corresponding to a loop* γ, *constructed* [17] *from the paths* α_j' *and* α_j''.

Let α_j be a path, separated (as in (6.11)) from the singular points $t_p, 1 \le p \le n$, and leading from the point x_0 to a point $x_j \in \left(\bigcap D_{t_j,\rho_j(\boldsymbol{t}^0)}\right) \setminus D_{t_j^0,\delta}$, where the intersection is taken over all $t_j \in D_{t_j^0,\delta}$. This path can be used for the coherent analytic continuations of the normalized by (6.18) solution to the neighborhoods of the point $t_j \in D_{t_j^0,\delta}$. The path α_j also generates a loop γ_{α_j} which goes around the disk neighborhood $D_{t_j^0,\delta}$. This loop does not depend on $\boldsymbol{t}$ and, for every $\boldsymbol{t} = (t_1, \ldots, t_n)$, $\boldsymbol{t} \in \mathsf{D}(\boldsymbol{t}^0, \delta)$, represents an element of the fundamental group $\pi(\overline{\mathbb{C}} \setminus \{t_1, \ldots, t_n\}; x_0)$. (Compare to the Definition 5.12). The loop γ_{α_j} is the composition of the path α_j, of the circular arc μ, starting and ending at the point x_j, and of the path α_j^{-1} (the path α_j^{-1} is the path α_j, traversed in the "opposite direction"). In more detail, μ is the circular arc of the radius $|x_j - t_j^0|$, centered at the point t_j^0, which starts at the point x_j, makes the full counterclockwise turn around the point t_j^0, and ends at the same point x_j (or, better to say, at the point $t_j^0 + e^{2\pi i}(x_j - t_j^0)$). Since $x_j \in \left(\bigcap D_{t_j,\rho_j(\boldsymbol{t}^0)}\right) \setminus D_{t_j^0,\delta}$, $\mu \subset D_{t_j,\rho_j(\boldsymbol{t}^0)} \setminus \{t_j\}$ for every $t_j \in D_{t_j^0,\delta}$. Since in $D_{t_j,\rho_j(\boldsymbol{t}^0)} \setminus \{t_j\}$ the representation (6.20) holds, Theorem 5.18 can be formulated in the following "$\boldsymbol{t}$-dependent" version:

---

[16]The similarity matrix is not unique: it can be multiplied by an arbitrary matrix commuting with $Q_j(\boldsymbol{t})$.

[17] Both paths,α_j' and α_j'', have the common starting point x_0. If their endpoints coincide, then the loop γ is just the composition $\gamma = (\alpha_j')^{-1}\alpha_j''$. In the general case, in which these endpoints may be different, $\gamma = (\alpha_j')^{-1}\nu\alpha_j''$, where ν is a path, connecting the endpoints.

Theorem 6.9. *Assume that the matrix $Q_j(\boldsymbol{t}^0)$ is non-resonant. Let α_j be a path leading from the initial point x_0 to a punctured neighborhood of the singular point t_j^0, and let γ_{α_j} be the loop generated by the path α_j (as described above).*

If $\boldsymbol{t} \in \mathsf{D}(\boldsymbol{t}^0, \delta)$, and $\delta > 0$ is small enough, then the monodromy matrix $M_{\gamma_{\alpha_j}}(\boldsymbol{t})$ for the equation (6.1)*, corresponding to the loop γ_{α_j}, and the exponent $A_{\alpha_j}(\boldsymbol{t})$ of the principal factor of the representation* (6.20) *of its solution $Y_{\alpha_j}(x, \boldsymbol{t})$ are related by the equality*

$$M_{\gamma_{\alpha_j}}(\boldsymbol{t}) = \exp\{2\pi i A_{\alpha_j}(\boldsymbol{t})\}. \tag{6.23}$$

The question arises, *whether the monodromy matrix $M_{\gamma_{\alpha_j}}(\boldsymbol{t})$ determines the exponent $A_{\alpha_j}(\boldsymbol{t})$ uniquely?* In particular, *does the equality $A_{\alpha_j}(\boldsymbol{t}') = A_{\alpha_j}(\boldsymbol{t}'')$ follow from the equality $M_{\gamma_{\alpha_j}}(\boldsymbol{t}') = M_{\gamma_{\alpha_j}}(\boldsymbol{t}'')$, if $\boldsymbol{t}', \boldsymbol{t}'' \in \mathsf{D}(\boldsymbol{t}^0, \delta)$, and $\delta \in$ is small enough?*

The latter question is closely related to the description of the set of solutions of the matrix equation (with respect to A):

$$\exp\{2\pi i A\} = M, \tag{6.24}$$

where M is a given $k \times k$ matrix. A description of the set of solutions of (6.24) can be found in [Gant], Chapter VIII, §8. The invertibility of M is the necessary and sufficient condition for the solvability of (6.24). Given $M, \det M \neq 0$, the set of solutions of (6.24) is always infinite. Our prime interest is in conditions under which a certain solution A_0 of the equation (6.24) is isolated. In general, non-isolated solutions may exist. For example, let $M = \operatorname{diag}[1,1]$, $A_0 = \operatorname{diag}[0,1]$, and let T be an *arbitrary* invertible matrix. Then $\exp\{2\pi i(T^{-1}A_0T)\} = M$. If T is non-diagonal matrix, then $T^{-1}A_0T \neq A_0$, but if T is close to I, then $T^{-1}A_0T$ is close to A_0.

Lemma 6.10. *Given a non-resonant matrix A_0, there exists $\varepsilon > 0$ such that from the conditions $\exp\{2\pi i A_0\} = \exp\{2\pi i A\}$ and $\|A - A_0\| < \varepsilon$ it follows that $A = A_0$.*

Proof. Since A_0 is non-resonant, there exists an open (not necessarily connected) set $\mathcal{O}, \mathcal{O} \subset \mathbb{C}$ containing the spectrum of A_0 such that the mapping $\zeta \mapsto \exp\{2\pi i\zeta\}$ is univalent (schlicht) in $\mathcal{O}$. If ε is small enough, the set $\mathcal{O}$ also contains the spectra of all matrices A satisfying the conditions $\|A - A_0\| < \varepsilon$. The set $\mathcal{G} = \exp\{2\pi i\mathcal{O}\}$ is an open set containing all the spectra of the matrices $\exp\{2\pi i A\}$ with $\|A - A_0\| < \varepsilon$, in particular, the spectrum of the matrix $\exp\{2\pi i A_0\}$. If $\frac{1}{2\pi i}\operatorname{Ln} w$, $\frac{1}{2\pi i}\operatorname{Ln} : \mathcal{G} \mapsto \mathcal{O}$, is a function, inverse to the function $\exp\{2\pi i\zeta\} : \mathcal{O} \mapsto \mathcal{G}$, that is $\frac{1}{2\pi i}\operatorname{Ln}(\exp\{2\pi i\zeta\}) \equiv \zeta \ \ \forall \zeta \in \mathcal{O}$, then $\frac{1}{2\pi i}\operatorname{Ln}(\exp\{2\pi i A\}) \equiv A$ for all matrices A whose spectra are contained in $\mathcal{O}$, in particular, $\forall\, A : \|A - A_0\| < \varepsilon$. □

Since the matrix $A_{\alpha_j}(\boldsymbol{t}^0)$ is similar to the matrix $Q_j(\boldsymbol{t}^0)$, the following result is the direct consequence of Lemma 6.10 and Theorem 6.9:

Theorem 6.11. *Let the matrix $Q_j(\boldsymbol{t}^0)$ be non-resonant, and let $A_{\alpha_j}(\boldsymbol{t})$ and $M_{\gamma_{\alpha_j}}(\boldsymbol{t})$ be the same as in the formulation of Theorem 6.9. If $\delta > 0$ is small enough, and if for some $\boldsymbol{t}', \boldsymbol{t}'' \in \mathsf{D}(\boldsymbol{t}^0, \delta)$, the equalities $M_{\gamma_{\alpha_j}}(\boldsymbol{t}') = M_{\gamma_{\alpha_j}}(\boldsymbol{t}'')$ hold, then the equalities $A_{\alpha_j}(\boldsymbol{t}') = A_{\alpha_j}(\boldsymbol{t}'')$ hold as well. In particular, if $M_{\gamma_{\alpha_j}}(\boldsymbol{t}) \equiv M_{\gamma_{\alpha_j}}(\boldsymbol{t}^0)$ in $\mathsf{D}(\boldsymbol{t}^0, \delta)$, then $A_{\alpha_j}(\boldsymbol{t}) \equiv A_{\alpha_j}(\boldsymbol{t}^0)$ in $\mathsf{D}(\boldsymbol{t}^0, \delta)$ as well.*

Remark 6.12. If the matrix $Q_j(\boldsymbol{t}^0)$ is resonant, then, trying to formulate a "$\boldsymbol{t}$-dependent" version of Theorem 5.16, we should seek the representation of the solution $Y_{\alpha_j}(\boldsymbol{t})$ near $\boldsymbol{t}^0$ in the form

$$Y_{\alpha_j}(x, \boldsymbol{t}) = H_{\alpha_j}(x, \boldsymbol{t})(x - t_j)^{Z_{\alpha_j}(\boldsymbol{t})}(x - t_j)^{A_{\alpha_j}(\boldsymbol{t})}, \quad x \in \operatorname{cov}(D_{t_j,\rho} \setminus \{t_j\}, x_j), \quad (6.25)$$

where the matrices $H_{\alpha_j}(x, \boldsymbol{t}), H^{-1}_{\alpha_j}(x, \boldsymbol{t})$ are holomorphic for $x \in D_{t_j,\rho(\boldsymbol{t}^0)}$ and $\boldsymbol{t} \in \mathsf{D}(\boldsymbol{t}^0, \rho_j(\boldsymbol{t}^0))$, the matrices $Z_{\alpha_j}(\boldsymbol{t}), A_{\alpha_j}(\boldsymbol{t})$ are holomorphic for $\boldsymbol{t} \in \mathsf{D}(\boldsymbol{t}^0, \rho_j(\boldsymbol{t}^0))$, and moreover the matrices $Z_{\alpha_j}(\boldsymbol{t})$ are diagonalizable, the eigenvalues of $Z_{\alpha_j}(\boldsymbol{t})$ are integer, the eigenvalues of $A_{\alpha_j}(\boldsymbol{t})$ and the eigenvalues of $Q_j(\boldsymbol{t})$ are related by equalities analogous to the equalities (5.29). However, if the matrix $Q_j(\boldsymbol{t}^0)$ is resonant, the representation of the form (6.25), with holomorphic $Z_{\alpha_j}(\boldsymbol{t}), A_{\alpha_j}(\boldsymbol{t})$, does not exist in general. The reason is that it is possible for the matrix $Q_j(\boldsymbol{t}^0)$ to be resonant, while all the matrices $Q_j(\boldsymbol{t})$ for $\boldsymbol{t} \neq \boldsymbol{t}^0$ are non-resonant. (The property of resonance is unstable under small perturbations). According to Theorem 5.15, the solution $Y_{\alpha_j}(\boldsymbol{t})$ must be of the form (6.20) for $\boldsymbol{t} \neq \boldsymbol{t}^0$. According to Theorem 5.16, it must be of the form (6.25) for $\boldsymbol{t} = \boldsymbol{t}^0$, where in general $Z_{\alpha_j}(\boldsymbol{t}^0) \neq 0$. However, such a bifurcation is incompatible with the holomorphy of $Z_{\alpha_j}(\boldsymbol{t})$.

7. Isoprincipal and isomonodromic deformations

Consider a Fuchsian equation (6.1), where $k \times k$ matrices $Q_p(\boldsymbol{t}), 1 \leq p \leq n$, are holomorphic in $\boldsymbol{t}$ for $\boldsymbol{t}$ from a domain $\boldsymbol{\mathcal{D}}$, $\boldsymbol{\mathcal{D}} \subset \mathbb{C}^n_*$. The equation (6.1) is considered as a differential equation with respect to the variable x, and $\boldsymbol{t} = (t_1, \ldots, t_n)$ is considered as a parameter. Recall that we use the terminology "*the deformation of Fuchsian equation*" for a Fuchsian equation depending on a parameter [18].

Definition 7.1. Let (6.1) be a deformation of Fuchsian equation, where $k \times k$ matrices $Q_p(\boldsymbol{t}), 1 \leq p \leq n$, are holomorphic in a domain $\boldsymbol{\mathcal{D}}, \boldsymbol{\mathcal{D}} \subset \mathbb{C}^n_*$, and let $\boldsymbol{t}^0 = (t^0_1, \ldots, t^0_n)$ be a point of $\boldsymbol{\mathcal{D}}$. The deformation is said to be isomonodromic with respect to a distinguished point x_0 at the pole loci $\boldsymbol{t}^0$ if for every loop γ with the distinguished point x_0, $\gamma \subset \mathbb{C} \setminus \{t^0_1, \ldots, t^0_n\}$, there exists $\delta > 0$, $\delta = \delta(\boldsymbol{t}^0, \gamma)$, such that the monodromy matrix $M_\gamma(\boldsymbol{t})$ of the deformation (6.1) along the loop γ does not depend on $\boldsymbol{t}$ in $\mathsf{D}(\boldsymbol{t}^0, \delta)$: $M_\gamma(\boldsymbol{t}) \equiv M_\gamma(\boldsymbol{t}^0)\, \forall\, \boldsymbol{t} \in \mathsf{D}(\boldsymbol{t}^0, \delta)$. (Of course, we assume that $\delta < \operatorname{dist}(\boldsymbol{t}^0, \gamma)$, so that $\gamma \subset \mathbb{C} \setminus \{t_1, \ldots, t_n\}$ for $\boldsymbol{t} \in \mathsf{D}(\boldsymbol{t}^0, \delta)$, and the monodromy matrix $M_\gamma(\boldsymbol{t})$ is well defined for such $\boldsymbol{t}$). The deformation is said to

[18] We consider only such deformations, where the loci $\boldsymbol{t} = (t_1, \ldots, t_n)$ serve as a parameter, and the dependence on the parameter is holomorphic.

be isomonodromic with respect to x_0 in $\boldsymbol{\mathcal{D}}$ if it is isomonodromic with respect to x_0 at every pole loci $\boldsymbol{t} \in \boldsymbol{\mathcal{D}}$.

Remark 7.2. To check that the deformation (6.1) is isomonodromic at $\boldsymbol{t}^0$, there is no need to examine the monodromy matrix $M_\gamma(\boldsymbol{t})$ for all loops γ. It is enough to choose loops $\gamma_j, j = 1, \dots, n, \gamma_j \subset \mathbb{C} \setminus \{t_1^0, \dots, t_n^0\}$, generating [19]the fundamental group $\pi(\mathbb{C} \setminus \{t_1^0, \dots, t_n^0\}, x_0)$, and to check the monodromy matrices $M_{\gamma_j}(\boldsymbol{t})$ do not depend on $\boldsymbol{t}$ for $\boldsymbol{t} \in \mathsf{D}(\boldsymbol{t}^0, \delta)$, $\delta > 0$ is small enough.

Remark 7.3. If γ is a loop with the distinguished point x_0, and $t_j \notin \gamma, j = 1, \dots, n$, then the monodromy matrix $M_\gamma(.)$ is holomorphic at the point $\boldsymbol{t} = (t_1, \dots, t_n)$. From the uniqueness theorem for holomorphic functions and from elementary topological considerations it follows that if the deformation (6.1) is isomonodromic at some $\boldsymbol{t} \in \boldsymbol{\mathcal{D}}$, then it is isomonodromic at every $\boldsymbol{t} \in \boldsymbol{\mathcal{D}}$.

Definition 7.4. Deformation (6.1) is said to be *isospectral*, if for every j, $j = 1, \dots, n$, the spectrum of the matrix $Q_j(\boldsymbol{t})$ does not depend on $\boldsymbol{t}$ in $\boldsymbol{\mathcal{D}}$: $\operatorname{spec} Q_j(\boldsymbol{t}') = \operatorname{spec} Q_j(\boldsymbol{t}'')$ (the sets coincides "with multiplicities") for every $\boldsymbol{t}', \boldsymbol{t}'' \in \boldsymbol{\mathcal{D}}$.

Theorem 7.5. *If a deformation* (6.1) *is isomonodromic with respect to a distinguished point, then it is also isospectral.*

Proof. For given j, the coefficients of the characteristic polynomials of the matrix $Q_j(\boldsymbol{t})$ are holomorphic with respect to $\boldsymbol{t}$ in $\boldsymbol{\mathcal{D}}$. Therefore, it is enough to prove that for some $\delta > 0$, this polynomial does not depend on $\boldsymbol{t}$ for $\boldsymbol{t} \in \mathsf{D}(\boldsymbol{t}^0, \delta)$ or, what is the same, that the spectrum $\operatorname{spec} Q_j(\boldsymbol{t})$ does not depend on $\boldsymbol{t}$ for $\boldsymbol{t} \in \mathsf{D}(\boldsymbol{t}^0, \delta)$. Let γ_j be a loop with the distinguished point x_0 wich makes one turn counterclockwise around the singular point t_j^0 and makes no turns around other singular points t_p^0, $p = 1, \dots, n, p \neq j$, as it was described in the formulation of Theorem 5.19. The monodromy matrix $M_{\gamma_j}(\boldsymbol{t})$ does not depend on $\boldsymbol{t}$ for $\boldsymbol{t}$ which are close to $\boldsymbol{t}^0$ (In general, $M_{\gamma_j}(\boldsymbol{t}) = M_{\gamma_j}(\boldsymbol{t}^0)$ if it is possible to pass from $\boldsymbol{t}^0$ to $\boldsymbol{t}$ in $\boldsymbol{\mathcal{D}}$ so that no t_p crosses the loop γ_j). According to Theorem 5.19, the set $\exp\{\operatorname{spec} Q_j(\boldsymbol{t})\}$ does not depend on $\boldsymbol{t}$ in $\mathsf{D}(\boldsymbol{t}^0, \delta)$. However, on the one hand, the spectrum $\operatorname{spec} Q_j(\boldsymbol{t})$ depends on $\boldsymbol{t}$ continuously, and on the other hand, the mapping $\zeta \mapsto \exp\{\zeta\}$ is locally univalent (schlicht). Therefore, the set $\operatorname{spec} Q_j(\boldsymbol{t})$ does not depend on $\boldsymbol{t}$ in $\mathsf{D}(\boldsymbol{t}^0, \delta)$ if δ is small enough. □

Let (6.1) be an isomodromic with respect to x_0 deformation of Fuchsian equation, and let for a certain j, the matrix $Q_j(\boldsymbol{t}^0)$ be non-resonant [20]. Let α_j, $\alpha_j \subset \mathbb{C} \setminus \{t_1^0, \dots, t_n^0\}$, be a path leading from the point x_0 to a small neighborhood of the point t_j^0. According to Theorem 6.11, for $\boldsymbol{t} \in \mathsf{D}(\boldsymbol{t}^0, \delta)$, the solution $Y_{\alpha_j}(x, \boldsymbol{t})$, normalized at the distinguished point x_0 by the condition (6.12) and continued

[19]If $\boldsymbol{t} \in \mathsf{D}(\boldsymbol{t}^0, \delta)$ and $\delta > 0$ is small enough, then such loops γ_j satisfy the condition $\gamma_j \subset \mathbb{C} \setminus \{t_1, \dots, t_n\}$ and generate the fundamental group $\pi(\mathbb{C} \setminus \{t_1, \dots, t_n\}, x_0)$. See Figure 4.

[20]According to Theorem 7.5, the matrix $Q_j(\boldsymbol{t})$ in non-resonant for every $\boldsymbol{t} \in \boldsymbol{\mathcal{D}}$.

analytically to the neighborhood of t_j^0, is representable near t_j in the form

$$Y_{\alpha_j}(x,\boldsymbol{t}) = H_{\alpha_j}(x,\boldsymbol{t})(x-t_j)^{A_{\alpha_j}}, \quad x \in D_{t_j,\rho}, \boldsymbol{t} \in \mathsf{D}(\boldsymbol{t}^0,\delta), \tag{7.1}$$

where δ and ρ are some positive numbers, the *regular factor* $H(x,\boldsymbol{t})$ is holomorphic in $x,\boldsymbol{t}$ and invertible for $x \in D_{t_j,\rho}$, $\boldsymbol{t} \in \mathsf{D}(\boldsymbol{t}^0,\delta)$, and the exponent A_{α_j} of the *principal factor* $(x-t_j)^{A_{\alpha_j}}$ ***does not depend on*** $\boldsymbol{t}$. Thus it is natural to give the following definition:

Definition 7.6. The deformation (6.1) is said to be **isoprincipal with respect to a distinguished point** x_0 **in the narrow sense** at the pole loci $\boldsymbol{t}^0 = (t_1^0,\dots,t_n^0)$, ($\boldsymbol{t}^0 \in \boldsymbol{\mathcal{D}}$), if for every $j=1,\dots,n$ and for some paths $\alpha_1,\dots,\alpha_n$ leading from x_0 to neighborhoods of the points $t_1^0,\dots,t_n^0$ respectively, the solution $Y_{\alpha_j}(x,\boldsymbol{t})$ normalized at x_0 by (6.12) is representable in the form (7.1) near t_j, where the matrices A_{α_j} do not depend on $\boldsymbol{t}$ for $\boldsymbol{t} \in \mathsf{D}(\boldsymbol{t}^0,\delta)$, $\delta > 0$ is small enough.

We emphasize that we do not require in this definition that the matrices $Q_j(\boldsymbol{t}^0)$ are non-resonant. Of course, in order for this definition to be correct, we must check that it does not depend on the choice of the paths α_j. This can be done easily. Below (Lemma 7.13) we prove this independence in a more general situation. Let us summarize the results in the form of two lemmas:

Lemma 7.7. *In the non-resonant case (i.e., if all the matrices $Q_j(\boldsymbol{t}^0)$ are non-resonant), every isomonodromic deformation is isoprincipal in the narrow sense as well.*

This lemma is the immediate consequence of Theorem 6.11. The converse is true always, without any assumption of non-resonance.

Lemma 7.8. *If the deformation* (6.1) *is isoprincipal (in the "narrow sense") at some $\boldsymbol{t}^0$ (no assumptions on the matrices $Q_j(\boldsymbol{t}^0), j=1,\dots,n$ are made), then this deformation is isomonodromic as well.*

Proof. If the solution $Y_{\alpha_j}(x,\boldsymbol{t})$ of (6.1) is representable in the form (7.1) with a constant A_{α_j}, then $M_{\gamma_{\alpha_j}}(\boldsymbol{t}) = \exp\{2\pi i A_{\alpha_j}\}$, where $M_{\gamma_{\alpha_j}}$ is the monodromy matrix along the loop γ_{α_j} generated by the path α_j. Therefore, $M_{\gamma_{\alpha_j}}$ does not depend on $\boldsymbol{t}$. □

Remark 7.9. If a deformation (6.1) is isomonodromic, but some of the matrices $Q_j(\boldsymbol{t})$ are not non-resonant, then this deformation may be not isoprincipal. As we shall see later, there is a rich class of deformations of Fuchsian equations all fundamental solutions $Y(x,\boldsymbol{t})$ of which are rational functions of x. Thus, such deformations are isomonodromic. (Their monodromies are trivial). However, most of these deformations are not isoprincipal.

From Lemmas 7.7 and 7.8 it follows that in the non-resonant case the classes of isomonodromic deformations and of isoprincipal (in the "narrow sense") deformations coincide. This is a useful observation. On the one hand, it is the property of "isoprincipalness" that implies the Schlesinger system for the matrix functions

$Q_p(\boldsymbol{t})$ appearing in the deformation. (We prove this implication below). On the other hand, there are methods, which are based on solving the so called Riemann-Hilbert problem, that allow to construct deformations of Fuchsian equations with the prescribed monodromy, but not with the prescribed principal factors of the solutions Y_{α_j}. In particular, such methods allow (under non-resonance condition and certain other restrictions) to construct directly isomonodromic (but not iso-principal) deformations. The proof of the fact that the resulting $Q_j(\boldsymbol{t})$ satisfy the Schlesinger system uses essentially the equivalency of these two classes of deformations in the non-resonant case.

However, the above given "narrow" definition of the isoprincipalness, which is quite natural in the non-resonant case, seems to be a little bit artificial and too restrictive in the general case. Indeed, in the general case the solutions $Y_{\alpha_j}(\boldsymbol{t})$ of the equation (6.1) are of the form (6.25), where exponents $Z_{\alpha_j}(\boldsymbol{t})$ and $A_{\alpha_j}(\boldsymbol{t})$ depend on $\boldsymbol{t}$ in general and posses some additional properties. In particular, the matrix $A_{\alpha_j}(\boldsymbol{t})$ is non-resonant, the matrix $Z_{\alpha_j}(\boldsymbol{t})$ is diagonalizable, with integer eigenvalues. (These exponents may be non-holomorphic, and even discontinuous functions of $\boldsymbol{t}$. See Remark 6.12). If these exponents are constant then we have good reason to call such a deformation isoprincipal. However, we have no need to go into detail. All that is important for us is that the principal factor $(x-t_j)^{Z_{\alpha_j}} \cdot (x-t_j)^{A_{\alpha_j}}$ depends on the difference $x - t_j$ only if the exponents Z_{α_j} and A_{α_j} do not depend on $\boldsymbol{t}$.

Definition 7.10. The deformation (6.1) is said to be isoprincipal with respect to a distinguished point x_0 (in the wide sense) at a pole loci $\boldsymbol{t}^0 = (t_1^0, \dots, t_n^0)$, ($\boldsymbol{t}^0 \in \boldsymbol{\mathcal{D}}$), if for every $j = 1, \dots, n$ and for some paths $\alpha_1, \dots, \alpha_n$ leading from x_0 to neighborhoods of the points $t_1^0, \dots, t_n^0$ respectively, the solution $Y_{\alpha_j}(x, \boldsymbol{t})$ normalized at x_0 by (6.12) is representable in the form

$$Y_{\alpha_j}(x, \boldsymbol{t}) = H_{\alpha_j}(x, \boldsymbol{t}) E_{\alpha_j}(x - t_j), \quad x \in D_{t_j,\rho} \setminus \{t_j\}, \boldsymbol{t} \in \mathsf{D}(\boldsymbol{t}^0, \delta), \tag{7.2}$$

where the matrix functions $H_{\alpha_j}(x, \boldsymbol{t})$, $\left(H_{\alpha_j}(x, \boldsymbol{t})\right)^{-1}$ are holomorphic in $x, \boldsymbol{t}$ for $x \in D_{t_j,\rho}, \boldsymbol{t} \in \mathsf{D}(\boldsymbol{t}^0, \delta)$, ($\rho$ and δ are some positive numbers), and $E_{\alpha_j}(\zeta), j = 1, \dots, n$, are matrix functions possessing the properties:

1. $E_{\alpha_j}(\zeta)$ are holomorphic on the Riemann surface of logarithm [21] and take invertible values there;
2. $E_{\alpha_j}(\zeta)$ do not depend on $\boldsymbol{t}$ and x.

Remark 7.11. Of course, to formulate Definition 7.10 more accurately, we have to describe the geometric configuration more accurately. For every j, we have to choose the point x_j according to (6.9), assuming that δ is small enough, in particular, (6.8) is satisfied. (The value $\rho_j(\boldsymbol{t}^0)$ is defined according to (6.7)). By such a choice of x_j, the condition (6.10) is satisfied. Then we have to choose the arguments $\arg(x_j - t_j^0)$ somehow, (6.14), and then $\arg(x_j - t_j)$ for $t_j \in D_{t_j^0,\delta}$

[21] The Riemann surface of logarithm is the universal covering $\mathrm{cov}(\mathbb{C} \setminus \{0\})$. Considered as functions defined in $\mathbb{C} \setminus \{0\}$, $E_{\alpha_j}(\zeta)$ are multi-valued in general.

according to (6.15). A point of the universal covering $\mathrm{cov}(D_{t_j,\rho_j(\boldsymbol{t}^0)} \setminus \{t_j\}, x_j)$ is a pair (x, μ), where x is a point from the punctured disk $D_{t_j,\rho_j(\boldsymbol{t}^0)}) \setminus \{t_j\}$, and μ is a path which starts at the point x_j and ends at the point x, $\mu \in D_{t_j,\rho_j(\boldsymbol{t}^0)}) \setminus \{t_j\}$. The path μ, together with the choice of $\arg(x_j - t_j)$, determines the value of $\arg(x-t_j)$, and hence, the value of $E_{\alpha_j}(x-t_j)$. On the other hand, the pair $(x, \mu) \in \mathrm{cov}(D_{t_j,\rho_j(\boldsymbol{t}^0)} \setminus \{t_j\}, x_j)$ determines the value of the analytic continuation Y_{α_j} along the path α_j at this point (x, μ). By definition, the value $Y_{\alpha_j}((x, \mu), \boldsymbol{t})$ is the result of the analytic continuation of the solution normalized at x_0 by (6.18) along the composed path $\mu \cdot \alpha_j$ from a neighborhood of x_0 to the point x. Since both values, $Y_{\alpha_j}(x, \boldsymbol{t})$ and $E_{\alpha_j}(x-t_j)$ are determined for every choice of the path μ (these values may depend on μ), the value $H_{\alpha_j}(x, \boldsymbol{t}) \stackrel{\text{def}}{=} Y_{\alpha_j}(x, \boldsymbol{t})\big(E_{\alpha_j}(x - t_j)\big)^{-1}$ is determined for $x \in D_{t_j,\rho_j(\boldsymbol{t}^0)} \setminus \{t_j\}$, and it may depend on μ. The definition 7.10 means firstly, that the value $H_{\alpha_j}(x, \boldsymbol{t})$ is determined only by the point $x \in D_{t_j,\rho_j(\boldsymbol{t}^0)} \setminus \{t_j\}$ itself, but does not depend on the path μ leading to this point, thus the function $H_{\alpha_j}(x, \boldsymbol{t})$ as a function of x is single-valued in the punctured disk $D_{t_j,\rho_j(\boldsymbol{t}^0)} \setminus \{t_j\}$. Secondly, this definition means that the singularity $x = t_j$ of the function $H_{\alpha_j}(x, \boldsymbol{t})$ is removable, that is there exists the limit $\lim_{x\to t_j, x\neq t_j} H_{\alpha_j}(x, \boldsymbol{t}) \stackrel{\text{def}}{=} H_{\alpha_j}(t_j, \boldsymbol{t})$. Thirdly, the definition 7.10 means that the value $H_{\alpha_j}(t_j, \boldsymbol{t})$ is an invertible matrix. (For $x \neq t_j$, the values $H_{\alpha_j}(t_j, \boldsymbol{t})$ are invertible, because both the values $Y_{\alpha_j}(x, \boldsymbol{t})$ and $E_{\alpha_j}(x - t_j)$ are invertible). So, the accurate definition of the isoprincipal deformation is rather bulky.

Below (Lemma 7.13) we prove that the notion of isoprincipal deformation is well defined, that is the property of the deformation to be isoprincipal at some point does not depend on the choice of the paths α_j.

Theorem 7.12. *If* (6.1) *is a deformation of Fuchsian equation which is isoprincipal with respect to a distinguished point in the wide sense (i.e, in the sense of Definition* 7.10) *at a pole loci* $\boldsymbol{t}^0 = (t_1^0, \dots, t_n^0)$, $(\boldsymbol{t}^0 \in \mathcal{D})$, *then the matrices* $E_{\alpha_j}(\zeta), j = 1, \dots, n$, *representing the principle factors, must satisfy the monodromic relations of the form*

$$E_{\alpha_j}(\zeta e^{2\pi i}) = E_{\alpha_j}(\zeta) M_{\alpha_j}, \quad \forall \zeta \neq 0, \tag{7.3}$$

where M_{α_j} *are constant (not depending on* ζ*) invertible matrices. For the monodromy matrix* $M_{\gamma_{\alpha_j}}(\boldsymbol{t})$ *of the equation* (6.1) *along the loop* γ_{α_j} *generated by the paths* α_j, *(see Figure 3), the equality*

$$M_{\gamma_{\alpha_j}}(\boldsymbol{t}) = M_{\alpha_j}, \quad \boldsymbol{t} \in \mathsf{D}(\boldsymbol{t}^0, \delta), \ \ j = 1, \dots, n. \tag{7.4}$$

holds. In particular, this deformation is isomonodromic with respect to x_0.

Proof. For fixed $\boldsymbol{t} \in \mathsf{D}(\boldsymbol{t}^0, \delta)$, and for $x \in D_{t,\rho} \setminus \{t_j\}$, consider two matrix functions of x: $Y_{\alpha_j}(x, \boldsymbol{t})$ and $Y_{\alpha_j}(t_j + (x - t_j)e^{2\pi i}, \boldsymbol{t})$. Here x and $t_j + (x - t_j)e^{2\pi i}$ are considered as points running over the universal covering $\mathrm{cov}(D_{t_j,\rho} \setminus \{t_j\}, x_j)$ of the punctured disk $D_{t_j,\rho} \setminus \{t_j\}$ with the distinguished point x_j. These two points of the universal covering have the same projections in the complex plane. Therefore, the matrices $Y_{\alpha_j}(x,\boldsymbol{t})$ and $Y_{\alpha_j}(t_j+(x-t_j)e^{2\pi i},\boldsymbol{t})$ can be considered as the values of two different solutions of the same linear differential equation at the same point. From $\det Y(x,\boldsymbol{t})_{|x=x_j}\neq 0$ (see (6.12)) it follows that $\det Y_{\alpha_j}(x,\boldsymbol{t})\neq 0$, $\det Y_{\alpha_j}(t_j+(x-t_j)e^{2\pi i},\boldsymbol{t})\neq 0$. Hence, these solutions are "proportional" as functions of x:

$$Y_{\alpha_j}(t_j+(x-t_j)e^{2\pi i},\boldsymbol{t})=Y_{\alpha_j}(x,\boldsymbol{t})M_{\alpha_j}(\boldsymbol{t}),\quad x\in D_{t_j,\rho}\setminus\{t_j\},\boldsymbol{t}\in \mathsf{D}(\boldsymbol{t}^0,\rho),\tag{7.5}$$

where $M_{\alpha_j}(\boldsymbol{t})$ are constant (with respect to x) non-degenerate matrices, which may depend on $\boldsymbol{t}$. Since the factor H_{α_j} is uni-valued ($H_{\alpha_j}(t_j+(x-t_j)e^{2\pi i},\boldsymbol{t})=H_{\alpha_j}(x,\boldsymbol{t})$), from (7.2) and (7.5) it follows that

$$E_{\alpha_j}((x-t_j)e^{2\pi i})=E_{\alpha_j}(x-t_j)M_{\alpha_j}(\boldsymbol{t})\tag{7.6}$$

Since x is an arbitrary point of $D_{t_j,\rho}\setminus\{t_j\}$, $x-t_j$ is an arbitrary point of $\mathbb{C}\setminus\{0\}$. Since the function E_{α_j} does not depend on $\boldsymbol{t}$, the matrix M_{α_j} does not depend on $\boldsymbol{t}$ as well. Moreover, the monodromic relation (7.3) holds at lest for $\zeta:0<|\zeta|<\rho$. However, the function E_{α_j} is holomorphic in the whole Riemann surface of logarithm. Therefore, the relation (7.3) holds for all $\zeta:|\zeta|>0$.

The expression for the monodromy matrix $M_{\gamma_{\alpha_j}}(\boldsymbol{t})$ along the loop γ_{α_j} was already obtained (see (5.19)):

$$M_{\gamma_{\alpha_j}}(\boldsymbol{t})=Y_{x_j,\alpha_j}(x_j,\boldsymbol{t})^{-1}\cdot Y_{x_j,\alpha_j}(t_j+(x_j-t_j)e^{2\pi i},\boldsymbol{t}).\tag{7.7}$$

The equality (7.4) is the consequence of (7.5) and (7.7). □

Let for a certain p, α_p be a path leading from the point x_0 to a neighborhood of the point t_p. Let α'_p be another path leading from x_0 to the same neighborhood. Let Y_{α_p} and $Y_{\alpha'_p}$ be the analytic continuations of the normalized at x_0 solution of (6.1) along the paths α_p and α'_p respectively. Then the solutions Y_{α_p} and $Y_{\alpha'_p}$ are related by the relation $Y_{\alpha'_p}(x,\boldsymbol{t})=Y_{\alpha_p}(x,\boldsymbol{t})\cdot M_\gamma(\boldsymbol{t})$, $x\in D_{t_p,\rho}\setminus\{t_j\}$, where $M_\gamma(\boldsymbol{t})$ is the monodromy matrix along the loop γ, constructed from the paths α_p and α'_p. (See footnote 17). If the deformation (6.1) is isoprincipal in the sense of the definition 7.10 by a certain choice of the paths $\alpha_j, j=1,\dots,n$, then, according to Theorem 7.12, this deformation is isomonodromic. In particular, the monodromy matrix $M_\gamma(\boldsymbol{t})$ in (7.8) does not depend on $\boldsymbol{t}$: $M_\gamma(\boldsymbol{t})=M_\gamma$, thus

$$Y_{\alpha'_p}(x,\boldsymbol{t})=Y_{\alpha_p}(x,\boldsymbol{t})\cdot M_\gamma,\quad x\in D_{t_p,\rho}\setminus\{t_j\},\tag{7.8}$$

Therefore, the solution $Y_{\alpha'_p}$ is representable in the form

$$Y_{\alpha'_p}(x,\boldsymbol{t})=H_{\alpha'_p}(x,\boldsymbol{t})\cdot E_{\alpha'_p}(x-t_j),\quad x\in D_{t_j,\rho}\setminus\{t_j\},\boldsymbol{t}\in\mathsf{D}(t_j,\rho),\tag{7.9}$$

with $H_{\alpha'_p}(x,\boldsymbol{t})=H_{\alpha_p}(x,\boldsymbol{t})M_\gamma$, $E_{\alpha'_p}(\zeta)=M_\gamma^{-1}E_{\alpha_p}(\zeta)M_\gamma$. Since the matrices $E_{\alpha_p}(\zeta)$ and M_γ do not depend on $\boldsymbol{t}$, the matrix $E_{\alpha'_p}(\zeta)$ does not depend on $\boldsymbol{t}$ as well. So, we prove the following

Lemma 7.13. *The notion of the isoprincipal deformation at a point is well defined. If the deformation* (6.1) *is isoprincipal at the point* $\boldsymbol{t}^0$ *for some choice of the paths*

α_j, $j = 1, \ldots, n$ leading from x_0 to neighborhoods of the singularities t_j^0, then this deformation is isoprincipal at this point for any other choice of such paths.

Definition 7.14. The deformation (6.1), where the coefficients $Q_p(\boldsymbol{t}), p = 1, \ldots, n$, are holomorphic for $\boldsymbol{t} \in \mathcal{D}$, $\mathcal{D}$ is an open set in $\mathbb{C}_*^n$, is said to be isoprincipal (in the wide sense) in $\mathcal{D}$ if it is isoprincipal at every point $\boldsymbol{t} \in \mathcal{D}$.

Theorem 7.15. *If the deformation* (6.1) *is isoprincipal with respect to x_0 at some $\boldsymbol{t}^0$, $\boldsymbol{t}^0 \in \mathcal{D}$, and the open set $\mathcal{D}$ is connected, then this deformation is isoprincipal with respect to x_0 in $\mathcal{D}$ (i.e., at every $\boldsymbol{t} \in \mathcal{D}$).*

We omit proof of this theorem.

8. Isoprincipalness of a deformation implies the Schlesinger system for its coefficients

Lemma 8.1. *Let the deformation* (6.1) *be isoprincipal with respect to a distinguished point x_0 (in the wide sense) at some pole loci $\boldsymbol{t}^0$, $\boldsymbol{t}^0 \in \mathcal{D}$, and let $Y(x, \boldsymbol{t})$, $x \in cov(\overline{\mathbb{C}} \setminus \{t_1, \ldots, t_n\}), \boldsymbol{t} \in \mathcal{D}$, be the solution of the differential equation* (6.1) *which is normalized by the condition* (6.12) *at the distinguished point x_0.*

Then the equalities

$$\frac{\partial Y(x, \boldsymbol{t})}{\partial t_k} = Q_k(\boldsymbol{t}) \left(\frac{1}{x_0 - t_k} - \frac{1}{x - t_k} \right) Y(x, \boldsymbol{t}), \quad k = 1, \ldots, n \tag{8.1}$$

are satisfied. In particular, if $x_0 = \infty$, then the equations (4.1b) *are satisfied.*

Proof. The key to the proof are the (trivial, but nevertheless very useful) equalities

$$\frac{d}{dx} E_{\alpha_j}(x - t_j) = -\frac{\partial}{\partial t_j} E_{\alpha_j}(x - t_j), \quad j = 1, \ldots, n, \tag{8.2}$$

which express the fact that the functions $E_{\alpha_j}(x - t_j)$ depend only on the difference of the arguments x and t_j. Expressing the logarithmic derivative $\frac{dY}{dx} Y^{-1}$ from the differential equation (6.1):

$$\frac{dY}{dx} Y^{-1} = \sum_p \frac{Q_p(\boldsymbol{t})}{x - t_p},$$

we derive its behavior in a neighborhood of the singular point t_j:

$$\frac{dY}{dx}(x, \boldsymbol{t}) Y^{-1}(x, \boldsymbol{t}) = \\ = \frac{Q_j(\boldsymbol{t})}{x - t_j} + (\text{a function of } x, \text{ holomorphic with respect to } x), \\ x \in D_{t_j, \rho}, \quad j = 1, \ldots n. \tag{8.3}$$

($D_{t_j, \rho} = \{z \in \mathbb{C} : |z - t_j| < \rho\}$, ρ is a positive number). We restrict our consideration to $\boldsymbol{t} \in \mathsf{D}(\boldsymbol{t}^0, \delta)$, where $\delta > 0$ is small enough, so that the representations

(7.2) of the solution $Y(x,\boldsymbol{t})$ hold in the appropriate neighborhoods $D_{t_j,\rho}\setminus\{t_j\}$ of the singular points t_j. Using the representation (7.2), we see that

$$\frac{dY_{\alpha_j}}{dx}(x,\boldsymbol{t})Y_{\alpha_j}^{-1}(x,\boldsymbol{t}) = H_{\alpha_j}(x,\boldsymbol{t})\left(\frac{d}{dx}E_{\alpha_j}(x-t_j)E_{\alpha_j}^{-1}(x-t_j)\right)H_{\alpha_j}^{-1}(x,\boldsymbol{t}) + \left(\frac{d}{dx}H_{\alpha_j}(x,\boldsymbol{t})\right)H_{\alpha_j}^{-1}(x,\boldsymbol{t}). \quad (8.4)$$

Since the function $H_{\alpha_j}(x,\boldsymbol{t})$ is holomorphic and invertible in an entire (non-punctured) neighborhood of the singular point t_j, comparing (8.3) and (8.4), we obtain

$$H_{\alpha_j}(x,\boldsymbol{t})\left(\frac{d}{dx}E_{\alpha_j}(x-t_j)\left(E_{\alpha_j}(x-t_j)\right)^{-1}\right)H_{\alpha_j}(x,\boldsymbol{t})^{-1} = \frac{Q_j}{x-t_j} + (\text{a function of } x, \text{ holomorphic with respect to } x), \quad x\in D_{t_j,\rho},\ \ j=1,\dots n. \quad (8.5)$$

Let us investigate the logarithmic derivatives $\dfrac{\partial Y}{\partial t_k}(x,\boldsymbol{t})Y^{-1}(x,\boldsymbol{t})$, as functions of x. First of all we remark that for fixed $\boldsymbol{t}$ these function are single-valued and holomorphic with respect to x in the domain $\mathbb{C}\setminus\{t_1,\dots,t_n\}$. Indeed, let $\mathcal{O}$ be a simply connected open set in $\bar{\mathbb{C}}$ separated from the points $t_1^0,\dots,t_n^0$: $\operatorname{dist}(\mathcal{O},\{t_1^0,\dots,t_n^0\})>0$. Let $Y_1(x,\boldsymbol{t}),Y_2(x,\boldsymbol{t})$ be two branches of the solution $Y(x,\boldsymbol{t})$ in $\mathcal{O}$, that is the functions $Y_1(x,\boldsymbol{t}),Y_2(x,\boldsymbol{t})$ are obtained by the analytic continuation of the solution of (6.1) along two different paths γ_1,γ_2 leading from x_0 to $\mathcal{O}$. These branches $Y_1(x,\boldsymbol{t}),Y_2(x,\boldsymbol{t})$ must satisfy the relation $Y_2(x,\boldsymbol{t})=Y_1(x,\boldsymbol{t})M_\gamma(\boldsymbol{t})$, where $M_\gamma(\boldsymbol{t})$ is the monodromy matrix corresponding to the loop γ constructing from the paths γ_1 and γ_2. In general, this monodromy matrix $M_\gamma(\boldsymbol{t})$ may depend on $\boldsymbol{t}$. However, according to Theorem 7.12, the considered deformation is isomonodromic. In particular, the matrix $M_\gamma(\boldsymbol{t})$ does not depend on $\boldsymbol{t}$, that is $Y_2(x,\boldsymbol{t})=Y_1(x,\boldsymbol{t})M_\gamma$, where M_γ does not depend on $\boldsymbol{t}$ and is invertible. (M_γ may depend on the choice of the branches Y_1 and Y_2.) Differentiating the last equality with respect to t_k, we obtain $\dfrac{\partial Y_2}{\partial t_k}=\dfrac{\partial Y_1}{\partial t_k}M_\gamma$. From the last two equalities it follows that $\dfrac{\partial Y_1}{\partial t_k}Y_1^{-1}=\dfrac{\partial Y_2}{\partial t_k}Y_2^{-1}$, that is the logarithmic derivative $\dfrac{\partial Y}{\partial t_k}Y^{-1}$ is a single-valued holomorphic function of x in $\overline{\mathbb{C}}\setminus\{t_1,\dots,t_n\}$. Using the representation (7.2) in a neighborhood of the singular point t_j, we see that for $j\neq k$

$$\frac{\partial Y}{\partial t_k}((x,\boldsymbol{t}))Y^{-1}((x,\boldsymbol{t})) = \frac{\partial H_{\alpha_j}(x,\boldsymbol{t})}{\partial t_k}H_{\alpha_j}^{-1}(x,\boldsymbol{t}), \quad x\in D_{t_j,\rho}.$$

(For $j\neq k$, the second factor $E_{\alpha_j}(x-t_j)$ in (7.2) does not depend on t_k.) Since the function $H_{\alpha_j}(x,\boldsymbol{t})$ is holomorphic with respect to $(x,\boldsymbol{t})$ for $x\in D_{t_j,\rho}, \boldsymbol{t}\in \mathsf{D}(\boldsymbol{t}^0,\delta)$, so is the function $\frac{\partial Y}{\partial t_k}((x,\boldsymbol{t}))Y^{-1}((x,\boldsymbol{t}))$ for $j\neq k$.

For $j = k$, we have to take into account the factor $E_{\alpha_j}(x - t_j)$:

$$\frac{\partial Y}{\partial t_k}((x,\boldsymbol{t}))Y^{-1}((x,\boldsymbol{t})) =$$

$$= H_{\alpha_k}(x,\boldsymbol{t})\left(\frac{\partial}{\partial t_k}E_{\alpha_k}(x-t_k)E^{-1}_{\alpha_k}(x-t_k)\right)H^{-1}_{\alpha_k}(x,\boldsymbol{t}) + \frac{\partial H_{\alpha_k}(x,\boldsymbol{t})}{\partial t_k}H^{-1}_{\alpha_k}(x,\boldsymbol{t}), \quad x \in D_{t_k,\rho}.$$

The function $\frac{\partial H_{\alpha_k}(x,\boldsymbol{t})}{\partial t_k}H^{-1}_{\alpha_k}(x,\boldsymbol{t})$ is holomorphic for $x \in D_{t_j,\rho}$. Taking into account (8.2) and (8.5), we come to the conclusion:

$$\frac{\partial Y}{\partial t_k}((x,\boldsymbol{t}))Y^{-1}((x,\boldsymbol{t})) = -\frac{Q_k}{x-t_k} + \text{(a function holomorphic with respect to } x), \quad x \in D_{t_k,\rho}. \tag{8.6}$$

Thus, the function $\frac{\partial Y}{\partial t_k}((x,\boldsymbol{t}))Y^{-1}((x,\boldsymbol{t}))$, considered as a function of the variable x, is holomorphic in the entire extended complex plane $\overline{\mathbb{C}}$, except the point t_k. Near t_k this function behaves as it described in (8.6). Therefore,

$$\frac{\partial Y}{\partial t_k}((x,\boldsymbol{t}))Y^{-1}((x,\boldsymbol{t})) = -\frac{Q_k}{x-t_k} + C_k(\boldsymbol{t}),$$

where $C_k(\boldsymbol{t})$ does not depend on x. To determine $C_k(\boldsymbol{t})$, we use the normalizing condition (6.12). Since $Y(x_0,\boldsymbol{t}) \equiv I$, $\frac{\partial Y(x_0,\boldsymbol{t})}{\partial t_k} = 0$, i.e., $-\frac{Q_k(\boldsymbol{t})}{x-t_k} + C_k(\boldsymbol{t}) = 0$ at $x = x_0$. Thus, (8.1) holds. □

Theorem 8.2. *Let a $k \times k$ matrix functions $Q_j(\boldsymbol{t})$, $\boldsymbol{t} = (t_1,\dots,t_n), j = 1,\dots,n$, are holomorphic for $\boldsymbol{t} \in \boldsymbol{\mathcal{D}}$, where $\boldsymbol{\mathcal{D}} \subset \mathbb{C}^n_*$ is an open connected set, and satisfying the condition*

$$\sum_{1\le j\le n} Q_j(\boldsymbol{t}) = 0, \quad \boldsymbol{t} \in \boldsymbol{\mathcal{D}}.$$

Assume that the deformation

$$\frac{dY}{dx} = \left(\sum_{1\le j\le n}\frac{Q_j(\boldsymbol{t})}{x-t_j}\right)Y$$

is isoprincipal (in the wide sense) with respect to the distinguished point $x_0 = \infty$ at least for one pole loci $\boldsymbol{t}^0 \in \boldsymbol{\mathcal{D}}$.

Then the matrix functions $Q_j(\boldsymbol{t}), j = 1,\dots,n$, satisfy the Schlesinger system (1.1) for all $\boldsymbol{t} \in \boldsymbol{\mathcal{D}}$.

Proof. Let $Y(x,\boldsymbol{t})$ be the solution of the differential equation (6.1) satisfying the normalizing condition $Y(x,\boldsymbol{t})_{x=\infty} = I$. According to Lemma 8.1, the equations (4.1b) are satisfied for $x \in \overline{\mathbb{C}} \setminus \{t_1,\dots,t_n\}, \boldsymbol{t} \in \mathsf{D}(\boldsymbol{t}^0,\delta)$. The equation (4.1a) is the same as the equation (6.1). According to Theorem 4.1, the matrix functions $Q_j(\boldsymbol{t})$,

$\boldsymbol{t} = (t_1, \dots, t_n), j = 1, \dots, n$ satisfy the Schlesinger system (1.1) for $\boldsymbol{t} \in \mathsf{D}(\boldsymbol{t}^0, \delta)$. Since matrix functions in both left and right-hand sides of (1.1) are holomorphic in $\boldsymbol{\mathcal{D}}$ and coincide in $\mathsf{D}(\boldsymbol{t}^0, \delta)$, they coincide everywhere in $\boldsymbol{\mathcal{D}}$. □

9. Construction of the isoprincipal deformations as a tool for solving the Schlesinger system

Theorem 8.2 opens a way for constructing solutions of the Schlesinger system. The Cauchy problem for this system can be formulated as follows.

Given $\boldsymbol{t}^0 \in \mathbb{C}_*^n$ and $k \times k$ matrices $Q_1^0, \dots, Q_n^0$, one needs to find matrix functions $Q_1(\boldsymbol{t}), \dots, Q_n(\boldsymbol{t})$ of the variable $\boldsymbol{t}$ satisfying the Schlesinger system (1.1) and the initial condition

$$Q_j(\boldsymbol{t}^0) = Q_j^0, \quad j = 1, \dots, n. \tag{9.1}$$

If we succeed in constructing a family of Fuchsian equations

$$\frac{dY}{dx} = \Big(\sum_{1 \le j \le n} \frac{Q_j(\boldsymbol{t})}{x - t_j} \Big) Y, \tag{9.2}$$

enumerated by the pole loci $\boldsymbol{t} = (t_1, \dots, t_n)$, with holomorphic coefficients $Q_1(\boldsymbol{t})$, $\dots, Q_n(\boldsymbol{t})$ (a holomorphic deformation) such that *this holomorphic deformation is isoprincipal with respect to* $x_0 = \infty$, then, according to Theorem 8.2, these coefficients will satisfy the Schlesinger system. In order for the initial conditions (9.1) to be satisfied, the family must contain the equation

$$\frac{dY}{dx} = \Big(\sum_{1 \le j \le n} \frac{Q_j^0}{x - t_j^0} \Big) Y, \tag{9.3}$$

which corresponds to the parameter value $\boldsymbol{t} = \boldsymbol{t}^0$.

If all the matrices $Q_1^0, \dots, Q_n^0$ are non-resonant (and hence, according to Theorems 7.12 and 7.5, all the matrices $Q_1(\boldsymbol{t}), \dots, Q_n(\boldsymbol{t})$ are non-resonant as well), this deformation is isoprincipal if, and only if, it is isomonodromic (Theorem 7.12 and Lemma 7.7). Therefore, if all the matrices $Q_1^0, \dots, Q_n^0$ are non-resonant, the Cauchy problem with initial conditions (9.1) is reduced to the problem of constructing the *isomonodromic* deformation (9.2) which in particular contains the equation (9.3). An approach to constructing such isomonodromic deformation lies in using *the Riemann-Hilbert problem.*

The Riemann-Hilbert problem can be formulated as follows.

Given $\boldsymbol{t} = (t_1, \dots, t_n) \in \mathbb{C}_*^n$, loops $\gamma_1(\boldsymbol{t}), \dots, \gamma_n(\boldsymbol{t})$ with the distinguished point x_0 that form the geometric configuration as it is plotted in Figure 1, and matrices $M_1, \dots, M_n$ satisfying the condition [22] *$M_{\pi(1)} \cdots \cdots M_{\pi(n)} = I$. One needs to construct*

[22] $\{\pi_1, \dots, \pi_n\}$ is a permutation of the indices $\{1, \dots, n\}$ which is determined by the geometric configuration of the paths $\alpha_1, \dots, \alpha_n$. For the configuration plotted in Figure 1 the permutation is trivial: $\pi_k = k$.

the Fuchsian differential equation (9.2) with poles $t_1, \ldots, t_n$ *whose monodromy matrices take the prescribed values*

$$M_{\gamma_j(\boldsymbol{t})} = M_j, \quad j = 1, \ldots, n. \tag{9.4}$$

An approach for solving the Riemann-Hilbert problem was developed by J. Plemelj in 1908, and, a little later but independently, by G.D. Birkhoff (see [Pleml], [Plem2], [Birk]). The first step of this approach requires to solve a factorization problem for a matrix function on a closed contour. The contour passes through points $t_1, \ldots, t_n$, and the matrix function is piece-wise constant, constructed from the monodromy matrices $M_1, \ldots, M_n$. This factorization problem can be reduced to a singular linear matrix integral equation on the contour. On the second step the solution of the factorization problem undergoes certain additional transformations in order to obtain the solution of the original Riemann-Hilbert problem. The second step is not completely painless: there are reefs here. In fact, not long ago A. A. Bolibruch has discovered examples of data for which the Riemann-Hilbert problem has no solution (see [AnBo], [Bol1], [Bol2], [Bol3]).

In order to apply the Riemann-Hilbert problem for the construction of the isomonodromic deformation, first we have to determine for the equation (9.3) (whose coefficients – matrices $Q_1^0, \ldots, Q_n^0$ – are the initial values in the Cauchy problem for the Schlesinger system) the monodromy matrices $M_1^0, \ldots, M_n^0$, corresponding to the loops $\gamma_1(\boldsymbol{t}^0), \ldots, \gamma_n(\boldsymbol{t}^0)$. Then we have to solve the Riemann-Hilbert problem for the following data: an arbitrary point $\boldsymbol{t} = (t_1, \ldots, t_n) \in \mathbb{C}_*^n$, the appropriate loops $\gamma_1(\boldsymbol{t}), \ldots, \gamma_n(\boldsymbol{t})$, and the matrices $M_1^0, \ldots, M_n^0$. As a result we shall obtain the matrices $Q_1(\boldsymbol{t}), \ldots, Q_n(\boldsymbol{t})$. Repeating this procedure for every $\boldsymbol{t}$, (and, therefore, solving the infinite family of the Riemann-Hilbert problems) we construct matrix *functions* $Q_1(\boldsymbol{t}), \ldots, Q_n(\boldsymbol{t})$. If we succeed in performing this construction in such a way that these matrix functions are holomorphic then we obtain the desired isomonodromic deformation. Finally, if the initial data are non-resonant, we obtain the solution of the Cauchy problem for the Schlesinger system. Thus solving the Cauchy problem for a *non-linear* system is reduced to solving a family of *linear* problems.

However, solving the Schlesinger system in the non-resonant case is not the ultimate goal of the present manuscript. We would like to consider the case which is, in a certain sense, just the opposite. More precisely, we consider the initial values $Q_1^0, \ldots, Q_n^0$ such that fundamental solutions of the Fuchsian system (9.3) are rational matrix functions of x in general position. Of course, a rational function is uni-valued in the complex plane, hence the monodromy in this case is trivial: $M_1 = \ldots = M_n = I$. In this case the spectrum of each matrix Q_j^0 is either $\{0, 1\}$ or $\{0, -1\}$ and $\operatorname{rank}(Q_j^0) = 1$. In particular, all the matrices $Q_1^0, \ldots, Q_n^0$ are resonant. Therefore, the considerations above, concerning the isomonodromic deformations, are not applicable in this case. Moreover, any family of Fuchsian systems with rational solutions is isomonodromic (the monodromies are trivial). Thus there are "too many" isomonodromic deformations here, and only one of them leads to the solution of the Schlesinger system. This is the isoprincipal deformation.

Let us recall Theorem 5.16 and, in particular, representation (5.28). If system (5.1) has a generic rational solution $Y(x)$ then in (5.28) we have $A_j = 0$, and hence $Y(x)$ is locally representable in a neighborhood of $x = t_j$ as

$$Y(x) = H_j(x)(x - t_j)^{Z_j}, \tag{9.5}$$

with $H_j(x)$ holomorphic and invertible there, and

$$Z_j^2 = \pm Z_j. \tag{9.6}$$

Furthermore, under condition (9.6) the factor $(x - t_j)^{Z_j}$ has the following simple form:

$$(x - t_j)^{Z_j} = \begin{cases} I - Z_j + (x - t_j)Z_j, \text{ if } Z_j^2 = Z_j, \\ I + Z_j - \dfrac{1}{x - t_j}Z_j, \text{ if } Z_j^2 = -Z_j. \end{cases} \tag{9.7}$$

Thus the factorization (9.5) takes the form

$$Y(x) = H_j(x)\Big(I - Z_j + (x - t_j)Z_j\Big), \quad \text{if} \quad Z_j^2 = Z_j, \tag{9.8}$$

$$Y(x) = H_j(x)\Big(I + Z_j - \frac{1}{x - t_j}Z_j\Big), \quad \text{if} \quad Z_j^2 = -Z_j. \tag{9.9}$$

Factorizations of the form (9.8), (9.9) are traditional in the theory of rational matrix functions. The factors (9.7) are just the principal factors of the rational solution $Y(x)$ at the singular point t_j. If $Y(x, \boldsymbol{t})$ is a family of rational solutions for a family of Fuchsian systems which gives an isoprincipal deformation then for every $\boldsymbol{t} = (t_1, \dots, t_n)$ at each singular point $x = t_j$ $Y(x, \boldsymbol{t})$ must admit the factorization

$$Y(x, \boldsymbol{t}) = H_j(x, \boldsymbol{t})\Big(I - Z_j + (x - t_j)Z_j\Big), \quad \text{if} \quad Z_j^2 = Z_j, \tag{9.10}$$

$$Y(x, \boldsymbol{t}) = H_j(x, \boldsymbol{t})\Big(I + Z_j - \frac{1}{x - t_j}Z_j\Big), \quad \text{if} \quad Z_j^2 = -Z_j, \tag{9.11}$$

where Z_j do not depend on $\boldsymbol{t}$, and $H_j(x, \boldsymbol{t})$ are holomorphic and invertible at $x = t_j$. The realization theory for rational matrix functions allows to construct families of Fuchsian systems whose solutions $Y(x, \boldsymbol{t})$ admit the factorizations (9.10), (9.11). Development of the realization theory appropriate for this goal originated with L.A. Sakhnovich [Sakhn]. This topic was further developed in [GKRM]. In [Kats1] and [Kats2] this theory was adapted for applications concerning the Schlesinger system. The forthcoming second part of the present manuscript will be dedicated to detailed construction of rational solutions of the Schlesinger system, based on the realization theory for rational matrix functions.

Appendix

The representation of a solution of a Fuchsian equation in a neighborhood of a singular point in the non-resonant case

Proof of Proposition 6.3. Choose a positive δ so small that the closure of the polydisk $\mathsf{D}(\boldsymbol{t}^0, \delta)$ is contained in the set $\boldsymbol{\mathcal{D}}$, where the matrix functions $Q_p(\boldsymbol{t}), p = 1, \dots, n$, are defined and holomorphic. We also impose the condition $\delta < \frac{1}{2}\rho_j(\boldsymbol{t}^0)$. So, if $|t_j - t_j^0| < \delta$, and $|t_p - t_p^0| < \delta, p \neq j$, then $|t_j - t_p| > \rho_j(\boldsymbol{t}^0)$. Since the matrix $Q_j(\boldsymbol{t}^0)$ is non-resonant, choosing, if necessary, a smaller δ, we can also assume that the matrix $Q_j(\boldsymbol{t})$ is non-resonant for all $\boldsymbol{t} \subset \mathsf{D}(\boldsymbol{t}^0, \delta)$. Moreover, if $\delta > 0$ is small enough, then the differences of eigenvalues $\lambda_p(\boldsymbol{t}) - \lambda_q(\boldsymbol{t})$ of the matrix $Q_j(\boldsymbol{t})$ are separated from non-zero integers:

$$|\lambda_p(\boldsymbol{t}) - \lambda_q(\boldsymbol{t}) - m| \geq \varepsilon > 0 \quad \forall\, p, q = 1, \dots, k, \quad \forall\, m \in \mathbb{Z} \setminus \{0\},\ \forall\, \boldsymbol{t} \in \mathsf{D}(\boldsymbol{t}^0, \delta), \tag{A.1}$$

where $\varepsilon > 0$ does not depend on $\boldsymbol{t} \in \mathsf{D}(\boldsymbol{t}^0, \delta)$, $m \in \mathbb{Z} \setminus \{0\}$.

Changing variable

$$x \to x + t_j, \tag{A.2}$$

we reduce the differential equation (6.1) to the form

$$\frac{dZ(x, \boldsymbol{t})}{dx} = \left(\frac{Q_j(\boldsymbol{t})}{r} + \Phi(x, \boldsymbol{t})\right) Z(x, \boldsymbol{t}), \tag{A.3}$$

where

$$Z(x, \boldsymbol{t}) = Y(x + t_j, \boldsymbol{t}), \quad \Phi(x, \boldsymbol{t}) = \sum_{\substack{1 \leq p \leq n \\ p \neq j}} \frac{Q_p(\boldsymbol{t})}{x + t_j - t_p}. \tag{A.4}$$

In view of the relation $|t_j - t_p| > \rho_j(\boldsymbol{t}^0)$, the function $\Phi(x, \boldsymbol{t})$ admits expansion of the form

$$\Phi(x, \boldsymbol{t}) = \sum_{0 \leq r < \infty} \Phi_r(\boldsymbol{t}) x^r, \quad x \in \mathbb{C},\ |x| < \rho_j(\boldsymbol{t}^0), \tag{A.5}$$

where $\Phi_r(\boldsymbol{t})$ are $k \times k$ matrix functions, which are holomorphic with respect to $\boldsymbol{t}$ for $\boldsymbol{t} \in \mathsf{D}(\boldsymbol{t}^0, \delta)$ and admit the estimate

$$\|\Phi_r(\boldsymbol{t})\| \leq C_1 \cdot (\rho_j(\boldsymbol{t}^0))^{-r}, \quad \boldsymbol{t} \in \mathsf{D}(\boldsymbol{t}^0, \delta),\ 0 \leq r < \infty, \tag{A.6}$$

where $C_1 < \infty$ does not depend on $\boldsymbol{t}$ and r.

We seek a normalized solution of (A.3), corresponding to the singular point $x = 0$, in the form

$$Z(x) = \Psi(x, \boldsymbol{t}) x^{Q_j(\boldsymbol{t})}, \tag{A.7}$$

where $\Psi(x, \boldsymbol{t})$ is power series in x:

$$\Psi(x, \boldsymbol{t}) = \sum_{0 \leq r < \infty} \Psi_r(\boldsymbol{t}) x^r, \quad \text{with} \quad \Psi_0(\boldsymbol{t}) = I. \tag{A.8}$$

Substituting the expressions (A.7)–(A.8) and (A.5) into (A.3) and using the differentiation rule

$$\frac{d}{dx}x^{Q_j(t)} = \frac{Q_j(t)}{x}x^{Q_j(t)}, \tag{A.9}$$

we obtain for $r = 0, 1, 2, \dots$ the equalities

$$(r+1)\Psi_{r+1}(t) + \Psi_{r+1}(t)Q_j(t) - Q_j(t)\Psi_{r+1}(t) = \sum_{\substack{l+m=r\\ l\geq 0, m\geq 0}} \Phi_l(t)\Psi_m(t), \tag{A.10}$$

relating the given sequence $\{\Phi_r(t)\}_{0\leq r<\infty}$ and the sequence $\{\Psi_r(t)\}_{0\leq r<\infty}$, which has to be found. The relations (A.10) express the fact that (A.7)–(A.8) is a *formal solution* of the differential equation (6.1). If the series (A.8) converge in a disk $\{x \in \mathbb{C} : |x| < \rho\}$, with $\rho > 0$, then (A.7), where $\Psi(x,t)$ is the sum of these series, is an actual solution of (6.1) in the punctured disk $\{x \in \mathbb{C} : |x| < \rho\} \setminus \{0\}$.

Since the constant term $\Psi_0(t)$ is given: $\Psi_0(t) = I$, the equations (A.10) can be considered as a recursive system for successive determining of $\Psi_1(t), \Psi_2(t), \dots$, etc. The r-th equation of the system (A.10) is of the form

$$\lambda X - [Q, X] = Y, \tag{A.11}$$

where $[Q,X] = QX - XQ$ – the commutator of the matruces Q and X. In (A.10),

$$\lambda = r+1, \quad Q = Q_j(t), \quad X = \Psi_{r+1}(t), \quad Y = \sum_{\substack{l+m=r\\ l\geq 0, m\geq 0}} \Phi_l(t)\Psi_m(t). \tag{A.12}$$

Now we interrupt the proof of Proposition 6.3 to discuss the matrix equation (A.11).

Lemma A.1. *Let $\mathfrak{M}_k$ be the set of all $k \times k$ matrices with complex entries. Given a matrix $Q \in \mathfrak{M}_k$, let us associate with this matrix the operator* ad_Q :

$$\mathrm{ad}_Q X = QX - XQ, \quad \mathrm{ad}_Q : \mathfrak{M}_k \to \mathfrak{M}_k. \tag{A.13}$$

Let $\lambda_1(Q), \dots, \lambda_k(Q)$ be the set of all eigenvalues of the matrix Q (enumerated with multiplicities). Then the set of all eigenvalues (enumerated with multiplicities) of the operator ad_Q *is the set of all the differences $\{\lambda_p(Q) - \lambda_q(Q)\}_{1\leq p,q\leq k}$.*

Proof. Assume first that all eigenvalues of the matrix Q are pairwise different. Let u_p and v_q be, respectively, eigenvector columns and eigenvector rows of the matrix Q:

$$Qu_p = \lambda_p u_p, \quad v_q Q = \lambda_q v_q, \quad 1 \leq p, q \leq k.$$

Then the matrices $X_{p,q} \stackrel{\text{def}}{=} u_p \otimes v_q$ are eigenvectors of the operator ad_Q:

$$\mathrm{ad}_Q X_{p,q} = (\lambda_p - \lambda_q)X_{p,q}, \quad 1 \leq p, q \leq k.$$

Thus, the numbers $\lambda_p(Q) - \lambda_q(Q)$ are eigenvalues of the operator ad_Q. Since the eigenvalues of the matrix Q are pairwise different, the vectors $\{u_p\}_{1\leq p\leq k}$ and $\{v_q\}_{1\leq q\leq k}$ form bases in the spaces of all k-vector columns and k-vector rows respectively. Since dimension of the space $\mathfrak{M}_k$ is equal to k^2, the set of matrices $\{X_{p,q}\}_{1\leq p,q\leq k}$ is a basis of the space $\mathfrak{M}_k$. Therefore, the set $\{\lambda_p(Q)-\lambda_q(Q)\}_{1\leq p,q\leq k}$

is the set of *all* eigenvalues of the operator ad_Q (enumerated with multiplicities). This result can be extended to general matrices Q, i.e., to matrices whose eigenvalues are not necessarily pairwise different. Such an extension can be done using the approximation reasoning. □

From the above result it follows that if λ satisfies the condition $\lambda \neq \lambda_p(Q)-\lambda_q(Q)$, $1 \le p,q \le k$, then the equation (Q, X, Y are $k \times k$ matrices)

$$\lambda X - \mathrm{ad}_Q X = Y \tag{A.14}$$

is solvable with respect to X for every Y. However, for what follows one needs not only the solvability of this equation but also an estimate for its solution.

Lemma A.2. *Let $\mathcal{L}$ be a K-dimensional vector space equipped with a norm $\|.\|_{\mathcal{L}}$, and let $\mathcal{A}$ be a linear operator acting in $\mathcal{L}$. Assume that the eigenvalues $\lambda_m(\mathcal{A})$ of the operator $\mathcal{A}$ are separated from zero:*

$$|\lambda_m(\mathcal{A})| \ge \varepsilon, \quad 1 \le m \le K. \tag{A.15}$$

Then the inverse operator $\mathcal{A}^{-1}$ admits the estimate from above:

$$\|\mathcal{A}^{-1}\| \le 2^K \|\mathcal{A}\|^{K-1}\varepsilon^{-K}, \tag{A.16}$$

where $\|\mathcal{A}\| = \sup \|\mathcal{A}x\|_{\mathcal{L}}, \|\mathcal{A}^{-1}\| = \sup\|\mathcal{A}^{-1}x\|_{\mathcal{L}}$, and sup*'s are taken over all $x \in \mathcal{L}$ such that $\|x\|_{\mathcal{L}} \le 1$.*

Proof. Let $\chi(\zeta) = \det(\zeta\mathcal{I} - \mathcal{A})$ be the characteristic polynomial of the operator $\mathcal{A}$,

$$\chi(\zeta) = \zeta^K - \sigma_1\zeta^{K-1} + \cdots + (-1)^{K-1}\sigma_{K-1}\zeta + (-1)^K\sigma_K,$$

where $\sigma_l, 1 \le l \le K$, denotes the l-th elementary symmetric polynomial of in $\lambda_1(\mathcal{A}), \ldots, \lambda_K(\mathcal{A})$. According to Hamilton-Cayley theorem, $\chi(\mathcal{A}) = 0$, hence

$$\mathcal{A}^{-1} = (-1)^{K-1}\frac{1}{\sigma_K}\mathcal{A}^{K-1} + (-1)^{K-2}\frac{\sigma_1}{\sigma_K}\mathcal{A}^{K-2} + \cdots + \frac{\sigma_{K-1}}{\sigma_K}\mathcal{I}. \tag{A.17}$$

From (A.17) it follows that

$$\|\mathcal{A}^{-1}\| \le \left|\frac{\sigma_0}{\sigma_K}\right| \|\mathcal{A}\|^{K-1} + \left|\frac{\sigma_1}{\sigma_K}\right| \|\mathcal{A}\|^{K-2} + \cdots + \left|\frac{\sigma_{K-1}}{\sigma_K}\right|. \quad (\sigma_0 \stackrel{\mathrm{def}}{=} 1)$$

It is clear that the fractions $\dfrac{\sigma_l}{\sigma_K}$ are the elementary symmetric polynomials in the inverse eigenvalues $\dfrac{1}{\lambda_1(\mathcal{A})}, \ldots, \dfrac{1}{\lambda_K(\mathcal{A})}$. From (A.15) we conclude that

$$\left|\frac{\sigma_l}{\sigma_K}\right| \le C_K^l \varepsilon^{l-K}, \quad \text{where } C_K^l \text{ are binomial coefficients: } C_K^l = \frac{K!}{l!(K-l)!}.$$

Therefore,

$$\|\mathcal{A}^{-1}\| \le C_K^0\varepsilon^{-K}\|\mathcal{A}\|^{K-1} + C_K^1\varepsilon^{1-K}\|\mathcal{A}\|^{K-2} + \cdots + C_K^{K-1}\varepsilon^{-1}\|\mathcal{A}\|^0.$$

Using the binomial formula, we rewrite the previous inequality as

$$\|\mathcal{A}^{-1}\| \le \|\mathcal{A}\|^{-1}\left(\left(1 + \varepsilon^{-1}\|\mathcal{A}\|\right)^K - 1\right), \tag{A.18}$$

Since $|\lambda_m(\mathcal{A})| \le \|\mathcal{A}\|$ for any eigenvalue $\lambda_m(\mathcal{A})$, inequality (A.15) implies that $1 \le \varepsilon^{-1}\|\mathcal{A}\|$, and hence $1+\varepsilon^{-1}\|\mathcal{A}\| \le 2\varepsilon^{-1}\|\mathcal{A}\|$. Weakening the inequality (A.18), we come to the more rough, but more simple inequality (A.16). □

Lemma A.3. *Let Q be a $k\times k$ matrix, with eigenvalues $\lambda_1,\dots,\lambda_k$, enumerated with multiplicities. Let λ be a complex number, and let for some $\varepsilon>0$ the inequalities*

$$|\lambda-(\lambda_p-\lambda_q)|>\varepsilon,\quad 1\le p,q\le k, \tag{A.19}$$

hold true. Then for every Y, the equation (A.14) (or, what is the same, the equation (A.11)) is solvable with respect to X, and the estimate

$$\|X\| \le 2^{k^2}\varepsilon^{-k^2}(|\lambda|+2\|Q\|)^{k^2-1}\|Y\|, \tag{A.20}$$

holds, where $\|.\|$ is an arbitrary norm on $\mathfrak{M}_k$ with the property: $\|AB\|\le\|A\|\|B\|$ for every $A,B\in\mathfrak{M}_k$.

Proof. We use Lemma A.2 for the operator $\mathcal{A}=\lambda\mathcal{I}-\mathrm{ad}_Q$, acting on the space $\mathfrak{M}_k$. It is clear that $\|\mathcal{A}\|\le|\lambda|+2\|Q\|$. According to Lemma A.1, the set $\{\lambda-(\lambda_p-\lambda_q)\}_{1\le p,q\le k}$ is the set of all eigenvalues of $\mathcal{A}$ (enumerated with multiplicities). The inequality (A.19) provides an estimate of these eigenvalues from below. Applying Lemma A.1 to this operator $\mathcal{A}$ (and taking into account that $\dim\mathfrak{M}_k=k^2$), we obtain the estimate (A.20). □

The following lemma gives an estimate, which is more precise than (A.20) for large λ.

Lemma A.4. *Let Q be a $k\times k$ matrix, with eigenvalues $\lambda_1,\dots,\lambda_k$, enumerated with multiplicities. Let λ be a complex number which is different from the numbers $\lambda_p-\lambda_q, 1\le p,q\le k$, and let the estimates for $\|Q\|$ from above and for $|\lambda-(\lambda_p-\lambda_q)|$ from below be given:*

$$|\lambda-(\lambda_p-\lambda_q)|\ge\varepsilon,\quad 1\le p,q\le k,\qquad \|Q\|\le\mu,\quad (\varepsilon>0,\ \mu<\infty). \tag{A.21}$$

Then for every $Y\in\mathfrak{M}_k$, the equation (A.14) *(or, what is the same, the equation* (A.11)) *is solvable with respect to $X\in\mathfrak{M}_k$, and the estimate*

$$\|X\|\le\frac{C(\varepsilon,\mu,k)}{1+|\lambda|}\|Y\|, \tag{A.22}$$

holds, where $\|.\|$ is an arbitrary norm on $\mathfrak{M}_k$ with the property: $\|AB\|\le\|A\|\|B\|$ for every $A,B\in\mathfrak{M}_k$, and $C(\varepsilon,\mu,k)<\infty$ is a constant depending only on ε,μ and k.

Proof. If $\lambda X-\mathrm{ad}_QX=Y$, then $|\lambda|\|X\|\le\|\mathrm{ad}_QX\|+\|Y\|\le2\|Q\|\|X\|+\|Y\|\le 2\mu\|X\|+\|Y\|$. Thus,

$$\|X\|\le\frac{1}{|\lambda|-2\mu}\|Y\|,\quad \text{for } |\lambda|>2\mu. \tag{A.23}$$

On the other hand, if we assume that $|\lambda|\le3\mu$ and replace in the inequality (A.20) $|\lambda|$ and $\|Q\|$ by the larger values 3μ and μ respectively, we come to the inequality

$$\|X\|\le10^{k^2}\varepsilon^{-k^2}\mu^{k^2-1}\|Y\|,\quad \text{for } |\lambda|\le3\mu. \tag{A.24}$$

Unifying the inequalities (A.23)and (A.24), we come to the inequality (A.22) with a suitable constant $C(\varepsilon, \mu, k)$ (which, of course, can be found explicitly). □

Now, after we have investigated the matrix equation (A.11), we resume the proof of Proposition 6.3. We apply Lemma A.3 to the equality (A.10) considered as an equation of the form (A.11), with X, Y, Q, λ defined in (A.12). The number $\max_{t\in D(t^0)} \|Q_j(t)\|$ serves as μ, the number ε is taken from (A.1). Substituting these X, λ and the estimate

$$\|Y\| \le \sum_{0\le m\le r} \|\Phi_{r-m}(t)\|\|\Psi_m(t)\| \tag{A.25}$$

for $\|Y\|$ into (A.22), we obtain, for $t \in D(t^0, \delta)$, $r = 0, 1, 2, \ldots$ the estimate

$$\|\Psi_{r+1}(t)\| \le \frac{C(\varepsilon, \mu, k)}{r+2} \sum_{0\le m\le r} \|\Phi_{r-m}(t)\|\|\Psi_m(t)\|. \tag{A.26}$$

Substituting the upper estimate (A.6) for $\|\Phi_{r-m}(t)\|$ into (A.26), we obtain:

$$\|\Psi_{r+1}(t)\|\rho^{r+1} \le \frac{C_2}{r+2} \sum_{0\le m\le r} \|\Psi_m(t)\|\rho^m, \quad \forall t \in D(t^0, \delta), \quad r = 0, 1, 2, \ldots, \tag{A.27}$$

where

$$C_2 = \rho \cdot C_1 \cdot C(\varepsilon, \mu, k), \tag{A.28}$$

$\rho = \rho_j(t^0)$ is defined in (6.7), C_1 and $C(\varepsilon, \mu, k)$ are the values from (A.6) and (A.22) respectively. It is important to note that the value $C_2 < \infty$ does not depend on r and on $t \in D(t^0)$.

Lemma A.5. *Let $\{a_r\}_{0\le r<\infty}$ be a sequence of non-negative numbers, satisfying the conditions*

$$a_{r+1} \le \frac{d}{r+2} \sum_{0\le m\le r} a_m, \quad r = 0, 1, 2, \ldots; \quad a_0 = 1, \tag{A.29}$$

where $d, 0 < d < \infty$, does not depend on r.

Then the sequence $\{a_r\}_{0\le r<\infty}$ admits the upper estimate

$$a_r \le (r+1)^{d-1}, \quad r = 0, 1, 2, \ldots. \tag{A.30}$$

Proof. For $r = 0$, the assertion (A.30) is true: $a_0 = 1$. Assume that the inequalities (A.30) hold for $r = 0, 1, \ldots, m$. Then, according to (A.30), the inequality

$$a_{m+1} \le \frac{d}{m+2} \sum_{0\le r\le m} (r+1)^{d-1} \tag{A.31}$$

holds. If $d \ge 1$, then

$$(r+1)^{d-1} \le \int_{r+1}^{r+2} t^{d-1} dt, \quad \sum_{0\le r\le m} (r+1)^{d-1} \le \int_{1}^{m+2} t^{d-1} dt \le \frac{(m+2)^d}{d},$$

and the inequality (A.30) holds for $r = m+1$. If $0 < d < 1$, then

$$(r+1)^{d-1} \leq \int\limits_{r}^{r+1} t^{d-1}dt, \quad \sum_{0\leq r\leq m} (r+1)^{d-1} \leq \int\limits_{0}^{m+1} t^{d-1}dt \leq \frac{(m+2)^d}{d},$$

and the inequality (A.30) holds for $r = m+1$ as well. By induction, the inequality (A.30) holds for all r. □

Applying Lemma A.5 to (A.27), with $a_r = \|\Psi_r(\boldsymbol{t})\|\rho^r$, we obtain the estimate

$$\|\Psi_r(\boldsymbol{t})\| \leq (r+1)^d \rho^{-r}, \quad r = 0,1,2,\dots, \quad \forall \boldsymbol{t} \in \mathsf{D}(\boldsymbol{t}^0, \delta), \tag{A.32}$$

where $\rho = \rho_j(\boldsymbol{t}^0)$ is defined in (6.7), $d = C_2$ from (A.28), $d < \infty$ does not depend on r and on $\boldsymbol{t} \in \mathsf{D}(\boldsymbol{t}^0, \delta)$.

From the estimate (A.32) it follows, that the power series (A.8) converge locally uniformly in the open circle $\{x : |x| < \rho\}$, and the convergence is uniform with respect to the parameter $\boldsymbol{t} \in \mathsf{D}(\boldsymbol{t}^0, \delta)$. Therefore, the sum $\Psi(x, \boldsymbol{t})$ of these power series is a function, holomorphic with respect to $x, \boldsymbol{t}$ for $\{x : |x| < \rho_j(\boldsymbol{t}^0)\}$, $\boldsymbol{t} \in \mathsf{D}(\boldsymbol{t}^0, \delta)$. Let us prove that the function $\Psi^{-1}(x, \boldsymbol{t})$ is holomorphic for these [23] $x, \boldsymbol{t}$ as well. Since $\Psi_0(\boldsymbol{t}) = I$, we can consider the *formal* power series

$$\Psi^{\text{inv}}(x, \boldsymbol{t}) = \sum_{0\leq r<\infty} \Psi_r^{\text{inv}}(\boldsymbol{t}) x^r, \tag{A.33}$$

which are *inverse* to the series (A.8), i.e.,

$$\begin{aligned}\Big(\sum_{0\leq r<\infty} \Psi_r(\boldsymbol{t})x^r\Big) &\cdot \Big(\sum_{0\leq r<\infty} \Psi_r^{\text{inv}}(\boldsymbol{t})x^r\Big) \\ &= \Big(\sum_{0\leq r<\infty} \Psi_r^{\text{inv}}(\boldsymbol{t})x^r\Big) \cdot \Big(\sum_{0\leq r<\infty} \Psi_r(\boldsymbol{t})x^r\Big) = I.\end{aligned} \tag{A.34}$$

It is clear that

$$\Psi_0^{\text{inv}}(\boldsymbol{t}) = I. \tag{A.35}$$

The expressions (A.3), (A.7), (A.8) and (A.34) imply that the product

$$Z^{\text{inv}}(x, \boldsymbol{t}) \stackrel{\text{def}}{=} x^{-Q_j(\boldsymbol{t})} \Psi^{\text{inv}}(x, \boldsymbol{t}) \tag{A.36}$$

is a formal solution of the differential equation

$$\frac{dZ^{\text{inv}}}{dx} = -Z^{\text{inv}}\Big(\frac{Q_j(\boldsymbol{t})}{x} + \Phi(x, \boldsymbol{t})\Big), \tag{A.37}$$

[23] Of course, in view of the condition $\Psi_0(\boldsymbol{t}) = I$, the function $\Psi^{-1}(x, \boldsymbol{t})$ is holomorphic for x which are small enough. However, it is very easy to prove the holomorphy of the function Ψ^{-1} in the whole disc $\{x : |x| < \rho\}$. Such a proof can be done "for free", using the previously established estimates.

As before (see (A.10)), substituting the expression (A.36) into (A.37), we obtain the infinite recursive system

$$(r+1)\Psi^{\text{inv}}_{r+1}(\boldsymbol{t}) + \Psi^{\text{inv}}_{r+1}(\boldsymbol{t})Q_j(\boldsymbol{t}) - Q_j(\boldsymbol{t})\Psi^{\text{inv}}_{r+1}(\boldsymbol{t}) = -\sum_{\substack{l+m=r\\ l\geq 0, m\geq 0}} \Psi^{\text{inv}}_m(\boldsymbol{t})\Phi_l(\boldsymbol{t}). \quad \text{(A.38)}$$

Like the r-th equation of the system (A.10), the r-th equation of the system (A.38) is of the form (A.11), with *the same* λ and Q as in (A.10): $\lambda = r+1, Q = Q_j(\boldsymbol{t})$. The only difference is that the right-hand side of (A.11) is $Y = \sum \Phi_l(\boldsymbol{t})\Psi_m(\boldsymbol{t})$ for (A.10), and $Y = -\sum \Psi^{\text{inv}}_m(\boldsymbol{t})\Phi_l(\boldsymbol{t})$ for (A.38). Because λ and Q for both matrix equations, corresponding to the systems (A.10) and (A.38), are the same, the estimates (A.22) for both equations hold with the same constant $C(\varepsilon,\mu,k)$ (Of course, ε,μ,k are the same). Instead of the estimate (A.25), the estimate

$$\|Y\| \leq \sum_{0\leq m\leq r} \|\Phi_{r-m}(\boldsymbol{t})\|\|\Psi^{\text{inv}}_m(\boldsymbol{t})\| \quad \text{(A.39)}$$

for the right-hand side of (A.11) should be used now. Finally, we come to the recursive system of inequalities

$$\|\Psi^{\text{inv}}_{r+1}(\boldsymbol{t})\| \leq \frac{C(\varepsilon,\mu,k)}{r+2} \sum_{0\leq m\leq r} \|\Phi_{r-m}(\boldsymbol{t})\|\|\Psi^{\text{inv}}_m(\boldsymbol{t})\|, \quad \forall \boldsymbol{t} \in \mathsf{D}(\boldsymbol{t}^0,\delta). \quad \text{(A.40)}$$

with respect to $\|\Psi^{\text{inv}}_m(\boldsymbol{t})\|$. However, the latter system of inequalities coincides with the system (A.26) of inequalities with respect to $\|\Psi_m(\boldsymbol{t})\|$. Since $\|\Psi^{\text{inv}}_0(\boldsymbol{t})\| = \|\Psi_0(\boldsymbol{t})\| = \|I\|$, the estimates for $\|\Psi_m(\boldsymbol{t})\|$ and $\|\Psi^{\text{inv}}_m(\boldsymbol{t})\|$, which follow from (A.26) and (A.40), coincide. Thus, the estimates

$$\|\Psi^{\text{inv}}_r(\boldsymbol{t})\| \leq (r+1)^d\rho^{-r}, \quad r = 0,1,2,\ldots, \quad \forall \boldsymbol{t} \in \mathsf{D}(\boldsymbol{t}^0,\delta), \quad \text{(A.41)}$$

hold with the same ρ and d that in (A.32). Therefore, the series (A.33) converge locally uniformly in the disk $\{x : |x| < \rho_j(\boldsymbol{t})\}$ and represent the matrix *function* $\Psi^{\text{inv}}(x,\boldsymbol{t})$ which is holomorphic with respect to $x, \boldsymbol{t}$ for $x : |x| < \rho_j(\boldsymbol{t}^0), \boldsymbol{t} \in \mathsf{D}(\boldsymbol{t}^0,\delta)$. The identities (A.34) for formal power series imply the identities

$$\Psi(x,\boldsymbol{t})\Psi^{\text{inv}}(x,\boldsymbol{t}) = \Psi^{\text{inv}}(x,\boldsymbol{t})\Psi(x,\boldsymbol{t}) = I \quad \forall\ x : |x| < \rho_j(\boldsymbol{t}^0), \boldsymbol{t} \in \mathsf{D}(\boldsymbol{t}^0,\delta) \quad \text{(A.42)}$$

for the appropriate holomorphic matrix functions. Thus, the values of the matrix function $\Psi(x,\boldsymbol{t})$ for $x : |x| < \rho_j(\boldsymbol{t}^0)$, $\boldsymbol{t} \in \mathsf{D}(\boldsymbol{t}^0,\delta)$ are invertible, and $\Psi^{-1}(x,\boldsymbol{t}) = \Psi^{\text{inv}}(x,\boldsymbol{t})$.

Performing the change of variables $x \to x - t_j$, inverse to the change (A.2), we obtain the representation (6.3) for the local solution $Y_j(x,\boldsymbol{t})$, with $H_j(x,\boldsymbol{t})) = \Psi(x - t_j, \boldsymbol{t})$. The properties of the function $H_j(x,\boldsymbol{t})$, stated in the formulation of Proposition 6.3, express the properties of $\Psi(x,\boldsymbol{t})$, as established above. This completes the proof of Proposition 6.3. □

References

[AKNS] ABLOWITZ, M., KAUP, D., NEWELL, A. and H. SEGUR. *Nonlinear-evolution equations of physical significance.* Phys. Rev. Let., **32**:2, (1973), 125–127.

[AnBo] ANOSOV, D.V. and A.A. BOLIBRUCH. *The Riemann-Hilbert Problem.* (Aspects of Mathematics, Vol. E 22), Vieveg, Braunschweig · Wiesbaden, 1994, i–ix + 190 pp.

[Arn] АРНОЛЬД, В.И. *Кольцо когомологий группы крашеных кос.* Матем. заметки, **5**:2, (1969) , c. 227–231. Engl. transl.: ARNOL'D, V.I. *The cohomology ring of the colored braid group.* Math. Notes, **5**, pp. 138–140 (1969).

[Birk] BIRKHOFF, G. D. *The generalized Riemann problem for linear differential equations and the allied problems for linear difference and q-difference equations.* Proc. Amer. Acad. Arts and Sci., **49** (1913)

[Bol1] БОЛИБРУХ, А. А. *Проблема Римана-Гильберта.* Успехи матем. наук, **45**:2 (1990), c. 3–47 . Engl. transl.: BOLIBRUCH, A.A. *The Riemann-Hilbert problem.* Russian Math. Surveys, **45**:2 (1990), pp. 1–47.

[Bol2] БОЛИБРУХ, А. А. *21-я проблема Гильберта для Фуксовых линейных систем.* Труды матем. института им. В.А. Стеклова, **206**. Москва, Наука, 1994., 160 c. Engl. transl.: BOLIBRUCH, A.A. *Hilbert's twenty first problem for Fuchsian systems.* Proc. Steklov Inst. Math, **5** (**206**), 1995, viii + 145 pp.

[Bol3] БОЛИБРУХ, А. А. *Фуксовы Дифференциальные Уравнения и Голоморфные Расслоения.* (In Russian): BOLIBRUCH, A.A. *Fuchsian Differencial Equations and Holomorphic Bundles.* МЦНМО (Издательство Московского Центра непрерывного математического образования.), Moscow, 2000, 120 pp.

[BoDoVe] BOUTET DE MONVEL, L., A.DOUADY and J.-L. VERDIER – editors. *Mathématique et Physique. Séminaire de l'Ecole Normale Supériore 1979–1982.* (Progress in Mathematics, vol. **37**). Birkhäuser, Boston · Basel · Stuttgart, 1983 .

[CoLe] CODDINGTON, E.A. and N. LEVINSON. *Theory of Ordinary Differential Equations.* McGraw Hill, New York·Toronto·London, 1955.

[DIKZ] DEIFT, P., ITS, A., KAPAEV, A. and X. ZHOU : *On the Algebro-Geometric Integration of the Schlesinger Equations.* Commun. Math. Phys., **203** (1999), pp. 613–633.

[Dick] DICKEY, L.A. *Soliton Equations and Hamiltonian Systems.* World Scientific. Singapure·New Jersey·London·Hong Kong, First Edition (Advanced Series in Mathematical Physics, Vol. 12) – 1991, ix+310 pp.; Second Edition (Advanced Series in Mathematical Physics, Vol. 26) – xii+420 pp., 2003.

[DJKM1] DATE, E., M. JIMBO, M. KASHIWARA, and T. MIWA. *Transformation groups for soliton equations.* Proc. Japan. Acad. Ser. **A** Math. Sci.: I. **53**:1 (1977), pp. 6–10; II. **53**:5 (1977), pp. 147–152; III. **53**:5 (1977), pp. 153–158; IV. **53**:6 (1977), pp. 183–185; V. **53**:7 (1977), pp. 219–224; VI. **54**:1 (1978), pp. 1–5; VII. **54**:2 (1978), pp. 36–41.

[DJKM2] DATE, E., M. JIMBO, M. KASHIWARA, and T. MIWA. *Solitons, τ functions and Euclidean Lie algebras.* Pp. 261–278 in: *Mathématique and Physique. Séminaire de l'Ecole Normale Supérieure 1979–1982.* BOUTET DE MONVEL, L., A. DOUADY, and J.-L. VERDIER – eds. (Progress in Math., **37**). Birkhäuser, Boston·Basel·Stuttgart, 1983

[FuL] FUCHS, L. *Gesammelte Mathematische Werke. Band 3.* (Herausgegeben by R. FUCHS und L SCHLESINGER). Mayer & Müller, Berlin, 1909.

[FuL1] FUCHS, L. *Zur Theorie der linearen Differentialgleichungen.* Sitzungsberichte der K. preuss. Akademie der Wissenschaften zu Berlin. Einleitung und No. 1–7, 1888, S. 1115–1126; No. 8–15, 1888, S.1273–1290; No. 16–21, 1889, S. 713–726; No. 22–31, 1890, S. 21–38. Reprinted in: [FuL], S. 1–68.

[FuL2] FUCHS, L. *Über lineare Differentialgleichungen, welche von Parametern unabhängige substitutionsgruppen besitzen.* Sitzungsberichte der K. preuss. Akademie der Wissenschaften zu Berlin. 1892, S. 157–176. Reprinted in: [FuL], S. 117–139.

[FuL3] FUCHS, L. *Über lineare Differentialgleichungen, welche von Parametern unabhängige Substitutionsgruppen besitzen.* Sitzungsberichte der K. preuss. Akademie der Wissenschaften zu Berlin. Einleitung und No. 1–4, 1893, S. 975–988; No. 5–8, 1894,, S. 1117–1127. Reprinted in: [FuL], S. 169–195.

[FuL4] FUCHS, L. *Über die Abhängigkeit der Lösungen einer linearen differentialgleichung von den in den Coefficienten auftretenden Parametren.* Sitzungsberichte der K. preuss. Akademie der Wissenschaften zu Berlin. 1895, S. 905–920. Reprinted in: [FuL], S. 201–217.

[FuR] FUCHS, R. *Sur quelquea équations différentielles linéares du second ordre.* Compt. Rend. de l'Académie des Sciences, Paris. **141** (1905), pp. 555–558.

[Gant] ГАНТМАХЕР, Ф. Р. *Теория матриц. 2-е издание.* Наука, 1966, 575 с. (In Russian). English transl.: GANTMACHER, F.R.. THE THEORY OF MATRICES. Chelsea, New York, 1959, 1960.

[Gar] GARNIER, R. *Sur une classe d'équations différentielles dont les intégrales générales ont leurs points critiques fixes.* Compt. Rend. de l'Académie des Sciences, Paris. **151** (1910), pp. 205–208.

[GKRM] GOHBERG, I., M.A. KAASHOEK, L. LEHRER and L. RODMAN. *Minimal divisors of rational matrix functions with prescribed zero and pole structures,* pp. 241–275 in: *Topics in Operator Theory, Systems and Networks.* (DYM, H. and I. GOHBERG- ed.) Operator Theory: Advances and Applications, **OT 12**, Birkhäuser, Basel Boston Stuttgart, 1984.

[Ha] HARNAD, J., *Dual isomonodromic tau functions and determinants of integrable Fredholm operators.* In: *Random Matrix Models and Their Applications,* (Mathematical Sciences Research Institute Publications, **40**), BLEHER, P. and A. ITS – editors. Cambridge Univ. Press 2001.

[HaI] HARNAD, J. and A. ITS. *Integrable Fredholm operators and dual isomonodromic deformations.* Commun. Math. Phys., **226** (2002), pp. 497–530.

[Hart] HARTMAN, PH. *Ordinary Differential Equations.* Wiley, New York·London·Sydney, 1964. Russian transl.: ХАРТМАН, Ф. *Обыкновенные Дифференциальные уравнения.* Москва, Мир, 1970, 720 с.

[HsSi] HSIEH, P., and Y. SIBIYA. *Basic Theory of Ordinary Differential Equations.* Springer-Verlag, New York·Berlin·Heidelberg, 1999, xi+468 pp.

[Huk] HUKUHARA, M. *Ordinary Differential Equations.* (In Japanese). Iwanami-Zensho **116**, Iwanami-Shoten, 1950.

[IKSY] IWASAKI, K., KIMURA, H., SHIMOMURA, SH. and M. YOSHIDA . *From Gauss to Painlevé. A Modern Theory of Special Functions.* (Aspects of mathematics: E , Vol. **16**). Vieweg, Braunschweig, 1991.

[JMMS] JIMBO, M., MIWA, T., MÔRI, Y. and M. SATO. *Density matrix of an impenetrable Bose gas and the fifth Painlevé transcendent.* Physica **1D** (1980), pp. 80–158.

[JMU] JIMBO, M., MIWA, T. and K. UENO. *Monodromy preserving deformation of linear ordinary differential equations with rational coefficients. I. General theory and τ-function.* Physica **2D** (1981), pp. 306–352.

[JM1] JIMBO, M. and T MIWA. *Monodromy preserving deformation of linear ordinary differential equations with rational coefficients. II.* Physica **2D**, (1981), pp. 407–448.

[JM2] JIMBO, M. and T MIWA. *Monodromy preserving deformation of linear ordinary differential equations with rational coefficients. III.* Physica **4D**, (1981), pp. 26–46.

[Kats1] KATSNELSON, V. *Fuchsian differential systems related to rational matrix fuctions in general position and the joint system realization* , pp. 117–143 in: Israel Mathematical Conference Proceedings, Vol. **11** (1997), *Proceedings of the Ashkelon Workshop on Complex Function Theory (May 1996)*, ZALCMAN, L. – editor.

[Kats2] KATSNELSON, V. *Right and left joint system representation of a rational matrix function in general position (System representation theory for dummies)* , pp. 337–400 in: *Operator theory, system theory and related topics. (The Moshe Livšic anniversary volume. Proceedings of the Conference on Operator Theory held at Ben-Gurion University of the Negev, Beer-Sheva and in Rehovot, June 29–July 4, 1997)*, ALPAY, D. and V. VINNIKOV – editors. (Operator Theory: Advances and Applications, vol. **123**), Birkhäuser Verlag, Basel, 2001.

[KiKo] KITAEV, A.V. and D.A. KOROTKIN. *On solution of the Schlesinger Equations in Terms of Θ-functions.* Intern. Math. Research Notes, 1998, No. **17**, pp. 877–905.

[KoMa] КОРОТКИН, Д. А. и В. Б. МАТВЕЕВ. *О тэта-функциональных решениях системы Шлезингера и уравнения Эрнста.* Функциональный анализ и его приложения , **34**:4 (2000), с. 18–34 (Russian). English transl.: KOROTKIN, D.A. and V.B. MATVEEV. *On theta function solutions of the Schlesinger system and the Ernst equation.* Funk. Anal. Appl., **34**:4 , (2000), pp. 252–264.

[Mah] MAHOUX, G. *Introduction to the theory of isomonodromic deformations of linear ordinary differential equations with rational coefficients.* Pp. 35–76 in: *The Painlevé Property. One Century Later.* CONTE R. – editor. (CRM Series in Mathematical Physics.) Springer-Verlag, New York·Berlin·Heidelberg, 1999.

[MSW] MASON, S.J., M.A. SINGER and N.M.J. WOODHOUSE. *Tau function and the twistor theory of integrable systems.* Journal of Geometry and Physics, **32** (2000), pp.397–430.

[Miw] MIWA, T. *Painlevé property of monodromy preserving deformation equations and analyticity of τ functions.* Publ. RIMS Kyoto Univ., **17** (1981), pp. 703–721.

[MJD] MIWA, T., M. JIMBO, and E. DATE. *Solitons. Differential equations, symmetries and infinite-dimensional algebras.* (Cambridge Tracts in Mathematics, **135**.) Cambridge University Press, Cambridge, 2000. x+108 pp.

[New] NEWELL, A.C. *Solitons in Mathematics and Physics.* (CBMS-NSF Regional Conference Series in Applied Mathematics, **48**.) SIAM, Philadelphia, PA, 1985. xvi+244 pp. Russian transl.: НЬЮЭЛЛ, А. *Солитоны в Математике и Физике.* Мир, 1989, 326 сс.

[Palm1] PALMER, J. *Determinants of Cauchy-Riemann operators as τ-functions.* Acta Applicandae Math, **18** (1990), pp. 199–223.

[Palm2] PALMER, J. *Deformation analysis of matrix models.* Physica **98D** (1994), pp. 166–185.

[Plem1] PLEMELJ, J. *Riemannsche Funktionenscharen mit gegebener Monodromiegruppe.* (*Riemann families with prescribed monodromy group*). Monatshefte für Math. und Phys., **XIX**, (1908), pp. 211–245 .

[Plem2] PLEMELJ, J. *Problems in the Sense of Riemann and Klein.* Intersience Publishers. A division of J. Wiley & Sons Inc., New York · London · Sidney , 1964 , 175 pp.

[Ra] RASCH, G. *Zur Theorie und Anwendung des Productintegrales.* Journ. für die reine und angew. Math., **171** (1934), pp.65–119.

[Sakhn] САХНОВИЧ, Л.А. *О факторизации передаточной оператор-функции.* Доклады АН СССР, **226**:4 (1976), с. 781–784. Engl. transl.: SAKHNOVICH, L.A., *On the factorization of an operator-valued transfer function.* Soviet. Math. Dokl. **17** (1970), pp. 203–207.

[Sa] SATO, M. *The KP hierarchy and infinite-dimensional Grassman manifolds.* Proc. of Symposia in Pure Math., **49**:1 (1989), pp. 51–66.

[SMJ1] SATO, M., MIWA, T and M. JIMBO. *Aspects of holonomic quantum fields. Isomonodromic deformations and Ising model.* Pp. 429–491 in *Complex Analysis, Microlocal Calculus and Relativistic Quantum Theory. Proceeding of the Colloquium held at Les Houches, Centre de Physique, September 1979,* IAGOLNITZER, D. – editor. (Lectures Notes in Physics, Vol. **126**). Springer-Verlag, Berlin·Heidelberg·New York, 1980.

[SMJ2] SATO, M., MIWA, T and M. JIMBO. *Holonomic quantum fields. II.* Publ. RIMS Kyoto Univ. **15** (1979), pp. 201–278.

[SaSa] SATO M. and Y. SATO. *Soliton equations as dynamical systems on infinite-dimensional Grassmanin manifolds.* Pp. 259–271 in: *Nonlinear Partial Differential Equations in Applied Science; Proceedings of The U.S.-Japan Seminar, Tokyo, 1982,* (Lect. Notes in Num. Appl. Anal., **5**), 1982, FUJITA, H., P. LAX, and G. STRANG – eds.

[Sch1] SCHLESINGER, L. *Über die Lösungen gewisser linearer Differentialgleichungen als Funktionen der singulären Punkte.* Journal für reine und angew. Math, **129** (1905), pp. 287–294.

[Sch2] SCHLESINGER, L. *Vorlesungen über lineare Differentialgleichungen.* Leipzig und Berlin, 1908.

[Sch3] SCHLESINGER, L. *Über eine Klasse von Differentialsystemen beliebiger Ordnung mit festen kritischen Punkten*. Journal für reine und angew. Math, **141** (1912), pp. 96–145.

[SeWi] SEGAL, G. and G. WILSON. *Loop groups and equations of KdV type*. Publ. Mathem. IHES, № **61** (1985), pp. 5–65 . Reprinted in [TeUh], pp. 403–466. Russian transl.: СИГАЛ, Г. и ДЖ. ВИЛЬСОН. *Группа петель и уравнения типа КдФ*. Стр. 379–442 в книге: ПРЕССЛИ, Э. и Г. СИГАЛ. *Группы Петель*. Москва, Мир, 1990.

[Sib] SIBUYA, Y. *Linear Differential Equations in the Complex Domain: Problems of Analytic Continuation*. (Translations of Math. Monographs, **82**), Amer. Math. Soc., Providence, Rhode Island, 1990, iv+267.

[TeUh] TERNG, CH. L. and K. UHLENBECK – editors. *Surveys in Differential Geometry: Integral Systems. (Lectures in Geometry and Topology, sponsored by Lehigh University's Journal of Differential Geometry.)* International Press, Boston, 1998.

[WNTB] WU, T.T., B. MCCOY, C. TRACY and E. BAROUCH. *Spin-spin correlation functions for the two-dimensional Ising model: Exact theory in the scaling region*. Phys. Rev. **B**, **13**:1 (1976), 316–374.

[ZaSh] ЗАХАРОВ, В.Е. и А.Б. ШАБАТ. *Точная теория двумерной самофокусировки и одномерной модуляции волн в нелинейной среде*. Журн. Эксперим. Теор. Физ., **61**:1, (1971), 118–134 (In Russian). English transl.: ZAKHAROV, V.E. and A.B. SHABAT (=A.B. ŠABAT). *Exact theory of two-dimensional self-focusing and one-dimensional self-modulation of waves in non-linear media*. Soviet Physics JETP, **34**:1 (1972), pp. 62–69.

V. Katsnelson
Department of Mathematics
the Weizmann Institute of Science
Rehovot, 76100, Israel
e-mail: `victor.katsnelson@weizmann.ac.il`

D. Volok
Department of Mathematics
Ben-Gurion University of the Negev
Beer-Sheva, 84105, Israel
e-mail: `volok@math.bgu.ac.il`

Operator Theory:
Advances and Applications, Vol. 149, 349–372

A Generalization of the $\tan 2\Theta$ Theorem

Vadim Kostrykin, Konstantin A. Makarov, and
Alexander K. Motovilov

Abstract. Let $\mathbf{A}$ be a bounded self-adjoint operator on a separable Hilbert space $\mathfrak{H}$ and $\mathfrak{H}_0 \subset \mathfrak{H}$ a closed invariant subspace of $\mathbf{A}$. Assuming that

$$\sup \operatorname{spec}(A_0) \le \inf \operatorname{spec}(A_1),$$

where A_0 and A_1 are restrictions of $\mathbf{A}$ onto the subspaces $\mathfrak{H}_0$ and $\mathfrak{H}_1 = \mathfrak{H}_0^\perp$, respectively, we study the variation of the invariant subspace $\mathfrak{H}_0$ under bounded self-adjoint perturbations $\mathbf{V}$ that are off-diagonal with respect to the decomposition $\mathfrak{H} = \mathfrak{H}_0 \oplus \mathfrak{H}_1$. We obtain sharp two-sided estimates on the norm of the difference of the orthogonal projections onto invariant subspaces of the operators $\mathbf{A}$ and $\mathbf{B} = \mathbf{A} + \mathbf{V}$. These results extend the celebrated Davis-Kahan $\tan 2\Theta$ Theorem. On this basis we also prove new existence and uniqueness theorems for contractive solutions to the operator Riccati equation, thus, extending recent results of Adamyan, Langer, and Tretter.

1. Introduction

Given a self-adjoint bounded operator $\mathbf{A}$ and a closed invariant subspace $\mathfrak{H}_0 \subset \mathfrak{H}$ of $\mathbf{A}$ we set $A_i = \mathbf{A}|_{\mathfrak{H}_i}$, $i = 0, 1$ with $\mathfrak{H}_1 = \mathfrak{H} \ominus \mathfrak{H}_0$. Assuming that the perturbation $\mathbf{V}$ is off-diagonal with respect to the orthogonal decomposition $\mathfrak{H} = \mathfrak{H}_0 \oplus \mathfrak{H}_1$ consider the 2×2 self-adjoint operator matrix

$$\mathbf{B} = \mathbf{A} + \mathbf{V} = \begin{pmatrix} A_0 & V \\ V^* & A_1 \end{pmatrix},$$

where V is a bounded operator from $\mathfrak{H}_1$ to $\mathfrak{H}_0$.

In the 1970 paper [8] Davis and Kahan proved that if

$$\sup \operatorname{spec}(A_0) < \inf \operatorname{spec}(A_1), \tag{1.1}$$

then the difference of the spectral projections

$$P = \mathsf{E}_{\mathbf{A}}\big((-\infty, \sup \operatorname{spec}(A_0)]\big) \quad \text{and} \quad Q = \mathsf{E}_{\mathbf{B}}\big((-\infty, \sup \operatorname{spec}(A_0)]\big)$$

for the operators $\mathbf{A}$ and $\mathbf{B}$, respectively, corresponding to the interval

$$(-\infty, \sup \operatorname{spec}(A_0)]$$

Received by the editors February 1, 2003. Submitted by M. Klaus.
Mathematics Subject Classification (2000). Primary 47A15, 47A55, 47A62; Secondary 47A53.
Key words and phrases. Perturbation theory, spectral subspaces, graph subspaces, operator Riccati equation.

admits the estimate

$$\|P - Q\| \le \sin\left(\frac{1}{2}\arctan\frac{2\|V\|}{d}\right) < \frac{\sqrt{2}}{2}, \tag{1.2}$$

where

$$d = \operatorname{dist}(\operatorname{spec}(A_0), \operatorname{spec}(A_1)).$$

Estimate (1.2) can be equivalently expressed as the $\tan 2\Theta$ Theorem:

$$\|\tan 2\Theta\| \le \frac{2\|V\|}{d}, \qquad \operatorname{spec}(\Theta) \subset [0, \pi/4),$$

where Θ is the operator angle between the subspaces $\operatorname{Ran} P$ and $\operatorname{Ran} Q$ (see, e.g., [10]).

By known results on graph subspaces (see, e.g., [3], [4], [6], [10]) estimate (1.2) in particular implies that the Riccati equation

$$A_1X - XA_0 - XVX + V^* = 0 \tag{1.3}$$

has a contractive solution $X : \mathfrak{H}_0 \to \mathfrak{H}_1$ satisfying the norm estimate

$$\|X\| = \frac{\|P - Q\|}{\sqrt{1 - \|P - Q\|^2}} \le \tan\left(\frac{1}{2}\arctan\frac{2\|V\|}{d}\right) < 1. \tag{1.4}$$

Moreover, the graph of X, i.e., the subspace $\mathcal{G}(\mathfrak{H}_0, X) := \{x \oplus Xx | x \in \mathfrak{H}_0\}$, coincides with the spectral subspace $\operatorname{Ran} \mathsf{E}_{\mathbf{B}}\big((-\infty, \sup\operatorname{spec}(A_0)]\big)$ of the operator $\mathbf{B}$.

Independently of the work of Davis and Kahan the existence of a unique contractive solution to the Riccati equation under condition (1.1) has been proven by Adamyan and Langer in [1], where the operators A_0 and A_1 were allowed to be semibounded. In a recent paper by Adamyan, Langer, and Tretter [2] the existence result has been extended to the case where the spectra of A_0 and A_1 intersect at one point $\lambda \in \mathbb{R}$, that is,

$$\sup\operatorname{spec}(A_0) = \inf\operatorname{spec}(A_1) = \lambda,$$

provided that at least one of the following conditions

$$\operatorname{Ker}(A_0 - \lambda) = \{0\}, \quad \operatorname{Ker}(A_1 - \lambda) \cap \operatorname{Ker} V = \{0\} \tag{1.5}$$

or

$$\operatorname{Ker}(A_1 - \lambda) = \{0\}, \quad \operatorname{Ker}(A_0 - \lambda) \cap \operatorname{Ker} V^* = \{0\} \tag{1.6}$$

holds. In this case the Riccati equation (1.3) has been proven to have a unique contractive solution X, which appears to be a strict contraction. The graph of X, as above, coincides with the spectral subspace $\operatorname{Ran} \mathsf{E}_{\mathbf{B}}((-\infty, \lambda))$ of the operator $\mathbf{B}$.

Conditions (1.5) and (1.6) are rather restrictive. In particular, in this case λ may be an eigenvalue neither for both A_0 and A_1 nor for $\mathbf{B}$.

The main goal of the present article is to drop conditions (1.5) and (1.6) and to carry out the analysis under the only assumption that

$$\sup\operatorname{spec}(A_0) \le \lambda \le \inf\operatorname{spec}(A_1). \tag{1.7}$$

Below we will prove (see Theorem 2.4) that under hypothesis (1.7) the **B**-invariant subspace

$$\mathfrak{Q} = \operatorname{Ran} \mathsf{E}_{\mathbf{B}}((-\infty, \lambda)) \oplus (\operatorname{Ker}(A_0 - \lambda) \cap \operatorname{Ker} V^*) \tag{1.8}$$

is the graph of a contractive operator $X : \mathfrak{H}_0 \to \mathfrak{H}_1$. Moreover, the norm of the operator X satisfies the lower bound

$$\|X\| \geq \frac{\delta}{\|V\|},$$

where

$$\delta = \max\{\inf \operatorname{spec}(\mathbf{A}) - \inf \operatorname{spec}(\mathbf{B}), \sup \operatorname{spec}(\mathbf{B}) - \sup \operatorname{spec}(\mathbf{A})\} \geq 0$$

is the maximal shift of the edges of the spectrum of the operator **A** under the perturbation **V**. These results can be stated equivalently as the two-sided estimate

$$\frac{\delta}{\sqrt{\delta^2 + \|V\|^2}} \leq \|P - Q\| \leq \frac{\sqrt{2}}{2}, \tag{1.9}$$

where P and Q are orthogonal projections in $\mathfrak{H}$ onto the subspaces $\mathfrak{H}_0$ and $\mathfrak{Q}$, respectively. Notice that the for the subspace $\mathfrak{Q}$ to be a spectral subspace of **B** it is necessary and sufficient that

$$\text{either} \quad \operatorname{Ker}(A_0 - \lambda) \cap \operatorname{Ker} V^* = \{0\} \quad \text{or} \quad \operatorname{Ker}(A_1 - \lambda) \cap \operatorname{Ker} V = \{0\},$$

and then, necessarily,

$$\text{either} \quad \mathfrak{Q} = \operatorname{Ran} \mathsf{E}_{\mathbf{B}}((-\infty, \lambda)) \quad \text{or} \quad \mathfrak{Q} = \operatorname{Ran} \mathsf{E}_{\mathbf{B}}((-\infty, \lambda]).$$

The fact that the subspace $\mathfrak{Q}$ is a graph of a contractive operator X means that the Riccati equation (1.3) has a contractive solution. In contrast to the case studied in [2], the solution X is in general (under hypothesis (1.7)) neither necessarily strictly contractive nor unique in the set of all contractive solutions to the Riccati equation. Moreover, if the invariant subspace $\mathfrak{Q}$ is not a spectral subspace of the operator **B**, then X is a non-isolated point (in the operator norm topology) of the set of all solutions of (1.3).

Below we will prove (see Theorem 4.1) that under assumption (1.7) the operator X is the unique solution to the Riccati equation (1.3) within the class of bounded linear operators from $\mathfrak{H}_0$ to $\mathfrak{H}_1$ satisfying the additional requirements that

$$\operatorname{Ker}(A_0 - \lambda) \cap \operatorname{Ker} V^* \subset \operatorname{Ker} X \quad \text{and} \quad \operatorname{spec}(A_0 + VX) \subset (-\infty, \lambda].$$

Furthermore, we formulate and prove necessary and sufficient conditions for a contractive solution to be a unique contractive solution (see Theorem 4.3 or/and Theorem 5.4). The solution X satisfying $\mathcal{G}(\mathfrak{H}_0, X) = \mathfrak{Q}$ is shown to be the unique contractive solution to (1.3) if and only if it is strictly contractive (see Corollary 4.4). Note that in the case where the contractive solution is non-unique but its graph is a spectral subspace for the operator **B**, a complete description of the set of all contractive solutions to the Riccati equation can be given by means of Theorem 6.2 in [10].

The techniques developed in the present work to prove that the subspace $\mathfrak{Q}$ (1.8) is the graph of a contractive operator (Theorem 2.4), are an extension of the geometric ideas of Davis and Kahan [7], [8]. The existence and uniqueness results (Theorems 4.1, 4.3, and 5.4) are obtained in the framework of the geometric approach of our recent paper [10]. The previously known results by Davis and Kahan [7], [8], Adamyan and Langer [1], and Adamyan, Langer, and Tretter [2] appear to be their direct corollaries.

A few words about the notations used throughout the paper. Given a linear operator A on a Hilbert space $\mathfrak{K}$, by $\operatorname{spec}(A)$ we denote the spectrum of A. If not explicitly stated otherwise, $\mathfrak{N}^\perp$ denotes the orthogonal complement in $\mathfrak{K}$ of a subspace $\mathfrak{N} \subset \mathfrak{K}$, i.e., $\mathfrak{N}^\perp = \mathfrak{K} \ominus \mathfrak{N}$. The identity operator on $\mathfrak{K}$ is denoted by $I_{\mathfrak{K}}$. The notation $\mathcal{B}(\mathfrak{K}, \mathfrak{L})$ is used for the set of bounded operators from the Hilbert space $\mathfrak{K}$ to the Hilbert space $\mathfrak{L}$. Finally, we write $\mathcal{B}(\mathfrak{K}) = \mathcal{B}(\mathfrak{K}, \mathfrak{K})$.

2. Upper bound

Throughout the whole work we adopt the following hypothesis.

Hypothesis 2.1. *Assume that the separable Hilbert space $\mathfrak{H}$ is decomposed into the orthogonal sum of two subspaces*

$$\mathfrak{H} = \mathfrak{H}_0 \oplus \mathfrak{H}_1. \tag{2.1}$$

Assume, in addition, that $\mathbf{B}$ is a self-adjoint operator on $\mathfrak{H}$ represented with respect to the decomposition (2.1) *as a 2×2 operator block matrix*

$$\mathbf{B} = \begin{pmatrix} A_0 & V \\ V^* & A_1 \end{pmatrix},$$

where A_i are bounded self-adjoint operators in $\mathfrak{H}_i$, $i = 0, 1$, while V is a bounded operator from $\mathfrak{H}_1$ to $\mathfrak{H}_0$. More explicitly, $\mathbf{B} = \mathbf{A} + \mathbf{V}$, where $\mathbf{A}$ is the bounded diagonal self-adjoint operator,

$$\mathbf{A} = \begin{pmatrix} A_0 & 0 \\ 0 & A_1 \end{pmatrix},$$

and the operator $\mathbf{V} = \mathbf{V}^*$ is an off-diagonal bounded operator

$$\mathbf{V} = \begin{pmatrix} 0 & V \\ V^* & 0 \end{pmatrix}.$$

Moreover, assume that

$$\sup \operatorname{spec}(A_0) \le \lambda \le \inf \operatorname{spec}(A_1) \tag{2.2}$$

for some $\lambda \in \mathbb{R}$.

If, under Hypothesis 2.1, λ is a multiple eigenvalue of the operator $\mathbf{B}$, then $\mathbf{B}$ has infinitely many invariant subspaces $\mathfrak{L}_\mathbf{B}$ such that

$$\operatorname{Ran} \mathsf{E}_\mathbf{B}((-\infty, \lambda)) \subsetneq \mathfrak{L}_\mathbf{B} \subsetneq \operatorname{Ran} \mathsf{E}_\mathbf{B}((-\infty, \lambda]) \tag{2.3}$$

that are necessarily not spectral subspaces.

A criterion for a **B**-invariant subspace $\mathfrak{L}_{\mathbf{B}}$ satisfying (2.3) to be a graph subspace associated with the decomposition $\mathfrak{H} = \mathfrak{H}_0 \oplus \mathfrak{H}_1$ is given by the following theorem.

Theorem 2.2. *Assume Hypothesis* 2.1. *Then*

$$\operatorname{Ker}(\mathbf{B} - \lambda) = (\operatorname{Ker}(A_0 - \lambda) \cap \operatorname{Ker} V^*) \oplus (\operatorname{Ker}(A_1 - \lambda) \cap \operatorname{Ker} V). \tag{2.4}$$

Moreover, given a subspace $\mathfrak{N} \subset \operatorname{Ker}(\mathbf{B} - \lambda)$, *the subspace*

$$\mathfrak{L}_{\mathbf{B}} := \operatorname{Ran} \mathsf{E}_{\mathbf{B}}((-\infty, \lambda)) \oplus \mathfrak{N} \tag{2.5}$$

is a graph subspace associated with the subspace $\mathfrak{H}_0$ *in the decomposition* $\mathfrak{H} = \mathfrak{H}_0 \oplus \mathfrak{H}_1$ *if and only if* $\mathfrak{N}$ *is a graph subspace associated with the subspace* $\operatorname{Ker}(A_0 - \lambda) \cap \operatorname{Ker} V^*$ *in the decomposition* (2.4).

Proof. Without loss of generality we may set $\lambda = 0$. The inclusion

$$(\operatorname{Ker} A_0 \cap \operatorname{Ker} V^*) \oplus (\operatorname{Ker} A_1 \cap \operatorname{Ker} V) \subset \operatorname{Ker} \mathbf{B} \tag{2.6}$$

is obvious. In order to prove the opposite inclusion assume that $\begin{pmatrix} x \\ y \end{pmatrix} \in \operatorname{Ker} \mathbf{B}$ with $x \in \mathfrak{H}_0$ and $y \in \mathfrak{H}_1$, i.e.,

$$A_0 x + V y = 0 \qquad \text{and} \qquad V^* x + A_1 y = 0. \tag{2.7}$$

Suppose that $x \notin \operatorname{Ker} A_0$. Then $(x, A_0 x) < 0$ and therefore $(x, Vy) > 0$ using the first equation in (2.7). From the second of equations (2.7) it follows that $(y, A_1 y) < 0$, which is in a contradiction with (2.2). Thus $x \in \operatorname{Ker} A_0$. Similarly one proves that $y \in \operatorname{Ker} A_1$. By using (2.7) it follows that $x \in \operatorname{Ker} V^*$ and $y \in \operatorname{Ker} V$ which together with (2.6) proves (2.4).

In order to prove the second statement of the theorem notice that by Theorem 5.5 in the Appendix the subspace $\mathfrak{N} \subset \operatorname{Ker} \mathbf{B}$ is a graph subspace associated with the subspace $\operatorname{Ker} A_0 \cap \operatorname{Ker} V^*$ in the decomposition (2.4) (recall that we assumed that $\lambda = 0$) if and only if

$$(\operatorname{Ker} A_0 \cap \operatorname{Ker} V^*) \cap (\operatorname{Ker} \mathbf{B} \ominus \mathfrak{N}) = (\operatorname{Ker} A_1 \cap \operatorname{Ker} V) \cap \mathfrak{N} = \{0\}.$$

Again from Theorem 5.5 it follows that the subspace $\operatorname{Ran} \mathsf{E}_{\mathbf{B}}((-\infty, 0)) \oplus \mathfrak{N}$ is a graph subspace associated with the subspace $\mathfrak{H}_0$ in the decomposition $\mathfrak{H} = \mathfrak{H}_0 \oplus \mathfrak{H}_1$ if and only if

$$\mathfrak{H}_0 \cap \big(\operatorname{Ran} \mathsf{E}_{\mathbf{B}}((-\infty, 0)) \oplus \mathfrak{N}\big)^{\perp} = \mathfrak{H}_1 \cap (\operatorname{Ran} \mathsf{E}_{\mathbf{B}}((-\infty, 0)) \oplus \mathfrak{N}) = \{0\}. \tag{2.8}$$

Therefore, to complete the proof it is sufficient to establish the following equalities

$$\mathfrak{H}_0 \cap \big(\operatorname{Ran} \mathsf{E}_{\mathbf{B}}((-\infty, 0)) \oplus \mathfrak{N}\big)^{\perp} = (\operatorname{Ker} A_0 \cap \operatorname{Ker} V^*) \cap (\operatorname{Ker} \mathbf{B} \ominus \mathfrak{N}) \tag{2.9}$$

and

$$\mathfrak{H}_1 \cap \big(\operatorname{Ran} \mathsf{E}_{\mathbf{B}}((-\infty, 0)) \oplus \mathfrak{N}\big) = (\operatorname{Ker} A_1 \cap \operatorname{Ker} V) \cap \mathfrak{N}. \tag{2.10}$$

First, we prove that the left-hand side of (2.9) is a subset of the right-hand side of (2.9), i.e.,

$$\mathfrak{H}_0 \cap \big(\operatorname{Ran} \mathsf{E}_{\mathbf{B}}((-\infty, 0)) \oplus \mathfrak{N}\big)^{\perp} \subset (\operatorname{Ker} A_0 \cap \operatorname{Ker} V^*) \cap (\operatorname{Ker} \mathbf{B} \ominus \mathfrak{N}). \tag{2.11}$$

Let $x \in \mathfrak{H}_0$ and $x \perp \operatorname{Ran} \mathsf{E}_{\mathbf{B}}((-\infty,0)) \oplus \mathfrak{N}$. Clearly,

$$(x, \mathbf{B}x) \geq 0 \quad \text{for} \quad x \perp \operatorname{Ran} \mathsf{E}_{\mathbf{B}}((-\infty,0)) \oplus \mathfrak{N}.$$

Moreover,

$$(x, \mathbf{B}x) = (x, \mathbf{A}x) \leq 0 \quad \text{for all} \quad x \in \mathfrak{H}_0,$$

since the operator matrix $\mathbf{V}$ is off-diagonal and the subspace $\mathfrak{H}_0$ is $\mathbf{A}$-invariant. Hence,

$$(x, \mathbf{A}x) = (x, \mathbf{B}x) = 0 \quad \text{for } x \in \mathfrak{H}_0 \cap \left(\operatorname{Ran} \mathsf{E}_{\mathbf{B}}((-\infty,0)) \oplus \mathfrak{N}\right)^{\perp},$$

which implies $x \in \operatorname{Ker} \mathbf{B} \ominus \mathfrak{N}$ by the variational principle. Therefore, the inclusion (2.11) holds.

The opposite inclusion

$$\mathfrak{H}_0 \cap \left(\operatorname{Ran} \mathsf{E}_{\mathbf{B}}((-\infty,0)) \oplus \mathfrak{N}\right)^{\perp} \supset (\operatorname{Ker} A_0 \cap \operatorname{Ker} V^*) \cap (\operatorname{Ker} \mathbf{B} \ominus \mathfrak{N})$$

is obvious, which proves (2.9).

The equality (2.10) is proven in a similar way. □

Remark 2.3. As it follows from (2.4) the closed $\mathbf{B}$-invariant subspace

$$\mathfrak{Q} = \operatorname{Ran} \mathsf{E}_{\mathbf{B}}((-\infty,\lambda)) \oplus \left(\operatorname{Ran} \mathsf{E}_{\mathbf{B}}(\{\lambda\}) \cap \mathfrak{H}_0\right) \subset \mathfrak{H}. \tag{2.12}$$

is a spectral subspace for the operator $\mathbf{B}$ if and only if

$$\text{either} \quad \operatorname{Ker}(A_0 - \lambda) \cap \operatorname{Ker} V^* = \{0\} \quad \text{or} \quad \operatorname{Ker}(A_1 - \lambda) \cap \operatorname{Ker} V = \{0\}.$$

The following theorem characterizes the subspace $\mathfrak{Q}$ as the graph of some contractive operator X from $\mathfrak{H}_0$ to $\mathfrak{H}_1$, that is, $\mathfrak{Q} = \mathcal{G}(\mathfrak{H}_0, X)$ where

$$\mathcal{G}(\mathfrak{H}_0, X) = \{x \oplus Xx | x \in \mathfrak{H}_0\}.$$

Theorem 2.4. *Assume Hypothesis* 2.1. *Then:*

(i) *The* $\mathbf{B}$*-invariant subspace* $\mathfrak{Q}$ *is a graph subspace* $\mathcal{G}(\mathfrak{H}_0, X)$ *associated with the subspace* $\mathfrak{H}_0$ *in decomposition* (2.1), *where the operator* X *is a contraction,* $\|X\| \leq 1$, *with the properties*

$$\begin{aligned} &\operatorname{Ker}(A_0 - \lambda) \cap \operatorname{Ker} V^* \subset \operatorname{Ker} X, \\ &\operatorname{Ker}(A_1 - \lambda) \cap \operatorname{Ker} V \subset \operatorname{Ker} X^*. \end{aligned} \tag{2.13}$$

(ii) *If either* $\operatorname{Ker}(A_0 - \lambda) = \{0\}$ *or* $\operatorname{Ker}(A_1 - \lambda) = \{0\}$, *then* X *is a strict contraction, i.e.,*

$$\|Xf\| < \|f\|, \quad f \neq 0.$$

(iii) *If*

$$d = \operatorname{dist}(\operatorname{spec}(A_0), \operatorname{spec}(A_1)) > 0, \tag{2.14}$$

then X *is a uniform contraction satisfying the estimate*

$$\|X\| \leq \tan\left(\frac{1}{2} \arctan \frac{2\|V\|}{d}\right) < 1. \tag{2.15}$$

Remark 2.5. In Section 4 we will establish necessary and sufficient conditions guaranteeing that the operator X referred to in Theorem 2.4 is a strict contraction (see Corollary 4.4 below). The claim (ii) of Theorem 2.4 will then appear to be a corollary of this more general result.

Proof of Theorem 2.4. Without loss of generality we assume that $\lambda = 0$.

(i) Step 1. First we prove the assertion under the additional assumption that $\operatorname{Ker} \mathbf{B} = \{0\}$.

By Theorem 2.2 the subspace $\mathfrak{Q}$ is a graph subspace associated with the decomposition $\mathfrak{H} = \mathfrak{H}_0 \oplus \mathfrak{H}_1$, i.e.,

$$\mathfrak{Q} = \mathcal{G}(\mathfrak{H}_0, X), \tag{2.16}$$

where X is a (possibly unbounded) densely defined closed operator from $\mathfrak{H}_0$ to $\mathfrak{H}_1$. Let $X = U|X|$ be the polar decomposition for X, where $U : \mathfrak{H}_0 \to \mathfrak{H}_1$ is a partial isometry with the initial subspace $(\operatorname{Ker} X)^\perp$ and the final subspace $\overline{\operatorname{Ran} X}$ and $|X| = (X^*X)^{1/2}$, the absolute value of X.

First we show that

$$\operatorname{spec}_{\mathrm{pp}}(|X|) \subset [0, 1], \tag{2.17}$$

where $\operatorname{spec}_{\mathrm{pp}}(|X|)$ denotes the set of all eigenvalues of $|X|$.

Let $0 \neq \mu \in \operatorname{spec}_{\mathrm{pp}}(|X|)$ and f be an eigenvector of $|X|$ corresponding to the eigenvalue μ, i.e.,

$$|X|f = \mu f, \quad 0 \neq f \in \operatorname{Dom}(|X|). \tag{2.18}$$

By (2.16)

$$F = f \oplus Xf = f \oplus \mu Uf \in \mathfrak{Q} \tag{2.19}$$

using (2.18). Since $f \perp \operatorname{Ker} |X| = \operatorname{Ker} X$, the element f belongs to the initial subspace of the isometry U. Moreover,

$$Uf = \mu^{-1} U|X|f \in \operatorname{Ran} X \subset \overline{\operatorname{Ran} X},$$

i.e., the element Uf belongs to the final subspace of U and hence $U^*Uf = f$, which, in particular, proves that $Uf \in \operatorname{Dom}(X^*)$. Therefore,

$$G = (-X^*Uf) \oplus Uf = (-\mu f) \oplus Uf \in \mathfrak{Q}^\perp. \tag{2.20}$$

Using (2.19), (2.20), and the hypothesis $\operatorname{Ker} \mathbf{B} = \{0\}$ one obtains the following two strict inequalities

$$0 > (F, \mathbf{B}F) = (f, A_0 f) + 2\mu \operatorname{Re}(V^*f, Uf) + \mu^2 (Uf, A_1 Uf), \tag{2.21}$$

$$0 < (G, \mathbf{B}G) = \mu^2 (f, A_0 f) - 2\mu \operatorname{Re}(V^*f, Uf) + (Uf, A_1 Uf). \tag{2.22}$$

If $\mu > 0$ satisfies (2.21) and (2.22), then necessarily $\mu \leq 1$. In order to see that we subtract (2.22) from (2.21) getting the inequality

$$(1 - \mu^2)\big((Uf, A_1 Uf) - (f, A_0 f)\big) > 4\mu \operatorname{Re}(V^*f, Uf). \tag{2.23}$$

Since $A_0 \leq 0$ and $A_1 \geq 0$, equation (2.23) implies

$$\operatorname{Re}(V^*f, Uf) < 0 \quad \text{for} \quad \mu > 1$$

which contradicts the orthogonality of the elements F and $\mathbf{B}G$:

$$(F, \mathbf{B}G) = \mu\big((Uf, A_1 Uf) - (f, A_0 f)\big) + (1-\mu^2)\mathrm{Re}(V^* f, Uf) = 0. \tag{2.24}$$

Hence, (2.17) is proven.

Our next goal is to prove that the operator X is a contraction.

Let $\{\mathcal{P}_n^{(0)}\}_{n\in\mathbb{N}}$ and $\{\mathcal{P}_n^{(1)}\}_{n\in\mathbb{N}}$ be two sequences of finite-dimensional orthogonal projections such that $\operatorname{Ran}\mathcal{P}_n^{(0)} \subset \mathfrak{H}_0$, $\operatorname{Ran}\mathcal{P}_n^{(1)} \subset \mathfrak{H}_1$, and

$$\operatorname*{s-lim}_{n\to\infty} \mathcal{P}_n^{(0)} = P, \qquad \operatorname*{s-lim}_{n\to\infty} \mathcal{P}_n^{(1)} = P^\perp, \tag{2.25}$$

where P is the orthogonal projection from $\mathfrak{H}$ onto $\mathfrak{H}_0$ and

$$\operatorname*{s-lim}_{n\to\infty} \mathsf{E}_{\mathbf{A}_n}(\{0\}) = \mathsf{E}_{\mathbf{A}}(\{0\}), \tag{2.26}$$

$$\operatorname*{s-lim}_{n\to\infty} \mathsf{E}_{\mathbf{A}_n}((-\infty,0)) = \mathsf{E}_{\mathbf{A}}((-\infty,0)), \tag{2.27}$$

where

$$\mathbf{A}_n = \begin{pmatrix} \mathcal{P}_n^{(0)} A_0 \mathcal{P}_n^{(0)} & 0 \\ 0 & \mathcal{P}_n^{(1)} A_1 \mathcal{P}_n^{(1)} \end{pmatrix}$$

are the corresponding finite-dimensional truncations of the operator $\mathbf{A}$. The existence of such sequences can easily be shown by splitting off the subspaces $\operatorname{Ker} A_0$ and $\operatorname{Ker} A_1$.

Introducing the finite rank operators

$$\mathbf{V}_n = \begin{pmatrix} 0 & \mathcal{P}_n^{(0)} V \mathcal{P}_n^{(1)} \\ \mathcal{P}_n^{(1)} V^* \mathcal{P}_n^{(0)} & 0 \end{pmatrix}$$

one concludes (see, e.g., Theorem I.5.2 in [5]) that

$$\operatorname*{s-lim}_{n\to\infty} (\mathbf{A}_n + \mathbf{V}_n) = \mathbf{B}. \tag{2.28}$$

Since $\operatorname{Ker}\mathbf{B} = \{0\}$, (2.28) implies (see, e.g., Theorem VIII.24 in [12])

$$\begin{aligned} \operatorname*{s-lim}_{n\to\infty} \mathsf{E}_{\mathbf{A}_n+\mathbf{V}_n}((-\infty,0)) &= \mathsf{E}_{\mathbf{B}}((-\infty,0)), \\ \operatorname*{s-lim}_{n\to\infty} \mathsf{E}_{\mathbf{A}_n+\mathbf{V}_n}((0,\infty)) &= \mathsf{E}_{\mathbf{B}}((0,\infty)), \end{aligned} \tag{2.29}$$

and hence

$$\operatorname*{s-lim}_{n\to\infty} \mathsf{E}_{\mathbf{A}_n+\mathbf{V}_n}(\{0\}) = \mathsf{E}_{\mathbf{B}}(\{0\}) = 0. \tag{2.30}$$

Let $\widehat{\mathbf{A}}_n$ and $\widehat{\mathbf{V}}_n$ denote the parts of the operators $\mathbf{A}_n$ and $\mathbf{V}_n$ associated with their invariant finite-dimensional subspace $\widehat{\mathfrak{H}}^{(n)} = \mathfrak{H}_0^{(n)} \oplus \mathfrak{H}_1^{(n)}$, where $\mathfrak{H}_0^{(n)} = \operatorname{Ran}\mathcal{P}_n^{(0)}$ and $\mathfrak{H}_1^{(n)} = \operatorname{Ran}\mathcal{P}_n^{(1)}$. By Theorem 2.2 the subspace (of the finite-dimensional Hilbert space $\widehat{\mathfrak{H}}_n$)

$$\operatorname{Ran}\mathsf{E}_{\widehat{\mathbf{A}}_n+\widehat{\mathbf{V}}_n}((-\infty,0)) \oplus \big(\operatorname{Ker}(\widehat{\mathbf{A}}_n + \widehat{\mathbf{V}}_n) \cap \mathfrak{H}_0^{(n)}\big) \subset \mathfrak{H}^{(n)}$$

is a graph subspace

$$\mathcal{G}\big(\operatorname{Ran}\mathsf{E}_{\widehat{\mathbf{A}}_n}((-\infty,0)) \oplus (\operatorname{Ker}(\widehat{\mathbf{A}}_n) \cap \mathfrak{H}_0^{(n)}), X_n\big)$$

for some $X_n \in \mathcal{B}(\mathfrak{H}_0^{(n)}, \mathfrak{H}_1^{(n)})$, $n \in \mathbb{N}$. Since X_n is of finite rank, $\|X_n\| \le 1$ by (2.17).

Applying Theorem 5.6 in the Appendix one arrives at the inequality

$$\|\mathsf{E}_{\widehat{\mathbf{A}}_n+\widehat{\mathbf{V}}_n}((-\infty,0))\oplus\widehat{\mathcal{S}}^{(n)} - \mathsf{E}_{\widehat{\mathbf{A}}_n}((-\infty,0))\oplus\widehat{\mathcal{T}}^{(n)}\| = \frac{\|X_n\|}{\sqrt{1+\|X_n\|^2}} \le \frac{\sqrt{2}}{2}, \quad (2.31)$$

where $\widehat{\mathcal{S}}^{(n)}$ and $\widehat{\mathcal{T}}^{(n)}$ are the orthogonal projections in $\mathfrak{H}^{(n)}$ onto the subspaces $\operatorname{Ker}(\widehat{\mathbf{A}}_n + \widehat{\mathbf{V}}_n) \cap \mathfrak{H}_0^{(n)}$ and $\operatorname{Ker}(\widehat{\mathbf{\Lambda}}_n) \cap \mathfrak{H}_0^{(n)}$, respectively.

The subspaces $\operatorname{Ran}(\widehat{\mathcal{S}}^{(n)})$ and $\operatorname{Ran}(\widehat{\mathcal{T}}^{(n)})$ of the space $\mathfrak{H}^{(n)}$ are naturally imbedded into the total Hilbert space $\mathfrak{H}$. Denoting by $\mathcal{S}^{(n)}$ and $\mathcal{T}^{(n)}$ the corresponding orthogonal projections in $\mathfrak{H}$ onto these subspaces (2.31) yields the estimate

$$\|\mathsf{E}_{\mathbf{A}_n+\mathbf{V}_n}((-\infty,0)) \oplus \mathcal{S}^{(n)} - \mathsf{E}_{\mathbf{A}_n}((-\infty,0)) \oplus \mathcal{T}^{(n)}\| \le \frac{\sqrt{2}}{2}. \quad (2.32)$$

From (2.26) it follows that

$$\operatorname*{s-lim}_{n\to\infty} \mathcal{T}^{(n)} = 0. \quad (2.33)$$

Meanwhile, by (2.30)

$$\operatorname*{s-lim}_{n\to\infty} \mathcal{S}^{(n)} = 0. \quad (2.34)$$

Combining (2.32)–(2.34) and passing to the limit $n \to \infty$, by the lower semicontinuity of the spectrum (see, e.g., [9], Sec. VIII.1.2) one concludes that

$$\|P - Q\| = \frac{\|X\|}{\sqrt{1+\|X\|^2}} \le \frac{\sqrt{2}}{2},$$

where Q is the orthogonal projection in $\mathfrak{H}$ onto the subspace $\mathfrak{Q}$ (2.12). This proves that the operator X is a contraction. The proof of (i) under the additional assumption that $\operatorname{Ker}\mathbf{B} = \{0\}$ is complete.

Step 2. Assume now that $\operatorname{Ker}\mathbf{B}$ is not necessarily trivial. From Theorem 2.2 it follows that the subspace $\operatorname{Ker}\mathbf{B}$ is $\mathbf{A}$-invariant. Denote by $\widehat{\mathbf{A}}$ and $\widehat{\mathbf{B}}$ the corresponding parts of the operators $\mathbf{A}$ and $\mathbf{B}$ associated with the reducing subspace $\widehat{\mathfrak{H}} = \operatorname{Ran}\mathsf{E}_{\mathbf{B}}(\mathbb{R}\setminus\{0\})$. Clearly, the operator $\widehat{\mathbf{B}}$ is an off-diagonal perturbation of the diagonal operator matrix $\widehat{\mathbf{A}}$ with respect to the decomposition $\widehat{\mathfrak{H}} = \widehat{\mathfrak{H}}_0 \oplus \widehat{\mathfrak{H}}_1$, where

$$\widehat{\mathfrak{H}}_0 := \mathfrak{H}_0 \ominus \big(\operatorname{Ran}\mathsf{E}_{\mathbf{B}}(\{0\}) \cap \mathfrak{H}_0\big),$$
$$\widehat{\mathfrak{H}}_1 := \mathfrak{H}_1 \ominus \big(\operatorname{Ran}\mathsf{E}_{\mathbf{B}}(\{0\}) \cap \mathfrak{H}_1\big),$$

and $\operatorname{Ker}\widehat{\mathbf{B}} = \{0\}$. Moreover, Hypothesis 2.1 is satisfied with the replacements $\mathfrak{H} \longrightarrow \widehat{\mathfrak{H}}$, $\mathfrak{H}_0 \longrightarrow \widehat{\mathfrak{H}}_0$, $\mathfrak{H}_1 \longrightarrow \widehat{\mathfrak{H}}_1$, and $\mathbf{A} \longrightarrow \widehat{\mathbf{A}}$, $\mathbf{B} \longrightarrow \widehat{\mathbf{B}}$.

By the first part of the proof the subspace $\operatorname{Ran}\mathsf{E}_{\mathbf{B}}\big((-\infty,0)\big)$, naturally imbedded into the Hilbert space $\widehat{\mathfrak{H}}$, is the graph of a contraction $\widehat{X}$,

$$\widehat{X} : \widehat{\mathfrak{H}}_0 \to \widehat{\mathfrak{H}}_1. \quad (2.35)$$

Clearly, the **B**-invariant subspace

$$\operatorname{Ran} \mathsf{E}_{\mathbf{B}}\big((-\infty,0)\big) \oplus \big(\operatorname{Ran} \mathsf{E}_{\mathbf{B}}(\{0\}) \cap \mathfrak{H}_0\big)$$

is a graph subspace $\mathcal{G}(\mathfrak{H}_0, X)$ associated with the subspace $\mathfrak{H}_0$ in decomposition (2.1) where the operator X is given by

$$Xf = \begin{cases} \widehat{X}f & \text{if} \quad f \in \operatorname{Ran} \mathsf{E}_{\mathbf{B}}((-\infty,0)) \text{ (naturally imbedded into } \widehat{\mathfrak{H}}), \\ 0 & \text{if} \quad f \in \operatorname{Ran} \mathsf{E}_{\mathbf{B}}(\{0\}) \cap \mathfrak{H}_0 = \operatorname{Ker} A_0 \cap \operatorname{Ker} V^*. \end{cases} \tag{2.36}$$

Since $\widehat{X}$ is a contraction by hypothesis, the operator X is also a contraction satisfying the properties (2.13) (with $\lambda = 0$):

$$\operatorname{Ker} A_0 \cap \operatorname{Ker} V^* \subset \operatorname{Ker} X \quad \text{and} \quad \operatorname{Ran} \mathsf{E}_{\mathbf{B}}(\{0\}) \cap \mathfrak{H}_1 = \operatorname{Ker} A_1 \cap \operatorname{Ker} V \subset \operatorname{Ker} X^*$$

using (2.35) and (2.36). The proof of (i) is complete.

(ii) If at least one of the subspaces Ker A_0 or Ker A_1 is trivial, then (in the notations above) we have that

$$(Uf, A_1 Uf) - (f, A_0 f) > 0,$$

and, therefore, equality (2.24) cannot be satisfied for $\mu = 1$. Hence $\mu = 1$ is not a singular number of the contraction X which proves that the operator X is a strict contraction, i.e.,

$$\|Xf\| < \|f\|, \quad \text{for any} \quad f \neq 0.$$

The proof of (ii) is complete.

(iii) Under Hypotheses 2.1 (with $\lambda = 0$) the fact that the spectra of the operators A_0 and A_1 are separated, i.e., $d = \operatorname{dist}\{\operatorname{spec}(A_0), \operatorname{spec}(A_1)\} > 0$, means that at least one of the subspaces Ker A_0 and Ker A_1 is trivial. Therefore, the following estimate holds

$$d\|f\|^2 < \big((Uf, A_1 Uf) - (f, A_0 f)\big) \tag{2.37}$$

and, hence, from (2.24) one derives the inequality

$$d < \frac{\mu^2 - 1}{\mu} \operatorname{Re}(V^* f, Uf) \leq \frac{1 - \mu^2}{\mu} \|V\|,$$

which proves that the operator X does not have singular values outside the interval $[0, \nu]$, where

$$\nu = \tan\left(\frac{1}{2} \arctan \frac{2\|V\|}{d}\right) < 1.$$

Using the same strategy as in the proof of (ii) one arrives to the conclusion that X is a uniform contraction satisfying the norm estimate

$$\|X\| \leq \tan\left(\frac{1}{2} \arctan \frac{2\|V\|}{d}\right) < 1,$$

which proves the upper bound (2.15). □

Remark 2.6. The operator X referred to in Theorem 2.4 is a contractive solution to the Riccati equation

$$A_1X - XA_0 - XVX + V^* = 0 \tag{2.38}$$

with the property that

$$\operatorname{spec}(A_0 + VX) \subset (-\infty, \lambda], \tag{2.39}$$

since the operator A_0+VX is similar to the part of $\mathbf{B}$ associated with the subspace $\mathfrak{Q}$ (see, e.g., [3]) and $\sup(\operatorname{spec}(\mathbf{B}|_{\mathfrak{Q}})) \leq \lambda$ by definition (2.12) of the invariant subspace $\mathfrak{Q}$. The similarity of $A_0 + VX$ and $\mathbf{B}|_{\mathfrak{Q}}$ can also be seen directly from the identity

$$(x+Xx, \mathbf{B}(x+Xx)) = (x, (A_0+X^*V^*)(I+X^*X)x) = (x, (I+X^*X)(A_0+VX)x)$$

valid for any $x \in \mathfrak{H}_0$.

The result of Theorem 5.6 in the Appendix shows that Theorem 2.4 admits the following equivalent formulation in terms of the corresponding spectral projections.

Theorem 2.7. *Assume Hypothesis* 2.1. *Denote by Q the orthogonal projection in $\mathfrak{H}$ onto the subspace $\mathfrak{Q}$* (2.12) *and by P the orthogonal projection onto $\mathfrak{H}_0$. Then:*

(i) $\|P - Q\| \leq \sqrt{2}/2$,

(ii) *If either* $\operatorname{Ker}(A_0 - \lambda) = \{0\}$ *or* $\operatorname{Ker}(A_1 - \lambda) = \{0\}$, *then*

$$\pm\frac{\sqrt{2}}{2} \notin \operatorname{spec}_{\mathrm{pp}}(P - Q).$$

(iii) *If*

$$d = \operatorname{dist}(\operatorname{spec}(A_0), \operatorname{spec}(A_1)) > 0, \tag{2.40}$$

then

$$\|P - Q\| \leq \sin\left(\frac{1}{2}\arctan\frac{2\|V\|}{d}\right) < \frac{\sqrt{2}}{2}. \tag{2.41}$$

Remark 2.8. Note that if (2.40) holds, then the interval $(\sup\operatorname{spec}(A_0), \inf\operatorname{spec}(A_1))$ belongs to the resolvent set of the operator $\mathbf{B}$. Although this fact is well known (see [1], [8]) we present a particularly simple and short alternative proof.

Given $\lambda \in (\sup\operatorname{spec}(A_0), \inf\operatorname{spec}(A_1))$, the subspace $\operatorname{Ran}\mathsf{E}_{\mathbf{B}}((-\infty, \lambda))$ is a graph subspace $\mathcal{G}(\mathfrak{H}_0, X_\lambda)$ where X_λ is a strictly contractive solution to the Riccati equation (2.38). By a uniqueness result (see Corollary 6.4 (i) in [10]) the solution X_λ does not depend on

$$\lambda \in (\sup\operatorname{spec}(A_0), \inf\operatorname{spec}(A_1)).$$

Therefore, $\mathsf{E}_{\mathbf{B}}\big((\sup\operatorname{spec}(A_0), \inf\operatorname{spec}(A_1))\big) = 0$ which proves the claim.

As a by-product of our considerations we also get the following important properties of the subspaces $\operatorname{Ker}(I_{\mathfrak{H}_0} - X^*X)$ and $\operatorname{Ker}(I_{\mathfrak{H}_1} - XX^*)$. They will be used in Sections 4 and 5 below.

Lemma 2.9. *Let X be the operator referred to in Theorem 2.4. Then*

$$\operatorname{Ker}(I_{\mathfrak{H}_0} - X^*X) \subset \operatorname{Ker}(A_0 - \lambda), \tag{2.42}$$

$$\operatorname{Ran} X|_{\operatorname{Ker}(I_{\mathfrak{H}_0} - X^*X)} \subset \operatorname{Ker}(A_1 - \lambda), \tag{2.43}$$

and

$$\operatorname{Ker}(I_{\mathfrak{H}_0} - X^*X) \subset \operatorname{Ker}(XVX - V^*). \tag{2.44}$$

Moreover, the subspace $\operatorname{Ker}(I_{\mathfrak{H}_0} - X^*X)$ *reduces both the operators VX and VV^*.*

Similarly,

$$\operatorname{Ker}(I_{\mathfrak{H}_1} - XX^*) \subset \operatorname{Ker}(A_1 - \lambda), \tag{2.45}$$

$$\operatorname{Ran} X^*|_{\operatorname{Ker}(I_{\mathfrak{H}_1} - XX^*)} \subset \operatorname{Ker}(A_0 - \lambda), \tag{2.46}$$

$$\operatorname{Ker}(I_{\mathfrak{H}_1} - XX^*) \subset \operatorname{Ker}(X^*V^*X^* - V), \tag{2.47}$$

and the subspace $\operatorname{Ker}(I_{\mathfrak{H}_1} - XX^*)$ *reduces both the operators X^*V^* and V^*V.*

Proof. Let $0 \neq f \in \operatorname{Ker}(I_{\mathfrak{H}_0} - X^*X)$, that is, $|X|f = f$, where $|X| = (X^*X)^{1/2}$. Then the elements $f \oplus Uf$ and $\mathbf{B}((-f) \oplus Uf)$ are orthogonal, where U is the partial isometry from the polar decomposition $X = U|X|$ (see the proof of Theorem 2.4 part (i)). This means that

$$(Uf, (A_1 - \lambda)Uf) = (f, (A_0 - \lambda)f), \tag{2.48}$$

which implies (2.42) and (2.43) (under Hypothesis 2.1).

In order to prove (2.44) and that the subspace $\operatorname{Ker}(I_{\mathfrak{H}_0} - X^*X)$ is both VX- and VV^*-invariant we proceed as follows.

By Remark 2.6 the operator X solves the Riccati equation

$$A_1X - XA_0 - XVX + V^* = 0 \tag{2.49}$$

and, hence,

$$X^*A_1 - A_0X^* - X^*V^*X^* + V = 0, \tag{2.50}$$

which in particular implies that

$$X^*A_1X - A_0X^*X - X^*V^*X^*X + VX = 0. \tag{2.51}$$

For any $f \in \operatorname{Ker}(I_{\mathfrak{H}_0} - X^*X)$ inclusions (2.42) and (2.43) yield

$$(A_1X - XA_0)f = (X^*A_1X - A_0)f = (X^*A_1X - A_0X^*X)f = 0.$$

Thus, from (2.49) and (2.51) it follows that

$$V^*f = XVXf, \qquad f \in \operatorname{Ker}(I_{\mathfrak{H}_0} - X^*X), \tag{2.52}$$

which proves (2.44) and the representation

$$VXf = X^*V^*X^*Xf = X^*V^*f, \qquad f \in \operatorname{Ker}(I_{\mathfrak{H}_0} - X^*X). \tag{2.53}$$

Combining (2.52) and (2.53) proves that

$$(I_{\mathfrak{H}_0} - X^*X)VXf = VXf - X^*XVXf = VXf - X^*V^*f = 0$$

for any $f \in \operatorname{Ker}(I_{\mathfrak{H}_0} - X^*X)$. That is, the subspace $\operatorname{Ker}(I_{\mathfrak{H}_0} - X^*X)$ is VX-invariant. From (2.53) it follows that $\operatorname{Ker}(I_{\mathfrak{H}_0} - X^*X)$ is also X^*V^*-invariant and, hence, $\operatorname{Ker}(I_{\mathfrak{H}_0} - X^*X)$ reduces the operator VX.

Equality (2.52) implies that

$$VV^*f = VXVXf, \qquad f \in \operatorname{Ker}(I_{\mathfrak{H}_0} - X^*X) \tag{2.54}$$

which proves that $\operatorname{Ker}(I_{\mathfrak{H}_0} - X^*X)$ is VV^*-invariant, since $\operatorname{Ker}(I_{\mathfrak{H}_0} - X^*X)$ is already proven to be a VX-invariant subspace. Since VV^* is self-adjoint, the subspace $\operatorname{Ker}(I_{\mathfrak{H}_0} - X^*X)$ reduces VV^*. The proof of (2.45), (2.46), and (2.47) is similar. □

Remark 2.10. It follows from Lemma 2.9 that the multiplicity m of the singular value $\mu = 1$ of the operator X satisfies the inequality

$$m \leq \min\{\dim \operatorname{Ker}(A_0 - \lambda), \dim \operatorname{Ker}(A_1 - \lambda)\}. \tag{2.55}$$

Equivalently,

$$\begin{aligned} \dim \operatorname{Ker}\left(Q - P - \frac{\sqrt{2}}{2}\right) &= \dim \operatorname{Ker}\left(Q - P + \frac{\sqrt{2}}{2}\right) \\ &\leq \min\{\dim \operatorname{Ker}(A_0 - \lambda), \dim \operatorname{Ker}(A_1 - \lambda)\}, \end{aligned} \tag{2.56}$$

where P and Q are orthogonal projections in $\mathfrak{H}$ onto the subspaces $\mathfrak{H}_0$ and $\mathfrak{Q}$ (2.12), respectively. The subspaces $\operatorname{Ker}(I_{\mathfrak{H}_0} - XX^*)$ and $\operatorname{Ker}(I_{\mathfrak{H}_1} - X^*X)$ will be studied in Section 5 below.

3. Lower bound

In this section we derive the lower bound on the norm of the difference of the orthogonal projections onto the $\mathbf{A}$-invariant subspace $\mathfrak{H}_0$ and the $\mathbf{B}$-invariant subspace $\mathfrak{Q}$ given by (2.12).

Theorem 3.1. *Assume Hypothesis* 2.1. *Let δ_- (δ_+) denote the shift of the bottom (top, respectively) of the spectrum of the operator $\mathbf{A}$ under the perturbation $\mathbf{V}$, i.e.,*

$$\delta_- = \inf \operatorname{spec}(\mathbf{A}) - \inf \operatorname{spec}(\mathbf{B}), \qquad \delta_+ = \sup \operatorname{spec}(\mathbf{B}) - \sup \operatorname{spec}(\mathbf{A}).$$

Then the solution X to the Riccati equation referred to in Theorem 2.4 *satisfies the lower bound*

$$\|X\| \geq \frac{\delta}{\|V\|}, \tag{3.1}$$

where

$$\delta = \max\{\delta_-, \delta_+\} \leq \|V\|.$$

Equivalently,

$$\|Q - P\| \geq \frac{\delta}{\sqrt{\delta^2 + \|V\|^2}}, \tag{3.2}$$

where P and Q are orthogonal projections onto the subspace $\mathfrak{H}_0$ and $\mathfrak{Q}$ (2.12), *respectively.*

Remark 3.2. From a general perturbation theory for off-diagonal perturbations it follows that $\delta_- \geq 0$ and $\delta_+ \geq 0$. For the proof of this fact we refer to [11].

Proof of Theorem 3.1. From Theorem 2.4 it follows that $\operatorname{Ran} \mathsf{E}_{\mathbf{B}}((-\infty, \lambda)) \oplus (\mathfrak{N}_0 \cap \operatorname{Ker} V^*)$ is the graph subspace $\mathcal{G}(\mathfrak{H}_0, X)$ associated with the subspace $\mathfrak{H}_0$ in the decomposition $\mathfrak{H} = \mathfrak{H}_0 \oplus \mathfrak{H}_1$ and by Remark 2.6 the operator X solves the Riccati equation (2.38). It is well known (see, e.g., [3]) that in this case

$$\operatorname{spec}(\mathbf{B}) = \operatorname{spec}(A_0 + VX) \cup \operatorname{spec}(A_1 - X^*V^*) \tag{3.3}$$

and, hence,

$$\inf \operatorname{spec}(\mathbf{B}) \geq \inf \operatorname{spec}(\mathbf{A}) - \|V\| \|X\| \tag{3.4}$$

and

$$\sup \operatorname{spec}(\mathbf{B}) \leq \sup \operatorname{spec}(\mathbf{A}) + \|V\| \|X\|, \tag{3.5}$$

which proves the lower bounds (3.1) and (3.2) using Theorem 5.6 in the Appendix. □

Remark 3.3. Let $\mathfrak{H}_0 = \mathfrak{H}_1 = \mathbb{C}$, let A_0 and A_1 be real numbers with $A_1 > A_0$, and let $V \in \mathbb{C}$. Then

$$\operatorname{spec}\begin{pmatrix} A_0 & V \\ V^* & A_1 \end{pmatrix}$$

$$= \left\{ A_0 - \|V\| \tan\left(\frac{1}{2} \arctan \frac{2\|V\|}{A_1 - A_0}\right), A_1 + \|V\| \tan\left(\frac{1}{2} \arctan \frac{2\|V\|}{A_1 - A_0}\right) \right\},$$

which can easily be seen by solving the characteristic equation

$$(A_0 - \lambda)(A_1 - \lambda) - \|V\|^2 = 0.$$

The upper bounds (2.15) and (2.41) give the exact value of the norms $\|X\|$ and $\|P - Q\|$, respectively, and, hence, the bounds (2.15) and (2.41) are sharp. In this case

$$\delta_- = \delta_+ = \|V\| \tan\left(\frac{1}{2} \arctan \frac{2\|V\|}{A_1 - A_0}\right)$$

and

$$\|P - Q\| = \sin\left(\frac{1}{2} \arctan \frac{2\|V\|}{A_1 - A_0}\right),$$

which shows that estimates (3.2) and (3.1) are also sharp.

4. Riccati equation: Uniqueness

Under Hypothesis 2.1, Theorem 2.4 and Remark 2.6 guarantee the existence of a contractive solution to the Riccati equation

$$A_1 X - X A_0 - X V X + V^* = 0 \tag{4.1}$$

with the properties

$$\begin{aligned} &\text{(i)} && \operatorname{Ker}(A_0 - \lambda) \cap \operatorname{Ker} V^* \subset \operatorname{Ker} X, \\ &\text{(ii)} && \operatorname{spec}(A_0 + VX) \subset (-\infty, \lambda]. \end{aligned} \tag{4.2}$$

If, in addition,

$$d = \operatorname{dist}(\operatorname{spec}(A_0), \operatorname{spec}(A_1)) > 0,$$

then X is a unique contractive solution to the Riccati equation (see Corollary 6.4 (i) in [10]; cf. [2]).

The following uniqueness result shows that there is no other solution to the Riccati equation (4.1) with properties (4.2).

Theorem 4.1. *Assume Hypothesis* 2.1. *Then a contractive solution to the Riccati equation* (4.1) *satisfying the properties* (4.2) *is unique.*

Proof. Exactly the same reasoning as in the proof of Theorem 2.4 (i) allows to conclude that without loss of generality one may assume that $\operatorname{Ker}(\mathbf{B} - \lambda) = 0$.

In this case the spectral subspace $\operatorname{Ran} \mathsf{E}_{\mathbf{B}}((-\infty, \lambda))$ is the graph of a contraction X which solves (4.1). Assume that $Y \in \mathcal{B}(\mathfrak{H}_0, \mathfrak{H}_1)$ is a (not necessarily contractive) solution to (4.1) different from X. Since the graph of X is a spectral subspace of $\mathbf{B}$, one concludes that the orthogonal projections onto the graphs $\mathcal{G}(\mathfrak{H}_0, X)$ and $\mathcal{G}(\mathfrak{H}_0, Y)$ of the operators X and Y commute.

We claim that $\mathcal{G}(\mathfrak{H}_0, X)^\perp \cap \mathcal{G}(\mathfrak{H}_0, Y)$ is nontrivial. To show this we set

$$\mathfrak{L} := \mathfrak{H}_0 \ominus \operatorname{Ran} P|_{\mathcal{G}(\mathfrak{H}_0, X) \cap \mathcal{G}(\mathfrak{H}_0, Y)},$$

where P is the orthogonal projection in $\mathfrak{H}$ onto the subspace $\mathfrak{H}_0$. Any $z \in \mathcal{G}(\mathfrak{H}_0, X) \cap \mathcal{G}(\mathfrak{H}_0, Y)$ admits the representations

$$z = x \oplus Xx \quad \text{for some} \quad x \in \mathfrak{L}^\perp,$$

$$z = y \oplus Yy \quad \text{for some} \quad y \in \mathfrak{L}^\perp,$$

where $\mathfrak{L}^\perp := \mathfrak{H}_0 \ominus \mathfrak{L}$. Obviously, $x = y = Pz$, and, therefore, $XPz = YPz$ for any $z \in \mathcal{G}(\mathfrak{H}_0, X) \cap \mathcal{G}(\mathfrak{H}_0, Y)$. Thus, $Y|_{\mathfrak{L}^\perp} = X|_{\mathfrak{L}^\perp}$ and

$$\mathcal{G}(\mathfrak{H}_0, X) \cap \mathcal{G}(\mathfrak{H}_0, Y) = \{x + Xx \mid x \in \mathfrak{L}^\perp\}.$$

Hence, $X \neq Y$ if and only if the subspace $\mathfrak{L}$ is nontrivial.

Note that

$$(x_0 \oplus Yx_0, x \oplus Yx) = 0, \qquad x := (I_{\mathfrak{H}_0} + Y^*Y)^{-1} y$$

for any $x_0 \in \mathfrak{L}^\perp$ and any $y \in \mathfrak{L}$. Hence,

$$\mathcal{G}(\mathfrak{H}_0, Y) \ominus (\mathcal{G}(\mathfrak{H}_0, X) \cap \mathcal{G}(\mathfrak{H}_0, Y)) = \{x + Yx \mid x = (I_{\mathfrak{H}_0} + Y^*Y)^{-1} y,\ y \in \mathfrak{L}\}. \quad (4.3)$$

Since the orthogonal projections onto the graph subspaces $\mathcal{G}(\mathfrak{H}_0, X)$ and $\mathcal{G}(\mathfrak{H}_0, Y)$ commute, by (4.3) we conclude that the subspace

$$\mathcal{G}(\mathfrak{H}_0, X)^\perp \cap \mathcal{G}(\mathfrak{H}_0, Y) = \mathcal{G}(\mathfrak{H}_0, Y) \ominus (\mathcal{G}(\mathfrak{H}_0, X) \cap \mathcal{G}(\mathfrak{H}_0, Y))$$

is nontrivial.

For any $z \in \mathfrak{H}$, $z \neq 0$ such that $z \in \mathcal{G}(\mathfrak{H}_0, Y)$ and $z \perp \mathcal{G}(\mathfrak{H}_0, X)$ we have

$$(z, \mathbf{B}z) > \lambda. \quad (4.4)$$

Therefore, for the operator $A_0 + VY$ the condition (ii) does not hold, since the spectrum of $A_0 + VY$ coincides with that of the restriction of $\mathbf{B}$ onto its invariant subspace $\mathcal{G}(\mathfrak{H}_0, Y)$ and by (4.4) the operator $\mathbf{B}|_{\mathcal{G}(\mathfrak{H}_0,Y)}$ has points of the spectrum to the right of the point λ. The proof is complete. □

Remark 4.2. If $\lambda \in \mathbb{R}$ is a multiple eigenvalue of the operator $\mathbf{B}$ and both $\operatorname{Ker}(\mathbf{B}-\lambda) \cap \mathfrak{H}_0 \neq \{0\}$ and $\operatorname{Ker}(\mathbf{B}-\lambda) \cap \mathfrak{H}_1 \neq \{0\}$, it follows from Theorem 2.2 that the Riccati equation (4.1) has uncountably many bounded solutions (even if $\mathfrak{H}$ is finite-dimensional). This can also be seen directly. Let

$$T : \operatorname{Ker}(A_0 - \lambda) \cap \operatorname{Ker} V^* \longrightarrow \operatorname{Ker}(A_1 - \lambda) \cap \operatorname{Ker} V$$

be any bounded operator acting from $\operatorname{Ker}(A_0-\lambda)\cap\operatorname{Ker} V^*$ to $\operatorname{Ker}(A_1-\lambda)\cap\operatorname{Ker} V$. The bounded operator $X \in \mathcal{B}(\mathfrak{H}_0, \mathfrak{H}_1)$ defined by

$$\widetilde{X}f = \begin{cases} Tf & \text{if } f \in \operatorname{Ker}(A_0 - \lambda) \cap \operatorname{Ker} V^* \\ 0 & \text{if } f \in \operatorname{Ker}(A_0 - \lambda) \ominus (\operatorname{Ker}(A_0 - \lambda) \cap \operatorname{Ker} V^*) \end{cases} \tag{4.5}$$

satisfies the equation

$$((A_1 - \lambda I_{\mathfrak{H}_1})\widetilde{X} - \widetilde{X}(A_0 - \lambda I_{\mathfrak{H}_0}) - \widetilde{X}V\widetilde{X} + V^*)f = 0 \quad \text{for all} \quad f \in \mathfrak{H}_0, \tag{4.6}$$

and, thus, it is also a solution to (4.1). If $\dim \operatorname{Ker}(A_0 - \lambda) \cap \operatorname{Ker} V^* = \infty$, $\dim \operatorname{Ker}(A_1 - \lambda) \cap \operatorname{Ker} V = \infty$, and T is a closed densely defined unbounded operator from $\operatorname{Ker}(A_0 - \lambda) \cap \operatorname{Ker} V^*$ to $\operatorname{Ker}(A_1 - \lambda) \cap \operatorname{Ker} V$, then the operator $\widetilde{X}$ defined by (4.5) is an unbounded solution to the Riccati equation (4.1) in the sense of Definition 4.2 in [10].

Our next result is the following uniqueness criterion.

Theorem 4.3. *Assume Hypothesis* 2.1. *A contractive solution X to the Riccati equation* (4.1) *is a unique contractive solution if and only if the graph of X is a spectral subspace of the operator* $\mathbf{B}$ *and $\mu = 1$ is not an eigenvalue of the operator $|X|$, the absolute value of X. In this case, X is a strict contraction and*

$$\text{either} \quad \mathsf{E}_{\mathbf{B}}((-\infty, \lambda)) = \mathcal{G}(\mathfrak{H}_0, X) \quad \text{or} \quad \mathsf{E}_{\mathbf{B}}((\lambda, +\infty)) = \mathcal{G}(\mathfrak{H}_1, -X^*).$$

Proof. "If Part". Since $\mu = 1$ is not an eigenvalue of the contraction X, it follows that X is a strict contraction. Then the claim follows from Corollary 6.4 in [10].

"Only If Part". Assume that X is the unique contractive solution to the Riccati equation. Then X coincides with the operator referred to in Theorem 2.4. We need to prove that the graph of X is a spectral subspace of $\mathbf{B}$ and that $\operatorname{Ker}(I_{\mathfrak{H}_0} - X^*X) = \{0\}$.

We will prove these statements by reduction to a contradiction. If the graph of X is not a spectral subspace of $\mathbf{B}$, then by Remark 2.3 and Theorem 2.4 both $\operatorname{Ker}(A_0 - \lambda) \cap \operatorname{Ker} V^*$ and $\operatorname{Ker}(A_1 - \lambda) \cap \operatorname{Ker} V$ are nontrivial. Let T be any contractive operator from $\operatorname{Ker}(A_0 - \lambda) \cap \operatorname{Ker} V^*$ to $\operatorname{Ker}(A_1 - \lambda) \cap \operatorname{Ker} V$ with $\operatorname{Ker} T \neq \{0\}$. Then the operator $\widetilde{X}$ defined by (4.5) is also a contractive solution to the Riccati equation (see Remark 4.2). Since by Theorem 2.4 $\operatorname{Ker}(A_0 - \lambda) \cap \operatorname{Ker} V^* \subset \operatorname{Ker} X$, where X is the contractive solution to the Riccati equation referred to in Theorem 2.4, the contractive solution $\widetilde{X}$ to the Riccati equation is different from X by construction. A contradiction.

Assume now that $\mu = 1$ is an eigenvalue of $|X|$, that is, $\operatorname{Ker}(I_{\mathfrak{H}_0} - X^*X)$ is nontrivial. By Lemma 2.9

$$\operatorname{Ker}(I_{\mathfrak{H}_0} - X^*X) \subseteq \operatorname{Ker}(XVX - V^*) \tag{4.7}$$

and $\operatorname{Ker}(I_{\mathfrak{H}_0} - X^*X)$ reduces both A_0 and VX. Applying Theorem 6.2 in [10] we conclude that the Riccati equation (4.1) has a contractive solution Y such that $\operatorname{Ker}(I_{\mathfrak{H}_0} - X^*X) = \operatorname{Ker}(I_{\mathfrak{H}_0} + Y^*X)$. This solution necessarily differs from X which contradicts the hypothesis that X is a unique contractive solution.

From Theorem 2.2 it follows now that the graph of X is the spectral subspace of the operator $\mathbf{B}$ and

$$\mathcal{G}(\mathfrak{H}_0, X) = \begin{cases} \mathsf{E}_{\mathbf{B}}((-\infty,\lambda)) & \text{if} \quad \operatorname{Ker}(A_0-\lambda)\cap\operatorname{Ker}V^* = \{0\} \\ \mathsf{E}_{\mathbf{B}}((-\infty,\lambda]) & \text{if} \quad \operatorname{Ker}(A_1-\lambda)\cap\operatorname{Ker}V = \{0\} \end{cases},$$

which proves the remaining statement of the theorem. □

As an immediate consequence of Theorem 4.3 we get the following results.

Corollary 4.4. *Let $X : \mathfrak{H}_0 \to \mathfrak{H}_1$ be the solution to the Riccati equation referred to in Theorem* 2.4. *It is the unique contractive solution if and only if it is strictly contractive.*

Remark 4.5. Let X be an arbitrary strictly contractive solution to the Riccati equation (4.1). In general this solution need not be the unique contractive solution or an isolated point of the set of all solutions (cf. Remark 4.2).

Corollary 4.6. *Let $X : \mathfrak{H}_0 \to \mathfrak{H}_1$ be the solution to the Riccati equation referred to in Theorem* 2.4. *It is an isolated point (in the operator norm topology) in the set of all solutions to the Riccati equation* (4.1) *if and only if either* $\operatorname{Ker}(A_0-\lambda)\cap\operatorname{Ker}V^* = \{0\}$ *or* $\operatorname{Ker}(A_1-\lambda)\cap\operatorname{Ker}V = \{0\}$.

Proof. By Remark 2.3 and Theorem 2.4 the graph $\mathcal{G}(\mathfrak{H}_0, X)$ is associated with a spectral subspace of the operator $\mathbf{B}$ if and only if either $\operatorname{Ker}(A_0-\lambda)\cap\operatorname{Ker}V^* = \{0\}$ or $\operatorname{Ker}(A_1-\lambda)\cap\operatorname{Ker}V = \{0\}$. Now the claim follows from Theorem 5.3 in [10]. □

Remark 4.7. If either $\operatorname{Ker}(A_0-\lambda)\cap\operatorname{Ker}V^* = \{0\}$ or $\operatorname{Ker}(A_1-\lambda)\cap\operatorname{Ker}V = \{0\}$ holds, Theorem 6.2 in [10] allows us to construct *all* contractive solutions to the Riccati equation from that referred to in Theorem 2.4.

5. More on the subspaces $\operatorname{Ker}(I_{\mathfrak{H}_0} - X^*X)$ and $\operatorname{Ker}(I_{\mathfrak{H}_1} - XX^*)$

The main goal of this section is to prove the fact that the subspace $\operatorname{Ker}(I_{\mathfrak{H}_0} - X^*X)$ associated with the operator X referred to in Theorem 2.4 admits an intrinsic description as the maximal VV^*-invariant subspace $\mathfrak{K}_0 \subset \mathfrak{H}_0$ with the properties

$$\begin{aligned} &\mathfrak{K}_0 \subset \operatorname{Ker}(A_0-\lambda)\cap\overline{\operatorname{Ran}V}, \\ &\operatorname{Ran}V^*|_{\mathfrak{K}_0} \subset \operatorname{Ker}(A_1-\lambda) \end{aligned} \tag{5.1}$$

(see Theorem 5.3 below). Similarly, the subspace $\mathrm{Ker}\,(I_{\mathfrak{H}_1} - XX^*)$ can be characterized as the maximal V^*V-invariant subspace with the properties

$$\begin{aligned} \mathfrak{K}_1 &\subset \mathrm{Ker}\,(A_1 - \lambda) \cap \overline{\mathrm{Ran}\,V^*}, \\ \mathrm{Ran}\,V|_{\mathfrak{K}_1} &\subset \mathrm{Ker}\,(A_0 - \lambda). \end{aligned} \tag{5.2}$$

We start with the observation that the maximal subspaces $\mathfrak{K}_0$ and $\mathfrak{K}_1$ with indicated properties do exist and admit a constructive description.

Lemma 5.1. *The subspaces*

$$\begin{aligned} \mathfrak{K}_0 :=&\mathrm{closure}\big\{x \in \mathrm{Ker}\,(A_0 - \lambda) \cap \mathrm{Ran}\,V \,\big| \\ &V^*(VV^*)^n x \in \mathrm{Ker}\,(A_1 - \lambda),\ (VV^*)^n x \in \mathrm{Ker}\,(A_0 - \lambda) \text{ for any } n \in \mathbb{N}_0\big\} \end{aligned} \tag{5.3}$$

and

$$\begin{aligned} \mathfrak{K}_1 :=&\mathrm{closure}\big\{x \in \mathrm{Ker}\,(A_1 - \lambda) \cap \mathrm{Ran}\,V^* \,\big| \\ &V(V^*V)^n x \in \mathrm{Ker}\,(A_0 - \lambda),\ (V^*V)^n x \in \mathrm{Ker}\,(A_1 - \lambda) \text{ for any } n \in \mathbb{N}_0\big\} \end{aligned} \tag{5.4}$$

reduce the operators VV^* and V^*V, respectively. Moreover, the subspaces $\mathfrak{K}_0$ and $\mathfrak{K}_1$ satisfy the properties (5.1) *and* (5.2), *respectively.*

The subspace $\mathfrak{K}_0$ is maximal in the sense that if $\mathfrak{L}_0$ is any other VV^*-invariant subspace with the properties (5.1), *then $\mathfrak{L}_0 \subset \mathfrak{K}_0$. Analogously, the subspace $\mathfrak{K}_1$ is maximal in the sense that if $\mathfrak{L}_1$ is any other V^*V-invariant subspace with the properties* (5.2), *then $\mathfrak{L}_1 \subset \mathfrak{K}_1$.*

Proof. Clearly, the subspace $\mathfrak{K}_0$ is invariant under the operator VV^* and, therefore, $\mathfrak{K}_0$ reduces VV^*, since VV^* is self-adjoint. It follows from (5.3) that (5.1) holds.

Now, let $\mathfrak{L}_0$ be an arbitrary closed subspace of $\mathrm{Ker}\,(A_0 - \lambda) \cap \overline{\mathrm{Ran}\,V}$ invariant under VV^* such that $V^*\mathfrak{L} \subset \mathrm{Ker}\,(A_1 - \lambda)$. Then $\mathfrak{L} \subset \mathfrak{K}_0$. Indeed, since $\mathfrak{L}$ is invariant under VV^*, we have $(VV^*)^n\mathfrak{L} \subset \mathfrak{L} \subset \mathrm{Ker}\,(A_0 - \lambda)$ for any $n \in \mathbb{N}$. Hence, $V^*(VV^*)^n\mathfrak{L} \subset V^*\mathfrak{L} \subset \mathrm{Ker}\,(A_1 - \lambda)$ for any $n \in \mathbb{N}_0$ and one concludes that $\mathfrak{L} \subset \mathfrak{K}_0$.

The maximality of the subspace $\mathfrak{K}_1$ is proven in a similar way. □

Lemma 5.2. *The subspaces $\mathfrak{K}_0$ and $\mathfrak{K}_1$ satisfy the properties that*

$$\overline{\mathrm{Ran}\,V^*|_{\mathfrak{K}_0}} = \mathfrak{K}_1, \qquad \overline{\mathrm{Ran}\,V|_{\mathfrak{K}_1}} = \mathfrak{K}_0, \tag{5.5}$$

$$\mathrm{Ran}\,V^*|_{\mathfrak{H}_0 \ominus \mathfrak{K}_0} \subset \mathfrak{H}_1 \ominus \mathfrak{K}_1, \qquad \mathrm{Ran}\,V|_{\mathfrak{H}_1 \ominus \mathfrak{K}_1} \subset \mathfrak{H}_0 \ominus \mathfrak{K}_0. \tag{5.6}$$

Proof. Equations (5.5) follow from the explicit description (5.3) and (5.4) of the subspaces $\mathfrak{K}_0$ and $\mathfrak{K}_1$, respectively.

Let $x \in \mathfrak{H}_0 \ominus \mathfrak{K}_0$ be arbitrary. Choose an arbitrary $y \in \mathfrak{K}_1$ and consider

$$(y, V^*x) = (Vy, x).$$

Since, by (5.5), $Vy \in \mathfrak{K}_0$ we have $(Vy, x) = 0$. Thus, $V^*x \in \mathfrak{H}_1 \ominus \mathfrak{K}_1$ which proves the first inclusion in (5.6). The second inclusion in (5.6) is proven similarly. □

Theorem 5.3. *Assume Hypothesis* 2.1. *Let* $X : \mathfrak{H}_0 \to \mathfrak{H}_1$ *be the solution to the Riccati equation* (4.1) *referred to in Theorem* 4.1. *Then*

$$\operatorname{Ker}(I_{\mathfrak{H}_0} - X^*X) = \mathfrak{K}_0 \quad \textit{and} \quad \operatorname{Ker}(I_{\mathfrak{H}_1} - XX^*) = \mathfrak{K}_1. \tag{5.7}$$

Moreover, $\overline{\operatorname{Ran} X|_{\mathfrak{K}_0}} = \mathfrak{K}_1$ *and*

$$X|_{\mathfrak{K}_0} = -\widehat{S}, \qquad \widehat{S} : \mathfrak{K}_0 \to \mathfrak{K}_1,$$

where $\widehat{S} = S|_{\mathfrak{K}_0}$ *with* $S : \mathfrak{H}_0 \to \mathfrak{H}_1$ *being the partial isometry with initial space* $\overline{\operatorname{Ran} V}$ *and final space* $\overline{\operatorname{Ran} V^*}$ *defined by the polar decomposition* $V^* = S(VV^*)^{1/2}$. *In particular,* $\mathfrak{K}_1 = \mathcal{G}(\mathfrak{K}_0, \widehat{S})$.

Proof. Without loss of generality we assume that $\lambda = 0$.

First, we will prove the inclusion

$$\operatorname{Ker}(I_{\mathfrak{H}_0} - X^*X) \subset \mathfrak{K}_0. \tag{5.8}$$

It is sufficient to establish that

(a) $\operatorname{Ker}(I_{\mathfrak{H}_0} - X^*X)$ reduces VV^*,
(b) $\operatorname{Ker}(I_{\mathfrak{H}_0} - X^*X) \subset \overline{\operatorname{Ran} V^*} \cap \operatorname{Ker} A_0$,
(c) $V^* \operatorname{Ker}(I_{\mathfrak{H}_0} - X^*X) \subset \operatorname{Ker} A_1$.

The statement (a) follows from Lemma 2.9.

In order to see that (b) holds note that if $z \in \operatorname{Ker} V^* \cap \operatorname{Ker} A_0$, then $Xz = 0$, since $\operatorname{Ker} V^* \cap \operatorname{Ker} A_0 \subset \operatorname{Ker} X$ by Theorem 2.4. Therefore, $\operatorname{Ker} V^* \cap \operatorname{Ker} A_0 \perp \operatorname{Ker}(I_{\mathfrak{H}_0} - X^*X)$ since

$$(z, x) = (z, X^*Xx) = (Xz, Xx) = 0 \quad \text{for any} \quad x \in \operatorname{Ker}(I_{\mathfrak{H}_0} - X^*X),$$

which proves (b) taking into account that $\operatorname{Ker}(I_{\mathfrak{H}_0} - X^*X) \subset \operatorname{Ker} A_0$ (see Lemma 2.9).

To prove (c) we proceed as follows. If $x \in \operatorname{Ker}(I_{\mathfrak{H}_0} - X^*X)$, then $A_0 x = 0$, since $\operatorname{Ker}(I_{\mathfrak{H}_0} - X^*X) \subset \operatorname{Ker} A_0$. The same reasoning as in the proof of Lemma 2.9 shows that (see (2.52))

$$V^*x = XVXx \qquad \text{and} \qquad VX = X^*V^*x.$$

Therefore, $(I_{\mathfrak{H}_1} - XX^*)V^*x = V^*x - XX^*V^*x = V^*x - XVXx = 0$, that is, $V^*x \in \operatorname{Ker}(I_{\mathfrak{H}_1} - XX^*)$. By Lemma 2.9 $\operatorname{Ker}(I_{\mathfrak{H}_1} - XX^*) \subset \operatorname{Ker} A_1$, which completes the proof of (c).

The inclusion $\operatorname{Ker}(I_{\mathfrak{H}_1} - XX^*) \subset \mathfrak{K}_1$ is proven similarly. Hence, we have established that

$$\operatorname{Ker}(I_{\mathfrak{H}_0} - X^*X) \subset \mathfrak{K}_0 \quad \text{and} \quad \operatorname{Ker}(I_{\mathfrak{H}_1} - XX^*) \subset \mathfrak{K}_1. \tag{5.9}$$

Now we turn to the proof of the opposite inclusions

$$\mathfrak{K}_0 \subset \operatorname{Ker}(I_{\mathfrak{H}_0} - X^*X) \qquad \text{and} \qquad \mathfrak{K}_1 \subset \operatorname{Ker}(I_{\mathfrak{H}_1} - XX^*). \tag{5.10}$$

Clearly, the subspaces $\widehat{\mathfrak{H}}_0 = \mathfrak{H}_0 \ominus \mathfrak{K}_0$ and $\widehat{\mathfrak{H}}_1 = \mathfrak{H}_1 \ominus \mathfrak{K}_1$ reduce the operators A_0 and A_1, respectively, since $\mathfrak{K}_0 \subset \operatorname{Ker} A_0$, $\mathfrak{K}_1 \subset \operatorname{Ker} A_1$, and A_0, A_1 are self-adjoint. Denote by $\widehat{A}_0$ and $\widehat{A}_1$ the corresponding parts of the operators A_0 and A_1 associated with these subspaces:

$$\widehat{A}_0 = A_0|_{\widehat{\mathfrak{H}}_0} \quad \text{and} \quad \widehat{A}_1 = A_1|_{\widehat{\mathfrak{H}}_1}.$$

Since by Lemma 5.2 $\operatorname{Ran} V|_{\widehat{\mathfrak{H}}_1} \subset \widehat{\mathfrak{H}}_0$, the restriction $\widehat{V}$ of the operator V onto $\mathfrak{H}_1$ is a map from $\widehat{\mathfrak{H}}_1$ to $\widehat{\mathfrak{H}}_0$. By Theorem 4.1 the Riccati equation

$$\widehat{A}_1\widehat{X} - \widehat{X}\widehat{A}_0 - \widehat{X}\widehat{V}\widehat{X} + \widehat{V}^* = 0 \tag{5.11}$$

has a unique solution $\widehat{X}$ satisfying $\operatorname{Ker} \widehat{A}_0 \cap \operatorname{Ker} \widehat{V}^* \subset \operatorname{Ker} \widehat{X}$ and $\operatorname{spec}(\widehat{A}_0+\widehat{V}\widehat{X}) \subset (-\infty, 0]$.

Let $S : \mathfrak{H}_0 \to \mathfrak{H}_1$ be a partial isometry with the initial subspace $\overline{\operatorname{Ran} V}$ and final subspace $\overline{\operatorname{Ran} V^*}$ defined by the polar decomposition $V^* = S(VV^*)^{1/2}$. From Lemma 5.2 it follows that $\operatorname{Ran} S|_{\mathfrak{K}_0} = \mathfrak{K}_1$.

Let $\widehat{S} : \mathfrak{K}_0 \to \mathfrak{K}_1$ be the restriction of S onto $\mathfrak{K}_0$: $\widehat{S} = S|_{\mathfrak{K}_0}$. Define the operator $Y : \mathfrak{H}_0 \to \mathfrak{H}_1$ by the following rule

$$Yx = \begin{cases} \widehat{X}x, & x \in \widehat{\mathfrak{H}}_0 \\ -\widehat{S}x, & x \in \mathfrak{K}_0 \end{cases}.$$

Since $\widehat{S}$ maps $\mathfrak{K}_0$ onto $\mathfrak{K}_1$ isometrically, one immediately concludes that

$$\mathfrak{K}_0 \subset \operatorname{Ker}(I - Y^*Y) \quad \text{and} \quad \mathfrak{K}_1 \subset \operatorname{Ker}(I - YY^*). \tag{5.12}$$

We claim that the operator Y solves the Riccati equation

$$A_1Y - YA_0 - YVY + V^* = 0. \tag{5.13}$$

Indeed, if $x \in \widehat{\mathfrak{H}}_0$ then

$$(A_1Y - YA_0 - YVY + V^*)x = 0 \tag{5.14}$$

as a consequence of (5.11). If $x \in \mathfrak{K}_0$, then $A_0x = 0$ (recall that we assumed that $\lambda = 0$). Moreover, $\widehat{S}x \in \mathfrak{K}_1$ and hence $A_1\widehat{S}x = 0$ resulting in

$$(A_1Y - YA_0 - YVY + V^*)x = (-YVY + V^*)x = (-\widehat{S}V\widehat{S} + V^*)x = 0,$$

where we have used the fact that $V\widehat{S}x \in \mathfrak{K}_0$ and the equality

$$\widehat{S}V\widehat{S}x = \widehat{S}(VV^*)^{1/2}\widehat{S}^*\widehat{S}x = \widehat{S}(VV^*)^{1/2}x = V^*x.$$

Therefore, Y solves the Riccati equation (5.14).

Our next claim is that

$$\operatorname{Ker} A_0 \cap \operatorname{Ker} V^* \subset \operatorname{Ker} Y. \tag{5.15}$$

Since $\mathfrak{K}_0 \subset \operatorname{Ker} A_0$, by Lemma 5.2 one concludes that the subspace $\operatorname{Ker} \widehat{A}_0 \cap \operatorname{Ker} \widehat{V}^*$, naturally imbedded into $\mathfrak{H}_0$, coincides with $\operatorname{Ker} A_0 \cap \operatorname{Ker} V^*$. One also concludes that the subspace $\operatorname{Ker} \widehat{X}$ naturally imbedded into $\mathfrak{H}_0$ coincides with

Ker Y by the definition of the operator Y. Therefore, (5.15) follows from the inclusion Ker $\widehat{A}_0 \cap$ Ker $\widehat{V}^* \subset$ Ker $\widehat{X}$.

Finally, observe that

$$VS|_{\mathfrak{K}_0} = (VV^*)^{1/2}S^*S|_{\mathfrak{K}_0} = (VV^*)^{1/2}|_{\mathfrak{K}_0} \geq 0$$

and

$$\operatorname{spec}(\widehat{A}_0 + \widehat{V}\widehat{X}) \subset (-\infty, 0].$$

Since the operator $A_0 + VY$ is diagonal with respect to the decomposition $\mathfrak{H}_0 = \widehat{\mathfrak{H}}_0 \oplus \mathfrak{K}_0$,

$$A_0 + VY = (\widehat{A}_0 + \widehat{V}\widehat{X}) \oplus (-VS|_{\mathfrak{K}_0}),$$

one infers that

$$\operatorname{spec}(A_0 + VY) \subset (-\infty, 0]. \tag{5.16}$$

Combining (5.13), (5.15), and (5.16) proves that the operator Y coincides with X using the uniqueness result of Theorem 4.1. Thus, (5.10) follows from (5.12).

Combining (5.10) and (5.9) proves (5.7).

The remaining statement of the theorem follows from the definition of the operator Y and the fact that $X = Y$. □

By Theorem 5.3 the uniqueness criterion (Theorem 4.3) admits the following equivalent purely geometric formulation.

Theorem 5.4. *Assume Hypothesis* 2.1 *and let X be the solution to the Riccati equation*

$$A_1X - XA_0 - XVX + V^* = 0$$

referred to in Theorem 2.4. Let $\mathfrak{K}_o$ and $\mathfrak{K}_1$ be the subspaces given by (5.3) *and* (5.4), *respectively. Then X is the unique contractive solution if and only if*

(i) *either* Ker $(A_0 - \lambda) \cap$ Ker V^* *or* Ker $(A_1 - \lambda) \cap$ Ker V *are trivial*

and

(ii) *either $\mathfrak{K}_0$ or $\mathfrak{K}_1$ (and hence both) are trivial.*

The solution X is strictly contractive.

Appendix: Two Subspaces

Here we collect some facts about two closed subspaces of a separable Hilbert space which are used in the body of the paper. Their comprehensive presentation with proofs as well as some further results and comments on the history of the problem can be found in [10].

Let (P, Q) be an ordered pair of orthogonal projections in the separable Hilbert space $\mathfrak{H}$. Denote

$$\begin{aligned}
\mathfrak{M}_{pq} &:= \{f \in \mathfrak{H} \,|\, Pf = pf,\ Qf = qf\}, \quad p, q = 0, 1,\\
\mathfrak{M}'_0 &:= \operatorname{Ran} P \ominus (\mathfrak{M}_{10} \oplus \mathfrak{M}_{11}),\\
\mathfrak{M}'_1 &:= \operatorname{Ran} P^\perp \ominus (\mathfrak{M}_{00} \oplus \mathfrak{M}_{01}),\\
\mathfrak{M}' &:= \mathfrak{M}'_0 \oplus \mathfrak{M}'_1,\\
P' &:= P|_{\mathfrak{M}'},\\
Q' &:= Q|_{\mathfrak{M}'}.
\end{aligned}$$

The space $\mathfrak{H}$ admits the canonical orthogonal decomposition

$$\mathfrak{H} = \mathfrak{M}_{00} \oplus \mathfrak{M}_{01} \oplus \mathfrak{M}_{10} \oplus \mathfrak{M}_{11} \oplus \mathfrak{M}'. \tag{A.1}$$

The following theorem provides a criterion for the subspace $\operatorname{Ran} Q$ to be a graph subspace associated with the subspace $\operatorname{Ran} P$.

Theorem 5.5. *Let P and Q be orthogonal projections in a Hilbert space $\mathfrak{H}$. The subspace* $\operatorname{Ran} Q$ *is a graph subspace* $\mathcal{G}(\operatorname{Ran} P, X)$ *associated with some closed densely defined (possibly unbounded) operator X :* $\operatorname{Ran} P \to \operatorname{Ran} P^\perp$ *with* $\operatorname{Dom}(X) \subset \operatorname{Ran} P$ *if and only if the subspaces $\mathfrak{M}_{01}(P,Q)$ and $\mathfrak{M}_{10}(P,Q)$ in the canonical decomposition* (A.1) *of the Hilbert space $\mathfrak{H}$ are trivial, i.e.,*

$$\mathfrak{M}_{01}(P,Q) = \mathfrak{M}_{10}(P,Q) = \{0\}. \tag{A.2}$$

For a given orthogonal projection P the correspondence between the closed subspaces $\operatorname{Ran} Q$ *satisfying* (A.2) *and closed densely defined operators X :* Ran $\to$ $\operatorname{Ran} P^\perp$ *is one-to-one.*

The subspaces $\mathfrak{M}_{11}$ and $\mathfrak{M}_{00}$ have a simple description in terms of the operator X: $\mathfrak{M}_{11} = \operatorname{Ker} X$ and $\mathfrak{M}_{00} = \operatorname{Ker} X^*$.

Note that $\mathfrak{M}_{01}(P,Q) = \mathfrak{M}_{10}(P,Q) = \{0\}$ if $\|P-Q\| < 1$. Moreover, Theorem 5.5 has the following corollary.

Theorem 5.6. *Let P and Q be orthogonal projections in a Hilbert space $\mathfrak{H}$. Then the inequality $\|P-Q\| < 1$ holds true if and only if* $\operatorname{Ran} Q$ *is a graph subspace associated with the subspace* $\operatorname{Ran} P$ *and some bounded operator* $X \in \mathcal{B}(\operatorname{Ran} P, \operatorname{Ran} P^\perp)$, *that is,* $\operatorname{Ran} Q = \mathcal{G}(\operatorname{Ran} P, X)$. *In this case*

$$\|X\| = \frac{\|P-Q\|}{\sqrt{1-\|P-Q\|^2}} \tag{A.3}$$

and

$$\|P-Q\| = \frac{\|X\|}{\sqrt{1+\|X\|^2}}. \tag{A.4}$$

Acknowledgments

V. Kostrykin is grateful to V. Enss, A. Knauf, and R. Schrader for useful discussions. A.K. Motovilov acknowledges the great hospitality and financial support by the Department of Mathematics, University of Missouri-Columbia, MO, USA. He was also supported in part by the Russian Foundation for Basic Research within Project RFBR 01-01-00958.

References

[1] V. Adamyan and H. Langer, *Spectral properties of a class of rational operator-valued functions*, J. Operator Theory **33** (1995), 259–277.

[2] V. Adamyan, H. Langer, and C. Tretter, *Existence and uniqueness of contractive solutions of some Riccati equations*, J. Funct. Anal. **179** (2001), 448–473.

[3] S. Albeverio, K.A. Makarov, and A.K. Motovilov, *Graph subspaces and the spectral shift function*, Canad. J. Math. **55** (2003), 449–503. (`arXiv:math.SP/0105142`)

[4] C. Apostol, C. Foias, and N. Salinas, *On stable invariant subspaces*, Integr. Equat. Oper. Theory **8** (1985), 721–750.

[5] M.Sh. Birman and M.Z. Solomjak, *Spectral Theory of Self-Adjoint Operators in Hilbert Space*, D. Reidel, Dordrecht, 1987.

[6] J. Daughtry, *Isolated solutions of quadratic matrix equations*, Linear Algebra Appl. **21** (1978), 89–94.

[7] C. Davis, *The rotation of eigenvectors by a perturbation. I and II*, J. Math. Anal. Appl. **6** (1963), 159–173; **11** (1965), 20–27.

[8] C. Davis and W.M. Kahan, *The rotation of eigenvectors by a perturbation. III*, SIAM J. Numer. Anal. **7** (1970), 1–46.

[9] T. Kato, *Perturbation Theory for Linear Operators*, Springer-Verlag, Berlin, 1966.

[10] V. Kostrykin, K.A. Makarov, and A.K. Motovilov, *Existence and uniqueness of solutions to the operator Riccati equation. A geometric approach*, pp. 181–198 in Yu. Karpeshina, G. Stolz, R. Weikard, Y. Zeng (Eds.), "Advances in Differential Equations and Mathematical Physics", Contemporary Mathematics **327**, Amer. Math. Soc., 2003. (`arXiv:math.SP/0207125`)

[11] V. Kostrykin, K.A. Makarov, and A.K. Motovilov, *Perturbation of spectra and spectral subspaces*, preprint `arXiv:math.SP/0306025` (2003).

[12] M. Reed and B. Simon, *Methods of Mathematical Physics, I: Fuctional Analysis*, Academic Press, New York, 1972.

Vadim Kostrykin
Fraunhofer-Institut für Lasertechnik
Steinbachstraße 15
D-52074 Aachen, Germany
e-mail: kostrykin@ilt.fraunhofer.de, kostrykin@t-online.de

Konstantin A. Makarov
Department of Mathematics
University of Missouri
Columbia, MO 65211, USA
e-mail: makarov@math.missouri.edu

Alexander K. Motovilov
Department of Mathematics
University of Missouri
Columbia, MO 65211, USA
Permanent Address:
Joint Institute for Nuclear Research
141980 Dubna, Moscow Region, Russia
e-mail: motovilv@thsun1.jinr.ru

Operator Theory:
Advances and Applications, Vol. 149, 373–385

Partly Free Algebras From Directed Graphs

David W. Kribs and Stephen C. Power

Abstract. We say that a nonselfadjoint operator algebra is *partly free* if it contains a free semigroup algebra. Motivation for such algebras occurs in the setting of what we call *free semigroupoid algebras*. These are the weak operator topology closed algebras generated by the left regular representations of semigroupoids associated with finite or countable directed graphs. We expand our analysis of partly free algebras from previous work and obtain a graph-theoretic characterization of when a free semigroupoid algebra with countable graph is partly free. This analysis carries over to norm closed *quiver algebras*. We also discuss new examples for the countable graph case.

Every finite or countable directed graph G recursively generates a Fock space Hilbert space and a family of partial isometries. These operators are of Cuntz-Krieger-Toeplitz type and also arise through the left regular representations of free semigroupoids determined by directed graphs. This was initially discovered by Muhly [23], and, in the case of finite graphs, more recent work with Solel [24] considered the norm closed algebras generated by these representations, which they called *quiver algebras*. In [19], we developed a structure theory for the weak operator topology closed algebras $\mathfrak{L}_G$ generated by the left regular representations coming from both finite and countable directed graphs; we called these algebras *free semigroupoid algebras*. In doing so, we found a unifying framework for a number of classes of algebras which appear in the literature, including; noncommutative analytic Toeplitz algebras $\mathfrak{L}_n$ [2, 10, 11, 20, 25, 26] (the prototypical free semigroup algebras), the classical analytic Toeplitz algebra H^∞ [14, 16], and certain finite-dimensional digraph algebras [19]. But this approach gives rise to a diverse collection of new examples which include finite-dimensional algebras, algebras with free behaviour, algebras which can be represented as matrix function algebras, and examples which mix these possibilities. The general theme of our work in [19] was a marriage of simple graph-theoretic properties with properties of the operator algebra. Furthermore, our technical analyses were chiefly spatial in nature; for instance, we proved the graph is a complete unitary invariant of both the free semigroupoid algebra and the quiver algebra.

Received by the editors November 25, 2003; revised June 11, 2003. Submitted by J. A. Ball.
Mathematics Subject Classification (2000). 47L55, 47L75.
Key words and phrases. Hilbert space, Fock space, partial isometry, directed graph, free semigroupoid, partly free operator algebra.
First author partially supported by a Canadian NSERC Post-doctoral Fellowship.

In the next section we give a short introduction to free semigroupoid algebras and discuss a number of examples. The second section contains an expanded analysis of a subclass called *partly free algebras*, which are characterized by containment of a copy of a free semigroup algebra. Specifically, we extend analysis from [19] to the case of countable graphs, obtaining a new graph-theoretic description of when a free semigroupoid algebra is partly free. In fact, we find that it is unusual for an algebra coming from a countable graph to not be partly free. We present a number of new illustrative examples for the countable graph case. Our analysis works equally well for quiver algebras, and we obtain a graph-theoretic condition for when these norm closed algebras are partly free.

1. Free Semigroupoid Algebras

Let G be a finite or countable directed graph, with edge set $E(G)$ and vertex set $V(G)$. Let $\mathbb{F}^+(G)$ be the *free semigroupoid* determined by G; that is, $\mathbb{F}^+(G)$ consists of the vertices, which act as units, and all allowable finite paths in G, with the natural operations of concatenation of allowable paths. Given a path w in $\mathbb{F}^+(G)$ we write $w = ywx$ when the initial and final vertices of w are, respectively, x and y. Let $\mathcal{H}_G = \ell^2(\mathbb{F}^+(G))$ be the Hilbert space with orthonormal basis indexed by elements of $\mathbb{F}^+(G)$. For each edge $e \in E(G)$ and vertex $x \in V(G)$, we may define partial isometries and projections on $\mathcal{H}_G$ by the following actions on basis vectors:

$$L_e\xi_w = \begin{cases} \xi_{ew} & \text{if } ew \in \mathbb{F}^+(G) \\ 0 & \text{otherwise} \end{cases}$$

and

$$L_x\xi_w = \begin{cases} \xi_{xw} = \xi_w & \text{if } w = xw \in \mathbb{F}^+(G) \\ 0 & \text{otherwise} \end{cases}$$

These operators may thus be regarded as 'partial creation operators' acting on a generalized Fock space Hilbert space. There is an equivalent tree perspective [19] which gives an appealing visual interpretation of the actions of these operators. The family $\{L_e, L_x\}$ also arises through the left regular representation $\lambda_G : \mathbb{F}^+(G) \to \mathcal{B}(\mathcal{H}_G)$, with $\lambda_G(e) = L_e$, and $\lambda_G(x) = L_x$. The associated *free semigroupoid algebra* is the weak operator topology closed algebra generated by this family,

$$\begin{aligned} \mathfrak{L}_G &= \text{WOT–Alg}\,\{L_e, L_x : e \in E(G), x \in V(G)\} \\ &= \text{WOT–Alg}\,\{\lambda_G(w) : w \in \mathbb{F}^+(G)\}. \end{aligned}$$

Remark 1.1. In the case of finite graphs, Muhly and Solel [23, 24] considered the norm closed algebras $\mathcal{A}_G$ generated by such a family, calling them *quiver algebras*. For both finite and countable graphs, we considered the classification problem for $\mathcal{A}_G$ in [19], and in Section 2.1 we derive partly free conditions for $\mathcal{A}_G$. Recently the C*-algebras generated by families of partial isometries associated with directed graphs have been studied heavily. The set of generators for these algebras are sometimes referred to as Cuntz-Krieger E-families (for instance see [4, 15, 21, 22]). On the other hand, the generators of free semigroupoid algebras are of Cuntz-Krieger-Toeplitz type in the sense that the C*-algebra generated by a family $\{L_e\}$ is generally the extension of a Cuntz-Krieger C*-algebra by the compact operators.

There is also a right regular representation $\rho_G : \mathbb{F}^+(G) \to \mathcal{B}(\mathcal{H}_G)$ determined by G, which yields partial isometries $\rho_G(w) \equiv R_{w'}$ for $w \in \mathbb{F}^+(G)$ acting on $\mathcal{H}_G$ by the equations $R_{w'}\xi_v = \xi_{vw}$, where w' is the word w in reverse order, with similar conditions. The corresponding algebra is

$$\begin{aligned} \mathfrak{R}_G &= \text{WOT–Alg}\,\{R_e, R_x : e \in E(G), x \in V(G)\} \\ &= \text{WOT–Alg}\,\{\rho_G(w) : w \in \mathbb{F}^+(G)\}. \end{aligned}$$

Given edges $e, f \in E(G)$, observe that $L_e R_f \xi_w = \xi_{ewf} = R_f L_e \xi_w$, for all $w \in \mathbb{F}^+(G)$, so that $L_e R_f = R_f L_e$, and similarly for the vertex projections. In fact, the commutant of $\mathfrak{R}_G$ coincides with $\mathfrak{L}_G = \mathfrak{R}_G'$ [19]. Further, the commutant of $\mathfrak{L}_G$ coincides with $\mathfrak{R}_G$, and thus $\mathfrak{L}_G$ is its own second commutant.

A useful ingredient in the proof is the observation that the algebras $\mathfrak{R}_G$ and $\mathfrak{L}_{G^t}$ are naturally unitarily equivalent, where G^t is the *transpose directed graph* obtained from G by reversing the directions of all edges. An important technical device obtained here is the existence of Fourier expansions for elements of $\mathfrak{L}_G$. Specifically, if $A \in \mathfrak{L}_G$ and $x \in V(G)$, then $A\xi_x = \sum_{w=wx} a_w \xi_w$ with $a_w \in \mathbb{C}$, and Cesaro type sums associated with the formal sum $\sum_{w\in\mathbb{F}^+(G)} a_w L_w$, converge in the strong operator topology to A. We write, $A \sim \sum_{w\in\mathbb{F}^+(G)} a_w L_w$. As a notational convenience, for projections determined by vertices $x \in V(G)$, we put $P_x \equiv L_x$ and $Q_x \equiv R_x$.

In [19] we proved that G is a complete unitary invariant of both $\mathfrak{L}_G$ and $\mathcal{A}_G$. Thus, different directed graphs really do yield different algebras. We finish this section by discussing a number of examples from simple graphs.

Examples 1.2. (i) The algebra generated by the graph with a single vertex x and single loop edge $e = xex$ is unitarily equivalent to the classical analytic Toeplitz algebra H^∞ [14, 16]. Indeed, the Hilbert space in this case may be naturally identified with the Hardy space H^2 of the unit disc, and under this identification L_e is easily seen to be unitarily equivalent to the unilateral shift U_+.

(ii) The noncommutative analytic Toeplitz algebras $\mathfrak{L}_n$, $n \geq 2$ [2, 9, 10, 11, 20, 25, 26], the fundamental examples of free semigroup algebras, arise from the graphs with a single vertex and n distinct loop edges. For instance, in the case $n = 2$ with loop edges $e = xex \neq f = xfx$, the Hilbert space is identified with unrestricted 2-variable Fock space $\mathcal{H}_2$. The operators L_e, L_f are equivalent to the natural creation operators on $\mathcal{H}_2$, also known as the Cuntz-Toeplitz isometries. Further, $P_x = I$, and thus $\mathfrak{L}_G \simeq \mathfrak{L}_2$.

(iii) As an example of a simple matrix function algebra, we may consider the graph G with vertices x, y and edges $e = xex$, $f = yfx$. Then $\mathfrak{L}_G$ is generated by $\{L_e, L_f, P_x, P_y\}$. If we make the natural identifications $\mathcal{H}_G = P_x\mathcal{H}_G \oplus P_y\mathcal{H}_G \simeq$

$H^2 \oplus H^2$ (respecting word length), then

$$L_e \simeq \begin{bmatrix} U_+ & 0 \\ 0 & 0 \end{bmatrix} \quad L_f \simeq \begin{bmatrix} 0 & 0 \\ U_+ & 0 \end{bmatrix} \quad P_x \simeq \begin{bmatrix} I & 0 \\ 0 & 0 \end{bmatrix} \quad P_y \simeq \begin{bmatrix} 0 & 0 \\ 0 & I \end{bmatrix}.$$

Thus, $\mathfrak{L}_G$ is seen to be unitarily equivalent to

$$\mathfrak{L}_G \simeq \begin{bmatrix} H^\infty & 0 \\ H_0^\infty & \mathbb{C}I \end{bmatrix}$$

where H_0^∞ is the subalgebra of H^∞ functions h with $h(0) = 0$.

(iv) By simply adding a directed edge $g = xgy$ to the previous graph, we obtain a very different algebra $\mathfrak{L}_{G'}$. In fact, $\mathfrak{L}_{G'}$ is a unitally partly free algebra in the sense of the next section because it contains isometries with mutually orthogonal ranges; for instance, $U = L_e^2 + L_f L_g$ and $V = L_e L_g + L_g L_e$ are isometries which satisfy $U^*V = 0$.

(v) If G is a finite graph with no directed cycles, then the Fock space $\mathcal{H}_G$ is finite-dimensional and so too is $\mathfrak{L}_G$. As an example, consider the graph with three vertices and two edges, labelled x_1, x_2, x_3, e, f where $e = x_2 e x_1$, $f = x_3 f x_1$. Then the Fock space is spanned by the vectors $\{\xi_{x_1}, \xi_{x_2}, \xi_{x_3}, \xi_e, \xi_f\}$ and with this basis the general operator $X = \alpha L_{x_1} + \beta L_{x_2} + \gamma L_{x_3} + \lambda L_e + \mu L_f$ in $\mathfrak{L}_G$ is represented by the matrix

$$X \simeq \begin{bmatrix} \alpha & & & & \\ & \beta & & & \\ & & \gamma & & \\ \lambda & & & \beta & \\ \mu & & & & \gamma \end{bmatrix}.$$

Algebraically, $\mathfrak{L}_G$ is isometrically isomorphic to the so-called digraph algebra in $\mathcal{M}_3(\mathbb{C})$ consisting of the matrices

$$\begin{bmatrix} \alpha & 0 & 0 \\ \lambda & \beta & 0 \\ \mu & 0 & \gamma \end{bmatrix}.$$

Recall that a digraph algebra $\mathcal{A}(H)$ is a unital subalgebra of $\mathcal{M}_n(\mathbb{C})$ which is spanned by some of the standard matrix units of $\mathcal{M}_n(\mathbb{C})$. The graph H is transitive and reflexive and is such that the edges of H naturally label the relevant matrix units.

(vi) Let $n \geq 1$ and consider the *cycle graph* C_n which has n vertices $x_1, \ldots, x_n$ and n edges $e_n = x_1 e_n x_n$ and $e_k = x_{k+1} e_k x_k$ for $k = 0, \ldots, n-1$. The *cycle algebra* $\mathfrak{L}_{C_n}$ may be identified with the WOT-closed semicrossed product $\mathbb{C}^n \times_\beta^\sigma \mathbb{Z}_+$ associated with the cyclic shift automorphism β of $\mathbb{C}^n$ [13]. To see this identify $L_{x_i}\mathcal{H}_G$ with H^2 for each i in the natural way (respecting word length). Then $\mathcal{H}_G = L_{x_1}\mathcal{H}_G \oplus \ldots \oplus L_{x_n}\mathcal{H}_G \simeq \mathbb{C}^n \otimes H^2$ and the operator $\alpha_1 L_{e_1} + \ldots + \alpha_n L_{e_n}$ is identified with the operator matrix

$$\begin{bmatrix} 0 & & & & \alpha_n T_z \\ \alpha_1 T_z & 0 & & & \\ & \alpha_2 T_z & 0 & & \\ & & \ddots & \ddots & \\ & & & \alpha_{n-1} T_z & 0 \end{bmatrix}.$$

Writing $H^\infty(z^n)$ for the subalgebra of H^∞ arising from functions of the form $h(z^n)$ with h in H^∞, the algebra $\mathfrak{L}_{C_n}$ is readily identified with the matrix function algebra

$$\begin{bmatrix} H^\infty(z^n) & z^{n-1}H^\infty(z^n) & \dots & zH^\infty(z^n) \\ zH^\infty(z^n) & H^\infty(z^n) & & \vdots \\ \vdots & & \ddots & \\ z^{n-1}H^\infty(z^n) & \dots & & H^\infty(z^n) \end{bmatrix}.$$

This in turn is identifiable with the crossed product above. In fact, this matrix function algebra is the WOT-closed variant of the matrix function algebra $\mathcal{B}_n$ of De Alba and Peters [13] for the norm closed semicrossed product $\mathbb{C}^n \times_\beta^\sigma \mathbb{Z}_+$. Such identifications are the Toeplitz versions of the identification of the graph C*-algebra of C_n with $\mathcal{M}_n(\mathrm{C}(\mathbb{T}))$.

2. Partly Free Algebras

We say that a WOT-closed operator algebra $\mathfrak{A}$ is *partly free* if it contains the free semigroup algebra $\mathfrak{L}_2$ as a subalgebra in the sense of the following definition.

Definition 2.1. A WOT-closed algebra $\mathfrak{A}$ is *partly free* if there is an inclusion map $\mathfrak{L}_2 \hookrightarrow \mathfrak{A}$ which is the restriction of an injection between the generated von Neumann algebras. If the map can be chosen to be unital, then $\mathfrak{A}$ is said to be *unitally partly free.*

These notions parallel the requirement that a C*-algebra contain the Cuntz algebra $\mathcal{O}_2$, or that a discrete group contain a free group. Theorems 2.5 and 2.6 determine when the algebras $\mathfrak{L}_G$ are partly free and unitally partly free for both finite and countable directed graphs. We require the following structural result on partial isometries in $\mathfrak{L}_G$.

Lemma 2.2. [19] *The initial projections of partial isometries V in $\mathfrak{L}_G$ are sums of projections from $\{P_x : x \in V(G)\}$. Specifically,*

$$V^*V = \sum_{x\in\mathcal{I}} P_x \quad \textit{where} \quad \mathcal{I} = \{x \in V(G) : V\xi_x \neq 0\}.$$

In fact, we proved much more in [19]. All partial isometries V in $\mathfrak{L}_G$ satisfy a standard form, written as $V = \sum_{x\in\mathcal{I}} \oplus L_{\eta_x}$, where $\{\eta_x\}_{x\in\mathcal{I}}$ are unit *wandering* vectors for $\mathfrak{R}_G$ supported on distinct $Q_x\mathcal{H}_G$, and hence the initial projections satisfy $L_{\eta_x}^* L_{\eta_x} = P_x$.

The cycle algebras $\mathfrak{L}_{C_n}$, $1 \leq n < \infty$, arise in the general theory, as does the algebra $\mathfrak{L}_{C_\infty}$ generated by an infinite graph analogue of C_n. Let C_∞ be the directed graph with vertex set $\{x_k : k \geq 1\}$ indexed by the natural numbers, and edge set $\{e_k = x_{k+1}e_k x_k : k \geq 1\}$. In the finite graph case, the cycle algebras $\mathfrak{L}_{C_n}$ are the key examples of free semigroupoid algebras which are not partly free. However, $\mathfrak{L}_{C_\infty}$ is unitally partly free, and this gives an indication of how pervasive these algebras are in the countable graph case.

Lemma 2.3. *The algebras $\mathfrak{L}_{C_n}$, $1 \leq n < \infty$, are not partly free. However, $\mathfrak{L}_{C_\infty}$ is unitally partly free.*

Proof. We proved this result in [19] for $1 \leq n < \infty$ by showing $\mathfrak{L}_{C_n}$ does not contain pairs of partial isometries U, V which satisfy condition (iii) of Theorem 2.5. For $n = \infty$, we shall construct a pair of isometries U, V in $\mathfrak{L}_{C_\infty}$ with orthogonal ranges. For $k \geq 1$, let u_k and v_k be the unique finite paths in $\mathbb{F}^+(C_\infty)$ with $u_k = x_{2k}u_k x_k$ and $v_k = x_{2k+1}v_k x_k$. Then we may define U and V by

$$U = \sum_{k\geq 1} \oplus L_{u_k} \quad \text{and} \quad V = \sum_{k \geq 1} \oplus L_{v_k},$$

where the sums converge in the strong operator topology. A unital injection of $\mathfrak{L}_2$ into $\mathfrak{L}_{C_\infty}$ is then defined by mapping the two generators of $\mathfrak{L}_2$ to U and V. □

In fact it turns out there are two ways in which an algebra $\mathfrak{L}_G$ can be partly free; namely G must either contain a double-cycle or a proper infinite (directed) path, in the sense of the following definition.

Definition 2.4. A *cycle* at vertex x is a path of edges $w = xwx$ for which only the initial edge has source vertex x. The graph G contains a *double-cycle* if there are distinct cycles $w_i = xw_ix$, $i = 1, 2$, at some vertex x in G. By a *proper infinite (directed) path* in G, we mean an infinite word $\omega = \cdots e_{i_3}e_{i_2}e_{i_1}$ in the edges of G such that no edges are repeated and every finite segment corresponds to an allowable finite directed path in G.

A little thought shows that the only way a graph G fails to have either a double-cycle or such a proper infinite path is if either there are no infinite paths at all, or, if every infinite path in the graph becomes periodic; in this case say that G has the periodic path property. If, on the other hand, G does not have the periodic path property, we shall say that G has the *aperiodic path property* meaning that there exists an aperiodic infinite path. This in turn means that there exists a proper infinite path or a double-cycle.

We define the *saturation* at a vertex x in G to be the set, $\operatorname{sat}(x)$, that consists of x, together with all finite and infinite paths which start at x, and all vertices that are final vertices for paths starting at x. Consider a graph which is a downward directed tree in which every branch has finite length and different branches may intersect. Also perform the following surgery: add arbitrary cycles to some vertices without creating double-cycles. Note there can be infinite branchings. This 'aperiodic looped tree graph' does not have the aperiodic path property and furthermore such a graph is a 'fairly typical' graph without that property. In fact, G does not have the aperiodic path property if and only if the saturation graph for each vertex has this kind of structure.

For the sake of brevity in the next proof we shall assume the finite vertex case [19].

Theorem 2.5. *The following assertions are equivalent for a finite or countable directed graph G:*

(i) *G has the aperiodic path property.*
(ii) *$\mathfrak{L}_G$ is partly free.*
(iii) *There are nonzero partial isometries U, V in $\mathfrak{L}_G$ with*

$$U^*U = V^*V, \quad UU^* \leq U^*U, \quad VV^* \leq V^*V, \quad U^*V = 0.$$

Proof. Since von Neumann algebra isomorphisms are spatial, condition (iii) is a reformulation of (ii). Thus it suffices to prove the equivalence of (i) and (iii).

To see (iii) $\Rightarrow$ (i), first suppose U and V are partial isometries in $\mathfrak{L}_G$ satisfying (iii) with $U^*U = P_x = V^*V$ for some $x \in V(G)$. Then there is at least one cycle $w = xwx$ over x, for otherwise $U = P_xUP_x = 0$ or $U = P_x$ and $0 = P_xU = V$. Suppose w is the only cycle over x. Then the compression algebra $P_x\mathfrak{L}_GP_x$ forms the WOT-closed subalgebra of $\mathfrak{L}_G$ consisting of elements $X \in \mathfrak{L}_G$ with Fourier expansions of the form $X \sim \sum_{k\geq 0} a_k L_{w^k}$. Let $\mathcal{H}_x = P_x\mathcal{H}_G$. Then $P_x\mathfrak{L}_G|_{\mathcal{H}_x}$ is evidently unitarily equivalent to $\mathfrak{L}_{C_1} \simeq H^\infty$. But $P_x\mathfrak{L}_G|_{\mathcal{H}_x}$ contains a pair of nonzero partial isometries $U|_{\mathcal{H}_x} = P_xU|_{\mathcal{H}_x}$, $V|_{\mathcal{H}_x} = P_xV|_{\mathcal{H}_x}$ satisfying condition (iii), and this contradicts Lemma 2.3. Thus, we deduce the existence of at least two cycles over x, and hence G contains a double-cycle.

For the general case, let U and V be partial isometries in $\mathfrak{L}_G$ satisfying (iii) with $U^*U = V^*V = \sum_{x\in\mathcal{I}} P_x$ as in Lemma 2.2. Suppose, by way of contradiction, that (i) fails so that G has the structure discussed above. The vertices x in the index set $\mathcal{I}$ are scattered over this graph. Plainly, either there exists such an x with no other y in $\mathcal{I}$ in the saturation of x or there is a cycle C containing several such edges such that the saturation of C contains no further vertices x in $\mathcal{I}$. In the first case we can consider the compression of the given U and V to $\mathcal{H}_x$, and in the second case the compression to $\mathcal{H}_C = \sum_{x\in V(C)} P_x\mathcal{H}_G$. Observe that the saturations of x and C may be infinite. However, the conditions in (iii) allow us to see that these compressions are indeed partial isometries because the hypothesis implies that the compression is the same (modulo a zero summand) as the restriction to the subspace $\mathfrak{L}_G\mathcal{H}_x$ or $\mathfrak{L}_G\mathcal{H}_C$. Now we get our contradiction by arguing as in the previous paragraph. Indeed, the first case is trivial and in the second case the compression of the algebra is unitarily equivalent to a cycle graph algebra $\mathfrak{L}_{C_n}$.

Towards the implication (i) $\Rightarrow$ (iii), suppose ω is a proper infinite path in G. Let $\mathcal{J}$ be the set of all vertices in G for which the saturation at each $x \in \mathcal{J}$ contains part, and hence an entire tail, of ω. Let $\{x_k, y_k\} \to \{z_k\}$ be a two-to-one map from the countable set $\mathcal{J}$ onto itself, such that the vertices $x_k, y_k \in \mathcal{J}$ belong to the saturation of $z_k \in \mathcal{J}$. Thus, there are distinct finite paths $u_k = x_k u_k z_k$, $v_k = y_k v_k z_k$. By design, $U = \sum_k \oplus L_{u_k}$, $V = \sum_k \oplus L_{v_k}$ are partial isometries in $\mathfrak{L}_G$ with orthogonal ranges, initial projection $U^*U = P_{\mathcal{J}} = V^*V$, and $U = P_{\mathcal{J}} U P_{\mathcal{J}}$, $V = P_{\mathcal{J}} V P_{\mathcal{J}}$.

Next, suppose distinct cycles w_1, w_2 form a double-cycle over a vertex x in G. Let $\mathcal{J}$ be the set of all vertices for which the saturation at each vertex in $\mathcal{J}$ contains this double-cycle. Let us enumerate the (possibly finite) vertices of $\mathcal{J}$ as $\{x_k\}_k$, and let $r_k = x r_k x_k$ be a path from x_k to x. For each vertex x_k in $\mathcal{J}$ choose, without repeating any choices, two paths u_k, v_k amongst the set $\{w_1^m w_2 : m \geq 1\}$. Then again by design, $U = \sum_k \oplus L_{u_k r_k}$, $V = \sum_k \oplus L_{v_k r_k}$ are partial isometries in $\mathfrak{L}_G$ with orthogonal ranges, initial projection $U^*U = P_{\mathcal{J}} = V^*V$, and $U = P_{\mathcal{J}} U P_{\mathcal{J}}$, $V = P_{\mathcal{J}} V P_{\mathcal{J}}$. Therefore, in both cases we have shown that (i) $\Rightarrow$ (iii), and this completes the proof. □

It follows from this result that the free semigroupoid algebras coming from countable graphs are typically partly free. Indeed, the condition on a countable graph G which forces $\mathfrak{L}_G$ to not be partly free is quite restrictive as outlined in the previous discussion. Below we discuss a number of examples.

The following is the unital version of the previous theorem. We shall say G has the *uniform aperiodic path property* if the saturation at every vertex includes an aperiodic infinite path.

Theorem 2.6. *The following assertions are equivalent for a finite or countable directed graph G:*

(i) *G has the uniform aperiodic path property.*
(ii) *$\mathfrak{L}_G$ is unitally partly free.*
(iii) *There are isometries U, V in $\mathfrak{L}_G$ with*

$$U^*V = 0.$$

Proof. Once again, condition (iii) is a restatement of (ii), so it suffices to prove the equivalence of (i) and (iii).

For (iii) $\Rightarrow$ (i), the proof of Theorem 2.5 can be adapted to show that the saturation at every vertex in the index set of vertices $\mathcal{I}$, determining the initial projection for the partial isometries U, V, includes a double-cycle or a proper infinite path. Hence we may apply this argument in the current case with $\mathcal{I} = V(G)$ as we are dealing with isometries here, $U^*U = V^*V = I = \sum_{x \in V(G)} P_x$, and it follows that G has the uniform aperiodic path property. to the finite vertex case general proof for countable the finite vertex case.

To see (i) $\Rightarrow$ (iii), consider the last two paragraphs in the proof of Theorem 2.5. As G satisfies the uniform aperiodic path property, it follows that we may decompose the vertex set for G into disjoint subsets $V(G) = \cup_i \mathcal{J}_i$, where each $\mathcal{J}_i$ is obtained as in one of these two cases; double-cycles, or proper infinite paths. In either case we can define partial isometries U_i, V_i in $\mathfrak{L}_G$ with orthogonal ranges, $U_i = P_{\mathcal{J}_i} U_i P_{\mathcal{J}_i}$, $V_i = P_{\mathcal{J}_i} V_i P_{\mathcal{J}_i}$, and $U_i^* U_i = P_{\mathcal{J}_i} = V_i^* V_i$. Thus, the operators $U = \sum_i \oplus U_i$, $V = \sum_i \oplus V_i$ are isometries in $\mathfrak{L}_G$ with mutually orthogonal ranges. □

We next add to the short list of known hyper-reflexive algebras [3, 5, 7, 8, 10] by extending our result from [19] to the case of countable graphs. Given an operator algebra $\mathfrak{A}$, a measure of the distance to $\mathfrak{A}$ is given by

$$\beta_{\mathfrak{A}}(X) = \sup_{L \in \operatorname{Lat} \mathfrak{A}} ||P_L^{\perp} X P_L||,$$

where P_L is the projection onto the subspace L and $\operatorname{Lat} \mathfrak{A}$ is the lattice of invariant subspaces for $\mathfrak{A}$. Evidently, $\beta_{\mathfrak{A}}(X) \leq \operatorname{dist}(X, \mathfrak{A})$, and the algebra $\mathfrak{A}$ is said to be *hyper-reflexive* if there is a constant C such that $\operatorname{dist}(X, \mathfrak{A}) \leq C \beta_{\mathfrak{A}}(X)$ for all X.

The free semigroup algebras $\mathfrak{L}_n$ were proved to be hyper-reflexive by Davidson ($n = 1$ [8]) and Davidson and Pitts ($n \geq 2$ [10]). Furthermore, motivated by the $\mathfrak{L}_n$ case Bercovici [5] proved an algebra is hyper-reflexive with distant constant no greater than 3 whenever its commutant contains a pair of isometries with orthogonal ranges.

Corollary 2.7. *Let G be a finite or countable directed graph such that the transpose graph G^t satisfies the uniform aperiodic path property; equivalently, $\mathfrak{L}_{G^t}$ is unitally partly free. Then $\mathfrak{L}_G$ is hyper-reflexive with distant constant at most 3.*

Proof. This is a direct consequence of Bercovici's result [5] since $\mathfrak{L}_G' = \mathfrak{R}_G \simeq \mathfrak{L}_{G^t}$. □

Problem 2.8. Is $\mathfrak{L}_G$ hyper-reflexive for every directed graph G?

We next present some simple examples for the countable graph case. The focus will be on the new aspect discovered here; the relevance of proper infinite paths.

Examples 2.9. (i) Let G be the directed graph with vertices $\{x_k : k \in \mathbb{Z}\}$ indexed by the integers and directed edges $\{e_k = x_{k+1} e_k x_k : k \in \mathbb{Z}\}$. We could also add (possibly infinite) directed paths $w_k = x_k w_k$. Then every vertex saturation in G contains a proper infinite tail $\omega_k = \cdots e_{k+1} e_k$ for some $k \in \mathbb{Z}$. Thus G satisfies the uniform aperiodic path property, and $\mathfrak{L}_G$ is unitally partly free.

Consider the interesting special case that occurs when all the paths $w_k = x_k w_k x_k$ are loop edges. Evidently G^t and G are isomorphic, hence the commutant $\mathfrak{L}_G' = \mathfrak{R}_G \simeq \mathfrak{L}_{G^t} \simeq \mathfrak{L}_G$ is unitarily equivalent to $\mathfrak{L}_G$ and is also unitally partly free. As a variation of this case, instead let H be the subgraph $H = \{x_k, e_k, w_k : k \geq 1\}$. Then $\mathfrak{L}_H$ is unitally partly free, but $\mathfrak{L}_H' \simeq \mathfrak{L}_{H^t}$ is not even partly free.

(ii) A non-discrete example is given by the graph Q consisting of vertices $\{x_q : q \in \mathbb{Q}\}$ indexed by the rational numbers, and directed edges $e_{qp} = x_q e_{qp} x_p$ whenever $p \leq q$. This example satisfies the uniform aperiodic path property, in fact there is an abundance of infinite non-overlapping directed paths emanating from each vertex, so $\mathfrak{L}_Q$ is unitally partly free. However, notice that the quiver algebra $\mathcal{A}_G$ is not even partly free (see Section 2.1). Further note that Q^t is graph isomorphic to Q, thus $\mathfrak{L}_Q \simeq \mathfrak{L}_{Q^t} \simeq \mathfrak{L}'_Q$ is unitarily equivalent to its commutant.

(iii) The following example was suggested to us by Ken Davidson. For $n \geq 2$, let $\mathbb{F}_n^+$ be the unital free semigroup on n noncommuting letters, written as $\{1, 2, \ldots, n\}$, with unit ϕ. Let G_n be the doubly-bifurcating (sideways) infinite tree with vertices $\{x_w : w \in \mathbb{F}_n^+\}$ indexed by words in $\mathbb{F}_n^+$, and directed edges $\{e_{iw} = x_{iw} e_{iw} x_w : w \in \mathbb{F}_n^+, 1 \leq i \leq n\}$ determined by the directions $w \mapsto iw$. Then G_n satisfies the uniform aperiodic path property (observe that $G_1 = C_\infty$), and hence $\mathfrak{L}_{G_n}$ is unitally partly free. An interesting point here is that the graph G_n itself has the structure of the full Fock space Hilbert space $\ell^2(\mathbb{F}_n^+)$ traced out by its left creation operators. Whereas, the Fock space $\mathcal{H}_{G_n}$ consists of infinitely many disjoint infinite-dimensional components, indexed by elements of $\mathbb{F}_n^+$. The transpose graph G_n^t is quite different from G_n. In fact the commutant algebra $\mathfrak{L}_{G_n^t} \simeq \mathfrak{R}_{G_n} = \mathfrak{L}'_{G_n}$ is not partly free.

(iv) Let G be the directed graph with vertices $\{x_k : k \geq 1\}$ and edges $\{e_k = x_k e_k x_1 : k \geq 1\}$. This graph has no infinite paths or double-cycles, hence $\mathfrak{L}_G$ is not partly free. A variation of this example, turning the e_k into non-overlapping paths of length k that only intersect at vertex x_1, produces a non-partly free $\mathfrak{L}_G$ with graph containing non-overlapping finite paths of arbitrarily large length.

(v) Let $G = \{x_1, x_2, e_k = x_2 e_k x_1 : k \geq 1\}$ with e_k distinct edges. Then $\mathfrak{L}_G$ is not partly free and is unitarily equivalent to its commutant $\mathfrak{L}'_G \simeq \mathfrak{L}_{G^t}$.

(vi) Let G be the directed graph with vertices $\{x_k : k \in \mathbb{Z}\}$ and directed edges $\{e_k : k \in \mathbb{Z}\}$ where

$$e_k = \begin{cases} x_{2m+1} e_k x_{2m} & \text{if } k = 2m \\ x_{2m+1} e_k x_{2m+2} & \text{if } k = 2m+1. \end{cases}$$

Then $\mathfrak{L}_G$ is not partly free and is unitarily equivalent to its commutant $\mathfrak{L}'_G \simeq \mathfrak{L}_{G^t}$.

2.1. Partly Free Quiver Algebras

Using Theorems 2.5 and 2.6 we may readily deduce graph-theoretic conditions for quiver algebras $\mathcal{A}_G$ to be partly free. We require the following structural result for partial isometries in $\mathcal{A}_G$ for the countable graph case. An immediate consequence is that $\mathcal{A}_G$ only contains isometries when G has finitely many vertices.

Lemma 2.10. *Let G be a countable directed graph. If V is a partial isometry in $\mathcal{A}_G$, then its initial projection $V^*V = \sum_{x \in \mathcal{I}} P_x$ is the sum of only finitely many P_x.*

Proof. In fact, if V is a partial isometry in $\mathfrak{L}_G$ for which $\mathcal{I}$ is an infinite set, then $\mathrm{dist}(V, \mathcal{A}_G) \geq 1$. Indeed, let $q(L)$ belong to the set of polynomials $\mathcal{A} = \mathrm{Alg}\{L_w : w \in \mathbb{F}^+(G)\}$ in the L_e and $L_x = P_x$. As $\mathcal{I}$ is infinite, there is a $y \in \mathcal{I}$ such that $q(L)P_y = 0$. Hence $1 = ||VP_y|| = ||VP_y - q(L)P_y|| \leq ||V - q(L)||$, and this proves the claim because $\mathcal{A}$ is (norm) dense in $\mathcal{A}_G$. The lemma follows since $\mathfrak{L}_G$ contains $\mathcal{A}_G$. □

Let $\mathcal{A}_2$ be the quiver algebra generated by the graph with a single vertex and two distinct loop edges. This is the *noncommutative disc algebra* of Popescu [26, 27], also considered by Arias [1], and Muhly-Solel [23, 24]. Say that $\mathcal{A}_G$ is *partly free*, or *unitally partly free*, if the maps in Definition 2.1 are injections of $\mathcal{A}_2$ into $\mathcal{A}_G$, and are restrictions of injections of the generated C*-algebras.

We say that a graph G has the *double-cycle property* if G contains a double-cycle, and G satisfies the *uniform double-cycle property* when every vertex saturation $\mathrm{sat}(x)$ includes a double-cycle. Compare the following results with Theorems 2.5 and 2.6, and notice how the proper infinite path phenomena only arises in the WOT-closed case. In particular, there are many examples in the countable graph case for which $\mathfrak{L}_G$ is partly free, but $\mathcal{A}_G$ is not.

Theorem 2.11. *The following assertions are equivalent for a finite or countable directed graph G:*

(i) *G has the double-cycle property.*
(ii) *$\mathcal{A}_G$ is partly free.*
(iii) *There are nonzero partial isometries U, V in $\mathcal{A}_G$ with*

$$U^*U = V^*V, \quad UU^* \leq U^*U, \quad VV^* \leq V^*V, \quad U^*V = 0.$$

Proof. As in Theorem 2.5 it suffices to establish the equivalence of (i) and (iii). But (i) $\Rightarrow$ (iii) is clear since $\mathcal{A}_G$ will include $U = L_{w_1}$, $V = L_{w_2}$, where w_1, w_2 are distinct cycles over a common vertex, when (i) holds. On the other hand, as $\mathfrak{L}_G$ contains $\mathcal{A}_G$, condition (iii) implies G satisfies the aperiodic path property by Theorem 2.5. But recall from the proof of (iii) $\Rightarrow$ (i) in Theorem 2.5 that the proper infinite path part of this property can only occur when the initial vertex set $\mathcal{I}$ is infinite. Hence by Lemma 2.10, G contains a double-cycle and (i) holds. □

In the unital case, the graph can only have finitely many vertices.

Theorem 2.12. *The following assertions are equivalent for a finite or countable directed graph G:*

(i) *G has finitely many vertices and satisfies the uniform double-cycle property.*
(ii) *$\mathcal{A}_G$ is unitally partly free.*
(iii) *There are isometries U, V in $\mathcal{A}_G$ with*

$$U^*V = 0.$$

Proof. This result follows from Theorem 2.6 as the previous result follows from Theorem 2.5, other than the extra vertex condition on G. In particular, (iii) implies G satisfies the uniform infinite path property, but since U, V are isometries the initial vertex set in this case is the entire vertex set of G, that is $\mathcal{I} = V(G)$. Therefore, G can only have finitely many vertices by Lemma 2.11 and must satisfy the uniform double-cycle property. □

Acknowledgements

We would like to thank Ken Davidson for organizing a workshop on nonselfadjoint operator algebras at the Fields Institute in Toronto (July 2002), where the authors had a number of productive conversations. The first named author would also like to thank members of the Department of Mathematics at Purdue University for kind hospitality during preparation of this article.

Note Added in Proof

Problem 2.8 has been answered in the affirmative for all finite graphs [17]. The problem remains open for general $\mathfrak{L}_G$. We also mention that the ideal structure of $\mathfrak{L}_G$ has been analyzed in [18]. Further, Ephrem [15] has recently identified graph conditions which characterize when a Cuntz-Krieger graph C*-algebra has type I representation theory. Interestingly, the condition he obtains for finite graphs is equivalent to the graph not having any double-cycles.

References

[1] A. Arias, *Multipliers and representations of noncommutative disc algebras,* Houston J. Math., **25** (1999), 99–120.

[2] A. Arias, G. Popescu, *Factorization and reflexivity on Fock spaces*, Int. Equat. Oper. Th. **23** (1995), 268–286.

[3] W. Arveson, *Interpolation problems in nest algebras*, J. Func. Anal. **20** (1975), 208–233.

[4] T. Bates, J. Hong, I. Raeburn, W. Szymenski, *The ideal structure of the* C*-*algebras of infinite graphs*, e-print arxiv math.OA/0109142, preprint, 2001.

[5] H. Bercovici, *Hyper-reflexivity and the factorization of linear functionals*, J. Func. Anal. **158** (1998), 242–252.

[6] A. Beurling, *On two problems concerning linear transformations in Hilbert space,* Acta Math. **81** (1949), 239–255.

[7] E. Christensen, *Perturbations of operator algebras II*, Indiana U. Math. J. **26** (1977), 891–904.

[8] K.R. Davidson, *The distance to the analytic Toeplitz operators*, Illinois J. Math. **31** (1987), 265–273.

[9] K.R. Davidson, E. Katsoulis, D.R. Pitts, *The structure of free semigroup algebras,* J. reine angew. Math. **533** (2001), 99-125.

[10] K.R. Davidson, D.R. Pitts, *Invariant subspaces and hyper-reflexivity for free semigroup algebras*, Proc. London Math. Soc. **78** (1999), 401–430.

[11] K.R. Davidson, D.R. Pitts, *The algebraic structure of non-commutative analytic Toeplitz algebras*, Math. Ann. **311** (1998), 275–303.

[12] K.R. Davidson, *Nest Algebras*, Longman Scientific & Technical, London, 1988.

[13] L.M. De Alba, J. Peters, *Classification of semicrossed products of finite-dimensional* C*-*algebras*, Proc. Amer. Math. Soc. **95** (1985), 557–564.

[14] R. Douglas, *Banach algebra techniques in operator theory,* Springer-Verlag, New York, 1998.

[15] M. Ephrem, *Characterizing liminal and type I graph* C*-*algebras*, arXiv:math.OA/0211241, preprint, 2003.

[16] K. Hoffman, *Banach spaces of analytic functions*, Dover Publications Inc., New York, 1988.

[17] F. Jaeck, S.C. Power, *The semigroupoid algebras of finite graphs are hyper-reflexive*, in preparation, 2003.

[18] M.T. Jury, D.W. Kribs, *Ideal structure in free semigroupoid algebras from directed graphs*, preprint, 2003.

[19] D.W. Kribs, S.C. Power, *Free semigroupoid algebras*, preprint, 2002.

[20] D.W. Kribs, *Factoring in non-commutative analytic Toeplitz algebras*, J. Operator Theory **45** (2001), 175–193.

[21] A. Kumjian, D. Pask, I. Raeburn, *Cuntz-Krieger algebras of directed graphs*, Pacific J. Math **184** (1998), 161–174.

[22] A. Kumjian, D. Pask, I. Raeburn, J. Renault, *Graphs, Groupoids, and Cuntz-Krieger algebras*, J. Funct. Anal. **144** (1997), 505–541.

[23] P.S. Muhly, *A finite-dimensional introduction to operator algebra*, A. Katavolos (ed.), Operator Algebras and Applications, 313–354, Kluwer Academic Publishers, 1997.

[24] P.S. Muhly, B. Solel, *Tensor algebras, induced representations, and the Wold decomposition*, Can. J. Math. **51** (4), 1999, 850–880.

[25] G. Popescu, *Multi-analytic operators and some factorization theorems*, Indiana Univ. Math. J. **38** (1989), 693–710.

[26] G. Popescu, *Multi-analytic operators on Fock spaces*, Math. Ann. **303** (1995), 31–46.

[27] G. Popescu, *Noncommuting disc algebras and their representations*, Proc. Amer. Math. Soc. **124** (1996), 2137–2148.

[28] S.C. Power, *Approximately finitely acting operator algebras*, J. Func. Anal. **189** (2002), 409–469.

David W. Kribs
Department of Mathematics and Statistics
University of Guelph
Guelph, Ontario
Canada N1G 2W1
e-mail: `dkribs@uoguelph.ca`

Stephen C. Power
Department of Mathematics and Statistics
Lancaster University
Lancaster, England
UK LA1 4YW
e-mail: `s.power@lancaster.ac.uk`

Operator Theory:
Advances and Applications, Vol. 149, 387–406

Uniform Approximation by Solutions of Elliptic Equations and Seminormality in Higher Dimensions

Mircea Martin

Abstract. The first main result in this article is a quantitative Hartogs-Rosenthal theorem for elliptic operators on smooth Riemannian vector bundles. The second main result is a generalization of Putnam's inequality to higher dimensions that deals with Riesz transforms models of seminormal self-adjoint systems of Hilbert space operators. Both results share a common origin illustrated in the similar way they are derived from uniform estimates of integral operators.

1. Introduction

The goal of this article is to explain, in a several variable setting, the connection between quantitative Hartogs-Rosenthal type theorems concerning uniform approximation on compact sets by solutions of elliptic equations and Putnam type commutator inequalities for seminormal systems of Hilbert space operators. In the case of a single variable, our main results are related to an inequality in rational approximation due to Alexander, and a spectral area estimate for the norm of the self-commutator of a seminormal operator due to Putnam.

Specifically, Alexander's inequality [Al1], [Al2] states that if Ω is a compact set in the complex plane and $C(\Omega)$ is the Banach algebra of complex-valued continuous functions on Ω, then

$$\mathrm{dist}_{C(\Omega)}[\bar{z}, R(\Omega)] \leq \left[\frac{1}{\pi}\mathrm{area}(\Omega)\right]^{1/2}, \tag{1.1}$$

where $\bar{z}$ is the complex conjugate coordinate function, and $R(\Omega)$ stands for the uniform closure in $C(\Omega)$ of rational functions that are analytic on open neighborhoods of Ω. The connection with the classical Hartogs-Rosenthal theorem [HR] is obvious. The latter is just a qualitative consequence of inequality (1.1). Whenever

Received by the editors February 2, 2003. Submitted by M. Klaus.
Mathematics Subject Classification (2000). Primary 42B20, 47B20.
Key words and phrases. Elliptic operators, hyponormal operators, Putnam's inequality, Riesz transforms.

area(Ω) equals zero, from (1.1) it follows that the closed subalgebra $R(\Omega)$ contains $\bar{z}$ and because z is also in $R(\Omega)$, by the Stone-Weierstrass theorem one gets $R(\Omega) = C(\Omega)$.

In its turn, Putnam's theorem [Pm2] states that if T is a bounded Hilbert space operator and its self-commutator, $[T^*, T] = T^*T - TT^*$, is semidefinite, then

$$\|[T^*, T]\| \leq \frac{1}{\pi}\text{area}(\text{spec}\, T), \tag{1.2}$$

where $\text{spec}\, T$ denotes the spectrum of T.

The similarity of inequalities (1.1) and (1.2) is not accidental. It illustrates once more the deep connection between one-variable complex analysis and operator theory. In one direction, Axler and Shapiro [AS] proved that Putnam's theorem for a subnormal operator, that is, the restriction of a normal operator to an invariant subspace, is a consequence of Alexander's inequality. In the opposite direction, relying on (1.2) and other subtle properties of seminormal operators, Putinar [Pr] generalized Alexander's inequality for arbitrary smooth functions $u : \mathbb{C} \to \mathbb{C}$ by proving that

$$\text{dist}_{C(\Omega)}[u, R(\Omega)] \leq \frac{2}{\sqrt{\pi}}\|\bar{\partial}u\|_{\Omega,\infty}^{1/2}\|\bar{\partial}u\|_{\Omega,1}^{1/2}, \tag{1.3}$$

where $\bar{\partial} = \partial/\partial\bar{z}$ and $\|\cdot\|_{\Omega,p}$ stand for the standard norms in the Lebesgue spaces of complex-valued functions $L^p(\Omega)$, $p = \infty$ or $p = 1$.

In this article we are going to formulate and prove generalizations of inequalities (1.1), (1.2), and (1.3) to higher dimensions. The generalization of (1.1) and (1.3) is stated in Section 2 as Theorem A and the generalization of (1.2) is stated in Section 3 as Theorem B. In contrast to the previous comments, instead of deriving one theorem from another, we will give independent direct proofs. However, it is exactly in their proofs where the expected link could be found. Both theorems are derived from a single general result proved in Section 4 that is concerned with uniform estimates for integral operators.

In Section 2 we will assume that Ω is a compact set in a Riemannian manifold M and $\mathfrak{D}$ is an elliptic operator on a Riemannian vector bundle E over M. Theorem A provides an estimate in the Banach space of continuous sections from Ω to E of the distance from a smooth global section u to the solution subspace of $\mathfrak{D}$ on Ω. That estimate is in terms of $\|\mathfrak{D}u\|_{\Omega,\infty}$ and $\|\mathfrak{D}u\|_{\Omega,1}$, and involves a universal constant associated in an explicit way with $\mathfrak{D}$. As typical examples one may use the Euclidean Dirac operator on $\mathbb{R}^{m+1}$ applied to functions with values in the Clifford algebra associated with $\mathbb{R}^m$, or the $\bar{\partial}$ operator on $\mathbb{C}^n$ applied to differential forms of type $(p, n-1)$, $0 \leq p \leq n$. If $m = 1$, or $n = 1$, we recover the $\bar{\partial}$ operator on $\mathbb{C}$, as used in inequality (1.3).

In Section 3 we will start with a specific form of Putnam's theorem for the Hilbert transform model of pure seminormal operators and then introduce a natural multi-dimensional generalization of that model referred to as the Riesz transforms model. The generalized model involves operators of multiplication by the coordinate functions and the Riesz operators on a direct integral Hilbert space over a compact set Ω in $\mathbb{R}^n$, $n \geq 2$. Theorem B provides a norm estimate of the commutator matrix associated with such a model in terms of the primary parameter involved in that model and the n-dimensional volume of Ω.

The complete proofs of Theorems A and B are given in Section 4 based on a technical result that pinpoints their common origin.

Each section also includes several references and comments that indicate contributions and part of the work done over the years in both areas, approximation theory by solutions of elliptic equations and the theory of seminormal operators.

As a concluding remark and, at the same time, as an answer to a question raised by the reviewer of this article, we would like to point out that Theorem A can be used to set up a Putnam type inequality in higher dimensions for a subnormal tuple by extending in a natural way the approach developed by Axler and Shapiro [AS] in the one-dimensional case. Specifically, we apply Theorem A to the $\bar{\partial}$ operator on $\mathbb{C}^n$ acting on the space of differential forms of type $(0, n-1)$, and then use the resulting inequality to estimate the self-commutator matrix of a subnormal tuple defined in terms of multiplication operators by the complex coordinate functions on spaces of closed forms. For more details and a complete proof, we refer to [Mn5].

2. Uniform Approximation by Solutions of Elliptic Equations

The problem of uniform approximation on compact sets by solutions of elliptic equations has a rather long history. Early contributions addressing the problem in a general setting are due to Malgrange [Me] and Browder [B1], [B2], The main result in this section, Theorem A, is a quantitative Hartogs-Rosenthal type theorem. The classical result, already mentioned in Section 1, and some other means of studying rational approximation are presented, for instance, in the monograph by Gamelin [G]. More general qualitative results in this area are discussed in the monograph by Tarkhanov [T, Chapter 6]. Other additional references are included at the end of this section right after the statement of Theorem A.

2.1. The Setting

Suppose M is a smooth connected noncompact Riemannian manifold of dimension n equipped with the Riemannian volume measure and let E be a smooth real Riemannian or complex Hermitian vector bundle over M.

The space of all smooth sections or compactly supported smooth sections from M to E will be denoted by $\mathcal{E}(M, E)$ and $\mathcal{D}(M, E)$, respectively. The fiber of E at a point $x \in M$ is denoted in what follows by E_x. For a later use, we also introduce the vector bundle $\mathcal{L}(E) = \mathcal{L}(E, E)$ over $M \times M$ whose fiber at a point $(x, y) \in M \times M$ equals the space $\mathcal{L}(E_y, E_x)$ consisting of linear maps from E_y into E_x. The norm in E_x and the operator norm in $\mathcal{L}(E_y, E_x)$ are denoted by $\|\cdot\|_x$ and $\|\cdot\|_{x,y}$.

Further, let $\Omega \subseteq M$ be a compact set and denote by $\mathcal{C}(\Omega, E)$ the Banach space of all continuous sections from Ω to the restriction of the bundle E to Ω, with the uniform norm given by

$$\|u\|_{\Omega,\infty} = \sup_{x\in\Omega} \|u(x)\|_x, \qquad u \in \mathcal{C}(\Omega, E).$$

We also introduce the norm $\|\cdot\|_{\Omega,p}$, $1 \le p < \infty$, defined as

$$\|u\|_{\Omega,p} = \left(\int_\Omega \|u(x)\|_x^p dx \right)^{1/p}, \qquad u \in \mathcal{C}(\Omega, E).$$

Next, suppose that $\mathfrak{D} : \mathcal{E}(M, E) \to \mathcal{E}(M, E)$ is an elliptic differential operator of order α, $1 \le \alpha \le n-1$, such that its transpose, $\mathfrak{D}'$, has the unique continuation property. We associate to $\mathfrak{D}$ and each compact set $\Omega \subseteq M$ the solution subspace $\mathcal{S}_\mathfrak{D}(\Omega, E)$ of $\mathcal{C}(\Omega, E)$ defined as the uniform closure in $\mathcal{C}(\Omega, E)$ of the set consisting of restrictions to Ω of sections u from $\mathcal{E}(M, E)$ such that $\mathfrak{D}u = 0$ on certain open neighborhoods of Ω in M that may vary from a section to another.

For more details on the unique continuation property we refer to the monograph by Tarkhanov [T], where this property is called *the uniqueness condition for the Cauchy problem in the small*, in short, $(U)_S$. As pointed out there, if the transpose $\mathfrak{D}'$ of $\mathfrak{D}$ has property $(U)_S$, then $\mathfrak{D}$ has a right fundamental solution, i.e., a smooth section $\Phi : M \times M \setminus \Delta \to \mathcal{L}(E, E)$ where $\Delta = \{(x, x) : x \in M\}$, such that

(i) $\Phi(x, y) : E_y \to E_x$ is a linear map for every $(x, y) \in M \times M \setminus \Delta$;
(ii) the associated singular integral operator $\mathfrak{I} : \mathcal{D}(M, E) \to \mathcal{E}(M, E)$ given by

$$\mathfrak{I}u(x) = \lim_{\rho\downarrow 0} \int_{M\setminus B(x,\rho)} \Phi(x, y)u(y)d\mathrm{vol}(y), \qquad x \in M, \tag{2.1}$$

where $B(x, \rho)$ stands for the closed geodesic ball in M centered at $x \in M$ with radius $\rho > 0$ is well defined;
(iii) $\mathfrak{I}$ has the property

$$\mathfrak{D}\mathfrak{I}u = u, \qquad u \in \mathcal{D}(M, E). \tag{2.2}$$

In other words, $\mathfrak{I}$ is a right inverse to $\mathfrak{D}$.

2.2. Auxiliary Results

Suppose that $\mathfrak{D}$ and Ω are as in 2.1 and Φ is a right fundamental solution of $\mathfrak{D}$. Operator $\mathfrak{I}$ defined by (2.1) makes sense for more general sections u, not necessarily smooth. In particular, we can define an operator $\mathfrak{I}_\Omega : \mathcal{E}(M, E) \to \mathcal{C}(\Omega, E)$ by setting $\mathfrak{I}_\Omega(\cdot) = \mathfrak{I}(\chi_\Omega \cdot)|_\Omega$, with χ_Ω the characteristic function of Ω, that is,

$$\mathfrak{I}_\Omega u(x) = \int_\Omega \Phi(x, y)u(y)d\mathrm{vol}(y), \qquad u \in \mathcal{E}(M, E),\ x \in \Omega, \tag{2.3}$$

where the integral has the same meaning as in (2.1).

The next result is a preliminary step in estimating the distance to the solution subspace $\mathcal{S}_\mathcal{D}(\Omega, E)$.

Lemma 1. *If $u \in \mathcal{E}(M,E)$, then*

$$\operatorname{dist}_{C(\Omega,E)}[u, \mathcal{S}_{\mathfrak{D}}(\Omega,E)] \leq \|\mathfrak{I}_\Omega(\mathfrak{D}u)\|_{\Omega,\infty}. \tag{2.4}$$

Proof. Let χ_n, $n \geq 1$, be a sequence of smooth compactly supported functions on M taking values in $[0,1]$ with each χ_n equal to 1 on an open relatively compact neighborhood U_n of Ω in M, $\operatorname{supp}\chi_{n+1} \subseteq \operatorname{supp}\chi_n$, and such that

$$\bigcap_{n\geq 1} \operatorname{supp}\chi_n = \Omega. \tag{2.5}$$

Suppose $u \in \mathcal{E}(M,E)$ is given and let $u_n \in \mathcal{E}(M,E)$ be defined as

$$u_n = u - \mathfrak{I}(\chi_n \mathfrak{D}u), \qquad n \geq 1.$$

By (2.2) we get

$$\mathfrak{D}u_n = \mathfrak{D}u - \chi_n\mathfrak{D}u = (1-\chi_n)\mathfrak{D}u,$$

whence $\mathfrak{D}u_n|_{U_n} = 0$, that is, $u_n|_\Omega \in \mathcal{S}_{\mathfrak{D}}(\Omega,E)$. Consequently,

$$\operatorname{dist}_{C(\Omega,E)}[u, \mathcal{S}_{\mathfrak{D}}(\Omega,E)] \leq \|u-u_n\|_{\Omega,\infty} = \|\mathfrak{I}(\chi_n\mathfrak{D}u)\|_{\Omega,\infty}. \tag{2.6}$$

Inequality (2.4) follows from (2.6) and (2.5) by taking the limit. The proof is complete. □

Next we introduce the function $\omega : M \times (0,\infty) \to (0,\infty)$ defined as

$$\omega(x,t) = \operatorname{vol}\{y \in M : y \neq x,\ \|\Phi(x,y)\|_{x,y} \geq t\}, \tag{2.7}$$

and set

$$A(\mathfrak{D}) = \frac{\kappa}{\kappa-1}\sup_{x\in M}\sup_{t>0} t[\omega(x,t)]^{1/\kappa}, \tag{2.8}$$

where $\kappa = n/(n-\alpha)$. For the operator $\mathfrak{I}_\Omega$ we have the following estimate.

Lemma 2. *If $u \in \mathcal{E}(M,E)$, then*

$$\|\mathfrak{I}_\Omega u(x)\|_x \leq A(\mathfrak{D})\|u\|_{\Omega,\infty}^{1/\kappa}\|u\|_{\Omega,1}^{1-1/\kappa}, \qquad x \in \Omega. \tag{2.9}$$

The proof of Lemma 2 is given in Section 4.

2.3. Main Results

Combining Lemmas 1 and 2 we get the following estimate of the distance in $C(\Omega,E)$ from a smooth global section u to the solution subspace $\mathcal{S}_{\mathfrak{D}}(\Omega,E)$.

Theorem A. *Suppose $\mathfrak{D} : \mathcal{E}(M,E) \to \mathcal{E}(M,E)$ is an elliptic operator of order α, $1 \leq \alpha \leq n-1$, and its transpose $\mathfrak{D}'$ has the unique continuation property. If $\Omega \subseteq M$ is a compact set and $u \in \mathcal{E}(M,E)$, then*

$$\operatorname{dist}_{C(\Omega,E)}[u, \mathcal{S}_{\mathfrak{D}}(\Omega,E)] \leq A(\mathfrak{D})\|\mathfrak{D}u\|_{\Omega,\infty}^{1/\kappa}\|\mathfrak{D}u\|_{\Omega,1}^{1-1/\kappa}, \tag{2.10}$$

where $A(\mathfrak{D})$ is as in (2.8) and $\kappa = n/(n-\alpha)$.

Since the space consisting of restrictions to Ω of sections from $\mathcal{E}(M,E)$ is dense in $C(\Omega,E)$, from Theorem A we get the next Hartog-Rosenthal type theorem.

Corollary A. *If $\Omega \subseteq M$ is a compact set of volume zero, then $\mathcal{S}_{\mathfrak{D}}(\Omega,E) = C(\Omega,E)$.*

As basic example of differential operators that fall within the framework outlined above, we mention the $\bar{\partial}$ operator on $\mathbb{C}$ applied to complex-valued functions, the $\bar{\partial}$ operator on $\mathbb{C}^n$, $n \geq 2$, applied to differential forms of type $(p, n-1)$, $0 \leq p \leq n$, the Laplace operator on $\mathbb{R}^n$, $n \geq 3$, or the Dirac operator on $\mathbb{R}^{m+1}$, $m \geq 1$, applied to functions with values in a module over the Clifford algebra associated with $\mathbb{R}^m$.

The corresponding specific forms of Collorary A have been previously established using a different approach in each case by Hartogs and Rosenthal [HR], when $M = \mathbb{C}$, $E = M \times \mathbb{C}$, and $\mathfrak{D} = \bar{\partial}$; Aizenberg and Dautov [AD], when $M = \mathbb{C}^n$, $E = M \times \Lambda^{*,n-1}(\mathbb{C}^n)$, and $\mathfrak{D} = \bar{\partial}$; Weinstock [W], when $M = \mathbb{R}^n$, $E = M \times \mathbb{R}^k$, and $\mathfrak{D}$ is an elliptic operator with constant coefficients; and Bracks, Delanghe, and Sommen [BDS], when $M = \mathbb{R}^{m+1}$, $E = M \times \mathfrak{H}$, with $\mathfrak{H}$ a Clifford bundle, and $\mathfrak{D} = D$, the Dirac operator.

Specific forms of Theorem A that of course generalize the results referred to above have been proved by Alexander [Al1], [Al2], when $M = \mathbb{C}$, $E = M \times \mathbb{C}$, $\mathfrak{D} = \bar{\partial}$, and $u(z) = \bar{z}$, the complex conjugate coordinate function on $\mathbb{C}$; Khavinson [Kh1], [Kh2], and Gustafsson and Khavinson [GK], when $M = \mathbb{R}^n$, $E = M \times \mathbb{R}^k$, and $\mathfrak{D}$ is a second-order elliptic operator; Putinar [Pr], when $M = \mathbb{C}$, $E = M \times \mathbb{C}$, and $\mathfrak{D} = \bar{\partial}$; Martin [Mn2], [Mn3], when $M = \mathbb{R}^{m+1}$, $E = M \times \mathfrak{H}$, with $\mathfrak{H}$ a Clifford bundle, and $\mathfrak{D} = D$, the Euclidean Dirac operator; Martin and Szeptycki [MnSz], when $M = \mathbb{R}^n$, $E = M \times \mathbb{R}^k$, and $\mathfrak{D}$ is an elliptic operator with constant coefficients; Martin [Mn5], when $M = \mathbb{C}^n$, $E = M \times \Lambda^{*,n-1}(\mathbb{C}^n)$, and $\mathfrak{D} = \bar{\partial}$; and Martin [Mn6] when $\mathfrak{D}$ is the Dirac operator on a Clifford bundle.

A more refined version of Corollary A in the space of sections on Ω of class C^p, $p \leq \alpha - 1$, when $M = \mathbb{R}^n$, $E = \mathbb{R}^n \times \mathbb{R}^k$, and $\mathfrak{D}$ is a differential operator of order α with surjective symbol and constant coefficients was proved by Tarkhanov [T, Chapter 6]. This result can be also derived from Theorem A.

The case of differential operators on $\mathbb{R}^n$ with constant matrix coefficients deserves an additional comment. For such an operator $\mathfrak{D}$, the fundamental solution Φ is two-sided and translation invariant, so one has $\Phi(x, y) = \Phi_0(x - y)$, $x \neq y$, where Φ_0 is a smooth matrix-valued function on $\mathbb{R}^n_0 = \mathbb{R}^n \setminus \{0\}$, homogeneous of degree $\alpha - n$. Consequently, the corresponding function $\omega : \mathbb{R}^n \times (0, \infty) \to (0, \infty)$ defined by (2.7) has the next simple form,

$$\omega(x, t) = \lambda t^{-(n-\alpha)/n}, \qquad x \in \mathbb{R}^n, \ t \in (0, \infty), \tag{2.11}$$

with $\lambda = \mathrm{vol}(\Lambda)$, where $\Lambda \subseteq \mathbb{R}^n$ is given by

$$\Lambda = \mathrm{vol}\{x \in \mathbb{R}^n_0 : \|\Phi_0(x)\| \geq 1\} \cup \{0\}. \tag{2.12}$$

From (2.8), (2.11), and (2.12) we get that the constant $A(\mathfrak{D})$ used in (2.9) and (2.10) can be expressed as

$$A(\mathfrak{D}) = \frac{n}{\alpha} \cdot \lambda^{(n-\alpha)/n}. \tag{2.13}$$

More details and a proof of Theorem A in this case including also the observation that inequality (2.9) with $A(\mathfrak{D})$ as in (2.13) is sharp can be found in Martin and Szeptycki [MnSz].

3. Seminormality in Higher Dimension

The theory of seminormal, i.e., hyponormal or cohyponormal operators, originated in connection with pairs (X, Y) of self-adjoint operators satisfying the commutator condition $\sqrt{-1}[X, Y] \geq 0$, which in terms of the operator $T = X + \sqrt{-1}Y$ and its adjoint $T^* = X - \sqrt{-1}Y$ reduces to $[T^*, T] \geq 0$.

The early contributions to the field are presented in the monograph by Putnam [Pn1]. Subsequent monographs by Clancey [Cl], Xia [X2], Martin and Putinar [MnPr], and Conway [Co] give accounts of the development of the one-variable theory at later stages. Steps towards a multi-variable theory are indicated in Athavale [At]; Cho, Curto, Huruya, and Zelazko [CCHZ]; Curto [Cu]; Curto and Jian [CJ]; Curto, Muhly, and Xia [CMX]; Douglas, Paulsen, and Yan [DPY]; Martin [Mn1], [Mn4]; Martin and Salinas [MnSa]; McCullough and Paulsen [McCP]; and Xia [X3], to mention just a few of the contributors in this area.

Our goal in this section is twofold. After introducing a definition of seminormality for pairs (X, Y) of self-adjoint n-tuples $X = (X_1, X_2, \ldots, X_n)$ and $Y = (Y_1, Y_2, \ldots, Y_n)$ of Hilbert space operators we describe a singular integral model of such pairs referred to as the Riesz transforms model, and then formulate the main result, Theorem B, which is a substitute for Putnam's inequality in regard to that model.

3.1. Notation and Definitions

We let $\mathcal{H}$ be a complex infinite-dimensional Hilbert space and denote by $\mathcal{L}(\mathcal{H})$ the C^*-algebra of all bounded linear operators on $\mathcal{H}$. The adjoint of $T \in \mathcal{L}(\mathcal{H})$ is denoted by T^* and the real and imaginary parts of T are defined as $\mathrm{Re}(T) = (T + T^*)/2$, $\mathrm{Im}(T) = (T - T^*)/2\sqrt{-1}$.

If $T = (T_1, T_2, \ldots, T_n)$ is an n-tuple of operators in $\mathcal{L}(\mathcal{H})$, we introduce the *self-commutator* of T as the $n \times n$ operator matrix given by

$$\mathcal{C}(T) = ([T_i^*, T_j])_{i,j=1}^n, \tag{3.1}$$

and regarded as an operator on the space $\mathbb{C}^n \otimes \mathcal{H}$.

We next take the self-adjoint n-tuples $X = (X_1, X_2, \ldots, X_n)$ and $Y = (Y_1, Y_2, \ldots, Y_n)$, where $X_i = \mathrm{Re}(T_i)$ and $Y_i = \mathrm{Im}(T_i)$, $1 \leq i \leq n$, and introduce

$$\mathcal{C}(X, Y) = \sqrt{-1}([X_i, Y_j] + [X_j, Y_i])_{i,j=1}^n. \tag{3.2}$$

In the case $n = 1$, we obviously have $\mathcal{C}(T) = \mathcal{C}(X, Y) = -\mathcal{C}(T^*)$, so basically we are dealing with a single commutator. If $n \geq 2$, these properties are no longer true, and although the three commutators are related to each other, they produce different operators on $\mathbb{C}^n \otimes \mathcal{H}$. Specifically, $\mathcal{C}(T)$ and $\mathcal{C}(T^*)$ are in general linearly independent, $\mathcal{C}(T^*)$ is the formal transpose of $-\mathcal{C}(T)$, and

$$\mathcal{C}(X,Y) = \frac{1}{2}(\mathcal{C}(T) - \mathcal{C}(T^*)). \tag{3.3}$$

The terminology used below is the same as in [Mn1], [Mn4], and [MnSa].

Definition 1. An n-tuple T of operators on $\mathcal{H}$ is called *right*, or *left seminormal* provided $\mathcal{C}(T)$ or $\mathcal{C}(T^*)$ is semidefinite as an operator on $\mathbb{C}^n \otimes \mathcal{H}$. In the case $\mathcal{C}(T) \geq 0$, or $\mathcal{C}(T) \leq 0$, the n-tuple T is said to be *right hyponormal*, or *right cohyponormal*, respectively, and when $\mathcal{C}(T^*) \geq 0$, or $\mathcal{C}(T^*) \leq 0$, the n-tuple T is called *left cohyponormal*, or *left hyponormal*, respectively.

If $n = 1$, we recover the standard definitions of seminormal, hyponormal, and cohyponormal operators. The concepts of joint hyponormality and t-hyponormality introduced in [At] and [X3] correspond in our terminology to left hyponormality and right hyponormality, respectively.

Definition 2. A pair of self-adjoint n-tuples X and Y of operators on $\mathcal{H}$ is called *jointly seminormal* provided its commutator $\mathcal{C}(X,Y)$ is semidefinite as an operator on $\mathbb{C}^n \otimes \mathcal{H}$. In the case when $\mathcal{C}(X,Y) \geq 0$, or $\mathcal{C}(X,Y) \leq 0$, the pair (X,Y) is said to be *jointly hyponormal*, or *jointly cohyponormal*, respectively.

The adjoint $T^* = (T_1^*, T_2^*, \dots, T_n^*)$ of a right hyponormal, or cohyponormal, tuple T is left cohyponormal, or hyponormal. We note that the self-adjoint pair (X,Y) associated with an n-tuple T which is simultaneously left and right hyponormal, or left and right cohyponormal, turns out to be jointly hyponormal, or jointly cohyponormal, as it follows from (3.3).

3.2. The Riesz Transforms Model

We now introduce a singular integral model of jointly seminormal pairs of self-adjoint tuples of operators that generalizes to higher dimension the Hilbert transform model discovered by Xia [X1] and Pincus [Pi] for operators with one-dimensional self-commutators and afterwards set up in full generality by Kato [Ka] and Muhly [My].

We first recall that the Riesz transforms of a complex-valued function u in the Schwartz space $\mathcal{S}(\mathbb{R}^n, \mathbb{C})$ are defined by the principal value integrals

$$R_i u(x) = \frac{\Gamma((n+1)/2)}{\pi^{(n+1)/2}} \text{p.v.} \int_{\mathbb{R}^n} \frac{x_i - y_i}{|x-y|^{n+1}} u(y)dy, \qquad 1 \leq i \leq n,\ x \in \mathbb{R}^n. \tag{3.4}$$

For the ease of notation, we let k_i denote the kernels

$$k_i(x) = \frac{\Gamma((n+1)/2)}{\pi^{(n+1)/2}} \cdot \frac{x_i}{|x|^{n+1}}, \qquad 1 \leq i \leq n,\ x \in \mathbb{R}_0^n = \mathbb{R}^n \setminus \{0\}, \tag{3.5}$$

and write (3.4) as $R_i u(x) = k_i * u(x)$, $1 \leq i \leq n$.

In addition, we introduce the operators M_i on $\mathcal{S}(\mathbb{R}^n, \mathbb{C})$ defined as

$$M_i u(x) = x_i u(x), \qquad 1 \leq i \leq n,\ x = (x_1, x_2, \dots, x_n) \in \mathbb{R}^n. \tag{3.6}$$

A simple calculation shows that the commutators $[M_i, R_j]$, $1 \le i, j \le n$, are singular integral operators, namely,

$$[M_i, R_j]u(x) = k_{ij} * u(x), \qquad u \in \mathcal{S}(\mathbb{R}^n, \mathbb{C}),\ x \in \mathbb{R}^n, \tag{3.7}$$

where the kernels k_{ij} are given by

$$k_{ij}(x) = x_i k_j(x) = x_j k_i(x), \qquad x \in \mathbb{R}^n_0, \tag{3.8}$$

with k_i and k_j as in (3.5). In particular, we have $[M_i, R_j] = [M_j, R_i]$, $1 \le i, j \le n$.

We next form the commutator $n \times n$ matrix $\mathcal{C} = ([M_i, R_j])^n_{i,j=1}$, and consider the corresponding operator on the Schwartz space $\mathcal{S}(\mathbb{R}^n, \mathbb{C}^n)$ of $\mathbb{C}^n$-valued functions on $\mathbb{R}^n$.

Finally, on $\mathcal{S}(\mathbb{R}^n, \mathbb{C})$ we take the inner product inherited from the standard Lebesgue space $L^2(\mathbb{R}^n)$ of square integrable complex-valued functions on $\mathbb{R}^n$, and then convert $\mathbb{C}^n \otimes \mathcal{S}(\mathbb{R}^n, \mathbb{C})$ into an inner product space in the usual way. We are now in a position to formulate the following result.

Lemma 3. *The operator $\mathcal{C} = ([M_i, R_j])^n_{i,j=1}$ is positive semidefinite, that is,*

$$\langle \mathcal{C}u, u \rangle \ge 0 \tag{3.9}$$

for any $u \in \mathcal{S}(\mathbb{R}^n, \mathbb{C}^n) \equiv \mathbb{C}^n \otimes \mathcal{S}(\mathbb{R}^n, \mathbb{C})$.

A complete proof of Lemma 3 based on a Fourier transform argument can be found in Martin [Mn4]. To make a point, we just want to notice that by taking the Fourier transforms of the kernels (3.5) we get

$$(R_i u)\hat{}\,(\xi) = -\sqrt{-1}\xi_i |\xi|^{-1} \hat{u}(\xi), \qquad u \in \mathcal{S}(\mathbb{R}^n, \mathbb{C}),\ \xi \in \mathbb{R}^n_0,$$

for each $1 \le i \le n$, a set of equations that immediately leads to several basic properties of the Riesz transforms. For instance, one gets that each R_i, $1 \le i \le n$, extends to a skew-adjoint operator on $L^2(\mathbb{R}^n)$, such that $\sum_{i=1}^n R_i^* R_i$ equals the identity on $L^2(\mathbb{R}^n)$ and $\sum_{i=1}^n \|R_i u\|^2 = \|u\|^2$ for any $u \in L^2(\mathbb{R}^n)$. More properties of Riesz transforms are discussed in Stein's monographs [S1], [S2].

As far as Lemma 3 is concerned, we need the Fourier transforms of the kernels k_{ij} defined by (3.8). Some calculations made in [Mn4] show that

$$\hat{k}_{ij}(\xi) = \delta_{ij} |\xi|^{-1} - \xi_i \xi_j |\xi|^{-3}, \qquad 1 \le i, j \le n,\ \xi \in \mathbb{R}^n_0. \tag{3.10}$$

Lemma 3 proves essential in developing the Riesz transforms model of a jointly seminormal pair (X, Y).

The two self-adjoint n-tuples involved in that model consist of operators on a direct integral Hilbert space $\mathfrak{H}$. To introduce it suppose $\mathfrak{H}$ is an infinite-dimensional separable complex Hilbert space and

$$\{0\} = \mathfrak{H}_0 \subset \mathfrak{H}_1 \subset \mathfrak{H}_2 \subset \mathfrak{H}_3 \subset \cdots \subset \mathfrak{H}$$

is a chain of subspaces such that

(i) $\dim \mathfrak{H}_k = k$, $k \ge 0$,
(ii) $\bigcup_{k \ge 0} \mathfrak{H}_k$ spans $\mathfrak{H}$.

For every $x \in \mathbb{R}^n$, let $\mathfrak{H}(x)$ be one of the subspaces $\mathfrak{H}_k$, $k \geq 0$, such that

(iii) $\Omega_k = \{x \in \mathbb{R}^n : \mathfrak{H}(x) = \mathfrak{H}_k\}$, $k \geq 0$, are Borel sets, and

(iv) $\Omega = \bigcup_{k\geq 1} \Omega_k$ is compact.

Next we form the direct integral Hilbert space

$$\mathcal{H} = \int_{\mathbb{R}^n}^{\oplus} \mathfrak{H}(x)dx = \int_{\Omega}^{\oplus} \mathfrak{H}(x)dx$$

consisting of classes of weakly measurable functions $u \in L^2(\mathbb{R}^n, \mathfrak{H})$ such that

$$u(x) \in \mathfrak{H}(x) \qquad \text{a.e.}$$

Let $P_\Omega : \mathbb{R}^n \to \mathcal{L}(\mathfrak{H})$ be the projection-valued map that assigns to every $x \in \mathbb{R}^n$ the orthogonal projection onto $\mathfrak{H}(x)$.

Finally, we take $n+1$ operator-valued functions $a_1, a_2, \ldots, a_n, b \in L^\infty(\mathbb{R}^n, \mathcal{L}(\mathfrak{H}))$ such that

$$\begin{aligned} a_i(x) &= P_\Omega(x)a_i(x)P_\Omega(x) = a_i(x)^* \quad \text{a.e.}, \qquad 1 \leq i \leq n, \\ b(x) &= P_\Omega(x)b(x)P_\Omega(x) \quad \text{a.e.} \end{aligned} \tag{3.11}$$

We will refer to the function b as the *primary parameter*, and to the functions $a_1, a_2, \ldots, a_n$ as the *secondary parameters*. Based on (3.11) we define $n+1$ operators $A_1, A_2, \ldots, A_n, B \in \mathcal{L}(\mathcal{H})$ as

$$\begin{aligned} A_i u(x) &= a_i(x)u(x), \qquad u \in \mathcal{H},\ 1 \leq i \leq n,\ x \in \Omega, \\ Bu(x) &= b(x)u(x), \qquad u \in \mathcal{H},\ x \in \Omega. \end{aligned} \tag{3.12}$$

We now introduce the self-adjoint n-tuples $X = (X_1, X_2, \ldots, X_n)$ and $Y = (Y_1, Y_2, \ldots, Y_n)$ of operators on $\mathcal{H}$ by setting $X_i = M_i \mid \mathcal{H}$, $1 \leq i \leq n$, with M_i given by (3.6), and $Y_i = A_i - \sqrt{-1}B^* P_\Omega R_i P_\Omega B$, $1 \leq i \leq n$, where R_i are the Riesz operators defined by (3.4).

We claim that the so-defined pair (X, Y) is jointly hyponormal, that is, the corresponding $n \times n$ commutator matrix $\mathcal{C}(X, Y)$ given by (3.2) is positive semidefinite. To prove our claim we first observe that

$$[X_i, Y_j] = [X_j, Y_i] = -\sqrt{-1}B^* P_\Omega [M_i, R_j] P_\Omega B, \qquad 1 \leq i, j \leq n.$$

Substituting these equations into (3.2) we get

$$\mathcal{C}(X, Y) = 2\left(B^* P_\Omega [M_i, R_j] P_\Omega B\right)_{i,j=1}^n . \tag{3.13}$$

It remains to observe that whenever $u \in \mathbb{C}^n \otimes \mathcal{H} \subset \mathbb{C}^n \otimes L^2(\mathbb{R}^n, \mathfrak{H})$ we get $\langle \mathcal{C}(X, Y)u, u\rangle = 2\langle \mathcal{C}_\Omega Bu, Bu\rangle$, where $\mathcal{C}_\Omega$ is the compression of the operator $\mathcal{C}$ in Lemma 3 to $\mathbb{C}^n \otimes L^2(\Omega, \mathfrak{H})$. Since u can be approximated by restrictions to Ω of functions from $\mathbb{C}^n \otimes \mathcal{S}(\mathbb{R}^n, \mathbb{C}^n)$, Lemma 3 shows that $\mathcal{C}_\Omega \geq 0$, hence $\mathcal{C}(X, Y) \geq 0$.

The reader should be aware that in the case when $n = 1$ the previous model coincides with the singular integral model of pure hyponormal operators. If $n = 1$, there is just one Riesz operator, the Hilbert transform. For details, we refer to [Cl], [Ka], [My], [Ps], [PX], [PXX], and [X1], [X2].

3.3. Putnam-Type Inequalities

We will refer to (1.2) as the one-dimensional form of Putnam's inequality for a seminormal operator T. Set $X = \operatorname{Re}(T)$ and $Y = \operatorname{Im}(T)$. A well-known result due also to Putnam and proved, for instance, in [Pm1], states that the projections of $\operatorname{spec}(T)$ onto the real and imaginary axes are equal to $\operatorname{spec} X$ and $\operatorname{spec} Y$, respectively. Therefore, (1.2) leads to the Cartesian form of Putnam's inequality,

$$\|\mathcal{C}(X,Y)\| \leq \frac{2}{\pi}\|Y\| \cdot \operatorname{length}(\operatorname{spec} X), \tag{3.14}$$

where we used the projection property recalled above in conjunction with $\operatorname{spec} Y \subseteq [-\|Y\|, \|Y\|]$.

The original proof of (1.2) in Putnam [Pu2] uses the Cartesian form (3.14) as a starting point and Putnam's cut-down technique, so (1.2) and (3.14) can be derived one from the other.

There is also a Cartesian form of Putnam's inequality for the Hilbert transform model. Using the notation in 3.2, that amounts to

$$\|\mathcal{C}(X,Y)\| \leq \frac{2}{\pi} \operatorname*{ess\,sup}_{x\in\Omega} \|b(x)\|^2 \cdot \operatorname{length}(\Omega), \tag{3.15}$$

where b is the primary parameter of the model. The proof of (3.15) is straightforward, and (3.15) turns out to be a refined version of (3.14), because $\Omega = \operatorname{spec} X$ and $\operatorname{ess\,sup}_{x\in\Omega} \|b(x)\|^2 = \|B\|^2 \leq \|Y\|$. For more details we refer to [Cl, Chapter 2]. However, it is important to notice that every pure seminormal operator is unitarily equivalent to a Hilbert transform model, therefore (3.15) implies (3.14) and also (1.2) for arbitrary seminormal operators.

The next result generalizes (3.15) for the Riesz transforms model.

Lemma 4. *Let X and Y be the self-adjoint tuples of the Riesz transform model introduced in 3.2. Then*

$$\|\mathcal{C}(X,Y)\| \leq \frac{2\Gamma\left(\dfrac{n+1}{2}\right)}{\pi^{(n+1)/2}} \sup_{x\in\Omega} \int_\Omega \frac{\|b(y)\|^2_{\mathcal{L}(\mathfrak{H})}}{|x-y|^{n-1}} dy. \tag{3.16}$$

Proof. Assume that $u = (u_1, u_2, \ldots, u_n) \in \mathbb{C}^n \otimes \mathcal{H} \subseteq \mathbb{C}^n \otimes L^2(\Omega, \mathfrak{H})$ is fixed.

By (3.13), (3.7), (3.8), and (3.5) we get the explicit equation

$$\langle \mathcal{C}(X,Y)u, u\rangle$$
$$= \frac{2\Gamma\left(\dfrac{n+1}{2}\right)}{\pi^{(n+1)/2}} \sum_{i,j=1}^n \int_\Omega \int_\Omega \frac{\langle (x_j - y_j)b(y)u_j(y), (x_i - y_i)b(x)u_i(x)\rangle}{|x-y|^{n+1}} dy dx, \tag{3.17}$$

where the inner products in the double integral are considered in the Hilbert space $\mathfrak{H}$.

We clearly have

$$|\langle (x_j - y_j)b(y)u_j(y), (x_i - y_i)b(x)u_i(x)\rangle_{\mathfrak{H}}| \\ \leq |x_i - y_i||x_j - y_j|\|b(x)\|_{\mathcal{L}(\mathfrak{H})}\|b(y)\|_{\mathcal{L}(\mathfrak{H})}\|u_i(x)\|_{\mathfrak{H}}\|u_j(y)\|_{\mathfrak{H}}, \tag{3.18}$$

for all $(x, y) \in \Omega \times \Omega$ and $1 \leq i, j \leq n$. In addition, using Cauchy inequality we get

$$\sum_{i=1}^{n} |x_i - y_i|\|u_i(x)\|_{\mathfrak{H}} \leq |x - y|\|u(x)\|_{\mathbb{C}^n \otimes \mathfrak{H}}, \tag{3.19}$$

and similarly

$$\sum_{i=1}^{n} |x_j - y_j|\|u_j(y)\|_{\mathfrak{H}} \leq |x - y|\|u(y)\|_{\mathbb{C}^n \otimes \mathfrak{H}}. \tag{3.20}$$

Since the left-hand side in (3.17) is nonnegative, based on (3.18), (3.19), and (3.20) we obtain

$$\|\mathcal{C}(X, Y)\| \leq \frac{2\Gamma\left(\dfrac{n+1}{2}\right)}{\pi^{(n+1)/2}} \int_\Omega \int_\Omega \frac{\|b(x)\|\|b(y)\|\|u(x)\|\|u(y)\|}{|x - y|^{n-1}} dy dx, \tag{3.21}$$

where the first two norms are in $\mathcal{L}(\mathfrak{H})$ and the last two in $\mathbb{C}^n \otimes \mathfrak{H}$. Let us now define two functions $\varphi, \psi \in L^2(\Omega \times \Omega)$ by

$$\varphi(x, y) = \frac{\|b(y)\|\|u(x)\|}{|x - y|^{(n-1)/2}}, \quad \psi(x, y) = \frac{\|b(x)\|\|u(y)\|}{|x - y|^{(n-1)/2}}, \tag{3.22}$$

and write (3.21) as

$$\langle \mathcal{C}(X, Y)u, u\rangle \leq \frac{2\Gamma\left(\dfrac{n+1}{2}\right)}{\pi^{(n+1)/2}} \langle \varphi, \psi\rangle_{L^2(\Omega\times\Omega)}. \tag{3.23}$$

By Cauchy-Schwarz inequality and observing that $\psi(x, y) = \varphi(y, x)$, $(x, y) \in \Omega\times\Omega$, we have

$$\langle \varphi, \psi\rangle_{L^2(\Omega\times\Omega)} \leq \|\varphi\|\|\psi\| = \|\varphi\|^2_{L^2(\Omega\times\Omega)}. \tag{3.24}$$

Finally, from (3.22) we obviously get

$$\|\varphi\|^2_{L^2(\Omega\times\Omega)} = \int_\Omega \|u(x)\|^2 \left\{\int_\Omega \frac{\|b(y)\|^2}{|x - y|^{n-1}} dy\right\} dx, \tag{3.25}$$

and by (3.23), (3.24), and (3.25) we conclude that

$$\langle \mathcal{C}(X, Y)u, u\rangle \leq \frac{2\Gamma\left(\dfrac{n+1}{2}\right)}{\pi^{(n+1)/2}} \left\{\sup_{x\in\Omega} \int_\Omega \frac{\|b(y)\|^2}{|x - y|^{n-1}} dy\right\} \|u\|^2,$$

an inequality that completes the proof of Lemma 4. □

The last auxiliary result we need is an estimate of the Riesz potential of order 1 involved in (3.16). To formulate it, we let $\nu : \mathbb{R}^n \to [0, \infty)$ denote the function given by

$$\nu(y) = \|b(y)\|_{\mathcal{L}(\mathfrak{H})}, \qquad y \in \mathbb{R}^n. \tag{3.26}$$

Lemma 5. *With the notation introduced above we have*

$$\int_{\mathbb{R}^n} \frac{\nu^2(y)}{|x-y|^{n-1}} dy \leq n[\mathrm{vol}(\mathbb{B}^n)]^{(n-1)/n} \|\nu\|_\infty^{2(n-1)/n} \|\nu\|_2^{2/n}, \qquad x \in \mathbb{R}^n, \tag{3.27}$$

where $\mathbb{B}^n$ *is the unit ball in* $\mathbb{R}^n$.

The proof of Lemma 5 is given in Section 4.

We are now in a position to state the main result in this section.

Theorem B. *The Riesz transforms model produces a jointly hyponormal pair* (X, Y) *of self-adjoint tuples such that*

$$\|\mathcal{C}(X,Y)\| \leq A_n \|B\|_\infty^{2(n-1)/n} \|B\|_2^{2/n}, \tag{3.28}$$

with

$$A_n = \frac{2n\Gamma\left(\dfrac{n+1}{2}\right)}{\pi^{(n+1)/2}} [\mathrm{vol}(\mathbb{B}^n)]^{(n-1)/n},$$

and

$$\|B\|_\infty = \operatorname*{ess\,sup}_{x \in \Omega} \|b(x)\|, \quad \|B\|_2 = \left(\int_\Omega \|b(x)\|^2 dx \right)^{1/2},$$

where b *is the primary parameter.*

As a direct consequence, we get the next estimate similar to (3.15).

Corollary B. *With the same notation as above we have*

$$\|\mathcal{C}(X,Y)\| \leq A_n \|B\|_\infty^2 [\mathrm{vol}(\Omega)]^{1/n}. \tag{3.29}$$

4. Main Technical Result

The main result in this section will be used to complete the proofs of Theorems A and B.

4.1. The Setting

Suppose M is a measurable space equipped with a nonnegative measure and let $K : M \times M \to [-\infty, \infty]$ be a measurable function such that the set $\Sigma = \{(x, y) \in M \times M : |K(x, y)| = \infty\}$ has measure zero.

The spaces of p-integrable real-valued functions on M, $1 \leq p \leq \infty$, are denoted subsequently by $L^p(M, \mathbb{R})$, and their standard norms will be denoted by $\|\cdot\|_p$.

We begin by associating to K the integral operator $\mathfrak{I} = \mathfrak{I}_K$ given by

$$\mathfrak{I}u(x) = \int_M K(x,y)u(y)dy, \qquad x \in M, \tag{4.1}$$

and defined for measurable functions u such that the integral above exists for almost all $x \in M$.

Next, for every x in M and $0 < t < \infty$ we introduce the set $\Omega[x,t] = \Omega_K[x,t]$ defined as

$$\Omega[x,t] = \{y \in M : |K(x,y)| \geq t\} \tag{4.2}$$

and let $\omega = \omega_K : M \times (0,\infty) \to [0,\infty]$ be the function given by

$$\omega(x,t) = \text{measure}\, \Omega[x,t]. \tag{4.3}$$

In other words, $\omega(x,\cdot) : (0,\infty) \to [0,\infty]$ is the distribution function corresponding to the measurable function $K(x,\cdot) : M \to \mathbb{R}$. We will assume that $\omega(x,\cdot)$ has finite values on $(0,\infty)$.

4.2. Main Result

We are now in a position to state the main result of this section.

Main Lemma. *Suppose* $1 < \kappa < \infty$. *The following two statements are equivalent.*

(i) *For each* $x \in M$ *there exists a positive constant* $A(K,x)$ *such that*

$$|\mathfrak{I}_K u(x)| \leq A(K,x) \|u\|_\infty^{1/\kappa} \|u\|_1^{1-1/\kappa}, \tag{4.4}$$

for every $u \in L^\infty(M,\mathbb{R}) \cap L^1(M,\mathbb{R})$.

(ii) *There exists a measurable function* $\lambda : M \to [0,\infty)$ *such that*

$$\omega_K(x,t) \leq \lambda(x) t^{-\kappa}, \qquad x \in M,\ 0 < t < \infty. \tag{4.5}$$

Whenever (i) *or* (ii) *is true, and* $\lambda(x)$ *in* (4.5) *is defined as*

$$\lambda(x) = \sup_{t>0} t^\kappa \omega(x,t), \qquad x \in M, \tag{4.6}$$

a posible value of $A(K,x)$ *in* (4.4) *is given by*

$$A(K,x) = \frac{\kappa}{\kappa - 1} [\lambda(x)]^{1/\kappa}. \tag{4.7}$$

Moreover, at points $x \in M$ *where* (4.5) *is an equality for all* $t \in (0,\infty)$, *the value of* $A(K,x)$ *in* (4.7) *is the best constant in inequality* (4.4).

Proof of Main Lemma – Part 1. We first prove that (i) implies (ii). Suppose $x \in M$ and $\tau \in (0,\infty)$ are given such that $\omega(x,\tau) \neq 0$, and let χ_τ be the characteristic functions of $\Omega[x,\tau]$. We are going to apply (4.4) to the function $u(\cdot) = K(x,\cdot)|K(x,\cdot)|^{-1}\chi_\tau(\cdot)$. On the one hand we get

$$\mathfrak{I}_K u(x) = \int_{\Omega[x,\tau]} |K(x,y)| dy \geq \tau\omega(x,\tau) \tag{4.8}$$

and

$$\|u\|_1 = \omega(x,\tau). \tag{4.9}$$

On the other hand we observe that

$$\|u\|_\infty = 1. \tag{4.10}$$

Substituting (4.8), (4.9), and (4.10) in (4.4) we get

$$\tau\omega(x,\tau) \leq A(x)[\omega(x,\tau)]^{1-1/\kappa}, \tag{4.11}$$

where $A(x) = A(K,x)$.

To conclude the proof, we notice that (4.11) implies

$$\omega(x,\tau) \leq [A(x)]^{\kappa}\tau^{-\kappa}, \tag{4.12}$$

hence (4.6) defines a measurable function $\lambda : X \to [0,\infty)$ such that (4.5) is true.

Part 2. We now show that (ii) implies (i). Suppose $x \in M$ is given and let u be a function in $L^{\infty}(M,\mathbb{R}) \cap L^{1}(M,\mathbb{R})$. This assumption makes it possible to associate to u the decreasing function $\mu_{u,x} : (0,\infty) \to [0,\infty)$ given by

$$\mu_{u,x}(t) = \int_{\Omega[x,t]} |u(y)|dy. \tag{4.13}$$

When $u \equiv 1$ we clearly have $\mu_{u,x}(t) = \omega(x,t)$ and from now we will denote the distribution function $\omega(x,\cdot)$ by μ_x. Condition (4.5) can be rewritten as

$$\mu_x(t) \leq \lambda(x)t^{-\kappa}, \qquad 0 < t < \infty. \tag{4.14}$$

We next select $\tau \in (0,\infty)$ and introduce the inner and outer parts of K defined as

$$K_{\text{inn}}(x,y) = \begin{cases} |K(x,y)| - \tau, & \text{if } y \in \Omega[x,\tau] \\ 0, & \text{if } y \in M \setminus \Omega[x,\tau] \end{cases} \tag{4.15}$$

and

$$K_{\text{out}}(x,y) = \begin{cases} \tau, & \text{if } y \in \Omega[x,\tau] \\ |K(x,y)|, & \text{if } y \in M \setminus \Omega[x,\tau]. \end{cases} \tag{4.16}$$

Since $|K(x,y)| = K_{\text{inn}}(x,y) + K_{\text{out}}(x,y)$, we clearly have

$$|\mathfrak{J}u(x)| \leq \mathfrak{J}_{\text{inn}}|u|(x) + \mathfrak{J}_{\text{out}}|u|(x), \qquad x \in M, \tag{4.17}$$

where $\mathfrak{J}_{\text{inn}} = \mathfrak{J}_{K_{\text{inn}}}$ and $\mathfrak{J}_{\text{out}} = \mathfrak{J}_{K_{\text{out}}}$.

We claim that

$$\mathfrak{J}_{\text{inn}}|u|(x) \leq \frac{\lambda(x)\|u\|_{\infty}}{\kappa - 1} \cdot \tau^{-\kappa+1} \tag{4.18}$$

and

$$\mathfrak{J}_{\text{out}}|u|(x) \leq \|u\|_1\tau. \tag{4.19}$$

To prove inequality (4.18), we express $\mathfrak{J}_{\text{inn}}|u|(x)$ a Stieltjes integral associated with the decreasing function $\mu_{u,x}$ defined in (4.13), namely,

$$\begin{aligned}\mathfrak{J}_{\text{inn}}|u|(x) &= \int_{\Omega[x,\tau]} [|K(x,y)| - \tau]|u(y)|dy \\ &= -\int_{\tau}^{\infty} (t-\tau)d\mu_{u,x}(t).\end{aligned}$$

From (4.13) and (4.14) we get

$$\mu_{u,x}(t) \leq \mu_x(t)\|u\|_{\infty} \leq \lambda(x)\|u\|_{\infty}t^{-\kappa}, \tag{4.20}$$

whence

$$\lim_{t\to\infty} t\mu_{u,x}(t) = 0.$$

An integration by parts for the last integral yields

$$\mathfrak{I}_{\text{inn}}|u|(x) = \int_{\tau}^{\infty} \mu_{u,x}(t)dt,$$

which, in conjunction with (4.20), results in

$$\mathfrak{I}_{\text{inn}}|u|(x) \leq \lambda(x)\|u\|_{\infty} \int_{\tau}^{\infty} t^{-\kappa}dt,$$

an inequality equivalent to (4.18).

We next prove inequality (4.19). First we notice that

$$\mathfrak{I}_{\text{out}}|u|(x) \leq \|K_{\text{out}}(x,\cdot)\|_{\infty}\|u\|_1. \tag{4.21}$$

From (4.16) we easily get $\|K_{\text{out}}(x,\cdot)\|_{\infty} = \tau$, and (4.19) follows from (4.21). By (4.17), (4.18), and (4.19) we have

$$|\mathfrak{I}u(x)| \leq \frac{\lambda(x)\|u\|_{\infty}}{\kappa - 1} \cdot \tau^{-\kappa+1} + \|u\|_1 \cdot \tau. \tag{4.22}$$

To minimize the right side of (4.22) we choose

$$\tau = \left[\lambda(x) \cdot \frac{\|u\|_{\infty}}{\|u\|_1}\right]^{1/\kappa}.$$

A direct calculation shows that for this specific value of τ inequality (4.22) takes the form (4.4) with a constant $A(K,x)$ as in (4.7).

Part 3. We yet have to show that (4.7) provides the best value of $A(K,x)$ at points $x \in M$ where

$$\mu_x(t) = \lambda(x)t^{-\kappa}, \qquad 0 < t < \infty. \tag{4.23}$$

This can be done by indicating an extremal function $u : M \to \mathbb{R}$.

We suppose that $x \in X$ is given such that (4.23) holds true and observe that

$$\lambda(x) = \mu_x(1) = \text{measure}\,\Omega[x,1]. \tag{4.24}$$

Next we set

$$u(y) = \begin{cases} K(x,y) \cdot |K(x,y)|^{-1}, & \text{if } y \in \Omega[x,1] \\ 0, & \text{if } y \in M \setminus \Omega[x,1]. \end{cases}$$

The left side in inequality (4.4) becomes

$$|\mathfrak{I}_K u(x)| = \int_{\Omega[x,1]} |K(x,y)|dy,$$

whence, by using (4.23) we get

$$|\mathfrak{I}_K u(x)| = -\int_1^{\infty} t d\mu_x(t) = \kappa\lambda(x) \int_1^{\infty} t^{-\kappa}dt,$$

that is,

$$|\mathfrak{I}_K u(x)| = \frac{\kappa}{\kappa - 1} \cdot \lambda(x).$$

From (4.24) we have

$$\|u\|_1 = \int_{\Omega[x,1]} 1dy = \lambda(x),$$

and obviously

$$\|u\|_\infty = 1.$$

Substituting the last three equations in (4.4) with $A(K,x)$ as in (4.7) we get an equality, hence inequality (4.4) is sharp.

The proof of Main Lemma is complete. □

4.3. Completing the Proofs

We go back to Lemma 2 and Lemma 5.

Proof of Lemma 2. We start by observing that

$$\|\mathfrak{J}u(x)\|_x \leq \int_\Omega \|\Phi(x,y)\|_{x,y}\|u(y)\|_y d\mathrm{vol}(y),$$

and then apply the Main Lemma to the kernel

$$K(x,y) = \|\Phi(x,y)\|_{x,y}, \qquad (x,y) \in M \times M.$$

With $A(K,x)$ as in (4.7) and $A(\mathfrak{D})$ as in (2.8) we clearly have

$$A(\mathfrak{D}) = \sup_{x\in M} A(K,x),$$

and (2.9) follows from (4.4). □

Proof of Lemma 5. We take the kernel K on $\mathbb{R}^n \times \mathbb{R}^n$ given by

$$K(x,y) = \frac{1}{|x-y|^{n-1}}, \qquad (x,y) \in \mathbb{R}^n \times \mathbb{R}^n,\ x \neq y,$$

and notice that the corresponding function ω defined as in (4.3) has the from

$$\omega(x,t) = \lambda t^{-\kappa}, \qquad x \in \mathbb{R}^n,\ t \in (0,\infty),$$

with $\lambda = \mathrm{vol}(\mathbb{B}^n)$ and $\kappa = n/(n-1)$. Since by (4.7) we get

$$A(K,x) = n[\mathrm{vol}(\mathbb{B}^n)]^{(n-1)/n},$$

inequality (3.27) follows from (4.4). □

References

[AD] Aizenberg, L.A. and Dautov, Sh.A., *Differential Forms Orthogonal to Holomorphic Functions or Forms, and Their Properties*, Transl. Math. Monographs, 56, Amer. Math. Soc., Providence, RI, 1983.

[Al1] Alexander, H., Projections of polynomial hulls, *J. Funct. Anal.* **13**(1973), 13–19.

[Al2] Alexander, H., On the area of the spectrum of an element of a uniform algebra. In *Complex Approximation Proceedings, Quebec, July 3–8, 1978*, Birkhäuser, 1980, 3–12.

[At] A. Athavale, *On joint hyponormal operators*, Proc. Amer. Math. Soc., **103** (1988), 417–423.

[AS] S. Axler, and J.H. Shapiro, *Putnam's theorem, Alexander's spectral area estimate, and VMO*, Math. Ann. **271** (1985), 161–183.

[B1] Browder, F., Approximation by solutions of partial differential equations, *Amer. J. Math.* **84**(1962), 134–160.

[B2] Browder, F., Functional analysis and partial differential equations. II, *Math. Ann.* **145**(1961), 81–226.

[BDS] Brackx, F., Delanghe, R., and Sommen, F., *Clifford Analysis*, Pitman Research Notes in Mathematics Series, 76, 1982.

[Cl] K.F. Clancey, *Seminormal Operators*, Lecture Notes in Math., **742**, Springer Verlag, 1979.

[CCHZ] M. Cho, R.E. Curto, T. Huruya, and W. Zelazko, *Cartesian form of Putnam's inequality for doubly commuting n-tuples*, preprint, 1999.

[Co] J.B. Conway, *The Theory of Subnormal Operators*, Mathematical Surveys and Monographs, 36, American Mathematical Society, Providence, RI, 1991.

[Cu] R.E. Curto, *Joint hyponormality: a bridge between hyponormality and subnormality*, in Operator Theory, Operator Algebras and Applications (W.B. Arveson and R.G. Douglas, eds.); part 2, Proc. Sympos. Pure Math., **51** (1990), 69–91.

[CJ] R.E. Curto, and R. Jian, *A matricial identity involving the self-commutator of a commuting n-tuple*, Proc. Amer. Math. Soc., **121** (1994), 461–464.

[CMX] R.E. Curto, P.A. Muhly, and J. Xia, *Hyponormal pairs of commuting operators*, in Operator Theory: Advances and Applications, **35**, Birkhäuser Verlag, 1988, 1–22.

[DPY] R.G. Douglas, V. Paulsen, and K. Yan, *Operator theory and algebraic geometry*, Bull. Amer. Math. Soc., **20** (1988), 67–71.

[G] Gamelin, T.W., *Uniform Algebras*, Prentice Hall, 1969.

[GK] Gustafsson, B. and Khavinson, D., On approximation by harmonic vector fields, *Houston J. Math.* **20**(1994), 75–92.

[HR] Hartogs, F. and Rosenthal, A., Über Folgen analytischer Functionen, *Math. Ann.*, **104** (1931), 606–610.

[Ka] T. Kato, *Smooth operators and commutators*, Studia Math., **31** (1968), 531–546.

[Kh1] Khavinson, D., On uniform approximation by harmonic functions, *Mich. Math. J.* **34**(1987), 465–473.

[Kh2] Khavinson, D., Duality and uniform approximation by solutions of elliptic equations, *Operator Theory: Advances and Applications* **35**(1988), Birkhäuser Verlag, Basel, 129–141.

[Mn1] M. Martin, *Joint seminormality and Dirac operators*, Integral Equations Operator Theory, **30** (1998), 101–121.

[Mn2] M. Martin, *Higher-dimensional Ahlfors-Beurling inequalities in Clifford analysis*, Proc. Amer. Math. Soc., **126** (1998), 2863–2871.

[Mn3] Martin, M., Convolution and maximal operator inequalities in Clifford analysis. In *Clifford Algebras and Their Applications in Mathematical Physics, Vol. 2: Clifford Analysis*, Progress in Physics **9**, Birkhäuser Verlag, Basel, 2000, 95–113.

[Mn4] M. Martin, *Self-commutator inequalities in higher dimension*, Proc. Amer. Math. Soc., **130** (2002), 2971–2983.

[Mn5] M. Martin, *Uniform approximation by closed forms in several complex variables*, preprint, 2002.

[Mn6] M. Martin, *Spin geometry, Clifford analysis, and joint seminormality*, Plenary Speaker Talk at the Satellite Conference to ICM-2002 on Clifford Analysis, Macao, China, August 2002; preprint 2002.

[MnPr] M. Martin and M. Putinar, *Lectures on Hyponormal Operators*, Operator Theory: Advances and Applications, **39**, Birkhäuser Verlag, Basel, 1989.

[MnSa] M. Martin, and N. Salinas, *Weitzenböck type formulas and joint seminormality*, in Contemporary Math., Amer. Math. Soc., **212** (1998), 157–167.

[MnSz] M. Martin, and P. Szeptycki, *Sharp inequalities for convolution operators with homogeneous kernels and applications*, Indiana Univ. Math. J., **46** (1997), 975–988.

[McCP] S. McCullough and V. Paulsen, *A note on joint hyponormality*, Proc. Amer. Math. Soc., **107** (1989), 187–195.

[My] P.S. Muhly, *A note on commutators and singular integrals*, Proc. Amer. Math. Soc., **54** (1976), 117–121.

[Ps] J.D. Pincus, *Commutators and systems of singular integral equations*, Acta Math., **121** (1968), 219–249.

[PX] J.D. Pincus, and D. Xia, *The analytic model of a hyponormal operator with rank one self-commutator*, Integral Equations Operator Theory, **4** (1981), 134–150.

[PXX] J.D. Pincus, D. Xia, and J. Xia, *The analytic model of a hyponormal operator with rank one self-commutator*, Integral Equations Operator Theory, **7** (1984), 516–535.

[Pr] Putinar, M., Extreme hyponormal operators, *Operator Theory: Advances and Applications* **28**(1988), 249–265.

[Pm1] C.R. Putnam, *Commutation Properties of Hilbert Space Operators and Related Topics*, Springer-Verlag, Berlin, Heidelberg, New York, 1967.

[Pm2] C.R. Putnam, *An inequality for the area of hyponormal spectra*, Math. Z., **116** (1970), 323–330.

[S1] Stein, E.M., *Singular Integrals and Differentiability Properties of Functions*, Princeton Univ. Press, Princeton, NJ, 1970.

[S2] Stein, E.M., *Harmonic Analysis: Real-Variable Methods, Orthogonality, and Oscillatory Integrals*, Princeton Univ. Press, Princeton, NJ, 1993.

[T] Tarkhanov, N.N., *The Cauchy Problem for Solutions of Elliptic Equations*, Akademie Verlag, Berlin, 1995.

[W] Weinstock, B.M., Uniform approximations by solutions of elliptic equations, *Proc. Amer. Math. Soc.* **41**(1973), 513–517.

[X1] D. Xia, *On non-normal operators. I*, Chinese J. Math., **3** (1963), 232–246; *II*, Acta Math. Sinica, **21** (1987), 103–108.

[X2] D. Xia, *Spectral Theory of Hyponormal Operators*, Birkhäuser Verlag, Basel-Boston-Stuttgart, 1983.

[X3] D. Xia, *On some classes of hyponormal tuples of commuting operators*, in Operator Theory: Advances and Applications, **48**, Birkhäuser Verlag, 1990, pp. 423–448.

Mircea Martin
Department of Mathematics
Baker University
Baldwin City
66006 Kansas, USA
e-mail: mircea.martin@bakeru.edu

Operator Theory:
Advances and Applications, Vol. 149, 407–439

Direct and Inverse Scattering for Skewselfadjoint Hamiltonian Systems

Cornelis van der Mee

Abstract. In this article the direct and inverse scattering theory for skewselfadjoint Hamiltonian systems on the line is developed. The inverse scattering problem of recovering the skewselfadjoint matrix potential from the reflection coefficient is solved explicitly using state space methods if bound states are assumed absent.

1. Introduction

Consider the skewselfadjoint Hamiltonian system of differential equations

$$-iJ_{2n}\frac{dX(x,\lambda)}{dx} - V(x)\,X(x,\lambda) = \lambda\,X(x,\lambda), \qquad x \in \mathbb{R}, \tag{1.1}$$

where

$$J_{2n} = \begin{bmatrix} I_n & 0 \\ 0 & -I_n \end{bmatrix}, \qquad V(x) = \begin{bmatrix} 0 & k(x) \\ -k(x)^\dagger & 0 \end{bmatrix}, \tag{1.2}$$

with I_n the identity matrix of order n, the $n \times n$ matrix function k has complex-valued entries belonging to $L^1(\mathbb{R})$, $\lambda \in \mathbb{R}$ is an eigenvalue parameter, and $\dagger$ denotes the matrix conjugate transpose. We call the function V the potential matrix, k the potential and the parameter λ the wavenumber. Note that $V(x)$ is a J_{2n}-selfadjoint $2n \times 2n$ matrix and satisfies

$$J_{2n}\,V(x) = -V(x)\,J_{2n}.$$

We can think of $X(x,\lambda)$ in (1.1) as either a column vector of $2n$ entries or as a $2n \times 2n$ matrix. For $\lambda \in \mathbb{R}$, we define the Jost solution from the left, $F_l(x,\lambda)$, and the Jost solution from the right, $F_r(x,\lambda)$, as the $2n \times 2n$ matrix solutions of (1.1) satisfying the boundary conditions

$$F_l(x,\lambda) = e^{i\lambda J_{2n}x}\left[I_{2n} + o(1)\right], \qquad x \to +\infty, \tag{1.3}$$

$$F_r(x,\lambda) = e^{i\lambda J_{2n}x}\left[I_{2n} + o(1)\right], \qquad x \to -\infty. \tag{1.4}$$

Received by the editors October 7, 2002; revised September 1, 2003. Submitted by M. Klaus.
Mathematics Subject Classification (2000). Primary 34A55, 81U40; Secondary 73D50.
Key words and phrases. Zakharov-Shabat system, AKNS system, canonical system, Marchenko equation.

Using (1.1), (1.3), and (1.4), we obtain

$$F_l(x,\lambda) = e^{i\lambda J_{2n}x} - iJ_{2n}\int_x^\infty dy\, e^{-i\lambda J_{2n}(y-x)}\,V(y)\,F_l(y,\lambda), \tag{1.5}$$

$$F_r(x,\lambda) = e^{i\lambda J_{2n}x} + iJ_{2n}\int_{-\infty}^x dy\, e^{i\lambda J_{2n}(x-y)}\,V(y)\,F_r(y,\lambda). \tag{1.6}$$

For a square matrix function $E(x)$, let us use $||E||_1$ to denote $\int_{-\infty}^\infty dx\,||E(x)||$, where $||\cdot||$ stands for the matrix norm defined by $||A|| = \sup\{||Av||_2 : \ ||v||_2 = 1\}$ and $||\cdot||_2$ is the Euclidean vector norm. Since the entries of $k(x)$ belong to $L^1(\mathbb{R})$, for each fixed $\lambda \in \mathbb{R}$ it follows by iteration that (1.5) and (1.6) are uniquely solvable and that $||F_l(x,\lambda)||$ and $||F_r(x,\lambda)||$ are bounded above by $e^{||k||_1}$. From (1.3)–(1.6) we get

$$F_l(x,\lambda) = e^{i\lambda J_{2n}x}\left[a_l(\lambda) + o(1)\right], \qquad x \to -\infty, \tag{1.7}$$

$$F_r(x,\lambda) = e^{i\lambda J_{2n}x}\left[a_r(\lambda) + o(1)\right], \qquad x \to +\infty, \tag{1.8}$$

where

$$a_l(\lambda) = I_{2n} - iJ_{2n}\int_{-\infty}^\infty dy\, e^{-i\lambda J_{2n}y}\,V(y)\,F_l(y,\lambda),$$

$$a_r(\lambda) = I_{2n} + iJ_{2n}\int_{-\infty}^\infty dy\, e^{-i\lambda J_{2n}y}\,V(y)\,F_r(y,\lambda).$$

In this article we solve the direct and inverse scattering problem for (1.1), where the inverse scattering problem consists of the determination of the potential $k(x)$ from either of the reflection coefficients $R(\lambda)$ and $L(\lambda)$, which are defined in (3.7) in terms of the matrices $a_l(\lambda)$ and $a_r(\lambda)$, plus suitable bound state data. In this paper we restrict ourselves to inverse scattering problems where there are no bound states.

Shabat [32] and Beals and Coifman [10, 11] considered the $n \times n$ matrix differential operator

$$d\varphi/dx = \lambda \boldsymbol{J}\varphi + q(x)\,\varphi,$$

where $\boldsymbol{J} = \operatorname{diag}\{\alpha_1,\ldots,\alpha_n\}$ with distinct complex α_j and $q(x)$ an $n \times n$ off-diagonal matrix with entries belonging to $L^1(\mathbb{R})$ or more restrictive classes, without requiring $q(x)$ to be selfadjoint. They proved that the inverse problem has a unique solution within a certain class of potentials for an open and dense set of scattering data. The solution of the inverse scattering problem for such linear systems is useful in solving the Cauchy problem for various nonlinear evolution equations. For details and further references, we refer the interested reader to [1, 12] and the references therein.

By putting $Z(x,\lambda) = \frac{1}{\sqrt{2}}[I_{2n} + i\boldsymbol{q}_{2n}]\,X(x,\lambda)$, where

$$\boldsymbol{q}_{2n} = \begin{bmatrix} 0 & I_n \\ I_n & 0 \end{bmatrix},$$

we can convert (1.1) into the massless Dirac equation of order $2n$. The direct and inverse scattering problems for the Dirac system on the half line were studied in [16]. The interested reader is referred to [16, 22, 23] and the references therein for more information on the Dirac system.

Melik-Adamyan [25, 26, 27], L.A. Sakhnovich [30, 31], and A.L. Sakhnovich [29] have studied the direct and inverse scattering problems for (1.1) on the half line. Alpay and Gohberg [3, 4, 5, 6] have applied state space methods to derive explicit expressions for the solution of the inverse problem for (1.1) on the half line from the general theory in [27] when the scattering data are rational functions and consist of either the spectral function of the differential operator $H = -iJ_{2n}\frac{d}{dx} - V(x)$ or a reflection function. Gohberg et al. [18, 19, 20] have solved a similar inverse problem when the scattering data consist of the spectral function of H and this function is rational.

In this article we develop a direct and inverse scattering theory for (1.1) when $k(x)$ has entries belonging to $L^1(\mathbb{R})$. Working within the framework established by Faddeev [15] and Deift and Trubowitz [14] for the Schrödinger equation on the line, we derive the analyticity and asymptotic properties of the Faddeev matrices and the scattering coefficients, employ them to derive a Riemann-Hilbert problem and various Marchenko integral equations, and recover the potential in terms of the solutions of the Marchenko equations. We prove the J_{2n}-unitarity of the scattering matrix and exploit this property to prove the unique solvability of the Marchenko equations. We also establish the unique canonical Wiener-Hopf factorization of the scattering matrix and show how the potential is obtained once the factors are known. After that, for rational reflection coefficients we present a procedure to compute explicitly the scattering matrix from a reflection coefficient, assuming there are no bound states. When the reflection coefficients are rational, we apply state space methods to solve the Marchenko equations and the inverse problem explicitly.

This article follows its predecessor [2], where the potential matrix $V(x)$ is selfadjoint, the scattering matrix is unitary, and there do not exist bound states. Here we are dealing with a more complicated but physically much more interesting problem, where the potential matrix $V(x)$ is J_{2n}-selfadjoint, the scattering matrix is J_{2n}-unitary, and bound states may exist. As a result, there are notable differences in the proof of many equations in this article from their counterparts in [2]. However, when deriving the Marchenko integral equations and applying state space methods to solve them we assume for simplicity that there are no bound states.

Let us discuss the organization of this article. In Section 2 we introduce the Faddeev matrices, obtain their analyticity properties, and analyze some other properties of the Faddeev matrices and the Jost solutions of (1.1). In Section 3 we define the scattering matrix $\boldsymbol{S}(\lambda)$ in terms of the spatial asymptotics of the Jost solutions, prove the J_{2n}-unitarity of $\boldsymbol{S}(\lambda)$, and obtain various properties of the scattering coefficients. In Section 4 we analyze the Fourier transforms of the Faddeev matrices and the scattering coefficients. We then go on, in Section 5, to derive a Riemann-Hilbert problem for the Faddeev matrices. In Section 6, we convert the Riemann-Hilbert problem into both coupled and uncoupled Marchenko integral equations and prove their unique solvability, assuming there are no bound states. In Section 7 we show how to construct $\boldsymbol{S}(\lambda)$ explicitly when one of the reflection coefficients is a rational function and bound states are absent. Finally, in Section 8 we give an explicit solution of the inverse scattering problem with rational reflection coefficients.

Let us give some definitions. By $\mathbb{C}^+$ and $\mathbb{C}^-$ we denote the open upper half and lower half complex planes, respectively. We will use the notation $L^j(I;\mathbb{C}^{p\times q})$ to denote the Banach space of all complex $p\times q$ matrix functions $z(\alpha)$ whose entries belong to $L^j(I)$, endowed with the norm $\left[\int_I d\alpha\, \|z(\alpha)\|^j\right]$; if $q=1$, we simply write $L^j(I;\mathbb{C}^p)$.

2. Scattering solutions

In this section we introduce the Faddeev matrices and study some of their properties. The results obtained here will be used later to establish various properties of the scattering matrix and to solve the inverse scattering problem by the Marchenko method.

Proposition 2.1. *Let $X(x,\lambda)$ and $Y(x,\lambda)$ be any two solutions of* (1.1). *Then, for real λ, $X(x,\lambda)^\dagger Y(x,\lambda)$ is independent of x.*

Proof. The result follows by differentiating $X(x,\lambda)^\dagger Y(x,\lambda)$ and using (1.1) together with the selfadjointness of $J_{2n}V(x)$ and J_{2n}. □

Proposition 2.2. *For $\lambda\in\mathbb{R}$, either Jost solution $F_l(x,\lambda)$ or $F_r(x,\lambda)$ forms a fundamental matrix of* (1.1) *and has determinant equal to one. Moreover, the matrices $a_l(\lambda)$ and $a_r(\lambda)$ appearing in* (1.7) *and* (1.8), *respectively, satisfy*

$$\det a_l(\lambda)=\det a_r(\lambda)=1. \tag{2.1}$$

Moreover, for $\lambda\in\mathbb{R}$, the Jost solutions satisfy

$$F_l(x,\lambda)=F_r(x,\lambda)\,a_l(\lambda), \tag{2.2}$$

$$F_r(x,\lambda)^\dagger\,F_l(x,\lambda)=a_r(\lambda)^\dagger=a_l(\lambda), \tag{2.3}$$

$$F_l(x,\lambda)^\dagger\,F_l(x,\lambda)=a_l(\lambda)^\dagger\,a_l(\lambda)=I_{2n}, \tag{2.4}$$

$$F_r(x,\lambda)^\dagger\,F_r(x,\lambda)=a_r(\lambda)^\dagger\,a_r(\lambda)=I_{2n}, \tag{2.5}$$

and hence

$$a_l(\lambda)\,a_r(\lambda)=a_r(\lambda)\,a_l(\lambda)=I_{2n}, \tag{2.6}$$

$$a_l(\lambda)^{-1}=a_l(\lambda)^\dagger,\qquad a_r(\lambda)^{-1}=a_r(\lambda)^\dagger. \tag{2.7}$$

In particular, $a_l(\lambda)$ and $a_r(\lambda)$ are unitary matrices.

Proof. From (1.1) it follows from [28] that

$$\frac{d\,[\det F_l(x,\lambda)]}{dx} = \big(\operatorname{tr}\{iJ_{2n}\,V(x) + i\lambda J_{2n}\}\big)\big(\det F_l(x,\lambda)\big),$$

where tr denotes the matrix trace. By (1.2), $iJ_{2n}\,V(x)+i\lambda J_{2n}$ has zero trace, and hence $\det F_l(x,\lambda)$ is independent of x and its value can be evaluated as $x\to+\infty$. Thus, we get $\det F_l(x,\lambda) = 1$, from which we also conclude that $F_l(x,\lambda)$ is a fundamental matrix of (1.1). Similarly, we find that $\det F_r(x,\lambda) = 1$ and $F_r(x,\lambda)$ is a fundamental matrix of (1.1). Then, from (1.2), (1.7), and (1.8) we obtain (2.1). Since either of $F_l(x,\lambda)$ and $F_r(x,\lambda)$ is a fundamental matrix of (1.1), with the help of (1.3) and (1.7), we get (2.2). Using Proposition 2.1, we obtain (2.3)–(2.5) by evaluating $F_r(x,\lambda)^\dagger\,F_l(x,\lambda)$, $F_l(x,\lambda)^\dagger\,F_l(x,\lambda)$, and $F_r(x,\lambda)^\dagger\,F_r(x,\lambda)$ as $x\to\pm\infty$. Then (2.6) and (2.7) readily follow. □

In terms of the Jost solutions, we define the Faddeev matrices $M_l(x,\lambda)$ and $M_r(x,\lambda)$ as

$$M_l(x,\lambda) = F_l(x,\lambda)\,e^{-i\lambda J_{2n}x}, \qquad M_r(x,\lambda) = F_r(x,\lambda)\,e^{-i\lambda J_{2n}x}. \tag{2.8}$$

From (1.3) and (1.4) we get

$$\begin{aligned} M_l(x,\lambda) &= I_{2n} + o(1), \qquad x\to+\infty,\\ M_r(x,\lambda) &= I_{2n} + o(1), \qquad x\to-\infty. \end{aligned}$$

Let us partition the Jost solutions and Faddeev matrices into $n\times n$ blocks as follows:

$$F_l(x,\lambda) = \begin{bmatrix} F_{l1}(x,\lambda) & F_{l2}(x,\lambda)\\ F_{l3}(x,\lambda) & F_{l4}(x,\lambda)\end{bmatrix}, \qquad F_r(x,\lambda) = \begin{bmatrix} F_{r1}(x,\lambda) & F_{r2}(x,\lambda)\\ F_{r3}(x,\lambda) & F_{r4}(x,\lambda)\end{bmatrix}, \tag{2.9}$$

$$M_l(x,\lambda) = \begin{bmatrix} M_{l1}(x,\lambda) & M_{l2}(x,\lambda)\\ M_{l3}(x,\lambda) & M_{l4}(x,\lambda)\end{bmatrix}, \quad M_r(x,\lambda) = \begin{bmatrix} M_{r1}(x,\lambda) & M_{r2}(x,\lambda)\\ M_{r3}(x,\lambda) & M_{r4}(x,\lambda)\end{bmatrix}. \tag{2.10}$$

We also define

$$\sigma_\pm(x) = \pm\int_x^{\pm\infty} dy\,||k(y)||. \tag{2.11}$$

Proposition 2.3. *Assume that the entries of $k(x)$ belong to $L^1(\mathbb{R})$. Then:*

(i) *For each fixed $x\in\mathbb{R}$, $\begin{bmatrix} M_{l1}(x,\lambda)\\ M_{l3}(x,\lambda)\end{bmatrix}$ can be extended to a matrix function that is continuous in $\lambda\in\overline{\mathbb{C}^+}$ and analytic in $\lambda\in\mathbb{C}^+$ and tends to $\begin{bmatrix} I_n\\ 0\end{bmatrix}$ as $\lambda\to\infty$ in $\overline{\mathbb{C}^+}$.*

(ii) *For all $\lambda\in\overline{\mathbb{C}^+}$, $M_{l1}(x,\lambda)$ and $M_{l3}(x,\lambda)$ are bounded by $e^{\sigma_+(x)}$ in the norm.*

(iii) *For each fixed* $x \in \mathbb{R}$, $\begin{bmatrix} M_{l2}(x,\lambda) \\ M_{l4}(x,\lambda) \end{bmatrix}$ *can be extended to a matrix function that is continuous in* $\lambda \in \overline{\mathbb{C}^-}$ *and analytic in* $\lambda \in \mathbb{C}^-$ *and tends to* $\begin{bmatrix} 0 \\ I_n \end{bmatrix}$ *as* $\lambda \to \infty$ *in* $\overline{\mathbb{C}^-}$.

(iv) *For all* $\lambda \in \overline{\mathbb{C}^-}$, $M_{l2}(x,\lambda)$ *and* $M_{l4}(x,\lambda)$ *are bounded by* $e^{\sigma_+(x)}$ *in the norm.*

Proof. Using (2.8) in (1.5), we obtain

$$M_l(x,\lambda) = I_{2n} - iJ_{2n} \int_x^\infty dy\, e^{-i\lambda J_{2n}(y-x)}\, V(y)\, M_l(y,\lambda)\, e^{i\lambda J_{2n}(y-x)}. \tag{2.12}$$

Iterating (2.12) once, we get the uncoupled systems

$$M_{l1}(x,\lambda) = I_n - \int_x^\infty dy \int_y^\infty dz\, e^{2i\lambda(z-y)}\, k(y)\, k(z)^\dagger\, M_{l1}(z,\lambda), \tag{2.13}$$

$$\begin{aligned} M_{l2}(x,\lambda) = &-i \int_x^\infty dy\, e^{-2i\lambda(y-x)}\, k(y) \\ &- \int_x^\infty dy \int_y^\infty dz\, e^{-2i\lambda(y-x)}\, k(y)\, k(z)^\dagger\, M_{l2}(z,\lambda), \end{aligned} \tag{2.14}$$

$$\begin{aligned} M_{l3}(x,\lambda) = &-i \int_x^\infty dy\, e^{2i\lambda(y-x)}\, k(y)^\dagger \\ &- \int_x^\infty dy \int_y^\infty dz\, e^{2i\lambda(y-x)}\, k(y)^\dagger\, k(z)\, M_{l3}(z,\lambda), \end{aligned} \tag{2.15}$$

$$M_{l4}(x,\lambda) = I_n - \int_x^\infty dy \int_y^\infty dz\, e^{-2i\lambda(z-y)}\, k(y)^\dagger\, k(z)\, M_{l4}(z,\lambda). \tag{2.16}$$

Iterating the Volterra integral equations (2.13) and (2.15), we prove that the series of iterates converge absolutely and uniformly in $\lambda \in \overline{\mathbb{C}^+}$, and we also get the estimate in (ii). Similarly, we prove that the series of iterates of (2.14) and (2.16) converge absolutely and uniformly in $\lambda \in \overline{\mathbb{C}^-}$ and that the estimate in (iv) holds. To prove the assertions concerning the large-λ limit we first consider $M_{l3}(x,\lambda)$. To deal with the first term on the right-hand side of (2.15) we define

$$\omega(\lambda) = \sup_{x\in\mathbb{R}} \left\| \int_x^\infty dy\, e^{2i\lambda(y-x)}\, k(y)^\dagger \right\|.$$

By approximating $k(y)$ by infinitely differentiable matrix functions of compact support (as in the proof of the Riemann-Lebesgue lemma) it follows that $\omega(\lambda) \to 0$ as $\lambda \to \infty$ in $\overline{\mathbb{C}^+}$. Iterating (2.15) we get $\|M_{l3}(x,\lambda)\| \le \omega(\lambda)\, e^{\sigma_+(x)}$, which implies that $\|M_{l3}(x,\lambda)\| \to 0$ as $\lambda \to \infty$ in $\overline{\mathbb{C}^+}$. Next we consider $M_{l1}(x,\lambda)$. Let $G_{l1}(x,\lambda) = M_{l1}(x,\lambda) - I_n$ and consider the following integral equation for $G_{l1}(x,\lambda)$ which follows from (2.13):

$$G_{l1}(x,\lambda) = H_{l1}(x,\lambda) - \int_x^\infty dy \int_y^\infty dz\, e^{2i\lambda(z-y)}\, k(y)\, k(z)^\dagger\, G_{l1}(z,\lambda),$$

where

$$H_{l1}(x,\lambda) = -\int_x^\infty dy \int_y^\infty dz\, e^{2i\lambda(z-y)}\, k(y)\, k(z)^\dagger.$$

Since $\|H_{l1}(x,\lambda)\| \le \nu(\lambda)\,\sigma_+(x)$, we conclude that $\|G_{l1}(x,\lambda)\| \to 0$ as $\lambda \to \infty$ in $\overline{\mathbb{C}^+}$. This proves the assertion of (i) regarding the limit $\lambda \to \infty$. The proof of the corresponding statement in (iii) is similar. □

We have a similar result for the Faddeev matrix $M_r(x,\lambda)$.

Proposition 2.4. *Assume that the entries of $k(x)$ belong to $L^1(\mathbb{R})$. Then:*

(i) *For each fixed $x \in \mathbb{R}$, $\begin{bmatrix} M_{r1}(x,\lambda) \\ M_{r3}(x,\lambda) \end{bmatrix}$ can be extended to a matrix function that is continuous in $\lambda \in \overline{\mathbb{C}^-}$ and analytic in $\lambda \in \mathbb{C}^-$ and tends to $\begin{bmatrix} I_n \\ 0 \end{bmatrix}$ as $\lambda \to \infty$ in $\overline{\mathbb{C}^-}$.*

(ii) *For all $\lambda \in \overline{\mathbb{C}^-}$, $M_{r1}(x,\lambda)$ and $M_{r3}(x,\lambda)$ are bounded by $e^{\sigma_-(x)}$ in the norm.*

(iii) *For each fixed $x \in \mathbb{R}$, $\begin{bmatrix} M_{r2}(x,\lambda) \\ M_{r4}(x,\lambda) \end{bmatrix}$ can be extended to a matrix function that is continuous in $\lambda \in \overline{\mathbb{C}^+}$ and analytic in $\lambda \in \mathbb{C}^+$ and tends to $\begin{bmatrix} 0 \\ I_n \end{bmatrix}$ as $\lambda \to \infty$ in $\overline{\mathbb{C}^+}$.*

(iv) *For all $\lambda \in \overline{\mathbb{C}^+}$, $M_{r2}(x,\lambda)$ and $M_{r4}(x,\lambda)$ are bounded by $e^{\sigma_-(x)}$ in the norm.*

Proof. Using (2.8) in (1.6), we obtain

$$M_r(x,\lambda) = I_{2n} + iJ_{2n}\int_{-\infty}^x dy\, e^{i\lambda J_{2n}(x-y)}\, V(y)\, M_r(y,\lambda)\, e^{-i\lambda J_{2n}(x-y)}. \tag{2.17}$$

Iterating (2.17) once, we obtain the four systems given by

$$M_{r1}(x,\lambda) = I_n - \int_{-\infty}^x dy \int_{-\infty}^y dz\, e^{-2i\lambda(y-z)}\, k(y)\, k(z)^\dagger\, M_{r1}(z,\lambda), \tag{2.18}$$

$$\begin{aligned} M_{r2}(x,\lambda) = {} & i\int_{-\infty}^x dy\, e^{2i\lambda(x-y)}\, k(y) \\ & - \int_{-\infty}^x dy \int_{-\infty}^y dz\, e^{2i\lambda(x-y)}\, k(y)\, k(z)^\dagger\, M_{r2}(z,\lambda), \end{aligned} \tag{2.19}$$

$$\begin{aligned} M_{r3}(x,\lambda) = {} & i\int_{-\infty}^x dy\, e^{-2i\lambda(x-y)}\, k(y)^\dagger \\ & - \int_{-\infty}^x dy \int_{-\infty}^y dz\, e^{-2i\lambda(x-y)}\, k(y)^\dagger\, k(z)\, M_{r3}(z,\lambda), \end{aligned} \tag{2.20}$$

$$M_{r4}(x,\lambda) = I_n - \int_{-\infty}^x dy \int_{-\infty}^y dz\, e^{2i\lambda(y-z)}\, k(y)^\dagger\, k(z)\, M_{r4}(z,\lambda). \tag{2.21}$$

Iterating (2.18)–(2.21) as in the proof of Proposition 2.3, we complete the proof. □

Let us write

$$a_l(\lambda) = \begin{bmatrix} a_{l1}(\lambda) & a_{l2}(\lambda) \\ a_{l3}(\lambda) & a_{l4}(\lambda) \end{bmatrix}, \qquad a_r(\lambda) = \begin{bmatrix} a_{r1}(\lambda) & a_{r2}(\lambda) \\ a_{r3}(\lambda) & a_{r4}(\lambda) \end{bmatrix}. \tag{2.22}$$

From (1.7), (1.8), and (2.8) we see that

$$\begin{bmatrix} a_{l1}(\lambda) & a_{l2}(\lambda) \\ a_{l3}(\lambda) & a_{l4}(\lambda) \end{bmatrix} = \lim_{x\to-\infty} \begin{bmatrix} M_{l1}(x,\lambda) & e^{-2i\lambda x}\, M_{l2}(x,\lambda) \\ e^{2i\lambda x}\, M_{l3}(x,\lambda) & M_{l4}(x,\lambda) \end{bmatrix}, \tag{2.23}$$

$$\begin{bmatrix} a_{r1}(\lambda) & a_{r2}(\lambda) \\ a_{r3}(\lambda) & a_{r4}(\lambda) \end{bmatrix} = \lim_{x\to+\infty} \begin{bmatrix} M_{r1}(x,\lambda) & e^{-2i\lambda x}\, M_{r2}(x,\lambda) \\ e^{2i\lambda x}\, M_{r3}(x,\lambda) & M_{r4}(x,\lambda) \end{bmatrix}. \tag{2.24}$$

Using (2.12), (2.17), (2.23), and (2.24) we find the integral representations

$$a_{l1}(\lambda) = I_n - i \int_{-\infty}^{\infty} dy\, k(y)\, M_{l3}(y,\lambda) \tag{2.25}$$

$$a_{l2}(\lambda) = -i \int_{-\infty}^{\infty} dy\, e^{-2i\lambda y}\, k(y)\, M_{l4}(y,\lambda), \tag{2.26}$$

$$a_{l3}(\lambda) = -i \int_{-\infty}^{\infty} dy\, e^{2i\lambda y}\, k(y)^\dagger\, M_{l1}(y,\lambda), \tag{2.27}$$

$$a_{l4}(\lambda) = I_n - i \int_{-\infty}^{\infty} dy\, k(y)^\dagger\, M_{l2}(y,\lambda), \tag{2.28}$$

$$a_{r1}(\lambda) = I_n + i \int_{-\infty}^{\infty} dy\, k(y)\, M_{r3}(y,\lambda), \tag{2.29}$$

$$a_{r2}(\lambda) = i \int_{-\infty}^{\infty} dy\, e^{-2i\lambda y}\, k(y)\, M_{r4}(y,\lambda), \tag{2.30}$$

$$a_{r3}(\lambda) = i \int_{-\infty}^{\infty} dy\, e^{2i\lambda y}\, k(y)^\dagger\, M_{r1}(y,\lambda), \tag{2.31}$$

$$a_{r4}(\lambda) = I_n + i \int_{-\infty}^{\infty} dy\, k(y)^\dagger\, M_{r2}(y,\lambda). \tag{2.32}$$

Proposition 2.5. *Assume that the entries of $k(x)$ belong to $L^1(\mathbb{R})$. Then:*

(i) *The matrices $a_{l1}(\lambda)$ and $a_{r4}(\lambda)$ are continuous in $\lambda \in \overline{\mathbb{C}^+}$ and analytic in $\lambda \in \mathbb{C}^+$ and tend to I_n as $\lambda \to \infty$ in $\overline{\mathbb{C}^+}$.*

(ii) *The matrices $a_{l4}(\lambda)$ and $a_{r1}(\lambda)$ are continuous in $\lambda \in \overline{\mathbb{C}^-}$ and analytic in $\lambda \in \mathbb{C}^-$ and tend to I_n as $\lambda \to \infty$ in $\overline{\mathbb{C}^-}$.*

(iii) *The matrices $a_{l2}(\lambda)$, $a_{l3}(\lambda)$, $a_{r2}(\lambda)$, and $a_{r3}(\lambda)$ are continuous in $\lambda \in \mathbb{R}$ and vanish as $\lambda \to \pm\infty$.*

(iv) *The matrices $a_{l2}(\lambda)$, $a_{l3}(\lambda)$, $a_{r2}(\lambda)$, and $a_{r3}(\lambda)$ satisfy*

$$a_{r2}(\lambda) = a_{l3}(\lambda)^\dagger, \qquad a_{r3}(\lambda) = a_{l2}(\lambda)^\dagger, \qquad \lambda \in \mathbb{R}. \tag{2.33}$$

Proof. Using Propositions 2.3 and 2.4 in (2.25)–(2.32), we get (i), (ii), and (iii). We obtain (iv) from (2.3). □

Using the notations of (2.9), let us form the following matrices:

$$f_+(x,\lambda) = \begin{bmatrix} F_{l1}(x,\lambda) & F_{r2}(x,\lambda) \\ F_{l3}(x,\lambda) & F_{r4}(x,\lambda) \end{bmatrix}, \quad f_-(x,\lambda) = \begin{bmatrix} F_{r1}(x,\lambda) & F_{l2}(x,\lambda) \\ F_{r3}(x,\lambda) & F_{l4}(x,\lambda) \end{bmatrix}. \tag{2.34}$$

Let an asterisk denote complex conjugation. From Propositions 2.3 and 2.4, it follows that $f_+(x,\lambda)$ is a solution of (1.1) that is continuous in $\lambda \in \overline{\mathbb{C}^+}$ and analytic in $\lambda \in \mathbb{C}^+$; similarly, $f_-(x,\lambda)$ is a solution of (1.1) that is continuous in $\lambda \in \overline{\mathbb{C}^-}$ and analytic in $\lambda \in \mathbb{C}^-$.

Proposition 2.6. *The $2n \times 2n$ matrix $f_-(x,\lambda^*)^\dagger f_+(x,\lambda)$ is independent of x for all $\lambda \in \overline{\mathbb{C}^+}$. Similarly, $f_+(x,\lambda^*)^\dagger f_-(x,\lambda)$ is independent of x for all $\lambda \in \overline{\mathbb{C}^-}$. We have*

$$f_-(x,\lambda^*)^\dagger f_+(x,\lambda) = \begin{bmatrix} a_{l1}(\lambda) & 0 \\ 0 & a_{r4}(\lambda) \end{bmatrix}, \qquad \lambda \in \overline{\mathbb{C}^+}. \tag{2.35}$$

Furthermore, $a_{l1}(\lambda)^\dagger$ and $a_{r4}(\lambda)^\dagger$ have analytic extensions to $\mathbb{C}^-$, $a_{r1}(\lambda)^\dagger$ and $a_{l4}(\lambda)^\dagger$ have analytic extensions to $\mathbb{C}^+$, and

$$a_{l1}(\lambda) = a_{r1}(\lambda^*)^\dagger, \qquad a_{r4}(\lambda) = a_{l4}(\lambda^*)^\dagger, \qquad \lambda \in \overline{\mathbb{C}^+}, \tag{2.36}$$

$$a_{r1}(\lambda) = a_{l1}(\lambda^*)^\dagger, \qquad a_{l4}(\lambda) = a_{r4}(\lambda^*)^\dagger, \qquad \lambda \in \overline{\mathbb{C}^-}. \tag{2.37}$$

Proof. Using (1.1), one can show that $f_\mp(x,\lambda^*)^\dagger f_\pm(x,\lambda)$ is independent of x for $\lambda \in \overline{\mathbb{C}^\pm}$. Evaluating it as $x \to \pm\infty$ and using (1.7) and (1.8) we get (2.35)–(2.37). □

As in the proof of Proposition 2.2 we find that $\det f_+(x,\lambda)$ is independent of x, and evaluating that determinant as $x \to \pm\infty$ we obtain

$$\det f_+(x,\lambda) = \det a_{l1}(\lambda) = \det a_{r4}(\lambda). \tag{2.38}$$

In analogy with (2.38), we get

$$\det f_-(x,\lambda) = \det a_{r1}(\lambda) = \det a_{l4}(\lambda), \tag{2.39}$$

Using (2.10) and (2.34), let us define

$$m_+(x,\lambda) = \begin{bmatrix} M_{l1}(x,\lambda) & M_{r2}(x,\lambda) \\ M_{l3}(x,\lambda) & M_{r4}(x,\lambda) \end{bmatrix} = f_+(x,\lambda)\, e^{-i\lambda J_{2n} x}, \tag{2.40}$$

$$m_-(x,\lambda) = \begin{bmatrix} M_{r1}(x,\lambda) & M_{l2}(x,\lambda) \\ M_{r3}(x,\lambda) & M_{l4}(x,\lambda) \end{bmatrix} = f_-(x,\lambda)\, e^{-i\lambda J_{2n} x}. \tag{2.41}$$

3. The scattering matrix

In this section we define and analyze the properties of the scattering coefficients of (1.1) when the entries of the potential $k(x)$ belong to $L^1(\mathbb{R})$.

We can write (2.6) as

$$a_{l1}(\lambda)\,a_{r1}(\lambda)+a_{l2}(\lambda)\,a_{r3}(\lambda)=I_n=a_{r1}(\lambda)\,a_{l1}(\lambda)+a_{r2}(\lambda)\,a_{l3}(\lambda), \tag{3.1}$$

$$a_{l1}(\lambda)\,a_{r2}(\lambda)+a_{l2}(\lambda)\,a_{r4}(\lambda)=0=a_{r1}(\lambda)\,a_{l2}(\lambda)+a_{r2}(\lambda)\,a_{l4}(\lambda), \tag{3.2}$$

$$a_{l3}(\lambda)\,a_{r1}(\lambda)+a_{l4}(\lambda)\,a_{r3}(\lambda)=0=a_{r3}(\lambda)\,a_{l1}(\lambda)+a_{r4}(\lambda)\,a_{l3}(\lambda), \tag{3.3}$$

$$a_{l3}(\lambda)\,a_{r2}(\lambda)+a_{l4}(\lambda)\,a_{r4}(\lambda)=I_n=a_{r3}(\lambda)\,a_{l2}(\lambda)+a_{r4}(\lambda)\,a_{l4}(\lambda). \tag{3.4}$$

For those real λ for which $a_{l1}(\lambda)$ and $a_{r4}(\lambda)$ are nonsingular, let us define the *transmission coefficients* $T_l(\lambda)$ from the left and $T_r(\lambda)$ from the right, and the *reflection coefficients* $R(\lambda)$ from the right and $L(\lambda)$ from the left, as follows:

$$T_l(\lambda)=a_{l1}(\lambda)^{-1},\qquad T_r(\lambda)=a_{r4}(\lambda)^{-1}, \tag{3.5}$$

$$R(\lambda)=a_{r2}(\lambda)\,a_{r4}(\lambda)^{-1},\qquad L(\lambda)=a_{l3}(\lambda)\,a_{l1}(\lambda)^{-1}. \tag{3.6}$$

From (3.2), (3.3), and (3.6) we get

$$R(\lambda)=-a_{l1}(\lambda)^{-1}\,a_{l2}(\lambda),\qquad L(\lambda)=-a_{r4}(\lambda)^{-1}\,a_{r3}(\lambda). \tag{3.7}$$

Note that using (2.3) and (3.1)–(3.7), we can express the matrices in (2.22) in terms of the scattering coefficients as follows

$$a_l(\lambda)=\begin{bmatrix} T_l(\lambda)^{-1} & -T_l(\lambda)^{-1}\,R(\lambda) \\ L(\lambda)\,T_l(\lambda)^{-1} & [T_r(\lambda)^\dagger]^{-1}\end{bmatrix}, \tag{3.8}$$

$$a_r(\lambda)=\begin{bmatrix} [T_l(\lambda)^\dagger]^{-1} & R(\lambda)\,T_r(\lambda)^{-1} \\ -T_r(\lambda)^{-1}\,L(\lambda) & T_r(\lambda)^{-1}\end{bmatrix}, \tag{3.9}$$

where the off-diagonal entries can be expressed in terms of $L(\lambda)$ or $R(\lambda)$ by using

$$L(\lambda)\,T_l(\lambda)^{-1}=[R(\lambda)\,T_r(\lambda)^{-1}]^\dagger, \tag{3.10}$$

which is immediate from (2.33).

The *scattering matrix* $\boldsymbol{S}(\lambda)$ associated with (1.1) is defined as follows:

$$\boldsymbol{S}(\lambda)=\begin{bmatrix} T_l(\lambda) & R(\lambda) \\ L(\lambda) & T_r(\lambda)\end{bmatrix}. \tag{3.11}$$

Theorem 3.1. *The scattering matrix $\boldsymbol{S}(\lambda)$ is continuous and J_{2n}-unitary, except at those $\lambda\in\mathbb{R}$ where $a_{l1}(\lambda)$ and $a_{r4}(\lambda)$ are singular. Further, it converges to I_{2n} as $\lambda\to\pm\infty$. Hence the scattering coefficients satisfy*

$$T_l(\lambda)\,T_l(\lambda)^\dagger-R(\lambda)\,R(\lambda)^\dagger=I_n=T_r(\lambda)^\dagger\,T_r(\lambda)-R(\lambda)^\dagger\,R(\lambda), \tag{3.12}$$

$$T_l(\lambda)^\dagger\,T_l(\lambda)-L(\lambda)^\dagger\,L(\lambda)=I_n=T_r(\lambda)\,T_r(\lambda)^\dagger-L(\lambda)\,L(\lambda)^\dagger, \tag{3.13}$$

$$T_r(\lambda)\,R(\lambda)^\dagger-L(\lambda)\,T_l(\lambda)^\dagger=0=T_r(\lambda)^\dagger\,L(\lambda)-R(\lambda)^\dagger\,T_l(\lambda). \tag{3.14}$$

Moreover, for those $\lambda\in\mathbb{R}$ where $a_{l1}(\lambda)$ and $a_{r4}(\lambda)$ are nonsingular, we have

$$\det T_l(\lambda)=\det T_r(\lambda), \tag{3.15}$$

$$\det\begin{bmatrix} I_n & -R(\lambda) \\ R(\lambda)^\dagger & I_n \end{bmatrix} = \det\begin{bmatrix} I_n & L(\lambda)^\dagger \\ -L(\lambda) & I_n \end{bmatrix} = |\det T_l(\lambda)|^2, \tag{3.16}$$

$$\det \boldsymbol{S}(\lambda) = \frac{\det T_l(\lambda)}{[\det T_l(\lambda)]^*}. \tag{3.17}$$

Proof. The continuity and the large-λ asymptotics follow from Proposition 2.5. Using (3.5)–(3.7) in (2.7), we get $\boldsymbol{S}(\lambda)\, J_{2n}\, \boldsymbol{S}(\lambda)^\dagger = J_{2n}$, from which (3.12)–(3.14) follow. Furthermore, from (2.38), (3.8), and (3.9) we obtain (3.15). Using (3.10), we can write (3.8) and (3.9) as

$$a_l(\lambda) = \begin{bmatrix} T_l(\lambda)^{-1} & 0 \\ 0 & [T_r(\lambda)^\dagger]^{-1} \end{bmatrix} \begin{bmatrix} I_n & -R(\lambda) \\ R(\lambda)^\dagger & I_n \end{bmatrix}, \tag{3.18}$$

$$a_r(\lambda) = \begin{bmatrix} [T_l(\lambda)^\dagger]^{-1} & 0 \\ 0 & T_r(\lambda)^{-1} \end{bmatrix} \begin{bmatrix} I_n & L(\lambda)^\dagger \\ -L(\lambda) & I_n \end{bmatrix}, \tag{3.19}$$

and hence, using (2.1), (3.15), (3.18), and (3.19), we get (3.16). Using (2.2), (2.34), (3.5), and (3.6) it follows that

$$f_-(x,\lambda) = f_+(x,\lambda)\, J_{2n}\, \boldsymbol{S}(\lambda)\, J_{2n}, \qquad \lambda \in \mathbb{R}. \tag{3.20}$$

Thus, from (3.5), (2.38), (2.39), (3.20), and $\det J_{2n} = (-1)^n$, we obtain (3.17). □

Corollary 3.2. *Suppose $a_{l1}(\lambda)$ and $a_{r4}(\lambda)$ are nonsingular for all $\lambda \in \overline{\mathbb{C}^+}$. Then the transmission coefficients $T_l(\lambda)$ and $T_r(\lambda)$ and their inverses $T_l(\lambda)^{-1}$ and $T_r(\lambda)^{-1}$ are continuous in $\lambda \in \overline{\mathbb{C}^+}$ and analytic in $\lambda \in \mathbb{C}^+$; these four matrices converge to I_n as $\lambda \to \infty$ in $\overline{\mathbb{C}^+}$. Similarly, the matrices $T_l(\lambda^*)^\dagger$ and $T_r(\lambda^*)^\dagger$ and their inverses $[T_l(\lambda^*)^\dagger]^{-1}$ and $[T_r(\lambda^*)^\dagger]^{-1}$ are continuous in $\lambda \in \overline{\mathbb{C}^-}$ and analytic in $\lambda \in \mathbb{C}^-$; these four matrices converge to I_n as $\lambda \to \infty$ in $\overline{\mathbb{C}^-}$.*

In general, $R(\lambda)$ and $L(\lambda)$ do not have analytic continuations off the real axis. In the special case when $k(x)$ vanishes on a half line, we have the following.

Proposition 3.3. *If $k(x)$ is supported in the right half line $\mathbb{R}^+$, then $a_{l3}(\lambda)$ extends to a function that is continuous on $\overline{\mathbb{C}^+}$, is analytic on $\mathbb{C}^+$, and vanishes as $\lambda \to \infty$ in $\overline{\mathbb{C}^+}$. Similarly, if $k(x)$ is supported in the left half line $\mathbb{R}^-$, then $a_{r2}(\lambda)$ extends to a function that is continuous on $\overline{\mathbb{C}^+}$, is analytic on $\mathbb{C}^+$, and vanishes as $\lambda \to \infty$ in $\overline{\mathbb{C}^+}$.*

Proof. If k has support in $\mathbb{R}^+$, then from (2.27) and Proposition 2.3 we see that $a_{l3}(\lambda)$ has an extension that is continuous in $\lambda \in \overline{\mathbb{C}^+}$, is analytic in $\lambda \in \mathbb{C}^+$, and converges to 0 as $\lambda \to \infty$ in $\overline{\mathbb{C}^+}$. In a similar manner, if k is supported in $\mathbb{R}^-$, using (2.30) and Proposition 2.3, we obtain that $a_{r2}(\lambda)$ extends to a function that is continuous on $\overline{\mathbb{C}^+}$, is analytic on $\mathbb{C}^+$, and vanishes as $\lambda \to \infty$ in $\overline{\mathbb{C}^+}$. □

4. Fourier transforms

Let $\mathcal{W}^q$ denote the Wiener algebra of all $q \times q$ matrix functions of the form

$$Z(\lambda) = Z_\infty + \int_{-\infty}^{\infty} d\alpha\, z(\alpha)\, e^{i\lambda\alpha}, \tag{4.1}$$

where $z(\alpha)$ is a $q \times q$ matrix function whose entries belong to $L^1(\mathbb{R})$ and $Z_\infty = Z(\pm\infty)$. Then $\mathcal{W}^q$ is a Banach algebra with a unit element and endowed with the norm

$$\|Z\|_{\mathcal{W}^q} = \|Z_\infty\| + \int_{-\infty}^{\infty} d\alpha\, \|z(\alpha)\|,$$

and its invertible elements are those $Z(\lambda)$ as in (4.1) for which Z_∞ and $Z(\lambda)$ are nonsingular matrices for all $\lambda \in \mathbb{R}$ (see, e.g., [17]). We will use $\mathcal{W}^q_\pm$ to denote the subalgebra of those functions $Z(\lambda)$ for which $z(\alpha)$ has support in $\mathbb{R}^\pm$ and $\mathcal{W}^q_{\pm,0}$ to denote the subalgebra of those functions $Z(\lambda)$ for which $Z_\infty = 0$ and $z(\alpha)$ has support in $\mathbb{R}^\pm$. Then, $\mathcal{W}^q = \mathcal{W}^q_+ \oplus \mathcal{W}^q_{-,0} = \mathcal{W}^q_{+,0} \oplus \mathcal{W}^q_-$.

In this section we prove that the matrix functions $M_l(x,\cdot)$, $M_r(x,\cdot)$, and $\boldsymbol{S}(\cdot)$ belong to $\mathcal{W}^{2n}$, and that $m_\pm(x,\cdot)$ belongs to $\mathcal{W}^{2n}_\pm$. Let us construct the L^1-matrix functions $b_\pm(x,\cdot)$, $B_l(x,\cdot)$, and $B_r(x,\cdot)$ such that

$$m_\pm(x,\lambda) = I_{2n} + \int_0^{\infty} d\alpha\, b_\pm(x,\alpha)\, e^{\pm i\lambda\alpha}, \tag{4.2}$$

$$\begin{cases} M_l(x,\lambda) = I_{2n} + \displaystyle\int_0^{\infty} d\alpha\, B_l(x,\alpha)\, e^{i\lambda J_{2n}\alpha}, \\ M_r(x,\lambda) = I_{2n} + \displaystyle\int_0^{\infty} d\alpha\, B_r(x,\alpha)\, e^{-i\lambda J_{2n}\alpha}. \end{cases} \tag{4.3}$$

Indeed, partitioning the matrix functions $B_l(x,\alpha)$ and $B_r(x,\alpha)$ in (4.3) into $n \times n$ blocks as

$$B_l(x,\alpha) = \begin{bmatrix} B_{l1}(x,\alpha) & B_{l2}(x,\alpha) \\ B_{l3}(x,\alpha) & B_{l4}(x,\alpha) \end{bmatrix},$$

$$B_r(x,\alpha) = \begin{bmatrix} B_{r1}(x,\alpha) & B_{r2}(x,\alpha) \\ B_{r3}(x,\alpha) & B_{r4}(x,\alpha) \end{bmatrix},$$

so that

$$b_+(x,\alpha) = \begin{bmatrix} B_{l1}(x,\alpha) & B_{r2}(x,\alpha) \\ B_{l3}(x,\alpha) & B_{r4}(x,\alpha) \end{bmatrix},$$

$$b_-(x,\alpha) = \begin{bmatrix} B_{r1}(x,\alpha) & B_{l2}(x,\alpha) \\ B_{r3}(x,\alpha) & B_{l4}(x,\alpha) \end{bmatrix}, \tag{4.4}$$

we apply (4.3) to (2.12) and (2.17), and derive the coupled integral equations for $\alpha > 0$

$$B_{l1}(x,\alpha) = -i \int_x^\infty dy\, k(y)\, B_{l3}(y,\alpha), \tag{4.5}$$

$$B_{l2}(x,\alpha) = -\frac{i}{2}k(x+\alpha/2) - i \int_x^{x+\alpha/2} dy\, k(y)\, B_{l4}(y,\alpha+2x-2y), \tag{4.6}$$

$$B_{l3}(x,\alpha) = -\frac{i}{2}k(x+\alpha/2)^\dagger - i \int_x^{x+\alpha/2} dy\, k(y)^\dagger\, B_{l1}(y,\alpha+2x-2y), \tag{4.7}$$

$$B_{l4}(x,\alpha) = -i \int_x^\infty dy\, k(y)^\dagger\, B_{l2}(y,\alpha), \tag{4.8}$$

$$B_{r1}(x,\alpha) = i \int_{-\infty}^x dy\, k(y)\, B_{r3}(y,\alpha), \tag{4.9}$$

$$B_{r2}(x,\alpha) = \frac{i}{2}k(x-\alpha/2) + i \int_{x-\alpha/2}^x dy\, k(y)\, B_{r4}(y,\alpha+2y-2x), \tag{4.10}$$

$$B_{r3}(x,\alpha) = \frac{i}{2}k(x-\alpha/2)^\dagger + i \int_{x-\alpha/2}^x dy\, k(y)^\dagger\, B_{l1}(y,\alpha+2y-2x), \tag{4.11}$$

$$B_{r4}(x,\alpha) = i \int_{-\infty}^x dy\, k(y)^\dagger\, B_{r2}(y,\alpha). \tag{4.12}$$

We first prove that, for each $x \in \mathbb{R}$, the four systems of integral equations (4.5) and (4.7), (4.6) and (4.8), (4.9) and (4.11), (4.10) and (4.12) have unique solutions with entries in $L^1(\mathbb{R}^+)$. Then for the matrix functions $m_\pm(x,\lambda)$, $M_l(x,\lambda)$, and $M_r(x,\lambda)$ defined in (4.2) and (4.3), we derive the integral relations (2.13)–(2.16) and (2.18)–(2.21). In this way we will have proved that $M_l(x,\cdot)$ and $M_r(x,\cdot)$ belong to $\mathcal{W}^{2n}$ and $m_\pm(x,\cdot)$ belongs to $\mathcal{W}^{2n}_\pm$.

Let us introduce the following mixed norm on the $2n \times 2n$ matrix functions $B(x,\alpha)$ depending on $(x,\alpha) \in \mathbb{R} \times \mathbb{R}^+$:

$$\|B(\cdot,\cdot)\|_{\infty,1} = \sup_{x\in\mathbb{R}} \|B(x,\cdot)\|_1. \tag{4.13}$$

The proof of the next result is identical to that of the analogous result in [2].

Theorem 4.1. *Assume that the entries of $k(x)$ belong to $L^1(\mathbb{R})$. Then, for each $x \in \mathbb{R}$, the four pairs of integral equations* (4.5) *and* (4.7), (4.6) *and* (4.8), (4.9) *and* (4.11), (4.10) *and* (4.12) *have unique solutions with finite mixed norm as defined in* (4.13). *Consequently, $m_+(x,\cdot)$ belongs to $\mathcal{W}^{2n}_+$, $m_-(x,\cdot)$ belongs to $\mathcal{W}^{2n}_-$, and $M_l(x,\cdot)$ and $M_r(x,\cdot)$ belong to $\mathcal{W}^{2n}$.*

The integral equations (4.5)–(4.12) allow us to derive the following relations for the potential $k(x)$:

$$k(x) = 2i\, B_{l2}(x,0^+) = -2i\, B_{r2}(x,0^+) = -2i\, B_{l3}(x,0^+)^\dagger = 2i\, B_{r3}(x,0^+)^\dagger. \tag{4.14}$$

Theorem 4.2. *The scattering coefficients* $a_{l2}(\lambda)$, $a_{l3}(\lambda)$, $a_{r2}(\lambda)$, *and* $a_{r3}(\lambda)$ *belong to* $\mathcal{W}^n$, *and vanish as* $\lambda \to \pm\infty$. *The scattering coefficients* $a_{l1}(\lambda)$, $a_{l4}(\lambda)$, $a_{r1}(\lambda)$, *and* $a_{r4}(\lambda)$ *belong to* $\mathcal{W}^n_+$, *and they converge to* I_n *as* $\lambda \to \infty$ *in* $\overline{\mathbb{C}^+}$.

Proof. Using (2.3) and (2.6)–(2.8) we get

$$a_l(\lambda) = e^{-i\lambda J_{2n}x} J_{2n} M_r(x,\lambda)^\dagger J_{2n} M_l(x,\lambda) e^{i\lambda J_{2n}x}, \tag{4.15}$$

$$a_r(\lambda) = e^{-i\lambda J_{2n}x} J_{2n} M_l(x,\lambda)^\dagger J_{2n} M_r(x,\lambda) e^{i\lambda J_{2n}x}. \tag{4.16}$$

From Theorem 4.1 we see that $M_l(x,\lambda)$ and $M_r(x,\lambda)$ belong to $\mathcal{W}^{2n}$. Using (4.15) and (4.16) at $x=0$, we can show that $a_l(\lambda)$ and $a_r(\lambda)$ are products of elements of $\mathcal{W}^{2n}$ and hence belong to $\mathcal{W}^{2n}$. □

If $a_{l1}(\lambda)$ and $a_{r4}(\lambda)$ are both nonsingular for any $\lambda \in \mathbb{R}$, then Theorem 4.2 and (3.6) show that the reflection coefficients $L(\lambda)$ and $R(\lambda)$ belong to $\mathcal{W}^n$. Theorem 4.2 and (3.5) show that in this case the transmission coefficients $T_l(\lambda)$ and $T_r(\lambda)$ belong to $\mathcal{W}^n$ as well. To prove that $T_l(\lambda)$ and $T_r(\lambda)$ belong to $\mathcal{W}^n_+$, one needs that $a_{l1}(\lambda)$ and $a_{r4}(\lambda)$ are nonsingular for all $\lambda \in \overline{\mathbb{C}^+}$.

5. Wiener-Hopf factorization

Using (2.40), (2.41), and (3.20), we obtain the Riemann-Hilbert problem

$$m_-(x,\lambda) = m_+(x,\lambda)\, \boldsymbol{G}(x,\lambda), \tag{5.1}$$

where $\boldsymbol{G}(x,\lambda)$ is the unitarily dilated scattering matrix given by

$$\boldsymbol{G}(x,\lambda) = e^{i\lambda J_{2n}x} J_{2n}\, \boldsymbol{S}(\lambda)\, J_{2n}\, e^{-i\lambda J_{2n}x} = \begin{bmatrix} T_l(\lambda) & -R(\lambda)\, e^{2i\lambda x} \\ -L(\lambda)\, e^{-2i\lambda x} & T_r(\lambda) \end{bmatrix}. \tag{5.2}$$

Here $\boldsymbol{G}(x,\lambda)$ is a J_{2n}-unitary matrix which is defined for those $\lambda \in \mathbb{R}$ where $a_{l1}(\lambda)$ and $a_{r4}(\lambda)$ are nonsingular.

Equation (5.1) can in principle be used to compute the potential from a reflection matrix. To do so, we first construct the scattering matrix $\boldsymbol{S}(\lambda)$ in terms of $L(\lambda)$ or $R(\lambda)$ alone. Indeed, given $R(\lambda)$ for $\lambda \in \mathbb{R}$ and assuming it to be continuous for $\lambda \in \mathbb{R}$, we first obtain the matrix function $T_{l0}(\lambda)$ which is continuous on $\overline{\mathbb{C}^+}$, is analytic on $\mathbb{C}^+$ and tends to I_n as $\lambda \to \infty$ in $\overline{\mathbb{C}^+}$, by performing the Wiener-Hopf factorization

$$T_{l0}(\lambda)\, T_{l0}(\lambda)^\dagger = I_n + R(\lambda)\, R(\lambda)^\dagger, \qquad \lambda \in \mathbb{R}, \tag{5.3}$$

in agreement with (3.12). In a similar way, the matrix function $T_{r0}(\lambda)$ which is continuous on $\overline{\mathbb{C}^+}$, is analytic on $\mathbb{C}^+$ and tends to I_n as $\lambda \to \infty$ in $\overline{\mathbb{C}^+}$, is constructed by performing the Wiener-Hopf factorization

$$T_{r0}(\lambda)^\dagger\, T_{r0}(\lambda) = I_n + R(\lambda)^\dagger\, R(\lambda), \qquad \lambda \in \mathbb{R}, \tag{5.4}$$

in agreement with (3.12). We then define the matrix function

$$L_0(\lambda) = T_{r0}(\lambda)\, R(\lambda)^\dagger\, [T_{l0}(\lambda)^\dagger]^{-1}, \qquad \lambda \in \mathbb{R}. \tag{5.5}$$

Let $T_{l0}(\lambda)$ and $T_{r0}(\lambda)$ be the $n \times n$ matrix functions that are continuous in $\overline{\mathbb{C}^+}$, are analytic in $\mathbb{C}^+$, are nonsingular in $\overline{\mathbb{C}^+}$, and tend to I_n as $\lambda \to \infty$ in $\overline{\mathbb{C}^+}$, such that (5.3) and (5.4) are satisfied, and let $L_0(\lambda)$ be the corresponding right-hand side of (5.5). Then $T_{l0}(\lambda)$, $T_{r0}(\lambda)$, $R(\lambda)$, and $L_0(\lambda)$ are the scattering coefficients if $a_{l1}(\lambda)$ and $a_{r4}(\lambda)$ are nonsingular for $\lambda \in \overline{\mathbb{C}^+}$ (i.e., in the absence of bound states). When both $a_{l1}(\lambda)$ and $a_{r4}(\lambda)$ are nonsingular for $\lambda \in \mathbb{R}$ and at least one of $a_{l1}(\lambda)$ and $a_{r4}(\lambda)$ is singular for some $\lambda \in \mathbb{C}^+$ (i.e., in the presence of bound states), there are rational matrix functions $B_l(\lambda)$ and $B_r(\lambda)$ that are unitary for $\lambda \in \mathbb{R}$, tend to I_n as $\lambda \to \infty$, and are analytic in $\mathbb{C}^-$, such that

$$T_l(\lambda) = T_{l0}(\lambda)B_l(\lambda), \qquad T_r(\lambda) = B_r(\lambda)T_{r0}(\lambda). \tag{5.6}$$

Then we easily see that

$$L(\lambda) = B_r(\lambda)L_0(\lambda)B_l(\lambda). \tag{5.7}$$

Moreover, (3.15) implies that

$$\det B_l(\lambda) = \det B_r(\lambda) = \prod_{j=1}^{\mathcal{N}} \left(\frac{\lambda + i\kappa_j}{\lambda - i\kappa_j} \right)^{p_j}, \tag{5.8}$$

where $i\kappa_1, \ldots, i\kappa_{\mathcal{N}}$ are the distinct poles of the transmission coefficients $T_l(\lambda)$ and $T_r(\lambda)$ in $\mathbb{C}^+$ and $p_1, \ldots, p_{\mathcal{N}}$ are the respective poles orders of $\det B_l(\lambda)$. Thus $T_l(\lambda)$ and $T_r(\lambda)$ necessarily have the same poles in $\mathbb{C}^+$ and their determinants have the same pole orders. However, except in the case $n = 1$ where $B_l(\lambda)$ and $B_r(\lambda)$ are scalar functions that both coincide with the right-hand side of (5.8), the transmission coefficients may be different and have different sets of partial pole orders at the same pole in $\mathbb{C}^+$.

6. The Marchenko method

In order to establish the connection between the Riemann-Hilbert problem (5.1) and the Marchenko integral equations, *we assume throughout this section that $a_{l1}(\lambda)$ and $a_{r4}(\lambda)$ are nonsingular for all $\lambda \in \mathbb{R}$*. This allows one to express the scattering coefficients in terms of their Fourier transforms as

$$R(\lambda) = \int_{-\infty}^{\infty} d\alpha\, \hat{R}(\alpha)\, e^{-i\lambda\alpha}, \qquad L(\lambda) = \int_{-\infty}^{\infty} d\alpha\, \hat{L}(\alpha)\, e^{-i\lambda\alpha}, \tag{6.1}$$

$$T_l(\lambda) = I_n + \int_{-\infty}^{\infty} d\alpha\, \nu_l(\alpha)\, e^{i\lambda\alpha}, \qquad T_r(\lambda) = I_n + \int_{-\infty}^{\infty} d\alpha\, \nu_r(\alpha)\, e^{i\lambda\alpha}. \tag{6.2}$$

Note that by Theorem 4.2, $\nu_l(\alpha)$ and $\nu_r(\alpha)$ vanish for $\alpha < 0$ and their entries belong to $L^1(\mathbb{R}^+)$, while the entries of $\hat{R}(\cdot)$ and $\hat{L}(\cdot)$ belong to $L^1(\mathbb{R})$. Let us define

$$g(x, \alpha) = \begin{bmatrix} 0 & -\hat{R}(2x + \alpha) \\ -\hat{L}(-2x + \alpha) & 0 \end{bmatrix}, \qquad \alpha > 0. \tag{6.3}$$

Theorem 6.1. *Suppose $a_{l1}(\lambda)$ and $a_{r4}(\lambda)$ are nonsingular for all $\lambda \in \overline{\mathbb{C}^+}$. Then for each $x \in \mathbb{R}$ the matrices $b_-(x,\cdot)$ and $b_+(x,\cdot)$ defined in (4.4) satisfy the $2n \times 2n$ systems of coupled Marchenko equations*

$$b_-(x,\alpha) = g(x,\alpha) + \int_0^\infty d\beta\, b_+(x,\beta)\, g(x,\alpha+\beta), \tag{6.4}$$

$$b_+(x,\alpha) = J_{2n}\, g(x,\alpha)^\dagger\, J_{2n} + \int_0^\infty d\beta\, b_-(x,\beta)\, J_{2n}\, g(x,\alpha+\beta)^\dagger\, J_{2n}, \tag{6.5}$$

where $\alpha > 0$.

Proof. Using (4.2), (5.1), and the fact that $b_+(x,\alpha) = b_-(x,\alpha) = 0$ for $\alpha < 0$, we get

$$m_+(x,\lambda)\,[\boldsymbol{G}(x,\lambda) - I_{2n}] = \int_{-\infty}^\infty d\alpha\, [b_-(x,\alpha) - b_+(x,-\alpha)]\, e^{-i\lambda\alpha}, \qquad \lambda \in \mathbb{R}. \tag{6.6}$$

Furthermore, from (5.2) we conclude that under the above assumptions

$$\boldsymbol{G}(x,\lambda) - I_{2n} = \int_{-\infty}^\infty d\alpha\, \boldsymbol{H}(\alpha)\, e^{i\lambda\alpha}, \qquad \lambda \in \mathbb{R}, \tag{6.7}$$

where

$$\boldsymbol{H}(\alpha) = \begin{bmatrix} \nu_l(\alpha) & -\hat{R}(2x-\alpha) \\ -\hat{L}(-2x-\alpha) & \nu_r(\alpha) \end{bmatrix}, \qquad \alpha \in \mathbb{R}. \tag{6.8}$$

The hypotheses of Theorem 6.1 imply that $\nu_l(\alpha)$ and $\nu_r(\alpha)$ are supported on $\alpha \in \mathbb{R}^+$. Upon writing

$$m_+(x,\lambda)\,[\boldsymbol{G}(x,\lambda) - I_{2n}] = [\boldsymbol{G}(x,\lambda) - I_{2n}] + [m_+(x,\lambda) - I_{2n}]\,[\boldsymbol{G}(x,\lambda) - I_{2n}],$$

by using (6.6) on the left-hand side, (4.2), (6.1)–(6.3), (6.7), and (6.8) on the right-hand side, together with the convolution theorem, we obtain (6.4). Similarly, using

$$\begin{aligned} m_-(x,\lambda)\, J_{2n}\, [\boldsymbol{G}(x,\lambda)^\dagger - I_{2n}]\, J_{2n} &= J_{2n}\, [\boldsymbol{G}(x,\lambda)^\dagger - I_{2n}]\, J_{2n} \\ &\quad + [m_-(x,\lambda) - I_{2n}]\, J_{2n}\, [\boldsymbol{G}(x,\lambda)^\dagger - I_{2n}]\, J_{2n}, \end{aligned}$$

we obtain (6.5). □

Using (6.4) in (6.5) and vice versa, we can uncouple these $2n \times 2n$ systems. Using the notations in (4.4), this leads to the uncoupled $n \times n$ Marchenko equations for $\alpha > 0$ given by

$$\begin{aligned} B_{l2}(x,\alpha) &= -\hat{R}(\alpha+2x) \\ &\quad - \int_0^\infty d\beta \int_0^\infty d\gamma\, B_{l2}(x,\gamma)\, \hat{R}(\beta+\gamma+2x)^\dagger\, \hat{R}(\alpha+\beta+2x), \end{aligned} \tag{6.9}$$

$$\begin{aligned} B_{l3}(x,\alpha) &= \hat{R}(\alpha+2x)^\dagger \\ &\quad - \int_0^\infty d\beta \int_0^\infty d\gamma\, B_{l3}(x,\gamma)\, \hat{R}(\beta+\gamma+2x)\, \hat{R}(\alpha+\beta+2x)^\dagger, \end{aligned} \tag{6.10}$$

$$B_{r2}(x,\alpha) = \hat{L}(\alpha - 2x)^{\dagger} - \int_0^{\infty} d\beta \int_0^{\infty} d\gamma\, B_{r2}(x,\gamma)\, \hat{L}(\beta+\gamma-2x)\, \hat{L}(\alpha+\beta-2x)^{\dagger}, \tag{6.11}$$

$$B_{r3}(x,\alpha) = -\hat{L}(\alpha - 2x) - \int_0^{\infty} d\beta \int_0^{\infty} d\gamma\, B_{r3}(x,\gamma)\, \hat{L}(\beta+\gamma-2x)^{\dagger}\, \hat{L}(\alpha+\beta-2x), \tag{6.12}$$

$$B_{l1}(x,\alpha) = -\int_0^{\infty} d\beta\, \hat{R}(\beta+2x)\, \hat{R}(\alpha+\beta+2x)^{\dagger} - \int_0^{\infty} d\beta \int_0^{\infty} d\gamma\, B_{l1}(x,\gamma)\, \hat{R}(\beta+\gamma+2x)\, \hat{R}(\alpha+\beta+2x)^{\dagger}, \tag{6.13}$$

$$B_{l4}(x,\alpha) = -\int_0^{\infty} d\beta\, \hat{R}(\beta+2x)^{\dagger}\, \hat{R}(\alpha+\beta+2x) - \int_0^{\infty} d\beta \int_0^{\infty} d\gamma\, B_{l4}(x,\gamma)\, \hat{R}(\beta+\gamma+2x)^{\dagger}\, \hat{R}(\alpha+\beta+2x), \tag{6.14}$$

$$B_{r1}(x,\alpha) = -\int_0^{\infty} d\beta\, \hat{L}(\beta-2x)^{\dagger}\, \hat{L}(\alpha+\beta-2x) - \int_0^{\infty} d\beta \int_0^{\infty} d\gamma\, B_{r1}(x,\gamma)\, \hat{L}(\beta+\gamma-2x)^{\dagger}\, \hat{L}(\alpha+\beta-2x), \tag{6.15}$$

$$B_{r4}(x,\alpha) = -\int_0^{\infty} d\beta\, \hat{L}(\beta-2x)\, \hat{L}(\alpha+\beta-2x)^{\dagger} - \int_0^{\infty} d\beta \int_0^{\infty} d\gamma\, B_{r4}(x,\gamma)\, \hat{L}(\beta+\gamma-2x)\, \hat{L}(\alpha+\beta-2x)^{\dagger}. \tag{6.16}$$

Theorem 6.2. *The coupled system of Marchenko integral equations* (6.4) *and* (6.5) *is uniquely solvable in* $L^1(\mathbb{R}^+;\mathbb{C}^{2n\times 2n})$. *The integral operator in each of the eight uncoupled Marchenko equations* (6.9)–(6.16) *is selfadjoint, and each of these eight equations is uniquely solvable in* $L^1(\mathbb{R}^+;\mathbb{C}^{n\times n})$.

Proof. The selfadjointness of the integral operators in (6.9)–(6.16) is clear. From (3.12), (3.13), and Corollary 3.2 it follows that

$$\sup_{\lambda\in\mathbb{R}} \|L(\lambda)\| < +\infty, \qquad \sup_{\lambda\in\mathbb{R}} \|R(\lambda)\| < +\infty. \tag{6.17}$$

Now observe that, for fixed $x \in \mathbb{R}$, the action of the integral operators with kernels $\hat{R}(\alpha+\beta+2x)$, $\hat{R}(\alpha+\beta+2x)^{\dagger}$, $\hat{L}(\alpha+\beta-2x)$, and $\hat{L}(\alpha+\beta-2x)^{\dagger}$ on $L^2(\mathbb{R}^+;\mathbb{C}^n)$ can be interpreted as follows: one imbeds $L^2(\mathbb{R}^+;\mathbb{C}^n)$ into $L^2(\mathbb{R};\mathbb{C}^n)$ isometrically, applies the sign flip $h(\beta) \mapsto h(-\beta)$, implements a convolution with an L^1-matrix function, and then projects orthogonally onto $L^2(\mathbb{R}^+;\mathbb{C}^n)$. Since the Fourier transforms of these matrix functions are bounded in $\lambda \in \mathbb{R}$, also these integral operators are bounded in $L^2(\mathbb{R};\mathbb{C}^n)$. Since the adjoints of all of these systems can be written in the form

$$(I + K^\dagger K)B = C$$

for some bounded operator K on a Hilbert function space, the system of equations (6.4) and (6.5) as well as each of the eight equations (6.9)–(6.16) are uniquely solvable on the direct sum of a suitable number of copies of $L^2(\mathbb{R}^+)$. Since, as a result of the integrability of $\hat{L}(\cdot)$ and $\hat{R}(\cdot)$, the integral operators are compact on both L^2 and L^1 (cf. Lemma XII 2.4 of [17], the proof for the L^2-case there can easily be adapted to cover the L^1-case), the system of equations (6.4) and (6.5) as well as each of the eight equations (6.9)–(6.16) are uniquely solvable on the direct sum of a suitable number of copies of $L^1(\mathbb{R}^+)$. □

From (4.14), we see that we can recover the potential $k(x)$ by solving any one of the four Marchenko equations (6.9)–(6.12).

The unique solvability of the Marchenko equations (6.9)–(6.16) has a number of other consequences. For example, if $R(\lambda)$ is analytic on $\mathbb{C}^+$, then $\hat{R}(\alpha)$ is supported on $\mathbb{R}^-$ and hence the right-hand sides in (6.9), (6.10), (6.13), and (6.14) vanish for $x > 0$. Since these equations are uniquely solvable, their solutions vanish as well and therefore $k(x) = 0$ for $x > 0$. On the other hand, if $L(\lambda)$ is analytic on $\mathbb{C}^+$, then $\hat{L}(\alpha)$ is supported on $\mathbb{R}^-$, and hence the right-hand sides in (6.11), (6.12), (6.15), and (6.16) vanish for $x < 0$. Since these equations are uniquely solvable, their solutions vanish as well, and therefore $k(x) = 0$ for $x < 0$. We have thus proved the converse of Proposition 3.3.

It remains to prove that the potential $k(x)$ obtained by the Marchenko method has entries in $L^1(\mathbb{R})$. To do so, we modify the inversion procedure as follows. We solve one of the Marchenko equations (6.9) and (6.10) for $x > 0$ and then employ (4.14) to compute $k(x)$ for $x > 0$. By the same token, we solve one of (6.11) and (6.12) for $x < 0$ and then use (4.14) to find $k(x)$ for $x < 0$. In fact, this procedure will be implemented in the case of rational reflection coefficients in Section 8.

We first derive the following partial characterization result.

Theorem 6.3. *Let $R(\lambda)$ be a matrix function in $\mathcal{W}^n$ such that*

$$\sup_{\lambda\in\mathbb{R}} \|R(\lambda)\| < +\infty, \qquad \int_0^\infty d\alpha \left(\|\hat{R}(\alpha)\| + \alpha\,\|\hat{R}(\alpha)\|^2\right) < +\infty, \tag{6.18}$$

where $\hat{R}(\alpha)$ is defined in (6.1). Then, for $x > 0$, the unique solutions $B_{l2}(x,\alpha)$ and $B_{l3}(x,\alpha)$ of (6.9) and (6.10), respectively, satisfy

$$\int_0^\infty dx\,\|B_{lj}(x,0^+)\| < +\infty, \qquad j = 2, 3.$$

In particular, the entries of $k(x) = 2i\, B_{l2}(x, 0^+)$ *and* $k(x) = -2i\, B_{l3}(x, 0^+)^\dagger$ *belong to* $L^1(\mathbb{R}^+)$. *Similarly, let* $L(\lambda)$ *be a matrix function in* $\mathcal{W}^n$ *such that*

$$\sup_{\lambda\in\mathbb{R}} \|L(\lambda)\| < +\infty, \qquad \int_{-\infty}^{0} d\alpha \left(\|\hat{L}(\alpha)\| - \alpha\, \|\hat{L}(\alpha)\|^2\right) < +\infty, \tag{6.19}$$

where $\hat{L}(\alpha)$ *is defined in* (6.1). *Then for* $x < 0$ *the unique solutions* $B_{r2}(x, \alpha)$ *and* $B_{r3}(x, \alpha)$ *of* (6.11) *and* (6.12), *respectively, satisfy*

$$\int_{-\infty}^{0} dx\, \|B_{rj}(x, 0^+)\| < +\infty, \qquad j = 2, 3.$$

In particular, the entries of $k(x) = -2i\, B_{r2}(x, 0^+)$ *and* $k(x) = 2i\, B_{r3}(x, 0^+)^\dagger$ *belong to* $L^1(\mathbb{R}^-)$.

Proof. We only prove the theorem for $x > 0$, as the proof for $x < 0$ is similar. Put

$$\hat{R}_\Delta(\alpha) = \begin{bmatrix} 0 & -\hat{R}(\alpha) \\ \hat{R}(\alpha)^\dagger & 0 \end{bmatrix},$$

and consider the integral equation

$$B_l(x, \alpha) - \int_0^\infty d\beta\, B_l(x, \beta)\, \hat{R}_\Delta(2x + \alpha + \beta) = \hat{R}_\Delta(2x + \alpha), \tag{6.20}$$

where $\alpha > 0$. This integral equation, which follows directly from (4.4) and (6.3) (6.5), has a unique solution in $L^1(\mathbb{R}^+; \mathbb{C}^{2n\times 2n})$ which coincides with the matrix function $B_l(x, \alpha)$ in (4.3). Moreover, the once iterated integral equation (6.20) has the form

$$(I + K^\dagger K)B = C$$

on $L^2(\mathbb{R}^+; \mathbb{C}^{2n\times 2n})$, which makes (6.20) uniquely solvable in $L^2(\mathbb{R}^+; \mathbb{C}^{2n\times 2n})$. Using the unique solvability of the equation obtained by taking the adjoint of the matrices on either side of (6.20) we obtain

$$\begin{aligned} \|B_l(x, \alpha)\| &\le \|\hat{R}_\Delta(2x + \alpha)\| + \left[\int_{2x+\alpha}^\infty d\gamma\, \|\hat{R}_\Delta(\gamma)\|^2\right]^{\frac{1}{2}} \left[\int_0^\infty d\beta\, \|B_l(x, \beta)\|^2\right]^{\frac{1}{2}} \\ &\le \|\hat{R}_\Delta(2x + \alpha)\| + C\left[\int_{2x+\alpha}^\infty d\gamma\, \|\hat{R}_\Delta(\gamma)\|^2\right]^{\frac{1}{2}} \left[\int_{2x}^\infty d\beta\, \|\hat{R}_\Delta(\beta)\|^2\right]^{\frac{1}{2}} \\ &\le \|\hat{R}_\Delta(2x + \alpha)\| + C\int_{2x}^\infty d\gamma\, \|\hat{R}_\Delta(\gamma)\|^2, \end{aligned}$$

where C is some constant, and therefore

$$\int_0^\infty dx\, \|B_l(x, \alpha)\| \le \int_0^\infty d\beta\, \|\hat{R}_\Delta(\beta)\| + \int_0^\infty dy\, y \|\hat{R}_\Delta(y)\|^2, \tag{6.21}$$

which is finite. $\square$

The natural conditions under which one would expect to be able to reconstruct a potential with L^1-entries for $x > 0$ are $R \in \mathcal{W}^n$ and

$$\sup_{\lambda\in\mathbb{R}} \|R(\lambda)\| < +\infty, \qquad \lim_{\lambda\to\pm\infty} \|R(\lambda)\| = 0. \tag{6.22}$$

However, evaluating the first iterate of (6.20) as $\alpha \to 0^+$, we get

$$B_l^{(1)}(x,0^+) = \int_{2x}^{\infty} d\beta\, \hat{R}_\Delta(\beta)^2 = \begin{bmatrix} \int_{2x}^{\infty} d\beta\, \hat{R}(\beta)\, \hat{R}(\beta)^\dagger & 0 \\ 0 & \int_{2x}^{\infty} d\beta\, \hat{R}(\beta)^\dagger\, \hat{R}(\beta) \end{bmatrix},$$

which strongly suggests that condition (6.18) is probably indispensable if the integral $\int_0^\infty dx\, \|B_l^{(1)}(x,0^+)\|$ is to be finite.

7. Construction of the scattering matrix

Throughout this section we assume that $R(\lambda)$ is a rational matrix function satisfying (6.22). We recall that then $R \in \mathcal{W}^n$ by the comments following the proof of Theorem 6.3. From the theory of transfer functions [7], since $R(\lambda) \to 0$ as $\lambda \to \pm\infty$, it follows that $R(\lambda)$ can be represented in the form

$$R(\lambda) = i\,\mathcal{C}(\lambda - i\mathcal{A})^{-1}\mathcal{B}, \qquad \lambda \in \mathbb{C}, \tag{7.1}$$

where $\mathcal{A}$, $\mathcal{B}$, and $\mathcal{C}$ are independent of λ and belong to $\mathbb{C}^{p\times p}$, $\mathbb{C}^{p\times n}$, and $\mathbb{C}^{n\times p}$, respectively, for some positive integer p. Here it is assumed that the order p of $\mathcal{A}$ is minimal, i.e., the realization (7.1) is minimal and hence unique up to similarity (cf. Theorems 6.1.4 and 6.1.5 in [24]).

Our goal is to construct $\boldsymbol{S}(\lambda)$ in terms of the matrices $\mathcal{A}$, $\mathcal{B}$, and $\mathcal{C}$ given in (7.1). Since $R(\lambda)$ is continuous for $\lambda \in \mathbb{R}$, from the minimality of the realization given in (7.1) it follows that $\mathcal{A}$ does not have any eigenvalues on the imaginary axis (cf. Theorem 6.2.2 of [24]). Using (7.1) in (5.3) and (5.4), we obtain

$$T_l(\lambda)\, T_l(\lambda^*)^\dagger = I_n + i\begin{bmatrix}\mathcal{C} & 0\end{bmatrix}(\lambda - i\mathcal{K}_l)^{-1}\begin{bmatrix}0\\ \mathcal{C}^\dagger\end{bmatrix}, \tag{7.2}$$

$$T_r(\lambda^*)^\dagger\, T_r(\lambda) = I_n + i\begin{bmatrix}0 & \mathcal{B}^\dagger\end{bmatrix}(\lambda - i\mathcal{K}_r)^{-1}\begin{bmatrix}\mathcal{B}\\ 0\end{bmatrix}, \tag{7.3}$$

where

$$\mathcal{K}_l = \begin{bmatrix}\mathcal{A} & -\mathcal{B}\mathcal{B}^\dagger\\ 0 & -\mathcal{A}^\dagger\end{bmatrix}, \qquad \mathcal{K}_r = \begin{bmatrix}\mathcal{A} & 0\\ -\mathcal{C}^\dagger\mathcal{C} & -\mathcal{A}^\dagger\end{bmatrix}. \tag{7.4}$$

Then $\mathcal{K}_l$ and $\mathcal{K}_r$ both have the set $\sigma(\mathcal{A}) \cup \{-\lambda^* : \lambda \in \sigma(\mathcal{A})\}$ as their spectrum ($\sigma(\mathcal{A})$ standing for the spectrum of $\mathcal{A}$), even though they need not have the same Jordan normal form. Note that the inverses of the right-hand sides in (7.2) and (7.3) can be written as

$$\left[T_l(\lambda^*)^\dagger\right]^{-1} T_l(\lambda)^{-1} = I_n - i\begin{bmatrix}\mathcal{C} & 0\end{bmatrix}(\lambda - i\mathcal{E})^{-1}\begin{bmatrix}0\\ \mathcal{C}^\dagger\end{bmatrix}, \tag{7.5}$$

$$T_r(\lambda)^{-1}\left[T_r(\lambda^*)^\dagger\right]^{-1} = I_n - i\begin{bmatrix}0 & \mathcal{B}^\dagger\end{bmatrix}(\lambda - i\mathcal{E})^{-1}\begin{bmatrix}\mathcal{B}\\0\end{bmatrix}, \tag{7.6}$$

where $\mathcal{E}$ is the "state characteristic matrix" given by

$$\mathcal{E} = \begin{bmatrix}\mathcal{A} & -\mathcal{B}\mathcal{B}^\dagger\\ -\mathcal{C}^\dagger\mathcal{C} & -\mathcal{A}^\dagger\end{bmatrix}, \tag{7.7}$$

which, apart from some factors $i = \sqrt{-1}$, has been used in [21]. We note that $\mathcal{K}_l$, $\mathcal{K}_r$, and $\mathcal{E}$ do not have eigenvalues on the imaginary axis. This follows from the invertibility of $I_n + R(\lambda)\,R(\lambda)^\dagger$ and Corollary 2.7 in [7]; for $\mathcal{K}_l$ and $\mathcal{K}_r$ this also follows immediately from the special form of the matrices $\mathcal{K}_l$ and $\mathcal{K}_r$ in (7.4) and the fact that $\mathcal{A}$ has no eigenvalues on the imaginary axis. Hence the matrices $(\lambda - i\mathcal{K}_l)^{-1}$, $(\lambda - i\mathcal{K}_r)^{-1}$, and $(\lambda - i\mathcal{E})^{-1}$ in (7.2), (7.3), (7.5), and (7.6) all exist for $\lambda \in \mathbb{R}$.

The following result is essential for obtaining explicit expressions for the factors $T_l(\lambda)$ and $T_r(\lambda)$ and their inverses.

Proposition 7.1. *Let $\mathcal{A}$, $\mathcal{B}$, and $\mathcal{C}$ be the matrices in the minimal realization given by* (7.1) *and consider the quadratic matrix equations*

$$\mathcal{A}\mathcal{X} + \mathcal{X}\mathcal{A}^\dagger = \mathcal{B}\mathcal{B}^\dagger - \mathcal{X}\mathcal{C}^\dagger\mathcal{C}\mathcal{X}, \tag{7.8}$$

$$\mathcal{A}^\dagger\mathcal{Y} + \mathcal{Y}\mathcal{A} = -\mathcal{C}^\dagger\mathcal{C} + \mathcal{Y}\mathcal{B}\mathcal{B}^\dagger\mathcal{Y}. \tag{7.9}$$

Then the spectrum of the matrix $\mathcal{E}$ given in (7.7) *is symmetric about the imaginary axis. Moreover, the spectral subspace $\mathcal{M}$ of $\mathcal{E}$ corresponding to its eigenvalues in the right half-plane is of the form*

$$\mathcal{M} = \left\{\begin{bmatrix}\mathcal{X}\\ I_p\end{bmatrix}u \,:\, u \in \mathbb{C}^p\right\}, \tag{7.10}$$

where $\mathcal{X}$ is a hermitian solution of (7.8), *and the spectral subspace $\mathcal{L}$ of $\mathcal{E}$ corresponding to its eigenvalues in the left half-plane is of the form*

$$\mathcal{L} = \left\{\begin{bmatrix}I_p\\ \mathcal{Y}\end{bmatrix}u \,:\, u \in \mathbb{C}^p\right\}, \tag{7.11}$$

where $\mathcal{Y}$ is a hermitian solution of (7.9). *The hermitian matrices $\mathcal{X}$ and $\mathcal{Y}$ are unique.*

Proof. The symmetry of the spectrum of $\mathcal{E}$ about the imaginary axis follows from the similarity $J_{2p}\boldsymbol{q}_{2p}\mathcal{E}\boldsymbol{q}_{2p}J_{2p} = -\mathcal{E}^\dagger$, where $\boldsymbol{q}_{2p}$ is defined by

$$\boldsymbol{q}_{2p} = \begin{bmatrix}0 & I_p\\ I_p & 0\end{bmatrix}.$$

The remaining assertions follow from Theorem 7.6.1 in [24] applied to the matrix $J_{2p}\mathcal{E}J_{2p}$ (to comply with the condition $D \geq 0$ there) and the Ξ-neutrality, where $\Xi = iJ_{2p}\boldsymbol{q}_{2p}$, of the spectral subspaces $\mathcal{M}$ and $\mathcal{L}$. Note that the spectral subspaces $\mathcal{M}$ and $\mathcal{L}$ both have dimension p, which is the order of $\mathcal{A}$, because $\mathcal{E}$ has no eigenvalues on the imaginary axis. Also note that the controllibility condition of Theorem 7.6.1 of [24] is satisfied as a result of the minimality of the realization in (7.1). Indeed, $\mathcal{C}^\dagger\mathcal{C}\mathcal{A}^j w = 0$ for $j \geq 0$ implies $\|\mathcal{C}\mathcal{A}^j w\|^2 = 0$ for $j \geq 0$, then $\mathcal{C}\mathcal{A}^j w = 0$ for $j \geq 0$, and then $w = 0$; similarly, one proves the other controllability statement. □

The nonlinear equations (7.8) and (7.9) are called state characteristic equations in [21] and (continuous algebraic) Riccati equations elsewhere in the literature (e.g., [24]). Since in the literature the term "hermitian" (instead of "selfadjoint") seems to have some tradition when referring to solutions of Riccati equations, we will use this terminology here.

The matrices $\mathcal{X}$ and $\mathcal{Y}$ used in Proposition 7.1 allow us to block diagonalize the matrix $\mathcal{E}$. Since the subspaces $\mathcal{L}$ and $\mathcal{M}$ have dimension p and $\mathcal{M} \cap \mathcal{L} = \{0\}$, the matrix Σ defined by

$$\Sigma = \begin{bmatrix} I_p & \mathcal{X} \\ \mathcal{Y} & I_p \end{bmatrix} \tag{7.12}$$

is nonsingular. Hence, both $I_p - \mathcal{X}\mathcal{Y}$ and $I_p - \mathcal{Y}\mathcal{X}$ are nonsingular, and

$$\Sigma^{-1} = \begin{bmatrix} (I_p - \mathcal{X}\mathcal{Y})^{-1} & -(I_p - \mathcal{X}\mathcal{Y})^{-1}\mathcal{X} \\ -(I_p - \mathcal{Y}\mathcal{X})^{-1}\mathcal{Y} & (I_p - \mathcal{Y}\mathcal{X})^{-1} \end{bmatrix}. \tag{7.13}$$

Theorem 7.2. *Let $\mathcal{A}$, $\mathcal{B}$, and $\mathcal{C}$ be the matrices in the minimal realization given by (7.1) and let $\mathcal{X}$ and $\mathcal{Y}$ be as in Proposition 7.1. Then*

$$\Sigma^{-1}\mathcal{E}\Sigma = \begin{bmatrix} \mathcal{E}_r & 0 \\ 0 & -\mathcal{E}_l^\dagger \end{bmatrix}, \tag{7.14}$$

where

$$\mathcal{E}_r = \mathcal{A} - \mathcal{B}\mathcal{B}^\dagger\mathcal{Y}, \qquad \mathcal{E}_l = \mathcal{A} + \mathcal{X}\mathcal{C}^\dagger\mathcal{C}. \tag{7.15}$$

Moreover, the matrices $\mathcal{E}_r$ and $\mathcal{E}_l$ have all their eigenvalues in the left half-plane and are related via the similarity transformation

$$\mathcal{E}_r = (I_p - \mathcal{X}\mathcal{Y})^{-1}\mathcal{E}_l\,(I_p - \mathcal{X}\mathcal{Y}). \tag{7.16}$$

Proof. The relations (7.14)–(7.16) follow by direct computation using (7.7)–(7.9), (7.12), and (7.13). The assertions about the spectra of $\mathcal{E}_r$ and $\mathcal{E}_l$ follow from (7.14) and Proposition 7.1 which imply that $\mathcal{E}|_{\mathcal{L}}$ is similar to $\mathcal{E}_r$ and $\mathcal{E}|_{\mathcal{M}}$ is similar to $-\mathcal{E}_l^\dagger$. □

In the following we also need representations of the form (7.10) and (7.11) for certain invariant subspaces of $\mathcal{K}_l$ and $\mathcal{K}_r$.

Proposition 7.3. *Let $\mathcal{A}$, $\mathcal{B}$, and $\mathcal{C}$ be the matrices in the minimal realization given by (7.1). Then the spectrum of $\mathcal{K}_l$ ($\mathcal{K}_r$) is symmetric about the imaginary axis. Moreover, the invariant spectral subspaces of $\mathcal{K}_l$ and $\mathcal{K}_r$ corresponding to the left and right half-planes all have dimension p. In the case of $\mathcal{K}_r$ both of the invariant subspaces are of the form*

$$\left\{ \begin{bmatrix} \widetilde{\mathcal{X}} \\ I_p \end{bmatrix} u : u \in \mathbb{C}^p \right\}, \tag{7.17}$$

where $\widetilde{\mathcal{X}}$ is a solution of the Riccati equation

$$\mathcal{A}\widetilde{\mathcal{X}} + \widetilde{\mathcal{X}}\mathcal{A}^\dagger = -\widetilde{\mathcal{X}}\mathcal{C}^\dagger\mathcal{C}\widetilde{\mathcal{X}}. \tag{7.18}$$

In the case of $\mathcal{K}_l$ both of the invariant subspaces are of the form

$$\left\{ \begin{bmatrix} I_p \\ \widetilde{\mathcal{Y}} \end{bmatrix} u : u \in \mathbb{C}^p \right\}, \tag{7.19}$$

where $\widetilde{\mathcal{Y}}$ is a hermitian solution of the Riccati equation

$$\mathcal{A}^\dagger\widetilde{\mathcal{Y}} + \widetilde{\mathcal{Y}}\mathcal{A} = \widetilde{\mathcal{Y}}\mathcal{B}\mathcal{B}^\dagger\widetilde{\mathcal{Y}}. \tag{7.20}$$

Proof. Apply Theorem 7.2.4 of [24] to $J_{2p}\mathcal{K}_l J_{2p}$ and $\boldsymbol{q}_{2p}\mathcal{K}_r\boldsymbol{q}_{2p}$. The symmetry of the spectrum about the imaginary axis follows as in the proof of Proposition 7.1 for $\mathcal{E}$. □

Before we can apply the main factorization result from [7] to (7.2) and (7.3), we need the following proposition based on the positive selfadjointness of the matrix functions in (7.2) and (7.3) for all $\lambda \in \mathbb{R}$.

Proposition 7.4. *Let $\mathcal{M}$ (resp. $\mathcal{L}$) be the invariant subspace of the matrix $\mathcal{E}$ given in (7.7) corresponding to the eigenvalues in the right and left half-plane, respectively, and let $\mathcal{N}$ (resp. $\mathcal{V}$) be the invariant subspace of $\mathcal{K}_r$ (resp. $\mathcal{K}_l$) corresponding to its eigenvalues in the right and left half-plane, respectively. Then*

$$\mathcal{L} \oplus \mathcal{N} = \mathbb{C}^{2p}, \qquad \mathcal{M} \oplus \mathcal{V} = \mathbb{C}^{2p}.$$

Proof. The above decompositions then follow from Theorem I 1.5 of [7], due to the existence of left and right canonical factorizations of a positive selfadjoint matrix function with respect to the imaginary line. □

Now let Π be the projection such that

$$\text{Im } \Pi = \mathcal{L}, \qquad \text{Ker } \Pi = \mathcal{N}, \tag{7.21}$$

and let $\mathcal{Q}$ be the projection such that

$$\text{Im } \mathcal{Q} = \mathcal{V}, \qquad \text{Ker } \mathcal{Q} = \mathcal{M}. \tag{7.22}$$

Applying Theorem 1.5 of [7] we can express the transmission coefficients in terms of the matrices appearing in (7.1) and the projections Π and $\mathcal{Q}$ as follows:

$$T_r(\lambda^*)^\dagger = I_n + i\begin{bmatrix} 0 & \mathcal{B}^\dagger \end{bmatrix}(\lambda - i\mathcal{K}_r)^{-1}(I_{2p} - \Pi)\begin{bmatrix} \mathcal{B} \\ 0 \end{bmatrix}, \tag{7.23}$$

$$T_r(\lambda) = I_n + i\begin{bmatrix} 0 & \mathcal{B}^\dagger \end{bmatrix}\Pi(\lambda - i\mathcal{K}_r)^{-1}\begin{bmatrix} \mathcal{B} \\ 0 \end{bmatrix}, \tag{7.24}$$

$$T_r(\lambda)^{-1} = I_n - i\begin{bmatrix} 0 & \mathcal{B}^\dagger \end{bmatrix}(\lambda - i\mathcal{E})^{-1}\Pi\begin{bmatrix} \mathcal{B} \\ 0 \end{bmatrix}, \tag{7.25}$$

$$[T_r(\lambda^*)^\dagger]^{-1} = I_n - i\begin{bmatrix} 0 & \mathcal{B}^\dagger \end{bmatrix}(I_{2p} - \Pi)(\lambda - i\mathcal{E})^{-1}\begin{bmatrix} \mathcal{B} \\ 0 \end{bmatrix}, \tag{7.26}$$

$$T_l(\lambda) = I_n + i\begin{bmatrix}\mathcal{C} & 0\end{bmatrix}(\lambda - i\mathcal{K}_l)^{-1}\mathcal{Q}\begin{bmatrix}0\\ \mathcal{C}^\dagger\end{bmatrix}, \tag{7.27}$$

$$T_l(\lambda^*)^\dagger = I_n + i\begin{bmatrix}\mathcal{C} & 0\end{bmatrix}(I_{2p} - \mathcal{Q})(\lambda - i\mathcal{K}_l)^{-1}\begin{bmatrix}0\\ \mathcal{C}^\dagger\end{bmatrix}, \tag{7.28}$$

$$[T_l(\lambda^*)^\dagger]^{-1} = I_n - i\begin{bmatrix}\mathcal{C} & 0\end{bmatrix}(\lambda - i\mathcal{E})^{-1}(I_{2p} - \mathcal{Q})\begin{bmatrix}0\\ \mathcal{C}^\dagger\end{bmatrix}, \tag{7.29}$$

$$T_l(\lambda)^{-1} = I_n - i\begin{bmatrix}\mathcal{C} & 0\end{bmatrix}\mathcal{Q}(\lambda - i\mathcal{E})^{-1}\begin{bmatrix}0\\ \mathcal{C}^\dagger\end{bmatrix}. \tag{7.30}$$

With the expressions (7.23)–(7.30) we have accomplished the desired canonical factorizations of the matrix functions on the right-hand sides of (7.2), (7.3), (7.5), and (7.6). Our next goal is to find more explicit representations for the projections Π and $\mathcal{Q}$ and for the invariant subspaces $\mathcal{N}$ and $\mathcal{V}$.

Proposition 7.5. *Let $\mathcal{X}$ and $\mathcal{Y}$ be as in Proposition 7.1 and let $\widetilde{\mathcal{X}} = \widetilde{\mathcal{X}}_+$ and $\widetilde{\mathcal{Y}} = \widetilde{\mathcal{Y}}_-$ be as in Proposition 7.3, where the subscript + (resp. −) indicates that the spectral subspaces given in (7.17) and (7.19) are those associated with the right (left) half-plane. Then the invariant subspaces $\mathcal{N}$ and $\mathcal{V}$ and the projections Π and $\mathcal{Q}$ can be written as*

$$\mathcal{N} = \left\{\begin{bmatrix}\widetilde{\mathcal{X}}_+\\ I_p\end{bmatrix}u : u \in \mathbb{C}^p\right\}, \qquad \mathcal{V} = \left\{\begin{bmatrix}I_p\\ \widetilde{\mathcal{Y}}_-\end{bmatrix}u : u \in \mathbb{C}^p\right\}, \tag{7.31}$$

$$\Pi = \begin{bmatrix}(I_p - \widetilde{\mathcal{X}}_+\mathcal{Y})^{-1} & -(I_p - \widetilde{\mathcal{X}}_+\mathcal{Y})^{-1}\widetilde{\mathcal{X}}_+\\ \mathcal{Y}(I_p - \widetilde{\mathcal{X}}_+\mathcal{Y})^{-1} & -\mathcal{Y}(I_p - \widetilde{\mathcal{X}}_+\mathcal{Y})^{-1}\widetilde{\mathcal{X}}_+\end{bmatrix}, \tag{7.32}$$

$$\mathcal{Q} = \begin{bmatrix}(I_p - \mathcal{X}\widetilde{\mathcal{Y}}_-)^{-1} & -(I_p - \mathcal{X}\widetilde{\mathcal{Y}}_-)^{-1}\mathcal{X}\\ \widetilde{\mathcal{Y}}_-(I_p - \mathcal{X}\widetilde{\mathcal{Y}}_-)^{-1} & -\widetilde{\mathcal{Y}}_-(I_p - \mathcal{X}\widetilde{\mathcal{Y}}_-)^{-1}\mathcal{X}\end{bmatrix}. \tag{7.33}$$

Furthermore, if $\mathcal{A}$ has all its eigenvalues in the set $\Gamma^\#$, then $\widetilde{\mathcal{X}}_+ = 0$ and

$$\mathcal{N} = \{0\} \oplus \mathbb{C}^p, \qquad \mathcal{V} = \mathbb{C}^p \oplus \{0\}, \tag{7.34}$$

$$\Pi = \begin{bmatrix}I_p & 0\\ \mathcal{Y} & 0\end{bmatrix}, \qquad \mathcal{Q} = \begin{bmatrix}I_p & -\mathcal{X}\\ 0 & 0\end{bmatrix}. \tag{7.35}$$

Proof. First, (7.31) is immediate from (7.17), (7.19), (7.21), and (7.22). Then (7.32) and (7.33) follow from (7.21), (7.22), and Proposition 7.1. If $\mathcal{A}$ has all its eigenvalues in the right half-plane, then $\widetilde{\mathcal{X}}_+ = \widetilde{\mathcal{Y}}_- = 0$, by the particular form of $\mathcal{K}_l$ and $\mathcal{K}_r$ in (7.4), and so (7.34) and (7.35) follow from (7.31)–(7.33). □

In order to find more explicit expressions for Π and $\mathcal{Q}$ when $\mathcal{A}$ has at least one eigenvalue in Γ, we employ suitable similarity transformations which bring the images of $\mathcal{K}_l$ and $\mathcal{K}_r$ in a form amenable to the same treatment as if $\mathcal{A}$ had only eigenvalues in the left half-plane. To set up these similarity transformations it is convenient to choose a basis such that $\mathcal{A}$, $\mathcal{B}$, and $\mathcal{C}$ are partitioned as

$$\mathcal{A} = \begin{bmatrix}\mathcal{A}_- & 0\\ 0 & \mathcal{A}_+\end{bmatrix}, \qquad \mathcal{B} = \begin{bmatrix}\mathcal{B}_-\\ \mathcal{B}_+\end{bmatrix}, \qquad \mathcal{C} = \begin{bmatrix}\mathcal{C}_- & \mathcal{C}_+\end{bmatrix}. \tag{7.36}$$

Here $\mathcal{A}_+$ ($\mathcal{A}_-$) has all its eigenvalues in right (left) half-plane and we denote its order by p_+ (p_-), so that $p_+ + p_- = p$. Moreover, $\mathcal{B}_+$, $\mathcal{B}_-$, $\mathcal{C}_+$, and $\mathcal{C}_-$ are $p_+ \times n$, $p_- \times n$, $n \times p_+$, and $n \times p_-$ matrices, respectively. Now put

$$\Phi_l = \begin{bmatrix} I_{p_-} & 0 & 0 & 0 \\ 0 & 0 & 0 & I_{p_+} \\ 0 & 0 & I_{p_-} & 0 \\ 0 & -I_{p_+} & 0 & P_1 \end{bmatrix}, \qquad \Phi_r = \begin{bmatrix} I_{p_-} & 0 & 0 & 0 \\ 0 & P_2 & 0 & I_{p_+} \\ 0 & 0 & I_{p_-} & 0 \\ 0 & -I_{p_+} & 0 & 0 \end{bmatrix}, \tag{7.37}$$

where P_1 and P_2 are the unique solutions of the equations (cf. Theorem I 4.1 of [17], Theorem VII 2.4 of [13])

$$\mathcal{A}_+ P_1 + P_1 \mathcal{A}_+^\dagger = \mathcal{B}_+ \mathcal{B}_+^\dagger, \tag{7.38}$$

$$P_2 \mathcal{A}_+ + \mathcal{A}_+^\dagger P_2 = \mathcal{C}_+^\dagger \mathcal{C}_+. \tag{7.39}$$

In fact, we have

$$P_1 = \int_0^\infty dt\, e^{-t\mathcal{A}_+} \mathcal{B}_+ \mathcal{B}_+^\dagger e^{-t\mathcal{A}_+^\dagger}, \qquad P_2 = \int_0^\infty dt\, e^{-t\mathcal{A}_+^\dagger} \mathcal{C}_+^\dagger \mathcal{C}_+ e^{-t\mathcal{A}_+}, \tag{7.40}$$

so that P_1 and P_2 are positive selfadjoint. Then, we easily compute

$$\Omega_1 = \Phi_l \mathcal{K}_l \Phi_l^{-1} = \begin{bmatrix} \Omega_3 & -\Omega_5 \Omega_5^\dagger \\ 0 & -\Omega_3^\dagger \end{bmatrix}, \ \Omega_2 = \Phi_r \mathcal{K}_r \Phi_r^{-1} = \begin{bmatrix} \Omega_4 & 0 \\ -\Omega_6^\dagger \Omega_6 & -\Omega_4^\dagger \end{bmatrix}, \tag{7.41}$$

where

$$\Omega_3 = \begin{bmatrix} \mathcal{A}_- & -\mathcal{B}_- \mathcal{B}_+^\dagger \\ 0 & -\mathcal{A}_+^\dagger \end{bmatrix}, \ \Omega_4 = \begin{bmatrix} \mathcal{A}_- & 0 \\ -\mathcal{C}_+^\dagger \mathcal{C}_- & -\mathcal{A}_+^\dagger \end{bmatrix}, \ \Omega_5 = \begin{bmatrix} \mathcal{B}_- \\ 0 \end{bmatrix}, \ \Omega_6 = \begin{bmatrix} \mathcal{C}_- & 0 \end{bmatrix}.$$

Note that all the eigenvalues of Ω_1 and Ω_2 lie in the open left half-plane. Therefore, in analogy to (7.34), (7.35), and Proposition 7.1, the projection operators $\mathcal{Q}$ and Π are such that

$$\operatorname{Im} \mathcal{Q} = \Phi_l^{-1} \left[\mathbb{C}^p \oplus \{0\} \right], \qquad \operatorname{Ker} \mathcal{Q} = \operatorname{Im} \begin{bmatrix} \mathcal{X} \\ I_p \end{bmatrix}, \tag{7.42}$$

$$\operatorname{Ker} \Pi = \Phi_r^{-1} \left[\{0\} \oplus \mathbb{C}^p \right], \qquad \operatorname{Im} \Pi = \operatorname{Im} \begin{bmatrix} I_p \\ \mathcal{Y} \end{bmatrix}. \tag{7.43}$$

Let us partition the inverses of Φ_l and Φ_r defined in (7.37) into $p \times p$ blocks as

$$\Phi_l^{-1} = \begin{bmatrix} \Lambda_{l1} & \Lambda_{l2} \\ \Lambda_{l3} & \Lambda_{l4} \end{bmatrix}, \qquad \Phi_r^{-1} = \begin{bmatrix} \Lambda_{r1} & \Lambda_{r2} \\ \Lambda_{r3} & \Lambda_{r4} \end{bmatrix}. \tag{7.44}$$

Note that

$$\Phi_l^{-1} = \boldsymbol{q}_{2p} \Phi_l \boldsymbol{q}_{2p}, \qquad \Phi_r^{-1} = \boldsymbol{q}_{2p} \Phi_r \boldsymbol{q}_{2p}, \tag{7.45}$$

so that the entries of Φ_l^{-1} and Φ_r^{-1} are readily available from (7.3).

Proposition 7.6. *Suppose a basis is chosen such that the matrices $\mathcal{A}$, $\mathcal{B}$, and $\mathcal{C}$ in the realization* (7.1) *have the form indicated in* (7.36). *Then the matrices $\widetilde{\mathcal{X}}_+$, $\widetilde{\mathcal{Y}}_-$, Π, and $\mathcal{Q}$ in Proposition* 7.5 *can be expressed as*

$$\widetilde{\mathcal{Y}}_- = \Lambda_{l3}\Lambda_{l1}^{-1} = \begin{bmatrix} 0 & 0 \\ 0 & P_1^{-1} \end{bmatrix}, \qquad \widetilde{\mathcal{X}}_+ = \Lambda_{r2}\Lambda_{r4}^{-1} = \begin{bmatrix} 0 & 0 \\ 0 & -P_2^{-1} \end{bmatrix}, \tag{7.46}$$

$$\mathcal{Q} = \begin{bmatrix} \Lambda_{l1}(\Lambda_{l1} - \mathcal{X}\Lambda_{l3})^{-1} & -\Lambda_{l1}(\Lambda_{l1} - \mathcal{X}\Lambda_{l3})^{-1}\mathcal{X} \\ \Lambda_{l3}(\Lambda_{l1} - \mathcal{X}\Lambda_{l3})^{-1} & -\Lambda_{l3}(\Lambda_{l1} - \mathcal{X}\Lambda_{l3})^{-1}\mathcal{X} \end{bmatrix}, \tag{7.47}$$

$$I_{2p} - \Pi = \begin{bmatrix} -\Lambda_{r2}(\Lambda_{r4} - \mathcal{Y}\Lambda_{r2})^{-1}\mathcal{Y} & \Lambda_{r2}(\Lambda_{r4} - \mathcal{Y}\Lambda_{r2})^{-1} \\ -\Lambda_{r4}(\Lambda_{r4} - \mathcal{Y}\Lambda_{r2})^{-1}\mathcal{Y} & \Lambda_{r4}(\Lambda_{r4} - \mathcal{Y}\Lambda_{r2})^{-1} \end{bmatrix}. \tag{7.48}$$

Proof. It follows from (7.37), (7.42), (7.44), and (7.45) that

$$\mathcal{V} = \operatorname{Im} \mathcal{Q} = \left\{ \begin{bmatrix} \Lambda_{l1} \\ \Lambda_{l3} \end{bmatrix} u : u \in \mathbb{C}^p \right\}.$$

Now (7.31), (7.37), (7.44), and (7.45) imply (7.46) for $\widetilde{\mathcal{Y}}_-$. Similarly, by (7.31), (7.37), and (7.43)–(7.45), we have

$$\mathcal{N} = \operatorname{Ker} \Pi = \left\{ \begin{bmatrix} \Lambda_{r2} \\ \Lambda_{r4} \end{bmatrix} u : u \in \mathbb{C}^p \right\},$$

and so, by comparison with (7.31), we obtain (7.46) for $\widetilde{\mathcal{X}}_+$. Then (7.47) and (7.48) follow on using (7.46) in (7.32) and (7.33). In the derivation of (7.48) we have also used the identity $(I_p - \widetilde{\mathcal{X}}_+\mathcal{Y})^{-1}\widetilde{\mathcal{X}}_+ = \widetilde{\mathcal{X}}_+(I_p - \mathcal{Y}\widetilde{\mathcal{X}}_+)^{-1}$. □

Note that in (7.48) we have stated the result for $I_{2p} - \Pi$ rather than Π because we will only need the former. By using (7.36) and (7.37) one easily verifies that $\widetilde{\mathcal{X}}_+$ and $\widetilde{\mathcal{Y}}_-$ given in (7.46) satisfy (7.18) and (7.20), respectively.

In order to use the results of Proposition 7.6 in (7.23)–(7.30) we need some additional notation. We decompose the solution $\mathcal{X}$ of (7.8) as

$$\mathcal{X} = \begin{bmatrix} \mathcal{X}_1 & \mathcal{X}_2 \\ \mathcal{X}_3 & \mathcal{X}_4 \end{bmatrix},$$

so that $\mathcal{X}_1$ and $\mathcal{X}_4$ are selfadjoint and have orders p_- and p_+, respectively, and $\mathcal{X}_2^\dagger = \mathcal{X}_3$. We denote by P_3 the unique solution of the equation

$$\Omega_4 P_3 + P_3 \mathcal{E}_l^\dagger = -\Lambda_{r4} \mathcal{B}\mathcal{B}^\dagger, \tag{7.49}$$

which is given by

$$P_3 = \int_0^\infty dt\, e^{t\Omega_4} \Lambda_{r4} \mathcal{B}\mathcal{B}^\dagger e^{t\mathcal{E}_l^\dagger}, \tag{7.50}$$

and we define P_4 to be the unique (and generally nonsquare matrix) solution of the equation

$$\Omega_4 P_4 + P_4 \mathcal{A}_-^\dagger = -\Lambda_{r4} \mathcal{B}\mathcal{B}_-^\dagger,$$

given by

$$P_4 = \int_0^\infty dt\, e^{t\Omega_4} \Lambda_{r4} \mathcal{B}\mathcal{B}_-^\dagger\, e^{t\mathcal{A}_-^\dagger}. \tag{7.51}$$

Note that in contrast to the solutions P_1 and P_2 of (7.38) and (7.39), respectively, the matrices P_3 and P_4 are in general not selfadjoint. Furthermore, we let

$$\mathcal{J}_l = (\Lambda_{l1} - \Lambda_{l3}\mathcal{X})^{-1}\Lambda_{l1}, \qquad \mathcal{J}_r = \mathcal{Y}(\Lambda_{r4} - \Lambda_{r2}\mathcal{Y})^{-1}, \tag{7.52}$$

and introduce the matrices

$$\widetilde{\mathcal{A}} = \begin{bmatrix} \widetilde{\mathcal{A}}_+ & 0 \\ 0 & \widetilde{\mathcal{A}}_- \end{bmatrix}, \qquad \widetilde{\mathcal{B}} = \begin{bmatrix} \widetilde{\mathcal{B}}_+ \\ \widetilde{\mathcal{B}}_- \end{bmatrix}, \qquad \widetilde{\mathcal{C}} = \begin{bmatrix} \widetilde{\mathbb{C}}_+ & \widetilde{\mathcal{C}}_- \end{bmatrix}, \tag{7.53}$$

where

$$\widetilde{\mathcal{A}}_+ = \begin{bmatrix} -\mathcal{E}_l^\dagger & 0 \\ 0 & -\mathcal{A}_-^\dagger \end{bmatrix}, \qquad \widetilde{\mathcal{A}}_- = \begin{bmatrix} -\mathcal{A}_\Gamma^\dagger & 0 \\ -\Lambda_{r4}\mathcal{B}\mathcal{B}_+^\dagger & \Omega_4 \end{bmatrix}, \tag{7.54}$$

$$\widetilde{\mathcal{B}}_+ = \begin{bmatrix} \mathcal{J}_l\, \mathcal{C}^\dagger \\ 0 \end{bmatrix}, \qquad \widetilde{\mathcal{B}}_- = \begin{bmatrix} (P_1 - \mathcal{X}_4)^{-1}(\mathcal{X}_3\, \mathcal{C}_-^\dagger + \mathcal{X}_4\, \mathcal{C}_+^\dagger) \\ -P_3 \mathcal{J}_l\, \mathcal{C}^\dagger \end{bmatrix}, \tag{7.55}$$

$$\widetilde{\mathbb{C}}_+ = \begin{bmatrix} -\mathcal{B}^\dagger(I_p + \mathcal{J}_r P_3) & \mathcal{B}^\dagger \mathcal{J}_r P_4 + \mathcal{B}_-^\dagger \end{bmatrix}, \qquad \widetilde{\mathcal{C}}_- = \begin{bmatrix} \mathcal{B}_+^\dagger & -\mathcal{B}^\dagger \mathcal{J}_r \end{bmatrix}. \tag{7.56}$$

We mention that $\widetilde{\mathcal{A}}$, $\widetilde{\mathcal{B}}$, and $\widetilde{\mathcal{C}}$ are $3p \times 3p$, $3p \times n$, and $n \times 3p$ matrices, respectively. Moreover, $\widetilde{\mathcal{A}}_+$, $\widetilde{\mathcal{A}}_-$, $\widetilde{\mathcal{B}}_+$, $\widetilde{\mathcal{B}}_-$, $\widetilde{\mathcal{C}}_+$ and $\widetilde{\mathcal{C}}_-$ are $(p_- + p) \times (p_- + p)$, $(p_+ + p) \times (p_+ + p)$, $(p_- + p) \times n$, $(p_+ + p) \times n$, $n \times (p_- + p)$, and $n \times (p_+ + p)$ matrices, respectively.

Next we present the main result of this section, expressing the scattering matrix in terms of the quantities defined above in connection with the similarity transformations induced by Φ_l and Φ_r.

Theorem 7.7. *Let $R(\lambda)$ be a rational reflection coefficient satisfying* (6.22). *Then the remaining entries of the scattering matrix* (3.11) *are given by*

$$T_l(\lambda) = I_n - i\,\mathcal{C}\Lambda_{l1}(\lambda - i\Omega_3)^{-1}(\Lambda_{l1} - \mathcal{X}\Lambda_{l3})^{-1}\mathcal{X}\mathcal{C}^\dagger, \tag{7.57}$$

$$T_r(\lambda) = I_n + i\,\mathcal{B}^\dagger \mathcal{J}_r(\lambda - i\Omega_4)^{-1}\Lambda_{r4}\mathcal{B}, \tag{7.58}$$

$$L(\lambda) = i\,\widetilde{\mathcal{C}}(\lambda - i\widetilde{\mathcal{A}})^{-1}\widetilde{\mathcal{B}}. \tag{7.59}$$

In the special case when $\mathcal{A}$ has all its eigenvalues in the left half-plane, these expressions simplify to

$$T_l(\lambda) = I_n - i\,\mathcal{C}(\lambda - i\mathcal{A})^{-1}\mathcal{X}\mathcal{C}^\dagger, \tag{7.60}$$

$$T_r(\lambda) = I_n + i\,\mathcal{B}^\dagger\mathcal{Y}(\lambda - i\mathcal{A})^{-1}\mathcal{B}, \tag{7.61}$$

$$L(\lambda) = -i\,\mathcal{B}^\dagger\mathcal{Y}(\lambda - i\mathcal{A})^{-1}\mathcal{X}\mathcal{C}^\dagger - i\,\mathcal{B}^\dagger(I_p - \mathcal{Y}\mathcal{X})(\lambda + i\mathcal{E}_l^\dagger)^{-1}\mathcal{C}^\dagger. \tag{7.62}$$

Proof. Using (7.27), (7.37), (7.41), (7.44), (7.45), (7.47), and the equality

$$\Phi_l \mathcal{Q} \begin{bmatrix} 0 \\ \mathcal{C}^\dagger \end{bmatrix} = -\begin{bmatrix} I_p \\ 0 \end{bmatrix} (\Lambda_{l1} - \mathcal{X}\Lambda_{l3})^{-1}\mathcal{X}\mathcal{C}^\dagger,$$

we obtain (7.57). From (7.23), (7.37), (7.41), (7.44), (7.45), (7.48), as well as the identity

$$\Phi_r(I_{2p} - \Pi) \begin{bmatrix} \mathcal{B} \\ 0 \end{bmatrix} = -\begin{bmatrix} 0 \\ I_p \end{bmatrix} (\Lambda_{r4} - \mathcal{Y}\Lambda_{r2})^{-1}\mathcal{Y}\mathcal{B},$$

it follows that

$$T_r(\lambda^*)^\dagger = I_n - i\,\mathcal{B}^\dagger\Lambda_{r4}(\lambda + i\Omega_4^\dagger)^{-1}(\Lambda_{r4} - \mathcal{Y}\Lambda_{r2})^{-1}\mathcal{Y}\mathcal{B}.$$

Now (7.58) follows by taking the adjoint and using (7.52). Note that Λ_{r2}, Λ_{r4}, and $\mathcal{Y}$ are hermitian.

With the help of (7.14), (7.26), (7.47), and (7.52), we derive

$$\left[T_l(\lambda^*)^\dagger\right]^{-1} = I_n - i\,\mathcal{C}\mathcal{X}(\lambda + i\mathcal{E}_l^\dagger)^{-1}\mathcal{J}_l\,\mathcal{C}^\dagger. \tag{7.63}$$

From (7.1), (7.63), and using

$$I_p - \mathcal{J}_l = -\Lambda_{l3}(\Lambda_{l1} - \mathcal{X}\Lambda_{l3})^{-1}\mathcal{X},$$
$$-i\,\mathcal{C}^\dagger\mathcal{C}\mathcal{X} = (\lambda + i\mathcal{A}^\dagger) - (\lambda + i\mathcal{E}_l^\dagger),$$

we get

$$\begin{aligned} R(\lambda^*)^\dagger[T_l(\lambda^*)^\dagger]^{-1} &= i\mathcal{B}^\dagger(\lambda + i\mathcal{A}^\dagger)^{-1}\Lambda_{l3}(\Lambda_{l1} - \mathcal{X}\Lambda_{l3})^{-1}\mathcal{X}\mathcal{C}^\dagger \\ &\quad - i\mathcal{B}^\dagger(\lambda + i\mathcal{E}_l^\dagger)^{-1}\mathcal{J}_l\,\mathcal{C}^\dagger. \end{aligned} \tag{7.64}$$

Using (5.5), (7.57), (7.64), and some standard results on realizations (Chapter 1 of [7]), we obtain

$$L(\lambda) = i\,\Omega_8(\lambda - i\Omega_7)^{-1}\Omega_9, \tag{7.65}$$

where

$$\Omega_7 = \begin{bmatrix} \Omega_4 & \Lambda_{r4}\mathcal{B}\mathcal{B}^\dagger & -\Lambda_{r4}\mathcal{B}\mathcal{B}^\dagger \\ 0 & -\mathcal{A}^\dagger & 0 \\ 0 & 0 & -\mathcal{E}_l^\dagger \end{bmatrix},$$

$$\Omega_8 = \mathcal{B}^\dagger\begin{bmatrix}\mathcal{J}_r & I_p & -I_p\end{bmatrix}, \qquad \Omega_9 = \begin{bmatrix} 0 \\ \Lambda_{l3}(\Lambda_{l1} - \mathcal{X}\Lambda_{l3})^{-1}\mathcal{X} \\ \mathcal{J}_l \end{bmatrix}\mathcal{C}^\dagger.$$

To bring $L(\lambda)$ into the form (7.59) we use a similarity transformation. Let

$$\Psi = \begin{bmatrix} 0 & 0 & 0 & I_p \\ 0 & I_{p_-} & 0 & 0 \\ 0 & 0 & I_{p_+} & 0 \\ -I_p & P_4 & 0 & P_3 \end{bmatrix}, \qquad \Psi^{-1} = \begin{bmatrix} P_3 & P_4 & 0 & -I_p \\ 0 & I_{p_-} & 0 & 0 \\ 0 & 0 & I_{p_+} & 0 \\ I_p & 0 & 0 & 0 \end{bmatrix},$$

where P_3 and P_4 have been defined in (7.50) and (7.51). Then it is straightforward to verify that

$$\widetilde{\mathcal{A}} = \Psi\Omega_7\Psi^{-1}, \qquad \widetilde{\mathcal{C}} = \Omega_8\Psi^{-1}, \qquad \widetilde{\mathcal{B}} = \Psi\Omega_9, \tag{7.66}$$

where $\widetilde{\mathcal{A}}$, $\widetilde{\mathcal{B}}$, and $\widetilde{\mathcal{C}}$ are the matrices defined in (7.53). Using (7.66) in (7.65), together with the fact that for one of the blocks of Ω_9 we can write

$$\begin{aligned} \Lambda_{l3}(\Lambda_{l1} - \mathcal{X}\Lambda_{l3})^{-1}\mathcal{X} &= \begin{bmatrix} 0 & 0 \\ 0 & I_{p_+} \end{bmatrix}\begin{bmatrix} I_{p_-} & -\mathcal{X}_2 \\ 0 & P_1 - \mathcal{X}_4 \end{bmatrix}^{-1}\begin{bmatrix} \mathcal{X}_1 & \mathcal{X}_2 \\ \mathcal{X}_3 & \mathcal{X}_4 \end{bmatrix} \\ &= \begin{bmatrix} 0 & 0 \\ (P_1 - \mathcal{X}_4)^{-1}\mathcal{X}_3 & (P_1 - \mathcal{X}_4)^{-1}\mathcal{X}_4 \end{bmatrix}, \end{aligned}$$

we obtain (7.59) with the matrices (7.54)–(7.56). The expressions (7.60)–(7.62) can be obtained from (5.5) and (7.23)–(7.30) by using the special forms (7.35) for Π and $\mathcal{Q}$, or by obvious reductions from (7.57)–(7.59). The details are omitted. □

8. Inverse problem with rational scattering matrices

Let $R(\lambda)$ have the form (7.1) for certain matrices $\mathcal{A}$, $\mathcal{B}$, and $\mathcal{C}$, where $\mathcal{A}$ has minimal order and hence does not have zero or purely imaginary eigenvalues. Then

$$R(\lambda) = i\,\mathcal{C}(\lambda - i\mathcal{A})^{-1}\mathcal{B} = -\int_{-\infty}^{\infty} dt\, e^{-i\lambda t}\,\mathcal{C}\,E(t;-\mathcal{A})\,\mathcal{B}, \tag{8.1}$$

where

$$E(t;-\mathcal{A}) = \begin{cases} e^{-t\mathcal{A}}P_{\mathcal{A}}^{(+)} = \dfrac{1}{2\pi i}\displaystyle\int_{\Gamma_+} dz\, e^{-tz}\,(z-\mathcal{A})^{-1}, & t>0, \\ -e^{-t\mathcal{A}}P_{\mathcal{A}}^{(-)} = -\dfrac{1}{2\pi i}\displaystyle\int_{\Gamma_-} dz\, e^{-tz}\,(z-\mathcal{A})^{-1}, & t<0, \end{cases} \tag{8.2}$$

is the bisemigroup generated by $\mathcal{A}$ (cf. [8, 9]). Here Γ_+ and Γ_- are the positively oriented simple Jordan contours in the right and left half-planes enclosing all of the eigenvalues of $\mathcal{A}$ in the open right and left half-planes, respectively, and $P_{\mathcal{A}}^{(+)}$ and $P_{\mathcal{A}}^{(-)}$ are the spectral projections of $\mathcal{A}$ corresponding to its eigenvalues in the right and left half-planes, respectively.

Our strategy for reconstructing $k(x)$ from $R(\lambda)$ is as follows. When $x>0$ we will solve the Marchenko equation (6.10) by using $R(\lambda)$ as the input, and when $x<0$ we will solve the Marchenko equation (6.11) by using $L(\lambda)$ as given in (7.59). Then we use (4.14) to determine $k(x)$. Throughout we assume the absence of poles of $T_l(\lambda)$, $T_r(\lambda)$, $T_l(\lambda)^{-1}$, and $T_r(\lambda)^{-1}$ in the upper half-plane.

First consider (6.10) with

$$\hat{R}(t) = -\mathcal{C}\,E(t;-\mathcal{A})\,\mathcal{B}, \qquad \hat{R}(t)^\dagger = -\mathcal{B}^\dagger\,E(t;-\mathcal{A}^\dagger)\,\mathcal{C}^\dagger,$$

which are obtained from (6.1) and (8.1). Introducing the positive selfadjoint $p\times p$ matrices

$$\mathcal{D}_1 = \int_0^\infty dt\, E(t;-\mathcal{A})\,\mathcal{B}\,\mathcal{B}^\dagger E(t;-\mathcal{A}^\dagger),\quad \mathcal{D}_2 = \int_0^\infty dt\, E(t;-\mathcal{A}^\dagger)\,\mathcal{C}^\dagger\,\mathcal{C}\,E(t;-\mathcal{A}),$$

and assuming $x>0$, we obtain for the hermitian integral kernel in (6.10)

$$\int_0^\infty d\beta\,\hat{R}(\gamma+\beta+2x)\,\hat{R}(\alpha+\beta+2x)^\dagger = \mathcal{C}\,E(\gamma+2x;-\mathcal{A})\,\mathcal{D}_1\,E(\alpha+2x;-\mathcal{A}^\dagger)\,\mathcal{C}^\dagger.$$

The unique solution of the separable integral equation (6.10) is then given by

$$B_{l3}(x,\alpha) = -\mathcal{B}^\dagger\left[I_p + E(2x;-\mathcal{A}^\dagger)\,\mathcal{D}_2\,E(2x;-\mathcal{A})\,\mathcal{D}_1\right]^{-1}E(\alpha+2x;-\mathcal{A}^\dagger)\,\mathcal{C}^\dagger, \tag{8.3}$$

where the inverse exists because of the unique solvability of (6.10). For later use we note that, by (7.36) and (7.40), we have

$$\mathcal{D}_1 = \begin{bmatrix} 0 & 0 \\ 0 & P_1 \end{bmatrix}, \qquad \mathcal{D}_2 = \begin{bmatrix} 0 & 0 \\ 0 & P_2 \end{bmatrix}. \tag{8.4}$$

When $x < 0$ we start from $L(\lambda)$ as given in (7.59), so that

$$\hat{L}(t) = -\widetilde{\mathcal{C}}\, E(t; -\widetilde{\mathcal{A}})\, \widetilde{\mathcal{B}}, \qquad \hat{L}(t)^\dagger = -\widetilde{\mathcal{B}}^\dagger\, E(t; -\widetilde{\mathcal{A}}^\dagger)\, \widetilde{\mathcal{C}}^\dagger.$$

Proceeding as in the derivation of (8.3), we obtain

$$B_{r2}(x,\alpha) = -\widetilde{\mathcal{B}}^\dagger \left[I_{3p} + E(-2x; -\widetilde{\mathcal{A}}^\dagger)\, \mathcal{D}_4\, E(-2x; -\widetilde{\mathcal{A}})\, \mathcal{D}_3\right]^{-1} E(\alpha - 2x; -\widetilde{\mathcal{A}}^\dagger)\, \widetilde{\mathcal{C}}^\dagger, \tag{8.5}$$

where the inverse exists because of the unique solvability of (6.11). Here $\mathcal{D}_3$ and $\mathcal{D}_4$ are the positive selfadjoint matrices given by

$$\mathcal{D}_3 = \int_0^\infty dt\, E(t; -\widetilde{\mathcal{A}})\, \widetilde{\mathcal{B}}\, \widetilde{\mathcal{B}}^\dagger\, E(t; -\widetilde{\mathcal{A}}^\dagger), \; \mathcal{D}_4 = \int_0^\infty dt\, E(t; -\widetilde{\mathcal{A}}^\dagger)\, \widetilde{\mathcal{C}}^\dagger\, \widetilde{\mathcal{C}}\, E(t; -\widetilde{\mathcal{A}}),$$

which, by means of (7.53)–(7.56), can be written as

$$\mathcal{D}_3 = \begin{bmatrix} P_5 & 0 & 0 & 0 \\ 0 & 0 & 0 & 0 \\ 0 & 0 & 0 & 0 \\ 0 & 0 & 0 & 0 \end{bmatrix}, \qquad \mathcal{D}_4 = \begin{bmatrix} P_6 & P_7 & 0 & 0 \\ P_8 & P_9 & 0 & 0 \\ 0 & 0 & 0 & 0 \\ 0 & 0 & 0 & 0 \end{bmatrix},$$

where

$$P_5 = \int_0^\infty dt\, e^{t\mathcal{E}_l^\dagger} \mathcal{J}_l \mathcal{C}^\dagger \mathcal{C} \mathcal{J}_l^\dagger e^{t\mathcal{E}_l}, \; P_6 = \int_0^\infty dt\, e^{t\mathcal{E}_l} (I_p + P_3^\dagger \mathcal{J}_r^\dagger) \mathcal{B}\mathcal{B}^\dagger (I_p + \mathcal{J}_r P_3) e^{t\mathcal{E}_l^\dagger}, \tag{8.6}$$

and P_7, P_8, and P_9 are irrelevant because they will not contribute to $k(x)$, as we will see.

Now we are ready to prove the main result of this section. Again we first state the general result and then specialize it to the particular case when $\mathcal{A}$ has all its eigenvalues in the left half-plane or, equivalently, when $R(\lambda)$ is analytic in $\mathbb{C}^+$.

Theorem 8.1. *Suppose that $R(\lambda)$ satisfies (6.22) and is given by the minimal representation (7.1) in a basis where (7.36) holds. Then the matrix potential $k(x)$ in (1.2) is given by*

$$k(x) = \begin{cases} 2i\, \mathcal{C}_+\, e^{-2x\mathcal{A}_+} \left[I_{p_+} + P_1\, e^{-2x\mathcal{A}_+^\dagger}\, P_2\, e^{-2x\mathcal{A}_+}\right]^{-1} \mathcal{B}_+, & x > 0, \\ 2i\, \mathcal{C}\mathcal{J}_l^\dagger \left[I_p + e^{-2x\mathcal{E}_l}\, P_6\, e^{-2x\mathcal{E}_l^\dagger}\, P_5\right]^{-1} e^{-2x\mathcal{E}_l} (I_p + P_3^\dagger \mathcal{J}_r^\dagger) \mathcal{B}, & x < 0. \end{cases} \tag{8.7}$$

If $R(\lambda)$ is analytic in $\mathbb{C}^+$, then

$$k(x) = \begin{cases} 0, & x > 0, \\ 2i\, \mathcal{C}\left[I_p - e^{-2x\mathcal{E}_l} \mathcal{X}\, e^{-2x\mathcal{E}_r^\dagger}\, \mathcal{Y}\right]^{-1} e^{-2x\mathcal{E}_l}\, (I_p - \mathcal{X}\mathcal{Y})\, \mathcal{B}, & x < 0, \end{cases} \tag{8.8}$$

where $\mathcal{X}$ *and* $\mathcal{Y}$ *are the unique solutions of* (7.8) *and* (7.9), *respectively. Moreover, if* $R(\lambda)$ *is analytic in* $\mathbb{C}^+$, *then the jump in the potential at* $x = 0$ *is given by*

$$k(0^+) - k(0^-) = -k(0^-) = -2i\,\mathcal{C}\,\mathcal{B}. \tag{8.9}$$

Proof. The representation (8.7) for $k(x)$ is a direct consequence of (4.14) and (8.3)–(8.6). Thus we need only establish the simplifications that occur when $R(\lambda)$ is analytic in $\mathbb{C}^+$. In this case $\mathcal{A}$ has all its eigenvalues in the left half-plane so that from (8.2) we get $E(t; -\mathcal{A}) = 0$ for $t > 0$. Thus $k(x) = 0$ for $x > 0$. For $x < 0$, starting with (7.36), we can simplify the expression in (8.7) by deleting the blocks associated with the spectrum of $\mathcal{A}$ in the right half-plane. This reduction is implemented by the following substitutions: $\Lambda_{l1} \mapsto I_p$, $\Lambda_{l3} \mapsto 0$, and hence

$$\mathcal{J}_l^\dagger \mapsto I_p, \tag{8.10}$$

by (7.52). Similarly, $\Lambda_{r2} \mapsto 0$, $\Lambda_{r4} \mapsto I_p$, and hence $\mathcal{J}_r^\dagger \mapsto \mathcal{Y}$. Since $\Omega_4 \mapsto \mathcal{A}$, the solution to (7.49) becomes $P_3 = -\mathcal{X}$ and thus

$$I_p + P_3^\dagger \mathcal{J}_r^\dagger \mapsto I_p - \mathcal{X}\mathcal{Y}. \tag{8.11}$$

Furthermore we can compute P_5 and P_6 in (8.6). We observe that P_5 and P_6 are solutions to the following Riccati equations:

$$P_5\mathcal{E}_l + \mathcal{E}_l^\dagger P_5 = -\mathcal{C}^\dagger\mathcal{C}, \tag{8.12}$$

$$P_6\mathcal{E}_l^\dagger + \mathcal{E}_l P_6 = -(I_p - \mathcal{X}\mathcal{Y})\mathcal{B}\mathcal{B}^\dagger(I_p - \mathcal{Y}\mathcal{X}). \tag{8.13}$$

First note the identity

$$\mathcal{E}_l^\dagger\mathcal{Y} + \mathcal{Y}\mathcal{E}_r = -\mathcal{C}^\dagger\mathcal{C}(I_p - \mathcal{X}\mathcal{Y}), \tag{8.14}$$

which follows from (7.9) and (7.15). On multiplying (8.14) from the right by $(I_p - \mathcal{X}\mathcal{Y})^{-1}$, using (7.16), and comparing the result with (8.12), we find that

$$P_5 = \mathcal{Y}(I_p - \mathcal{X}\mathcal{Y})^{-1} = (I_p - \mathcal{Y}\mathcal{X})^{-1}\mathcal{Y}. \tag{8.15}$$

Similarly, on multiplying the identity

$$\mathcal{E}_r\mathcal{X} + \mathcal{X}\mathcal{E}_l^\dagger = \mathcal{B}\mathcal{B}^\dagger(I_p - \mathcal{Y}\mathcal{X}),$$

which follows from (7.8) and (7.15), from the left by $I_p - \mathcal{X}\mathcal{Y}$, using (7.16), and comparing the result with (8.13), we obtain

$$P_6 = -\mathcal{X}(I_p - \mathcal{Y}\mathcal{X}). \tag{8.16}$$

Since, by (7.16) and its adjoint,

$$(I_p - \mathcal{Y}\mathcal{X})e^{-2x\mathcal{E}_l^\dagger} = e^{-2x\mathcal{E}_r^\dagger}(I_p - \mathcal{Y}\mathcal{X}),$$

the result (8.8) for $x < 0$ follows from inserting (8.10), (8.11), (8.15), (8.16) in (8.7). Finally, letting $x \to 0$ from below gives $k(0^-) = -2i\,\mathcal{C}\mathcal{B}$, and hence (8.9) follows. □

Acknowledgments

The research was supported in part by MIUR under COFIN grant No. 2002014121 and by INdAM.

References

[1] M.J. Ablowitz, P.A. and Clarkson: "Solitons, nonlinear evolution equations and inverse scattering," London Math. Soc. Lecture Note Series 149, Cambridge Univ. Press, Cambridge, 1991.

[2] T. Aktosun, M. Klaus, and C. Van der Mee: "Direct and inverse scattering for self-adjoint Hamiltonian systems on the line," Integral Equations and Operator Theory 38, 129–178 (2000).

[3] D. Alpay and I. Gohberg: "Inverse spectral problems for differential operators with rational scattering matrix functions," J. Differential Equations 118, 1–19 (1995).

[4] D. Alpay and I. Gohberg: Inverse scattering problem for differential operators with rational scattering matrix functions. In: Böttcher, A. and Gohberg, I. (eds.): "Singular integral operators and related topics (Tel Aviv, 1995)," Birkhäuser OT 90, Basel and Boston, 1996, pp. 1–18.

[5] D. Alpay ánd I. Gohberg: Potential associated with rational weights. In: Gohberg, I. and Lyubich, Yu. (eds.): "New results in operator theory and its applications," Birkhäuser OT 98, Basel and Boston, 1997, pp. 23–40.

[6] D. Alpay and I. Gohberg: "State space method for inverse spectral problems," Progr. Systems Control Theory 22, 1–16 (1997).

[7] H. Bart, I. Gohberg, and M.A. Kaashoek: "Minimal factorization of matrix and operator functions," Birkhäuser OT 1, Basel and Boston, 1979.

[8] H. Bart, I. Gohberg, and M.A. Kaashoek: "Exponentially dichotomous operators and inverse Fourier transforms," Report 8511/M, Econometric Institute, Erasmus University of Rotterdam, 1985.

[9] H. Bart, I. Gohberg, and M.A. Kaashoek: "Wiener-Hopf factorization, inverse Fourier transforms and exponentially dichotomous operators," J. Funct. Anal. 68, 1–42 (1986).

[10] R. Beals and R.R. Coifman: "Scattering and inverse scattering for first order systems," Comm. Pure Appl. Math. 37, 39–90 (1984).

[11] R. Beals and R.R. Coifman: "Scattering and inverse scattering for first-order systems: II," Inverse Problems 3, 577–593 (1987).

[12] R. Beals, P. Deift, and C. Tomei: "Direct and inverse scattering on the line," Math. Surveys and Monographs, 28, Amer. Math. Soc., Providence, 1988.

[13] R. Bhatia: "Matrix analysis," Graduate Texts in Mathematics 169, Springer, New York, 1997.

[14] P. Deift and E. Trubowitz: "Inverse scattering on the line," Comm. Pure Appl. Math. 32, 121–251 (1979).

[15] L.D. Faddeev: "Properties of the S-matrix of the one-dimensional Schrödinger equation," Amer. Math. Soc. Transl. (Ser. 2), 65, 139–166 (1967) [Trudy Mat. Inst. Steklova 73, 314–336 (1964) (Russian)].

[16] M.G. Gasymov: "The inverse scattering problem for a system of Dirac equations of order $2n$," Trans. Moscow Math. Soc. 19, 41–119 (1968) [*Trudy Moscov. Mat. Obšč.* 19, 41–112 (1968) (Russian)].

[17] I. Gohberg, S. Goldberg, and M.A. Kaashoek: "Classes of linear operators. Vol. I," Birkhäuser OT 49, Basel and Boston, 1990.

[18] I. Gohberg, M.A. Kaashoek, and A.L. Sakhnovich: "Canonical systems with rational spectral densities: Explicit formulas and applications," Math. Nachr. 149, 93–125 (1998).

[19] I. Gohberg, M.A. Kaashoek, and A.L. Sakhnovich: "Pseudo-canonical systems with rational Weyl functions: Explicit formulas and applications," J. Differential Equations 146, 375–398 (1998).

[20] I. Gohberg, M.A. Kaashoek, and A.L. Sakhnovich: Canonical systems on the line with rational spectral densities: Explicit formulas. In: Adamayan, V.M., Gohberg, I., Gorbachuk, M.L., and Gorbachuk, V.I. (eds.): "Differential Operators and Related Topics (Odessa, 1997), Vol. I," Birkhäuser OT 117, Basel and Boston, 2000, pp. 127–139.

[21] I. Gohberg and S. Rubinstein: Proper contractions and their unitary minimal completions. In: Gohberg, I. (ed.): "Topics in interpolation theory of rational matrix-valued functions," Birkhäuser OT 33, Basel and Boston, 1988, pp. 223–247.

[22] B. Grébert: "Inverse scattering for the Dirac operator on the real line," Inverse Problems 8, 787–807 (1992).

[23] D.B. Hinton, A.K. Jordan, M. Klaus, J.K. and Shaw: "Inverse scattering on the line for a Dirac system," J. Math. Phys. 32, 3015–3030 (1991).

[24] P. Lancaster and L. Rodman: "Algebraic Riccati equations," Oxford University Press, New York, 1995.

[25] F.È. Melik-Adamyan: "On the properties of the S-matrix of canonical differential equations on the whole axis," Akad. Nauk Armjan. SSR Dokl. 58, 199–205 (1974) [Russian].

[26] F.È. Melik-Adamyan: "Canonical differential operators in a Hilbert space," Izv. Akad. Nauk Armjan. SSR Ser. Mat. 12, 10–31, 85 (1977) [Russian].

[27] F.È. Melik-Adamyan: "On a class of canonical differential operators," Soviet J. Contemporary Math. Anal. 24, 48–69 (1989) [Izv. Akad. Nauk Armyan. SSR Ser. Mat. 24, 570–592, 620 (1989) (Russian)]

[28] I.G. Petrovski: "Ordinary differential equations," Prentice-Hall, Englewood Cliffs, 1966.

[29] A.L. Sakhnovich: "Nonlinear Schrödinger equation on a semi-axis and an inverse problem associated with it," Ukr. Math. J. 42, 316–323 (1990) [Ukrain. Mat. Zh. 42, 356–363 (1990) (Russian)].

[30] L.A. Sakhnovich: "Factorization problems and operator identities," Russ. Math. Surv. 41, 1–64 (1986) [Usp. Mat. Nauk 41, 4–55, 240 (1986) (Russian)].

[31] L.A. Sakhnovich: "The method of operator identities and problems in analysis," St. Petersburg Math. J. 5, 1–69 (1994) [Algebra i Analiz 5, 3–80 (1993) (Russian)].

[32] A.B. Shabat: "An inverse scattering problem," Differential Equations 15, 1299–1307 (1980) [Differ. Urav. 15, 1824–1834, 1918 (1979) (Russian)].

Cornelis van der Mee
Dipartimento di Matematica e Informatica
Università di Cagliari, Viale Merello 92
I-09123 Cagliari, Italy
e-mail: `cornelis@bugs.unica.it`

Operator Theory:
Advances and Applications, Vol. 149, 441–465

Factorization of Block Triangular Matrix Functions in Wiener Algebras on Ordered Abelian Groups

Cornelis V.M. van der Mee, Leiba Rodman, Ilya M. Spitkovsky, and Hugo J. Woerdeman

Abstract. The notion of Wiener-Hopf type factorization is introduced in the abstract framework of Wiener algebras of matrix-valued functions on connected compact abelian groups. Factorizations of 2×2 block triangular matrix functions with elementary functions on the main diagonal are studied in detail. A conjecture is formulated concerning characterization of dual groups with the property that every invertible matrix function in a Wiener algebra admits a factorization. Applications of factorization are given to systems of difference equations and orthogonal families of functions.

1. Wiener Algebras

Let G be a (multiplicative) connected compact abelian group and let Γ be its (additive) character group. Recall that Γ consists of all continuous homomorphisms of G into the group of unimodular complex numbers. Since G is compact, Γ is discrete. In applications, often Γ is an additive subgroup of $\mathbb{R}$, the group of real numbers, or of $\mathbb{R}^k$, and G is the Bohr compactification of Γ. The group G can be also thought of as the character group of Γ, an observation that will be often used.

The group G has a unique invariant measure ν satisfying $\nu(G) = 1$, while Γ is equipped with the discrete topology and the (translation invariant) counting measure. It is well-known [31] that, because G is connected, Γ can be made into a linearly ordered group. So let $\preceq$ be a linear order such that $(\Gamma, \preceq)$ is an ordered group, i. e., if $x, y, z \in \Gamma$ and $x \preceq y$, then $x + z \preceq y + z$. *Throughout the paper it will be assumed that Γ is ordered with a fixed linear order $\preceq$.* The notation $\prec$, $\succeq$, $\succ$,

Received by the editors April 14, 2003; revised August 30, 2003. Submitted by J. A. Ball.
Mathematics Subject Classification (2000). Primary 47A68. Secondary 43A17, 42C05.
Key words and phrases. Wiener-Hopf factorization, Wiener algebras, linearly ordered groups, difference equations, orthogonal polynomials.
The research of CVMvdM leading to this article was supported by INdAM and MIUR under grant No. 2002014121. The research of LR, IMS, and HJW was partially supported by NSF grant DMS-9988579. The research of IMS and HJW was also partially supported by the Semester Research Grants from the College of William & Mary.

max, min (with obvious meaning) will also be used. We put $\Gamma_+ = \{x \in \Gamma : x \succeq 0\}$ and $\Gamma_- = \{x \in \Gamma : x \preceq 0\}$.

For any nonempty set M, let $\ell^1(M)$ stand for the complex Banach space of all complex-valued M-indexed sequences $\boldsymbol{x} = \{x_j\}_{j\in M}$ having at most countably many nonzero terms that are finite with respect to the norm

$$\|\boldsymbol{x}\|_1 = \sum_{j\in M} |x_j|.$$

Then $\ell^1(\Gamma)$ is a commutative Banach algebra with unit element with respect to the convolution product $(\boldsymbol{x} * \boldsymbol{y})_j = \sum_{k\in\Gamma} x_k\, y_{j-k}$. Further, $\ell^1(\Gamma_+)$ and $\ell^1(\Gamma_-)$ are closed subalgebras of $\ell^1(\Gamma)$ containing the unit element.

Given $a = \{a_j\}_{j\in\Gamma} \in \ell^1(\Gamma)$, by the *symbol* of a we mean the complex-valued continuous function $\hat{a}$ on G defined by

$$\hat{a}(g) = \sum_{j\in\Gamma} a_j \langle j, g\rangle, \qquad g \in G, \tag{1}$$

where $\langle j, g\rangle$ stands for the action of the character $j \in \Gamma$ on the group element $g \in G$ (thus, $\langle j, g\rangle$ is a unimodular complex number), or, by Pontryagin duality, of the character $g \in G$ on the group element $j \in \Gamma$. The set $\{j \in \Gamma : a_j \neq 0\}$ will be called the *Fourier spectrum* of $\hat{a}$ given by (1). Since Γ is written additively, we have

$$\langle \alpha + \beta, g\rangle = \langle \alpha, g\rangle \cdot \langle \beta, g\rangle, \qquad \alpha, \beta \in \Gamma, \quad g \in G,$$

$$\langle \alpha, gh\rangle = \langle \alpha, g\rangle \cdot \langle \alpha, h\rangle, \qquad \alpha \in \Gamma, \quad g, h \in G.$$

We will use the shorthand notation e_α for the function

$$e_\alpha(g) = \langle \alpha, g\rangle, \qquad g \in G. \tag{2}$$

Thus, $e_{\alpha+\beta} = e_\alpha e_\beta$, $\alpha, \beta \in \Gamma$.

The entries a_j can be retrieved from the symbol $\hat{a}$ as follows:

$$a_j = \int_G \hat{a}(g) e_j(g)\, d\nu(g), \qquad j \in \Gamma.$$

The set of all symbols of elements $a \in \ell^1(\Gamma)$ forms an algebra $W(G)$ of continuous functions on G. The algebra $W(G)$ (with pointwise multiplication and addition) is isomorphic to $\ell^1(\Gamma)$. Denote by $W(G)_+$ (resp., $W(G)_-$) the algebra of symbols of elements in $\ell^1(\Gamma_+)$ (resp., $\ell^1(\Gamma_-)$).

For every Banach algebra $\mathcal{A}$ with identity element we denote its group of invertible elements by $\mathcal{G}(\mathcal{A})$. We have the following result:

Theorem 1. *Let G be a compact abelian group with the character group Γ, and let $W(G)^{n\times n}$ be the corresponding Wiener algebra of $n \times n$ matrix functions. Then $\hat{A} \in \mathcal{G}(W(G)^{n\times n})$ if and only if $\hat{A}(g) \in \mathcal{G}(\mathbb{C}^{n\times n})$ for every $g \in G$.*

This is an immediate consequence of Theorem A.1 in [21] (also proved in [1]). To rephrase the theorem in terms of this result, in the Banach algebra $\mathcal{A} = W(G)^{n\times n}$ we distinguish the closed Banach subalgebra $\mathcal{Z} = \{\hat{a}\, I_n : \hat{a} \in W(G)^{1\times 1}\}$ contained in the center of $\mathcal{A}$ and the algebra $\mathcal{F}$ of constant $n \times n$ matrix functions. Then the algebraic tensor product $\mathcal{Z} \otimes \mathcal{F}$ is dense in $\mathcal{A}$ and the multiplicative functionals on $\mathcal{Z}$ are the evaluation maps $\hat{a} \mapsto \hat{a}(g)$, where $g \in G$. Further, the multiplicative projectors (in the terminology of [21]) turn out to be the evaluation maps $\hat{A} \mapsto \hat{A}(g)$, where $g \in G$. Thus the theorem is immediate from Theorem A.1 in [21], which completes the proof.

We now consider the discrete abelian subgroup Γ' of Γ and denote its character group by G'. Then we introduce the annihilator

$$\Lambda = \{g \in G : \langle j, g\rangle = 1 \text{ for all } j \in \Gamma'\}, \tag{3}$$

which is a closed subgroup of G and hence a compact group. According to Theorem 2.1.2 in [31], we have $G' \simeq (G/\Lambda)$.

Let us now introduce the natural projection $\pi : G \to (G/\Lambda)$. We observe that the above theorem also applies to $W(G')^{n\times n}$. Given $A \in \ell^1(\Gamma)^{n\times n}$ with its Fourier spectrum restricted to Γ' (i.e., $A_j = 0$ for $j \in \Gamma \setminus \Gamma'$), we have two symbol definitions:

$$\hat{A}_\Gamma(g) = \sum_{j\in\Gamma'} A_j \langle j, g\rangle, \qquad g \in G,$$

$$\hat{A}_{\Gamma'}(g) = \sum_{j\in\Gamma'} A_j \langle j, g\rangle, \qquad g \in G',$$

where we have taken into account that $A_j = 0$ for $j \in \Gamma \setminus \Gamma'$. The latter can be replaced by

$$\hat{A}_{\Gamma'}([g]) = \sum_{j\in\Gamma'} A_j \langle j, g\rangle, \qquad [g] \in (G/\Lambda),$$

where $[g] = \pi(g)$ for $g \in G$. Obviously, $\langle j, g\rangle$ only depends on $[g] = \pi(g)$ if $j \in \Gamma'$. (If $[g_1] = [g_2]$, then $g_1 g_2^{-1} \in \Lambda$ and hence $\langle j, g_1 g_2^{-1}\rangle = 1$ for all $j \in \Gamma'$, which implies the statement.) Thus the two symbol definitions are equivalent in the sense that the value of "the" symbol $\hat{A}$ on $g \in G$ only depends on $[g] = \pi(g)$. The following result is now obtained.

Theorem 2. *Let Γ' be a subgroup of the discrete abelian group Γ, let G and G' be the character groups of Γ and Γ', respectively, and let Λ be defined by* (3)*. If $\hat{A} \in W(G)^{n\times n}$ is an element which has all of its Fourier spectrum within Γ', then $\hat{A} \in \mathcal{G}(W(G')^{n\times n})$ if and only if $\hat{A}(g) \in \mathcal{G}(\mathbb{C}^{n\times n})$ for every $g \in G$.*

For the proof note that according to Theorem 1, we need to show that $\hat{A}_{\Gamma'}([g]) \in \mathcal{G}(\mathbb{C}^{n\times n})$ for every $[g] \in (G/\Lambda)$ if and only if $\hat{A}(g) \in \mathcal{G}(\mathbb{C}^{n\times n})$ for every $g \in G$. By the paragraph preceding Theorem 2, the statement is clear, since under the hypotheses on $\hat{A}$ the values $\hat{A}(g)$ of the symbol $\hat{A}$ depend only on $[g] \in (G/\Lambda)$.

We now consider factorizations, and begin with scalar-valued functions. A linearly ordered group Γ is called *Archimedean* if for every $u, v \in \Gamma$, $u, v \succ 0$, there exists a positive integer n such that $nu \succeq v$.

Proposition 3. *Assume that Γ is Archimedean, or that Γ is finitely generated. If $a \in \ell^1(\Gamma)$ and $\hat{a}(g) \neq 0$ for every $g \in G$, then $\hat{a}$ admits a factorization*

$$\hat{a}(g) = \hat{a}_+(g) e_j(g) \hat{a}_-(g), \qquad g \in G, \tag{4}$$

where $a_+ \in \mathcal{G}(\ell^1(\Gamma_+))$, $a_- \in \mathcal{G}(\ell^1(\Gamma_-))$, and $j \in \Gamma$. The element $j = j(\hat{a})$ is uniquely determined by $\hat{a}$.

Using the terminology of the classical case $\Gamma = \mathbb{Z}$, G the unit circle, we call the element $j(\hat{a})$ the *winding number* of $\hat{a}$. In the case $\Gamma = \mathbb{R}$ and G is the Bohr compactification of $\mathbb{R}$, the winding number is known as the *mean motion*. For the particular case $\Gamma = \mathbb{R}^k$, Proposition 3 was proved in [25] by elementary means; see also [30], where the case when Γ is a subgroup of $\mathbb{R}^k$ is studied. For the case $\Gamma = \mathbb{Z}^d$ ($\mathbb{Z}$ stands for the group of integers) and G the d-torus, Proposition 3 was proved in [11]. The general situation of Proposition 3 is easily reduced to the cases just mentioned in view of Hölder's theorem (see, e.g., [15]) which asserts that every linearly ordered Archimedean abelian group is order isomorphic to a subgroup of $\mathbb{R}$ (with the natural linear order induced from $\mathbb{R}$).

The concept of factorization as in Proposition 3 extends to $n \times n$ matrix functions in $(W(G))^{n\times n}$. A *(left) factorization* of $A(g) \in (W(G))^{n\times n}$ is a representation of the form

$$A(g) = A_+(g) \left(\operatorname{diag}\left(e_{j_1}(g), \dots, e_{j_n}(g)\right)\right) A_-(g), \quad g \in G, \tag{5}$$

where $A_+ \in \mathcal{G}((W(G)_+)^{n\times n})$, $A_- \in \mathcal{G}((W(G)_-)^{n\times n})$, and $j_1, \dots, j_n \in \Gamma$. Here and elsewhere we use $\operatorname{diag}(x_1, \dots, x_n)$ to denote the $n \times n$ diagonal matrix with $x_1, \dots, x_n$ on the main diagonal, in that order. The elements j_k are uniquely defined (if ordered $j_1 \preceq j_2 \preceq \dots \preceq j_n$); this can be proved by a standard argument (see [18, Theorem VIII.1.1]). The elements $j_1, \dots, j_n$ in (5) are called the *(left) factorization indices* of A. Analogously, by a *right factorization* of $A(g) \in (W(G))^{n\times n}$ we mean a representation of the form

$$A(g) = A_-(g) \left(\operatorname{diag}\left(e_{j_1}(g), \dots, e_{j_n}(g)\right)\right) A_+(g), \quad g \in G, \tag{6}$$

where $A_+ \in \mathcal{G}((W(G)_+)^{n\times n})$, $A_- \in \mathcal{G}((W(G)_-)^{n\times n})$, and $j_1, \dots, j_n \in \Gamma$. Unless stated otherwise, all notions involving factorization will pertain to left factorization.

If all factorization indices are zero, the factorization is called *canonical*. If a factorization of A exists, the function A is called *factorizable*. For $\Gamma = \mathbb{Z}$ and G the unit circle, the definitions and the results are classical [19], [18], [7]; many results have been generalized to $\Gamma = \mathbb{R}^k$ (see [4] and references there), and Γ a subgroup of $\mathbb{R}^k$ (see [27], [30]).

We postpone development of a comprehensive theory of factorization to another occasion. In this paper we will be concerned mainly with factorization of block triangular 2×2 matrix functions of the form

$$A(g) = \begin{bmatrix} e_\lambda(g)I_p & 0 \\ c_1 e_{\alpha_1}(g) + \cdots + c_m e_{\alpha_m}(g) & e_{-\lambda}(g)I_q \end{bmatrix}, \qquad g \in G, \tag{7}$$

where $\lambda, \alpha_j \in \Gamma$, and $c_j \in \mathbb{C}^{q\times p}$, applications of factorizations of this kind of functions, and related problems. Here and elsewhere, we use I_k (or I if k is clear from context) to denote the $k \times k$ identity matrix, and $0_{p\times q}$ (or 0) to denote the $p \times q$ zero matrix.

We immediately observe, as in the case of $\Gamma = \mathbb{R}$ [4], that by applying elementary row (resp., column) operations that do not change the factorizability property and (in case a factorization exists) the indices, one can eliminate terms in (7) with $\alpha_k \succeq \lambda$ and with $\alpha_k \preceq -\lambda$. In particular, when $\lambda \preceq 0$, the function (7) is factorizable with indices $\pm\lambda$ (p and q times, respectively). Thus, we often assume in the sequel that

$$\lambda \succ 0, \qquad -\lambda \prec \alpha_1 \prec \ldots \prec \alpha_m \prec \lambda. \tag{8}$$

Note also that for A defined by (7), A^{-1} has finite Fourier spectrum.

Recently, there is an interest in special cases of left and right factorizations (5), (6) in which the factors $A_\pm$ and their inverses $A_\pm^{-1}$ have finite Fourier spectrum ([17], [5]). If this happens we say that the factorization is *finite*. The function $A \in (W(G))^{n\times n}$ will be called *finitely factorizable* if it admits a finite factorization.

Proposition 4. *If A is given as in* (7), *and if one of the four functions A_+, A_+^{-1}, A_-, A_-^{-1} in a factorization* (5) *has finite Fourier spectrum, then the factorization* (5) *is finite. An analogous statement applies to factorization* (6).

Proof. Say, A_+ has finite Fourier spectrum. By taking determinants in (5), we easily see that A_+ and A_- have constant nonzero determinants. Thus, A_+^{-1} has finite Fourier spectrum, as follows from the formula

$$A_+^{-1} = \frac{\operatorname{adj}(A_+)}{\det(A_+)},$$

where $\operatorname{adj}(X)$ stands for the algebraic adjoint of a matrix X, and it suffices to show that the Fourier spectrum of A_-^{-1} is finite. To this end, rewrite (5) in the form

$$AA_-^{-1} = A_+ \operatorname{diag}(e_{j_1}, \ldots, e_{j_n})$$

and observe that the right-hand side, and hence also the left-hand side, has finite Fourier spectrum. Using the special structure of A (7), it is easy to see that A_-^{-1} has finite Fourier spectrum. □

To conclude the introduction we describe the contents of the paper. In Section 2 the Portuguese transformation (known in the case of $\Gamma = \mathbb{R}$, see, e.g., [4]) is described in the context of abstract groups. In Section 3 we focus on factorization of matrices of the form (7) whose off-diagonal blocks are binomials ($m = 2$). As we shall see, even in this case factorization does not always exist without additional hypotheses on the order. Problems of factorizability vs. invertibility are treated in

Section 4. There, known examples [22], [4] of nonfactorable functions of the form (7) with off-diagonal trinomials in the real case ($\Gamma = \mathbb{R}$) play an essential role. Applications of factorization to systems of differential equations and to orthogonal functions are given in the last two sections.

2. Portuguese transformation

The Portuguese transformation is the main tool to prove factorizability and compute indices of 2×2 matrix functions of the form (7). In explicit form, under the assumption that the matrices c_k commute, it appeared in [3], although for some particular cases it was employed earlier [34], [32]. Without the commutativity hypotheses, the Portuguese transformation was given in [26] (where it was called the BKST transformation), and a thorough exposition, from the viewpoint of corona theorems, of the Portuguese transformation is found in [4]. All previous works on the Portuguese transformation were restricted to the case $\Gamma = \mathbb{R}$.

In this section, we give formulas for the Portuguese transformation in the abstract setting of ordered abelian groups. Since corona theorems are not generally available in this setting, we will use the algebraic approach employed in [26] which does not utilize results of corona type.

Consider (7), and assume that $p = q$ and that (8) holds. Furthermore, assume that the matrix c_1 is invertible. Then, renaming terms in the off-diagonal entry of (7), and replacing m by $m+1$, we re-write (7) in the form

$$A(g) = \begin{bmatrix} e_\lambda(g) I_p & 0 \\ a e_{-\nu}(g)(I - \sum_{k=1}^{m} b_k e_{\gamma_k}(g)) & e_{-\lambda}(g) I_p \end{bmatrix}, \qquad g \in G. \tag{9}$$

Thus, $\nu = -\alpha_1$, $\gamma_j = \alpha_{j+1} + \nu$ $(j = 1, 2, \cdots, m)$, $0 \prec \gamma_1 \prec \cdots \prec \gamma_m \prec \lambda + \nu$, and $b_1, \dots b_m$ are nonzero $p \times p$ matrices. We make another assumption:

$$(\aleph) \qquad n\gamma_1 \succ \lambda + \nu \quad \text{for some positive integer } n.$$

Denote by $\mathbb{Z}^+$ the set of nonnegative integers, and for any $N = (n_1, \dots, n_m) \in (\mathbb{Z}^+)^m$ define

$$y_N(A) = \sum b_{j_1} b_{j_2} \dots b_{j_w}, \tag{10}$$

where $w = n_1 + \cdots + n_m$, and the sum in (10) is taken over all ordered w-tuples of integers $(j_1, \dots, j_w)$ exactly n_k of which are equal to k, for $k = 1, \dots, m$ (if $N = (0, \dots, 0)$, we let $y_N(A) = I$). Finally, define

$$f(g) = \sum y_{(n_1,\dots,n_m)}(A)(g) a^{-1} e_{-\lambda + n_1\gamma_1 + \cdots + n_m\gamma_m}(g), \qquad g \in G, \tag{11}$$

where the sum is taken over all $(n_1, \dots, n_m) \in (\mathbb{Z}^+)^m$ for which

$$-\nu \prec -\lambda + n_1\gamma_1 + \cdots + n_m\gamma_m \prec \nu. \tag{12}$$

If the set of $(n_1, \dots, n_m)$'s satisfying (12) is empty (for example, this would be the case when $\nu \prec 0$), then we take f to be the zero function. Condition $(\aleph)$ guarantees that the sum in (11) is finite.

Theorem 5. *Under the conditions in the previous paragraph, there exist* $A_+ \in \mathcal{G}((W(G)_+)^{n\times n})$ *and* $A_- \in \mathcal{G}((W(G)_-)^{n\times n})$ *with finite Fourier spectra such that the equality*

$$A = A_+ \, B \, A_-$$

holds, where

$$B = \begin{bmatrix} e_\nu I_p & 0 \\ f & e_{-\nu} I_p \end{bmatrix}.$$

In particular, the matrix functions A *and* B *are factorizable (resp., finitely factorizable) only simultaneously, and in case they are factorizable, they have the same factorization indices.*

Theorem 5 is proved by using the calculations given in [26, Section 2]. The formulas for $A_\pm$ (which we will not reproduce) are also given there.

Note that Assumption ($\aleph$) is valid for the function B. Thus, one can apply the Portuguese transformation again to B provided the matrix coefficient of the lowest term in the bottom left corner of B is invertible (analogously to the condition of invertibility of c_1). This condition is always satisfied if $p = 1$. Once the formulas of Theorem 5 are in place, one can repeat without difficulty many results on factorization that depend on the Portuguese transformation. We refer the reader to [4, Chapters 14 and 15] and references therein. Here, we state just one such result.

Theorem 6. *Let* A *have the form (7), where* $0 \prec \alpha_1 \prec \dots \prec \alpha_m$, *and let the matrix* c_1 *be invertible. Assume that Condition* ($\aleph$) *holds. Then* A *is finitely factorizable, and the factorization indices are* $\pm\alpha_1$ *(*p *times each).*

Observe that the invertibility of c_1 is essential in Theorem 6; in [26] an example was given (with $\Gamma = \mathbb{R}$) of a non-factorizable function of the form (7) with $m = 4$ and $0 \preceq \alpha_1 \prec \alpha_2 \prec \alpha_3 \prec \alpha_4$.

3. Off diagonal binomials

We prove in this section the following result:

Theorem 7. *Let* A *have the form*

$$A(g) = \begin{bmatrix} e_\lambda(g) I_p & 0 \\ c_1 e_\sigma(g) - c_2 e_\mu(g) & e_{-\lambda}(g) I_q \end{bmatrix}, \qquad g \in G, \tag{13}$$

where $\lambda, \sigma, \mu \in \Gamma$, $-\lambda \prec \sigma \prec \mu \prec \lambda$, *and* $c_j \in \mathbb{C}^{q\times p}$ *(the cases when one or both of* c_1 *and* c_2 *are equal to zero are not excluded). Assume that the following condition holds:*

$$n(\mu - \sigma) \succeq \max\{\lambda - \sigma, \lambda + \mu\} \quad \text{for some positive integer } n. \tag{14}$$

Then A *admits a finite factorization.*

For the case $\Gamma = \mathbb{R}$, Theorem 7 was proved in [23] (in this case (14) is automatically satisfied). We adapt the approach of [23] to prove Theorem 7 in full generality.

Proof. Let r and s be the nonnegative integers having the following properties:

$$\lambda - \sigma - (s+1)(\mu - \sigma) \prec 0 \preceq \lambda - \sigma - s(\mu - \sigma), \tag{15}$$

$$\lambda + \mu - (r+1)(\mu - \sigma) \prec 0 \preceq \lambda + \mu - r(\mu - \sigma). \tag{16}$$

In view of (14), s and r are correctly defined.

Consider first the case when $p = q$ and both matrices c_1 and c_2 are invertible. We distinguish between three possibilities:

$$s > r \Leftrightarrow \mu + \lambda - s(\mu - \sigma) \prec 0; \tag{17}$$

$$s < r \Leftrightarrow \lambda - \sigma - r(\mu - \sigma) \prec 0; \tag{18}$$

$$s = r \Leftrightarrow \begin{cases} \lambda - \sigma - (s+1)(\mu - \sigma) \prec 0 \preceq \lambda - \sigma - s(\mu - \sigma), \\ \lambda + \mu - (s+1)(\mu - \sigma) \prec 0 \preceq \lambda + \mu - s(\mu - \sigma). \end{cases} \tag{19}$$

Then the following formulas (which can be verified in a straightforward manner) give a factorization $A = A_+ \Lambda A_-$ of A. In the case that (17) holds:

$$A_+ = \begin{bmatrix} -\sum_{j=0}^{s-1} (c_2^{-1} c_1)^j c_2^{-1} e_{\lambda - \mu - j(\mu - \sigma)} & I \\ I & 0 \end{bmatrix}, \quad \Lambda = \begin{bmatrix} e_\mu I & 0 \\ 0 & e_{-\mu} I \end{bmatrix},$$

$$A_- = \begin{bmatrix} c_1 e_{\sigma - \mu} - c_2 & e_{-\mu - \lambda} I \\ (c_2^{-1} c_1)^s e_{\mu + \lambda - s(\mu - \sigma)} & \sum_{j=0}^{s-1} (c_2^{-1} c_1)^j c_2^{-1} e_{-j(\mu - \sigma)} \end{bmatrix}.$$

Clearly, $A_+ \in \mathcal{G}((W(G)_+)^{2p \times 2p})$, $A_- \in (W(G)_-)^{2p \times 2p}$, and since (use Schur complements)

$$\begin{aligned} \pm \det(A_-) &= (e_{-\mu - \lambda})^p \det \Big((c_2^{-1} c_1)^s e_{\mu + \lambda - s(\mu - \sigma)} \\ &\quad - e_{\mu + \lambda} \left(\sum_{j=0}^{s-1} (c_2^{-1} c_1)^j c_2^{-1} e_{-j(\mu - \sigma)} \right) (c_1 e_{\sigma - \mu} - c_2) \Big) \\ &= (e_{-\mu - \lambda})^p \det (e_{\mu + \lambda} I) = 1 \end{aligned}$$

it follows that $A_-^{-1} \in (W(G)_-)^{2p \times 2p}$. In the case that (18) holds:

$$A_+ = \begin{bmatrix} e_{\lambda - \sigma} I & -c_1^{-1} \sum_{j=0}^{r-1} (c_2 c_1^{-1})^j e_{j(\mu - \sigma)} \\ -c_2 e_{\mu - \sigma} + c_1 & (c_2 c_1^{-1})^r e_{r(\mu - \sigma) - \lambda + \sigma} \end{bmatrix},$$

$$\Lambda = \begin{bmatrix} e_\sigma I & 0 \\ 0 & e_{-\sigma} I \end{bmatrix}, \quad A_- = \begin{bmatrix} I & c_1^{-1} \sum_{j=0}^{r-1} (c_2 c_1^{-1})^j e_{j(\mu - \sigma) - \sigma - \lambda} \\ 0 & I \end{bmatrix}.$$

Again, clearly $A_+ \in (W(G)_+)^{2p\times 2p}$, $A_- \in \mathcal{G}((W(G)_-)^{2p\times 2p})$, and since

$$\det(A_+) = (e_{\lambda-\sigma})^p \det\left(e_{-\lambda+\sigma}(-c_2e_{\mu-\sigma}+c_1)\left(c_1^{-1}\sum_{j=0}^{r-1}(c_2c_1^{-1})^j e_{j(\mu-\sigma)}\right)\right.$$
$$\left. + \quad (c_2c_1^{-1})^r e_{r(\mu-\sigma)-\lambda+\sigma}\right) = 1$$

we also have $A_+ \in \mathcal{G}((W(G)_+)^{2p\times 2p})$. In the case that (19) holds:

$$A_+ = \begin{bmatrix} I + \sum_{j=1}^{s}(c_1^{-1}c_2)^j e_{j(\mu-\sigma)} & -\sum_{j=0}^{s-1}(c_2^{-1}c_1)^j c_2^{-1} e_{\lambda-\mu-j(\mu-\sigma)} \\ -c_2(c_1^{-1}c_2)^s e_{s(\mu-\sigma)-\lambda+\mu} & I \end{bmatrix},$$

$$\Lambda = \begin{bmatrix} e_{\lambda-s(\mu-\sigma)}I & 0 \\ 0 & e_{-\lambda+s(\mu-\sigma)}I \end{bmatrix},$$

$$A_- = \begin{bmatrix} (c_2^{-1}c_1)^s & e_{s(\mu-\sigma)-\mu-\lambda}\sum_{j=0}^{s-1}(c_2^{-1}c_1)^j c_2^{-1} e_{-j(\mu-\sigma)} \\ c_1 e_{\sigma+\lambda-s(\mu-\sigma)} & e_{-s(\mu-\sigma)}I + \sum_{j=0}^{s-1}(c_2c_1^{-1})^{s-j} e_{-j(\mu-\sigma)} \end{bmatrix}.$$

It is easily verified that $A_\pm \in (W(G)_\pm)^{2p\times 2p}$ (follows from (19)). Similarly, $A_\pm^{-1} \in (W(G)_\pm)^{2p\times 2p}$. This concludes the proof in the case when $p = q$ and both c_1 and c_2 are invertible.

Consider now the general case. Applying the transformation

$$A \mapsto \begin{bmatrix} T_1 & 0 \\ 0 & T_2 \end{bmatrix} A \begin{bmatrix} T_1^{-1} & 0 \\ 0 & T_2^{-1} \end{bmatrix},$$

where T_1 and T_2 are invertible matrices of sizes $p \times p$ and $q \times q$, respectively, the matrices c_1 and c_2 are replaced by $T_2c_1T_1^{-1}$ and $T_2c_2T_1^{-1}$, respectively. Now use the well-known canonical form (also known as Kronecker form) for pairs of rectangular matrices under the transformation $(X, Y) \mapsto (TXS, TYS)$, where T and S are invertible matrices of appropriate sizes, see, e.g., [16], or [20, Appendix]. Ignoring the zero blocks in the canonical form of (c_1, c_2) (trivial case), and blocks where both matrices are invertible (this case was taken care of already), we are left with the following situations to consider:

(a) c_1 and c_2 are of size $k \times (k+1)$ (so $q = k$, $p = k+1$) of the form

$$c_1 = \begin{bmatrix} I_k & 0_{k\times 1} \end{bmatrix}, \quad c_2 = \begin{bmatrix} 0_{k\times 1} & I_k \end{bmatrix}.$$

(b) c_1 and c_2 are of size $(k+1) \times k$ (so $q = k+1$, $p = k$) of the form

$$c_1 = \begin{bmatrix} I_k \\ 0_{1\times k} \end{bmatrix}, \quad c_2 = \begin{bmatrix} 0_{1\times k} \\ I_k \end{bmatrix}.$$

(c) c_1 is the $k \times k$ upper triangular nilpotent Jordan block, denoted by V_k, and $c_2 = I_k$ (so $p = q = k$).

(d) $c_1 = I_k$, and $c_2 = V_k$ (so $p = q = k$).

Let J_k be the $k \times k$ matrix with 1's along the top-right to the left-bottom diagonal and zeros in all other positions. If $A(g) = [a_{i,j}(g)]_{i,j=1}^n \in (W(G))^{n\times n}$, then A^* will denote the matrix function defined by $[\overline{a_{j,i}(g)}]_{i,j=1}^n$; clearly, $A^* \in (W(G))^{n\times n}$, and if $A \in (W(G)_\pm)^{n\times n}$, then $A^* \in (W(G)_\mp)^{n\times n}$. The transformation

$$A \mapsto \begin{bmatrix} 0 & J_k \\ J_k & 0 \end{bmatrix} A^* \begin{bmatrix} 0 & J_k \\ J_k & 0 \end{bmatrix}$$

transforms the case (c) to the case (d). The transformation

$$A \mapsto \begin{bmatrix} 0 & J_{k+1} \\ J_k & 0 \end{bmatrix} A^* \begin{bmatrix} 0 & J_k \\ J_{k+1} & 0 \end{bmatrix}$$

transforms the case (b) to the case (a). Thus, it will suffice to consider the cases (a) and (d).

We will need the nonnegative integers s and r introduced by (15) and (16), respectively, as well as two additional nonnegative integers r' and s' defined as follows: $s' = s$ if $n(\mu-\sigma) \neq \lambda - \sigma$ for any positive integer n, and $s' = s-1$ otherwise; $r' = r$ if $n(\mu-\sigma) \neq \lambda+\mu$ for any positive integer n, and $r' = r-1$ otherwise.

Consider the case (d) first:

$$A = \begin{bmatrix} e_\lambda I_k & 0 \\ e_\sigma I_k - e_\mu V_k & e_{-\lambda} I_k \end{bmatrix}. \tag{20}$$

Let

$$B_+ = \begin{bmatrix} I_k - e_{\mu-\sigma} V_k & -e_{\lambda-\sigma} I_k \\ 0 & \sum_{j=0}^{k-1} e_{j(\mu-\sigma)} V_k^j \end{bmatrix}, \quad B_- = \begin{bmatrix} \sum_{j=0}^{r-1} e_{j(\mu-\sigma)-\lambda-\sigma} V_k^j & I_k \\ -I_k & 0 \end{bmatrix}.$$

Clearly, $B_\pm \in \mathcal{G}((W(G)_\pm)^{n\times n})$. Using the property that $V_k^k = 0$, we obtain

$$\begin{aligned} B_+ A B_- &= \begin{bmatrix} e_{-\sigma} I_k & 0 \\ -\sum_{j=r}^{k-1} e_{j(\mu-\sigma)-\lambda} V_k^j & e_\sigma I_k \end{bmatrix} \\ &= \begin{cases} \begin{bmatrix} e_{-\sigma} I_k & 0 \\ 0 & e_\sigma I_k \end{bmatrix} & \text{if } r \geq k; \\ \begin{bmatrix} e_{-\sigma} I_r & 0 & 0 \\ 0 & H_{k-r} & 0 \\ 0 & 0 & e_\sigma I_r \end{bmatrix} & \text{if } 1 \leq r < k; \end{cases} \end{aligned}$$

where

$$H_{k-r} = \begin{bmatrix} e_{-\sigma} I_{k-r} & 0 \\ -\sum_{j=0}^{k-r-1} e_{(j+r)(\mu-\sigma)-\lambda} V_{k-r}^j & e_\sigma I_{k-r} \end{bmatrix}.$$

If $s' < r < k$, then since we have $\lambda - \mu \preceq (r-1)(\mu-\sigma)$ (this relation is easily obtained from $(r+1)(\mu-\sigma) \succ \lambda+\mu$), it follows that H_{k-r} admits a factorization of the form

$$H_{k-r} = \begin{bmatrix} I_{k-r} & 0 \\ -\sum_{j=0}^{k-r-1} e_{(j+r-1)(\mu-\sigma)-\lambda+\mu} V_{k-r}^j & I_{k-r} \end{bmatrix} \begin{bmatrix} e_{-\sigma} I_{k-r} & 0 \\ 0 & e_\sigma I_{k-r} \end{bmatrix}.$$

If $r \le \min\{s', k-1\}$, then let

$$C_+ = \begin{bmatrix} \sum_{j=0}^{k-r-1} e_{j(\mu-\sigma)} V_{k-r}^j & e_{\lambda-\sigma-r(\mu-\sigma)} I_{k-r} \\ 0 & I_{k-r} - e_{\mu-\sigma} V_{k-r} \end{bmatrix},$$

$$C_- = \begin{bmatrix} e_{\lambda+\sigma-r(\mu-\sigma)} I_{k-r} & -I_{k-r} \\ I_{k-r} & 0 \end{bmatrix}.$$

We have $C_\pm \in \mathcal{G}((W(G)_\pm)^{(k-r)\times(k-r)})$; to verify that $C_- \in (W(G)_-)^{(k-r)\times(k-r)}$ we need to verify the relation $\lambda + \sigma \preceq r(\mu - \sigma)$, which in turn follows easily from $(r+1)(\mu-\sigma) \succ \lambda + \mu$. Furthermore,

$$C_+ H_{k-r} C_- = \begin{bmatrix} e_{\lambda-r(\mu-\sigma)} I_{k-r} & 0 \\ -e_\mu V_{k-r} & e_{-\lambda+r(\mu-\sigma)} I_{k-r} \end{bmatrix}$$

$$= \begin{cases} \begin{bmatrix} e_{\lambda-r(\mu-\sigma)} & 0 \\ 0 & e_{-\lambda+r(\mu-\sigma)} \end{bmatrix} & \text{if } s' \ge r = k-1; \\ \begin{bmatrix} e_{\lambda-r(\mu-\sigma)} & 0 & 0 \\ 0 & \widetilde{H}_{k-r-1} & 0 \\ 0 & 0 & e_{-\lambda+r(\mu-\sigma)} \end{bmatrix} & \text{if } 1 \le r \le \min\{s', k-2\}, \end{cases} \tag{21}$$

where

$$\widetilde{H}_{k-r-1} = \begin{bmatrix} e_{\lambda-r(\mu-\sigma)} I_{k-r-1} & 0 \\ -e_\mu I_{k-r-1} & e_{-\lambda+r(\mu-\sigma)} I_{k-r-1} \end{bmatrix}. \tag{22}$$

Factorizability of $\widetilde{H}_{k-r-1}$ follows from Theorem 6, or could be verified directly.

Finally, consider the case (a):

$$A = \begin{bmatrix} e_\lambda I_k & 0 & 0 \\ 0 & e_\lambda & 0 \\ e_\sigma I_k - e_\mu V_k & h & e_{-\lambda} I_k \end{bmatrix}, \quad \text{where} \quad h = \begin{bmatrix} 0_{(k-1)\times 1} \\ -e_\mu \end{bmatrix}.$$

Let

$$B_+ = \begin{bmatrix} I_k - e_{\mu-\sigma} V_k & b & -e_{\lambda-\sigma} I_k \\ 0 & 1 & 0 \\ 0 & 0 & \sum_{j=0}^{k-1} e_{j(\mu-\sigma)} V_k^j \end{bmatrix}, \quad \text{where} \quad b = \begin{bmatrix} 0_{(k-1)\times 1} \\ -e_{\mu-\sigma} \end{bmatrix},$$

$$B_- = \begin{bmatrix} \sum_{j=0}^{r-1} e_{j(\mu-\sigma)-\lambda-\sigma} V_k^j & 0 & I_k \\ 0 & 1 & 0 \\ -I_k & 0 & 0 \end{bmatrix}.$$

Clearly, $B_+ \in \mathcal{G}((W(G)_+)^{(2k+1)\times(2k+1)})$ and (because of $(r-1)(\mu-\sigma)-\lambda-\sigma \preceq 0$, which follows from (16)) we have $B_- \in \mathcal{G}((W(G)_-)^{(2k+1)\times(2k+1)})$. A computation shows that

$$\Phi_0 := B_+ A B_- = \begin{bmatrix} e_{-\sigma} I_k & 0 & 0 \\ 0 & e_\lambda & 0 \\ -\sum_{j=r}^{k-1} e_{j(\mu-\sigma)-\lambda} V_k^j & h_k & e_\sigma I_k \end{bmatrix},$$

where

$$(h_k)^T = \begin{bmatrix} -e_{(k-1)(\mu-\sigma)+\mu} & \cdots & -e_{(\mu-\sigma)+\mu} & -e_\mu \end{bmatrix}. \tag{23}$$

Let us define for $j = 0, 1, \ldots, \min\{k, s'\} - 1$ the auxiliary matrices

$$R_{+,k-j} = \begin{bmatrix} 1 & 0 & e_{\lambda-\mu-j(\mu-\sigma)} \\ 0 & I_{k-j-1} & h_{k-j-1}e_{-\sigma} \\ 0 & 0 & 1 \end{bmatrix},$$

$$R_{-,k-j} = \begin{bmatrix} e_{\sigma-\mu} & 0 & -1 \\ 0 & I_{k-j-1} & 0 \\ 1 & 0 & 0 \end{bmatrix}, \quad R_{k-j} = \begin{bmatrix} e_{\lambda-j(\mu-\sigma)} & 0 \\ h_{k-j} & e_{\sigma}I_{k-j} \end{bmatrix}.$$

Clearly, $R_{-,k-j} \in \mathcal{G}((W(G)_-)^{(k-j+1)\times(k-j+1)})$. By (15), we easily obtain $R_{+,k-j} \in \mathcal{G}((W(G)_+)^{(k-j+1)\times(k-j+1)})$. We also have the recurrence relations

$$R_{+,k-j}R_{k-j}R_{-,k-j} = \begin{bmatrix} R_{k-j-1} & 0 \\ 0 & e_{\mu} \end{bmatrix}, \quad R_0 = e_{\lambda-k(\mu-\sigma)}, \tag{24}$$

for $j = 0, \ldots, \min\{k, s'\} - 1$.

If $k \le \min\{r, s'\}$, then $\Phi_0 = \operatorname{diag}(e_{-\sigma}I_k, R_k)$, and applying consecutively (24) for $j = 0, \ldots, k-1$, we obtain a factorization $A = A_+\Lambda A_-$ with $\Lambda = \operatorname{diag}(e_{-\sigma}I_k, e_{\lambda-k(\mu-\sigma)}, e_{\mu}I_k)$. If $s' < k \le r$, then again $\Phi_0 = \operatorname{diag}(e_{-\sigma}I_k, R_k)$. Applying (24) for $j = 0, 1, \ldots, s'-1$, we reduce R_k to $\operatorname{diag}(R_{k-s'}, e_{\mu}I_{s'})$, and in view of the inequality $\lambda - \mu \preceq s'(\mu - \sigma)$ the equation

$$R_{k-s'} = \begin{bmatrix} 1 & 0 \\ h_{k-s'}e_{s'(\mu-\sigma)-\lambda} & I_{k-s'} \end{bmatrix} \begin{bmatrix} e_{\lambda-s'(\mu-\sigma)} & 0 \\ 0 & e_{\sigma}I_{k-s'} \end{bmatrix}$$

represents a factorization of $R_{k-s'}$. If $s' < r < k$, then in view of the inequalities

$$\lambda - \mu \preceq s'(\mu - \sigma) \preceq (r-1)(\mu - \sigma)$$

(which follow from (15)), we have

$$A_+ := \begin{bmatrix} I_k & 0 & 0 \\ 0 & 1 & 0 \\ \sum_{j=r}^{k-1} e_{j(\mu-\sigma)-\lambda+\sigma}V_k^j & 0 & I_k \end{bmatrix} \in \mathcal{G}((W(G)_+)^{(2k+1)\times(2k+1)}),$$

and since $A_+\Phi_0 = \operatorname{diag}(e_{-\sigma}I_k, R_k)$, we have reduced the proof of factorizability of A to that of R_k. Now repeat the arguments given above for the case $k \le \min\{r, s'\}$ to obtain the factorizability of R_k.

It remains to consider the case $r \le \min\{k-1, s'\}$. Then

$$\lambda - \mu \succ (r-1)(\mu - \sigma).$$

We have the recurrence relations (which can be verified using (24)):

$$\Phi_{j+1} = \begin{bmatrix} I_k & 0 & 0 \\ 0 & R_{+,k-j} & 0 \\ 0 & 0 & I_j \end{bmatrix} \Phi_j \begin{bmatrix} I_k & 0 & 0 \\ 0 & R_{-,k-j} & 0 \\ 0 & 0 & I_j \end{bmatrix}, \quad j = 0, 1, \ldots, r-2,$$

where

$$\Phi_j = \operatorname{diag}(e_{-\sigma}I_r, F_j, e_{\mu}I_j), \qquad j = 1, \ldots, r-1,$$

and where

$$F_j = \begin{bmatrix} e_{-\sigma}I_{k-r} & 0 & 0 \\ 0 & e_{\lambda-j(\mu-\sigma)} & 0 \\ f_j & h_{k-j} & e_\sigma I_{k-j} \end{bmatrix}, \ f_j = \begin{bmatrix} -\sum_{\ell=0}^{k-r-1} e_{(\ell+r)(\mu-\sigma)-\lambda} V_{k-r}^{\ell} \\ 0_{(r-j)\times(k-r)} \end{bmatrix}.$$

Thus, it suffices to construct a factorization for F_{r-1}. Note that

$$\begin{aligned} F_{r-1} &= \begin{bmatrix} e_{-\sigma}I_{k-r+1} & 0 \\ -\sum_{j=0}^{k-r} e_{(j+r)(\mu-\sigma)-\lambda} V_{k-r+1}^{j} & e_\sigma I_{k-r+1} \end{bmatrix} \\ &\times \begin{bmatrix} I_{k-r} & 0 & 0 \\ 0 & e_{\lambda+\sigma-(r-1)(\mu-\sigma)} & 0 \\ 0 & 0 & I_{k-r+1} \end{bmatrix}. \end{aligned}$$

Letting

$$C_+ = \begin{bmatrix} \sum_{j=0}^{k-r} e_{j(\mu-\sigma)} V_{k-r+1}^{j} & e_{\lambda-\sigma-r(\mu-\sigma)} I_{k-r+1} \\ 0 & I_{k-r+1} - e_{\mu-\sigma} V_{k-r+1} \end{bmatrix},$$

$$C_- = \begin{bmatrix} e_{\lambda+\sigma-r(\mu-\sigma)} I_{k-r} & 0 & -I_{k-r} & 0 \\ 0 & e_{\sigma-\mu} & 0 & -1 \\ I_{k-r} & 0 & 0 & 0 \\ 0 & 1 & 0 & 0 \end{bmatrix},$$

we have (in view of (15) and (16))

$$C_\pm \in \mathcal{G}((W(G)_\pm)^{(2k-2r+2)\times(2k-2r+2)}),$$

and we obtain analogously to (21) that

$$\begin{aligned} C_+F_{r-1}C_- &= \left[\begin{array}{c|cc} e_{\lambda-r(\mu-\sigma)} I_{k-r+1} & \multicolumn{2}{c}{0} \\ \hline -e_\mu V_{k-r+1} & e_{r(\mu-\sigma)-\lambda} I_{k-r} & 0 \\ & 0 & e_\mu \end{array}\right] \\ &= \operatorname{diag}\left(e_{\lambda-r(\mu-\sigma)}, \widetilde{H}_{k-r}, e_\mu\right), \end{aligned}$$

where $\widetilde{H}_{k-r}$ has the form (22) (with size of blocks $(k-r)\times(k-r)$). The factorizability of $\widetilde{H}_{k-r}$ follows from Theorem 6. □

Using the proof of Theorem 7, one can deduce formulas for the factorization indices, and in particular, necessary and sufficient conditions for canonical factorization. We omit the details, as they are long and cumbersome.

Once formulas for the factors in the proof of Theorem 7 are available, we may use them to derive additional results. For example:

Theorem 8. *Let A be given by* (13), *where $p = q$, $-\lambda \prec \sigma \prec \mu \prec \lambda$, and $c_j \in \mathbb{C}^{p\times p}$. Assume that one of the two conditions below is satisfied:*

(1) *$n(\mu-\sigma) \prec \lambda - \mu$ for all positive integers n, the matrix c_2 is invertible, and the spectral radius of $c_2^{-1}c_1$ is less than one;*
(2) *$n(\mu-\sigma) \prec \lambda + \sigma$ for all positive integers n, the matrix c_1 is invertible, and the spectral radius of $c_2c_1^{-1}$ is less than one.*

Then A is factorizable.

Proof. If (1) holds, the factorization is given by the formulas

$$A_+ = \begin{bmatrix} -\sum_{j=0}^{\infty}(c_2^{-1}c_1)^j c_2^{-1} e_{\lambda-\mu-j(\mu-\sigma)} & I \\ I & 0 \end{bmatrix}, \quad \Lambda = \begin{bmatrix} e_\mu I & 0 \\ 0 & e_{-\mu} I \end{bmatrix},$$

$$A_- = \begin{bmatrix} c_1 e_{\sigma-\mu} - c_2 & e_{-\mu-\lambda} I \\ 0 & \sum_{j=0}^{\infty}(c_2^{-1}c_1)^j c_2^{-1} e_{-j(\mu-\sigma)} \end{bmatrix}.$$

If (2) holds, the factorization is given by formulas

$$A_+ = \begin{bmatrix} e_{\lambda-\sigma} I & -c_1^{-1}\sum_{j=0}^{\infty}(c_2 c_1^{-1})^j e_{j(\mu-\sigma)} \\ -c_2 e_{\mu-\sigma} + c_1 & 0 \end{bmatrix},$$

$$\Lambda = \begin{bmatrix} e_\sigma I & 0 \\ 0 & e_{-\sigma} I \end{bmatrix}, \quad A_- = \begin{bmatrix} I & c_1^{-1}\sum_{j=0}^{\infty}(c_2 c_1^{-1})^j e_{j(\mu-\sigma)-\sigma-\lambda} \\ 0 & I \end{bmatrix}. \qquad \square$$

An interesting case occurs when both inequalities in conditions (1) and (2) of Theorem 8 are satisfied. We then have:

Theorem 9. *Let A be given as in Theorem 8, and assume that*

$$n(\mu - \sigma) \prec \min\{\lambda - \mu, \lambda + \sigma\} \quad \textit{for all integers} \quad n > 0. \tag{25}$$

If the matrix c_1 is invertible and the spectrum of $c_1^{-1}c_2$ does not intersect the unit circle, or if c_2 is invertible and the spectrum of $c_2^{-1}c_1$ does not intersect the unit circle, then A admits a factorization. Moreover, the factorization indices belong to the set $\{\pm\sigma, \pm\mu\}$.

Proof. Say c_1 is invertible. Using a transformation $c_j \mapsto Sc_jT$, $j = 1, 2$, for suitable invertible S and T, we can assume without loss of generality that $c_1 = I$ and c_2 is in the Jordan form. Now apply the result of Theorem 8 to each Jordan block of c_2 separately, to obtain Theorem 9. $\square$

The condition on the spectrum of $c_1^{-1}c_2$ (or of $c_2^{-1}c_1$) in Theorem 9 is essential, as the following example shows: Let $\Gamma = \mathbb{Z}^2$ with the lexicographical order, G is the 2-torus, and

$$A = \begin{bmatrix} z & 0 \\ 1 - w & z^{-1} \end{bmatrix} \tag{26}$$

(here z is the first variable, w is the second variable; so z corresponds to $e_{(1,0)}$, w corresponds to $e_{(0,1)}$, and the lexicographical order is such that $(1,0) \succ (0,n)$ for every integer n). The function (26) is not factorizable. Indeed, if it was, then considering A as a function of z only, with w a parameter, we would obtain that the factorization indices of A are independent of $w \in \mathbb{T}$ (the unit circle), whereas in fact the factorization indices are zeros if $w \neq 1$, and they are ± 1 if $w = 1$:

$$\begin{bmatrix} z & 0 \\ 1 - w & z^{-1} \end{bmatrix} = \begin{bmatrix} 1 & (1-w)^{-1}z \\ 0 & 1 \end{bmatrix} \begin{bmatrix} 0 & -(1-w)^{-1} \\ 1 - w & z^{-1} \end{bmatrix}, \quad w \neq 1$$

is a canonical factorization.

4. Invertibility vs factorizability

Clearly, a necessary condition for the existence of a factorization is that $\det A(g) \neq 0$ for every $g \in G$. For the case $\Gamma = \mathbb{Z}$, G is the unit circle, this condition is well known to be also sufficient for the existence of a factorization. An example (a 2×2 triangular matrix function of the form (7)) shows that in general this condition is not sufficient (see Section 15.1 in [4]). Motivated by this example, we formulate a conjecture.

Conjecture 10. *Assume $n \geq 2$. Every function $A \in \mathcal{G}(W(G)^{n \times n})$ admits a factorization if and only if Γ is (isomorphic to) a subgroup of the additive group of rational numbers $\mathbb{Q}$.*

For $n = 1$ the statement of the conjecture is false (see Proposition 3).

Toward this conjecture, we prove the following result:

Theorem 11. *If Γ is not isomorphic to a subgroup of $\mathbb{Q}$, then there exists a 2×2 matrix function of the form*

$$A(g) = \begin{bmatrix} e_\lambda(g) & 0 \\ c_1 e_{\alpha_1}(g) + c_2 + c_3 e_{\alpha_3}(g) & e_{-\lambda}(g) \end{bmatrix}, \qquad g \in G, \tag{27}$$

where $\lambda, \alpha_1, \alpha_2, \alpha_3 \in \Gamma$, and $c_1, c_2, c_3 \in \mathbb{C}$, which does not admit a finite factorization.

Conversely, if Γ is isomorphic to a subgroup of $\mathbb{Q}$, then every function $A \in \mathcal{G}(W(G)^{n \times n})$ such that its Fourier spectrum is contained in a finitely generated subgroup, admits a factorization.

We need several preliminary results for the proof of Theorem 11. A subset S of $\mathbb{R}^k$ (the vectors in $\mathbb{R}^k$ are understood as column vectors) is called a *halfspace* if it has the following properties (i) - (iv): (i) $\mathbb{R}^k = S \cup (-S)$; (ii) $S \cap (-S) = \{0\}$; (iii) if $x, y \in S$ then $x + y \in S$; (iv) if $x \in S$ and α is a nonnegative real number, then $\alpha x \in S$. Clearly, every halfspace S in $\mathbb{R}^k$ induces a linear order $\preceq$ on $\mathbb{R}^k$ (there are linear orders on $\mathbb{R}^k$ which are not induced by any halfspace): The linear order $\preceq_S$ induced by a halfspace S is defined by the property that $x \preceq_S y$ if and only if $y - x \in S$. A standard example of a halfspace is given by

$$E_k = \{(x_1, \ldots, x_k)^T \in \mathbb{R}^k : x_1 = x_2 = \cdots = x_{j-1} = 0,\ x_j \neq 0 \Rightarrow x_j > 0\} \cup \{0\}, \tag{28}$$

where the superscript ^{T} denotes the transpose.

All halfspaces in $\mathbb{R}^k$ are described as follows:

Lemma 12. *A set $S \subset \mathbb{R}^k$ is a halfspace if and only if there exists a real invertible $k \times k$ matrix A such that*

$$S = AE_k \stackrel{\text{def}}{=} \{Ax \ : \ x \in E_k\}. \tag{29}$$

For a proof see [28], [12], for example.

It is often convenient to extend linear orders with respect to which countable additive subgroups of $\mathbb{R}^k$ are ordered groups, to so-called *term orders* on $\mathbb{R}^k$, i.e., linear orders $\preceq$ on $\mathbb{R}^k$ having the properties $x \preceq y \Rightarrow x + z \preceq y + z$ and $(x \preceq y$ and $c \in \mathbb{R}^+) \Rightarrow cx \preceq cy$ (note that the term orders are in one-to-one correspondence with halfspaces). That this is always possible follows from the next lemma [6]:

Lemma 13. *If $\preceq$ is a linear order on $\mathbb{Z}^k$, then there exists a term order $\preceq_{\mathbb{R}^k}$ on $\mathbb{R}^k$, and an order preserving (with respect to $\preceq$ and $\preceq_{\mathbb{R}^k}$) one-to-one group homomorphism from $\mathbb{Z}^k$ into $\mathbb{R}^k$.*

We now prove the following auxiliary result.

Lemma 14. *Let Γ_0 be a countable subgroup of $\mathbb{R}^k$, and let $\preceq_0$ be a term order on $\mathbb{R}^k$. Then for every selection of two finite subsets Ω_+, Ω_- of Γ_0 such that Ω_+ is positive and Ω_- is negative with respect to $\preceq_0$, there exists a term order $\preceq$ on $\mathbb{R}^k$ with the following properties:*

(a) *Γ_0 is Archimedean with respect to $\preceq$;*
(b) *Ω_+ is positive and Ω_- is negative with respect to $\preceq$.*

Proof. Since term orders are in one-to-one correspondence with halfspaces, for every term order $\preceq_0$ there exists a subspace of $\mathbb{R}^k$ of dimension $k-1$ which divides $\mathbb{R}^k$ in a set of positive points and a set of negative points (with respect to $\preceq_0$). Let $\mathcal{V}$ be the set of all $(k-1)$-dimensional subspaces V of $\mathbb{R}^k$ such that no point of Ω_+ can be continuously connected to a point of Ω_- without crossing V. Then $\mathcal{V}$ is open in the gap topology (see, e.g., [20, Chapter 13] for definition and properties of the gap topology), while the $(k-1)$-dimensional subspaces of $\mathbb{R}^k$ that have an empty intersection with $\Gamma_0 \setminus \{0\}$, form a dense subset of the metric space of linear subspaces of $\mathbb{R}^k$ with the gap topology. Thus there exists $V \in \mathcal{V}$ such that $V \cap \Gamma_0 = \{0\}$. Now let $\preceq$ be a term order on $\mathbb{R}^k$ which orders points first by their signed distance from V (with points in Ω_+ having positive distances to V) and then by applying any term order on V to the orthogonal projections of points onto V. Since no two different points of Γ_0 can have the same signed distance to V, we see that the signed distance from V is an order preserving group homomorphic embedding of $(\Gamma, \preceq|_{\Gamma_0})$ into $\mathbb{R}$. Consequently, Γ_0 is Archimedean with respect to $\preceq$. □

Proposition 15. *Consider $\mathbb{R}^k$ with a term order $\preceq$. Let $\nu, \delta \in \mathbb{R}^k$ be positive with respect to this order, and assume that the subgroup of $\mathbb{R}^k$ generated by ν and δ is not cyclic. Then none of the functions*

$$\begin{bmatrix} e_{\nu+\delta}(g) & 0 \\ c_1 e_{-\nu}(g) + c_2 + c_3 e_\delta(g) & e_{-\nu-\delta}(g) \end{bmatrix}, \qquad g \in G, \quad \mathbb{R}^k \text{ the dual of } G, \tag{30}$$

admit a finite factorization, with respect to $\Gamma = \mathbb{R}^k$ with the term order $\preceq$, for any triple of nonzero complex numbers (c_1, c_2, c_3) satisfying

$$(\log|c_3|)\nu + (\log|c_1|)\delta = (\log|c_2|)(\nu + \delta). \tag{31}$$

Note that (31) holds, in particular, when $|c_3| = |c_2| = |c_1|$. Note also that the subgroup generated by ν and δ is non-cyclic if and only if an equality $m\nu = n\delta$, $m, n \in \mathbb{Z}$ implies $m = n = 0$.

Proof. The case $k = 1$ is known: [22], also [4, Section 15.1]. In fact, in this case a stronger property holds: the matrix function (30) does not admit any factorization, finite or not.

Consider now the general case. Arguing by contradiction, assume that a function of the form (30) admits a finite factorization with respect to $\preceq$:

$$\begin{bmatrix} e_{\nu+\delta} & 0 \\ c_1 e_{-\nu} + c_2 + c_3 e_\delta & e_{-\nu-\delta} \end{bmatrix} = A_+ \begin{bmatrix} e_{\mu_1} & 0 \\ 0 & e_{\mu_2} \end{bmatrix} A_-, \tag{32}$$

where $A_\pm \in \mathcal{G}((W(G)_\pm)^{n\times n})$, $\{\mu_1, \mu_2\} \subset \mathbb{R}^k$, and the Fourier spectrum of $A_\pm$ is finite. By taking determinants in (32) we see that actually $\mu_2 = -\mu_1$ and $\det(A_\pm)$ is a nonzero constant, so without loss of generality we may assume that $\det(A_\pm) = 1$. Let Γ_0 be the (countable) subgroup of $\mathbb{R}^k$ generated by ν, δ, μ_1, and the Fourier spectra of A_+ and A_-, and let G_0 be the group for which Γ_0 is the dual. By Lemma 14 there is a term order $\preceq_0$ on $\mathbb{R}^k$ with respect to which Γ_0 is Archimedean and such that the same formula (32) is a factorization with respect to $\preceq_0$. By Hölder's theorem [31], Γ_0 is $\preceq_0$-order isomorphic to a subgroup of $\mathbb{R}$. Such an isomorphism would transform (32) to a factorization of a function of the form (30) where $k = 1$, a contradiction with the known case [22], [4, Section 15.1]. □

It remains an open question whether functions (30) admit a non-finite factorization in the case $k > 1$.

The next lemma is well-known in abstract group theory, see [24, Chapter VIII, §30], for example.

Lemma 16. *An abstract group Ω is isomorphic to a subgroup of $\mathbb{Q}$ if and only if Ω is torsion-free (every nonzero element has infinite order) and every finitely generated subgroup of Ω is cyclic.*

Proof of Theorem 11. Let Γ be a discrete ordered abelian group that is not isomorphic to a subgroup of $\mathbb{Q}$. By Lemma 16 there exist $\nu_0, \delta_0 \in \Gamma$ that are positive elements and generate the noncyclic subgroup Γ_0 of Γ. Say, $\nu_0 \succ \delta_0 \succ 0$. By Proposition 15 the functions

$$A = \begin{bmatrix} e_{\nu_0+\delta_0} & 0 \\ c_1 e_{-\nu_0} + c_2 + c_3 e_{\delta_0} & e_{-\nu_0-\delta_0} \end{bmatrix}, \tag{33}$$

where

$$c_1, c_2, c_3 \neq 0, \quad (\log|c_3|)\nu + (\log|c_1|)\delta = (\log|c_2|)(\nu+\delta),$$

are not finitely factorizable with respect to Γ_0 (to make Proposition 15 applicable, identify Γ_0 with $\mathbb{Z}^2$ and use Lemma 13).

It remains to prove that the functions (33) are not finitely factorizable with respect to Γ, and with the factors $A_\pm$ having finite spectrum. Suppose one of such functions is:

$$A = A_+ \begin{bmatrix} e_{j_1} & 0 \\ 0 & e_{j_2} \end{bmatrix} A_-, \tag{34}$$

for some c_1, c_2, c_3 as in (33), where $A_\pm \in \mathcal{G}((W(G)_\pm)^{2p\times 2p})$, the Fourier spectra of $A_\pm$ are finite, and $j_1, j_2 \in \Gamma$. Comparing determinants in (33) and in (34), we see that in fact $j_1 + j_2 = 0$.

Let Γ' be a finitely generated subgroup of Γ that contains j_1, ν_0, δ_0, and the Fourier spectra of A_+ and A_-, and let G' be the dual group of Γ'. By Theorems 1 and 2, $A_\pm^{-1} \in W(G')^{2p\times 2p}$. It now follows that $A_\pm \in \mathcal{G}((W(G')_\pm)^{2p\times 2p}$, and (34) is actually a factorization with respect to Γ'. Identifying Γ' with $\mathbb{Z}^m$ for some m, we see that by Lemma 13, Γ' is order isomorphic to a subgroup of $\mathbb{R}^m$, and transforming (34) with the help of such an order isomorphism, we obtain a contradiction with Proposition 15.

The converse statement follows from Lemma 16 and the well-known result that the converse statement holds for $\Gamma = \mathbb{Z}$. Indeed, every finitely generated subgroup of $\mathbb{Q}$ is cyclic. □

5. Systems of difference equations

As an application of factorization we consider systems of difference equations on a general discrete ordered abelian group $(\Gamma, \preceq)$ of the form

$$\sum_{s\in\Gamma_+} A(t-s)x(s) = b(t), \qquad t \in \Gamma_+, \tag{35}$$

where $A(t)$ is an $n\times n$ matrix function whose spectrum is a subset of Γ and for which $\sum_{t\in\Gamma} \|A(t)\| < \infty$, and $b(t)$ is an $n\times 1$ matrix function whose spectrum is a subset of Γ_+ and for which $\sum_{t\in\Gamma_+} \|b(t)\| < \infty$. We seek solutions x which are $n\times 1$ matrix functions whose spectrum is a subset of Γ_+ and for which $\sum_{t\in\Gamma_+} \|x(t)\| < \infty$. Let us now define

$$b(t) = \sum_{s\in\Gamma_+} A(t-s)x(s), \qquad t \in \Gamma_- \setminus \{0\}.$$

Then (35) is true for every $t \in \Gamma$, while $x(t)$ for $t \in \Gamma_+$ and $b(t)$ for $t \in \Gamma_- \setminus \{0\}$ are the unknowns. Putting

$$\hat{x}(g) = \textstyle\sum_{t\in\Gamma_+} x(t)\langle t, g\rangle, \qquad \hat{u}(g) = -\textstyle\sum_{t\in\Gamma_-\setminus\{0\}} b(t)\langle t, g\rangle,$$

$$\hat{b}(g) = \textstyle\sum_{t\in\Gamma_+} b(t)\langle t, g\rangle, \qquad \hat{A}(g) = \textstyle\sum_{t\in\Gamma} A(t)\langle t, g\rangle,$$

we obtain

$$\hat{A}(g)\hat{x}(g) + \hat{u}(g) = \hat{b}(g), \qquad g \in G, \tag{36}$$

where $\hat{u}(g)$ belongs to $(W(G))^{n\times n}$ and has only spectrum in $\Gamma_- \setminus \{0\}$.

By standard methods (such as [18] for $\Gamma = \mathbb{Z}$) one can prove that if $\hat{A}(g)$ has a right canonical factorization

$$\hat{A}(g) = \hat{A}_-(g)\hat{A}_+(g), \qquad g \in G,$$

where

$$\hat{A}_+(g)^{-1} = \sum_{t \succeq 0} \alpha_+(t)e_t(g), \quad \hat{A}_-(g)^{-1} = \sum_{t \preceq 0} \alpha_-(t)e_t(g),$$

then (35) has a unique solution which is given by

$$x(t) = \sum_{u \succeq 0} F(t,u)b(u), \qquad t \in \Gamma_+, \tag{37}$$

where

$$F(t,u) = \sum_{0 \preceq v \preceq \min(t,u)} \alpha_+(t-v)\alpha_-(v-u).$$

Let us now consider (35) in the case when $\hat{A}(g)$ has a right, not necessarily right canonical, factorization. In the statement below, we denote by $\sigma(\cdot)$ the Fourier spectrum, and use the shorthand $[\alpha,\beta)$ for the set $\{\gamma \in \Gamma : \alpha \preceq \gamma \prec \beta\}$, with the obvious modification for the set $(\alpha,\beta]$.

Theorem 17. *Suppose $\hat{A}(g)$ has the right factorization*

$$\hat{A}(g) = \hat{A}_-(g)\operatorname{diag}(e_{j_1}(g),\ldots,e_{j_n}(g))\hat{A}_+(g), \qquad g \in G,$$

where $j_1 \preceq \ldots \preceq j_n$. Put $P_r = \operatorname{row}(\delta_{ir})_{i=1}^n$, where δ_{ir} is the Kronecker symbol. Then (35) has at least one solution if and only if

$$\sigma\left(P_r\hat{A}_-(g)^{-1}\hat{b}(g)\right) \cap [0,j_r) = \emptyset \text{ whenever } j_r \succ 0. \tag{38}$$

The solutions of the homogeneous version of (36) are exactly those $\hat{x}(g) \in (W(G)_+)^{n\times 1}$ for which

$$\sigma\left(P_r\hat{A}_+(g)\hat{x}(g)\right) \subseteq [j_r,0) \text{ whenever } j_r \prec 0. \tag{39}$$

Thus all solutions are unique if and only if $0 \preceq j_1 \preceq \ldots \preceq j_n$.

Proof. From (36) we immediately have

$$e_{j_r}(g)P_r\hat{A}_+(g)\hat{x}(g) + P_r\hat{A}_-(g)^{-1}\hat{u}(g) = P_r\hat{A}_-(g)^{-1}\hat{b}(g). \tag{40}$$

Then for $j_r \succ 0$ condition (38) is clear, because the first term on the left-hand side of (39) has its spectrum in $[j_r,\infty)$ and the second term in $(-\infty,0)$. Thus for $j_r \succ 0$ the right-hand side of (40) should not have any spectrum in $[0,j_r)$ in order for a solution to exist. If (38) holds, then one of the solutions of (36) is delivered by $\hat{x}(g) = \hat{A}_+(g)^{-1}\operatorname{diag}(e_{-j_1}(g),\ldots,e_{-j_n}(g))\hat{A}_-(g)^{-1}\hat{b}(g)$.

To prove the uniqueness statement, we write (36) with $\hat{b}(g) \equiv 0$ in the form

$$e_{j_r}(g)P_r\hat{A}_+(g)\hat{x}(g) = -P_r\hat{A}_-(g)^{-1}\hat{u}(g),$$

where the left-hand side has its spectrum in $[j_r,\infty)$ and the right-hand side in $\{\mu \in \Gamma : \mu \prec 0\}$. □

6. Orthogonal families of functions

In this section we show how factorization can be used to determine orthogonal families of functions. Let $f = \sum_{j\in\Gamma} f_j e_j \in W(G)$, where e_j is defined by (2), be such that

$$f(g) \geq \epsilon > 0 \quad \text{for every } g \in G. \tag{41}$$

We let $L^2(f\nu)$ be the weighted L^2 Hilbert space with respect to the normalized invariant measure ν on G with the weight f.

Theorem 18. *Assume that $f \in W(G)$ satisfies (41). For $\mu \succeq 0$ let*

$$Q_\mu f = \sum_{j\in\Gamma,\ -\mu\preceq j\preceq\mu} f_j e_j.$$

Assume that for every $\mu \succeq 0$ there exists a factorization

$$\begin{bmatrix} Q_\mu f & e_{-\mu} \\ e_\mu & 0 \end{bmatrix} = \begin{bmatrix} a_\mu & b_\mu \\ c_\mu & d_\mu \end{bmatrix} \begin{bmatrix} p_\mu & q_\mu \\ r_\mu & s_\mu \end{bmatrix}, \tag{42}$$

where

$$\begin{bmatrix} a_\mu & b_\mu \\ c_\mu & d_\mu \end{bmatrix} \in \mathcal{G}((W(G)_+)^{2\times 2}), \qquad \begin{bmatrix} p_\mu & q_\mu \\ r_\mu & s_\mu \end{bmatrix} \in \mathcal{G}((W(G)_-)^{2\times 2})$$

with the additional properties that 0 is not in the Fourier spectrum of β_μ and $\alpha_\mu \in \mathcal{G}(W(G)_+)$, where α_μ and β_μ are taken from

$$\begin{bmatrix} \alpha_\mu & \beta_\mu \\ \gamma_\mu & \delta_\mu \end{bmatrix} = \begin{bmatrix} a_\mu & b_\mu \\ c_\mu & d_\mu \end{bmatrix}^{-1}.$$

Then the family

$$\pi_\mu(g) := e_\mu(g)\overline{\alpha_\mu(g)}, \quad g \in G, \qquad \mu \succeq 0$$

forms an orthogonal set in $L^2(f\nu)$. In addition, the Fourier spectrum of π_μ lies in $\{g \in \Gamma \ :\ 0 \preceq g \preceq \mu\}$ and contains μ.

Note that factorization (42) is, upon a row interchange, and except for the additional properties, a canonical factorization of a triangular 2×2 matrix function of the form (7).

Proof. We will use in the proof the fact that $\langle e_\lambda, e_\mu\rangle = 0$ if $\lambda \neq \mu$; this orthogonality is a result of the unitarity of the Fourier transform from $\ell^2(\Gamma)$ onto $L^2(G)$ [31].

From the factorization (42) it follows that

$$\alpha_\mu(Q_\mu f) = p_\mu - e_\mu\beta_\mu(z),$$

which has Fourier spectrum in $\{g \in \Gamma \ :\ g \preceq 0 \text{ or } \mu \prec g\}$. Here we used that 0 is not in the Fourier spectrum of β. Thus $e_\mu(Q_\mu f)\alpha_\mu^*$ has Fourier spectrum in $\{g \in \Gamma \ :\ g \prec 0 \text{ or } \mu \preceq g\}$. Here and elsewhere α_μ^* is a shorthand notation for the function $\overline{\alpha_\mu(g)}$, $g \in G$. Thus for $0 \preceq l \prec \mu$ we have that

$$\langle e_l, e_\mu(Q_\mu f)\alpha_\mu^*\rangle = 0,$$

where $\langle \, , \rangle$ is the inner product in $L^2(\nu)$. In addition, since $\alpha_\mu = e_\mu q_\mu$, it follows that α_μ has Fourier spectrum in $\{g \in \Gamma \ : \ 0 \preceq g \preceq \mu\}$, and since α_μ is invertible it cannot be the zero element in $W(G)$. From the location of the Fourier spectrum of α_μ it follows that $e_\mu(f - (Q_\mu f)(z))\alpha_\mu^*$ has Fourier spectrum in $\{g \in \Gamma \ : \ g \prec 0 \text{ or } \mu \prec g\}$. Thus it follows that

$$0 = \langle e_l, e_\mu (Q_\mu f)\alpha_\mu^* \rangle = \langle e_l, e_\mu f \alpha_\mu^* \rangle, \ 0 \preceq l \prec \mu.$$

Thus $e_\mu \alpha_\mu^*$ is orthogonal to e_l, $0 \preceq l \prec \mu$, in $L^2(f\nu)$. Moreover, the Fourier spectrum of $e_\mu\alpha_\mu^*$ lies in $\{g \in \Gamma \ : \ 0 \preceq g \preceq \mu\}$ and must contain μ (because α_μ is assumed to be invertible). □

In the case that Γ is a subgroup of $\mathbb{R}^k$ the existence of a factorization (42) with the additional properties described in Theorem 18 follows from the results in [29] (combine Theorem 2.22 with Lemma 4.9; see also the proof of Theorem 4.11). The very special case that $\Gamma = \mathbb{Z}$ and $G = \mathbb{T}$, the unit circle, follows of course directly as a particular case. We state the result.

Theorem 19. *Let $f(z) = \sum_{j=-\infty}^{\infty} f_j z^j \in W(\mathbb{T})$ be such that $f(z) > 0$, $z \in \mathbb{T}$. For each nonnegative integer n, put*

$$Q_n f(z) = \sum_{j=-n}^{n} f_j z^j.$$

Then there exist factorizations

$$\begin{bmatrix} Q_n f(z) & z^{-n} \\ z^n & 0 \end{bmatrix} = \begin{bmatrix} a & b \\ c & d \end{bmatrix} \begin{bmatrix} p & q \\ r & s \end{bmatrix}, \tag{43}$$

where

$$\begin{bmatrix} a & b \\ c & d \end{bmatrix}, \begin{bmatrix} \alpha & \beta \\ \gamma & \delta \end{bmatrix} := \begin{bmatrix} a & b \\ c & d \end{bmatrix}^{-1}$$

are analytic in $\overline{\mathbb{D}}$, the closure of the unit disk,

$$\begin{bmatrix} p & q \\ r & s \end{bmatrix}, \begin{bmatrix} p & q \\ r & s \end{bmatrix}^{-1}$$

are analytic in $(\mathbb{C} \cup \{\infty\}) \setminus \overline{\mathbb{D}}$, and in addition $\beta(0) = 0$ and α^{-1} is analytic in $\overline{\mathbb{D}}$. Moreover, if we define

$$\pi_n(z) := z^n \overline{\alpha(1/\overline{z})},$$

then the family π_n, $n = 0, 1, 2, \ldots$, consists of orthogonal polynomials with respect to the inner product on $L_2(\mathbb{T})$ with weight $f(z)$.

While the last statement of this theorem is a direct consequence of Theorem 18, it is interesting to make the connection to the results in [13], [14], [9], and [10]. In these papers a machinery has been developed where Riemann-Hilbert problems are used to solve various asymptotics problems in the area of orthogonal polynomials and their applications. This approach has been very successful and continues to be. The book [8] may be consulted for further recent developments in this area.

We end this paper with a proof of Theorem 19 based on results in [13] and [9] specialized to the case of the unit circle.

Alternative proof of Theorem 19. Consider the following Riemann-Hilbert problem: find a 2×2 matrix-valued function $Y(z)$ such that

$$\begin{cases} Y(z), & \text{is analytic in } \mathbb{C} \setminus \mathbb{T}; \\ Y_+(z) = Y_-(z) \begin{bmatrix} 1 & z^{-n} f(z) \\ 0 & 1 \end{bmatrix}, & z \in \mathbb{T} \\ Y(z) \begin{bmatrix} z^{-n} & 0 \\ 0 & z^n \end{bmatrix} = (I + O(\frac{1}{|z|})), & \text{uniformly as } z \to \infty. \end{cases}$$

For the definition of Y_+ and Y_- we require that $Y(z)$ is continuous up to the boundary in each component of $\mathbb{C} \setminus \mathbb{T}$, and we define

$$Y_+(z) = \lim_{w \to z,\ w \in \mathbb{D}} Y(w) \quad \text{and} \quad Y_-(z) = \lim_{w \to z,\ w \in \mathbb{C} \setminus \mathbb{D}} Y(w).$$

By [13], [9] (see also [2, Lemma 4.1], where the result is written out for the case of a circle) it follows that this problem has a unique solution $Y(z) = (Y_{ij})_{i,j=1}^2$, that $Y_{11}(z)$ is the nth monic orthogonal polynomial with respect to the weight $f(e^{i\theta})\frac{d\theta}{2\pi}$ on the unit circle, and that $Y_{21}(z)$ is a polynomial of degree $n-1$. Further note that $\det Y(z)$ has no jump on the circle, and that $\det Y(z) \to 1$ as $z \to \infty$. Thus $\det Y(z) \equiv 1$. Let now

$$R_1(z) = \overline{Y(\frac{1}{\bar{z}})} \begin{bmatrix} z^n & 0 \\ 0 & z^{-n} \end{bmatrix}, z \in \mathbb{D}, \qquad \text{and} \qquad R_2(z) = \overline{Y(\frac{1}{\bar{z}})} \begin{bmatrix} 0 & 1 \\ 1 & 0 \end{bmatrix}, z \in \mathbb{C} \setminus \overline{\mathbb{D}},$$

and extend them continuously to the boundary (which is possible since Y_+ and Y_- are well-defined). Then for $|z| = 1$ we have that

$$R_2(z) = \overline{Y_+(\frac{1}{\bar{z}})} \begin{bmatrix} 0 & 1 \\ 1 & 0 \end{bmatrix} = \overline{Y_-(\frac{1}{\bar{z}})} \begin{bmatrix} 1 & z^n f(z) \\ 0 & 1 \end{bmatrix} \begin{bmatrix} 0 & 1 \\ 1 & 0 \end{bmatrix} =$$

$$R_1(z) \begin{bmatrix} z^{-n} & 0 \\ 0 & z^n \end{bmatrix} \begin{bmatrix} 1 & z^n f(z) \\ 0 & 1 \end{bmatrix} \begin{bmatrix} 0 & 1 \\ 1 & 0 \end{bmatrix} = R_1(z) \begin{bmatrix} f(z) & z^{-n} \\ z^n & 0 \end{bmatrix}.$$

Furthermore, as $Y_{21}(z)$ is a polynomial of degree $n-1$, it follows that $(R_1)_{21}(0) = 0$. Next, let

$$\begin{bmatrix} \alpha & \beta \\ \gamma & \delta \end{bmatrix} = R_1(z) \begin{bmatrix} 1 & \sum_{j=1}^{\infty} f_{j+n} z^j \\ 0 & 1 \end{bmatrix}, \begin{bmatrix} p & q \\ r & s \end{bmatrix} = R_2(z) \begin{bmatrix} 1 & 0 \\ -\sum_{j=-\infty}^{-1} f_{j-n} z^j & 1 \end{bmatrix}.$$

Then (43) is satisfied. Moreover, note that $\pi(z) = Y_{11}(z)$. By the results of Szegő [33], π has all its zeros in the open unit disk, so α^{-1} is analytic in $\overline{\mathbb{D}}$. In order to get that $\beta(0) = 0$, one may multiply

$$\begin{bmatrix} \alpha & \beta \\ \gamma & \delta \end{bmatrix}$$

with an appropriate constant matrix on the left (e.g., the inverse of its value at 0). □

Finally, we remark that the formulas for the factors given in the first line of the proof of Theorem 2.18 in [29] are in direct correspondence with the formulas for the solution to the Riemann-Hilbert problem given in [13], [9] (see [2, Lemma 4.1] where it is explicitly written out for the circle case).

Acknowledgments

We wish to thank Professor Arno Kuijlaars for useful discussions leading to the proof of Theorem 19 based on the Riemann-Hilbert problem approach. We also thank the anonymous referee for insightful suggestions that led to improved exposition in several places.

References

[1] G.R. Allan, *One-sided inverses in Banach algebras of holomorphic vector-valued functions*, J. London Math. Soc. **42**, 463–470 (1967).

[2] J. Baik, P. Deift and K. Johansson, *On the distribution of the length of the longest increasing subsequence of random permutations*, J. Amer. Math. Soc. **12** (1999), no. 4, 1119–1178.

[3] M.A. Bastos, Yu.I. Karlovich, I.M. Spitkovsky, and P.M. Tishin, *On a new algorithm of almost periodic factorization.* In: I. Gohberg, R. Mennicken, and C. Tretter (eds.), *Recent Progress in Operator Theory (Regensburg, 1995)*, Birkhäuser OT 103, Basel and Boston, 1998, pp. 53–74.

[4] A. Böttcher, Yu.I. Karlovich, and I.M. Spitkovsky, *Convolution Operators and Factorization of Almost Periodic Matrix Functions*, Birkhäuser OT 131, Basel and Boston, 2002.

[5] M.C. Câmara and A.F. dos Santos, private communication.

[6] L. Cerlienco and M. Mureddu, *Rappresentazione matriciale degli ordini l.c. su* $\mathbb{R}^n$ *e su* $\mathbb{N}^n$, Rend. Sem. Fac. Sc. Univ. Cagliari 66, 49–68 (1996).

[7] K.F. Clancey and I. Gohberg, *Factorization of Matrix Functions and Singular Integral Operators*, Birkhäuser OT 3, Basel and Boston, 1981.

[8] P. Deift, *Orthogonal Polynomials and Random Matrices: A Riemann-Hilbert Approach*, Courant Lecture Notes in Mathematics, **3**, 1999.

[9] P. Deift, T. Kriecherbauer, K.T.-R. McLaughlin, S. Venakides, and X. Zhou, *Uniform asymptotics for polynomials orthogonal with respect to varying exponential weights and applications to universality questions in random matrix theory*, Comm. Pure Appl. Math. **52** (1999), 1335–1425.

[10] P. Deift, T. Kriecherbauer, K.T.-R. McLaughlin, S. Venakides, and X. Zhou, *Strong asymptotics of orthogonal polynomials with respect to exponential weights*, Comm. Pure Appl. Math. **52** (1999), 1491–1552.

[11] T. Ehrhardt and C. van der Mee, *Canonical factorization of continuous functions on the d-torus*, Proc. Amer. Math. Soc. 131, 801–813 (2002).

[12] J. Erdős, *On the structure of ordered real vector spaces*, Publ. Math. Debrecen 4, 334–343 (1956).

[13] A.S. Fokas, A.R. Its, and A.V. Kitaev, *Discrete Painlevé equations and their appearance in quantum gravity*, Comm. Math. Phys. **142** (1991), 313–344.

[14] A.S. Fokas, A.R. Its, and A.V. Kitaev, *The isomonodromy approach to matrix models in 2D quantum gravity*, Comm. Math. Phys. **147** (1992), 395–430.

[15] L. Fuchs, *Partially Ordered Algebraic Systems*, Pergamon Press, Oxford, 1963.

[16] F.R. Gantmacher, *Applications of the Theory of Matrices*, Interscience Publishers, New York, 1959. (Translation from Russian.)

[17] J.S. Geronimo and H.J. Woerdeman, *Positive extensions, Riesz-Féjer factorization, and autoregressive filters in two variables*, Annals of Mathematics, to appear.

[18] I.C. Gohberg and I.A. Feldman, *Convolution Equations and Projection Methods for their Solution*, Transl. Math. Monographs 41, Amer. Math. Soc., Providence, R. I., 1974.

[19] I.C. Gohberg and M.G. Krein, *Systems of integral equations on a half line with kernels depending on the difference of arguments*, Amer. Math. Soc. Transl. (2)14, 217–287 (1960).

[20] I. Gohberg, P. Lancaster, and L. Rodman, *Invariant Subspaces of Matrices with Applications*, Wiley – Interscience, New York, etc, 1986.

[21] I.C. Gohberg and Yu. Leiterer, *Factorization of operator functions with respect to a contour. II. Canonical factorization of operator functions close to the identity*, Math. Nachr. **54**, 41–74 (1972). (Russian)

[22] Yu. Karlovich and I.M. Spitkovsky, *Factorization of almost periodic matrix functions*, J. Math. Anal. Appl. 193, 209–232 (1995).

[23] Yu.I. Karlovich and I.M. Spitkovsky, *Factorization of almost periodic matrix functions and (semi)-Fredholmness of some convolution type equations*, No. 4421-85 dep., VINITI, Moscow, 1985. (Russian).

[24] A.G. Kurosh, *The theory of groups*, Volumes I and II, Chelsea Publishing Co., New York 1960.

[25] A.I. Perov and A.V. Kibenko, *A theorem on the argument of an almost periodic function of several variables*, Litovskii Matematicheskii Sbornik, 7, 505–508 (1967). (Russian)

[26] D. Quint, L. Rodman, and I.M. Spitkovsky, *New cases of almost periodic factorization of triangular matrix functions*, Michigan Math. J. 45(1), 73–102 (1998).

[27] L. Rodman, I.M. Spitkovsky, and H.J. Woerdeman, *Carathéodory-Toeplitz and Nehari problems for matrix-valued almost periodic functions*, Trans. Amer. Math. Soc. 350, 2185–2227 (1998).

[28] L. Rodman, I.M. Spitkovsky, and H.J. Woerdeman, *Factorization of almost periodic matrix functions of several variables and Toeplitz operators.* In: H. Bart, I. Gohberg, and A.C.M. Ran (eds.), *Operator Theory and Analysis* (Amsterdam, 1997), Birkhäuser OT 122, Basel and Boston, 2001, pp. 385–416.

[29] L. Rodman, I.M. Spitkovsky, and H.J. Woerdeman, *Abstract band method via factorization, positive and band extensions of multivariable almost periodic matrix functions, and spectral estimation*, Memoirs of Amer. Math. Soc. 160, No. 762 (2002).

[30] L. Rodman, I.M. Spitkovsky, and H.J. Woerdeman, *Noncanonical factorizations of almost periodic multivariable matrix functions*, Operator Theory: Advances and Applications, 142 (2003), 311–344.

[31] W. Rudin, *Fourier Analysis on Groups*, John Wiley, New York, 1962.

[32] I.M. Spitkovsky and P.M. Tishin, *Factorization of New Classes of Almost-periodic Matrix Functions*, Reports of the extended sessions of a seminar of the I. N. Vekua

Institute for Applied Mathematics, Tbilisi University Press 3(1), 170–173 (1989). (Russian)

[33] G. Szegő, *Orthogonal polynomials*, fourth ed., American Mathematical Society, Providence, R.I., 1975, American Mathematical Society, Colloquium Publications, Vol. XXIII.

[34] P.M. Tishin, *Factorization and Formulas for the Partial Indices of certain classes of Matrix Functions*, Candidate dissertation, Odessa State University, Odessa, 1990. (Russian)

Cornelis V.M. van der Mee
Dipartimento di Matematica e Informatica
Università di Cagliari
Viale Merello 92
I-09123 Cagliari, Italy
e-mail: cornelis@bugs.unica.it

Leiba Rodman
Department of Mathematics
College of William and Mary
P. O. Box 8795
Williamsburg, VA 23187-8795, USA
e-mail: lxrodm@math.wm.edu

Ilya M. Spitkovsky
Department of Mathematics
College of William and Mary
P. O. Box 8795
Williamsburg, VA 23187-8795, USA
e-mail: ilya@math.wm.edu

Hugo J. Woerdeman
Department of Mathematics
College of William and Mary
P. O. Box 8795
Williamsburg, VA 23187-8795, USA
and
Department of Mathematics
Katholieke Universiteit Leuven
Celestijnenlaan 200B
B-3001 Heverlee, Belgium
e-mail: hugo@math.wm.edu

Operator Theory:
Advances and Applications, Vol. 149, 467–486

Semidefinite Invariant Subspaces: Degenerate Inner Products

Christian Mehl, André C.M. Ran, and Leiba Rodman

Abstract. The paper concerns several classes of matrices in possibly degenerate indefinite inner products, such as expansive, dissipative, normal and plus matrices. The main results concern existence of invariant maximal semidefinite subspaces for matrices in these classes.

1. Introduction

The theory and applications of semidefinite invariant subspaces for certain classes of operators in indefinite inner product spaces, both finite and infinite-dimensional, is well-developed by now (see, e.g., the monographs [1], [2], [4], [8], [14], [17]). However, most results in this area are available under the additional hypothesis that the indefinite inner product is regular, i.e., the only vector orthogonal to the whole space is the zero vector. At the same time, nonregular, or degenerate, indefinite inner products do appear in various applications (see [3], [16], [21]). Because of this, and of general mathematical interest, it is worthwhile to develop a more general theory of classes of operators and semidefinite invariant subspaces in indefinite inner products that does not presuppose regularity. Some work in this direction already exists (see [22]). In the present paper, we continue work in this direction, with emphasis on semidefinite invariant subspaces.

We confine ourselves to finite-dimensional spaces, as proofs of several of our main results depend on finite-dimensionality, although many statements in Section 2 can be extended to some infinite-dimensional indefinite inner product spaces.

Throughout the paper, we consider the vector space $\mathbb{F}^n$, where $\mathbb{F}$ is the real field or the complex field. We fix the indefinite inner product $[\cdot, \cdot]$ determined by a not necessarily invertible Hermitian (or symmetric in the real case) $n \times n$ matrix H via the formula

$$[x,y] = \langle Hx, y\rangle, \qquad x,y \in \mathbb{F}^n.$$

(Here, $\langle\cdot,\cdot\rangle$ denotes the standard inner product.) By $i_+(H)$ (respectively, $i_-(H)$, $i_0(H)$) we denote the number of positive (respectively, negative, zero) eigenvalues (counted with multiplicities) of H.

Received by the editors December 23, 2002; revised January 8, 2003. Submitted by J. A. Ball.
Mathematics Subject Classification (2000). 15A63, 15A57.
Key words and phrases. Degenerate inner products, semidefinite invariant subspaces, expansive matrices, plus-matrices, dissipative matrices, normal matrices.

The rest of the paper is organized as follows. Section 2 is preliminary and reviews the theory of semidefinite subspaces. Many results there may be well known but not easily found in the literature. In Sections 3 to 6, we then discuss the existence of maximal semidefinite subspaces that are invariant for matrices from various classes with respect to the indefinite inner product, namely:

a) expansive matrices A: $[Ax, Ax] \geq [x, x]$, $x \in \mathbb{F}^n$;
b) plus-matrices A: $[Ax, Ax] \geq 0$ for all $x \in \mathbb{F}^n$ such that $[x, x] \geq 0$;
c) dissipative matrices A: $\mathrm{Im}[Ax, x] \geq 0$, $x \in \mathbb{F}^n$ (in the complex case);
d) normal matrices A: $A^*HA = HAH^\dagger A^*H$, where $H^\dagger$ denotes the Moore-Penrose generalized inverse of H.

We note that in each case it is easy to check that if A is from one of the classes of matrices in a)–d) with respect to the inner product induced by H, then $P^{-1}AP$ is from the corresponding class with respect to the inner product induced by P^*HP, provided that P is nonsingular.

We start investigating expansive matrices in Section 3. The key result is that for an H-expansive matrix $A \in \mathbb{F}^{n\times n}$ and an A-invariant H-nonnegative subspace $\mathcal{M}_0$ there exists an A-invariant H-nonnegative subspace $\mathcal{M} \supseteq \mathcal{M}_0$ such that $\dim \mathcal{M} = i_+(H) + i_0(H)$. This result is the basis for analogous results for plus-matrices, described in Section 4, and for dissipative matrices, described in Section 5. In Section 6 we discuss H-normal matrices. We first prove that in a nondegenerate indefinite inner product space any H-normal matrix has an invariant maximal H-nonnegative subspace. Our proof is constructive, and in that way it differs from the one given in [2]. The second main result of the section concerns spaces with a degenerate indefinite inner product. We show that if N is H-normal, then there is an H-nonnegative N-invariant subspace $\mathcal{M}$ with $\dim \mathcal{M} = i_+(H) + i_0(H)$.

We shall use the following notations in the sequel: $\mathbb{N}$ is the set of positive numbers; $e_k = \begin{bmatrix} 0 & \cdots & 0 & 1 & 0 & \cdots & 0 \end{bmatrix}^T \in \mathbb{F}^n$ denotes the kth standard unit vector (with 1 in the kth position) - the dimension n is understood from context; $\mathrm{Span}\{x_1, \ldots, x_k\}$ is the subspace spanned by the vectors $x_1, \ldots, x_k$; $\mathrm{Im}\, X$ is the column space of a matrix X; for a complex number z, $\mathrm{Im}\, z$ is the imaginary part of z; Z_n is the $n \times n$ matrix with ones on the upper right - lower left diagonal and zeros elsewhere; $\mathcal{J}_n(\lambda)$ denotes a Jordan block of size n associated with the eigenvalue λ; I_p and 0_p stand for the $p \times p$ identity and the $p \times p$ zero matrix (if p is clear from context, it will often be omitted); $A_1 \oplus \cdots \oplus A_k$ is a block diagonal matrix with diagonal blocks $A_1, \ldots, A_k$ (in that order); A^T is the transpose of a matrix A; $A \geq 0$ stands for positive semidefiniteness of a matrix A; and $\|X\|$ denotes the operator norm (largest singular value) of a matrix X.

2. Semidefinite subspaces

In this section we develop a general framework to handle various types of semidefinite subspaces. The class of H-*nonnegative* subspaces is defined by

$$\mathcal{S}_{\geq 0}(H) = \Big\{V : V \subseteq \mathbb{F}^n \text{ subspace such that } [x,x] \geq 0 \text{ for all } x \in V\backslash\{0\}\Big\}. \quad (2.1)$$

Analogously, one defines $\mathcal{S}_{\eta\ 0}(H)$, where η is one of the signs $>$, $\leq$, $<$, $=$, and the elements of these classes are called H-*positive*, H-*nonpositive*, H-*negative*, H-*neutral* subspaces, respectively. Clearly, if a subspace $V \subseteq \mathbb{F}^n$ belongs to one of these classes, then so does every subspace of V. Observe that by default (nonexistence of nonzero vectors) the zero subspace is simultaneously H-positive, H-negative, and H-neutral. The classes $\mathcal{S}_{\eta\ 0}(H)$ are naturally transformed under congruence $H \mapsto S^*HS$, where $S \in \mathbb{F}^{n\times n}$ is invertible. Namely,

$$\mathcal{M} \in \mathcal{S}_{\eta\ 0}(H) \text{ if and only if } S^{-1}\mathcal{M} \stackrel{\text{def}}{=} \{S^{-1}x | \, x \in \mathcal{M}\} \in \mathcal{S}_{\eta\ 0}(S^*HS). \quad (2.2)$$

As a consequence, we obtain the following useful lemma (whose elementary proof is omitted).

Lemma 2.1. *Let $\mathcal{M}$ be H-positive, respectively, H-negative, with dimension d. Then there exists a nonsingular matrix $S \in \mathbb{F}^{n\times n}$ such that $S^{-1}\mathcal{M} = Span\{e_1, \ldots, e_d\}$ and $S^*HS = H_{11} \oplus H_{22}$, where $H_{11} \in \mathbb{F}^{d\times d}$ is positive definite, respectively, negative definite.*

A subspace $\mathcal{M} \in \mathcal{S}_{\geq 0}(H)$ is called *maximal H-nonnegative* if there is no larger subspace in the set $\mathcal{S}_{\geq 0}(H)$. Analogously, maximal H-positive, H-nonpositive, H-negative, and H-neutral subspaces are defined.

Proposition 2.2. *Let $\mathbb{F} = \mathbb{C}$ or $\mathbb{F} = \mathbb{R}$. Then*

(a) *$\mathcal{M} \in \mathcal{S}_{\geq 0}(H)$ is maximal H-nonnegative if and only if* $\dim \mathcal{M} = i_+(H) + i_0(H)$;

(b) *$\mathcal{M} \in \mathcal{S}_{> 0}(H)$ is maximal H-positive if and only if* $\dim \mathcal{M} = i_+(H)$;

(c) *$\mathcal{M} \in \mathcal{S}_{\leq 0}(H)$ is maximal H-nonpositive if and only if* $\dim \mathcal{M} = i_-(H) + i_0(H)$;

(d) *$\mathcal{M} \in \mathcal{S}_{< 0}(H)$ is maximal H-negative if and only if* $\dim \mathcal{M} = i_-(H)$;

(e) *$\mathcal{M} \in \mathcal{S}_{=0}(H)$ is maximal H-neutral if and only if* $\dim \mathcal{M} = \min(i_+(H), i_-(H)) + i_0(H)$.

The "if" parts of Proposition 2.2 follow from the interlacing inequalities for eigenvalues of compressions of Hermitian matrices to a subspace (see, e.g., [13, Theorem 4.3.8] or [18, Chapter 8]). We omit the proofs of "only if" parts, noting only that Lemma 2.1 may be used for the proof of "only if" in (b), and the "only if" of (a) can be obtained from (b) by considering $H + \epsilon I$, where $\epsilon > 0$, and letting $\epsilon \longrightarrow 0$.

In the following, we will discuss the existence of maximal nonnegative subspaces that contain a given nonnegative subspace. Therefore, a description of nonnegative and maximal nonnegative subspaces in terms of the range of certain matrices is needed. Note that, setting $i_+(H) = p$, $i_-(H) = q$, and $i_0(H) = \nu$, we can assume that H has the form

$$H = I_p \oplus -I_q \oplus 0_\nu. \tag{2.3}$$

Indeed, by applying a congruence $H \mapsto S^*HS$ for a suitable invertible matrix S, and simultaneously transforming $\mathcal{M}_0 \mapsto S^{-1}\mathcal{M}_0$, the form (2.3) can always be achieved. Then, H-nonnegative subspaces can be conveniently described (see, e.g., [4] for the case of invertible H):

Lemma 2.3. *Let H be given by (2.3). Then a subspace $\mathcal{M} \subseteq \mathbb{F}^n$ of dimension $d > 0$ is H-nonnegative if and only if $\mathcal{M}$ has the form*

$$\mathcal{M} = \operatorname{Im} \begin{bmatrix} P & 0 \\ K & 0 \\ Y & X \end{bmatrix}, \tag{2.4}$$

where $P \in \mathbb{F}^{d\times r}$, $K \in \mathbb{F}^{q\times r}$, and $\begin{bmatrix} Y & X \end{bmatrix} \in \mathbb{F}^{\nu\times d}$ satisfy $P^*P = I_r$, $\|K\| \leq 1$, $X^*X = I_{d-r}$, and $X^*Y = 0$ for some r with $0 \leq r \leq d$.

Proof. It is easy to see that every subspace of the form (2.4) is H-nonnegative.

Conversely, let $\mathcal{M} \in \mathcal{S}_{\geq 0}(H)$, dim $\mathcal{M} = d$. Let $f_1, \ldots, f_d$ be a basis of $\mathcal{M}$, and partition

$$f_j = \begin{bmatrix} f_{1j} \\ f_{2j} \\ f_{3j} \end{bmatrix}, \quad f_{1j} \in \mathbb{F}^p, \; f_{2j} \in \mathbb{F}^q, \; f_{3j} \in \mathbb{F}^\nu.$$

We note that the vectors $[f_{1j}, f_{3j}]^T$, $j = 1, \ldots, d$, are linearly independent; otherwise, $\mathcal{M}$ would contain a nonzero vector of the form $\left[0, f^T, 0\right]^T$, $f \in \mathbb{F}^q$, a contradiction with $\mathcal{M}$ being H-nonnegative.

Let r be the dimension of the span of the vectors $f_{1j}, j = 1, \ldots, d$. Without loss of generality, we may assume that $f_{11}, \ldots, f_{1r}$ are linearly independent. By subtracting from some of the f_j, $j > r$, linear combinations of $f_1, \ldots, f_r$, we may moreover assume without loss of generality that $f_{1j} = 0$ for $j > r$. It then follows that also $f_{2j} = 0$ for $j > r$, as otherwise $\mathcal{M}$ contains H-negative vectors.

Put $F = [f_{11} \cdots f_{1r}]$. Let $P = FT$, where $T \in \mathbb{F}^{r\times r}$ is chosen so that $P^*P = I_e$. As F^*F is invertible such a choice of T is possible. Now we take $K = [f_{21} \cdots f_{2r}]\,T$.

Define $X_1 = [f_{3,r+1} \cdots f_{3d}]$ and $Y = [f_{31} \cdots f_{3r}]\,T$. Then

$$\mathcal{M} = \operatorname{Im}\,[f_1 \cdots f_d] = \operatorname{Im} \begin{bmatrix} P & 0 \\ K & 0 \\ Y & X_1 \end{bmatrix}.$$

It follows that rank $X_1 = d - r$. Then there is a matrix S such that $X_1 = XS$ with $X^*X = I_{d-r}$. So

$$\mathcal{M} = \operatorname{Im} \begin{bmatrix} P & 0 \\ K & 0 \\ Y & X \end{bmatrix}.$$

Next, we show that we can take Y in such a way that $X^*Y = 0$. Indeed, the latter condition simply means that Im X and Im Y are orthogonal. This can be achieved by observing that adding to Y a matrix of the form XW, where W is arbitrary, does not change $\mathcal{M}$. Then $X^*Y + X^*XW = X^*Y + W = 0$ if we take $W = -X^*Y$. Finally, a straightforward calculation shows that $\|K\| \leq 1$. □

Note that the representation (2.4) for a given $\mathcal{M}$ is not unique. We have the following result.

Lemma 2.4. *If*

$$\operatorname{Im} \begin{bmatrix} P_1 & 0 \\ K_1 & 0 \\ Y_1 & X_1 \end{bmatrix} = \operatorname{Im} \begin{bmatrix} P_2 & 0 \\ K_2 & 0 \\ Y_2 & X_2 \end{bmatrix}, \tag{2.5}$$

where P_1 and P_2 are $p \times r_1$ and $p \times r_2$ matrices such that $P_1^*P_1 = I$ and $P_2^*P_2 = I$, and $X_1^*X_1 = I$, $X_2^*X_2 = I$, $X_i^*Y_i = 0$ for $i = 1, 2$, then $r_1 = r_2$ and $P_1 = P_2W$, $Y_1 = Y_2W$, $K_1 = K_2W$ for some unitary $d \times d$ matrix W, and $X_1 = X_2V$ for some unitary V. Such unitary matrices W and V are unique.

Proof. We have

$$\begin{bmatrix} P_1 & 0 \\ K_1 & 0 \\ Y_1 & X_1 \end{bmatrix} = \begin{bmatrix} P_2 & 0 \\ K_2 & 0 \\ Y_2 & X_2 \end{bmatrix} Q, \quad Q = \begin{bmatrix} Q_1 & Q_2 \\ Q_3 & Q_4 \end{bmatrix}, \quad Q_1 \in \mathbb{F}^{r_2 \times r_1}, \tag{2.6}$$

for some invertible matrix Q which is easily seen to be unique. Conditions $P_1^*P_1 = I$ and $P_2^*P_2 = I$ imply $Q_1^*Q_1 = I$; in particular, $r_2 \geq r_1$. Reversing the roles of the matrices in the left and the right-hand sides of (2.6), we obtain also $r_1 \geq r_2$. Thus, $r_1 = r_2$, and Q_1 is unitary. Next, we obtain from (2.6):

$$Q^*(Z_2 \oplus I)Q = Z_1 \oplus I,$$

where $Z_j = I + K_j^*K_j + Y_j^*Y_j$, $j = 1, 2$. In particular,

$$Q_1^*Z_2Q_1 = Z_1 - Q_3^*Q_3, \tag{2.7}$$

and since upon subtracting a positive semidefinite matrix from a positive definite matrix the eigenvalues can only decrease, we obtain that $\lambda_k(Z_2) \leq \lambda_k(Z_1)$, $k = 1, 2 \ldots, r_1$, where $\lambda_k(X)$ stand for the eigenvalues of a positive definite matrix X, in the non-increasing order. Reversing the roles of Z_1 and Z_2 we obtain also $\lambda_k(Z_1) \leq \lambda_k(Z_2)$, and hence Z_1 and Z_2 have the same eigenvalues. Now (2.7) implies $Q_3 = 0$. Analogously $Q_2 = 0$. Now from (2.6) we have $X_1 = X_2Q_4$, and (as for Q_1) we show that Q_4 is unitary. □

In particular Lemma 2.4 applies to an H-nonnegative subspace of the form (2.5). However, for the validity of the lemma, it is not necessary that $\|K_1\| \leq 1$, $\|K_2\| \leq 1$.

Corollary 2.5. *Every maximal H-nonnegative subspace $\mathcal{M}$ can be uniquely written in the form*

$$\mathcal{M} = \operatorname{Im} \begin{bmatrix} I_p & 0 \\ K & 0 \\ 0 & I_\nu \end{bmatrix}, \tag{2.8}$$

where $K \in \mathbb{F}^{q\times p}$ satisfies $\|K\| \le 1$. Conversely, every subspace of the form (2.8) is maximal H-nonnegative.

Indeed, the uniqueness of (2.8) follows from Lemma 2.4. Existence of (2.8) follows from Lemma 2.3 in which P is unitary, because in view of Proposition 2.2(a) the dimension of $\mathcal{M}$ is equal to $p+\nu$.

Next, we express containment of an H-nonnegative subspace in a maximal such subspace, in terms of the representation (2.4).

Lemma 2.6. *Let*

$$\mathcal{M}_0 = \operatorname{Im} \begin{bmatrix} P_0 & 0 \\ K_0 & 0 \\ Y_0 & X_0 \end{bmatrix} \in \mathcal{S}_{\ge 0}(H), \quad \mathcal{M} = \operatorname{Im} \begin{bmatrix} I_p & 0 \\ K & 0 \\ 0 & I_\nu \end{bmatrix} \in \mathcal{S}_{\ge 0}(H),$$

where $P_0 \in \mathbb{F}^{p\times r}$, $K_0 \in \mathbb{F}^{q\times r}$, $K \in \mathbb{F}^{q\times p}$ are such that $P_0^*P_0 = I$ and $\|K_0\| \le 1$, $\|K\| \le 1$, and $X_0^*X_0 = I$, $X_0^*Y_0 = 0$. (In particular, $\mathcal{M}$ is maximal H-nonnegative). Then $\mathcal{M}_0 \subseteq \mathcal{M}$ if and only if $K_0 = KP_0$.

Proof. If $K_0 = KP_0$, then obviously $\begin{bmatrix} P_0 \\ K_0 \end{bmatrix} = \begin{bmatrix} I \\ K \end{bmatrix} P_0$, and therefore $\mathcal{M}_0 \subseteq \mathcal{M}$. Conversely, if $\mathcal{M}_0 \subseteq \mathcal{M}$, then there exist matrices B_{11}, B_{12}, B_{21}, B_{22} such that

$$\begin{bmatrix} P_0 & 0 \\ K_0 & 0 \\ Y_0 & X_0 \end{bmatrix} = \begin{bmatrix} I & 0 \\ K & 0 \\ 0 & I \end{bmatrix} \begin{bmatrix} B_{11} & B_{12} \\ B_{21} & B_{22} \end{bmatrix}.$$

It is immediate from this equality that in fact $B_{12} = 0$, $B_{22} = X_0$, $B_{21} = Y_0$ and $B_{11} = P_0$. It then follows that $K_0 = KB_{11} = KP_0$. □

We will need also the following description of H-neutral subspaces:

Lemma 2.7. *Let H be invertible and let $\mathcal{M}$ be H-neutral with dimension d. Then $2d \le n$ and there exists a nonsingular $P \in \mathbb{F}^{n\times n}$ such that*

$$P^{-1}\mathcal{M} = \operatorname{Span}\{e_1,\dots,e_d\} \qquad \text{and} \qquad P^*HP = \begin{bmatrix} 0 & I_d & 0 \\ I_d & 0 & 0 \\ 0 & 0 & H_{33} \end{bmatrix}.$$

Proof. The result follows from the properties of skewly-linked neutral subspaces (see, for example, [14, Lemma 3.1], or [7]). For completeness, we offer an independent proof. By equation (2.2), we may assume that $\mathcal{M} = \operatorname{Span}\{e_1,\dots,e_d\}$. Since H is nonsingular, we obtain from Proposition 2.2 that $2d \le n$. Partition

$$H = \begin{bmatrix} H_{11} & H_{12} & H_{13} \\ H_{12}^* & H_{22} & H_{23} \\ H_{13}^* & H_{23}^* & H_{33} \end{bmatrix}.$$

Then $H_{11} = 0$ and hence, $\begin{bmatrix} H_{12} & H_{13} \end{bmatrix}$ is of full row rank. But then, there exists matrices $Q_1 \in \mathbb{F}^{d\times d}$ and $Q_2 \in \mathbb{F}^{(n-2d)\times(n-2d)}$ such that $Q_1 \begin{bmatrix} H_{12} & H_{13} \end{bmatrix} Q_2 = \begin{bmatrix} I_d & 0 \end{bmatrix}$. Setting $P_1 = Q_1 \oplus Q_2$, we obtain

$$P_1^* H P_1 = \begin{bmatrix} 0 & I_d & 0 \\ I_d & \hat{H}_{22} & \hat{H}_{23} \\ 0 & \hat{H}_{23}^* & \hat{H}_{33}^* \end{bmatrix}, \quad P_1^{-1}\mathcal{M} = \mathcal{M}.$$

Setting furthermore

$$P = P_1 \begin{bmatrix} I & -\frac{1}{2}\hat{H}_{22} & -\hat{H}_{23} \\ 0 & I & 0 \\ 0 & 0 & I \end{bmatrix},$$

we obtain that $P^{-1}\mathcal{M} = \mathcal{M}$ and P^*HP has the desired form. □

3. Expansive matrices

In this section we focus on H-expansive matrices. Recall that a matrix $A \in \mathbb{F}^{n\times n}$ is called *H-expansive* if $[Ax, Ax] \geq [x, x]$ for all $x \in \mathbb{F}^n$, or, equivalently, if $A^*HA - H \geq 0$. It turns out that the kernel $\operatorname{Ker} A := \{x \in \mathbb{F}^n : Ax = 0\}$ of an H-expansive matrix A is H-negative (as long as H is invertible). To prove this, we need the following auxiliary result.

Lemma 3.1. *Let $A \in \mathbb{F}^{n\times n}$ be H-expansive. Then there exists a nonsingular matrix $S \in \mathbb{F}^{n\times n}$ such that*

$$S^*HS = H_1 \oplus (-I_{p_2}) \oplus 0_{p_3}, \quad S^*A^*HAS = M_1 \oplus 0_{p_2} \oplus 0_{p_3}, \tag{3.1}$$

where $M_1, H_1 \in \mathbb{F}^{p_1\times p_1}$ are nonsingular and $p_1, p_2, p_3 \in \mathbb{N} \cup \{0\}$.

Proof. By the well-known canonical forms under congruence for pairs of Hermitian matrices or pairs of symmetric matrices in the real case (see, e.g., [24], [25]), we may assume that H and A^*HA have the forms

$$H = H_1 \oplus H_2 \oplus \cdots \oplus H_m, \quad A^*HA = M_1 \oplus M_2 \oplus \cdots \oplus M_m,$$

where M_j and H_j have the same size, H_1 and M_1 are nonsingular, and M_j and H_j, $j > 1$, are blocks of one of the following types:

type 1: $H_j = \varepsilon Z_p$ and $M_j = \varepsilon Z_p \mathcal{J}_p(0)$ for some $p \in \mathbb{N}$, $\varepsilon = \pm 1$;

type 2: $p \in \mathbb{N} \cup \{0\}$,

$$H_j = \begin{bmatrix} 0 & 0 & I_p \\ 0 & 0 & 0 \\ I_p & 0 & 0 \end{bmatrix} \in \mathbb{F}^{(2p+1)\times(2p+1)}, \quad M_j = \begin{bmatrix} 0 & 0 & 0 \\ 0 & 0 & I_p \\ 0 & I_p & 0 \end{bmatrix} \in \mathbb{F}^{(2p+1)\times(2p+1)},$$

type 3: $H_j = \varepsilon Z_p \mathcal{J}_p(0)$ and $M_j = \varepsilon Z_p$ for some $p \in \mathbb{N}$, $\varepsilon = \pm 1$.

Clearly, since A is H-expansive, each block $M_j - H_j$ is positive semidefinite. It is easy to check that this is possible if and only if $p = 1$ and $\varepsilon = -1$ if H_j and M_j are of type 1, $p = 1$ and $\varepsilon = 1$ if H_j and M_j are of type 3, and $p = 0$ if H_j

and M_j are of type 2. But then, after eventually permuting some blocks, H and A^*HA have the forms

$$H = H_1 \oplus 0_{p_2} \oplus (-I_{p_3}) \oplus 0_{p_4}, \quad A^*HA = M_1 \oplus 0_{p_2} \oplus 0_{p_3} \oplus I_{p_4}.$$

Note that $M_1 - H_1$ is still positive semidefinite, i.e., the number of positive eigenvalues of M_1 is larger or equal to the number of positive eigenvalues of H_1 (see, e.g., 7.7.4 in [13]). From the well-known fact that the number of positive (negative) eigenvalues of A^*HA is always less or equal to the number of positive (negative, respectively) eigenvalues of H (see, e.g., 4.5.11 in [13]), it follows that blocks of type 3 cannot occur and hence, A^*HA and H have forms as in (3.1). □

Corollary 3.2. *Let H be invertible and let $A \in \mathbb{F}^{n\times n}$ be H-expansive. Then* Ker A *is H-negative.*

Proof. If A is nonsingular, Ker A is H-negative by definition. Otherwise, let $y \in$ Ker A. Then $y \in$ Ker A^*HA. Since H is invertible, it follows immediately from Lemma 3.1 and equation (3.1) that $y^*Hy < 0$. □

Proposition 3.3. *Let $A \in \mathbb{F}^{n\times n}$ be H-expansive. Then* Ker H *is A-invariant.*

Proof. Applying a transformation of the form $(A, H) \mapsto (S^{-1}AS, S^*HS)$, we may assume that H and A^*HA have the forms as in (3.1). Applying one more transformation on M_1 and H_1, we may furthermore assume that

$$H = I_{p_1} \oplus -I_{p_2} \oplus -I_{p_3} \oplus 0, \quad A^*HA = \begin{bmatrix} M_{11} & M_{12} \\ M_{12}^* & M_{22} \end{bmatrix} \oplus 0,$$

where $p_1, p_2, p_3 \in \mathbb{N} \cup \{0\}$, $M_{11} \in \mathbb{F}^{p_1\times p_1}$, $M_{12} \in \mathbb{F}^{p_1\times p_2}$, $M_{22} \in \mathbb{F}^{p_2\times p_2}$. Let A be partitioned conformably:

$$A = \begin{bmatrix} A_{11} & A_{12} & A_{13} & A_{14} \\ A_{21} & A_{22} & A_{23} & A_{24} \\ A_{31} & A_{32} & A_{33} & A_{34} \\ A_{41} & A_{42} & A_{43} & A_{44} \end{bmatrix}.$$

With $A^*HA - H$ also $M_{11} - I_{p_1} = A_{11}^*A_{11} - A_{21}^*A_{21} - A_{31}^*A_{31} - I_{p_1}$ must be positive semidefinite. This is possible only if $A_{11}^*A_{11}$ is positive definite, i.e., only if A_{11} is nonsingular. Next, we show $A_{14} = 0$. To see this, assume that A_{14} is not zero. Then there exists a matrix $P \in \mathbb{F}^{p_4\times p_1}$, $p_4 = n - p_1 - p_2 - p_3$, such that $A_{11} - PA_{14}$ is singular. (For example, if $v \in \mathbb{F}^{p_4}$ is such that $A_{14}v \neq 0$ choose P such that $PA_{14}v = A_{11}v$.) Applying the transformation $(A, H) \mapsto (\mathcal{P}^{-1}A\mathcal{P}, \mathcal{P}^*H\mathcal{P})$ with

$$\mathcal{P} = \begin{bmatrix} I_{p_1} & 0 & 0 & 0 \\ 0 & I_{p_2} & 0 & 0 \\ 0 & 0 & I_{p_3} & 0 \\ P & 0 & 0 & I_{p_4} \end{bmatrix},$$

we find that the $(1,1)$-block of $\mathcal{P}^{-1}A\mathcal{P}$ is $A_{11} - PA_{14}$, whereas $\mathcal{P}^*H\mathcal{P} = H$ and $\mathcal{P}^*A^*HA\mathcal{P} = A^*HA$. This contradicts the fact just mentioned that the $(1,1)$-block of $\mathcal{P}^{-1}A\mathcal{P}$ must be nonsingular. But $A_{14} = 0$ implies that the $(4,4)$-block of A^*HA has the form $-A_{24}^*A_{24} - A_{34}^*A_{34} = 0$. This is possible only if $A_{24} = 0$ and $A_{34} = 0$. Hence, Ker H is A-invariant. □

The key result in this section is the following theorem.

Theorem 3.4. *Let $A \in \mathbb{F}^{n\times n}$ be H-expansive and let $\mathcal{M}_0 \subseteq \mathbb{F}^n$ be an A-invariant subspace which is H-nonnegative. Then there exists an H-nonnegative A-invariant subspace $\mathcal{M} \supseteq \mathcal{M}_0$ such that* $\dim \mathcal{M} = i_+(H) + i_0(H)$.

Proof. For the case when H is invertible, the proof follows by a well-known argument that originated with M. G. Kreĭn [15] using the Schauder's fixed point theorem [11, Section 106] and the representation of $\mathcal{M}_0$ according to Lemma 2.3:

$$\mathcal{M}_0 = \operatorname{Im} \begin{bmatrix} P_0 \\ K_0 \end{bmatrix},$$

where $P_0^*P_0 = I$, $\|K_0\| \leq 1$; and see [2, Section 3.3] for an application of a fixed point theorem in a more general Krein space context.

Thus, consider the case that H is singular. Without loss of generality, we may assume that H has the form (2.3), i.e., $H = H_1 \oplus 0$, where $H_1 = I_p \oplus (-I_q)$. Then Proposition 3.3 implies that A takes the form

$$A = \begin{bmatrix} A_{11} & 0 \\ A_{21} & A_{22} \end{bmatrix},$$

where $A_{11} \in \mathbb{F}^{(p+q)\times(p+q)}$ is easily seen to be H_1-expansive. Represent $\mathcal{M}_0$ according to Lemma 2.3:

$$\mathcal{M}_0 = \operatorname{Im} \begin{bmatrix} Q_0 & 0 \\ Y_0 & X_0 \end{bmatrix}, \quad Q_0 = \begin{bmatrix} P_0 \\ K_0 \end{bmatrix},$$

where $P_0^*P_0 = I_p$, $\|K_0\| \leq 1$, $X_0^*X_0 = I$, and $X_0^*Y_0 = 0$. Then $\widetilde{\mathcal{M}_0} = \operatorname{Im} Q_0$ is H_1-nonnegative and A_{11}-invariant. By the part already proved, there exists an $i_+(H)$-dimensional, H_1-nonnegative, and A_{11}-invariant subspace $\widetilde{\mathcal{M}}$ that contains $\mathcal{M}_0$. Let $\widetilde{\mathcal{M}} = \operatorname{Im} Q$ for some matrix Q of appropriately chosen dimension, i.e., $Q_0 = QW_0$ for some matrix W_0. Then $\mathcal{M} := \operatorname{Im}(Q \oplus I_\nu)$ is H-nonnegative and A-invariant. Furthermore, it contains $\mathcal{M}_0$, and has dimension $i_+(H) + i_0(H)$. □

Obviously, every H-isometric matrix is H-expansive (recall that a matrix A is called *H-isometric* if $[Ax, Ax] = [x, x]$ for every $x \in \mathbb{F}^n$), Thus, as an important corollary of Theorem 3.4, we obtain:

Theorem 3.5. *Let $A \in \mathbb{F}^{n\times n}$ be an H-isometry, and let $\mathcal{M}_0 \subseteq \mathbb{F}^n$ be an A-invariant H-nonnegative, respectively, H-nonpositive, subspace. Then there exists an A-invariant H-nonnegative, respectively, H-nonpositive, subspace $\mathcal{M}$ such that $\mathcal{M} \supseteq \mathcal{M}_0$ and* $\dim \mathcal{M} = i_+(H)$, *respectively,* $\dim \mathcal{M} = i_-(H)$.

The part of Theorem 3.5 concerning H-nonpositive subspaces follows by noticing that A is also expansive with respect to $-H$, and applying Theorem 3.4 with H replaced by $-H$.

4. Plus-matrices

Recall that a matrix $A \in \mathbb{F}^{n\times n}$ is called a *plus-matrix* if $[Ax, Ax] \geq 0$ for every $x \in \mathbb{F}^n$ such that $[x, x] \geq 0$. The following lemma is well known in the complex case (and can be proved by using the Toeplitz-Hausdorff theorem on convexity of the numerical range [1], and see [6] for a proof in the real case).

Lemma 4.1. *If A is a plus-matrix, then there exists a nonnegative number k such that $[Ax, Ax] \geq k[x, x]$ for all $x \in \mathbb{F}^n$.*

Denote by $k(A)$ the smallest value of $k \geq 0$ for which Lemma 4.1 holds.

Theorem 4.2. *Let A be a plus matrix such that* Im *A is not H-nonnegative, and let $\mathcal{M}_0 \subseteq \mathbb{F}^n$ be an A-invariant subspace which is H-nonnegative. Then there exists H-nonnegative A-invariant subspace $\mathcal{M} \supseteq \mathcal{M}_0$ such that dim $\mathcal{M} = i_+(H) + i_0(H)$.*

Proof. By Lemma 4.1, we have $[Ax, Ax] \geq k(A)[x, x]$ for all $x \in \mathbb{F}^n$. Since Im A is not H-nonnegative, it follows that $k(A) > 0$. Scaling A, if necessary, we can assume that $k(A) = 1$. Then A is H-expansive, and the result follows from Theorem 3.4. □

The hypothesis that Im A is not H-nonnegative is essential in Theorem 3.4. The following example illustrates this.

Example 4.3. Let p and q are real numbers such that $4p + q^2 > 0$. Furthermore, let

$$A = \begin{bmatrix} 0 & 1 & 0 & p \\ -1 & 0 & 1 & q \\ 0 & 0 & 0 & 0 \\ 0 & 0 & 0 & 0 \end{bmatrix}, \quad H = \begin{bmatrix} 0 & 0 & 1 & 0 \\ 0 & 0 & 0 & 1 \\ 1 & 0 & 0 & 0 \\ 0 & 1 & 0 & 0 \end{bmatrix}, \quad \mathcal{M}_0 = \text{Span}\left\{\begin{bmatrix} 1 \\ 0 \\ 1 \\ 0 \end{bmatrix}\right\}.$$

Then, A is a plus matrix; in fact, $[Ax, Ax] = 0$ for every $x \in \mathbb{F}^4$. Moreover, $A\mathcal{M}_0 = \{0\}$ which implies, in particular, that $\mathcal{M}_0$ is A-invariant. The subspace $\mathcal{M}_0$ is H-positive. On the other hand, Ker $A = \text{Span}\{[1, 0, 1, 0]^T, [0, -p, -q, 1]^T\}$ is not H-nonnegative. If $F = \mathbb{R}$, then the only two-dimensional A-invariant subspaces are Ker A and Im A, and the latter does not contain $\mathcal{M}_0$. If $F = \mathbb{C}$, then, besides Ker A, there are two two-dimensional A-invariant subspaces that contain $\mathcal{M}_0$, namely, Span $\{[1\ 0\ 1\ 0]^T,\ \ [1\ \pm i\ 0\ 0]^T\}$. But

$$\begin{bmatrix} 1 & \mp i & 0 & 0 \\ 1 & 0 & 1 & 0 \end{bmatrix} H \begin{bmatrix} 1 & 1 \\ \pm i & 0 \\ 0 & 1 \\ 0 & 0 \end{bmatrix} = \begin{bmatrix} 0 & 1 \\ 1 & 2 \end{bmatrix},$$

so neither of these two subspaces is H-nonnegative. □

Theorem 4.4. *Let $A \in \mathbb{F}^{n\times n}$ be a plus-matrix. Then there exists an H-nonnegative A-invariant subspace $\mathcal{M}$ such that* dim $\mathcal{M} = i_+(H) + i_0(H)$.

Proof. By Lemma 4.1 we have $[Ax, Ax] \geq k(A)[x, x]$ for all $x \in \mathbb{F}^n$ and $k(A) \geq 0$. If $k(A) = 0$, then the range of A is an H-nonnegative subspace, and we are done. Indeed, any maximal H-nonnegative subspace $\mathcal{M}$ containing the range of A is A-invariant. Otherwise, apply the previous theorem with $\mathcal{M}_0 = \{0\}$. □

5. Dissipative matrices

Let us consider the case $\mathbb{F} = \mathbb{C}$ first. A matrix $B \in \mathbb{C}^{n\times n}$ is called *H-dissipative* if $\operatorname{Im}[Bx, x] \geq 0$ for every $x \in \mathbb{C}^n$. The dissipativity condition can be easily interpreted in terms of positive definiteness: A matrix B is H-dissipative if and only if

$$i(HB - B^*H) \leq 0. \tag{5.1}$$

Lemma 5.1.

(a) *Let A be H-expansive, and let $w, \eta \in \mathbb{C}$ be such that w has positive imaginary part, $|\eta| = 1$, and η is not an eigenvalue of A. Then the matrix*

$$B = (wA - \overline{w}\eta I)(A - \eta I)^{-1} \tag{5.2}$$

is H-dissipative.

(b) *Let B be H-dissipative, and let $w, \eta \in \mathbb{C}$ be such that $|\eta| = 1$, w has positive imaginary part, and is not an eigenvalue of B. Then the matrix*

$$A = \eta(B - wI)(B - wI)^{-1} \tag{5.3}$$

is H-expansive.

The proof is obtained by elementary algebraic manipulations and therefore is omitted.

Theorem 5.2. ($\mathbb{F} = \mathbb{C}$) *Let $B \in \mathbb{C}^{n\times n}$ be H-dissipative, and let $\mathcal{M}_0 \subseteq \mathbb{C}^n$ be a B-invariant H-nonnegative, respectively, H-nonpositive subspace. Then there exists a B-invariant maximal H-nonnegative, respectively, maximal H-nonpositive subspace $\mathcal{M}$ such that $\mathcal{M} \supseteq \mathcal{M}_0$.*

Proof. Assume $\mathcal{M}_0$ is H-nonnegative. Let A be given by (5.3). Then A is H-expansive. Note also that $\mathcal{M}_0$ is A-invariant, because A is a function of B. By Theorem 3.4, there exists an A-invariant subspace $\mathcal{M}$ which is maximal H-nonnegative and contains $\mathcal{M}_0$. Since B is a function of A (given by formula (5.2)) $\mathcal{M}$ is also B-invariant. This proves the part of Theorem 5.2 for H-nonnegative subspaces. If $\mathcal{M}_0$ is H-nonpositive, apply the already proved part of Theorem 5.2 to the $(-H)$-dissipative matrix $-B$. □

Recall that a matrix A is called H-skewadjoint if $HA = -A^*H$. Since every H-selfadjoint matrix is obviously H-dissipative and since every H-skewadjoint matrix is just an H-selfadjoint matrix multiplied with the imaginary unit, we immediately obtain the following corollary:

Corollary 5.3. ($\mathbb{F} = \mathbb{C}$) *Let $B \in \mathbb{C}^{n\times n}$ be H-selfadjoint or H-skewadjoint, and let $\mathcal{M}_0 \subseteq \mathbb{C}^n$ be a B-invariant H-nonnegative, respectively, H-nonpositive subspace. Then there exists a B-invariant maximal H-nonnegative, respectively, maximal H-nonpositive subspace $\mathcal{M}$ such that $\mathcal{M} \supseteq \mathcal{M}_0$.*

Theorem 5.2 can also be extended to the real case. But first, we have to modify the definition for H-dissipative matrices, since (5.1) for real matrices B and H implies that B is H-selfadjoint, and not every real H-selfadjoint matrix has an invariant subspace that is maximal H-nonnegative, as easy examples show. Thus, for the case $\mathbb{F} = \mathbb{R}$, let us call $C \in \mathbb{R}^{n\times n}$ *real H-dissipative* if $HC+C^T H \leq 0$, i.e., if $-iC$ is H-dissipative. (Note that this corresponds exactly to the definition of dissipative matrices (or operators) in the sense of [23].) Clearly, in the complex case the theory of real H-dissipative matrices is essentially the same as the theory of H-dissipative matrices, but in the real case, the two definitions lead to fundamentally different classes of matrices.

Theorem 5.4. ($\mathbb{F} = \mathbb{R}$) *Let $C \in \mathbb{R}^{n\times n}$ be real H-dissipative, and let $\mathcal{M}_0 \subseteq \mathbb{R}^n$ be a C-invariant H-nonnegative, respectively, H-nonpositive subspace. Then there exists a C-invariant maximal H-nonnegative, respectively, maximal H-nonpositive subspace $\mathcal{M}$ such that $\mathcal{M} \supseteq \mathcal{M}_0$.*

Proof. Let $\varrho < 0$ such that ϱ is neither an eigenvalue of C, nor of $-C$. Then the real matrix $A = (C+\varrho I)(C-\varrho I)^{-1}$ is H-expansive by a simple computation. The rest of the proof proceeds analogously to the proof of Theorem 5.2. □

Since any H-skewadjoint matrix is real H-dissipative, we immediately obtain the following corollary:

Corollary 5.5. ($\mathbb{F} = \mathbb{R}$) *Let $C \in \mathbb{R}^{n\times n}$ be H-skewadjoint, and let $\mathcal{M}_0 \subseteq \mathbb{R}^n$ be a C-invariant H-nonnegative (respectively, H-nonpositive) subspace. Then there exists a C-invariant maximal H-nonnegative (respectively, maximal H-nonpositive) subspace $\mathcal{M}$ such that $\mathcal{M} \supseteq \mathcal{M}_0$.*

6. Normal matrices

First, let H be invertible. Then for $M \in \mathbb{F}^{n\times n}$, the matrix $M^{[*]_H} := H^{-1}M^*H$, or $M^{[*]}$ if there is no risk of confusion, is called the *H-adjoint* of M. This is the unique matrix satisfying

$$[Mx, y] = [x, M^{[*]}y] \quad \text{for all } x, y \in \mathbb{F}^n. \tag{6.1}$$

A matrix $N \in \mathbb{F}^{n\times n}$ is called *H-normal* if and only if N and $N^{[*]}$ commute, i.e., if and only if $H^{-1}N^*HN = NH^{-1}N^*H$. If H is singular, then for $M \in \mathbb{F}^{n\times n}$ an "H-adjoint", i.e., a matrix $M^{[*]}$ satisfying (6.1) need not exist, and if it exists, it need not be unique. Therefore, we have to modify the definition of H-normal matrices in this case. Noting that H-normality is equivalent to the identity $N^*HN = HNH^{-1}N^*H$, we say that a matrix $N \in \mathbb{F}^{n\times n}$ is *H-normal* if and only if

$$N^*HN = HNH^{\dagger}N^*H,$$

where $H^\dagger$ denotes the Moore-Penrose generalized inverse of H. (This definition has been introduced in [19] and used subsequently also in [5].) In the case $\mathbb{F} = \mathbb{C}$ and H invertible, it is well known that there exists a nonsingular matrix $P \in \mathbb{C}^{n\times n}$ such that

$$P^{-1}NP = N_1 \oplus \cdots \oplus N_m, \quad P^*HP = H_1 \oplus \cdots \oplus H_m, \tag{6.2}$$

where, for each j, N_j and H_j have the same size and N_j has at most two distinct eigenvalues, see [10], and [12] for a corresponding result in the real case. This is no longer true in the case that H is singular as the following example illustrates:

Example 6.1. Let

$$N = \begin{bmatrix} -1 & 0 & 0 \\ 0 & 0 & 0 \\ 0 & 0 & 3 \end{bmatrix}, \quad H = \begin{bmatrix} 0 & 1 & 0 \\ 1 & -1 & 1 \\ 0 & 1 & 0 \end{bmatrix}, \quad H^\dagger = \frac{1}{16}\begin{bmatrix} 3 & 4 & 5 \\ 4 & 0 & 12 \\ 5 & 12 & 3 \end{bmatrix}.$$

Then it is easy to check that A is H-normal, in fact, $N^*HN = HNH^\dagger N^*H = 0$. Suppose, a decomposition as in (6.2) with $m > 1$ exists. Note that the eigenvectors of N associated with the eigenvalues -1 and 3 are H-neutral and that the eigenvector associated with the eigenvalue 0 is H-negative. Then the fact that Ker H is not N-invariant and that $i_+(H) = i_-(H) = i_0(H) = 1$ implies that the only possible decomposition as in (6.2) with $m > 1$ is

$$P^{-1}NP = N_1 \oplus \begin{bmatrix} 0 \end{bmatrix}, \quad P^*HP = \begin{bmatrix} 1 & 0 \\ 0 & 0 \end{bmatrix} \oplus \begin{bmatrix} -1 \end{bmatrix}$$

for some nonsingular matrix $P \in \mathbb{F}^{3\times 3}$ (up to congruent forms for the blocks of P^*HP). This contradicts the fact that the subspace spanned by the eigenvectors of N associated with the eigenvalues -1 and 3 is H-neutral.

The question whether every H-normal matrix has an invariant subspace that is maximal H-nonnegative has a negative answer in the case $\mathbb{F} = \mathbb{R}$ as can be seen from the example

$$A = \begin{bmatrix} \alpha & \beta \\ -\beta & \alpha \end{bmatrix}, \quad H = \begin{bmatrix} 0 & 1 \\ 1 & 0 \end{bmatrix},$$

where α, β are real and $\beta \neq 0$. The answer, however, is affirmative in the case $\mathbb{F} = \mathbb{C}$ (and invertible H), see [2, Corollary 3.4.12]. For the sake of completeness, we present an independent proof for this result.

Theorem 6.2. *Let H be invertible and let $N \in \mathbb{C}^{n\times n}$ be H-normal. Then there exists an H-nonnegative N-invariant subspace $\mathcal{M}$ such that* $\dim \mathcal{M} = i_+(H)$.

Proof. Let $N = A + S$ be the decomposition of N into its H-selfadjoint part A and its H-skewadjoint part S, i.e., $A = 1/2(N + N^{[*]})$ and $S = 1/2(N - N^{[*]})$. Then it is easy to check that N is H-normal if and only if A and S commute. We will now show by induction on n that A and S have a common invariant subspace that is maximal H-nonnegative. Clearly, this subspace is also N-invariant. The case $n = 1$ is clear, since either the zero space $\{0\}$ or the full space $\mathbb{C}$ is maximal

H-nonnegative. Next, let $n > 1$. Since A and S commute, they have a common eigenvector $v \in \mathbb{C}^n$.

Case (1): Span$\{v\}$ is not H-neutral, i.e., Span$\{v\}$ is H-positive or H-negative. By Lemma 2.1, we may assume that

$$A = \begin{bmatrix} A_{11} & A_{12} \\ 0 & A_{22} \end{bmatrix}, \quad S = \begin{bmatrix} S_{11} & S_{12} \\ 0 & S_{22} \end{bmatrix}, \quad H = \begin{bmatrix} H_{11} & 0 \\ 0 & H_{22} \end{bmatrix},$$

where $A_{11}, S_{11}, H_{11} \in \mathbb{C}$, $H_{11} \neq 0$. Since A is H-selfadjoint and S is H-skewadjoint, i.e., $A^*H = HA$ and $-S^*H = HS$, we obtain that $A_{12} = S_{12} = 0$. Then the result follows by using the induction hypothesis on A_{11}, S_{11} and A_{22}, S_{22}.

Case (2): Span$\{v\}$ is H-neutral. By Lemma 2.7, we may assume that

$$A = \begin{bmatrix} A_{11} & A_{12} & A_{13} \\ 0 & A_{22} & A_{23} \\ 0 & A_{32} & A_{33} \end{bmatrix} S = \begin{bmatrix} S_{11} & S_{12} & S_{13} \\ 0 & S_{22} & S_{23} \\ 0 & S_{32} & S_{33} \end{bmatrix} H = \begin{bmatrix} 0 & I_d & 0 \\ I_d & 0 & 0 \\ 0 & 0 & H_{33} \end{bmatrix}.$$

Then the fact that A is H-selfadjoint and S is H-skewadjoint implies $A_{23} = S_{23} = 0$ as it can easily be seen from the identities $A^*H = HA$ and $-S^*H = HS$. Also A_{33} is H_{33}-selfadjoint, and S_{33} is H_{33}-skewadjoint. From $A_{23} = S_{23} = 0$ one easily sees by considering the $3,3$-block entry in $SA = AS$, that also $A_{33}S_{33} = S_{33}A_{33}$. Then, by the induction hypothesis, A_{33} and S_{33} have a common H-nonnegative invariant subspace $\widetilde{\mathcal{M}}$ of dimension $\tilde{d} = i_+(H_{33})$. Writing $\widetilde{\mathcal{M}} = \operatorname{Im} X$ for some matrix $X \in \mathbb{C}^{n \times \tilde{d}}$ of full column rank, we find that

$$\mathcal{M} = \operatorname{Im} \begin{bmatrix} I_d & 0 \\ 0 & 0 \\ 0 & X \end{bmatrix}$$

is both A- and S-invariant, H-nonnegative, and has dimension $i_+(H) = d + i_+(H_{33})$. □

Next, we consider the question if every H-normal matrix $N \in \mathbb{C}^{n \times n}$ has an invariant subspace that is maximal H-nonnegative and that contains a given H-nonnegative N-invariant subspace $\mathcal{M}_0$. (Note that, in general, $\mathcal{M}_0$ is invariant neither for the H-selfadjoint part A, nor for the H-skewadjoint part S of N.) In its full generality, this question is an open problem, but for the special case that $\mathcal{M}_0$ is neutral, we can give an affirmative answer, even for a larger class than the class of normal matrices. For this, let us introduce the following notation. If $\mathcal{M} \subseteq \mathbb{C}^n$ is a subspace, then

$$\mathcal{M}^{[\perp]} := \{v \in \mathbb{C}^n : [v, w] = 0 \text{ for all } w \in \mathcal{M}\}.$$

Theorem 6.3. *Let H be invertible, let $X \in \mathbb{C}^{n \times n}$, and let $\mathcal{M}_0$ be an H-neutral X-invariant subspace. If $\mathcal{M}_0$ is $(XX^{[*]} - X^{[*]}X)$-invariant and if $\mathcal{M}_0^{[\perp]} \cap (H^{-1}\mathcal{M}_0)^{[\perp]}$ is $(XX^{[*]} - X^{[*]}X)$-neutral, then there exists an H-nonnegative X-invariant subspace $\mathcal{M}$ of dimension $i_+(H)$ such that $\mathcal{M}_0 \subseteq \mathcal{M} \subseteq \mathcal{M}_0^{[\perp]}$.*

Proof. If $d := \dim \mathcal{M}_0 = 0$, then $\mathcal{M}_0^{[\perp]} \cap (H^{-1}\mathcal{M}_0)^{[\perp]} = \mathbb{C}^n$. Since this space is $(XX^{[*]} - X^{[*]}X)$-neutral by assumption, it follows that $(XX^{[*]} - X^{[*]}X) = 0$, i.e., X is H-normal. Then the result follows from Theorem 6.2. If $n = 2d$, then $\mathcal{M}_0$ is already maximal H-nonnegative. The remainder of the proof now proceeds by induction on n. The case $n = 1$ is clear, since necessarily $\mathcal{M}_0 = \{0\}$. Thus, let $n > 1$. By the above, we may assume $0 < d < n/2$. By Lemma 2.7, we may moreover assume that $\mathcal{M}_0 = \text{Span}\{e_1, \ldots, e_d\}$ and

$$H = \begin{bmatrix} 0 & I_d & 0 \\ I_d & 0 & 0 \\ 0 & 0 & H_{33} \end{bmatrix}.$$

Partitioning X conformably, we obtain that

$$X = \begin{bmatrix} X_{11} & X_{12} & X_{13} \\ 0 & X_{22} & X_{23} \\ 0 & X_{32} & X_{33} \end{bmatrix}, \quad X^{[*]} = \begin{bmatrix} X_{22}^* & X_{12}^* & X_{32}^* H_{33} \\ 0 & X_{11}^* & 0 \\ H_{33}^{-1} X_{23}^* & H_{33}^{-1} X_{13}^* & X_{33}^{[*]} \end{bmatrix}, \tag{6.3}$$

Note that

$$\mathcal{M}_0^{[\perp]} = \text{Im} \begin{bmatrix} I_d & 0 \\ 0 & 0 \\ 0 & I_{n-2d} \end{bmatrix}, \quad H^{-1}\mathcal{M}_0^{[\perp]} = \text{Im} \begin{bmatrix} 0 & 0 \\ I_d & 0 \\ 0 & I_{n-2d} \end{bmatrix}.$$

Thus, the fact that $\mathcal{M}_0$ is $(XX^{[*]} - X^{[*]}X)$-invariant and that $\mathcal{M}_0^{[\perp]} \cap (H^{-1}\mathcal{M}_0)^{[\perp]}$ is $(XX^{[*]} - X^{[*]}X)$-neutral implies that $XX^{[*]} - X^{[*]}X$ has the pattern

$$XX^{[*]} - X^{[*]}X = \begin{bmatrix} * & * & * \\ 0 & * & * \\ 0 & * & 0 \end{bmatrix}$$

if it is partitioned conformably with X. Computing the $(2,1)$-, $(3,1)$-, and $(3,3)$-block of $XX^{[*]} - X^{[*]}X$, we obtain the identities

$$X_{23} H_{33}^{-1} X_{23}^* = 0, \tag{6.4}$$

$$X_{33} H_{33}^{-1} X_{23}^* = H_{33}^{-1} X_{23}^* X_{11}, \tag{6.5}$$

$$X_{33} X_{33}^{[*]} - X_{33}^{[*]} X_{33} = H_{33}^{-1} X_{23}^* X_{13} + H_{33}^{-1} X_{13}^* X_{23}. \tag{6.6}$$

Let us consider the subspace $\widetilde{\mathcal{M}}_0 = \text{Im}\, H_{33}^{-1} X_{23}^*$. Then $\widetilde{\mathcal{M}}_0$ is H_{33}-neutral (because of (6.4)), X_{33}-invariant (because of (6.5)), and $(X_{33}X_{33}^{[*]} - X_{33}^{[*]}X_{33})$-invariant, because by (6.6) we have that

$$(X_{33}X_{33}^{[*]} - X_{33}^{[*]}X_{33}) H_{33}^{-1} X_{23}^* = H_{33}^{-1} X_{23}^* X_{13} H_{33}^{-1} X_{23}^* + \underbrace{H_{33}^{-1} X_{13}^* X_{23} H_{33}^{-1} X_{23}^*}_{= 0 \text{ by } (6.4)}$$

Next, we show that $\widetilde{\mathcal{M}}_0^{[\perp]} \cap (H_{33}^{-1}\widetilde{\mathcal{M}}_0)^{[\perp]}$ is $(X_{33}X_{33}^{[*]} - X_{33}^{[*]}X_{33})$-neutral. For this, let B be a matrix such that $\widetilde{\mathcal{M}}_0^{[\perp]} \cap (H_{33}^{-1}\widetilde{\mathcal{M}}_0)^{[\perp]} = \operatorname{Im} B$. Then

$$B^*H_{33}^{-1}X_{23}^* = B^*H_{33}(H_{33}^{-1}H_{33}^{-1}X_{23}^*) = 0, \text{ because } \operatorname{Im} B \subseteq (H_{33}^{-1}\widetilde{\mathcal{M}}_0)^{[\perp]}, \quad (6.7)$$

$$\text{and } X_{23}B = (H_{33}^{-1}X_{23}^*)^*H_{33}B = 0, \text{ because } \operatorname{Im} B \subseteq \widetilde{\mathcal{M}}_0^{[\perp]}. \quad (6.8)$$

Using (6.6), we obtain

$$B^*\left(X_{33}X_{33}^{[*]} - X_{33}^{[*]}X_{33}\right)B = B^*H_{33}^{-1}X_{23}^*X_{13}B + B^*H_{33}^{-1}X_{13}^*X_{23}B = 0,$$

by (6.7) and (6.8), and thus, $\operatorname{Im} B$ is $(X_{33}X_{33}^{[*]} - X_{33}^{[*]}X_{33})$-neutral. By the induction hypothesis, there exists an H_{33}-nonnegative X_{33}-invariant subspace $\widetilde{\mathcal{M}}$ of dimension $i_+(H_{33})$ such that $\widetilde{\mathcal{M}}_0 \subseteq \widetilde{\mathcal{M}} \subseteq \widetilde{\mathcal{M}}_0^{[\perp]}$. Let C be a matrix such that $\widetilde{\mathcal{M}} = \operatorname{Im} C$. Since $\widetilde{\mathcal{M}} \subseteq \widetilde{\mathcal{M}}_0^{[\perp]}$, we obtain in particular that $X_{23}C = (H_{33}^{-1}X_{23}^*)^*H_{33}C = 0$, and since $\widetilde{\mathcal{M}}$ is X_{33}-invariant, there exists a matrix $Y \in \mathbb{C}^{(n-2d)\times(n-2d)}$ such that $X_{33}C = CY$. Now choose

$$\mathcal{M} = \operatorname{Im} \begin{bmatrix} I_d & 0 \\ 0 & 0 \\ 0 & C \end{bmatrix}.$$

Clearly, $\mathcal{M}$ is nonnegative with dimension $i_+(H) = d + i_+(H_{33})$ and $\mathcal{M}_0 \subseteq \mathcal{M} \subseteq \mathcal{M}_0^{[\perp]}$. Moreover, we obtain

$$\begin{bmatrix} X_{11} & X_{12} & X_{13} \\ 0 & X_{22} & X_{23} \\ 0 & X_{32} & X_{33} \end{bmatrix} \begin{bmatrix} I_d & 0 \\ 0 & 0 \\ 0 & C \end{bmatrix} = \begin{bmatrix} I_d & 0 \\ 0 & 0 \\ 0 & C \end{bmatrix} \begin{bmatrix} X_{11} & X_{13}C \\ 0 & Y \end{bmatrix},$$

i.e., $\mathcal{M}$ is X-invariant. This concludes the proof. □

If $XX^{[*]} - X^{[*]}X = 0$, then clearly any subspace is $(XX^{[*]} - X^{[*]}X)$-invariant and $(XX^{[*]} - X^{[*]}X)$-neutral. Hence, we immediately obtain the following corollary.

Corollary 6.4. *Let H be invertible, let $N \in \mathbb{C}^{n\times n}$ be H-normal, and let $\mathcal{M}_0$ be an H-neutral N-invariant subspace. Then there exists an H-nonnegative N-invariant subspace $\mathcal{M}$ of dimension $i_+(H)$ such that $\mathcal{M}_0 \subseteq \mathcal{M} \subseteq \mathcal{M}_0^{[\perp]}$.*

Note that a direct proof of Corollary 6.4 cannot proceed completely analogously to the proof of Theorem 6.3, because X_{33} in (6.3) need not be H_{33}-normal.

Let us now consider the case that H is not necessarily invertible. As we can see from Example 6.1, the kernel of H need not be invariant for an H-normal matrix $N \in \mathbb{C}^{n\times n}$. Therefore, to generalize Theorem 6.2 to the case of singular H, we need some preparations. Let us start with a simple form for H-normal matrices.

Theorem 6.5. *Let $N \in \mathbb{C}^{n\times n}$ be H-normal. Then there exists a nonsingular matrix $P \in \mathbb{C}^{n\times n}$ such that*

$$P^{-1}NP = \begin{bmatrix} A_1 & 0 & 0 & 0 & 0 \\ 0 & A_2 & 0 & 0 & B_2 \\ 0 & 0 & A_3 & 0 & B_3 \\ 0 & 0 & 0 & A_4 & 0 \\ C_1 & C_2 & C_3 & C_4 & D \end{bmatrix}, \quad P^*HP = \begin{bmatrix} H_1 & 0 & 0 & 0 & 0 \\ 0 & H_2 & 0 & 0 & 0 \\ 0 & 0 & 0 & I_m & 0 \\ 0 & 0 & I_m & 0 & 0 \\ 0 & 0 & 0 & 0 & 0 \end{bmatrix}, \tag{6.9}$$

where $A_1, H_1 \in \mathbb{C}^{n_1\times n_1}$, $A_2, H_2 \in \mathbb{C}^{n_2\times n_2}$, $A_3, A_4 \in \mathbb{C}^{m\times m}$, and the other blocks have corresponding sizes. Moreover, A_1, H_1, H_2, A_3 are nonsingular, A_2, A_4 are nilpotent, A_1 is H_1-normal, A_2 is H_2-normal, A_3 and A_4^* commute, and $A_2^*H_2B_2 = 0$, $B_2^*H_2B_2 = 0$, $A_4^*B_3 = 0$.

Proof. Without loss of generality, we may assume that

$$H = \begin{bmatrix} \tilde{H} & 0 \\ 0 & 0 \end{bmatrix}, \quad N = \begin{bmatrix} A & B \\ C & D \end{bmatrix},$$

where $\tilde{H}$ is nonsingular and N is partitioned conformably with H. Since N is H-normal, we obtain that

$$\begin{bmatrix} A^*\tilde{H}A & A^*\tilde{H}B \\ B^*\tilde{H}A & B^*\tilde{H}B \end{bmatrix} = N^*HN = HNH^\dagger N^*H = \begin{bmatrix} \tilde{H}A\tilde{H}^{-1}A^*\tilde{H} & 0 \\ 0 & 0 \end{bmatrix}. \tag{6.10}$$

In particular, A is $\tilde{H}$-normal. Thus, since $\tilde{H}$ is invertible, there exists a nonsingular matrix Q such that

$$Q^{-1}AQ = A_{11} \oplus \cdots \oplus A_{kk}, \quad Q^*\tilde{H}Q = H_{11} \oplus \cdots \oplus H_{kk},$$

where, for each j, A_{jj} and H_{jj} have the same sizes, and either A_{jj} has only one eigenvalue, or A_{jj} has two distinct eigenvalues and

$$A_{jj} = \begin{bmatrix} A_{jj1} & 0 \\ 0 & A_{jj2} \end{bmatrix}, \quad H_{jj} = \begin{bmatrix} 0 & I_{p_j} \\ I_{p_j} & 0 \end{bmatrix},$$

where both $A_{jj1} \in \mathbb{C}^{p_j\times p_j}$ and $A_{jj2} \in \mathbb{C}^{p_j\times p_j}$ have only one eigenvalue. (For a proof, see [20], for example.) If A_{jj} is singular and has two distinct eigenvalues, then one of the blocks A_{jj1}, A_{jj2} must be nilpotent. Clearly, we may assume that in this case always A_{jj2} is nilpotent, applying a permutation otherwise. Let us group together all nonsingular blocks, all singular blocks that have only one eigenvalue, and all singular blocks that have two distinct eigenvalues. Thus, after applying an appropriate block permutation, we may assume that N and H have the forms

$$N = \begin{bmatrix} A_1 & 0 & 0 & 0 & B_1 \\ 0 & A_2 & 0 & 0 & B_2 \\ 0 & 0 & A_3 & 0 & B_3 \\ 0 & 0 & 0 & A_4 & B_4 \\ C_1 & C_2 & C_3 & C_4 & D \end{bmatrix}, \quad P^*HP = \begin{bmatrix} H_1 & 0 & 0 & 0 & 0 \\ 0 & H_2 & 0 & 0 & 0 \\ 0 & 0 & 0 & I_m & 0 \\ 0 & 0 & I_m & 0 & 0 \\ 0 & 0 & 0 & 0 & 0 \end{bmatrix},$$

where $A_1, H_1 \in \mathbb{C}^{n_1 \times n_1}$ are nonsingular, $A_2, H_2 \in \mathbb{C}^{n_2 \times n_2}$, H_2 is nonsingular, A_2 is nilpotent, $A_3 \in \mathbb{C}^{m \times m}$ is nonsingular, and $A_4 \in \mathbb{C}^{m \times m}$ is nilpotent. The fact that A is $\tilde{H}$-normal implies furthermore that A_1 is H_1-normal, A_2 is H_2-normal, and $A_3 A_4^* = A_4^* A_3$. Finally, equation (6.10) implies $B_1 = 0$, $B_4 = 0$, $A_2^* H_2 B_2 = 0$, $B_2^* H_2 B_2 = 0$, and $A_4^* B_3 = 0$. □

We are now able to generalize Theorem 6.2 to the case that H is singular.

Theorem 6.6. *Let $N \in \mathbb{C}^{n \times n}$ be H-normal. Then there exists an H-nonnegative N-invariant subspace $\mathcal{M}$ such that* $\dim \mathcal{M} = i_+(H) + i_0(H)$.

Proof. Without loss of generality, we may assume that N and H are in the simple form (6.9). Using the same notation as in Theorem 6.5, let

$$\mathcal{M}_0 = \operatorname{Im} \begin{bmatrix} B_2 & A_2 B_2 & \dots & A_2^{n-1} B_2 \end{bmatrix},$$

i.e., $\mathcal{M}_0$ is the controllable subspace of the pair (A, B) (for basic properties of controllable subspaces see, e.g., [9, Section 2.8] or [17, Chapter 4]). Then $\mathcal{M}_0$ is A_2-invariant and contains Im B_2, see, for example, [17, Proposition 4.1.2]. Next, we show that $\mathcal{M}_0$ is H_2-neutral. Therefore, it is sufficient to prove that

$$B_2^* (A_2^*)^i H_2 A_2^j B_2 = 0 \quad \text{for all } i, j = 0, \dots, n-1. \tag{6.11}$$

Since by Theorem 6.5 we have $B_2^* H_2 B_2 = 0$ (which covers the case $i = j = 0$), equation (6.11) is guaranteed if $A_2^* H_2 A_2^j B_2 = 0$ for $j = 0, \dots, n-1$ which we will prove by induction on j. For $j = 0$, this follows directly from Theorem 6.5. If $j > 0$, then we have

$$A_2^* H_2 A_2^j B_2 = (A_2^* H_2 A_2) A_2^{j-1} B_2 = H_2 A_2 H_2^{-1} A_2^* H_2 A_2^{j-1} B_2 = 0,$$

because A_2 is H_2-normal and because of the induction hypothesis. Then, applying Corollary 6.4 on A_2 and $\mathcal{M}_0$, there exists an H_2-nonnegative A_2-invariant subspace $\mathcal{M}_2 \supseteq \mathcal{M}_0$ of dimension $i_+(H_2)$. Furthermore, there exists an H_1-nonnegative A_1-invariant subspace $\mathcal{M}_1$ of dimension $i_+(H_1)$ by Theorem 6.2. Let $\mathcal{M}_1 = \operatorname{Im} X_1$ and $\mathcal{M}_2 = \operatorname{Im} X_2$ for some matrices X_1, X_2 of appropriate dimensions and set

$$\mathcal{M} = \operatorname{Im} \begin{bmatrix} X_1 & 0 & 0 & 0 \\ 0 & X_2 & 0 & 0 \\ 0 & 0 & I_m & 0 \\ 0 & 0 & 0 & 0 \\ 0 & 0 & 0 & I_\nu \end{bmatrix} \subseteq \mathbb{C}^n,$$

where $\nu = n - n_1 - n_2 - 2m$. Then $\mathcal{M}$ is H-nonnegative and $\dim \mathcal{M} = i_+(H) + i_0(H)$. Moreover, the fact that Im $B_2 \subseteq \mathcal{M}_0 \subseteq \mathcal{M}_2 = \operatorname{Im} X_2$ implies that $\mathcal{M}$ is N-invariant. This concludes the proof. □

Acknowledgment

The research of L. Rodman is partially supported by NSF Grant DMS-9988579.

References

[1] T. Ando. *Linear operators on Kreĭn spaces.* Hokkaido University, Research Institute of Applied Electricity, Division of Applied Mathematics, Sapporo, Japan, 1979.

[2] T.Ya. Azizov and I.S. Iohvidov. *Linear Operators in Spaces with an Indefinite Metric.* John Wiley and Sons, Ltd., Chichester, 1989. (Translated from Russian.)

[3] P. Binding and R. Hryniv. Full and partial range completeness. *Oper. Theory Adv. Appl.* 130:121–133, 2002. Birkhäuser, Basel.

[4] J. Bognár. *Indefinite inner product spaces.* Springer-Verlag, New York-Heidelberg, 1974.

[5] V. Bolotnikov, C.K. Li, P. Meade, C. Mehl, and L. Rodman. Shells of matrices in indefinite inner product spaces *Electronic J. of Linear Algebra,* 9:67–92, 2002.

[6] Y. Bolshakov, C.V.M. van der Mee, A.C.M. Ran, B. Reichstein, and L. Rodman. Polar decompositions in finite-dimensional indefinite scalar product spaces: special cases and applications. *Recent developments in operator theory and its applications. Oper. Theory Adv. Appl.* (I. Gohberg, P. Lancaster, P. N. Shivakumar, eds.) 87:61–94, 1996, Birkhäuser, Basel. Errata, *Integral Equations and Operator Theory,* 17:497–501, 1997.

[7] Y. Bolshakov, C.V.M. van der Mee, A.C.M. Ran, B. Reichstein, and L. Rodman. Extensions of isometries in finite-dimensional indefinite scalar product spaces and polar decompositions. *SIAM J. of Matrix Analysis and Applications,* 18:752–774, 1997.

[8] I. Gohberg, P. Lancaster, and L. Rodman. *Matrices and Indefinite Scalar Products.* Birkhäuser Verlag, Basel, Boston, Stuttgart, 1983.

[9] I. Gohberg, P. Lancaster, and L. Rodman. *Invariant Subpaces of Matrices with Applications.* Wiley Interscience, New York etc., 1986.

[10] I. Gohberg and B. Reichstein. On classification of normal matrices in an indefinite scalar product. *Integral Equations Operator Theory,* 13:364–394, 1990.

[11] H. Heuser. *Funktionalanalysis.* Teubner, Stuttgart, 1975.

[12] O. Holtz and V. Straus. On classification of normal operators in real spaces with indefinite scalar product. *Linear Algebra Appl.* 255:113–155, 1997.

[13] R. Horn and C.R. Johnson. *Matrix Analysis.* Cambridge University Press, Cambridge, 1985.

[14] I.S. Iohvidov, M.G. Kreĭn, and H. Langer. *Introduction to the spectral theory of operators in spaces with an indefinite metric.* Mathematical Research 9, Akademie-Verlag, Berlin, 1982.

[15] M.G. Kreĭn. On an application of the fixed point principle in the theory of linear transformations of spaces with an indefinite metric. (Russian) *Uspehi Matem. Nauk* (N.S.) 5, 1950, no. 2(36), 180–190.

[16] P. Lancaster, A.S. Markus, and P. Zizler. The order of neutrality for linear operators on inner product spaces. *Linear Algebra Appl.* 259:25–29, 1997.

[17] P. Lancaster and L. Rodman. *Algebraic Riccati Equations.* Clarendon Press, Oxford, 1995.

[18] P. Lancaster and M. Tismenetsky. *Theory of Matrices with Applications,* 2-nd ed. Academic Press, Orlando, FL, 1985.

[19] C.K. Li, N.K. Tsing, and F. Uhlig. Numerical ranges of an operator on an indefinite inner product space. *Electron. J. Linear Algebra* 1:1–17, 1996.

[20] B. Lins, P. Meade, C. Mehl, and L. Rodman. Normal matrices and polar decompositions in indefinite inner products. *Linear and Multilinear Algebra*, 49:45–89, 2001.

[21] A. Luger. A factorization of regular generalized Nevanlinna functions. *Integral Equations Operator Theory* 43:326–345, 2002.

[22] C. Mehl and L. Rodman. Symmetric matrices with respect to sesquilinear forms. *Linear Algebra Appl.* 349:55–75, 2002.

[23] R.S. Phillips. Dissipative operators and hyperbolic systems of partial differential equations. *Trans. Amer. Math. Soc.* 90:193–254, 1959.

[24] R. Thompson. The characteristic polynomial of a principal subpencil of a Hermitian matrix pencil. *Linear Algebra Appl.*, 14:135–177, 1976.

[25] R. Thompson. Pencils of complex and real symmetric and skew matrices. *Linear Algebra Appl.*, 147:323–371, 1991.

Christian Mehl
Fakultät II; Institut für Mathematik
Technische Universität Berlin
D-10623 Berlin, Germany
e-mail: `mehl@math.tu-berlin.de`

André C.M. Ran
Afdeling Wiskunde
Faculteit der Exacte Wetenschappen
Vrije Universiteit Amsterdam
De Boelelaan 1081a
NL-1081 HV Amsterdam, The Netherlands
e-mail: `ran@cs.vu.nl`

Leiba Rodman
College of William and Mary
Department of Mathematics
P.O. Box 8795
Williamsburg, VA 23187-8795, USA
e-mail: `lxrodm@math.wm.edu`

Operator Theory:
Advances and Applications, Vol. 149, 487–491

On The Dual Spectral Set Conjecture

Steen Pedersen

Abstract. Let a, b, c, d be real numbers with $a \neq 0$ and $c \neq 0$. Suppose $a\mathbb{Z} + b$ and $c\mathbb{Z} + d$ are disjoint. Let $\Lambda = (a\mathbb{Z} + b) \cup (c\mathbb{Z} + d)$. It is shown that if the exponentials $e_\lambda(x) = e^{i2\pi\lambda x}$, $\lambda \in \Lambda$, form an orthogonal basis for some bounded measurable set Ω then the translates $\Omega + \lambda$, $\lambda \in \Lambda$, tile the real line for some bounded measurable set Ω'.

1. Introduction

Let $e_\lambda(x) = e(\lambda x) = e^{2\pi i\lambda x}$. A pair (Ω, Λ) consisting of a set $\Omega \subset \mathbb{R}$ of finite non-zero measure and a discrete set $\Lambda \subset \mathbb{R}$ is called a *spectral pair* if

$$\mathcal{E}_\Lambda = \{e_\lambda : \lambda \in \Lambda\}$$

is an orthogonal basis for $L^2(\Omega)$. We say Ω is a *spectral set* if (Ω, Λ) is a spectral pair for some Λ, and that Λ is a *spectrum* if (Ω, Λ) is a spectral pair for some Ω. Simple examples of spectral pairs include:

- $\Omega = [0, 1]$, $\Lambda = \mathbb{Z}$ (Fourier series)
- $\Omega = \{0, 1\} + \left[0, \frac{1}{2}\right) = \left[0, \frac{1}{2}\right) \cup \left[1, \frac{3}{2}\right)$, $\Lambda = \left\{0, \frac{1}{2}\right\} + 2\mathbb{Z} = 2\mathbb{Z} \cup \left(\frac{1}{2} + 2\mathbb{Z}\right)$
- $\Omega = \left\{0, \frac{3}{2}\right\} + \left[0, \frac{1}{2}\right)$, $\Lambda = \mathbb{Z}$ or $\Lambda = \left\{0, \frac{1}{3}\right\} + 2\mathbb{Z}$

If λ is a real number then we write

$$\Omega + \lambda = \{\omega + \lambda : \omega \in \Omega\}.$$

The pair (Ω, Λ) is a *tiling pair* if

$$\bigcup_{\lambda \in \Lambda} (\Omega + \lambda) = \mathbb{R}$$

where the union is disjoint. Disjointness and equality are up to sets of measure zero. is an orthogonal basis for $L^2(\Omega)$. We say Ω is a *tile* if (Ω, Λ) is a tiling pair for some Λ, and that Λ is a *tiling set* if (Ω, Λ) is a tiling pair for some Ω. What we call a tile is called a *translational* tile by some authors.

Received by the editors December 20, 2002; revised August 8, 2003. Submitted by L. Rodman.
Mathematics Subject Classification (2000). Primary 42A99, 42C99, 51M04, 52C99.
Key words and phrases. Fourier basis, non-harmonic Fourier series, tiling, spectral set, spectral pair.
This is a version of the author's talk at IWOTA 2002. The author wishes to thank David Larson and Joe Ball for the invitation to participate in IWOTA 2002.

Simple examples of tiling pairs include:

- $\Omega = [0,1]$, $\Lambda = \mathbb{Z}$
- $\Omega = \{0,1\} + [0,\frac{1}{2})$, $\Lambda = \{0,\frac{1}{2}\} + 2\mathbb{Z}$
- $\Omega = \{0,\frac{3}{2}\} + [0,\frac{1}{2})$, $\Lambda = \mathbb{Z}$ or $\Lambda = \{0,\frac{1}{2},1\} + 3\mathbb{Z}$

Connections between spectral pairs, tiling pairs, and wavelets are explored in [10], for example.

Remark 1.1. Exponential bases and tilings by translations make sense for domains in $\mathbb{R}^d$ for $d \geq 1$. In this note we will only consider $d = 1$.

For the domain

$$\Omega = [0,\tfrac{1}{2}) \cup [\tfrac{3}{2},2)$$

the possible spectra (containing 0) are

$$\Lambda = 2\mathbb{Z} \cup (2\mathbb{Z} + \tfrac{p}{3}),\ p \text{ odd}.$$

The possible tiling sets (containing 0) are

$$\Lambda = 3\mathbb{Z} \cup (3\mathbb{Z} + \tfrac{r}{2}) \cup (3\mathbb{Z} + \tfrac{s}{2}),$$

where r, s are integers so that

$$\{0,3\} + \{0,r,s\} = \{0,r,s,3,3+r,3+s\}$$

is a complete set of remainders modulo 6. For example, $r = 1$ and $s = 2$.

Conjecture 1.2 (Spectral Set Conjecture, Fuglede 1974, see [1]). *Ω is a spectral set if and only if Ω is a tile.*

Stated in terms of spectral pairs this says that (Ω, Λ) is a spectral pair for some Λ if and only if (Ω, Λ') is a tiling pair for some Λ'.

Conjecture 1.3 (Dual Spectral Set Conjecture, Jorgensen-Pedersen 1999, see [4]). *Λ is a spectrum if and only if Λ is a tiling set.*

Stated in terms of spectral pairs this says that (Ω, Λ) is a spectral pair for some Ω if and only if (Ω', Λ) is a tiling pair for some Ω'.

Remark 1.4. As we saw above the domain $\Omega = [0,\frac{1}{2}) \cup [\frac{3}{2},2)$ only has spectra consisting of two lattices while any tiling set consists of three lattices. This shows that the flexibiity $\Omega \neq \Omega'$ is needed in the Dual Spectral Set Conjecture.

Remark 1.5. These conjectures also make sense in dimension $d > 1$, however, there they are much harder. The restriction $d = 1$ offers some hope, in part because all spectra are unions of translated lattices. This is not true in higher dimensions.

The Dual Spectral Set Conjecture is verified for many sets in [7].

2. Formulation of our main result

Suppose Ω is the union of two intervals

$$\Omega = [0, r) \cup [a, a+1-r), 0 < r < 1, r \leq a.$$

In [6] Łaba showed that Ω is a spectral set if and only if $a - r \in \mathbb{Z}$ or $r = \frac{1}{2}$ and $2a \in \mathbb{Z}$ if and only if Ω is a tile. The spectra for Ω are $\Lambda = \mathbb{Z}$ or $\Lambda = \{0, \frac{p}{n}\} + 2\mathbb{Z}$, where p is odd and $n = 2a$.

We seek a dual version of this result. Consider an interval $\Omega = [a, a+r)$. The collection of all spectra for Ω is $\Lambda_x = \frac{1}{r}\mathbb{Z} + x$, $x \in \mathbb{R}$. So the "dual" of two intervals is two lattices

$$\Lambda = a\mathbb{Z} \cup (b\mathbb{Z} + c),$$

where we can think of a^{-1} as the length of the first interval and b^{-1} as the length of the second interval.

Given that Ω is the union of two intervals we were able to find all spectra. The only non-trivial possibility is

$$\Omega = \left[0, \tfrac{1}{2}\right) \cup \left[\tfrac{n}{2}, \tfrac{n+1}{2}\right) \text{ and } \Lambda = 2\mathbb{Z} \cup \left(\tfrac{p}{n} + 2\mathbb{Z}\right)$$

where n is an integer and p is odd. Conversely, starting with a spectrum and trying to determine all the corresponding spectral sets is more difficult. For example, given $\Lambda = \mathbb{Z}$, i.e., $p = n = 1$ in the notation above, we cannot conclude that Ω is an interval. We can only conclude that

$$\Omega = \bigcup_{n \in \mathbb{Z}} (n + I_n)$$

up to sets of measure zero where

$$[0, 1) = \bigcup_{n \in \mathbb{Z}} I_n$$

is a measure-theoretic partition of $[0, 1)$. Despite this difficulty we will establish

Theorem 2.1. *Suppose $\Lambda = a\mathbb{Z} \cup (b\mathbb{Z} + c)$ for some $a, b, c \in \mathbb{R}$ with $a \neq 0$ and $b \neq 0$. If Λ is a spectrum (for some bounded Ω) then Λ is a tiling set (for some bounded Ω').*

The converse was established after the meeting, see [9] for details.

3. Proof of Theorem 2.1

Replacing Ω by $a\Omega$ and Λ by $\frac{1}{a}\Lambda$ we may assume $a = 1$. Hence

$$\Lambda = \mathbb{Z} \cup (b\mathbb{Z} + r).$$

The Fourier transform

$$\widehat{\chi_\Omega}(\xi) = \int_\Omega e(t\xi)\, dt$$

is continuous and $\widehat{\chi_\Omega}(0) = |\Omega|$, so $\widehat{\chi_\Omega}(\xi) \neq 0$ for small ξ. Since

$$\widehat{\chi_\Omega}(\lambda' - \lambda) = \langle e_\lambda | e_{\lambda'} \rangle = 0$$

for $\lambda \neq \lambda'$ in Λ it follows that $\Lambda - \Lambda$ cannot contain small non-zero numbers. However,

$$\{k + bl + r : k, l \in \mathbb{Z}\} \subset \Lambda - \Lambda,$$

so we can conclude that b is rational. Suppose $b = \frac{p}{q}$ with $\gcd(p,q) = 1$. After a change of scale we may assume

$$\Lambda = p\mathbb{Z} \cup (q\mathbb{Z} + r),$$

where $\gcd(p,q) = 1$, $0 < p$, and $0 \leq r < q$.

Since $p\mathbb{Z} \subset \Lambda$ the exponentials $\mathcal{E}_{p\mathbb{Z}}$ are orthogonal in $L^2(\Omega)$, hence $p' = p|\Omega|$ is an integer ([3], Theorem 6.2). Similarly, $q' = q|\Omega|$ is an integer. Since $p'q = pq'$ and $\gcd(p,q) = 1$ it follows that p divides p'. Thus $|\Omega|$ is an integer.

Suppose $r \in \mathbb{Z}$. Then

$$\mathbb{Z} = \{pm + qn : m, n \in \mathbb{Z}\} \subset \Lambda - \Lambda$$

since $\gcd(p,q) = 1$. So $\widehat{\chi_\Omega}(k) = 0$ for all $k \in \mathbb{Z}\backslash\{0\}$. Hence $\mathcal{E}_\mathbb{Z}$ is orthogonal in $L^2(\Omega)$ and $\mathcal{E}_\Lambda \subseteq \mathcal{E}_\mathbb{Z}$. But $\mathcal{E}_\Lambda$ is a maximal orthogonal set, hence $\Lambda = \mathbb{Z}$, corresponding to $p = q = 1$. Note $\Lambda = \mathbb{Z}$ is a tiling set for $\Omega' = [0,1)$.

Suppose $r \notin \mathbb{Z}$. It follows from [5] that

$$\lim_{R\to\infty} \frac{\#\Lambda \cap (-R, R)}{2R} = |\Omega|.$$

Since $\Lambda = p\mathbb{Z} \cup (q\mathbb{Z} + r)$ and r is not an integer it follows that

$$\frac{1}{p} + \frac{1}{q} = |\Omega|.$$

But $|\Omega|$ is an integer and $\gcd(p,q) = 1$, so we must have $p = q = 1$. Hence

$$\Lambda = \mathbb{Z} \cup (\mathbb{Z} + r), 0 \leq r < 1.$$

By [2] there exists an integer n such that

$$0 < |\Omega \cap (\Omega + n)| < |\Omega|$$

and for any such integer $e(nr) = -1$, so $r = \frac{p}{2n}$ where p is odd. Hence

$$\Lambda = \mathbb{Z} \cup \left(\mathbb{Z} + \tfrac{p}{2n}\right).$$

But this is a tiling set for

$$\Omega' = \bigcup_{j=0}^{n-1} \left[\tfrac{j}{n}, \tfrac{j}{n} + \tfrac{1}{2n}\right).$$

This completes the proof.

We conclude this note by considering some examples.

- We saw above that the domain
$$\Omega = \left[0, \tfrac{1}{2}\right) \cup \left[\tfrac{3}{2}, 2\right)$$
has spectra (containing 0)
$$\Lambda = 2\mathbb{Z} \cup \left(2\mathbb{Z} + \tfrac{p}{3}\right), \; p \text{ odd.}$$
After division by 2 we have
$$\Lambda = \mathbb{Z} \cup \left(\mathbb{Z} + \tfrac{p}{6}\right), \; p \text{ odd}$$
so $6 = 2n$ implies $n = 3$ and therefore
$$\Omega' = \left[0, \tfrac{1}{6}\right) \cup \left[\tfrac{2}{6}, \tfrac{3}{6}\right) \cup \left[\tfrac{4}{6}, \tfrac{5}{6}\right).$$
- It follows from the last part of the proof that $\mathbb{Z} \cup \left(\mathbb{Z} + \frac{1}{3}\right)$ is not a spectrum for any Ω, because 3 is not even. It is not hard to see that this Λ is not a tiling set for any Ω'. See also [8].

Acknowledgment

Many thanks to the audience for keeping the talk lively by asking numerous questions and to the referee for making constructive suggestions improving this note.

References

[1] Bent Fuglede, *Commuting self-adjoint partial differential operators and a group theoretic problem,* J. Funct. Anal. **16** (1974), 101–121.

[2] Palle E.T. Jorgensen and Steen Pedersen, *Harmonic analysis on tori,* Acta Appl. Math. **10** (1987), 87–99.

[3] Palle E.T. Jorgensen and Steen Pedersen, *Spectral theory for Borel sets in R^n of finite measure,* J. Funct. Anal. **107** (1992), 74–104.

[4] Palle E.T. Jorgensen and Steen Pedersen, *Spectral pairs in Cartesian coordinates,* J. Fourier Anal. Appl. **5** (1999), 285–302.

[5] Mihail N. Kolountzakis and Jeffrey C. Lagarias, *Structure of tilings of the line by a function,* Duke Math. J. **82** (1996), 653–678.

[6] Izabella Łaba, *Fuglede's conjecture for a union of two intervals,* Proc. Amer. Math. Soc. **129** (2001), 2965–2972.

[7] Jeffrey C. Lagarias, J. Reeds and Yang Wang, *Orthonormal bases of exponentials for the n-cube,* Duke Math. J. **103** (2000), 25–37.

[8] Jeffrey C. Lagarias and Yang Wang, *Tiling the line with translates of one tile,* Invent. Math. **124** (1996), 341–365.

[9] Steen Pedersen, *The dual spectral set conjecture,* preprint.

[10] Yang Wang, *Wavelets, tiling and spectral sets,* Duke Math. J. **114** (2002), 43–57.

Steen Pedersen, Department of Mathematics, Wright State University,
3640 Colonel Glenn Highway, Dayton, OH 45435, USA
e-mail: `steen@math.wright.edu`
URL: `http://www.wright.edu/~steen.pedersen/`

Operator Theory:
Advances and Applications, Vol. 149, 493–515

Stochastic Controllability of Linear Interest Rate Models

Mark A. Petersen, Helgard Raubenheimer, Frederik C. van der Walt, and Herrie F. van Rooy II

Abstract. We consider the controllability problem for the linear Heath-Jarrow-Morton-Musiela (HJMM) interest rate model that is realized by an infinite-dimensional stochastic differential equation (SDE). Although it is clear that interest rates are not generally controllable, the objective of our paper is nevertheless to establish necessary and sufficient conditions for the stochastic controllability of a special subclass of the aforementioned models. In this process we determine a control that transfers the said model from an arbitrary interest rate to any other interest rate in the state space of forward rate curves. Our method of solving this problem involves a consideration of the deterministic and stochastic controllability operators related to the aforementioned SDE and their resolvents and a regulator problem associated with the minimum energy principle. In this regard, a formula for a minimizing control is given explicitly in terms of an invertible deterministic controllability operator. Also, we briefly comment on connections between the main results of the paper and the related Ho-Lee, Hull-White and Cox-Ingersoll-Ross interest rate models.

1. Introduction and preliminaries

By contrast to standard stock markets that consist of a finite number of traded assets, the study of bond markets is underpinned by the entire term structure of interest rates. Such variables are by nature infinite-dimensional and are generally not directly observable. Empirically it is necessary to devise curve fitting methods for the daily estimation of the term structure. The aim of this paper is to investigate how infinite-dimensional models of the term structure of interest rates can be controlled and regulated by analytic means. In particular, we study the system theoretic concept of controllability as it pertains to linear interest rate models of Heath-Jarrow-Morton-Musiela (HJMM) type (see [19], [28] and [9]).

Received by the editors March 7, 2003; revised November 10, 2003. Submitted by J. W. Helton.
Mathematics Subject Classification (2000). Primary 60G44, 90A09; Secondary 93B15.
Key words and phrases. Controllability, Heath-Jarrow-Morton-Musiela Interest Rate Models.
This research was generously supported by grants from the National Research Foundation of South Africa with Gun Numbers 2053343 and 2053080.

In the Heath-Jarrow-Morton (HJM) model the evolution of the forward rate curve is explained in terms of a stochastic structure. This model can be considered to be a unification in terms of the family of Itô processes indexed by the continuum of the maturities of all continuous interest rate models. The stochastic differential equations that arise in this case involve interest rates that may be regarded as a field of random variables that changes with respect to the parameters of time t and maturity T. The initial time moment is identified with zero and is fixed while the final time moment is denoted by T ($T > 0$) and is variable. Furthermore, we recall that a bond with maturity T is a default-free zero coupon bond with nominal value 1 and with price at time t being denoted by $p(t,T)$. We know that there is a one-to-one relation between the time t term structure of bond prices and the time t term structure of interest rates or forward curves $\{f(t,T) : T \geq t\}$ given by

$$p(t,T) = \exp\left\{ -\int_t^T f(t,s)ds \right\}.$$

In other words, $f(t,T)$ is the continuously compounded instantaneous forward rate for date T prevailing at time t. The forward curve contains all the necessary information for pricing bonds, swaps and forward rate agreements of all maturities. In addition, for central banks it is an important indicator for formulating monetary policy. Moreover, for time t and a fixed maturity T, the instantaneous forward rate $f(t,T)$ evolves, under a given measure, according to the diffusion process

$$\begin{aligned} df(t,T) &= \alpha(t,T)dt + \sigma(t,T)dW(t), \\ f(0,T) &= \widetilde{f}(0,T), \end{aligned}$$

with $T \longmapsto \widetilde{f}(0,T)$ the market instantaneous-forward curve rate at time $t = 0$ and where $W = (W_1, \ldots, W_m)$ is an m-dimensional Brownian motion, $\sigma(t,T) = (\sigma_1(t,T), \ldots, \sigma_m(t,T))$ is a vector of adapted processes and $\alpha(t,T)$ is itself an adapted process. Musiela (see [28] for more details) reparametrized the HJM model by substituting time of maturity T with the parameter time to maturity x which has a positive correlation with the idea of a forward rate curve as an infinite-dimensional state variable. This enabled an infinite number of stochastic equations to be represented as a single (linear or nonlinear) system (compare [12]).

Our motivation for considering the controllability of linear interest rate models is that it is of crucial importance that financial institutions are able to regulate the interest rate on a day-to-day basis. In particular, in countries with emerging economies interventionary mechanisms for adjusting interest rate levels are widely used because of factors like the relatively dramatic fluctuation in foreign exchange rates, interest rates and inflation. At present, in these countries, the effectiveness of the procedures and mechanisms that are used for the frequent amendment of interest rate levels vary greatly and to a large extent suffer from the lack of a sound mathematical basis. The topicality of issues such as the analysis and control of the systems dynamics of financial models in related research areas such as pension fund theory (see [31] and [33]), dividend payment strategies, maintenance of a prudential margin, insurance and reinsurance and incomplete markets is another motivating factor. Also, we know that controllability has a role to play in minimality that in turn has been shown in [4] (see also [5], [6]) to be of considerable consequence for linear interest rate models. In particular, it is known that such finite-factor term structure models are useful for practitioners. Of the two groups of practitioners in the fixed income market, namely the fund managers and the interest rate option traders, the latter have a special interest in the low dimensionality of the interest rate model since the number of factors usually equals the number of instruments one needs to hedge in the model. It is known that the daily adjustment of huge numbers of instruments becomes infeasible due to transaction costs. Also, we know that it is important to develop analytic methods to describe the term structure of interest rates in order to support and improve existing estimation methods (for, e.g., Nelson-Siegel, Svennson, smoothing splines) that are used by central banks. Furthermore, it is our opinion that analytic means of describing interest rate models is an effective tool in respect of the regulation and sensitivity analysis of such models.

Next, we provide a few preliminaries about notation and terminology. Dom A denotes the domain of the (bounded linear) operator A and $R(\lambda, A)$ is the notation used for its resolvent $(\lambda I - A)^{-1}$. If $\mathcal{X}$ and $\mathcal{Y}$ are Banach spaces then $\mathcal{L}(\mathcal{X},\mathcal{Y})$ denotes the space of all bounded linear operators from $\mathcal{X}$ to $\mathcal{Y}$. Furthermore, $\mathcal{H}$ and $\mathcal{U}_{ad}$ is the notation used for the separable Hilbert state space of forward rate curves and the admissible control set, respectively. Our choice of an appropriate control set (see Subsection 3.2) is heavily reliant on the work done by Bensoussan and Viot in [3]. Throughout the paper $\mathcal{U}$ will be an open convex subset of $\mathcal{H}$ that is separable. In fact, we will demonstrate in Subsection 3.2 that by making appropriate choices $\mathcal{U}_{ad} \subset \mathcal{U} \subset \mathcal{H}$. For $\mathcal{X}$ a separable Hilbert space, we denote the space of equivalence classes of all functions from $[0,T]$ to $\mathcal{X}$ that are Lebesque measurable and square integrable with respect to the Lebesque measure by $L_2([0,T];\mathcal{X})$. $L_2^{\mathcal{F}}([0,T];\mathcal{X})$ is the space of $\mathcal{F}_t$-adapted, $\mathcal{X}$-valued measurable processes $\varphi(t,\omega)$ on $[0,T]$ such that

$$\mathbf{E}\int_0^T ||\varphi(t,\omega)||^2 < \infty.$$

The notation $\Delta = \{(t,s) :\ 0 \le s \le t \le T\}$ is used for the triangular set over $[0,T]$. $\mathcal{B}_2(\Delta, \mathcal{L}(\mathcal{X},\mathcal{Y}))$ denotes the class of all $\mathcal{L}(\mathcal{X},\mathcal{Y})$-valued functions on Δ that are strongly measurable and square integrable with respect to the Lebesque measure on Δ. For $s \le r \le T$, we denote the controllability operator corresponding to a specific type of stochastic and deterministic system by Π_r^T and Γ_s^T, respectively.

There are several possible concepts of controllability that we can investigate for linear stochastic interest rate models. In the main, we have a choice of two classes of notions that both arise from various ideas about proximity in probability theory. The first category comes about by introducing a new state space that consists of the probability measures over the original state space. This results in the original stochastic system becoming an infinite-dimensional deterministic system to which all the classical theory about the controllability of such systems can be applied. The second class occurs when one views the original state space $\mathcal{H}$ as a system with random states. When we formulate the concept for this class we use statistical concepts to express the proximity of one state to another. In this paper, our strategy for investigating the stochastic controllability of interest rate models comes from the latter class. In particular, the notion of controllability of linear interest rates that we develop in this paper is based on the premise that the drift term can be influenced by a positive (control) function u from $[0, \infty)$ to the state space. In a follow-up study we will construct an example that illustrates that this is possible under certain bond market conditions.

The rest of the paper is organized in the following way. In Section 2 we present a brief description of the linear interest rate model in the HJMM framework that we will consider. Moreover, we make a specific choice for the state space of forward rate curves that is consistent with the economic nature of the investigation. In the third section, we construct an appropriate subclass of HJMM interest rate models whose members are candidates for controllability. This is accomplished by choosing a suitable partially observable stochastic model and admissible control set at the outset. We solve controllability problems for infinite-dimensional stochastic interest rate models in the fourth section. More specifically, we provide necessary and sufficient conditions for HJMM interest rate models to be approximately, completely and stochastically controllable. These results make use of methods that are related to the deterministic and stochastic controllability and resolvent operators associated with such stochastic models. The formulas for these controllability operators can be used to derive the required (minimizing) control explicitly. In addition, during the aforementioned procedure, we are required to solve a regulator problem related to the minimum energy principle. In Section 5 we comment briefly on the relationship between our aforegoing analysis and the Cox-Ingersoll-Ross, Ho-Lee and Hull-White interest rate models. The final section provides conclusive remarks and a discussion about possible future directions of research.

2. The HJMM interest rate model

In this section, we provide a brief description of the HJMM interest rate model (see [19], [28] and [9]). In addition, in Subsection 2.2 we decide on the most economic Hilbert space to be considered as a state space of forward rate curves.

2.1. Basic description

As was described in [19], the HJM interest rate model for the forward curve $x \mapsto r(t, x)$ is fixed by the structure of its volatility σ and the market price of risk. In this case, $r(t, x)$ is the notation used to denote the forward rate at time t with maturation date $t + x$. In this model, we consider a default free, frictionless bond market with perfectly divisible bonds on a filtered probability space

$(\Omega,\ \mathcal{F},\ \{\mathcal{F}_t\}_{t\geq 0},\ \mathbf{P})$. Furthermore, denote the price at time t of a zero coupon bond maturing at $t+x$ by $p(t,x)$, where x is time to maturity and

$$p(t,x) = \exp\{-y(t,x)\},$$

where the period yield $y(t,x)$ is defined by

$$y(t,x) = \int_0^x r(t,s)ds.$$

The expression $\int_0^x \cdot(t,s)ds$ denotes integration with respect to time to maturity x. Also, $r(t,x)$ the forward rate contracted at t maturing at $t+x$ has the form

$$r(t,x) = -\frac{\partial \log p(t,x)}{\partial x}.$$

Moreover, we denote the short rate by $R(t)$, where $R(t) = r(t,0)$. As is well-known the HJMM approach addresses the question of the modelling of the dynamics for the entire forward rate curve. Here the yield curve r is the state variable rather than the short rate R.

As regards notation, in the ensuing discussion the forward rate at time t with maturation date $t+x$, is simply denoted by $r(t)$. From [15], we know that every classical HJM model can more or less be realized as a stochastic differential equation (SDE) of the form

$$\begin{cases} dr(t) &= (Ar(t) + D(r(t)))\,dt + \sigma(r(t))dW(t), \\ r(0) &= r^*(0), \end{cases} \tag{2.1}$$

where W is an m-dimensional Wiener process, $\sigma(r(t))dW(t) = \sum_{j=1}^{m} \sigma_j(r(t))dW_j(t)$ and the initial curve $\{r^*(0,x):\ x \geq 0\}$ is interpreted as the observed forward rate curve. This equation evolves on some open convex subset $\mathcal{U}$ in a separable Hilbert state space $\mathcal{H}$ (to be specified in Subsection 2.2 below) of forward rate curves. More specifically, we have for $A,\ D \in \mathcal{L}(\mathcal{U},\mathcal{H})$ and σ that

$$A = \frac{\partial}{\partial x} : \text{Dom } A \subset \mathcal{H} \to \mathcal{H};\ D : \mathcal{U} \subset \mathcal{H} \to \mathcal{H};$$
$$\sigma = (\sigma_1,\ \sigma_2,\ \ldots,\ \sigma_m) : \mathcal{U} \subset \mathcal{H} \to \mathcal{H}^m,$$

respectively. Also, the model must be arbitrage-free which leads to the existence of an equivalent local martingale measure $\mathbf{Q} \sim \mathbf{P}$. In this case the drift term in the equation (2.1) can be written in terms of the volatility σ and be specified as

$$\frac{\partial}{\partial x}r(t) + D(r(t)) = \frac{\partial}{\partial x}r(t) + \sigma(r(t))\int_0^x \sigma(r(s))ds. \tag{2.2}$$

Here (2.1) is commonly referred to as the **HJMM equation** and (2.2) is called the **HJMM drift condition**. This means that the pricing formula for interest rate sensitive contingent claims only depend on σ. Furthermore, the deterministic counterpart of the stochastic linear HJMM interest rate model (2.1) may be represented as

$$\begin{cases} dr(t) &= (Ar(t) + D(r(t)))\,dt, \\ r(0) &= r^*(0). \end{cases} \tag{2.3}$$

2.2. State space

When choosing an appropriate state space for our subsequent analysis it is important to bear the following considerations in mind. In some instances, the linear operator $\frac{\partial}{\partial x}$ in the HJMM equation (2.1) may not be bounded which means that strong solutions may not exist and we have to make certain regularity assumptions. Our choice of state space corresponds with the one made in [8] (see also [15], [16], [17] and [18] for the construction of suitable Hilbert spaces for HJMM theory). We provide a brief outline of the key features of this space below. For each $\beta, \gamma \in \mathbf{R}$ the space $\mathcal{H}_{\beta,\gamma}$ of all infinitely differentiable functions $h : \mathbf{R} \to \mathbf{R}$ is defined by

$$\mathcal{H}_{\beta,\gamma} = \{h \in C^{\infty}[0,\infty) : ||h|| < \infty\}.$$

Here the norm is represented by

$$||h||^2_{\beta,\gamma} = \sum_{n=0}^{\infty} \beta^{-n} \int_0^{\infty} \left(\frac{d^n h(x)}{dx^n}\right)^2 e^{-\gamma x} dx$$

with corresponding inner product

$$\langle h, g \rangle = \sum_{n=0}^{\infty} \beta^{-n} \int_0^{\infty} \frac{d^n h(x)}{dx^n} . \frac{d^n g(x)}{dx^n} e^{-\gamma x} dx. \tag{2.4}$$

Since the results are uniform in terms of β and γ, we are allowed to omit the subindices β, γ, i.e., $\mathcal{H} = \mathcal{H}_{\beta,\gamma}$. Here $\mathcal{H}$ is a Hilbert space with respect to the inner product (2.4) (compare [8] and [15]) with every function in $\mathcal{H}$ being real analytic. We also have that the $\mathcal{H}$ described above is separable. Furthermore from [15] we know that it is preferable that $\mathcal{H}$ is continuously embedded in $C([0,\infty); \mathbf{R})$. In other words, for any choice of $x \in [0,\infty)$ the pointwise evaluation $r \longmapsto r(x)$ is a linear functional on $\mathcal{H}$ that is continuous. Furthermore, $\mathcal{H}$ contains the constant function 1. We also insist that the family of right shifts

$$S_t r = r(t+x) \text{ for } t \in [0,\infty) \tag{2.5}$$

forms a strongly continuous semigroup on $\mathcal{H}$ with generator $\frac{\partial}{\partial x}$. Furthermore, we may assume that the domain of $\frac{\partial}{\partial x}$ has the form

$$\left\{h \in \mathcal{H} \cap C^1([0,\infty); \mathbf{R}) : \frac{\partial}{\partial x} h \in \mathcal{H}\right\}.$$

The state space described above has all the properties needed to perform our analysis in the sequel in an economic setting. For the purposes of our discussion, however, we need to bear in mind that the stochastic differential equation (2.1) evolves on some open convex subset $\mathcal{U}$ of the state space $\mathcal{H}$ and not on the whole of $\mathcal{H}$. Furthermore, since every subset of a separable inner product space is separable we have that $\mathcal{U}$ described above is also separable.

3. Constructing a controllable HJMM interest rate model

In this section we investigate the possibility of influencing the HJMM interest rate model (2.1) by postmultiplying $D(r(t))$ in the drift term (2.2) by a positive function $u : [0, \infty) \to \mathcal{U}$, where $\mathcal{U}$ is some open convex subset of the state space $\mathcal{H}$ of forward rate curves.

In practice one is often not able to observe the interest rate directly. For this reason, we choose as a starting point a partially observable infinite-dimensional HJMM model which will be specified in Subsection 3.1. In addition, in Subsection 3.2 we describe the control sets that have meaning for our study. Finally, in this section, we construct the subclass of HJMM interest rate models that are amenable to stochastic controllability. We demonstrate that this subclass arises quite naturally from the aforementioned partially observable model.

3.1. Partially observable HJMM interest rate models

We commence our analysis with a time independent, partially observable infinite-dimensional model (compare [7]) that we represent by

$$\left\{\begin{array}{rcl} d\widetilde{r}(t) & = & (A\widetilde{r}(t) + D(r(t))u(t))\,dt + \sigma^{'}d\widetilde{W}(t), \\ dz(t) & = & C\widetilde{r}(t) + Fd\widehat{W}(t), \\ r(0) & = & r^{*}(0), \end{array}\right. \tag{3.1}$$

where $D(r(t))$ in (2.1) acting on u that belongs to a yet to be defined admissible control set $\mathcal{U}_{ad} \subset \mathcal{U}$, $\sigma^{'}$, C and F are given by

$$D(r(t)) : \mathcal{U}_{ad} \subset \mathcal{H} \to \mathcal{H},\ \sigma^{'} : Q^{1/2}\mathcal{E} \to \mathcal{H},\ C : \mathcal{U} \subset \mathcal{H} \to \mathbf{R}^m,\ F : \mathbf{R}^m \to \mathbf{R}^m,$$

respectively. The process z usually stands for an observable quantity such as a stock price or an index or a combination of the two. For the sake of our analysis we will investigate the situation where $\widetilde{W}(t)$ is an m-dimensional Wiener process on some separable Hilbert space $\mathcal{E}$ with covariance operator $\widetilde{Q}$ and $\widehat{W}(t)$ is a vector-valued Wiener process on $\mathbf{R}^m$ with covariance operator Q. Also, the volatility $\sigma^{'}$ in (3.1) is a Hilbert Schmidt operator from $\widetilde{Q}^{1/2}\mathcal{E}$ into $\mathcal{H}$ with the Hilbert-Schmidt norm $||\cdot||_2$. Furthermore, we assume that $D(r(t))$ is a ounded linear operator from $\mathcal{U}_{ad}$ with generic element $u(t)$ to $\mathcal{H}$. Also, Q and F are invertible and Q^{-1}, F and $F^{-1} \in \mathcal{L}(\mathbf{R}^m)$.

For each $z \in L_2(\mathcal{Z}_t, \mathcal{U})$ in (3.1) from Theorem 5.6 of [25] we know that there exists a process $\psi(\cdot) \in L_2^{\mathcal{Z}}([0,T]; \mathcal{L}(\mathbf{R}^k, \mathbf{R}^n))$ with the property that $z \in L_2(\mathcal{Z}_t, \mathcal{U})$

is given by

$$z = \mathbf{E}(z|\mathcal{Z}_t) = \mathbf{E}z + \int_0^T \psi(s)dW(s). \tag{3.2}$$

3.2. Admissible control sets

The admissible control set $\mathcal{U}_{ad}$ that we choose has to be amenable to economic interpretation and $D(r(t))$ from (3.1) has to be well-defined on it. Furthermore, its elements have to be independent of the control process and share properties with the elements in some open convex subset $\mathcal{U}$ of the separable Hilbert state space $\mathcal{H}$ given in Subsection 2.2. We would also prefer that our controls satisfy some type of feedback control law. Because of these specifications, a suitable control set is chosen by mainly relying on the results of Bensoussan and Viot presented in [3]. There it is shown how stochastic processes like interest rates can be controlled in an infinite-dimensional setting.

The set-up of a suitable admissible control set $\mathcal{U}_{ad}$ may be described as follows. Consider a $\mathbf{Q}$-measurable function z on Ω to $L_2([0,T];\mathcal{U})$. For $0 < t \leq T$, denote by z_t the restriction of z to $[0,t] \times \Omega$ and by $\Sigma(z_t)$ denote the Σ-algebra generated by $z_t(s)$, $0 \leq s \leq t$. Let $\mathcal{U}_{z_t} = L_2(\Omega, \Sigma(z_t), \mathbf{Q}, \mathcal{U})$. The set

$$\int^{\oplus} \mathcal{U}_{z_t} dt = \{z \in L_2([0,T];\mathcal{U}) : z_t \in \mathcal{U}_{z_t} \text{ for almost every } t \in [0,T]\}$$

is the Hilbertian sum of subspaces $\mathcal{U}_{z_t}$. This sum may have a dependence on the control process which as in [3] may be avoided by fixing the class of admissible controls as

$$\mathcal{U}_{ad} = \int^{\oplus} \mathcal{U}^0_{z_t} dt \cap \int^{\oplus} \mathcal{U}_{z_t} dt,$$

where $z^0(\cdot)$ is the observation process under zero control. Furthermore, it was shown in [3] that actually $\mathcal{U}^0_{z_t} = \mathcal{U}_{z_t}$ and so its control set is independent of the choice of control. It follows that for $u \in \mathcal{U}_{ad}$ that the stochastic system (3.1) is well-defined and u is a feedback control that obeys laws of the type where $u(t) = \psi(t, z_t)$ is admissible for ψ being measurable, nonanticipative and satisfying a uniform Lipschitz condition.

Next, we introduce the Σ-algebras

$$\mathcal{Z}_t = \Sigma(z_t) \text{ and } \mathcal{Z}^0_t = \Sigma(z^0_t),$$

generated by $z_t(\cdot)$ and $z^0_t(\cdot)$, respectively. Also, we consider an important subclass $\mathcal{U}^g_{ad} \subset \mathcal{U}_{ad}$ (see [10] for more details) defined by a linear feedback on the observation

$$\mathcal{U}^g_{ad} = \left\{ \begin{array}{c} u(t) = w(t) + \int_0^t K(t,s)dz(s) : \\ w \in L_2([0,T];\mathcal{U}), \ K \in \mathcal{B}_2(\Delta, \mathcal{L}(\mathbf{R}^k, \mathcal{U})) \end{array} \right\} \subset \mathcal{U}. \tag{3.3}$$

We contend that this choice of the control u not only satisfies the requirements insisted upon above, but also transfers (3.1) from an arbitrary interest rate r_0 to an arbitrary element h of the open convex subset $\mathcal{U}$ of the state space $\mathcal{H}$ of forward rate curves in a natural way. This fact will be verified in Section 4 below.

3.3. An appropriate subclass of HJMM interest rate models

From [35] we know that if $u(\cdot) \in \mathcal{U}_{ad}$ then for $\widetilde{r}(t)$ from (3.1) the Kalman filter ([23])

$$r(t) = \mathbf{E}\{\widetilde{r}(t)|\mathcal{Z}_t\} = \mathbf{E}\{\widetilde{r}(t)|\mathcal{Z}_t^0\}$$

is the mild solution of the linear HJMM interest rate model

$$\begin{cases} dr(t) &= (Ar(t) + D(r(t))u(t))\,dt + \sigma(r(t))dW(t), \\ r(0) &= r^*(0), \end{cases} \tag{3.4}$$

where $u(t)$ is given as in (3.3) and

$$D(r(t)) = \sigma(r(t)) \int_0^x \sigma(r(s))ds \tag{3.5}$$

and

$$\sigma(r(\cdot)) = P(\cdot)C^*(FQF^*)^{-1}F, \tag{3.6}$$

with the linear operator $P(\cdot)$ satisfying the algebraic Riccati equation

$$\frac{d}{dt}\langle P(t)h,k\rangle - \langle P(t)h, A^*k\rangle - \langle A^*h, P(t)k\rangle - \langle \sigma'\widetilde{Q}\sigma'^*h, k\rangle$$
$$+\langle P(t)C^*(FQF^*)^{-1}CP(t)h,k\rangle = 0,\ \ P(0) = P_0,\ \ h,k \in \text{Dom } A^*.$$

We note that this special subclass of HJMM interest rate models for the forward curve $r(t)$ given by (3.4) is fixed by the structure of its volatility σ in (3.6) that, in turn, may be expressed in terms of the coefficients and covariance associated with a partially observable interest rate model of the type given by (3.1). Furthermore, we assume that P may be chosen in such a way that the volatility (3.6) is (locally) Lipschitz continuous in r. From [15] we have that in this case, σ can be shown to be (locally) bounded. An important consequence of this is that the linear operator $D(r(t))$ in (3.4) is (locally) bounded which is a prerequisite for the analysis in the sequel.

As was demonstrated in [12] and [15] developing a notion of solution for general infinite-dimensional stochastic systems can be a tricky business. For instance, it is well-known that we are able to distinguish between mild, weak and strong solutions of such systems. In particular, strong solutions are very seldom encountered in the context of interest rate models. Under the conditions specified in the previous paragraph (see [35]) for more details), we are able to write the mild solution $r(t; r_0, u)$ of (3.4) explicitly as

$$r(t; r_0, u) = S_t r^*(0) + \int_0^t S_{t-s}D(s)u(s)ds + \int_0^t S_{t-s}\sigma(r(s))dW(s). \tag{3.7}$$

In addition, we note that a deterministic counterpart of (3.4) may be given as

$$\begin{aligned} \frac{dr_d(t)}{dt} &= Ar_d(t) + D(r(t))v(t), \\ r_d(0) &= r_d^*(0), \end{aligned} \tag{3.8}$$

with a solution of the form

$$r_d(t) = S_t r_d^*(0) + \int_0^t S_{t-s} D(s) v(s) ds. \tag{3.9}$$

4. Controllability of linear HJMM interest rate models

In this section we investigate the controllability of linear HJMM interest rate models of the type given by (2.1) with stochastic volatility (3.6). Henceforth we shall refer to such models only as (2.1). Firstly, in Subsection 4.1, we construct a controllability operator for the stochastic interest rate model (3.4) that we investigate here. Next, we consider a linear regulator problem of minimizing a certain cost function related to such models (see Subsection 4.2). Also, we discuss the complete and approximate controllability of the aforementioned interest rate model in Subsection 4.3. Finally, we analyze the stochastic controllability of such HJMM interest rate models. The general problem that arises in this section may be stated as follows.

Can we find a control u that drives the linear stochastic HJMM interest rate model given by (3.4) *from an arbitrary interest rate r_0 to any other interest rate h contained in the open convex subset $\mathcal{U}$ of the separable state space $\mathcal{H}$ of forward rate curves ?*

Throughout our discussion, the techniques that we use to solve aspects of the aforementioned problem is based on the theoretical development in contributions on infinite-dimensional stochastic systems theory such as [2], [10], [12], [25], [26] and [35, 36]).

4.1. Controllability operators for linear interest rate models

Our strategy for solving the controllability problem for the linear HJMM interest rate model described above is based on an analysis of the stochastic controllability operator

$$\Pi_s^T : L_2(\mathcal{Z}_T, \mathcal{U}) \to L_2(\mathcal{Z}_T, \mathcal{U})$$

corresponding to (3.4) and its solution (3.7) given by

$$\Pi_s^T\{\cdot\} = \int_s^T S_{T-t} D(r(t)) D(r(t))^* S_{T-t}^* \mathbf{E}\{\cdot|\mathcal{Z}_t\} dt, \tag{4.1}$$

where S_t and $D(r(t))$ are given by (2.5) and (3.5), respectively. Moreover, the related deterministic controllability operator Γ_s^T corresponding to (3.8) and its solution (3.9) has the form

$$\Gamma_s^T\{\cdot\} = \int_s^T S_{T-t}D(r(t))D(r(t))^* S_{T-t}^* dt. \tag{4.2}$$

The controllability operators (4.1) and (4.2) will be shown to be closely related to the controllability of infinite-dimensional HJMM interest rate models in the following ways. Firstly, in Lemma 4.2, we identify an unique optimal control at which a cost function is minimized in terms of the deterministic controllability operator Γ_s^T given by (4.2) and its resolvent. In the case where Γ_s^T is invertible, an explicit formula for the control that drives the initial interest rate r_0 to any other $h \in \mathcal{U}$ is given (see Proposition 4.3). Furthermore, the equivalences between (small-time) approximate, complete and stochastic controllability of our (stochastic) HJMM interest rate model and its deterministic counterpart are established by using the aforementioned controllability operators and their resolvents.

Next, we state and prove an important lemma that provides a relationship between the stochastic controllability operator $\Pi_s^T : L_2(\mathcal{Z}_T,\mathcal{U}) \to L_2(\mathcal{Z}_T,\mathcal{U})$ corresponding to (3.4) given by (4.1) and the deterministic controllability operator Γ_s^T corresponding to (3.8) with the form (4.2) (compare Lemma 3 of [27]).

Lemma 4.1. (Existence of Optimal Process)
For each $z \in L_2(\mathcal{Z}_t,\mathcal{U})$ there exists a process $\psi(\cdot) \in L_2^{\mathcal{Z}}([0,T];\mathcal{L}(\mathbf{R}^k,\mathbf{R}^n))$ with the properties that

1. *the stochastic controllability operator*

$$\Pi_0^T z = \Gamma_0^T \mathbf{E}z + \int_0^T \Gamma_s^T \psi(s)dW(s); \tag{4.3}$$

2. *for all $\lambda > 0$, we have that*

$$R(\lambda_1, -\Pi_0^T)z = R(\lambda_1, -\Gamma_0^T)\mathbf{E}z + \int_0^T R(\lambda_1, -\Gamma_s^T)\psi(s)dW(s). \tag{4.4}$$

Proof. The proofs of these facts are as follows.

1. We note that a consideration of the definition of the stochastic controllability operator Π_s^T in (4.1), the stochastic version of Fubini's Theorem and (3.2) leads to the required representation

$$\begin{aligned}
\Pi_0^T z &= \int_0^T S_{T-t}D(r(t))D(r(t))^* S_{T-t}^* \mathbf{E}(z|\mathcal{Z}_t)dt \\
&= \int_0^T S_{T-t}D(r(t))D(r(t))^* S_{T-t}^* [\mathbf{E}z + \int_0^t \psi(s)dW(s)]dt \\
&= \Gamma_0^T \mathbf{E}z + \int_0^T S_{T-t}D(r(t))D(r(t))^* S_{T-t}^* \int_0^t \psi(s)dW(s)dt
\end{aligned}$$

$$
\begin{aligned}
&= \Gamma_0^T \mathbf{E}z + \int_0^T \int_0^t S_{T-t} D(r(t)) D(r(t))^* S_{T-t}^* \psi(s) dW(s) dt \\
&= \Gamma_0^T \mathbf{E}z + \int_0^T \int_s^T S_{T-t} D(r(t)) D(r(t))^* S_{T-t}^* dt \psi(s) dW(s) \\
&= \Gamma_0^T \mathbf{E}z + \int_0^T \Gamma_s^T \psi(s) dW(s).
\end{aligned}
$$

2. From the formulas (3.2) and (4.3) we may conclude that

$$
R(\lambda_1, -\Pi_0^T)^{-1} z = R(\lambda_1, -\Gamma_0^T)^{-1} \mathbf{E}z + \int_0^T R(\lambda_1, -\Gamma_s^T)^{-1} \psi(s) dW(s).
$$

Furthermore, from the definition of the inverse operator $R(\lambda_1, -\Pi_0^T)^{-1}$ it follows that

$$
z = R(\lambda_1, -\Pi_0^T) R(\lambda_1, -\Gamma_0^T)^{-1} \mathbf{E}z + \int_0^T R(\lambda_1, -\Pi_0^T) R(\lambda_1, -\Gamma_s^T)^{-1} \psi(s) dW(s).
$$

The formula in (4.4) follows when we compare this with equation (3.2). □

4.2. Regulator problem for linear HJMM interest rate models

We define a linear regulator problem related to the minimum energy principle. The problem is to minimize

$$
J(u) = \mathbf{E}||r(T; r_0, u) - h||^2 + \lambda \mathbf{E} \int_0^T ||u(t)||^2 dt, \tag{4.5}
$$

over all $u(\cdot) \in \mathcal{U}_{ad}$, where $r(T; r_0, u)$ is a stochastic interest rate; $h \in L_2(\mathcal{Z}_T, \mathcal{U})$ and $\lambda > 0$ are parameters; and h has the representation

$$
h = \mathbf{E}(h) + \int_0^T h(s) dW(s).
$$

Next, we investigate the existence of a unique optimal control $u^\lambda(\cdot) \in \mathcal{U}_{ad}$ at which the functional (4.5) takes on its minimum value (see Lemma 4 of [27]).

Lemma 4.2. (Existence of Optimal Control)
There exists a unique optimal control $u^\lambda(\cdot) \in \mathcal{U}_{ad}$ at which (4.5) takes on its minimum value and

$$
\begin{aligned}
u^\lambda(t) &= -D(r(t))^* S_{T-t}^* \{ R(\lambda, -\Gamma_0^T)(S_T r_0 - \mathbf{E}(h)) \\
&\quad + \int_0^t R(\lambda, -\Gamma_0^T)[S_{T-s}\sigma(r(s)) - h(s)] dW(s) \Big\};
\end{aligned} \tag{4.6}
$$

$$
\begin{aligned}
r(T, r_0, u^\lambda) - h &= \lambda R(\lambda, -\Gamma_0^T)(S_T r_0 - \mathbf{E}(h)) \\
&\quad + \int_0^t \lambda R(\lambda, -\Gamma_s^T)[S_{T-s}\sigma(r(s)) - h(s)] dW(s).
\end{aligned} \tag{4.7}
$$

Proof. The problem of minimizing the functional (4.5) has a unique solution $u^\lambda(\cdot) \in \mathcal{U}_{ad}$ which is completely characterized by the stochastic maximum principle (see for example [1] and [2]) and has the following form.

$$u^\lambda(t) = -\lambda^{-1} D(r(t))^* S_{T-t}^* \mathbf{E}\left\{r(T, r_0, u^\lambda) - h|\mathcal{Z}_t\right\}. \tag{4.8}$$

Using this in (3.7) we have that

$$\begin{aligned}
r(T, r_0, u^\lambda) &= S_T r_0 + \int_0^T S_{T-s} D(s) u^\lambda(s) ds + \int_0^T S_{T-s}\sigma(r(s))dW(s) \\
&= S_T r_0 + \int_0^T S_{T-s}\sigma(r(s))dW(s) \\
&\quad -\lambda^{-1}\int_0^T S_{T-s}D(s)D(s)^* S_{T-s}^* \mathbf{E}\{r(T, r_0, u^\lambda) - h|\mathcal{Z}_s\}ds \\
&= S_T r_0 + \int_0^T S_{T-s}\sigma(r(s))dW(s) - \lambda^{-1}\Pi_0^T(r(T, r_0, u^\lambda) - h).
\end{aligned}$$

Hence, it follows that

$$\begin{aligned}
\lambda r(T, r_0, u^\lambda) &= \lambda\left(S_T r_0 + \int_0^T S_{T-s}\sigma(r(s))dW(s) - \lambda^{-1}\Pi_0^T(r(T, r_0, u^\lambda) - h)\right) \\
&= \lambda\left(S_T r_0 + \int_0^T S_{T-s}\sigma(r(s))dW(s)\right) - \Pi_0^T(r(T, r_0, u^\lambda) - h)
\end{aligned}$$

which implies that

$$(\lambda I + \Pi_0^T) r(T, r_0, u^\lambda) = \lambda\left(S_T r_0 + \int_0^T S_{T-s}\sigma(r(s))dW(s)\right) + \Pi_0^T h.$$

Consequently, it follows from Lemma 4.1 that

$$\begin{aligned}
r(T, r_0, u^\lambda) - h &= \lambda(\lambda I + \Pi_0^T)^{-1}\left(S_T r_0 + \int_0^T S_{T-s}\sigma(r(s))dW(s)\right) \\
&\quad +(\lambda I + \Pi_0^T)^{-1}(\lambda I + \Pi_0^T - \lambda I)h - h \\
&= \lambda R(\lambda, -\Pi_0^T)\left(S_T r_0 + \int_0^T S_{T-s}\sigma(r(s))dW(s) - h\right) \\
&= \lambda R(\lambda, -\Gamma_0^T)\left(S_T r_0 - \mathbf{E}h\right) \\
&\quad + \int_0^T \lambda R(\lambda, -\Gamma_0^T)\left(S_{T-s}\sigma(r(s)) - h(s)\right) dW(s).
\end{aligned}$$

Thus (4.7) holds. Substituting (4.7) into (4.8) we obtain (4.6) in the following way.

$$\begin{aligned} u^\lambda(t) &= -\lambda^{-1} D(r(t))^* S^*_{T-t} \mathbf{E}\left\{ r(T, r_0, u^\lambda) - h | \mathcal{Z}_t \right\} \\ &= -\lambda^{-1} D(r(t))^* S^*_{T-t} \mathbf{E} \\ \times & \left\{ \lambda R(\lambda, -\Gamma_0^T)(S_T r_0 - \mathbf{E} h) \right. \\ & \left. + \int_0^T \lambda R(\lambda, -\Gamma_s^T)\left(S_{T-s}\sigma(r(s)) - h(s)\right) dW(s) | \mathcal{Z}_t \right\} \\ &= -D(r(t))^* S^*_{T-t} \left\{ R(\lambda, -\Gamma_0^T)(S_T r_0 - \mathbf{E} h) \right. \\ & \left. + \int_0^T R(\lambda, -\Gamma_s^T)\left(S_{T-s}\sigma(r(s)) - h(s)\right) dW(s) \right\}. \end{aligned}$$

This proves the lemma. □

The next proposition uses Lemma 4.1 above to provide an explicit formula for the control steering any initial interest rate r_0 to any other arbitrary interest rate h.

Proposition 4.3. (Control Formula)
Assume that for any $0 \le t \le T$ the deterministic controllability operator Γ_t^T is invertible. Then

1. *for any interest rate $r_0 \in \mathcal{U}$ and $h \in L_2(\mathcal{Z}_T, \mathcal{U})$ the control*

$$u^0(t) = -D(r(t))^* S^*_{T-t} \left\{ (\Gamma_0^T)^{-1}(S_T r_0 - \mathbf{E}(h)) + \int_0^t (\Gamma_0^T)^{-1}[S_{T-s}\sigma(r(s)) - h(s)] dW(s) \right\}$$

   drives r_0 to h at time T.
2. *$u^0(\cdot)$ can be picked from among all the controls $u(\cdot)$ that drive r_0 to h at time T, to minimize the integral*

$$\mathbf{E} \int_0^T ||u(t)||^2 dt.$$

Proof. The proof relies rather heavily on the arguments by Zabczyk in [36]. Also, an application of Lemma 4.1 to the aforementioned results is essential for the proof. □

4.3. Complete and approximate controllable HJMM interest rate models

The definition of complete and approximate controllability of the linear HJMM interest rate model is provided below. Suppose for these definitions that the reachability subspace

$$\mathcal{R}(t, r_0) = \{ r(t, r_0, u) : u \in \mathcal{U}_{ad} \}.$$

Definition 4.4. (Complete Controllability of the HJMM Interest Rate Model) The linear HJMM interest rate model (2.1) is **completely controllable** on $[0,T]$ if for (3.4) all the points in $L_2(\mathcal{Z}_T,\mathcal{U})$ can be reached from the initial interest rate r_0 at time T, i.e., if

$$\mathcal{R}(T,r_0) = L_2(\mathcal{Z}_T,\mathcal{U}).$$

Definition 4.5. (Approximate Controllability of HJMM Interest Rate Models) The linear HJMM interest rate model (2.1) is **approximately controllable** on $[0,T]$ if for (3.4)

$$\overline{\mathcal{R}(T,r_0)} = L_2(\mathcal{Z}_T,\mathcal{U}).$$

It has been shown in certain instances (see [1] and [26]) that the extension of the concepts of complete and approximate controllability from deterministic to stochastic control systems is not really very meaningful. In other words, it is a well-known fact that the (small time) complete and approximate controllability of certain linear stochastic systems at time T, and the (small time) complete and approximate controllability of deterministic systems are equivalent. The following lemma will be used to show that this is also true for the linear stochastic HJMM interest rate model (3.4) and its deterministic counterpart (3.8) under investigation in our paper.

Lemma 4.6. (Controllability of Stochastic and Deterministic Models) *The following statements are true for the HJMM interest rate model* (2.1) *and its corresponding deterministic model* (2.3).

1. *If the stochastic HJMM interest rate model in* (2.1) *is approximately controllable on* $[0,T]$ *then its deterministic counterpart* (2.3) *is approximately controllable on every* $[s,T]$, $0 \le s < T$.
2. *If the deterministic model* (2.3) *is small time approximately controllable on every* $[s,T]$, $0 \le s < T$ *then the stochastic HJMM interest rate model in* (2.1) *is small time approximately controllable on* $[0,T]$.

Proof.

1. Assume that the stochastic HJMM interest rate model in (2.1) is approximately controllable on $[0,T]$. Then

$$\mathbf{E}||\lambda R(\lambda, -\Pi_0^T)z||^2 \to 0.$$

From (4.7) in Lemma 4.1 we conclude that

$$\mathbf{E}||\lambda R(\lambda, -\Pi_0^T)z||^2 \tag{4.9}$$

$$= ||\lambda R(\lambda, -\Gamma_0^T)\mathbf{E}z||^2 + \mathbf{E}||\sum_{j=1}^{k}\int_0^T \lambda R(\lambda, -\Gamma_r^T)\psi_j(r)dW_j(r)||^2$$

$$= ||\lambda R(\lambda, -\Gamma_0^T)\mathbf{E}z||^2 + \mathbf{E}\sum_{j=1}^{k} D_j(t)\int_0^T ||\lambda R(\lambda, -\Gamma_s^T)\psi_j(s)||^2 ds \to 0$$

which for all $\psi(\cdot) \in L_2^{\mathcal{Z}}([0,T];\mathcal{L}(\mathbf{R}^k,\mathcal{U}))$ has the result that

$$\mathbf{E}\sum_{j=1}^{k} D_j(t)\int_0^T ||\lambda R(\lambda, -\Pi_s^T)\psi_j(s)||^2 ds \to 0.$$

This implies that a subsequence $\{\lambda_k\}$ exists such that for all $h \in \mathcal{U}$

$$||\lambda_k R(\lambda_k, -\Gamma_s^T)h||^2 \to 0, \text{ almost everywhere on } [0,T].$$

Because of the continuity of $R(\lambda, -\Gamma_s^T)$ this property holds for all $0 \leq s < T$. The latter means that the corresponding deterministic model (3.8) is approximately controllable on every $[s,T]$, $0 \leq s < T$.

2. Assume that the deterministic model (3.8) is small time controllable on every $[s,\tau]$, $0 \leq s < \tau$. Then

$$||\lambda R(\lambda, -\Gamma_s^\tau)h|| \to 0 \text{ as } \lambda \to 0^+.$$

Since by the Lebesque Dominated Convergence Theorem we have

$$\sum_{j=1}^{k} D_j(t)||\lambda R(\lambda, -\Pi_s^\tau)\psi_j(s)||^2 \leq ||\psi_j(s)||^2$$

it follows from (4.9) that

$$\mathbf{E}||\lambda R(\lambda, -\Pi_0^T)z||^2 \to 0 \text{ as } \lambda \to 0^+.$$

This means that the stochastic HJMM interest rate model in (2.1) is small time approximately controllable on $[0,T]$. □

We note that in the case where the right shifts $S_t = e^{At}$ we have that the generator is analytic and hence we have the following result.

Theorem 4.7. (Approximate and Small Time Approximate Controllability) *The linear stochastic HJMM interest rate model* (2.1) *is approximately controllable on* $[0,T]$ *if and only if it is small time approximately controllable.*

Proof. This result follows from the definitions of approximate and small time approximate controllability and Lemma 4.6. □

4.4. Stochastically controllable linear HJMM interest rate models

The concepts of approximate and complete controllability for stochastic systems referred to in Subsection 4.3 may be weakened in order to obtain the definition of stochastic controllability. In general, a stochastically controllable stochastic control system is a system that is able to reach an arbitrarily small neighbourhood of each point in the state space with a probability close to one.

Next, we define the stochastic controllability of linear HJMM interest rate models. For this, we have to introduce the sets $\mathcal{A}_\epsilon^p(t, r_0)$ given by

$$\mathcal{A}_\epsilon^p(t,r_0) = \{h \in \mathcal{U} : \exists u \in \mathcal{U}_{ad},\ \mathbf{Q}(||r(t,r_0,u) - h||^2 \leq \epsilon) \geq p)$$

and $\mathcal{A}(t, r_0)$, where

$$\mathcal{A}(T, r_0) = \cap_{\epsilon>0,\ 0\leq p\leq 1} \mathcal{A}_\epsilon^p(T, r_0).$$

Moreover, it is a well-known fact that $\overline{\mathcal{A}(T, r_0)} = \mathcal{A}(T, r_0)$. The following definition is an important one.

Definition 4.8. (Stochastic Controllability of HJMM Interest Rate Models) The linear stochastic HJMM interest rate model (2.1) is **stochastically controllable** if for (3.4)

$$\overline{\mathcal{A}(T, r_0)} = \mathcal{A}(T, r_0) = \mathcal{U}.$$

This means that if we are given an arbitrary $\epsilon > 0$ it is possible to steer from the initial interest rate r_0 to within a given distance $\sqrt{\epsilon}$ of all interest rates in the separable open convex subset $\mathcal{U}$ of the state space $\mathcal{H}$ of forward rate curves at time T with a probability close to one.

Theorem 4.9.
(Approximate and Stochastic Controllability of Interest Rate Models)
The linear stochastic HJMM interest rate model (2.1) *is approximately controllable on* $[0, T]$ *if and only if it is small time stochastically controllable with the control set* $\mathcal{U}_{ad}^g$.

Proof. In order to prove this theorem we consider the solution (3.7) of the interest rate model (3.4). If (2.1) is approximately controllable then by Lemma 4.6 it is approximately controllable on each $[s, T]$. In this case, we have that

$$\lambda R(\lambda, -\Gamma_s^T) \to 0 \text{ strongly as } \lambda \to 0^+.$$

Furthermore, Lemma 4.2 claims that for any fixed $h \in \mathcal{U}$ that is nonrandom there exists a Gaussian control

$$u^\lambda(t) = -D(r(t))^* S_{T-t}^*$$

$$\left\{R(\lambda, -\Gamma_0^T)(S_T r_0 - \mathbf{E}(h)) + \int_0^t R(\lambda, -\Gamma_0^T)[S_{T-s}\sigma(r(s)) - h(s)]dW(s)\right\}$$

in $\mathcal{U}_{ad}^g$ with the property that

$$r(T, r_0, u^\lambda) - h = \lambda R(\lambda, -\Gamma_0^T)(S_T r_0 - \mathbf{E}h) + \int_0^T \lambda R(\lambda, -\Gamma_s^T) S_{T-s}\sigma(r(s))dW(s).$$

From this we conclude that

$$\mathbf{E}||r(T, r_0, u^\lambda) - h||^2 \to 0 \text{ as } \lambda \to 0^+.$$

In this case, the small time stochastic controllability of the HJMM interest rate model in (2.1) with control set $\mathcal{U}_{ad}^g$ is a consequence of Chebychev's inequality.

Conversely, suppose that the HJMM interest rate model with stochastic volatility in (2.1) with control set $\mathcal{U}_{ad}^g$ is small time stochastically controllable. Let $h \in \mathcal{U}$ and consider

$$\{\epsilon_n : \epsilon_n > 0,\ \epsilon_n \to 0\} \text{ and } \{p_n : 0 \leq p_n \leq 1,\ p_n \to 1\}.$$

In this case, we are assured of the existence of a sequence $\{u^n\} \in \mathcal{U}_{ad}^g$ with the property that

$$\mathbf{Q}\{||r(T, r_0, u^n) - h||^2 \leq \epsilon\} \geq p_n.$$

From this we deduce that for any $\epsilon > 0$ there is a number N such that $0 < \epsilon_n < \epsilon^2$. Furthermore, we have that

$$\begin{aligned} \mathbf{Q}\{||r(T, r_0, u^n) - h||^2 > \epsilon\} &= 1 - \mathbf{Q}\{||r(T, r_0, u^n) - h||^2 < \epsilon\} \\ &\leq 1 - \mathbf{Q}\{||r(T, r_0, u^n) - h||^2 \leq \epsilon_n\} \\ &= \mathbf{Q}\{||r(T, r_0, u^n) - h||^2 > \epsilon_n\} \\ &\leq 1 - p_n, \end{aligned}$$

for all $n > N$. This inequality suggests that $r(T, r_0, u^n)$ converges to h in probability which in turn implies that for any $\epsilon > 0$, it follows that

$$\mathbf{Q}\{||r(T, r_0, u^n) - h|| > \epsilon\} \to 0, \text{ as } n \to \infty.$$

In this case, for all $x \in \mathcal{U}$ we have

$$\lim_{n \to \infty} \mathbf{E}e^{i\langle r(T, r_0, u^n), x\rangle} = \mathbf{E}e^{i\langle h, x\rangle}.$$

Moreover, since $r(T, r_0, u^n)$ is a Gaussian random variable (as the solution (3.7) of the stochastic model (3.4) corresponding to the Gaussian control u^n), from the aforementioned convergence of characteristic functions and the gaussian properties of $r(T, r_0, u^n)$ and h it follows that

$$\mathbf{E}e^{i\langle r(T, r_0, u^n), x\rangle} = e^{i\langle m_n, x\rangle - 1/2\langle \Lambda_n x, x\rangle}$$

and

$$\lim_{n \to \infty} e^{i\langle m_n, x\rangle - 1/2\langle \Lambda_n x, x\rangle} = e^{i\langle h, x\rangle}.$$

As a result we may deduce that for all $x \in \mathcal{U}$ we have

$$\langle m_n, x\rangle \to \langle h, x\rangle \text{ and } \langle \Lambda_n x, x\rangle \to 0 \text{ as } n \to \infty,$$

where $m_n = \mathbf{E}r(T, r_0, u^n)$ and $\Lambda_n = \text{cov}\,(r(T, r_0, u^n))$. Convergence in the first instance results in the sequence $\{m_n\} \in \mathcal{U}$ converging weakly to h in $\mathcal{U}$. Mazur's Theorem implies that we can construct the sequence

$$h_n = \sum_{i=1}^{n} c_i^n m_i, \; c_i^n \geq 0, \; \sum_{i=1}^{n} c_i^n = 1, \; i = 1, \; 2, \; \ldots$$

of convex combinations of $m_i = \mathbf{E}r(T, r_0, u^i)$, $i = 1, \; 2, \; \ldots, \; n$ so that h_n converges to h in the strong topology of $\mathcal{U}$. Next, we write

$$\widetilde{u}^n = \sum_{i=1}^{n} c_i^n u_i, \; n = 1, \; 2, \; \ldots$$

It follows directly that $\widetilde{u}^n \in \mathcal{U}_{ad}$. Because of the affineness of (3.4), it follows that

$$h_n = \sum_{i=1}^{n} c_i^n \mathbf{E}r(T, r_0, u^i) = \mathbf{E}r(T, r_0, \sum_{i=1}^{n} c_i^n u_i) = \mathbf{E}r(T, r_0, \widetilde{u}^n).$$

Next, if $\widetilde{v}^n = \mathbf{E}\widetilde{u}^n \in \mathcal{V}$ then $h_n = y(T, r_0, \widetilde{u}^n)$ and as a result

$$\lim_{n\to\infty} ||y(T, r_0, \widetilde{u}^n) - h|| = \lim_{n\to\infty} ||h_n - h|| = 0.$$

Our conclusion is that the deterministic model (2.3) is small time approximately controllable. Moreover, by the equivalence in Lemma 4.6, we may conclude that the stochastic HJMM interest rate model (2.1) is also small time approximately controllable. □

5. Connections with related interest rate models

An interest rate model that is closely related to the discussion above is the Cox-Ingersoll-Ross ([11] and the references contained therein) model. Also, two of the interest rate models that we identify as having deterministic volatility and that can be discussed in our context are the Ho-Lee ([21]) and Hull-White ([22] and the references therein) models. A brief comment on their applicability in terms of the results developed above are given in this section.

5.1. Ho-Lee model

The Ho-Lee model is an interest rate model with time independent, deterministic volatility that may be studied in the context of the stochastic model (2.1) has volatility of the form

$$\sigma(x) = \sigma,$$

where σ is a scalar constant. In other words, we have only one driving Wiener process. In this case it is a simple matter to apply our main results to factors involving a drift term with

$$D(r(t)) = \sigma^2 x.$$

In particular, under certain bond market conditions we can quite easily compute the controllability operators and the (optimal) controls for the Ho-Lee interest rate model.

5.2. Hull-White model

The Hull-White interest rate model has time independent, deterministic volatility that is given by

$$\sigma(x) = \sigma e^{-cx},$$

where σ and c are scalar constants so that once again we only have one driving Wiener process. It is closely associated with the model (2.1) with

$$D(r(t)) = -\frac{\sigma^2}{c} e^{-cx}\{e^{-cx} - 1\}$$

and may therefore be analyzed in our context. Since the Hull-White interest rate model is similar to the Ho-Lee model this analysis will be similar in many respects.

5.3. Cox-Ingersoll-Ross model

The Cox-Ingersoll-Ross interest rate model is directly related to the stochastic system (2.1) and has time dependent volatility

$$\sigma(r(t)) = \sigma\sqrt{r(t)}.$$

6. Conclusions

In this paper we have discussed the stochastic controllability of linear interest rate models of HJMM type. In order to solve the problems related to this issue we adopted an approach which involved (infinite-dimensional) stochastic systems theoretic concepts. Necessary and sufficient conditions for the stochastic controllability of linear stochastic HJMM interest rate models is presented in our main result.

The results about controllability in this paper may be extended in several directions. For instance, the very topical question of the controllability of nonlinear interest rate models deserves some future attention. The present paper concentrates on the approximate, complete and stochastic controllability of stochastic interest rate models. However, equivalent notions of controllability for output processes may be interesting to investigate. Also, for instance, we could study the idea of dynamic path controllability for interest rate models (see [29]). Also, we are currently considering the practical implementation of the results determined in this paper (see the multi-authored conference paper [32]).

The use of stochastic control in other areas of current research is fairly widespread. For instance, the papers [13] and [14] are related to the very topical issue of pricing and hedging in incomplete markets which in recent times seems to have emerged as a rich source of stochastic control problems. Of course, since the appearance of the aforementioned literature much has happened. However, these references provide some useful clues to where the solutions to stochastic control problems appearing in such diverse financial scenarios as funding systems (in particular, pension funds; see, for instance, [30], [31] and [33]), insurance and reinsurance (see, for instance, [20]), option pricing and hedging, portfolio selection (see [24]), loan and debt management (see, for instance, [34]) and interest rate theory (see [32]) may lie.

References

[1] A.E. Bashirov and N.I. Mahmudov, *On Concepts of Controllability of Deterministic and Stochastic Systems.* SIAM Journal of Control and Optimization **37** (1999), 1808–1821.

[2] A.E. Bensoussan, *Stochastic Maximum Principle for Systems with Partial Information and Application to the Separation Principle.* Applied Stochastic Analysis (London, 1989), 157–172, Stochastics Monogr., 5, Gordon and Breach, New York, 1991.

[3] A.E. Bensoussan and M. Viot, *Optimal Control of Stochastic Linear Distributed Parameter Systems.* SIAM Journal of Control **13** (1975), 904–926.

[4] T. Björk and B.J. Christensen, *Interest Rate Dynamics and Consistent Forward Rate Curves.* Mathematical Finance **9** (1999), 323–348.

[5] T. Björk and A. Gombani, *Minimal Realization of Interest Rate Models.* Finance and Stochastics **3** (1997), 413–432.

[6] T. Björk and C. Landen, *On the Construction of Finite Dimensional Realizations for Nonlinear Forward Rate Models.* Finance and Stochastics **6** (2002), 303–331.

[7] T. Björk, C. Landen and L. Svensson, *Finite Dimensional Markovian Realizations for Stochastic Volatility Forward Rate Models.* Working Paper (2002).

[8] T. Björk and L. Svensson, *On the Existence of Finite Dimensional Realizations for Nonlinear Forward Rate Models.* Mathematical Finance **11** (2001), 205–243.

[9] A. Brace and M. Musiela, *A Multi-Factor Gauss Markov Implementation of Heath Jarrow and Morton.* Mathematical Finance **4** (1994), 563–576.

[10] R. Curtain and A. Ichikawa, *A Separation Principle for Stochastic Evolution Equations.* SIAM Journal of Control and Optimization **15** (1977), 367–383.

[11] J. Cox, J. Ingersoll and S. Ross, *A Theory of the Term Structure of Interest Rates.* Econometrica **53** (1985), 385–408.

[12] G. Da Prato and J. Zabczyk, *Stochastic Equations in Infinite Dimensions.* Cambridge University Press, Cambridge, 1992.

[13] N. El Karoui and M.C. Quenez, *Programmation Dynamique et Evaluation Des Actifs Contingents en Marche Incomplet.* C.R. Acad. Sci. Paris, Serie I **313** (1991), 851–854.

[14] N. El Karoui and M.C. Quenez, *Dynamic Programming and Pricing of Contingent Claims in an Incomplete Market.* SIAM J. Control and Optimization **33** (1995), 29–66.

[15] D. Filipović, *Consistency Problems for HJM Interest Rate Models*, Springer-Verlag, Berlin, 2001.

[16] D. Filipović and J. Teichmann, *Finite Dimensional Realizations for Stochastic Equations in the Heath-Jarrow-Morton Framework.* Working Paper, Technische Universität, Vienna.

[17] D. Filipović and J. Teichmann, Existence of Invariant Manifolds for Stochastic Equations in Infinite Dimension. Journal of Functional Analysis **197** (2003), 398–432.

[18] D. Filipović and J. Teichmann, *Regularity of Finite-Dimensional Realizations for Evolution Equations.* Journal of Functional Analysis **197** (2003), 433–446.

[19] D. Heath, R. Jarrow and A. Morton, *Bond Pricing and the Term Structure of Interest Rates: A New Methodology for Contingent Claims Valuation.* Econometrica **60** (1992), 77–105.

[20] C. Hipp and M. Taksar, *Stochastic Control for Optimal New Business.* Liber amicorum for Etienne De Vylder on the occasion of his 60th birthday, Insurance: Mathematics and Economics **26** (2000), 185–192.

[21] T. Ho and S. Lee, *Term Structure Movements and Pricing Interest Rate Contingent Claims.* Journal of Finance **41** (1986), 1011–1029.

[22] J. Hull and A. White, *Numerical Procedures for Implementing Term Structure Models II: Two-Factor Models.* The Journal of Derivatives **2** (1994), 37–47.

[23] R.E. Kalman, *Mathematical Description of Linear Dynamical Systems.* SIAM Journal on Control and Optimization **1** (1963), 152–192.

[24] I. Karatzas, J.P. Lehoczky and S.E. Shreve, *Optimal Portfolio and Consumption Decisions for a Small Investor on a Finite Horizon.* SIAM Journal of Control and Optimization **25** (1987), 1557–1586.

[25] R.S. Lipster and A.N. Shiryaev, *Statistics of Random Processes.* Springer-Verlag, New York, 1977.

[26] N.I. Mahmudov, *Controllability of Linear Stochastic Systems.* IEEE Transactions on Automatic Control **46** (2001), 724–731.

[27] N.I. Mahmudov and A. Denker, *On Controllability of Linear Stochastic Systems.* International Journal of Control **73** (2000), 144–151.

[28] M. Musiela, *Stochastic PDE's and Term Structure Models.* Mathematical Finance **4** (1994), 563–576.

[29] H. Nijmeijer, *On Dynamic Decoupling and Dynamic Path Controllability in Economic Systems.* Journal of Economic Dynamics and Control **13** (1989), 21–39.

[30] M. Parlar, *Stochastic Control of a Pension Fund Model with First-Order Markov-Dependent Parameters.* Optimal Control Application Methods **2** (1981), 175–189.

[31] M.A. Petersen, *On Optimal Control Problems in Funding Systems.* Electronic Proceedings of the First African Control Conference (AFCON 2003), University of Cape Town's Breakwater Campus, Cape Town, South Africa, 3 December 2003, pp. 240–247.

[32] M.A. Petersen, E. Letsaolo, H. Raubenheimer, N. Sepadi, F.C. van der Walt and H.F. van Rooy, *On the Controllability of Ho-Lee, Hull-White, Black-Karasinski and Cox-Ingersoll-Ross Interest Rate Models.* Electronic Proceedings of the First African Control Conference (AFCON 2003), University of Cape Town's Breakwater Campus, Cape Town, South Africa, 3 December 2003, pp. 228–233.

[33] M.A. Petersen, H. Raubenheimer and J.H. van Schuppen, *Control of Pension Funds with Optimal Stochastic Control Theory,* Mathematical Finance, Submitted.

[34] M. Taksar and X. Zhou, Optimal Risk and Dividend Control for a Company with a Debt Liability. The Interplay between Insurance, Finance and Control (Aarhus, 1997). Insurance Math. Econom. 22 (1998), no. 1, 105–122.

[35] J. Zabczyk, *Controllability of Stochastic Linear Systems.* Systems and Control Letters **1** (1991), 25–31.

[36] J. Zabczyk, *Mathematical Control Theory.* Birkhäuser, Boston/Basel/Berlin, 1992.

Mark A. Petersen
Department of Mathematics and Applied Mathematics
Potchefstroom University for CHE
Potchefstroom X 6001
South Africa
e-mail: wskmap@puknet.puk.ac.za

Helgard Raubenheimer
Centre for Business Mathematics and Informatics
North-West University (Potchefstroom Campus)
Potchefstroom X 6001
South Africa
e-mail: bwihr@puknet.puk.ac.za

Frederik C. van der Walt
Centre for Business Mathematics and Informatics
North-West University (Potchefstroom Campus)
Potchefstroom X 6001
South Africa
e-mail: bwifcvdw@puknet.puk.ac.za

Herrie F. van Rooy II
Centre for Business Mathematics and Informatics
North-West University (Potchefstroom Campus)
Potchefstroom X 6001
South Africa
e-mail: bwihfvr@puknet.puk.ac.za

Operator Theory:
Advances and Applications, Vol. 149, 517–523

On the Generalized Joint Eigenvector Expansion for Commuting Normal Operators

Artem Pulemyotov

Abstract. Let $A = (A_x)_{x\in X}$ be a family of commuting normal operators in a separable Hilbert space H_0. Obtaining the spectral expansion of A involves the construction of the corresponding joint resolution of identity E. The support, $\operatorname{supp} E$, is not, generally, a set of full E-measure. This causes numerous difficulties, in particular, when proving the projection spectral theorem, i.e., the main theorem about the expansion in generalized joint eigenvectors. In this review, we provide an example of a joint resolution of identity with an empty support and show $\operatorname{supp} E$ has a full outer measure under the conditions of the projection spectral theorem. This result can be used to simplify the proof of the theorem and to refine its assertions.

1. Introduction

Let A be a normal operator acting in a separable Hilbert space H_0. Let E be the resolution of identity (RI) of A. The following spectral integral representation takes place:

$$A = \int_{\mathbb{C}} \lambda dE(\lambda).$$

If an operator with a discrete spectrum $(\lambda_j)_{j=1}^{\infty}$ is considered, then

$$E(\alpha) = \sum_{\lambda_j \in \alpha} P(\lambda_j), \; A = \sum_{j=1}^{\infty} \lambda_j P(\lambda_j),$$

where $P(\lambda_j)$ is the projection onto the eigensubspace corresponding to λ_j. These formulas can be generalized to hold in the case of a continuous spectrum.

More precisely, let H_0 be a separable space, and let A be standardly connected (Section 2) with the rigging

$$H_- \supset H_0 \supset H_+ \supset D. \tag{1.1}$$

Here H_+ and H_- are Hilbert spaces of "test" and "generalized" vectors with a quasinuclear (i.e., Hilbert-Schmidt) embedding $H_+ \subset H_0$, D is a linear topological space embedded densely and continuously into H_+. The RI E, if understood as

Received by the editors November 18, 2002; revised February 17, 2003. Submitted by L. Rodman.
Mathematics Subject Classification (2000). 47A70, 47B15, 47B25.
Key words and phrases. Spectral theorem, support, joint resolution of identity, generalized eigenvector.

an operator-valued measure whose values are operators from H_+ to H_-, can be differentiated with respect to a certain scalar measure ρ. Namely, a derivative $(dE/d\rho)(\lambda) = P(\lambda) : H_+ \to H_-$ exists, and the integral representation

$$A = \int_{\mathbb{C}} \lambda P(\lambda) d\rho(\lambda)$$

holds (here A is understood to be restricted to $H_+ \cap \mathrm{Dom}(A)$). Moreover, if λ belongs to a certain set π of full ρ-measure then the range $\mathrm{Ran}(P(\lambda))$ consists of generalized eigenvectors $\xi \in H_-$ corresponding to λ, i.e., $A\xi = \lambda\xi$ in some generalized sense (Section 2). Note that π can now be chosen to satisfy the inclusion $\pi \subset \operatorname{supp} \rho$ with $\operatorname{supp} \rho$ denoting the support of the measure ρ. The described result constitutes the so-called BGKM-expansion for a single operator ([6], [7], [1], [8], [11], see also Chapter 2 of [2] and Chapter 3 of [5]).

It is possible to obtain an analogous expansion for an arbitrary family $A = (A_x)_{x\in X}$ of commuting (in the sense of RI's) normal operators standardly connected with the rigging (1.1). In order to do that, one has to construct the joint resolution of identity (JRI) E of the family A, i.e., a projection-valued measure on a σ-algebra of subsets of $\mathbb{C}^X$, where $\mathbb{C}^X$ is the space of all mappings $X \ni x \mapsto \lambda(x) \in \mathbb{C}$. The following integral representation takes place for the operators A_x:

$$A_x = \int_{\mathbb{C}^X} \lambda(x) dE(\lambda(\cdot)) \ (x \in X).$$

No essential difficulties arise when proving the possibility to differentiate E: there exists a derivative $(dE/d\rho)(\lambda(\cdot)) = P(\lambda(\cdot)) : H_+ \to H_-$ such that

$$A_x = \int_{\mathbb{C}^X} \lambda(x) P(\lambda(\cdot)) d\rho(\lambda(\cdot)) \ (x \in X). \tag{1.2}$$

It is more complicated to prove the second part of the following statement known as the projection spectral theorem ([3], see also [5]). *Representation (1.2) holds for the operators A_x, and there exists a set π of full ρ-measure (maybe, full outer ρ-measure) satisfying the following: if $\lambda(\cdot) \in \pi$ then the range* $\mathrm{Ran}(P(\lambda(\cdot)))$ *consists of generalized joint eigenvectors of A corresponding to the joint eigenvalue $\lambda(\cdot)$, i.e., $\xi \in H_-$ such that $A_x\xi = \lambda(x)\xi$ in some generalized sense for each $x \in X$.*

Note that the collection of all the joint eigenvalues of the family A forms, by definition, the generalized spectrum $g(A)$ of A.

Equality $A_x\xi = \lambda(x)\xi$ is not very hard to prove for a fixed x and $\xi \in \mathrm{Ran}(P(\lambda(\cdot)))$, where $\lambda(\cdot) \in \pi_x \subset \mathbb{C}^X$. Here π_x is a set of full ρ-measure depending on x. It would be natural to define $\pi = \bigcap_{x\in X} \pi_x$ to obtain this equality for all $x \in X$, $\lambda(\cdot) \in \pi$. When X is at most countable, $\pi = \bigcap_{x\in X} \pi_x$ does have the necessary properties. However, when X is not countable, the intersection $\bigcap_{x\in X} \pi_x$ might not have a full outer ρ-measure.

To construct π in the case of an uncountable X one has to involve the concept of the support of a measure. Proof of the equality $A_x\xi = \lambda(x)\xi$ is now based on the so-called "Separability Lemma", see [12]. If it is known that the measure ρ

is proper (i.e., it is known that $\operatorname{supp}\rho$ has a full outer ρ-measure), and so are its restrictions to some special sets, then it is of no difficulty to prove that π has a full outer ρ-measure. E.g., this situation takes place if all the operators A_x are bounded. However, examples exist showing that the support of a JRI might not have a full outer measure (example of a JRI with even an empty support is given further). This circumstance essentially complicates the proof of the projection spectral theorem for the case of unbounded operators.

It should also be noted that the inclusion $\pi \subset \operatorname{supp}\rho$ can not be obtained in the "unbounded" case using the arguments from [3] and [5].

This paper is devoted to the following result: ρ is proper under the conditions of the projection spectral theorem, and so is its arbitrary restriction (even if the operators under consideration are unbounded). This enables to prove the inclusion $\pi \subset \operatorname{supp}\rho$ and to simplify the construction of the set π in the "unbounded" case. We also provide examples of JRI's with empty supports.

The theory of generalized joint eigenvector expansions appears to be a powerful tool when working with several objects of infinite-dimensional analysis. Its applications to harmonic analysis of functions of infinitely many variables, to the theory of representations of commutative algebraic structures, to construction of spaces of test and generalized functions of infinitely many variables is shown in [5]. In particular, the well-known Wiener-Ito-Segal transform is a Fourier transform corresponding to the generalized eigenvector expansion of a commuting family of field operators in a Fock space. For the latest questions concerning these topics, see, e.g., [10], [4] and [9]

2. Preliminaries

To make it more convenient for the reader, we adduce the main notions and facts below. See Chapter 2 of [2] or Chapter 3 of [5] for details.

Let $A = (A_x)_{x\in X}$ be a family of commuting normal operators acting in a separable Hilbert space H_0. Suppose A to be connected with the quasinuclear rigging

$$H_- \supset H_0 \supset H_+ \supset D \tag{2.1}$$

in a standard way. The latter means that D is a linear topological space embedded densely and continuously into H_+, and the inclusions $D \subset \operatorname{Dom}(A_x)$, $A_x^* \restriction D, A_x \restriction D \in \mathcal{L}(D \to H_+)$ hold for each $x \in X$ (hereafter $\operatorname{Dom}(A_x)$ is the domain of A_x, $\mathcal{L}(D \to H_+)$ the space of bounded operators from D to H_+). We will now introduce some notations: $\mathbb{1}$ is the identity operator in H_0, O is the embedding of H_+ into H_0, O^+ is the embedding of H_0 into H_-.

Consider *cylindrical sets* in the Tychonoff product $\mathbb{C}^X$. A cylindrical set is defined by distinct points $x_1, \ldots, x_n \in X$ (the coordinates) and a base $\Delta \in \mathcal{B}(\mathbb{C}^n)$ as follows:

$$\operatorname{Cyl}(x_1, \ldots, x_n; \Delta) = \{\lambda(\cdot) \in \mathbb{C}^X \mid (\lambda(x_1), \ldots, \lambda(x_n)) \in \Delta\} \ (n \in \mathbb{N})$$

(hereafter $\mathcal{B}(R)$ denotes the Borel σ-algebra in R). The collection of all cylindrical sets is an algebra $\mathcal{C}(\mathbb{C}^X)$. Let $\mathcal{C}_\sigma(\mathbb{C}^X)$ be a σ-algebra generated by $\mathcal{C}(\mathbb{C}^X)$. Define a RI E^0 on the sets of the form $\mathrm{Cyl}(x_1,\ldots,x_n;\Delta_1\times\cdots\times\Delta_n)$ $(\Delta_1,\ldots,\Delta_n\in\mathcal{B}(\mathbb{C}))$ by the formula

$$E^0(\mathrm{Cyl}(x_1,\ldots,x_n;\Delta_1\times\cdots\times\Delta_n)) = E_{x_1}(\Delta_1)\ldots E_{x_n}(\Delta_n)$$

(hereafter E_x denotes the RI of the operator A_x). E^0 determines uniquely a RI E on $\mathcal{C}_\sigma(\mathbb{C}^X)$. E is called the *JRI* of the family A. Remark that a JRI can be constructed analogously if a family $(E_x)_{x\in X}$ of commuting RI's (instead of operators) is given.

The equality

$$O^+E(\alpha)O = \int_\alpha P(\lambda(\cdot))d\rho(\lambda(\cdot)) \ (\alpha\in\mathcal{C}_\sigma(\mathbb{C}^X)) \tag{2.2}$$

holds (here $\mathcal{C}_\sigma(\mathbb{C}^X)\ni\alpha\mapsto\rho(\alpha)=\mathrm{Tr}(O^+E(\alpha)O)$ is a scalar measure and $P(\lambda(\cdot)):H_+\to H_-$ is a defined ρ-almost everywhere operator-valued function). $P(\lambda(\cdot))$ is weakly measurable with respect to $\mathcal{C}_\sigma(\mathbb{C}^X)$, $\mathrm{Tr}(P(\lambda(\cdot)))=1$. ρ is called the *spectral measure* of the family A, $P(\lambda(\cdot))$ is called its *generalized projection.*

Suppose R to be a Hausdorff topological space with a fixed base $\Sigma\ni R$ of neighborhoods; $\mathcal{R}$ to be a σ-algebra generated by Σ. The *support* of a (scalar or operator-valued) measure θ on $\mathcal{R}$ is defined as the intersection of all closed sets of full measure from $\mathcal{R}$, i.e., closed $\alpha\in\mathcal{R}$ such that $\theta(\alpha)=\theta(R)$. Denote the support of θ by $\operatorname{supp}\theta$. Remark that $\operatorname{supp}\theta$ is a consistent characteristics of θ if every open set $\mathcal{R}\ni o\supset\operatorname{supp}\theta$ has a full θ-measure. In this case, θ is called a proper measure.

Let us fix a set $R'\subset R$ and topologize it with the relative topology of R. A collection Σ' of sets of the form $u'=u\cap R'$ $(u\in\Sigma)$ forms a base of neighborhoods in R'. Let $\mathcal{R}'$ be a σ-algebra generated by Σ'. The mapping $\mathcal{R}\ni\alpha\mapsto\alpha'=\alpha\cap R'$ carries the whole of $\mathcal{R}$ onto the whole of $\mathcal{R}'$. If R' has a full outer θ-measure (i.e., if any set $\mathcal{R}\ni v\supset R'$ has a full θ-measure), then θ determines a measure θ' on the σ-algebra $\mathcal{R}'$ according to the formula $\mathcal{R}'\ni\alpha'\mapsto\theta'(\alpha')=\theta(\alpha)$ with α satisfying $\alpha'=\alpha\cap R'$. θ' is called a *modification* of θ with respect to R' and denoted also by $\theta_{R'}$. If $R'\in\mathcal{R}$, then θ' equals to the restriction $\theta\restriction R'$. The equality $\operatorname{supp}\theta'=(\operatorname{supp}\theta)\cap R'$ takes place. Properness of θ' implies the properness of θ.

A nonzero vector $\xi\in H_-$ is called a *generalized joint eigenvector* of the family A with an eigenvalue $\lambda(\cdot)\in\mathbb{C}^X$ if it satisfies $(\xi,A_x^*\phi)_{H_0}=\lambda(x)(\xi,\phi)_{H_0}$ and $(\xi,A_x\phi)_{H_0}=\overline{\lambda(x)}(\xi,\phi)_{H_0}$ for each $x\in X$, $\phi\in D$. A collection of all family's eigenvalues is called a *generalized spectrum* of A and denoted by $g(A)$.

Speaking about the JRI E, we note E and ρ are absolutely continuous with respect to one another. The equality $\operatorname{supp}E=\operatorname{supp}\rho$ holds true. The restriction of ρ to a set from $\mathcal{C}_\sigma(\mathbb{C}^X)$ is proper if and only if the restriction of E to this set is proper. In particular, properness of ρ is equivalent to the properness of E.

3. Support of a JRI and the projection spectral theorem

Let us first give an example of a JRI with an empty support [13]. More examples can be found in [2], [5], [13].

Example 3.1. Assume $X = \Lambda_1 \sqcup \Lambda_2 \sqcup \{\xi\}$ with $\Lambda_1 = \Lambda_2 = \mathbb{R}$, ξ a particular abstract point. $(A_x)_{x\in X}$ act in $L_2(\mathbb{R}, d\mu)$ (μ is the Lebesgue measure) as operators of multiplication by the function $f_x(t)$. Let $f_\xi(t) = t$ and for $x_1 \in \Lambda_1$, $x_2 \in \Lambda_2$ let $f_{x_1}(t) = \frac{1}{t-x_1}\sin\frac{1}{t-x_1}$, $f_{x_2}(t) = \frac{1}{t-x_2}\cos\frac{1}{t-x_2}$. If E denotes the JRI of $(E_x)_{x\in X}$ then $\operatorname{supp} E = \emptyset$ (see [13] for the proof).

As we have noted in Section 1, one faces numerous difficulties when working with families of unbounded operators. In particular, the JRI of such a family is not, generally, proper. According to the following result [14], properness of the JRI and its restrictions does take place under the conditions of the projection spectral theorem.

Theorem 3.1. *Assume $A = (A_x)_{x\in X}$ to be a family of commuting normal operators connected with quasinuclear rigging (2.1) in a standard way. If D is a separable projective limit of Hilbert spaces, then the JRI E of the family A is proper. Moreover, the restriction $E \restriction \eta$ is proper for an arbitrary $\eta \in \mathcal{C}_\sigma(\mathbb{C}^X)$ satisfying $E(\eta) \neq 0$.*

Sketch of the Proof. We introduce a topology on the indices and show that the set $C(X)$ of continuous functions acting from X to $\mathbb{C}$ has a full outer E-measure. This lets us modify E with respect to $C(X)$. Properness of E follows the properness of the modification immediately. Remark that the topology we introduce on the indices must be a Hausdorff topology. Therefore, it is necessary to replace X with the factorset $\tilde{X} = X/\sim$ ($\sim$ is a certain equivalence).

The proof consists of four parts. Part I is devoted to the construction of a set γ of full outer E-measure having some particular properties (it should be mentioned that the arguments here are based on the proof of Theorem 3.2 in [3], see also Theorem 2.7 in Chapter 3 of [5]). We will show further that γ only contains functions continuous in some special sense.

After factorizing X in Part II, we consider an auxiliary RI $\tilde{E}$ on a certain σ-algebra of subsets of $\mathbb{C}^{\tilde{X}}$. After "projecting" E onto $\mathbb{C}^{\tilde{X}}$ by means of the mapping prj, we prove that E is proper if $\tilde{E}$ is.

Hausdorff topology is introduced on $\tilde{X}$ in Part III of the proof. Properties of γ stated in Part I imply the continuity of all functions from $\operatorname{prj}\gamma$. Moreover, $\operatorname{prj}\gamma$ has a full outer $\tilde{E}$-measure. This circumstance gives a possibility to modify $\tilde{E}$ with respect to the set $C(\tilde{X})$ of continuous functions. It suffices to prove the properness of the modification $\tilde{E}_{C(\tilde{X})}$ to obtain the properness of $\tilde{E}$. Properness of $\tilde{E}_{C(\tilde{X})}$ is proved in Part IV. □

Corollary 3.1. Let the assumptions of Theorem 3.1 be fulfilled. The spectral measure ρ of the family A is proper. Moreover, the restriction $\rho \restriction \eta$ is proper for an arbitrary $\eta \in \mathcal{C}_\sigma(\mathbb{C}^X)$ satisfying $\rho(\eta) \neq 0$.

With the aid of Theorem 3.1 we come up with a refined version of the projection spectral theorem [14]. Its previous version ([3] and [5]) did not state the formula $\pi \subset \operatorname{supp} \rho$ in the general case. This inclusion was only known to hold if the operators A_x (except for not more than a countable number of them) were bounded. Using Theorem 3.1 also makes the proof of the projection spectral theorem simpler. More precisely, it's now possible to avoid several difficult arguments connected to the compactification method used in the previously known version of the proof.

Theorem 3.2 (The projection spectral theorem). *Assume $A = (A_x)_{x \in X}$ to be a family of commuting normal operators connected with quasinuclear rigging (2.1) in a standard way. Let E be the JRI, ρ the spectral measure, $P(\lambda(\cdot))$ the generalized projection of A. If D is a separable projective limit of Hilbert spaces, then there exists a set $\pi \subset g(A) \cap \operatorname{supp} \rho$ of full outer ρ-measure such that the following holds: if $\lambda(\cdot) \in \pi$ then the range $\operatorname{Ran}(P(\lambda(\cdot)))$ consists of generalized joint eigenvectors of A corresponding to $\lambda(\cdot)$ and $\operatorname{Tr}(P(\lambda(\cdot))) = 1$.*

The representation

$$A_x = \int_\pi \lambda(x) dE_\pi(\lambda(\cdot)),$$
$$\operatorname{Dom}(A_x) = \left\{ f \in H_0 \,\middle|\, \int_\pi |\lambda(x)|^2 d(E_\pi(\lambda(\cdot))f, f)_{H_0} < \infty \right\} \quad (x \in X) \tag{3.1}$$

takes place (one needs to place a complex bar over $\lambda(x)$ to obtain a formula for A_x^*). If A_x, A_x^* are understood to be restricted to $\operatorname{Dom}(A_x) \cap H_+$, then $dE_\pi(\lambda(\cdot))$ can be replaced by $P(\lambda(\cdot)) d\rho_\pi(\lambda(\cdot))$.

Acknowledgments

I express my gratitude to Prof. Yu.M. Berezansky for setting the problem, for his support and productive discussions. I am also grateful to Prof. S.A. Ovsienko for supporting me and my research.

References

[1] Yu.M. Berezansky, On expansions in eigenfunctions of general selfadjoint differential operators, *Dokl. Akad. Nauk SSSR* **108**(3) (1956) 379–382.

[2] Yu.M. Berezansky, *Self-adjoint Operators in Spaces of Functions of Infinitely Many Variables* (AMS, 1986); (in Russian, Naukova Dumka, 1978).

[3] Yu.M. Berezansky, The Projection Spectral Theorem, *Russian Mathematical Surveys* **39**(4) (1984) 1–62.

[4] Yu.M. Berezansky, Poisson Measure as the Spectral Measure of Jacobi Field, *Infinite Dim. Anal., Quantum Prob. Related Topics* **1**(1) (2000) 121–139.

[5] Yu.M. Berezansky, Yu.G. Kondratiev, *Spectral Methods in Infinite-Dimensional Analysis* (Kluwer, 1995); (in Russian, Naukova Dumka, 1988).

[6] L. Gårding, Application of the Theory of Direct Integrals of Hilbert Spaces to some Integral and Differential Operators, *Lect. Ser. Inst. Fluid Dynam. a. Appl. Math.* **11** (1954).

[7] I.M. Gelfand, A.G. Kostyuchenko, On expansions in eigenfunctions of differential and other operators, *Dokl. Acad. Nauk SSSR* **103**(3) (1955) 349–352.

[8] G.I. Kats, On expansions in eigenfunctions of selfadjoint operators, *Dokl. Akad. Nauk SSSR* **119**(1) (1958) 19–22.

[9] Yu.G. Kondratiev, E.W. Lytvynov, Operators of Gamma White Noise Calculus, *Infinite Dim. Anal., Quantum Prob. Related Topics* **3**(3) (2000) 303–335.

[10] Yu.G. Kondratiev, J.L. Da Silva, L. Streit, G.F. Us, Analysis on Poisson and Gamma Spaces, *Infinite Dim. Anal., Quantum Prob. Related Topics* **1**(1) (1998) 91–117.

[11] K. Maurin, Entwicklung positiv definiter Kerne nach Eigendistributionen. Differenzierbarkeit der Spektralfunktion eines hypoelliptischen Operators, *Bull. Acad. Polon. Sci. Ser. math. astr. et phys.* **6**(3) (1958) 149–155.

[12] M.A. Naimark, *Normed Algebras* (Noordhoff, 1972); (in Russian, Nauka, 1968).

[13] A.D. Pulemyotov, On the Support of the Product of Resolutions of Identity, *Methods Func. Anal. Topology* **7**(2) (2001) 75–80.

[14] A.D. Pulemyotov, Support of a Joint Resolution of Identity and the Projection Spectral Theorem, submitted to *Infinite Dimensional Analysis, Quantum Probability and Related Topics.*

Artem Pulemyotov
Dept. of Algebra
Faculty of Mathematics and Mechanics
Kyiv National T. Shevchenko University
64 Volodymyrska Str.
Kyiv 01033, Ukraine
e-mail: pulen@i.kiev.ua

Operator Theory:
Advances and Applications, Vol. 149, 525–538

Dynamics and Stabilization of an Elastic Tape Moving Axially Between Two Sets of Rollers

David L. Russell

Abstract. We consider a thin elastic tape moving axially between two sets of rollers with speed $c > 0$ and subject to a positive tension. We derive linear and nonlinear equations of motion and demonstrate instability of the nominal equilibrium state for c sufficiently large in the case of the linear model. We then proceed to show that such instability can be overcome with active boundary control synthesized via feedback on boundary data. Well-posedness of the uncontrolled system and the closed loop controlled system are both demonstrated.

1. Setting; equations of motion

We consider a thin tape composed of an elastic material moving axially between two sets of rollers located at $x = 0$ and $x = 1$; the tape moves to the left with speed $c > 0$. We treat the tape as an Euler-Bernoulli beam with density per unit length $\rho > 0$ and bending moment $A > 0$. We suppose in addition that the strip is subject to a tension $\kappa > 0$. The equilibrium position of the tape is the x-axis $y = 0$ and displacements from that equilibrium are represented by $y = \eta(x,t)$. In deriving the equations of motion we will assume any needed smoothness properties of $\eta(x,t)$; we will consider questions of well-posedness and regularity in a later section.

Focusing attention on a material particle moving with the tape we see that its transverse velocity is $\frac{\partial \eta}{\partial t} - c\frac{\partial \eta}{\partial x}$. Using this in the standard kinetic energy expression the total energy may be taken to be

$$\mathcal{E}(\eta(\cdot,t)) = \frac{1}{2}\int_0^L E(x,t)\,dx,$$

$$E(x,t) = \rho\left(\left(\frac{\partial \eta}{\partial t} - c\frac{\partial \eta}{\partial x}\right)(x,t)\right)^2 + A\left(\frac{\partial^2 \eta}{\partial x^2}('')\right)^2 + \kappa\left(\frac{\partial \eta}{\partial x}('')\right)^2. \quad (1.1)$$

In some situations it is desirable to reinforce the restoring forces to ensure finite amplitude. For this purpose the quadratic energy $\mathcal{E}$ is modified by adding a quartic

Received by the editors December 30, 2002. Submitted by J. A. Ball.
Mathematics Subject Classification (2000). 93D15, 93C20, 74H45, 74H55, 74H60.
Key words and phrases. Axial motion, elastic tape, boundary control, distributed parameter system, stability .

term to the quadratic integrand to obtain

$$\begin{aligned}\tilde{E}(x,t) &= \rho\left(\left(\frac{\partial\eta}{\partial t} - c\frac{\partial\eta}{\partial x}\right)(x,t)\right)^2 \\ &+ A\left(\frac{\partial^2\eta}{\partial x^2}('')\right)^2 + \kappa\left(\frac{\partial\eta}{\partial x}('')\right)^2 + A_1\left(\frac{\partial\eta}{\partial x}('')\right)^4,\end{aligned}$$

where A_1 is a positive constant. Standard variational techniques then give, as the equation of motion, the nonlinear partial differential equation

$$\rho\left(\frac{\partial^2\eta}{\partial t^2} - 2c\frac{\partial^2\eta}{\partial t\partial x} + c^2\frac{\partial^2\eta}{\partial x^2}\right) + A\frac{\partial^4\eta}{\partial x^4} - \left(\kappa + 6A_1\left(\frac{\partial\eta}{\partial x}\right)^2\right)\frac{\partial^2\eta}{\partial x^2} = 0. \tag{1.2}$$

Equations of this type have been studied extensively, but in a somewhat different context, in [14], [15], [16]. In the present paper, however, we study only the linear case corresponding to $A_1 = 0$ and the quadratic energy (1.1), in which case (1.2) is replaced by the simpler linear partial differential equation

$$\rho\left(\frac{\partial^2\eta}{\partial t^2} - 2c\frac{\partial^2\eta}{\partial t\partial x}\right) + A\frac{\partial^4\eta}{\partial x^4} + \left(c^2\rho - \kappa\right)\frac{\partial^2\eta}{\partial x^2} = 0. \tag{1.3}$$

For fixed rollers the boundary conditions take the form

$$\eta(0,t) = \frac{\partial\eta}{\partial x}(0,t) = \eta(L,t) = \frac{\partial\eta}{\partial x}(L,t) = 0. \tag{1.4}$$

We do not specifically assume that the tape has a periodic structure, as it would, e.g., in the case of a band saw blade, nor do we assume the tape to be wrapped around the rollers at the ends, for which case it is known (see, e.g., [2]) that the radius of curvature of the rollers has a definite influence on system dynamics.

With the boundary conditions (1.4) $\frac{\partial^4\eta}{\partial x^4}$ and $-\frac{\partial^2\eta}{\partial x^2}$ are unbounded, positive self-adjoint operators and we are led to expect that for any given tension $\kappa > 0$ instabilities may arise for a sufficiently large speed $c > 0$. Indeed, this has been known for some time; see, e.g., [18], [19]. In §2., with the aid of the symbolic manipulation capabilities of MATLAB$^{(R)}$, we confirm this with the aid of a "prototype system", of which the above can be considered a perturbation, and show precisely how the bifurcation takes place. Then, in §3., we propose a family of closed loop boundary control mechanisms, realized mathematically in the form of modified boundary conditions, to overcome such instabilities. Since the feedback process involves only boundary data at the boundary point where the control operates, complications inherent in attempting to measure the entire distributed tape state are avoided; these results should be compared with those in [3]. Some suggestions are advanced for optimization within this class of closed loop controls. Finally, in §4., we discuss the well-posedness and regularity of both the uncontrolled system (1.3), (1.4) and the controlled system as developed in §3.

We should note that a considerable body of literature on axially moving continua is in existence. In addition to the papers noted above the reader is referred to the comprehensive study of energy and conserved functionals for such systems in [11] and the treatment of nonlinear dynamics and bifurcations in this context appearing in [10] and in [5]. A mechanical engineering study of a class of related problems for moving machinery belts appears in [1]. Any investigation of the history of the subject would need to include the 1965 study [8] of band saw vibrations. There are many more important contributions which we are unable to cite individually here. The corresponding problems for an axially moving elastic string have been widely researched; see, e.g., [17], [4], [7] and [20]. Online exploration of MathSciNet, Cambridge Scientific Abstracts, etc., with key words such as "axially moving ! (beam, strip, tape, string,...)" is highly recommended.

2. Instability for certain values of c and κ

Our purpose in this section is to prove the following result.

Theorem 2.1. *For $\sqrt{\frac{c^2\rho-\kappa}{A}}-\frac{2\pi}{L}>0$ and $\kappa>0$, both sufficiently small, the system (1.3), (1.4) has solutions growing exponentially in time and thus is unstable.*

Remark. Indeed, we will go further than the result stated and show how the "pitchfork" bifurcation to instability takes place as $\sqrt{\frac{c^2\rho-\kappa}{A}}$ passes through the value $\frac{2\pi}{L}$.

Preliminaries. Assuming a solution of the form $\eta(x,t) = e^{\lambda t}\hat{\eta}(x)$ we obtain the eigenvalue equation

$$\frac{A}{\rho}\frac{d^4\hat{\eta}}{dx^4} + \left(c^2 - \frac{\kappa}{\rho}\right)\frac{d^2\hat{\eta}}{dx^2} - 2c\lambda\frac{d\hat{\eta}}{dx} + \lambda^2 = 0 \tag{2.1}$$

along with the obvious counterparts of (1.4). We first normalize the length to unity by setting $x = L\tilde{x}$. This has the effect of replacing A by $\tilde{A} = \frac{A}{L^4}$, c by $\tilde{c} = \frac{c}{L}$ and κ by $\tilde{\kappa} = \frac{\kappa}{L^2}$ in (2.1). Renaming $\tilde{x}$ as x again, setting

$$\hat{c} = \frac{\tilde{c}\rho^{\frac{1}{2}}}{\tilde{A}^{\frac{1}{2}}},\quad \hat{\lambda} = \frac{\lambda\rho^{\frac{1}{2}}}{\tilde{A}^{\frac{1}{2}}},\quad \hat{\kappa} = \frac{\tilde{\kappa}}{\tilde{A}}$$

in (2.1) and then dropping the ^ for the sake of economy of notation, we have

$$\frac{d^4\hat{\eta}}{dx^4} + \left(c^2 - \kappa\right)\frac{d^2\hat{\eta}}{dx^2} - 2c\lambda\frac{d\hat{\eta}}{dx} + \lambda^2\hat{\eta} = 0. \tag{2.2}$$

Confining attention to cases where $c^2 > \kappa$, we now set $\tilde{c}^2 = c^2 - \kappa$ in (2.2), and then drop the ~, to obtain

$$\frac{d^4\hat{\eta}}{dx^4} + c^2\frac{d^2\hat{\eta}}{dx^2} - 2c\lambda\sqrt{1+\frac{\kappa}{c^2}}\frac{d\hat{\eta}}{dx} + \lambda^2 = 0. \tag{2.3}$$

With these changes the inequality in the statement of the theorem now becomes $c > 2\pi$. Since the theorem applies to cases where $\frac{\kappa}{c^2}$ is small, (2.3) can be interpreted as a perturbation of the prototype equation

$$\frac{d^4\hat{\eta}}{dx^4} + c^2\frac{d^2\hat{\eta}}{dx^2} - 2c\lambda\frac{d\hat{\eta}}{dx} + \lambda^2 = 0 \tag{2.4}$$

corresponding to a prototype partial differential equation

$$\frac{\partial^2 \eta}{\partial t^2} - 2c\frac{\partial^2 \eta}{\partial t \partial x} + c^2\frac{\partial^2 \eta}{\partial x^2} + \frac{\partial^4 \eta}{\partial x^4} = 0. \tag{2.5}$$

The characteristic equation for the differential equation (2.4) takes the form

$$r^4 + (c\,r - \lambda)^2 = 0. \tag{2.6}$$

Setting $\hat{r} = r - \frac{\lambda}{c}$ we have

$$\left(\hat{r} + \frac{\lambda}{c}\right)^4 + c^2\hat{r}^2 = 0 \Rightarrow \left(\hat{r} + \frac{\lambda}{c}\right)^2 = \pm\, i\, c\hat{r},$$

or

$$\hat{r}^2 + \left(\frac{2\lambda}{c} \pm i\,c\right)\hat{r} + \frac{\lambda^2}{c^2} = 0,$$

which gives us the two families of solutions

$$\hat{r} = -\frac{\lambda}{c} - \frac{i\,c}{2} \pm i\,c\sqrt{1 - \frac{4\lambda\, i}{c^2}}, \quad \hat{r} = -\frac{\lambda}{c} - \frac{i\,c}{2} \pm i\,c\sqrt{1 + \frac{4\lambda\, i}{c^2}}.$$

Going back to $r = \hat{r} + \frac{\lambda}{c}$ we have the four families of solutions of (2.6), parametrized by c and λ,

$$r_1(c,\lambda) = -\frac{i\,c}{2} + \frac{i\,c}{2}\sqrt{1 - \frac{4\lambda\, i}{c^2}}, \quad r_2(c,\lambda) = -\frac{i\,c}{2} - \frac{i\,c}{2}\sqrt{1 - \frac{4\lambda\, i}{c^2}},$$

$$r_3(c,\lambda) = \frac{i\,c}{2} + \frac{i\,c}{2}\sqrt{1 - \frac{4\lambda\, i}{c^2}}, \quad r_4(c,\lambda) = \frac{i\,c}{2} - \frac{i\,c}{2}\sqrt{1 - \frac{4\lambda\, i}{c^2}}. \tag{2.7}$$

Suppressing c and λ as arguments of the r_k we have, initially, the general solution of (2.6) in the form

$$\hat{\eta}(x) = b_1 e^{r_1 x} + b_2 e^{r_2 x} + b_3 e^{r_3 x} + b_4 e^{r_4 x}. \tag{2.8}$$

Taking B to be the four-dimensional vector with components b_k and applying the boundary conditions (cf. (1.4)) $\hat{\eta}(0) = 0,\ \hat{\eta}'(0) = 0,\ \hat{\eta}(1) = 0,\ \hat{\eta}'(1) = 0$, we have

$$\tilde{\mathbf{M}}B = 0, \quad \tilde{\mathbf{M}} = \tilde{\mathbf{M}}(c,\lambda) = \begin{pmatrix} 1 & 1 & 1 & 1 \\ r_1 & r_2 & r_3 & r_4 \\ e^{r_1} & e^{r_2} & e^{r_3} & e^{r_4} \\ r_1\, e^{r_1} & r_2\, e^{r_2} & r_3\, e^{r_3} & r_4\, e^{r_4} \end{pmatrix} \tag{2.9}$$

and the eigenvalue equation takes the form

$$\tilde{d}(c,\lambda) \equiv \det \tilde{\mathbf{M}}(c,\lambda) = 0. \tag{2.10}$$

Proposition 2.1. *The function $\tilde{d}(c,\lambda)$ is even with respect to λ.*

This is, of course, to be expected since replacement of λ by $-\lambda$ and x by $1-x$ leaves (2.4) invariant. However, a somewhat more satisfying proof is obtained from the readily observed relations $r_1(c,\lambda) = -r_4(c,-\lambda)$ and $r_2(c,\lambda) = -r_3(c,-\lambda)$. Forming $\tilde{\mathbf{M}}(c,-\lambda)$ and using these relations in all four columns, followed by multiplication of each column by the appropriate exponential e^{r_k} to remove terms e^{-r_k}

and then carrying out an even number of column interchanges and an even number of row interchanges, we return to the matrix $\tilde{\mathbf{M}}(c,\lambda)$ again. In the process we use the fact that the sum of the $r_k(c,\lambda)$ is identically zero.

Proof of Theorem 2.1. As we may readily ascertain, $\tilde{d}(c,0) \equiv 0$. This has no effect on determination of eigenvalues for the boundary value problem because the b_k obtained with $\lambda = 0$ are such that $\hat{\eta}(x) \equiv 0$; $\lambda = 0$ is not an eigenvalue. The problem is that both $r_1(c,\lambda)$ and $r_4(c,\lambda)$ reduce to 0 when $\lambda = 0$ and, in addition, $\lim_{\lambda\to 0}(r_1(c,\lambda) - r_4(c,\lambda))/\lambda = 0$. To remove 0 as an extraneous root of $\tilde{d}(c,\lambda) = 0$ while maintaining evenness with respect to λ we slightly modify our basis of solutions by replacing $e^{r_1 x}$, $e^{r_4 x}$, respectively, by

$$e^{r_1 x} + \lambda^{-2}\left(e^{r_1 x} - e^{r_4 x}\right),\ e^{r_4 x} - \lambda^{-2}\left(e^{r_1 x} - e^{r_4 x}\right).$$

Then the matrix $\tilde{\mathbf{M}}(c,\lambda)$ is replaced by

$$\mathbf{M}(c,\lambda) = \tilde{\mathbf{M}}(c,\lambda)\left(\mathbf{I} + \lambda^{-2}\mathbf{N}\right),\quad \mathbf{N} = \begin{pmatrix} 1 & 0 & 0 & -1 \\ 0 & 1 & 0 & 0 \\ 0 & 0 & 1 & 0 \\ -1 & 0 & 0 & 1 \end{pmatrix} \tag{2.11}$$

whose determinant will be denoted by $d(c,\lambda)$, readily seen to be an analytic function of c and λ in a neighborhood of the point $(\hat{c},0)$ for any $\hat{c} > 0$. Using MATLAB$^{(R)}$ symbolic manipulation we can see that near such a point we have

$$d(c,\lambda) = d_0(c) + d_2(c)\,\lambda^2 + \mathcal{O}(|\lambda|^4).$$

We verify easily (even without computer symbolic manipulation that $d(2\pi,0) = d_0(2\pi) = 0$. With symbolic manipulation we can go further and see that for c near 2π

$$d_0(c) = \frac{4}{\pi}(c-2\pi) - \frac{4}{\pi^2}(c-2\pi)^2 + \left(\frac{3}{\pi^3} - \frac{2}{3\pi}\right)(c-2\pi)^3 + \mathcal{O}\left(|c-2\pi|^4\right).$$

The corresponding expression for $d_2(c)$ takes the form

$$d_2(c) = -\frac{3}{2\pi^4} + \left(\frac{9}{2\pi^5} - \frac{1}{(56\pi - 64\pi^5)\,\pi^6}\right)(c-2\pi) + \mathcal{O}\left(|c-2\pi|^2\right).$$

Thus when $c = 2\pi$ we have $d(2\pi,0) = 0$ and $d(2\pi,\lambda) < 0$ for λ near 0. Fixing $\hat{\lambda} > 0$ such that $d(2\pi,\hat{\lambda}) < 0$ we will continue to have $d(c,\hat{\lambda}) < 0$ for c sufficiently near 2π. Since $\frac{d}{dc}d_0(2\pi) = \frac{4}{\pi}$ we have $d_0(c,0) > 0$ for $c - 2\pi > 0$ and small and $d_0(c,0) < 0$ for $c - 2\pi < 0$ and small. It follows that, for $c - 2\pi > 0$ and sufficiently small there will be a real number $\lambda_+(c)$with $0 < \lambda_+(c) < \hat{\lambda}$ such that $d(c,\lambda_+(c)) = d(c,-\lambda_+(c)) = 0$. A comparable analysis shows that there is a $\tilde{\lambda} > 0$ such that for $c - 2\pi < 0$ and sufficiently small there will be a real number $\lambda_-(c)$ with $0 < \lambda_-(c) < \tilde{\lambda}$ such that $d(c,i\,\lambda_-(c)) = d(c,-i\,\lambda_-(c)) = 0$. Since the exponential functions $e^{r_k(c,\lambda_\pm(c))}$ are distinct the corresponding non-zero solutions of $\tilde{\mathbf{M}}(c,\lambda_\pm(c))\,B = 0$ lead to non-zero solutions $\phi_\pm(c,x)$ of the eigenvalue value problem (2.4) and (1.4) for values of c as stipulated and it is easy to see that this remains true for the modified equation $\mathbf{M}(c, \lambda_\pm(c))\, B = 0$. These results imply the existence of a solution $e^{\lambda_+(c)t}\phi_+(c,x)$ of the prototype system (2.5), (1.4) growing exponentially in time and thus inconsistent with stability of that system.

Finally we note that for a fixed $c > 2\pi$ as described, solutions $\lambda_\pm(c,\kappa)$ of the equation $d(c,\kappa,\lambda)$ continuously approaching $\lambda_\pm(c)$ as $\kappa \downarrow 0$ will be obtained, where $d(c,\kappa,\lambda)$ is the determinant of the matrix $\mathbf{M}(c,\kappa,\lambda)$ replacing $\mathbf{M}(c,\lambda)$ for small values of $\kappa > 0$. Thus the instability conclusion extends to the original system (1.3), (1.4) and the proof is complete. □

3. A boundary stabilization procedure

Having shown that the system (1.3), (1.4) becomes unstable for certain values of the axial velocity parameter c, our objective now is to show that stability can be regained with the use of active boundary controls synthesized in terms of collocated boundary data. Our emphasis on use of boundary data alone for feedback purposes is, of course, prompted by the realization that data from the moving belt at points other than boundary points is ordinarily unavailable without the introduction of advanced measurement techniques unlikely to be economically viable in most application contexts.

A *controlled endpoint* of our moving tape system will be taken to be either $x = 0$ or $x = L$ with boundary conditions of the form

$$\frac{\partial^2 \eta}{\partial x^2} = u, \quad \frac{\partial^3 \eta}{\partial x^3} = v \tag{3.1}$$

replacing the boundary conditions (1.4) at the endpoint in question. In this article a *linear collocated boundary feedback relation* takes the form

$$u = j_1 \frac{\partial \eta}{\partial t} + j_2 \frac{\partial \eta}{\partial x} + j_3 \frac{\partial^2 \eta}{\partial t \partial x}, \quad v = k_1 \frac{\partial \eta}{\partial t} + k_2 \frac{\partial \eta}{\partial x} + k_3 \frac{\partial^2 \eta}{\partial t \partial x} + k_4 \frac{\partial^2 \eta}{\partial x^2} + k_5 \frac{\partial^3 \eta}{\partial t \partial x^2}, \tag{3.2}$$

where the j_m, k_m indicated are real coefficients and the right-hand sides of (3.2) are evaluated at the same endpoint as are the left-hand sides of (3.1). We then have the following result.

Theorem 3.1. *For $c^2\rho - \kappa > 0$ there exist feedback coefficients as in (3.2) active at the endpoint $x = 0$, with $x = L$ corresponding to a fixed endpoint (cf. (1.4)) for which the system energy (1.1) is non-increasing. There are no such feedback coefficients active at $x = L$ with $x = 0$ a fixed endpoint.*

Proof. We consider a stability analysis based on the energy expression (1.1) for the system (1.3), (1.4). We set

$$\zeta(x,t) = \frac{\partial \eta}{\partial t}(x,t) - c\frac{\partial \eta}{\partial x}(x,t); \tag{3.3}$$

then the expression for $E(x,t)$ in (1.1) becomes

$$E(x,t) = \rho\zeta(x,t)^2 + A\left(\frac{\partial^2\eta}{\partial x^2}(\text{"})\right)^2 + \kappa\left(\frac{\partial\eta}{\partial x}(\text{"})\right)^2. \tag{3.4}$$

Rewriting (3.3) and (1.3) we have the first-order system in t consisting of the equations

$$\frac{\partial\eta}{\partial t} = c\,\frac{\partial\eta}{\partial x} + \zeta, \tag{3.5}$$

$$\rho\,\frac{\partial\zeta}{\partial t} = -A\frac{\partial^4\eta}{\partial x^4} - \left(c^2\rho - \kappa\right)\frac{\partial^2\eta}{\partial x^2} + \rho c\,\frac{\partial\zeta}{\partial x}. \tag{3.6}$$

Then for sufficiently smooth solutions we have, using (1.1), (3.5) and (3.6),

$$\begin{aligned}
\frac{d\mathcal{E}}{dt} &= \int_0^L\left[\zeta\,\rho\frac{\partial\zeta}{\partial t} + A\frac{\partial^2\eta}{\partial x^2}\frac{\partial^3\eta}{\partial t\partial x^2} + \kappa\frac{\partial\eta}{\partial x}\frac{\partial^2\eta}{\partial t\partial x}\right]dx \\
&= \int_0^L\left[\zeta\left(-A\frac{\partial^4\eta}{\partial x^4} + \kappa\frac{\partial^2\eta}{\partial x^2} + \rho c\frac{\partial\zeta}{\partial x}\right) + \left(A\frac{\partial^4\eta}{\partial x^4} - \kappa\frac{\partial^2\eta}{\partial x^2}\right)\frac{\partial\eta}{\partial t}\right]dx \\
&+ \left[A\left(\frac{\partial^2\eta}{\partial t\partial x}\frac{\partial^2\eta}{\partial x^2} - \frac{\partial\eta}{\partial t}\frac{\partial^3\eta}{\partial x^3}\right) + \kappa\frac{\partial\eta}{\partial x}\frac{\partial\eta}{\partial t}\right]\Bigg|_0^L = \int_0^L c\left(A\frac{\partial^4\eta}{\partial x^4} - \kappa\frac{\partial^2\eta}{\partial x^2}\right)\frac{\partial\eta}{\partial x}\,dx \\
&+ \left[\frac{\rho c}{2}\zeta^2 + A\left(\frac{\partial^2\eta}{\partial t\partial x}\frac{\partial^2\eta}{\partial x^2} - \frac{\partial\eta}{\partial t}\frac{\partial^3\eta}{\partial x^3}\right) + \kappa\frac{\partial\eta}{\partial x}\left(\zeta + c\frac{\partial\eta}{\partial x}\right)\right]\Bigg|_0^L = \left[\frac{\rho c}{2}\left(\frac{\partial\eta}{\partial t} - c\frac{\partial\eta}{\partial x}\right)^2\right. \\
&\left. - \frac{\kappa c}{2}\left(\frac{\partial\eta}{\partial x}\right)^2 + \kappa\frac{\partial\eta}{\partial x}\frac{\partial\eta}{\partial t} + Ac\frac{\partial^3\eta}{\partial x^3}\frac{\partial\eta}{\partial x} - \frac{Ac}{2}\left(\frac{\partial^2\eta}{\partial x^2}\right)^2 + A\left(\frac{\partial^2\eta}{\partial t\partial x}\frac{\partial^2\eta}{\partial x^2} - \frac{\partial\eta}{\partial t}\frac{\partial^3\eta}{\partial x^3}\right)\right]\Bigg|_0^L
\end{aligned} \tag{3.7}$$

With

$$Y^* \equiv \left(\frac{\partial\eta}{\partial t}\ \ \frac{\partial\eta}{\partial x}\ \ \frac{\partial^3\eta}{\partial x^3}\ \ \frac{\partial^2\eta}{\partial x^2}\ \ \frac{\partial^2\eta}{\partial t\partial x}\right)$$

and Y the corresponding column vector, the boundary quadratic form (3.7) is $Y^*\mathbf{Q}Y$, wherein

$$\mathbf{Q} = \begin{pmatrix}
\frac{\rho c}{2} & \frac{\kappa}{2} - \frac{c^2\rho}{2} & -\frac{A}{2} & 0 & 0 \\
\frac{\kappa}{2} - \frac{c^2\rho}{2} & \frac{c^3\rho}{2} - \frac{c\kappa}{2} & \frac{Ac}{2} & 0 & 0 \\
-\frac{A}{2} & \frac{Ac}{2} & 0 & 0 & 0 \\
0 & 0 & 0 & -\frac{Ac}{2} & \frac{A}{2} \\
0 & 0 & 0 & \frac{A}{2} & 0
\end{pmatrix}.$$

The linear feedback relations ensuring that

$$Y(L,t)^*\mathbf{Q}\,Y(L,t) - Y(0,t)^*\mathbf{Q}\,Y(0,t) \le 0$$

is to be obtained by imposing boundary conditions at $x = 0$ or $x = L$, in the form of linear restrictions on the components of Y, leaving the other boundary point one of fixed type. The lower right-hand 2×2 block of $\mathbf{Q}$ has eigenvalues

$-\frac{Ac}{4} \pm \frac{Ac}{4}\sqrt{1+\frac{4}{c^2}}$, which are real with opposite sign. On the other hand the 3×3 upper left-hand block of $\mathbf{Q}$ has, when $c^2\rho > \kappa$, positive trace $\frac{c\rho}{2} + \frac{c^3\rho}{2} - \frac{c\kappa}{2}$ and negative determinant $-\frac{A^2 c\kappa}{8}$, implying that it has two positive eigenvalues and one negative eigenvalue. It follows that for $c^2\rho > \kappa$ the matrix $\mathbf{Q}$ has three positive eigenvalues, $\mu_1,\ \mu_2,\ \mu_3$, and two negative eigenvalues, $\mu_{-1},\ \mu_{-2}$ with corresponding mutually orthogonal eigenvectors $\Phi_1,\ \Phi_2,\ \Phi_3$ and $\Phi_{-1},\ \Phi_{-2}$.

To make $Y(L,t)^*\mathbf{Q}\,Y(L,t) < 0$ when $Y(L,t) \neq 0$ we would need to restrict $Y(L,t)$ to a subspace on which this form is negative. But, since $\mathbf{Q}$ has three positive eigenvalues, such a subspace cannot have dimension greater than 2, requiring the imposition of at least three linear conditions on $Y(L,t)$, which exceeds the number of boundary conditions we can impose at $x = L$. We conclude that we cannot find a linear feedback control in the form of boundary conditions based on the right-hand ("pitching end") $x = L$ stabilizing the system. So we leave the rollers at $x = L$ fixed, corresponding to

$$\eta(L,t) = \frac{\partial \eta}{\partial x}(L,t) = 0,$$

and work with the left-hand ("catching end") rollers at $x = 0$.

Since $-\mathbf{Q}$ has just two positive eigenvalues, $-\mu_{-1}$ and $-\mu_{-2}$, and three negative eigenvalues $-\mu_1,\ -\mu_2,\ -\mu_3$, subspaces on which $-Y(0,t)^*\mathbf{Q}\,Y(0,t) < 0$, lying within the negative cone of that quadratic form, are three-dimensional, requiring the imposition of just 2 linear conditions on $Y(0,t)$. We denote generic boundary conditions of this sort by

$$B_1^*\,Y(0,t) = 0, \ \ B_2^*\,Y(0,t) = 0; \tag{3.8}$$

they clearly include the case $B_1 = \Phi_{-1}$, $B_2 = \Phi_{-2}$, which we will refer to as "eigenvector boundary conditions". Further, with the boundary conditions $\eta(L,t) = \frac{\partial \eta}{\partial x}(L,t) = 0$ the form $Y(L,t)^*\mathbf{Q}\,Y(L,t)$ reduces to $-\frac{Ac}{2}\left(\frac{\partial^2 \eta}{\partial x^2}(L,t)\right)^2 \leq 0$. So with these "fixed roller" conditions at $x = L$ and appropriate boundary conditions – for example, the "eigenvector" boundary conditions just noted – at $x = 0$, we can achieve $\frac{d}{dt}\mathcal{E}(\eta(\cdot,t)) \leq 0$. This completes the proof. □

Additional work would be required, of course, to show that we can achieve an exponential rate of decay for the energy, corresponding to uniform exponential asymptotic stability for the closed loop system; we intend to undertake that study in a later article.

Study of the eigenvector boundary conditions is facilitated by the block diagonal structure of the matrix $\mathbf{Q}$. Let us take μ_{-1} to be the negative eigenvalue of the lower right 2×2 block of that matrix and μ_1 to be the positive eigenvalue of that block. The eigenvector boundary condition $\Phi_{-1}^* Y(0,t) = 0$ corresponding to the eigenvalue μ_{-1} leads to a boundary condition of the form $\frac{\partial^2 \eta}{\partial x^2}(0,t) = \frac{A}{2\mu_1}\frac{\partial^2 \eta}{\partial t \partial x}(0,t)$, which may be interpreted in terms of torque induced rotation of the roller assembly at $x = 0$. The eigenvector boundary condition associated with the 3×3 block may be interpreted in terms of lateral force induced vertical motion of the same roller assembly but its analytic expression as a linear equation involving $\frac{\partial\eta}{\partial t}(0,t)$, $\frac{\partial\eta}{\partial x}(0,t)$ and $\frac{\partial^3\eta}{\partial x^3}(0,t)$, although obtainable via symbolic manipulation, appears to be too complicated to be very enlightening. We note, however, that the structure of this 3×3 block ensures that the eigenvector boundary condition $\Phi^*_{-2}Y(0,t) = 0$, and more generally the associated boundary conditions $B_2^* Y(0,t) = 0$ of (3.8), are solvable for $\frac{\partial^3\eta}{\partial x^3}$ and thus consistent with (3.1), (3.2).

In a given application we may suppose boundary conditions at the "catching end" initially given in the open loop form (3.1). Equipped with an initial stabilizing feedback relation of the type expressed in (3.8), perhaps the eigenvector boundary conditions, a given quadratic performance criterion and initial state and disturbance statistics, one could undertake step by step performance optimization within the whole admissible class of boundary feedback relations at $x = 0$. Such a project would involve extension to distributed parameter systems of *restricted feedback* optimal control methods as treated, e.g., in [12] and elsewhere.

4. Semigroup framework for the fixed and boundary stabilized systems

Letting $c_1 = \frac{2c}{\rho}$, $A_1 = \frac{A}{\rho}$, $c_2 = c^2 - \frac{\kappa}{\rho}$ and now defining (note that this is different from the definition in the preceding section!) $\zeta = \frac{\partial\eta}{\partial t}$, (1.3) corresponds to the first-order system

$$\frac{\partial}{\partial t}\begin{pmatrix}\eta\\ \zeta\end{pmatrix} = \mathcal{D}\begin{pmatrix}\eta\\ \zeta\end{pmatrix} = \begin{pmatrix}0 & I\\ -A_1\frac{\partial^4}{\partial x^4} - c_2\frac{\partial^2}{\partial x^2} & c_1\frac{\partial}{\partial x}\end{pmatrix}\begin{pmatrix}\eta\\ \zeta\end{pmatrix} \equiv \begin{pmatrix}0 & I\\ -D_4 - D_2 & D_1\end{pmatrix}\begin{pmatrix}\eta\\ \zeta\end{pmatrix}. \tag{4.1}$$

The state space $\mathcal{Y}$ includes pairs $\eta \in H_0^2(0,L)$, $\zeta \in L^2(0,L)$, where $H_0^2(0,L)$ denotes the closed subspace of the Sobolev space $H^2(0,L)$ consisting of functions $y(x)$ such that $y(0) = y'(0) = y(L) = y'(L) = 0$. Initially we use the inner product and norm

$$\left\langle\begin{pmatrix}\eta\\ \zeta\end{pmatrix},\begin{pmatrix}\hat\eta\\ \hat\zeta\end{pmatrix}\right\rangle \equiv \int_0^L\left(A_1\frac{\partial^2\eta}{\partial x^2}\overline{\frac{\partial^2\hat\eta}{\partial x^2}} + \zeta\bar{\hat\zeta}\right)dx,\quad \left\|\begin{pmatrix}\eta\\ \zeta\end{pmatrix}\right\| \equiv \left\langle\begin{pmatrix}\eta\\ \zeta\end{pmatrix},\begin{pmatrix}\eta\\ \zeta\end{pmatrix}\right\rangle^{\frac{1}{2}} \tag{4.2}$$

corresponding to the standard energy inner product and norm for the Euler-Bernoulli beam. The domain of $\mathcal{D}$ consists of pairs $\eta \in H^4(0,L)$, $\zeta \in H^2(0,L)$, both satisfying the boundary conditions just indicated.

Theorem 4.1. *With the indicated domain, the operator $\mathcal{D}$ generates a strongly continuous semigroup in $\mathcal{Y}$.*

Proof. With the indicated boundary conditions $D_4 = A_1\frac{\partial^4}{\partial x^4}$ is a positive self-adjoint operator and we can form its positive self-adjoint square root $D_4^{\frac{1}{2}}$ whose domain consist of functions in $H^2(0,L)$ satisfying the indicated boundary conditions. Then making the change of coordinates

$$\eta = D_4^{-\frac{1}{2}}\omega$$

we obtain the modified system

$$\frac{\partial}{\partial t}\begin{pmatrix}\omega\\ \zeta\end{pmatrix} = \mathcal{D}_1\begin{pmatrix}\omega\\ \zeta\end{pmatrix} = \begin{pmatrix}0 & D_4^{\frac{1}{2}}\\ -D_4^{\frac{1}{2}} - D_2 D_4^{-\frac{1}{2}} & D_1\end{pmatrix}\begin{pmatrix}\omega\\ \zeta\end{pmatrix} \tag{4.3}$$

with state space $\mathcal{Y}_1 = L^2(0,L) \times L^2(0,L)$ endowed with the usual inner product and norm. The domain of $\mathcal{D}_1$ is then $H_0^2(0,L) \times H_0^2(0,L)$.

It is well known that

$$\mathcal{D}_0 \equiv \begin{pmatrix}0 & D_4^{\frac{1}{2}}\\ -D_4^{\frac{1}{2}} & 0\end{pmatrix}$$

generates a strongly continuous group of isometries in $\mathcal{Y}_1$ which is also, in this specialized isometric sense, a strongly continuous contraction semigroup, the antisymmetric operator $\mathcal{D}_0$ being, in a comparably specialized sense, dissipative. In $H_0^2(0,L)$ the operator $D_1 = c_1\frac{\partial}{\partial x}$ is also dissipative in the sense of being antisymmetric. From the work in [13] we know that with the indicated boundary conditions

$$\frac{\partial^2}{\partial x^2} = (I+P)\,D_4^{\frac{1}{2}},$$

where P is a bounded operator on $L^2(0,L)$. Thus the term $-D_2D_4^{-\frac{1}{2}}$ in (4.3) represents a bounded operator. Further, for $u \in H_0^2(0,L)$ and $\epsilon > 0$, we have

$$\int_0^L |\frac{\partial u}{\partial x}|^2 dx = -\int_0^L \frac{\partial^2 u}{\partial x^2}\bar{u}dx \le \frac{\epsilon^2}{2}\int_0^L |\frac{\partial^2 u}{\partial x^2}|^2 dx + \frac{1}{2\epsilon^2}\int_0^L |u|^2 dx$$

$$\le \frac{1}{2}\left(\epsilon\,(1+\|P\|)\,\|D_4^{\frac{1}{2}}u\| + \frac{1}{\epsilon}\|u\|\right)^2 .$$

Using this result with a standard result on perturbation of contraction semigroups (see, e.g., [9]) we conclude that

$$\mathcal{D}_2 = \begin{pmatrix}0 & D_4^{\frac{1}{2}}\\ -D_4^{\frac{1}{2}} & D_1\end{pmatrix}$$

generates a strongly continuous contraction semigroup on $\tilde{\mathcal{Y}}$; then the bounded perturbation $-D_2D_4^{-\frac{1}{2}}$ applied to $\mathcal{D}_2$ gives the operator $\mathcal{D}_1$ which is then seen, by the standard bounded perturbation theorem, to generate a C_0 semigroup, not in general a contraction semigroup, on $\mathcal{Y}_1$, from which we conclude that $\mathcal{D}$ generates a strongly continuous semigroup on $\mathcal{Y}$. This completes the proof. □

Remark. It is preferable, of course, to work with an inner product and norm equivalent to the energy expression (1.1). Since the transformation

$$\begin{pmatrix} \hat{\eta} \\ \hat{\zeta} \end{pmatrix} = \begin{pmatrix} I & 0 \\ -\frac{c}{\rho}\frac{\partial}{\partial x} & I \end{pmatrix} \begin{pmatrix} \eta \\ \zeta \end{pmatrix} \tag{4.4}$$

is bounded and boundedly invertible on $H_0^2(0,L) \times L^2(0,L)$ and since

$$\int_0^L \left(A \frac{\partial^2 \eta}{\partial x^2} \overline{\frac{\partial^2 \hat{\eta}}{\partial x^2}} \right) dx$$

is bounded above and below by

$$\int_0^L \left(A \frac{\partial^2 \eta}{\partial x^2} \overline{\frac{\partial^2 \hat{\eta}}{\partial x^2}} + \kappa \frac{\partial \eta}{\partial x} \overline{\frac{\partial \hat{\eta}}{\partial x}} \right) dx$$

on $H_0^2(0,L)$ we can achieve that objective, with no change in the semigroup generation properties of $\mathcal{D}$, by taking $\mathcal{Z}$ to consist of the same function pairs as those in $\mathcal{Y}$ but endowed with the inner product and norm

$$\left\langle \begin{pmatrix} \eta \\ \zeta \end{pmatrix}, \begin{pmatrix} \hat{\eta} \\ \hat{\zeta} \end{pmatrix} \right\rangle_{\mathcal{Z}} \equiv \int_0^L \left(A \frac{\partial^2 \eta}{\partial x^2} \overline{\frac{\partial^2 \hat{\eta}}{\partial x^2}} + \kappa \frac{\partial \eta}{\partial x} \overline{\frac{\partial \hat{\eta}}{\partial x}} + \rho \left(\zeta - c \frac{\partial \eta}{\partial x} \right) \overline{\left(\hat{\zeta} - c \frac{\partial \hat{\eta}}{\partial x} \right)} \right) dx$$

$$\left\| \begin{pmatrix} \eta \\ \zeta \end{pmatrix} \right\|_{\mathcal{Z}} \equiv \left\langle \begin{pmatrix} \eta \\ \zeta \end{pmatrix}, \begin{pmatrix} \eta \\ \zeta \end{pmatrix} \right\rangle_{\mathcal{Z}}^{\frac{1}{2}}. \tag{4.5}$$

Combining the above results with standard theorems on strongly continuous semigroups, including the regularity of solutions corresponding to initial states lying in the domain of the semigroup generator, we have the needed background for the results of §2.

Next we need to establish that the stabilizing boundary conditions identified in §3 result in a well-posed system, corresponding to a strongly continuous semigroup. First we define

$$\tilde{\mathcal{Y}} \equiv \left\{ \eta \in H^2(0,L),\ \zeta \in L^2(0,L) \,\middle|\, \eta(L) = \eta'(L) = 0 \right\}$$

with the inner product and norm (4.2), and we define $\tilde{\mathcal{Z}}$ to consist of the same functions but now equipped with the inner product and norm (4.5).

Theorem 4.2. *Let $\tilde{\mathcal{D}}$ have the same form as $\mathcal{D}$ in (4.1) but with domain modified from that of $\mathcal{D}$ by taking the boundary conditions at $x = 0$ to have the form indicated in (3.8). Then $\tilde{\mathcal{D}}$ generates a strongly continuous contraction semigroup in $\tilde{\mathcal{Y}}$ which is a contraction semigroup in $\tilde{\mathcal{Z}}$.*

Proof. The analysis of §3 is readily adapted to show that the operator $\tilde{\mathcal{D}}$ is *dissipative* in $\tilde{\mathcal{Z}}$ in the sense of that term as defined in [9] (p. 14). Applying the Lumer-Phillips theorem as developed there, we see that $\tilde{\mathcal{D}}$ generates a strongly continuous semigroup of contractions in $\tilde{\mathcal{Z}}$ if there is some $\lambda_0 > 0$ such that the range of $\tilde{\mathcal{D}} - \lambda_0 \mathcal{I}$, $\mathcal{I}$ denoting the identity operator, is all of $\tilde{\mathcal{Z}}$. In view of the boundedness of (4.4), it is sufficient to show that, given any $F \in \tilde{\mathcal{Y}}$ we can find Y in the domain of $\tilde{\mathcal{D}}$ in $\tilde{\mathcal{Y}}$ such that

$$\left(\tilde{\mathcal{D}} - \lambda_0 \mathcal{I}\right) Y = F. \tag{4.6}$$

Letting $Y = (\eta, \zeta)^*$, $F = (\phi, \psi)^*$, this becomes (cf. (4.1))

$$\zeta - \lambda_0 \eta = \phi, \quad -A_1 \frac{d^4\eta}{dx^4} - c_2 \frac{d^2\eta}{dx^2} + c_1 \frac{d\zeta}{dx} - \lambda_0 \zeta = \psi.$$

On substitution of the first equation into the second this results in

$$A_1 \frac{d^4\eta}{dx^4} + c_2 \frac{d^2\eta}{dx^2} - c_1 \lambda_0 \frac{d\eta}{dx} + \lambda_0^2 \eta + \psi + \lambda_0 \phi - c_1 \frac{d\phi}{dx} = 0. \tag{4.7}$$

Since $\phi \in H^2(0, L)$, $\psi \in L^2(0, L)$, retracing the redefinition of coefficients at the beginning of this section, (4.7) is equivalent to (cf. (2.2))

$$\frac{d^4\eta}{dx^4} + \left(c^2 - \frac{\kappa}{\rho}\right) \frac{d^2\eta}{dx^2} - 2c\lambda \frac{d\eta}{dx} + \lambda^2 \eta + f = 0, \quad f \in L^2(0, L). \tag{4.8}$$

Let $\hat{\eta}_1(x)$ and $\hat{\eta}_2(x)$ be the solutions of the homogeneous counterpart of (4.8) corresponding to initial conditions

$$\hat{\eta}_1(L) = \hat{\eta}_1'(L) = \hat{\eta}_1'''(L) = 0, \quad \hat{\eta}_1''(L) = 1,$$

$$\hat{\eta}_2(L) = \hat{\eta}_2'(L) = \hat{\eta}_2''(L) = 0, \quad \hat{\eta}_2'''(L) = 1,$$

respectively, let $\hat{Y}_1(x) = \begin{pmatrix} \hat{\eta}_1(x) \\ \lambda_0 \hat{\eta}_1(x) \end{pmatrix}$, $\hat{Y}_2(x) = \begin{pmatrix} \hat{\eta}_2(x) \\ \lambda_0 \hat{\eta}_2(x) \end{pmatrix}$, and (cf. (3.8)) define the 2×2 Gram matrix

$$\mathbf{G} = \begin{pmatrix} B_1^* \hat{Y}_1(0) & B_1^* \hat{Y}_2(0) \\ B_2^* \hat{Y}_1(0) & B_2^* \hat{Y}_2(0) \end{pmatrix}.$$

Then it will be familiar that (4.8) can fail to have a unique solution if and only if $\det \mathbf{G} = 0$. If $\det \mathbf{G} = 0$, however, there will be constants a_1, a_2, not both zero, such that

$$\hat{\eta}(x) \equiv a_1 \hat{\eta}_1(x) + a_2 \hat{\eta}_2(x),$$

which clearly cannot be identically zero, satisfies $\hat{\eta}(L) = \hat{\eta}'(L) = 0$ and (cf. (3.8)), with $\hat{Y}(x) = \begin{pmatrix} \hat{\eta}(x) \\ \lambda_0 \hat{\eta}(x) \end{pmatrix}$,

$$B_1^* \hat{Y}(0) = 0, \; B_2^* \hat{Y}(0) = 0. \tag{4.9}$$

(In other words, the Fredholm alternative applies to (4.6).) Then, since we have assumed that $\lambda_0 > 0$,

$$\eta(x, t) = e^{\lambda_0 t} \hat{\eta}(x)$$

is a solution of our controlled system growing exponentially with increasing t; the energy (1.1) then also increases exponentially with increasing t, contradicting the way in which B_1 and B_2 were selected in §3. It follows that (4.6) is uniquely solvable for all $f \in L^2(0, L)$ and we conclude that the "closed loop operator" $\tilde{\mathcal{D}}$ generates a strongly continuous semigroup on $\tilde{\mathcal{Y}}$; a contraction semigroup on $\tilde{\mathcal{Z}}$. □

References

[1] Abrate, S.: *Vibration of belts and belt drives*, Mech. and Mach. Thy. 27 (1992), pp. 645–659

[2] Asokanthan, S.F., and S.T. Ariaratnam: *Flexural instabilities in axially moving bands*, J. Vib. & Acoust. Trans. ASME, 116 (1994), pp. 275–279

[3] Cai, L.: *Active control of axially moving continua*, Proc. Soc. Ph.-Opt. Instr. Eng. 2620 (1995), pp. 780–785

[4] Chung, C.H., and C.A. Tan: *Active vibration control of the axially moving string by wave cancellation*, J. Vib. & Acoust. Trans. ASME, 117 (1995), pp. 49–55

[5] Galip Ulsey, A.: *Vibration and control in rotating or translating elastic systems*, J. Dyn. Syst., Meas. & Contr. 106 (1984), pp. 6–14

[6] Lagnese, J.E.: *Recent progress in exact boundary controllability and uniform stabilizability of thin beams and plates*, in *Distributed Parameter Control Systems*, G. Chen, E. B. Lee, W. Littman and L. Markus, Eds., Marcel Dekker, New York, 1991, pp. 61–111

[7] Lee, S.-Y., and C.D. Mote, Jr.: *Vibrational control of an axially moving string by boundary control*, J. Dyn. Syst., Meas. & Contr. 118 (1996), pp. 66–74

[8] Mote, C.D., Jr.: *A study of bandsaw vibrations*, J. Franklin Inst. 279 (1965), pp 430–444

[9] Pazy, A.: *Semigroups of Linear Operators and Applications to Partial Differential Equations*, Springer-Verlag, New York, 1983

[10] Pellicano, F. and F. Vestroni: *Nonlinear dynamics and bifurcations of an axially moving beam*, J. Vib. & Acoust. Trans. ASME, 122 (2000), pp. 21–30

[11] Renshaw, A.A., C.D. Rahn, J.A. Wickert and C.D. Mote: *Energy and conserved functionals for axially moving materials*, J. Vib. & Acoust. Trans. ASME, 120 (1998), pp. 634–636

[12] Russell, D.L.:*Mathematics of Finite Dimensional Control Systems: Theory and Design*, Marcel Dekker, New York, 1979

[13] Russell, D.L.: *On the positive square root of the fourth derivative operator*, Quarterly of Applied Mathematics, 46 (1988), pp. 751–773

[14] Russell, D.L.: *Freely Propagating Waves in a Supported Nonlinear Elastic Beam*, in *Nonsmooth/ Nonconvex Mechanics*, D.Y. Gao, R.W. Ogden and G. E. Stavroulakis, Eds., Kluwer Academic Publishers, 2000

[15] Russell, D.L.: *Forced Propagation of Steady State Waves in a Supported Elastic Beam*, Discrete and Continuous Dynamical Systems, 6(2001), pp. 337–348

[16] Russell, D.L. and L.W. White: *Static buckling in a supported nonlinear elastic beam*, in *Control of Nonlinear Distributed Parameter Systems*, G. Chen, I. Lasiecka and J. Zhou, Eds., Marcel Dekker, New York, 2001

[17] Shahruz, S.M., and D.A. Kurmaji: *Vibration suppression of a non-linear axially moving string by boundary control*, J. Sound Vib. 201 (1997), pp. 145–152

[18] Wickert, J.A.: *Non-linear vibration of a travelling tensioned beam*, Int. J. Non-Linear Mech. 27 (1992), pp. 503–517

[19] Wickert, J.A., and C.D. Mote, Jr.: *Classical vibrational analysis of axially moving continua*, J. Appl. Mech. 57 (1990), pp. 738–744

[20] Ying, S. and C.A. Tan: *Active vibration control of the axially moving string using space feedforward and feedback controllers*, J. Vib. & Acoust. Trans. ASME, 118 (1996), pp. 306–312

David L. Russell
Department of Mathematics
Virginia Polytechnic Institute and State University
Blacksburg, VA 24061-123, USA
e-mail: `russell@math.vt.edu`

Operator Theory:
Advances and Applications, Vol. 149, 539–551

Weyl-Titchmarsh Matrix Functions and Spectrum of Non-selfadjoint Dirac Type Equation

L. Sakhnovich

Abstract. In this paper we consider the matrix non-selfadjoint equations of the Dirac type with the corresponding boundary condition at $x = 0$. We investigate the connection between the spectrum of the equation and the singularities of the corresponding Weyl-Titchmarsh matrix function. Using the Weyl-Titchmarsh matrix function, we construct the Green matrix function.

1. Introduction

Let us consider the matrix equation

$$\frac{dY}{dx} = [izj + \xi(x)]\,Y, \quad 0 \leq x < \infty, \tag{1}$$

where $\xi(x)$, j are $n \times n$ matrices ($n = 2m$) and

$$j = \begin{bmatrix} I_m & 0 \\ 0 & -I_m \end{bmatrix}, \quad \xi(x) = \begin{bmatrix} 0 & q(x) \\ -q^*(x) & 0 \end{bmatrix}. \tag{2}$$

The matrix function $Y(x, z)$ has the form $Y(x, z) = col[Y_1(x, z), Y_2(x, z)]$, where $Y_1(x, z)$ and $Y_2(x, z)$ are $m \times m$ matrix functions. We associate with system (1) the following boundary condition

$$Y_1(0) = Y_2(0). \tag{3}$$

In our paper we investigate the connection between the spectrum of equation (1) and the singularities of the corresponding Weyl-Titchmarsh function (W-T function).

We suppose that

$$||\xi(x)|| \leq M \quad ,0 \leq x < \infty \tag{4}$$

Separately we consider the case ,when $||\xi(x)||\to 0$ as $x\to\infty$. For this case we proved the following assertions.

Received by the editors October 22, 2002; revised March 21, 2003. Submitted by L. Rodman.
Mathematics Subject Classification (2000). 34B20.
Key words and phrases. Differential systems, Green functions, Short-range potentials.

1. The set of non-real eigenvalues of the boundary problem (1),(3) coincides with the set of poles of the corresponding W-T matrix function $v(z)$ in the domain $Imz\neq 0$.
2. The set of poles of $v(z)(Imz\neq 0)$ is either finite or countable.
3. The set of poles of $v(z)(Imz\neq 0)$ is finite if the potential $\xi(x)$ is short- range, i.e., $\xi(x) = 0$, when $x > \ell$.

With the help of W-T matrix function we have constructed the Green matrix function of equation (1),(3). For the selfadjoint case the Green matrix function was constructed in the paper [6] (see also [8]).

We remark that the properties of W-T matrix functions of non-selfadjoint systems (1), (3) play an important role in the theory of the non-linear integrable equations (see [3], [7]). A number of results referring to the spectrum of the system (1), (3) is contained in the articles [2], [13].

2. Principal notions

Equation (1) can be reduced to the form

$$\frac{dU}{dx} = [iz\,J + B(x)]\,U, \qquad 0 \leq x < \infty, \tag{5}$$

where

$$U(x,z) = TY(x,z), B(x) = T\xi(x)\,T^*, \tag{6}$$

$$T = \frac{1}{\sqrt{2}}\begin{bmatrix} I_m & -I_m \\ I_m & I_m \end{bmatrix}, \quad Tj\,T^* = J = \begin{bmatrix} 0 & I_m \\ I_m & 0 \end{bmatrix}. \tag{7}$$

From (2) and (6) we deduce that

$$B(x) = \begin{bmatrix} a(x) & b(x) \\ -b(x) & -a(x) \end{bmatrix}, \tag{8}$$

where

$$a(x) = [-q(x) + q^*(x)]\,/2, \qquad b(x) = [q(x) + q^*(x)]\,/2.$$

We represent $U(x,z)$ in the form $U(x,z) = col[U_1(x,z), U_2(x,z)]$, where $U_1(x,z)$ and $U_2(x,z)$ are $m \times m$ matrix functions. The boundary condition (3) takes the form

$$U_1(0,z) = 0. \tag{9}$$

Definition 1. [9] The $m \times m$ holomorphic matrix function $v(z)$ is called the *Weyl-Titchmarsh matrix function* for system (5) if the inequality

$$\int_0^\infty [I_m, v^*(z)]\,W^*(x,z)\,W(x,z)\begin{bmatrix} I_m \\ v(z) \end{bmatrix} dx < \infty, \quad Imz < -M, \tag{10}$$

is fulfilled, where $W(x,z)$ is the $(2m) \times (2m)$ matrix solution of equation (5) and

$$W(0,z) = I_{2m}. \tag{11}$$

(Here $[I_m, v^*(z)]$ is the $m \times 2m$ block matrix.)

Remark 1. In (10) and in the sequel, given a positive semidefinite matrix A, we use the expression $A < \infty$ to mean $Tr\,A < \infty$. If A is a positive semi-definite matrix, we write $A \geq 0$.

Let us represent the matrix function $\mathfrak{A}(\ell, z) = W^*(\ell, \bar{z})$ in the block form

$$\mathfrak{A}(\ell, z) = \begin{bmatrix} a(\ell, z) & b(\ell, z) \\ c(\ell, z) & d(\ell, z) \end{bmatrix},$$

where all blocks are of order $m \times m$. Now we introduce the class of $m \times m$ matrix functions

$$\begin{aligned} v(\ell, z) = &[a(\ell, z)\,P(z) + b(\ell, z)\,Q(z)] \\ &[c(\ell, z)\,P(z) + d(\ell, z)\,Q(z)]^{-1}. \end{aligned} \tag{12}$$

Here the meromorphic $m \times m$ matrix functions $P(z)$ and $Q(z)$ are such that

$$\begin{cases} P^*(z)\,P(z) + Q^*(z)\,Q(z) > 0 \\ P^*(z)\,Q(z) + Q^*(z)\,P(z) \leq 0, \qquad Imz < -M. \end{cases} \tag{13}$$

Proposition 1. [9] *Let condition* (4) *be fulfilled. Then the limit*

$$v(z) = \lim_{\ell \to \infty} v(\ell, z) \tag{14}$$

exists, and the matrix function $v(z)$ *defined by relation* (14) *is the unique Weyl-Titchmarsh matrix function of the system* (1) *and* (3).

We note that the limit in the right side of (14) does not depend on the choice of $P(z)$ and $Q(z)$. We have the case of the limit Weyl point (see [9]).

3. Adjoint differential systems

Now we introduce the system

$$\frac{d\tilde{W}}{dx} = [iz\,J + B^\star(x)]\,\tilde{W}, \quad \tilde{W}(0, z) = I_n \tag{15}$$

adjoint to system (5). We start with the identity

$$\tilde{W}^\star(x, \bar{z}) J W(x, z) = J \tag{16}$$

which can be proved by differentiation and using the relations

$$W(0, z) = \tilde{W}(0, z) = I_n, \quad B(x)J = -JB(x) \tag{17}$$

It follows from(16) that

$$W(x, z) J \tilde{W}^\star(x, \bar{z}) = J \tag{18}$$

By $\tilde{v}(z)$ we denote the W-T matrix function of system (15). Let us represent $W(x, z)$ and $\tilde{W}(x, z)$ in the block form $W(x, z) = [Y_1(x, z), Y_2(x, z)]$, $\tilde{W}(x, z) = [\tilde{Y}_1(x, z), \tilde{Y}_2(x, z)]$, where $Y_k(x, z)$ and $\tilde{Y}_(x, z)$ are $2m \times m$ matrix functions ($k = 1, 2$). It follows from (10) that columns of matrix functions

$$Y(x, z) = Y_1(x, z) + Y_2(x, z) v(z) \tag{19}$$

and

$$\tilde{Y}(x,z) = \tilde{Y}_1(x,z) + \tilde{Y}_2(x,z)\tilde{v}(z) \tag{20}$$

belong to $L_n^2(0,\infty)$.

4. Green functions

Now we consider the equation

$$-iJ\frac{du}{dx} = [zI_n - iJB(x)]u + f(x), \quad 0 \le x < \infty, \tag{21}$$

where $u(x,z)$ and $f(x)$ are $n\times 1$ vector functions and

$$u(x,z) = col[u_1(x,z), u_2(x,z)].$$

Here $u_k(x,z)$ are $m\times 1$ vector functions. The boundary condition has the form

$$u_1(0,z) = 0 \tag{22}$$

Let us introduce the Green function $g(x,t,z)$:

$$g(x,t,z) = iY(x,z)\tilde{Y}_2^\star(t,\bar{z}), \quad t < x, \tag{23}$$

$$g(x,t,z) = iY_2(x,z)\tilde{Y}^\star(t,\bar{z}), \quad t > x. \tag{24}$$

The corresponding Green operator has the form

$$(Gf)(x) = \int_0^\infty g(x,t,z)f(t)dt, \tag{25}$$

where $f(t)\in L_n^2(0,\infty)$.

Remark 2. The integral in (25) converges absolutely because the columns of $\tilde{Y}(x,\bar{z})$ belong to $L_n^2(0,\infty)$.

We suppose further that relation

$$\tilde{v}^\star(\bar{z}) = -v(z) \tag{26}$$

is fulfilled. We shall show later that this relation is true if condition (4) holds.

Proposition 2. *If relation* (26) *is valid , then the function*

$$u(x,z) = (Gf)(x) \tag{27}$$

satisfies equation (21) *and boundary condition* (22).

Proof. It is easy to see that condition (22) is fulfilled. From relation (22) we deduce that

$$-iJ\frac{du}{dx} = [zI_n - iJB(x)]u + J[Y(x,z)\tilde{Y}_2^\star(x,\bar{z}) + Y_2(x,z)\tilde{Y}^\star(x,\bar{z})]f(x). \tag{28}$$

Using (26) and (28) we have

$$J[Y(x,z)\tilde{Y}_2^\star(x,\bar{z})+Y_2(x,z)\tilde{Y}^\star(x,\bar{z})] = JW(x,z)\times\begin{bmatrix} 0 & I_m \\ I_m & v(z)+\tilde{v}^\star(\bar{z})\end{bmatrix}\times\tilde{W}^\star(x,\bar{z}). \tag{29}$$

The right side of relation (29) can be written in the form

$$JW(x,z)J\tilde{W}^{\star}(x,\bar{z}) = I_n. \tag{30}$$

The proposition follows directly from (26) and (28)–(30). □

Lemma 4.1. *Let f(t) belongs to $L_n^2(0,A)$ and $f(t)=0$,when $t>A$. Then function $u(x)=(Gf)(x)$ belongs to $L_n^2(0,\infty)$.*

Proof. If $x>A$ then

$$u_A(x) = iY(x,z)\int_0^A \tilde{Y}_2^{\star}(t,\bar{z})f(t)dtY(x,z)D_A, \tag{31}$$

where D_A is $m\times 1$ matrix independent of x. Hence $u_A(x)$ belongs to $L_n^2(0,\infty)$ and the lemma is proved. □

Theorem 1. *Let the relations*

$$||q(x)||\leq M \quad ,0\leq x<\infty \tag{32}$$

and

$$\tilde{v}^{\star}(\bar{z}) = -v(z), \quad |Imz|>M \tag{33}$$

be fulfilled. Then G is a bounded operator in the space $L_n^2(0,\infty)$ for $|Imz|>M$.

Proof. We give a modification of a proof in [6] pertaining to the selfadjoint case. Let $\Delta_k = [0,t_k], 0<t_k<\infty$ and $t_k\to\infty$. We consider the vector functions $f(t)\in L_n^2(0,\infty)$ and f_k defined as $f_k(t)=f(t)$ for $0\leq t\leq t_k$ and$f_k(t)=0$ for $t>t_k$. If $0\leq x\leq t_k$, then we obtain the equality

$$u_k(x) = iY(x,z)\int_0^x \tilde{Y}_2^{\star}(t,\bar{z})f(t)dt + iY_2(x,z)\int_x^{t_k} \tilde{Y}^{\star}(t,\bar{z})f(t)dt. \tag{34}$$

It follows from (31) and (34) that

$$u(x,z)-u_k(x,z) = iY_2(x,z)\int_{t_k}^{\infty} \tilde{Y}^{\star}(t,\bar{z})f(t)dt. \tag{35}$$

Hence, the sequence $u_k(x,z)$ converges uniformly to $u(x,z)$ on compact intervals. If $x>t_k$, then in view of (31) we have

$$u_k(x,z) = Y(x,z)D_k, \tag{36}$$

i.e., $u_k(x,z)\in L_n^2(0,\infty)$ and

$$u_k^{\star}(x,z)Ju_k(x,z) = D_k^{\star}Y^{\star}(x,z)JY(x,z)D_k. \tag{37}$$

The vector function $Y(x,z)$ belongs to $L_n^2(0,\infty)$. Hence, there exists such a sequence $t_k\to\infty$ that

$$Y(t_k,z)\to 0. \tag{38}$$

We deduce from relations (36) and (38) that

$$u_k^{\star}(t_k,z)\to 0. \tag{39}$$

Using equation (28), we obtain

$$u_k^\star(x,z)(-iJ)u_k(x,z)I_0^{t_k} \quad + \quad (z-\bar{z})\int_0^{t_k} u_k^\star(x,z)u_k(x,z)dx \tag{40}$$

$$= \quad \int_0^{t_k}[u_k^\star(x,z)f_k(x) - f_k^\star(x)u_k^\star(x,z)]dx - I_A,$$

where $I_A = i\int_0^{t_k} u_k^\star(x,z)A(x)u_k(x,z)dx$ and $A(x) = JB(x) + B^\star(x)J$. Using (39) and (40) as $t_k{\to}\infty$ we have

$$\int_0^\infty [I_n + iA(x)/(z-\bar{z})]u_k(x,z)dx \tag{41}$$

$$= [1/(z-\bar{z})]\int_0^\infty [u_k^\star(x,z)f_k(x) - f_k^\star(x)u_k^\star(x,z)]dx. \tag{42}$$

If $|Imz| > M/(1-\delta), \quad 0 < \delta < 1$ then the inequality

$$[I_n + iA(x)/(z-\bar{z})]{\geq}\delta I_n \tag{43}$$

is true. It follows from (41), (42) and Cauchy inequality that

$$\int_0^\infty u_k^\star(x,z)u_k(x,z)dx{\leq}\frac{1}{(Imz\delta)^2}\int_0^\infty f_k^\star(x)f_k(x)dx. \tag{44}$$

The theorem is proved. □

Corollary 1. *Let conditions of Theorem* 1 *be fulfilled. Then the boundary problem* (21), (22) *has unique solution*

$$u(x) = (Gf)(x). \tag{45}$$

Proof. It follows directly from Theorem 1 that $u(x,z)$ defined by (44) satisfies (21) and (22). For uniqueness it is enough to prove that $u(x,z) = 0$, when $u(x,z){\in}L_n^2(0,\infty)$ and satisfies relations (21) and (22), where $f(x) = 0$. From $u(x,z){\in}L_n^2(0,\infty)$ we deduce that

$$u(x,z) = Y(x,z)D, \tag{46}$$

where D is $m\times 1$ vector independent of x. Using relations

$$Y_1(0,z) = \begin{bmatrix} I_m \\ 0 \end{bmatrix}, \qquad Y_2(0,z) = \begin{bmatrix} 0 \\ I_m \end{bmatrix}$$

and (19), (45) we have $u_1(0,z) = D$. Hence, in view of (22) we deduce that $D = 0$, i. e., $u(x,z) = 0$. The corollary is proved. □

5. Short-range potentials

Let us consider equation (1), (2), where $q(x)$ is a short-range potential, i.e.,

$$||q(x)||{\leq}M \quad (0{\leq}x<\infty), \quad q(x) = 0 \quad (x{\geq}\ell). \tag{47}$$

Definition 2. [9] The $m \times m$ holomorphic matrix function $\phi(z)$ is called Weyl-Titchmarsh matrix function for system (1) with boundary condition

$$y_2(0) = 0, \tag{48}$$

if the inequality

$$\int_0^\infty [\phi^*(z), I_m]\, w^*(x,z)\, w(x,z) \begin{bmatrix} \phi(z) \\ I_m \end{bmatrix} dx < \infty, \quad Imz < -M \tag{49}$$

is fulfilled, where $w(x,z)$ is the $(2m) \times (2m)$ matrix solution of equation (1) and

$$w(0,z) = I_{2m}. \tag{50}$$

In view of (48) we have

$$[w_{11}(\ell,z)\phi(z) + w_{12}(\ell,z)]e^{i(x-\ell)z} \in L_n^2(0,\infty), \tag{51}$$

$$[w_{21}(\ell,z)\phi(z) + w_{22}(\ell,z)]e^{-i(x-\ell)z} \in L_n^2(0,\infty). \tag{52}$$

It follows from (46) and (50) , (51) that

$$\phi(z) = -{w_{21}}^{-1}(\ell,z)w_{22}(\ell,z), \quad Imz > 0, \tag{53}$$

$$\phi(z) = -{w_{11}}^{-1}(\ell,z)w_{12}(\ell,z), \quad Imz < 0. \tag{54}$$

Lemma 5.1. *Let conditions* (46) *be fulfilled. Then the relations*

$$\phi(z) = -\phi^\star(\bar{z})^{-1} \quad , v(z) = -v^\star(\bar{z})^{-1} \tag{55}$$

are true.

Proof. We start with the identity

$$w^\star(x,z)w(x,\bar{z}) = I_n, \tag{56}$$

which can be proved by differentiation. It follows from (55) that

$$w(x,z)w^\star(x,\bar{z}) = I_n. \tag{57}$$

We deduce from (56) that

$${w_{11}}^{-1}(\ell,z)w_{12}(\ell,z) = -{w_{21}}^\star(\ell,\bar{z})w_{22}^\star(\ell,\bar{z})^{-1} \tag{58}$$

Now using (52), (53) and (57) we obtain the first relation of (54). The second relation of (54) we obtain from the first one and from equality (see [12]):

$$v(z) = [I_m + \phi(z)][\phi(z) - I_m]^{-1}. \tag{59}$$

The lemma is proved. □

Corollary 2. *Let conditions* (46) *be fulfilled. Then all singular points of* $\phi(z)$ *and* $v(z)$ $(Imz \neq 0)$ *are poles.*

Lemma 5.2. *Let conditions* (46) *be fulfilled. Then the relation*

$$v^\star(\bar{z}) = -\tilde{v}(z), \quad Imz \neq 0 \tag{60}$$

is true.

Proof. According to Proposition 1 we have

$$v(z) = \lim_{r\to\infty} W_{11}{}^{\star}(r,\bar{z})W_{12}^{\star}(r,\bar{z})^{-1}, \quad Imz < -M, \tag{61}$$

$$\tilde{v}(z) = \lim_{r\to\infty} \tilde{W}_{21}^{\star}(r,\bar{z})\tilde{W}_{22}^{\star}(r,\bar{z})^{-1}, \quad Imz < -M. \tag{62}$$

Using the relation $JB^{\star}(x)J = B(x)$ we deduce that

$$\tilde{W}(r,z) = JW(r,z)J. \tag{63}$$

It follows from (60)–(62) that

$$v^{-1}(z) = \tilde{v}(z). \tag{64}$$

Relations (54) and (63) imply equality (59) in the domain $Imz < -M$. This equality is valid in the domain $Imz\neq 0$ as the matrix functions $v(z)$ and $\tilde{v}(z)$ are meromorphic. The lemma is proved. □

Proposition 3. *In case* (46) *the number of poles of W-T matrix function* $v(z)$ *in the domain* $Imz\neq 0$ *is finite.*

Proof. The matrix function $v(z)$ has no poles in the domain $Imz < -M$ (see [9]). Using (59) we deduce that $v(z)$ has no poles in the domain $|Imz| > M$. Now we use the following representation (see [11]):

$$w(x,z) = e^{izxj} + \int_{-x}^{x} e^{izt}N(x,t)dt, \tag{65}$$

where $N(x,t)$ is bounded matrix function in the domain $0\leq|t|\leq x\leq \ell$. It follows from (64) that

$$w(\ell,z) = e^{iz\ell j}[1+o(1)], \quad z\to\infty, \tag{66}$$

where $|Imz| > M$. From (64) and (65) we obtain that

$$\phi(z)\to 0, \quad -M\leq Imz < 0, \quad z\to\infty, \tag{67}$$

$$\phi(z)\to 0, \quad 0 < Imz\leq M, \quad z\to\infty. \tag{68}$$

Hence the following relations

$$v(z)\to -I_m, \quad -M\leq Imz < 0, \quad z\to\infty, \tag{69}$$

$$v(z)\to I_m, \quad 0 < Imz\leq M, \quad z\to\infty \tag{70}$$

are true. In view of (66) and (67) there exists L such that all the poles of $v(z)$ are contained in the domain $|Imz|\leq M, |Rez|\leq L$. It follows from (52), (53) and (58) that the poles of the matrix function $v(z)$ coincide with the zeroes of the entire function $det[w_{22}(\ell,z) + w_{21}(\ell,z)]$ in the domain $Imz > 0$ and with the zeroes of the entire function $det[w_{11}(\ell,z) + w_{12}(\ell,z)]$ in the domain $Imz < 0$. Consequently $v(z)$ has a finite number of the poles. The proposition is proved. □

6. Bounded potentials

In this section we suppose that condition (4) is fulfilled.

Proposition 4. *Let $\phi(z)$ and $\phi_1(z)$ be W-T matrix functions for the cases $\xi(x)$ of the form (2) and $\xi_1(x) = \overline{\xi(x)}$ respectively. Then the relation $\phi_1(z) = \overline{\phi(-\bar{z})}$ is true.*

Proof. Let $w(x,z)$ satisfy relations (1) and (49). Then the matrix function $w_1(x,z) = \overline{w(x,-\bar{z})}$ satisfies the equation

$$\frac{dw_1(x,z)}{dx} = izjw_1(x,z) + \xi_1(x)w_1(x,z). \tag{71}$$

Using relation $w(x,z)col[\phi(z), I_m] \in L_m^2(0,\infty)$ we have

$$\overline{w(x,-\bar{z})}col[\overline{\phi(-\bar{z})}, I_m] \in L_m^2(0,\infty),$$

i. e., $\phi_1(z) = \overline{\phi(-\bar{z})}$. The proposition is proved. □

Corollary 3. *If $q(x) = \overline{q(x)}$ then $\phi(z) = \overline{\phi(-\bar{z})}$.*

Proposition 5. *Let $\phi(z)$ and $\phi_2(z)$ be W-T matrix functions for the cases $\xi(x)$ of the form (2) and*

$$\xi_2(x) = \begin{pmatrix} 0 & -\overline{q^\star(x)} \\ \overline{q(x)} & 0 \end{pmatrix} \tag{72}$$

respectively. Then the relation $\phi_2(z) = \overline{\phi(\bar{z})}^{-1}$ is true.

Proof. The matrix function

$$w_2(x,z) = J\overline{w(x,\bar{z})}J \tag{73}$$

satisfies the equation

$$\frac{dw_2(x,z)}{dx} = izjw_2(x,z) + \xi_2(x)w_2(x,z). \tag{74}$$

Using relation (72), we obtain that

$$w_2(x,z)col[I_m, \overline{\phi(\bar{z})}] \in L_m^2(0,\infty), \quad \text{i. e.,} \quad w_2(x,z)col[\overline{\phi(\bar{z})}^{-1}, I_m] \in L_m^2(0,\infty).$$

Hence $\phi_2(z) = \overline{\phi(\bar{z})}^{-1}$. The proposition is proved. □

We can reformulate the previous results for $\phi(z)$ in terms of $v(z)$.

Proposition 6. *Let $v(z)$ and $v_1(z)$ be W-T matrix functions for the cases $\xi(x)$ of the form (2) and $\xi_1(x) = \overline{\xi(x)}$ respectively. Then the relation $v_1(z) = \overline{v(-\bar{z})}$ is true.*

Corollary 4. *If $q(x) = \overline{q(x)}$ then $v(z) = \overline{v(-\bar{z})}$.*

Proposition 7. *Let $v(z)$ and $v_2(z)$ be W-T matrix functions for the cases $\xi(x)$ of the form (2) and*

$$\xi_2(x) = \begin{pmatrix} 0 & -\overline{q^\star(x)} \\ \overline{q(x)} & 0 \end{pmatrix} \tag{75}$$

respectively. Then the relation $v_2(z) = -\overline{v(\bar{z})}$ is true.

From Proposition 7 we obtain the following assertion.

Corollary 5. *Let condition* (46) *be fulfilled. Then there exists one and only one W-T matrix function $v(z)$ in the domain $|Imz| > M$.*

Proposition 8. *Let condition* (46) *be fulfilled. Then the following relation*

$$v^\star(\bar{z}) = -\tilde{v}(z), \quad |Imz| > M \tag{76}$$

is true.

Proof. Let us denote by $\phi_r(z)$ and $v_r(z)$ the W-T matrix functions in case $q_r(x) = q(x)$, $0 \le x \le r$ and $q_r(x) = 0$, $r < x$. It follows from (14), (49) and (52), (53) that $\lim_{r\to\infty} \phi_r(z) = \phi(z)$, $|Imz| > M$. Hence we have $\lim_{r\to\infty} v_r(z) = v(z)$, $|Imz| > M$. From the last relation and (59) we deduce the relation (75). The proposition is proved. □

Remark 3. It follows from (75) that Theorem 1 is valid if condition (32) is fulfilled.

7. The case when $\|\xi(x)\|\to 0, \quad x\to\infty$

We separately consider the case when

$$\|\xi(x)\| \le M \quad (0 \le x < \infty), \quad \|\xi(x)\|\to 0, \quad x\to\infty \tag{77}$$

Proposition 9. *Let condition* (76) *be fulfilled. Then all singular points of W-T matrix function $\phi(z)$ of system* (1) *in the half-plane $Imz < 0$ are poles.*

Proof. Let $\epsilon > 0$. Then there exists σ such that

$$\|\xi(x)\| < \epsilon, \quad |x| \ge \sigma. \tag{78}$$

Using Definition 2 we have

$$\int_\sigma^\infty \begin{bmatrix} \phi_\sigma^\star(x,z) & I_m \end{bmatrix} w_\sigma^\star(x,z) w_\sigma(x,z) \begin{bmatrix} \phi_\sigma(z) \\ I_m \end{bmatrix} dx < \infty, \tag{79}$$

where $Imz < -\epsilon$ and $w_\sigma(x,z)$ is the matrix solution of equation (1), which satisfies the condition

$$w_\sigma(\sigma, z) = I_m. \tag{80}$$

It follows from (1),(2) and (79) that

$$w_\sigma(x,z) = w(x,z) A^{-1}(\sigma, z), \tag{81}$$

where $A(\sigma, z) = w(\sigma, z)$. Comparing (10) and (78) we deduce the inequality

$$\int_0^\infty \Phi^\star(\sigma, z) w_\sigma^\star(x, z) w_\sigma(x, z) \Phi(\sigma, z) dx < \infty. \tag{82}$$

Here

$$\Phi(\sigma, z) = A^{-1}(\sigma, z) \begin{bmatrix} \phi_\sigma(z) \\ I_m \end{bmatrix}. \tag{83}$$

We represent $A^{-1}(\sigma, z)$ in the following block form

$$A^{-1}(\sigma, z) = \begin{bmatrix} b_{11}(\sigma, z) & b_{12}(\sigma, z) \\ b_{21}(\sigma, z) & b_{22}(\sigma, z) \end{bmatrix}, \tag{84}$$

where all blocks are of $m \times m$ order. Relations (10) and (81)–(83) imply the equality

$$\phi(z) = [b_{11}(\sigma, z)\phi_\sigma(z) + b_{12}(\sigma, z)] \times [b_{21}(\sigma, z)\phi_\sigma(z) + b_{22}(\sigma, z)]^{-1} \tag{85}$$

The matrix functions $b_{ij}(\sigma, z)$ are entire. Hence in view of (84) all singular points of $\phi(z)$ in the domain $Imz < -\epsilon$ are poles. As ϵ is arbitrary, the proposition is proved. □

Corollary 6. *Let conditions (76) be fulfilled. Then all singular points of the corresponding W-T matrix function $\phi(z)$ in the domain $Imz \neq 0$ are poles.*

Using (58) we can reformulate the previous result for $\phi(z)$ in terms of $v(z)$.

Proposition 10. *Let relations (76) be fulfilled. Then all singular points of the corresponding W-T matrix function $v(z)$ in the domain $Imz \neq 0$ are poles.*

8. Green function (continuation)

Let us consider equation (21) with boundary condition (22). Under conditions (76) we shall construct Green functions and Green operators in the domain $Imz \neq 0$. It follows from (21) that

$$u(x, z) = e^{izxJ}(c(z) + \int_0^x e^{iztJ}[B(t)u(t, z) + iJf(t)]dt), \tag{86}$$

where in view of (22) we have $c(z) = col[0, c_2(z)]$.

Setting (see (6) and (7)) $\tilde{u}(x, z) = T^\star u(x, z), \tilde{f}(t) = T^\star f(t), \tilde{c}(z) = T^\star c(z)$ we deduce from (85) that

$$\tilde{u}(x, z) = e^{izxj}\left(\tilde{c}(z) + \int_0^x e^{-iztj}[\xi(t)\tilde{u}(t, z) + ij\tilde{f}(t)]dt\right). \tag{87}$$

Since $\tilde{u}(x, z) \in L^2_{2m}(0, \infty)$ and $\tilde{c}(z) = col[c_2(z), c_2(z)]$ the relation

$$c_2(z) = \int_0^\infty e^{izt}[q^\star(t)\tilde{u}_1(t, z) - if_2(t)]dt, Imz > 0 \tag{88}$$

is true. In the domain $Imz < 0$ we have

$$c_2(z) = -\int_0^\infty e^{izt}[q(t)\tilde{u}_2(t, z) + if_1(t)]dt. \tag{89}$$

Hence, equation (86) takes the form ($Imz > 0$)

$$\tilde{u}_1(x,z) = \int_0^\infty e^{iz(x+t)}[q^\star(t)\tilde{u}_1(t,z) - if_2(t)]dt + \int_0^x e^{iz(x-t)}[q(t)\tilde{u}_2(t,z) + if_1(t)]dt, \tag{90}$$

and

$$\tilde{u}_2(x,z) = \int_x^\infty e^{iz(t-x)}[q^\star(t)\tilde{u}_1(t,z) - if_2(t)]dt \tag{91}$$

System (89) and (90) can be written in the following form

$$\tilde{u}(x,z) = T_1(z)\tilde{u}(x,z) + T_2(z)f, \quad Imz > 0, \tag{92}$$

where $T_1(z)$ is compact ,$T_2(z)$ is bounded. The representation of type (91) is valid in the domain $Imz < 0$ too. We denote by G_0 the set of z ($Imz\neq 0$), for which the operator $I - T_1(z)$ is not invertible. From representation (91) and Gohberg's theorem [4] we deduce the following result.

Proposition 11. *Let conditions* (76) *be fulfilled. Then the following assertions are true.*

1. *The set G_0 hasn't limit points in the domain $Imz\neq 0$.*
2. *If $z_0\in G_0$, then z_0 is a pole of $[I - T_1(z)]^{-1}$*
3. *The spectrum of boundary problem* (21) *and* (22)*, which belongs to the domain $Imz\neq 0$ coincides with the set G_0.*

With the help of Proposition 11 we obtain the following assertion.

Theorem 2. *Let conditions* (76) *be fulfilled. Then the non-real spectrum of boundary problem* (21) *and* (22) *coincides with the set of poles of $v(z)$ in the domain $Imz\neq 0$.*

References

[1] Ablowitz, M.J. and Segur, H., *Solitons and the Inverse Scattering Transform*, SIAM, Philadelphia, 1981.

[2] Bronski, J.C., Semiclassical Eigenvalue Distribution of the Zakharov-Shabat Eigenvalue Problem, *Physica*, D97 (1996) 376–397.

[3] Bullough, R.K. and Caudrei, P.J.(eds.), *Solitons, Topics in Current Physics*, Springer-Verlag, 1980.

[4] Gohberg, I., On linear operators analytically depending on a parameter, *Dokl. Akad. Nauk SSSR*, 78, No. 4 (1951) 629–631 (Russian).

[5] Hille, E., *Lectures on Ordinary Differential Equations*, Addison-Wesley, 1969.

[6] Hinton, D.B. and Shaw, J.K., Hamiltonian Systems of Limit Point or Limit Circle Type with Both Endpoints Singular, *Journal of Differential Equations*, 30 (1983) 444–464.

[7] Li, Y. and McLaughlin, D.W., Morse and Melnikov Functions for NLS PDE's, *Comm. Math. Phys.* 162, (1994) 175–214.

[8] Marchenko, V.A., *Sturm-Liouville Operators and Applications*, Birkhäuser, Basel, (1986).

[9] Sakhnovich, A.L., Nonlinear Schrödinger Equation on Semi-Axis and Inverse Problem Associated with it, *Ukrainskii Math. Zhurnal* 42, No. 3, (1990) 356–363 (Russian).

[10] Sakhnovich, L.A., *Spectral Theory of Canonical Differential Systems. Method of Operator Identities*, Operator Theory, Adv. and Appl., 107, Birkhäauser, (1999).

[11] Sakhnovich, L.A., Spectral Analysis of Volterra Operators Given in the space $L^2_m(,\ell)$, *Ukrainskii Math. Zhurnal*, 2 (1964), 259–268 (Russian).

[12] Sakhnovich, L.A., Weyl-Titchmarsh Matrix Functions for Matrix Dirac Type Equations (non-selfadjoint case), *Inverse Problems*, 18, (2002), 1525–1536.

[13] Tkachenko, V., Non-selfadjoint Dirac operators with finite-band spectra, *Integral Equations and Operator Theory*, 36, No.3, (2000) 325–348.

L. Sakhnovich
735 Crawford ave.
Brooklyn, 11223, New York, NY, USA
Affiliation:
Research Professor
Department of Mathematics
University of Connecticut at Storrs
e-mail: Lev.Sakhnovich@verizon.net

Operator Theory:
Advances and Applications, Vol. 149, 553–568

Operator Theory and the Corona Problem on the Bidisk

Tavan T. Trent

Abstract. We give five equivalent operator theory formulations for the corona problem for $H^\infty(D^2)$. Also, we give two sufficient and one necessary condition for the solution of the problem. Finally, we indicate the solution to the corona problem for the multiplier algebra of symmetric Fock space on the unit ball in $\mathbb{C}^n$.

It has become increasingly evident in recent years that the corona problem on the bidisk can be reformulated as an operator theory problem [Am], [AmMe], [AgMc2], [BT]. We will outline some of these equivalent forms. By the "corona problem on the bidisk" we mean the following:

P 1. Assume $f, g \in H^\infty(D^2)$ and $0 < \epsilon^2 \le |f(\underline{z})|^2 + |g(\underline{z})|^2 < 1$ for all $\underline{z} \in D \times D = D^2$. Do there exist $u, v \in H^\infty(D^2)$ with $fu + gv \equiv 1$?

Let $T_F = (T_f, T_g)$, where T_f denotes the analytic Toeplitz operator defined on $H^2(D^2)$ by multiplication by f, for $f \in H^\infty(D^2)$ and T_g is defined analogously. Here $H^2(D^2)$ denotes the Hardy space on the bidisk. If the answer to the corona problem above is "yes," then for any $h \in H^2(D^2)$

$$f(uh) + g(vh) = h,$$

i.e., T_F is onto. It is easy to check that in this case, $T_F T_F^*$ is bounded below. Conversely, if $T_F T_F^* \ge \delta^2 I$, then obviously T_F is onto.

Now Li [L] and Lin [Li], independently, showed that for $f, g \in H^\infty(D^2)$, if $0 < \epsilon^2 \le |f(\underline{z})|^2 + |g(\underline{z})|^2 \le 1$ for all $z \in D^2$, then T_F is onto and so $T_F T_F^*$ is bounded below. (See [Tre1] for an elementary proof and extension of Li and Lin's result in the Hilbert space case.)

Hence to solve the corona theorem above we must solve the "operator corona problem on the bidisk":

P 2. Assume $f, g \in H^\infty(D^2)$ and $T_f T_f^* + T_g T_g^* \ge \delta^2 I$ for some $\delta > 0$. Do there exist $u, v \in H^\infty(D^2)$ with $T_f T_u + T_g T_v = I$?

Received by the editors January 15, 2003; revised June 6, 2003. Submitted by J. W. Helton.
Mathematics Subject Classification (2000). 32A35, 47A48.
Key words and phrases. Corona theorem, bidisk, operator theory, Fock space.
Partially supported by NSF Grant DMS-0100294.

If we approached Carleson's corona theorem on D, the unit disk, in the preceding manner, then the resulting "operator corona theorem on the disk," (the analog of **P2** for D), can be solved in several ways: using the commutant lifting theorem [Sz-NF], [S], nest algebras and expectations [Ar1], Krein space and Beurling's theorem [BH], operator equations and expectations [KMT], control theory (transfer functions) [BT], and other ways.

We will discuss some of these approaches for the bidisk. But first note that Agler's work on the Nevanlinna-Pick problem on the bidisk [A1], [A2], suggests a direct approach. We have

Theorem 1. [BT, AgMc2]. *Let $f, g \in H^\infty(D^2)$ satisfy $T_F T_F^* = T_f T_f^* + T_g T_g^* \geq \epsilon^2 I$ for some $\epsilon > 0$. Then there exist $u, v \in H^\infty(D^2)$ with $fu + gv = 1$ if and only if there exists $\delta > 0$ and there exist $A \geq 0, B \geq 0$ with*

$$A \geq T_z A T_z^* \text{ and } B \geq T_w B T_w^* \tag{1}$$

and

$$T_f T_f^* + T_g T_g^* - \delta^2 I = A + B. \tag{2}$$

So to solve the corona problem on the bidisk, we consider the equivalent problem

P 3. Assume $f, g \in H^\infty(D^2)$ satisfying $T_F T_F^* \geq \epsilon^2 I$ for some $\epsilon > 0$. Do there exist $\delta > 0$, $A \geq 0$, $B \geq 0$ with

$$T_F T_F^* - \delta^2 I = A + B$$

with A, B satisfying (1)?

Next we'll examine the meaning of (1) and (2). Note that if $S = T_z$, $W = T_w$ and $f \in H^\infty(D^2)$, then $T_f = f(S, W)$. So what does (2) mean, i.e., what does $f(S,W)f(S,W)^* + g(S,W)g(S,W)^* - \delta^2 I = A + B$ mean, where A and B satisfy (1)?

More generally, if $G \in H^\infty(D^4)$, what does it mean to say that

$$G(S, W, S^*, W^*) = A + B$$

with A and B satisfying (1)? Let

$$G(z_1, z_2, z_3, z_4) = \sum_{j,k,l,m=0}^{\infty} a_{jklm} z_1^j z_2^k z_3^l z_4^m.$$

If $T_1 T_2 = T_2 T_1$, $\|T_i\| < 1$, we define

$$G(T_1, T_2, T_1^*, T_2^*) = \sum_{j,k,l,m=0}^{\infty} a_{jklm} T_1^j T_2^k T_1^{*l} T_2^{*m}.$$

Following Agler [A2], we show that

Theorem 2 (Agler). *Let $G \in H^\infty(D^4)$. Then*

$$G(S,W,S^*,W^*) = A+B \text{ and } A,B \text{ satisfy (1) } \text{ if and only if}$$
$$G(T_1,T_2,T_1^*,T_2^*) \geq 0 \quad \text{for all } \|T_i\|<1 \text{ with } T_1T_2 = T_2T_1.$$

Outline of proof. ($\Rightarrow$) Denote by $k_{\underline{z}}$ the reproducing kernel for $H^2(D^2)$ where $\underline{z} = (z_1, z_2)$. From $\langle G(S,W,S^*,W^*)k_{\underline{w}}, k_{\underline{z}}\rangle = \langle A\,k_{\underline{w}}, k_{\underline{z}}\rangle + \langle B\,k_{\underline{w}}, k_{\underline{z}}\rangle$, we deduce that $G(z_1,w_1,\overline{z}_2,\overline{w}_2) = (1-z_1\overline{z}_2)\,\alpha(\underline{z},\underline{w}) + (1-w_1\overline{w}_2)\,\beta(\underline{z},\underline{w})$ where

$$\alpha(\underline{z},\underline{w}) \stackrel{\text{def}}{=} \langle \sqrt{B-WBW^*}k_{\underline{w}}, \sqrt{B-WBW^*}k_{\underline{z}}\rangle$$

and

$$\beta(\underline{z},\underline{w}) \stackrel{\text{def}}{=} \langle \sqrt{A-ZAZ^*}k_{\underline{w}}, \sqrt{A-ZAZ^*}k_{\underline{z}}\rangle$$

are positive definite on $D^2 \times D^2$, analytic in $\underline{z}$, and co-analytic in $\underline{w}$. Thus $\alpha(\underline{z},\underline{w}) = \sum_{n=1}^\infty \alpha_n(\underline{z})\overline{\alpha_n(\underline{w})}$, where $\alpha_n \in Hol(D^2)$ for all n. (See, for example, Ball-Trent [BT].) Similarly for $\beta(\underline{z},\underline{w})$. Now a functional calculus argument shows that for $\|T_1\|<1$ with $T_1T_2 = T_2T_1$,

$$\begin{aligned} G(T_1,T_2,T_1^*,T_2^*) &= \sum_{n=1}^\infty [\alpha_n(T_1,T_2)(I-T_1T_1^*)\alpha_n(T_1,T_2)^* \\ &\quad + \beta_n(T_1,T_2)(I-T_2T_2^*)\beta_n(T_1,T_2)^*] \\ &\geq 0. \end{aligned}$$

($\Leftarrow$) Let $\underline{z}_n = (z_n, w_n)$ for $n = 1,2,\ldots$ be distinct points in D^2 with $\lim_{n\to\infty} \underline{z}_n = 0$. Let $\mathcal{M}_n = sp\{k_{\underline{z}_j}\}_{j=1}^n$ and P_n denote the projection onto $\mathcal{M}_n$. Then $P_n \nearrow I$ strongly. Thus to show that $G \triangleq G(S,W,S^*,W^*) = A+B$ with A,B positive semidefinite and satisfying (1), we need only show that $P_nG\,P_n = P_nA\,P_n + P_nB\,P_n = A_n + B_n$, where $A_n, B_n \in B(\mathcal{M}_n)$ and

$$A_n, B_n \geq 0 \text{ with } A_n \geq P_nS\,A_nS^*P_n \text{ and } B_n \geq P_nW\,B_nW^*P_n. \tag{3}$$

Let $\alpha_{jk} \triangleq \langle A_nk_{\underline{z}_k}, k_{\underline{z}_j}\rangle$, $\beta_{jk} \triangleq \langle B_nk_{\underline{z}_k}, k_{\underline{z}_j}\rangle$, and $\mathcal{A} = [\alpha_{jk}], \mathcal{B} = [\beta_{jk}]$ be matrices in $M(\mathbb{C}^n)$. Then (3) means that

$$\mathcal{A},\mathcal{B} \geq 0 \text{ with } \mathcal{A} \geq D_1\mathcal{A}\,D_1^*,\ \mathcal{B} \geq D_2\mathcal{B}\,D_2^* \tag{4}$$

where D_1 is the diagonal matrix with entries $(z_1, z_2, \ldots, z_n)$
and D_2 is the diagonal matrix with entries $(w_1, w_2, \ldots, w_n)$.

Also

$$\langle P_nG\,P_nk_{\underline{z}_k}, k_{\underline{z}_j}\rangle = \frac{G(z_j, w_j, \overline{z}_k, \overline{w}_k)}{(1-z_j\overline{z}_k)(1-w_j\overline{w}_k)}.$$

Let $g_{jk} = G(z_j, w_j, \overline{z}_k, \overline{w}_k)$ and $\mathcal{G} = [g_{jk}]$. Then $P_n G P_n = A_n + B_n$ satisfies (3) if and only if

$$g_{jk} = (a_{jk} - z_j a_{jk} \overline{z}_k) - w_j (a_{jk} - z_j a_{jk} \overline{z}_k) \overline{w}_k \\ + (b_{jk} - w_j b_{jk} \overline{w}_k) - z_j (b_{jk} - w_j b_{jk} \overline{w}_k) \overline{z}_k. \qquad (5)$$

That is

$$\mathcal{G} = [\mathcal{A} - D_1 \mathcal{A} D_1^*] - D_2 [\mathcal{A} - D_1 \mathcal{A} D_1^*] D_2^* \\ + [\mathcal{B} - D_2 \mathcal{B} D_2^*] - D_1 [\mathcal{B} - D_2 \mathcal{B} D_2^*] D_1^*.$$

Then for $R = \mathcal{A} - D_1 \mathcal{A} D_1^* \geq 0$ and $Q = \mathcal{B} - D_2 \mathcal{B} D_2^* \geq 0$, we must show that

$$\mathcal{G} = R - D_2 R D_2^* + Q - D_1 Q D_1^*$$

where $R, Q \geq 0$. [Notice that if $R \geq 0$ then, since $\|D_1\| < 1$, $R = \mathcal{A} - D_1 \mathcal{A} D_1^*$ with $\mathcal{A} = \sum_{n=1}^{\infty} D_1^n R D_1^{n*} \geq 0$.]

Let $\mathcal{C} = \{Q - D_1 Q D_1^* + R - D_2 R D_2^* : Q, R \in M_n(\mathbb{C}), Q \geq 0, R \geq 0\}$. Then $\mathcal{C}$ is convex and closed (check the diagonal entries first; since $\|D_1\| < 1$, a limit point of $\mathcal{C}$ will have bounded diagonal entries). Moreover, if $P \geq 0$ in $M_n(\mathbb{C})$, then $P \in \mathcal{C}$ since $P = A - D_1 A D_1^*$ where $A = \sum_{n=0}^{\infty} D_1^n P D_1^{*n} \geq 0$. We wish to show that $\mathcal{G} \in \mathcal{C}$.

Suppose that for $Q \in M_n(\mathbb{C})$, we have

$$Re\, tr[Q\, c] \geq 0 \quad \text{for all } c \in \mathcal{C}.$$

Then

$$0 \leq Re\, tr[Q\, c] = \frac{1}{2}\, tr[Q\, c + c\, Q^*] = \frac{1}{2}\, tr[(Q + Q^*)\, c],$$

so we may assume that $Q = Q^*$ and $tr\,[Q\, c] \geq 0$. Moreover, we see that $0 \leq tr[Q\, \underline{x} \otimes \overline{\underline{x}}] = \langle Q\, x, x \rangle$ for all $\underline{x} \in \mathbb{C}^n$ since $\underline{x} \otimes \overline{\underline{x}} \in \mathcal{C}$. Hence $Q \geq 0$. In addition, since $A - D_i A D_i^* \in \mathcal{C}$ for $A \geq 0$ we have $0 \leq tr[Q(A - D_i A D_i^*)] = tr[(Q - D_i^* Q D_i) A]$ for all $A \geq 0$ and, as above, we may conclude that $Q \geq 0$ and

$$Q \geq D_i^* Q\, D_i.$$

Thus if we define $\mathcal{D}_i$ by

$$\mathcal{D}_i(\underline{y}) = \begin{cases} \sqrt{Q}\, D_i \underline{x} & \text{for } \underline{y} = \sqrt{Q}\, \underline{x} \\ 0 & \text{for } \underline{y} \perp ran\, Q \end{cases}$$

we have that $\|\mathcal{D}_i\| < 1$ and $\mathcal{D}_1 \mathcal{D}_2 = \mathcal{D}_2 \mathcal{D}_1$.

By our hypothesis, $\overline{G}(\mathcal{D}_1^*, \mathcal{D}_2^*, \mathcal{D}_1, \mathcal{D}_2) \geq 0$. Let $\underline{1}$ denote the vector of all 1's in $\mathbb{C}^n$. Then

$$\begin{aligned} tr\,[Q\,\mathcal{G}] &= \sum_{s,t,l,m} g_{s,t,l,m}\; tr[Q\, D_1^s D_2^t\, \underline{1} \otimes \underline{1}\, D_1^{l*} D_2^{m*}] \\ &= \sum_{s,t,l,m} \overline{g}_{l,m,s,t} \langle \mathcal{D}_1^s \mathcal{D}_2^t \sqrt{Q}\,\underline{1},\ \mathcal{D}_1^l \mathcal{D}_2^m \sqrt{Q}\,\underline{1}\rangle \\ &= \langle \sum_{s,t,l,m} \overline{g}_{l,m,s,t} \mathcal{D}_1^{*l} \mathcal{D}_2^{*m} \mathcal{D}_1^s \mathcal{D}_2^t \sqrt{Q}\,\underline{1}, \sqrt{Q}\,\underline{1}\rangle \\ &= \langle \overline{G}(\mathcal{D}_1^*, \mathcal{D}_2^*, \mathcal{D}_1, \mathcal{D}_2) \sqrt{Q}\,\underline{1}, \sqrt{Q}\,\underline{1}\rangle \\ &\geq 0. \end{aligned}$$

Thus we deduce that $\mathcal{G} \in \mathcal{C}$. This completes the outline. □

A few remarks are in order. If $f, g \in H^\infty(D^2)$ and $f(S,W)\,f(S,W)^* + g(S,W)\,g(S,W)^* \leq I$, then applying reproducing kernels, we get that

$$|f(z,w)|^2 + |g(z,w)|^2 \leq 1.$$

So by von Neumann's inequality, we have

$$f(T_1,T_2)\,f(T_1,T_2)^* + g(T_1,T_2)\,g(T_1,T_2)^* \leq I$$

for all $\|T_i\| < 1$ with $T_1T_2 = T_2T_1$.

Thus by Theorem 2, an equivalent formulation of the corona problem on the bidisk is a "lower bound" von Neumann inequality. That is,

P 4. Assume $f, g \in H^\infty(D^2)$ and

$$f(S,W)\,f(S,W)^* + g(S,W)\,g(S,W)^* - \epsilon^2 I \geq 0. \tag{6}$$

for some $\epsilon > 0$. Does there exist $\delta > 0$ so that

$$f(T_1,T_2)\,f(T_1,T_2)^* + g(T_1,T_2)\,g(T_1,T_2)^* \geq \delta^2 I \tag{7}$$

for all $\|T_i\| < 1$ with $T_1T_2 = T_2T_1$?

Notice that for the unit disk **P4** is true, since T^* with $\|T\| < 1$ is unitarily equivalent to $S^*\,|_{\mathcal{M}_*}$, where $\mathcal{M}_*$ is an invariant subspace of S^* and S is the unilateral shift with infinite codimension.

Unfortunately, a pair of commuting strict contractions T_1, T_2 satisfies:

$$\begin{aligned} V^*T_1^*V &= S^*\,|_{\mathcal{M}_*} \\ V^*T_2^*V &= W^*\,|_{\mathcal{M}_*} \end{aligned}$$

where V is unitary and $\mathcal{M}_*$ is invariant for both S^* and W^* if and only if

$$I - T_1T_1^* - T_2T_2^* + T_1T_2T_1^*T_2^* \geq 0.\ \text{[A3,At]}$$

Thus it does not directly follow from model theory that (6) $\Longrightarrow$ (7).

Now we examine the commutant lifting (and transfer function), CLT, approach for the bidisk case. The basic idea for CLT is to use the hypothesis that $T_F T_F^* \geq \epsilon^2 I$ as follows:

Let

$$\mathcal{M}_* = range(T_F^*) \text{ and } X^* = (T_F T_F^*)^{-1} T_F \in \mathcal{B}(\mathcal{M}_*, \bigoplus_1^2 H^2(D^2)).$$

Then

$$S^* \mathcal{M}_* \subset \mathcal{M}_*$$
$$W^* \mathcal{M}_* \subset \mathcal{M}_*.$$

Also,

$$X^* S^* \mid_{\mathcal{M}_*} = S^{(2)*} X^*$$
$$X^* W^* \mid_{\mathcal{M}_*} = W^{(2)*} X^*.$$

Now if CLT applied, then there would exist a lifting

$$Y^* \in \mathcal{B}(H^2(D^2), \bigoplus_1^2 H^2(D^2))$$

and

$$Y^* S^* = S^{(2)*} Y^*$$
$$Y^* W^* = W^{(2)*} Y^*. \tag{8}$$

This would solve the problem because

$$(8) \implies Y = \begin{pmatrix} T_u \\ T_v \end{pmatrix} \text{ for } u, v \in H^\infty(D^2)$$

and $Y^* \mid_{\mathcal{M}_*} = X^*$ means that

$$\begin{pmatrix} T_u^* & T_v^* \end{pmatrix} \begin{pmatrix} T_f^* h \\ T_g^* h \end{pmatrix} = X^* \begin{pmatrix} T_f^* h \\ T_g^* h \end{pmatrix} = h$$

i.e.,

$$T_f T_u + T_g T_v = I.$$

But the work of [BLTT] shows that a necessary and sufficient condition for X^* to have such a lifting is that there exists an $M < \infty$ with $M^2 I - X X^* = A + B$ and A, B satisfy (1). Assume $M^2 I - X X^* = A + B$ and A, B satisfy (1). Then

$$M^2 I - T_F^* (T_F T_F^*)^{-2} T_F = A + B$$
$$\implies \quad T_F T_F^* - \frac{1}{M^2} I = \frac{T_F A T_F^* + T_F B T_F^*}{M^2} = A_1 + B_1.$$

and it is easy to check that A_1, B_1 satisfy (1). Thus the CLT and transfer function approaches both lead to **P4**.

We will now consider the Arveson [Ar1] and [KMT] approaches. Let $\mathcal{L}$ denote a lattice of subspaces, E, of a Hilbert space H and assume that $\{0\}$, H are in $\mathcal{L}$. We let $Alg(\mathcal{L})$ or $\mathbb{A}_\mathcal{L}$ denote $\{A \in B(H) : \ AE \subset E \ \text{ for all } \ E \in \mathcal{L}\}$. Then $\mathbb{A}_\mathcal{L}$

is a WOT closed subalgebra of $B(H)$. In the case that $\mathcal{L}$ is linearly ordered, we denote $\mathcal{L}$ by $\mathcal{N}$ and refer to $\mathcal{N}$ as a nest of subspaces and of $\mathbb{A}_{\mathcal{N}}$ as a nest algebra.

The following theorem can be found in [KMT] and was motivated by the special case in Arveson [Ar1]. See [D] for a discussion of Arveson's theorem using the distance formula for nest algebras. In [KMT] the distance formula was replaced by Douglas' lemma and a Hahn-Banach argument.

Theorem 3. [KMT] *Let $\mathbb{A}_{\mathcal{N}}$ be a nest algebra with nest $\mathcal{N}$. Let $X, Y \in \mathbb{A}_{\mathcal{N}}$. Then there exists $Z \in \mathbb{A}_{\mathcal{N}}$ with $\|Z\| = K$ and $ZX = Y$ if and only if $K^2 X^* P_{E^\perp} X \geq Y^* P_{E^\perp} Y$ for all $E \in \mathcal{N}$.*

Let $\mathbb{A}$ be the algebra of upper triangular operators on $H^2(D)$ with respect to the orthonormal basis $\{e^{int}\}_{n=0}^{\infty}$. Thus if $f \in H^\infty(D)$, then

$$T_f^* = \begin{bmatrix} \overline{f}_0 & \overline{f}_1 & \overline{f}_2 & \cdots \\ 0 & \overline{f}_0 & \overline{f}_1 & \ddots \\ 0 & 0 & \overline{f}_0 & \ddots \\ 0 & \vdots & & \ddots \end{bmatrix} \in \mathbb{A} \qquad \text{where} \quad f(z) = \sum_0^\infty f_n z^n \quad \text{and} \quad |z| < 1.$$

Let $E_n = sp\{e^{ijt}\}_{j=0}^{n}$. Clearly, $\mathcal{N} = \{E_n\} \cup \{\{0\}, H\}$ is a nest and $\mathbb{A} = \mathbb{A}_{\mathcal{N}}$. If $\mathcal{A} = \mathbb{A} \otimes M_2(\mathbb{C})$, i.e., 2×2 matrices with entries in $\mathbb{A}$, then $\mathcal{A}$ is also a nest algebra with nest $N = \{E \oplus E : E \in \mathcal{N}\}$.

If $f, g \in H^\infty(D)$, then

$$T_F^* = \begin{pmatrix} T_f^* & 0 \\ T_g^* & 0 \end{pmatrix} \in \mathcal{A}.$$

If we assume that

$$T_F T_F^* \geq \epsilon^2 \begin{pmatrix} I_{H^2(D)} & 0 \\ 0 & 0 \end{pmatrix}$$

then for all $n \geq 0$

$$S^n T_f T_f^* S^{*n} + S^n T_g T_g^* S^{*n} \geq \epsilon^2 S^n S^{*n}$$

or

$$T_f S^n S^{*n} T_f^* + T_g S^n S^{*n} T_g^* \geq \epsilon^2 S^n S^{*n}$$

or

$$T_f P_{E_n^\perp} T_f^* + T_g P_{E_n^\perp} T_g^* \geq \epsilon^2 P_{E_n^\perp}.$$

Finally,

$$\begin{pmatrix} T_f & T_g \\ 0 & 0 \end{pmatrix} \begin{pmatrix} P_{E_n^\perp} & 0 \\ 0 & P_{E_n^\perp} \end{pmatrix} \begin{pmatrix} T_f^* & 0 \\ T_g^* & 0 \end{pmatrix} \geq \epsilon^2 \begin{pmatrix} I & 0 \\ 0 & 0 \end{pmatrix} \begin{pmatrix} P_{E_n^\perp} & 0 \\ 0 & P_{E_n^\perp} \end{pmatrix} \begin{pmatrix} I & 0 \\ 0 & I \end{pmatrix}$$

$$T_F P_{(E_n \oplus E_n)}^\perp T_F^* \geq \epsilon^2 \begin{pmatrix} I & 0 \\ 0 & 0 \end{pmatrix} P_{(E_n \oplus E_n)}^\perp \begin{pmatrix} I & 0 \\ 0 & 0 \end{pmatrix}.$$

Thus by Arveson's theorem there exists $A^* \in \mathcal{A}$ such that

$$A^*T_F^* = \begin{pmatrix} I & 0 \\ 0 & 0 \end{pmatrix}$$

That is, there exist A^*, $B^* \in \mathbb{A}$ with $A^*T_f^* + B^*T_g^* = I$.

Let $U \in B(L^2(\partial D))$ denote the bilateral shift, $U(e^{int}) = e^{i(n+1)t}$ for $n \in Z$ and $t \in [0, 2\pi]$. Then for m any fixed Banach limit on $l^\infty(N)$, we define an expectation operator, $E(\cdot)$, from $B(L^2(\partial D))$ into $L^\infty(\partial D)$ as follows. For $A \in B(L^2(\partial D))$ we define $\langle E(A)h, k\rangle = m(\{\langle U^{n*}AU^n h, k\rangle\}_{n=0}^\infty)$ for $h, k \in L^2(\partial D)$. For details, see the very nice discussion in Davidson [D]. The expectation $E(\cdot)$ has the following properties: for $A \in B(L^2(\partial D))$ and $\varphi \in L^\infty(\partial D)$,

$$E(A^*) = E(A)^*,\ E(T_{E(A)}) = E(A),\ \|E(A)\|_\infty \leq \|A\|,\ E(T_\varphi) = \varphi,$$
$$E(AT_\varphi) = E(T_\varphi A) = \varphi\, E(A),\ E(\lambda A + B) = \lambda\, E(A) + E(B),$$

and if $A^* \in \mathbb{A}$ then $E(A)$ is a bounded analytic function on D.

Using $E(\cdot)$, we get another proof of the operator corona theorem as follows. By Arveson's theorem, since $T_fT_f^* + T_gT_g^* \geq \epsilon^2 I$, there exists $A^*, B^* \in \mathbb{A}$ with $A^*T_f^* + B^*T_g^* = I$. So applying $E(\cdot)$ to $T_fA + T_gB = I$, we get

$$f\, E(A) + g\, E(B) = 1,$$

where $E(A), E(B)$ are bounded analytic functions on D.

For the bidisk case, if $f, g \in H^\infty(D^2)$ with $T_FT_F^* \geq \epsilon^2 I_{H^2(D^2)}$, the relevant algebra is no longer a nest algebra, but rather a tensor product of two nest algebras. So $T_f^* \in \mathbb{A} \otimes \mathbb{A}$ ("upper triangular matrices with upper triangular entries" with respect to the usual orthonormal basis $\{e^{int}e^{ims}\}_{n,m=0}^\infty$ on $H^2(D^2)$). Let $\mathcal{A}_2 = (\mathbb{A} \otimes \mathbb{A}) \otimes M_2(\mathbb{C})$. Also, let

$$T_F^* = \begin{bmatrix} T_f^* & 0 \\ T_g^* & 0 \end{bmatrix} \in \mathcal{A}_2.$$

It can be verified that an expectation argument analogous to the one for $H^2(D)$ holds. [Repeat the previous argument for vector-valued, i.e., $H^2(D)$-valued, $H^2(D)$ space.] The conclusion is the following:

Assume

$$T_FT_F^* \geq \epsilon^2 \begin{pmatrix} I_{H^2(D^2)} & 0 \\ 0 & 0 \end{pmatrix}.$$

If there exist $A^*, B^* \in \mathbb{A} \otimes \mathbb{A}$ satisfying

$$A^*T_f^* + B^*T_g^* = I,$$

then there exist $u, v \in H^\infty(D^2)$ with

$$fu + gv = 1.$$

So the problem is to extend Arveson's theorem.

P 5. Suppose $X, Y \in \mathcal{A}_2$ with $k^2 X^* P_{\mathcal{E}^\perp} X \geq Y^* P_{\mathcal{E}^\perp} Y$ for some $k < \infty$ and all $\mathcal{E} \in Lat(\mathcal{A}_2)$. Does there exist a $Z \in \mathcal{A}_2$ with $ZX = Y$?

As we have mentioned, in Arveson's original proof of Theorem 3 the distance formula for nest algebras was used. That is, if $\mathbb{A}_\mathcal{N}$ is a nest algebra on H and if $T \in B(H)$, then $dis_{\|.\|}(T, \mathbb{A}_\mathcal{N}) = \sup\{\|P_{E^\perp} T P_E\| : E \in \mathcal{N}\}$.

For tensor products of nest algebras, it is not known whether $dis_{\|.\|}(T, \mathbb{A} \otimes \mathbb{A}) \leq K \sup\{\|P_{\mathcal{E}^\perp} T P_\mathcal{E}\| : \mathcal{E} \in Lat\, \mathbb{A} \otimes \mathbb{A}\}$ for some $K < \infty$. But it is trivial that

$$\sup\{\|P_{\mathcal{E}^\perp} T P_\mathcal{E}\| : \mathcal{E} \in Lat\, \mathbb{A} \otimes \mathbb{A}\} \leq dis_{\|.\|}(T, \mathbb{A} \otimes \mathbb{A}).$$

If the distance formula for tensor products of nest algebras could be established, there would still remain an obstacle for any version of a corona theorem.

Namely, one needs to show

P 6. Assume $T_F T_F^* \geq \epsilon^2 I$. Does there exist $\delta > 0$ so that $T_f P_{\mathcal{E}^\perp} T_f^* + T_g P_{\mathcal{E}^\perp} T_g^* \geq \delta^2 P_{\mathcal{E}^\perp}$ for all $\mathcal{E} \in Lat(\mathbb{A} \otimes \mathbb{A})$?

P6 is thus a necessary condition for the corona theorem to hold. It trivially follows in the one-dimensional case, since $P_{E_n} = S^n S^{*n}$ for S the unilateral shift. [Refer back to the argument following Theorem 3.]

To further understand the meaning of (2) and give two additional reformulations, we will need the following lemma which is a tweaking of Ando's theorem. (See Davidson [D] for a discussion of Ando's theorem.)

Lemma 1. *Assume that $T_1, T_2 \in B(\mathcal{H})$ with $T_1 T_2 = T_2 T_1$ and $\|T_i\| < 1$. Then there exists a Hilbert space $\mathcal{K} \supset \mathcal{H}$ and commuting unilateral shifts (of possibly ∞-multiplicity), S_i on $\mathcal{K}$ such that*

$$S_i^* \mathcal{H} \subset \mathcal{H}$$

and

$$S_i^* \mid_\mathcal{H} = T_i^* \qquad i = 1, 2.$$

Proof. We outline the proof. By the Nagy-Foias theory, we may assume $T_1^* = S^* \mid_{\mathcal{M}_*}$ where $\mathcal{M}_* \subseteq \mathcal{H}$ and $S \in B(\mathcal{H})$ is a unilateral shift of ∞-multiplicity. Then assume $T_2 \in B(\mathcal{M}_*)$, $\|T_2\| < 1$, and $T_2^* S^* \mid_{\mathcal{M}_*} = S^* \mid_{\mathcal{M}_*} T_2^*$. By the commutant lifting theorem (see [RR]), there exists a $B \in B(\mathcal{H})$ with $B^* S^* = S^* B^*$, $\|B\| = \|T_2\| < 1$, and

$$B^* \mid_{\mathcal{M}_*} = T_2^*.$$

We need one more factorization result. Namely, $I - B^* B$ satisfies $S^*(I - B^* B) S = (I - B^* B)$ and $I - B^* B$ is invertible since $\|B\| < 1$. Thus by the abstract Szegö theorem due to Lowdenslager (see [RR]), there exists $C \in B(\mathcal{H})$ with $CS = SC$

and $I - B^*B = C^*C$. We now define our unilateral shifts (of ∞-multiplicity) acting on $\overset{\infty}{\underset{1}{\oplus}}\mathcal{H}$. (Thus, $\mathcal{K} = \overset{\infty}{\underset{1}{\oplus}}\mathcal{H}$.)

$$\text{Let } S_1 = \begin{bmatrix} S & 0 & 0 & \dots \\ 0 & S & 0 & \dots \\ 0 & 0 & S & \\ \vdots & \vdots & & \ddots \end{bmatrix} \text{ and } S_2 = \begin{bmatrix} B & 0 & 0 & 0 & \dots \\ C & 0 & 0 & 0 & \dots \\ 0 & I & 0 & 0 & \dots \\ 0 & 0 & I & 0 & \ddots \\ \vdots & \vdots & & \ddots & \ddots \end{bmatrix}.$$

Then trivial computations show that $S_i^*S_i = I$. Also,

$$S_2^{*n} = \begin{bmatrix} B^{*n} & B^{*n-1}C^* & B^{*n-2}C^* & \dots & C^* & 0 & 0 & \dots \\ 0 & 0 & 0 & \dots & 0 & I & 0 & \dots \\ \vdots & \vdots & \vdots & \vdots & \vdots & \ddots & \ddots & \end{bmatrix}$$

Since $B^{*n} \overset{s}{\to} 0$, it is easy to check that $S_2^{*n} \overset{s}{\to} 0$. Now if $\mathcal{N}_* = \mathcal{M}_* \oplus 0 \oplus 0 \dots$ in $\overset{\infty}{\underset{1}{\oplus}}\mathcal{H}$, we have $S_1^*\mathcal{N}_* \subset \mathcal{N}_*$ and $S_2^*\mathcal{N}_* \subset \mathcal{N}_*$. Moreover,

$$S_1^* \mid_{\mathcal{N}_*} = S^* \mid_{\mathcal{M}_*} \oplus 0 \oplus \dots = T_1^* \oplus 0 \oplus \dots$$

and

$$S_2^* \mid_{\mathcal{N}_*} = B^* \mid_{\mathcal{M}_*} \oplus 0 \oplus \dots = T_2^* \oplus 0 \oplus \dots .$$

One can identify $\mathcal{N}_*$ with $\mathcal{M}_*$, and $T_i^* \oplus 0 \oplus \dots$ with $T_i^*, i = 1, 2$. This finishes the proof of the lemma. □

The lemma directly leads to:

P 7. If $f, g \in H^\infty(D^2)$ satisfy

$$\epsilon^2 I \le f(S,W)f(S,W)^* + g(S,W)g(S,W)^*,$$

where $S = M_z, W = M_w$ on $H^2(\partial D \times \partial D)$, does there exist $\delta > 0$ such that for all S_1 and S_2,

$$\delta^2 I \le f(S_1,S_2)\, f(S_1,S_2)^* + g(S_1,S_2)\, g(S_1,S_2)^*,$$

where S_i are unilateral shifts of ∞-multiplicity and $S_1S_2 = S_2S_1$?

We will further consider the meaning of

$$\epsilon^2 I \le T_fT_f^* + T_gT_g^* \le I \text{ for } f, g \in H^\infty(D^2). \tag{9}$$

From (9) it follows that for $z_i, w_i \in D$, $i = 1, 2$, we have

$$\begin{aligned} &f(z_1,w_1)f(z_2,w_2)^* + g(z_1,w_1)g(z_2,w_2)^* - \epsilon^2 \\ &\quad = ((1 - w_1\overline{w}_2) - z_1(1 - w_1\overline{w}_2)\overline{z}_2)\mathcal{A}(z_1,w_1)\mathcal{A}(z_1,w_2)^* \end{aligned} \tag{10}$$

where

$$\mathcal{A}(z_1,w_1)\mathcal{A}(z_2,w_2)^* = \langle (T_fT_f^* + T_gT_g^* - \epsilon^2 I)\, k_{z_2,w_2}, k_{z_1,w_1}\rangle.$$

The fact that positive definite forms have such a factorization follows from the relationship between positive definite forms, reproducing kernels, and orthonormal bases. A reference for this approach to such factorizations is given in [BT]. Assume that $\|T_i\| < 1$ and $T_1T_2 = T_2T_1$. Utilizing Lemma 1, we assume that $T_1^* = S^* |_{\mathcal{M}_*}$ and $T_2^* = B^* |_{\mathcal{M}_*}$, where S is a shift of ∞-multiplicity and $BS = SB$.

Now employing a Riesz-type functional calculus, used, for example, by Agler in [A3], for $z, w \in D$, we may replace z_1 by zI, z_2 by wI, w_1 by $B(z)$, and w_2 by $B(w)$ in (10), where $B(z)$ is the value of the analytic Toeplitz operator B at z. Thus we have

$$\begin{aligned} &f(zI, B(z))f(wI, B(w))^* + g(zI, B(z))g(wI, B(w))^* - \epsilon^2 I \\ &\quad = \mathcal{A}(zI, B(z))[I - B(z)B(w)^* - z(I - B(z)B(w)^*)\overline{w}]\mathcal{A}(wI, B(w))^*. \end{aligned} \tag{11}$$

But $BS = SB$ and $\|B\| < 1$, so for $k_w(z) = \frac{1}{1-\overline{w}z}$, the reproducing kernel for $H^2(D)$ and $\underline{x}, \underline{y} \in l^2$, we have

$$\langle (I - BB^*)k_w\underline{x}, k_z\underline{y}\rangle = \langle (I - B(z)B(w)^*)\underline{x}, \underline{y}\rangle k_w(z).$$

Since $I - BB^* = PP^*$ for $P \geq 0$, we have

$$I - B(z)B(w)^* = (1 - z\overline{w})\mathcal{P}(z)\mathcal{P}(w)^* \text{ for } \mathcal{P}(z)\underline{x} = P(k_z\underline{x}).$$

Finally, from (11), we derive that

$$\begin{aligned} &f(zI, B(z))f(wI, B(w))^* + g(zI, B(z))g(wI, B(w))^* - \epsilon^2 I \\ &\quad = (1 - z\overline{w})^2 \mathsf{A}(z)\mathsf{A}(w)^*, \end{aligned} \tag{12}$$

where $\mathsf{A}(z) = \mathcal{A}(zI, B(z))\mathcal{P}(z)$. Dividing by $(1 - z\overline{w})^2$ and using the fact that $k_w(z) = \frac{1}{(1-\overline{w}z)^2}$ is the reproducing kernel for Bergman space on the disk, $A^2(D)$, we conclude that

$$f(M_z, M_B)f(M_z, M_B)^* + g(M_z, M_B)g(M_z, M_B)^* - \epsilon^2 I \geq 0, \tag{13}$$

where M_z, M_B act on $\overset{\infty}{\underset{1}{\oplus}} A^2(D)$.

Let $F(z) = f(zI, B(z))$ and $G(z) = g(zI, B(z))$. We let A_F denote

$$f(M_z, M_B) \qquad \text{acting on} \qquad \overset{\infty}{\underset{1}{\oplus}} A^2(D)$$

and similarly, A_G denotes $g(M_z, M_B)$.

Then with this notation, (13) says that

$$(A_F, A_G)(A_F, A_G)^* \geq \epsilon^2 I.$$

This leads to the following sufficient problem.

P 8. Assume that

$$F(z) = [f_{ij}(z)]_{i,j=1}^{\infty} \quad \text{with} \quad f_{ij} \in H^{\infty}(D)$$

and that $A_F \in B(\overset{\infty}{\underset{1}{\oplus}} A^2(D))$, where A_F denotes multiplication by $F(z)$ on $\overset{\infty}{\underset{1}{\oplus}} A^2(D)$. Define A_G similarly. Does

$$A_F A_F^* + A_G A_G^* \geq \epsilon^2 I \text{ for some } \epsilon > 0$$

imply that for some $\delta > 0$

$$T_F T_F^* + T_G T_G^* \geq \delta^2 I,$$

where $T_F, T_G \in B(\overset{\infty}{\underset{1}{\oplus}} H^2(D))$?

Roughly speaking, the question is whether the invertibility of a certain operator acting on Bergman spaces entails its invertibility when acting on Hardy spaces.

If the answer to **P8** is "yes," we claim that the corona theorem on $H^{\infty}(D^2)$ follows. To see this, suppose that **P8** is true and there exist $f, g \in H^{\infty}(D^2)$, with $\epsilon^2 I \leq T_f T_f^* + T_g T_g^* \leq I$, but the corona theorem on the bidisk fails. Then by (7), there exist strict contractions $\{A_n\}_{n=1}^{\infty}$ and $\{B_n\}_{n=1}^{\infty}$ with $A_n B_n = B_n A_n$ and

$$f(A_n, B_n) f(A_n, B_n)^* + g(A_n, B_n) g(A_n, B_n)^* - \frac{1}{n^2} I \not\geq 0. \tag{14}$$

But by Lemma 1, Ando's theorem, there exist $\mathcal{M}_{n,*}$ with $S^* \mathcal{M}_{n,*} \subseteq \mathcal{M}_{n,*}$ and $\mathcal{B}_n \in B(\mathcal{M}_{n,*})$ with $\mathcal{B}_n^* S^* = S^* \mathcal{B}_n^*$ and $\mathcal{B}_n^* \mathcal{M}_{n,*} \subseteq \mathcal{M}_{n,*}$ such that $V^* A_n^* V = S^* \mid_{\mathcal{M}_{n,*}}$ and $V^* B_n^* V = \mathcal{B}_n^* \mid_{\mathcal{M}_{n,*}}$ for some unitary operator V. Recall that S denotes the unilateral shift of ∞-multiplicity.

Let

$$F_j(z) = f(zI, B_j(z)) \quad \text{and} \quad G_j(z) = g(zI, B_j(z)),$$

and further

$$F(z) = \begin{bmatrix} F_1(z) & 0 & \cdots \\ 0 & F_2(z) & \\ \vdots & & \ddots \end{bmatrix} \quad \text{and} \quad G(z) = \begin{bmatrix} G_1(z) & 0 & \cdots \\ 0 & G_2(z) & \\ \vdots & & \ddots \end{bmatrix}.$$

By our observations preceding **P8**, it follows that for

$$A_F = \begin{bmatrix} A_{F_1} & 0 & \cdots \\ 0 & A_{F_2} & \\ \vdots & & \ddots \end{bmatrix} \quad \text{and} \quad A_G = \begin{bmatrix} A_{G_1} & 0 & \cdots \\ 0 & A_{G_2} & \\ \vdots & & \ddots \end{bmatrix}$$

we have

$$A_{F_j} A_{F_j}^* + A_{G_j} A_{G_j}^* \geq \epsilon^2 I \quad \text{for all} \quad j = 1, 2, \ldots$$

and hence

$$A_F A_F^* + A_G A_G^* \geq \epsilon^2 I.$$

Now, if **P8** has "yes" for an answer then for some $\delta > 0$, we get

$$T_F T_F^* + T_G T_G^* \geq \delta^2 I$$

or

$$T_{F_n} T_{F_n}^* + T_{G_n} T_{G_n}^* \geq \delta^2 I \text{ for all } n = 1, 2, \ldots.$$

This contradicts (14). Thus, the corona theorem on $H^\infty(D^2)$ holds, if **P8** has an affirmative answer.

But if **P8** has a negative answer, this is also of interest. We note that in [Tr], Treil solved the Sz.-Nagy-Foias problem: that is, there exists $F(z) = [f_{ij}(z)]_{i,j=1}^\infty$, $f_{ij} \in H^\infty(D)$ with $\epsilon^2 I \leq F(z)F(z)^* \leq I$ for all $z \in D$ (so $T_F \in B(\bigoplus_1^\infty H^2(D))$ with $\|T_F\| \leq 1$), but there does not exist $G(z) = [g_{ij}(z)]_{i,j=1}^\infty$, $g_{ij} \in H^\infty(D)$ with T_G bounded and $T_F T_G = I$. Of course, we must not have $\delta^2 I \leq T_F T_F^*$ or CLT would give us an analytic G with $T_F T_G = I$. If the answer to **P8** is "no," this implies there exists a Treil counterexample with stronger properties. For more on the Treil counterexample see Amar and Menini [AmMe].

Although **P8** gives only a sufficient condition for the $H^\infty(D^2)$-corona problem, perhaps the fact that it is only a one-variable problem might make it more tractable.

For our last example, let μ denote a positive finite Borel measure on $\partial D \times \partial D$. For $f \in H^\infty(D^2)$, T_f^μ will denote the operator of multiplication by f on $P^2(\mu) \triangleq \{p(z,w) : p \text{ an analytic polynomial in z and w}\}^{-L^2(\mu)}$.

P 9. If $\epsilon^2 I \leq T_f T_f^* + T_g T_g^*$ for $f, g \in H^\infty(D^2)$ and some $\epsilon > 0$, does there exist $\delta > 0$ so that

$$\delta^2 I \leq T_f^\mu (T_f^\mu)^* + T_g^\mu (T_g^\mu)^*$$

for all positive finite Borel measures μ on $\partial D \times \partial D$?

Using a max-min argument, it can be shown that **P9** is equivalent to the $H^\infty(D^2)$ corona problem. For more on this, see [Am] and [AgMc1]. We believe that the investigation of this problem is of great general interest. We will outline the positive solution to **P9** in the case of $H^2(D)$. For a reference, see Nikolskii [N].

Let μ be a positive finite Borel measure on ∂D, then

$$\mu = w\, d\sigma + \mu_s \text{ where } w \in L^1(d\sigma) \text{ and } \mu_s \perp d\sigma.$$

It then follows that for $\mu_1 = w\, d\sigma$ and $\mu_2 = \mu_s$, we have

$$T_f^\mu = T_f^{\mu_1} \oplus T_f^{\mu_2}.$$

Now $T_f^{\mu_2}$ is normal, so if $T_f T_f^* + T_g T_g^* \geq \epsilon^2 I$, then $|f(e^{it})|^2 + |g(e^{it})|^2 \geq \epsilon^2$ for all $e^{it} \in \partial D$.

Thus, since $T_f^{\mu_2}$ and $T_g^{\mu_2}$ are normal,

$$\begin{aligned} T_f^{\mu_2}(T_f^{\mu_2})^* + T_g^{\mu_2}(T_g^{\mu_2})^* &= (T_f^{\mu_2})^* T_f^{\mu_2} + (T_g^{\mu_2})^* T_g^{\mu_2} \\ &= T^{\mu_2}_{|f|^2+|g|^2} \geq \epsilon^2 I. \end{aligned}$$

By Szegö's theorem, we have two cases. Either T^{μ_1} is normal and we may argue as above to conclude that

$$T_f^{\mu_1}(T_f^{\mu_1})^* + T_g^{\mu_1}(T_g^{\mu_1})^* \geq \epsilon^2 I,$$

or else $w = |F|^2$, where F is an outer function in $H^2(D)$. In this latter case, it is easily checked that for $\varphi \in H^\infty(D)$ and p an analytic polynomial

$$\|T_\varphi^{\mu_1 *}(p)\|^2_{\mu_1} = \|T_\varphi^*(F\,p)\|^2_\sigma.$$

But then

$$\|T_f^*(p)\|^2_\sigma + \|T_g^*(p)\|^2_\sigma \geq \epsilon^2 \|p\|^2_\sigma$$

leads to

$$\begin{aligned} \|T_f^{\mu_1 *}(p)\|^2_{\mu_1} + \|T_g^{\mu_1 *}(p)\|^2_{\mu_1} &= \|T_f^*(F\,p)\|^2_\sigma + \|T_g^*(F\,p)\|^2_\sigma \\ &\geq \epsilon^2 \|F\,p\|^2_\sigma = \epsilon^2 \|p\|^2_{\mu_1}. \end{aligned}$$

Thus,

$$T_f^{\mu_1}(T_f^{\mu_1})^* + T_g^{\mu_1}(T_g^{\mu_1})^* \geq \epsilon^2 I$$

in this case also. We conclude that **P9** is solved with $\delta = \epsilon$ for the case $H^2(D)$.

We end this paper with a positive corona theorem result. We let $\mathcal{F}^2(B_n)$ denote the symmetric Fock space on the n-ball, B_n, in $\mathbb{C}^n$. See Arveson [Ar2] and also Ortega and Fabrega [OF]. That is,

$$\mathcal{F}^2(B_n) = \{f : B_n \to \mathbb{C} \mid f \text{ is analytic and if } f(z_1, \ldots, z_n) =$$
$$\sum a_{j_1,\ldots,j_n} z_1^{j_1} \ldots z_n^{j_n} \text{ then } \|f\|^2_{\mathcal{F}^2(B_n)} \stackrel{def}{=} \sum |a_{j_1,\ldots,j_n}|^2 \frac{j_1! \ldots j_n!}{(j_1 + \cdots + j_n)!} < \infty\}.$$

It is well-known that the reproducing kernel for $\mathcal{F}^2(B_n)$ is $k_{\underline{w}}(\underline{z}) = \frac{1}{1-\langle \underline{w}, \underline{z}\rangle_{\mathbb{C}^n}}$ where $\underline{w}, \underline{z} \in B_n$ [BTV]. We let $\mathcal{M}(\mathcal{F}^2(B_n))$ denote the multipliers on $\mathcal{F}^2(B_n)$. That is,

$$\mathcal{M}(\mathcal{F}^2(B_n)) = \{\varphi \in \mathcal{F}^2(B_n) : \varphi f \in \mathcal{F}^2(B_n) \text{ for all } f \in \mathcal{F}^2(B_n)\}.$$

Although

$$\mathcal{M}(\mathcal{F}^2(B_n)) \subset H^\infty(B^n)$$

these spaces are not identical [Ar2]. Then the corona theorem for $\mathcal{M}(\mathcal{F}^2(B_n))$ holds.

Theorem 4. *Let* $f, g \in \mathcal{M}(\mathcal{F}^2(B_n))$ *with* $\epsilon^2 \leq |f(\underline{z})|^2 + |g(\underline{z})|^2 \leq 1$ *for all* $\underline{z} \in B_n$. *Then there exists* $u, v \in \mathcal{M}(\mathcal{F}^2(B_n))$ *such that* $fu + gv = 1$.

Proof. By Ortega and Fabrega [OF], there exists a $\delta > 0$ such that $\delta^2 I \leq M_f M_f^* + M_g M_g^*$, where M_f denotes multiplication by f on $\mathcal{F}^2(B_n)$. [M_f is a bounded operator by the closed graph theorem.] Now $\frac{1}{k_{\underline{w}}(\underline{z})} = 1 - \langle \underline{z}, \underline{w} \rangle$, so $k_{\underline{w}}(\underline{z})$ has "one-positive square." Then by [BTV], it follows that a CLT argument applies to show that there exists $u, v \in \mathcal{M}(\mathcal{F}^2(B_n))$ with $fu + gv = 1$. $\square$

Acknowledgements

The author would like to thank the reviewer for many helpful suggestions.

References

[A1] J. Agler, *Interpolation*, preprint, 1986.

[A2] ———, *On the representation of certain holomorphic functions defined on a polydisk*, Topics in Operator Theory: Ernst D. Hellinger Memorial Volume (L. de Branges, I. Gohberg, and J. Rovnyak, eds.) vol. 48, Birkhäuser Verlag, Basel, 1990, 47–66.

[A3] ———, *The Arveson extension theorem and coanalytic models*, Int. Equ Op. Theory **5** (1982), 608–631.

[AgMc1] J. Agler and J.W. McCarthy, *Complete Nevanlinna-Pick kernels*, J. Func. Analysis, **175** (2000), 111–124.

[AgMc2] ———, *Nevanlinna-Pick interpolation on the bidisk*, J. Reine Angew. Math. **506** (1999), 191–204.

[Am] E. Amar, *On the Toeplitz Corona problem*, preprint.

[AmMe] E. Amar and C. Menini, *A counterexample to the corona theorem for operators on $H^2(\mathbb{D}^n)$*, preprint.

[Ar1] W.B. Arveson, *Interpolation problems in nest algebras*, J. Func. Analysis, **3** (1975), 208–233.

[Ar2] ———, *Subalgebras of C^*-algebras III*, Acta Math. **181** (1998), 159–228.

[At] A. Athavale, *Model theory on the unit ball in $\mathbb{C}^m$*, J. Oper. Theory **27** (1992), 347–358.

[BH] J.A. Ball and J.W. Helton, *A Beurling-Lax theorem for the Lie group $U(m,n)$ which contains most classical interpolation theory*, J. Oper. Theory **9** (1983), 107–142.

[BLTT] J.A. Ball, W.S. Li, D. Timotin, and T.T. Trent, *A commutant lifting theorem on the polydisk: Interpolation problems for the bidisc*, Indiana Univ. J. **48** (1999), 653–675.

[BT] J.A. Ball and T. Trent, *Unitary colligations, reproducing kernel Hilbert spaces, and Nevanlinna-Pick interpolation in several variables*, J. Funct. Anal. **157** (1998), 1–61.

[BTV] J.A. Ball, T.T. Trent, and V. Vinnikov, *Interpolation and commutant lifting for multipliers on reproducing kernel Hilbert spaces*, Oper. Theory: Advances and Applications **122** (2001), 89–138.

[D] K.R. Davidson, *Nest algebras*, Pitman Research Notes in Mathematics, Longman Scientific and Technical, Essex, 1988.

[KMT] E.G. Katsoulis, R.L. Moore, and T.T. Trent, *Interpolation in nest algebras and applications to operator corona theorems*, J. Oper. Theory, **29** (1993), 115–123.

[L] S.-Y. Li, *Corona problems of several complex variables*, Madison Symposium of Complex Analysis, Contemporary Mathematics, vol. 137, Amer. Math. Soc., Providence, 1991.

[Li] K.C. Lin, *H^p corona theorem for the polydisk*, Trans. Amer. Math. Soc. **341** (1994), 371–375.

[N] N. K.Nikolskii, *Operators, Functions, and Systems: An Easy Reading. Volume 1: Hardy, Hankel, and Toeplitz*, Mathematical Surveys and Monographs, vol. 92, Amer. Math. Soc., Providence, 2002.

[OF] J.M. Ortega and J. Fabrega, *Pointwise multipliers and decomposition theorems in analytic Besov spaces*, Math. Z. **235** (2000), 53–81.

[RR] M. Rosenblum and J. Rovnyak, *Hardy Classes and Operator Theory*, Oxford University Press, New York, 1985.

[S] C.F. Schubert, *Corona theorem as operator theorem*, Proc. Amer. Math. Soc., **69** (1978), 73–76.

[Sz-NF] B. Sz-Nagy and C. Foias, *On contractions similar to isometries and Toeplitz Operators*, Ann. Acad. Sci. Fenn., series A I Math, **2** (1976), 553–564.

[T] V.A. Tolokonnikov, *The corona theorem in algebras of bounded analytic functions*, Amer. Math. Soc. Trans. **149** (1991), 61–93.

[Tr] S.R. Treil, *Angles between coinvariant subspaces and an operator-valued corona problem, a question of Szokefalvi-Nagy*, Soviet Math. Dokl. **38** (1989), 394–399.

[Tre1] T.T. Trent, *An H^2 corona theorem on the bidisk for infinitely many functions*, to appear in J. Linear Equations and its Applications.

Tavan T. Trent
Department of Mathematics
The University of Alabama
Box 870350
Tuscaloosa, AL 35487-0350, USA
e-mail: ttrent@gp.as.ua.edu

Operator Theory:
Advances and Applications, Vol. 149, 569–585

Factorization of Polynomials With Estimates of Norms

Alexander Volberg

Abstract. We want to factorize the polynomials with the estimate of norms and degrees of factors. This problem has connections with many different questions in Functional and Harmonic analysis including the Fejér-Riesz factorization, Peller's lemma on trace norm of finite Hankel matrix, some extension problems for finite Toeplitz matrices, and a recent result of Ferguson and Lacey on the weak factorization of analytic functions in the bidisc. We try to establish elementary relations between these questions. We give a certain elementary factorization theorem for polynomials of two variables.

0. Introduction

We want to give here an overview of what is known about the factorization of polynomials with estimates of norms, and about the relations of this problem with other problems of function theory and operator theory. This subject starts with relatively simple problems, but then, very naturally and very fast, hits on some very difficult ones.

We are interested in the following 2 questions.

Q.1 How to prove that a polynomial $Q(z) = Q_0 + \cdots + Q_n z^n$ of (even) degree n can be represented as

$$Q(z) = \Phi(z)^T \Psi(z), \tag{0.1}$$

where $\Phi(z) = \Phi_0 + \cdots + \Phi_{n_0} z^{n_0}$, $\Psi(z) = \Psi_0 + \cdots + \Psi_{n_1} z^{n_1}$, and where Φ, Ψ are polynomials with vector (columns) coefficients

1) n_0 and $n_1 = n/2$, (or, at worst, $\leq An$, A is absolute).
2) Φ_i, $\Psi_j \in \mathbb{C}^M$. We do not care about M, but we give information about it if available.
3) The main thing we care in (0.1) is

$$\|\Phi\|_2 \cdot \|\Psi\|_2 \leq A\|Q\|_1, \tag{0.2}$$

Received by the editors December 12, 2002; revised May 27, 2003. Submitted by L. Rodman.
Mathematics Subject Classification (2000). 30E20, 47B37, 47B40, 30D55.
Key words and phrases. Fejér-Riesz factorization theorem, trace class, weak factorization, Toeplitz matrices.

where

$$\|\Phi\|_2 = \left(\sum_{i=0}^{n_0} \|\Phi_i\|^2_{\mathbb{C}^M}\right)^{1/2}, \ \|\Psi\|_2 = \left(\sum_{j=0}^{n_1} \|\Psi_j\|^2_{\mathbb{C}^M}\right)^{1/2},$$

and $\| \ \|_{\mathbb{C}^M}$ is a usual Hilbert norm.

Notice that we can introduce scalar polynomials

$$\varphi_m = \sum_{i=0}^{n_0} \varphi_{im} z^i, \ \psi_m = \sum_{j=0}^{n_1} \psi_{jm} z^j,$$

and then (0.1), (0.2) give us weak factorization of $Q(z)$ in the sense

$$\begin{aligned} Q(z) &= \sum_{m=1}^{M} \varphi_m(z)\psi_m(z), \\ \sum_{m=1}^{M} &\|\psi_m\|_2 \|\psi_m\|_2 \leq A\|Q\|_1. \end{aligned} \tag{0.3}$$

Notice that the previous estimate of norms is stronger than this one.

We can immediately formulate a

Problem I. Is (0.3) possible with $M = 1$? I learned this question from Joaquim Bruna and Konstantin Dyakonov.

Q.2 How to prove that a polynomial ($z = (z_1, z_2)$) $Q(z) = \sum_{k,\ell=0}^{N} Q_{k,\ell} z_1^k z_2^\ell$ with frequencies in $[0, N] \times [0, N]$ with some (even) N can be represented as

$$Q(z) = \Phi(z)^T \Psi(z),$$

where Φ and Ψ are polynomials with frequencies in $[0, \frac{N}{2}] \times [0, \frac{N}{2}]$ (or at worst in $[0, AN] \times [0, AN]$, with absolute constant A), columns $\Phi_{k,\ell}, \Psi_{k,\ell}$ are in $\mathbb{C}^M$ and

$$\left(\sum_{k,\ell} \|\Phi_{k,\ell}\|^2_{\mathbb{C}^M}\right)^{1/2} \left(\sum_{k,\ell} \|\Psi_{k,\ell}\|^2_{\mathbb{C}^M}\right)^{1/2} \leq A\|Q\|_{L^1(\mathbb{T}^2)}. \tag{0.4}$$

Question 2 has a formally weaker version as a question of weak factorization of Q: does there exist a family of polynomials $\varphi_m(z)$, $\psi_m(z)$ with frequencies in

$$\left[0, \frac{N}{2}\right] \times \left[0, \frac{N}{2}\right]$$

such that

$$\begin{aligned} Q(z) &= \sum_{m=1}^{M} \varphi_m(z)\psi_m(z), \\ \sum_{m=1}^{M} &\|\varphi_m\|_{L^2(\mathbb{T}^2)} \, \|\psi_m\|_{L^2(\mathbb{T}^2)} \leq A\|Q\|_1 \, ? \end{aligned} \tag{0.5}$$

The relation between Question 2 and weak factorization question is not clear to us. The only clear thing is that the positive answer to Question 2 gives the positive answer to weak factorization question. In one variable there is no difference between the factorization with the help of polynomials with vector coefficients and weak factorization question.

Remark. We will see that in one variable case (0.3) (weak factorization) and (0.1), (0.2) (factorization through vector polynomials) constitute the same problem. However, the best constants we can get in (0.3) and (0.2) are different and the one in (0.2) is of course bigger. As far as we know both of these constants are unknown. They are related to some important problems in Functional Analysis, for example, to the best constant in the extension of finite Toeplitz matrix to a bounded infinite Toeplitz matrix. In two variable cases we do not know whether (0.4) holds in general. Here we are making an attempt to get an approach to (0.4) and (0.5) simpler than the one in [5]. We do not succeed, but maybe our approach reveals connections with some other problems.

Notice that in both Question 2 and the weak factorization question we emphasize the estimates on the degrees of the factor polynomials, given that the initial function is a polynomial of a fixed degree N.

A very important work of Ferguson and Lacey [5] provides the weak factorization in two variables (but they are not interested in the control of the degrees if the data is a polynomial).

Here is their result. Let $H^1(\mathbb{D}^2)$ denote functions from $L^1(\mathbb{T}^2)$ with frequencies only in the first quadrant:

$$\hat{f}(n,k) = 0, n < 0 \text{ or } k < 0\,.$$

Let $H^2(\mathbb{D}^2)$ denote functions from $L^2(\mathbb{T}^2)$ with frequencies only in the first quadrant.

Theorem (Ferguson-Lacey). Every function $f \in H^1(\mathbb{D}^2)$ can be written as

$$f = \sum_{i=1}^{\infty} \phi_i \psi_i, \quad \phi_i, \psi_i \in H^2(\mathbb{D}^2)\,,$$

and (with absolute constant C)

$$\sum_i \|\phi_i\|_{L^2(\mathbb{T}^2)} \|\psi_i\|_{L^2(\mathbb{T}^2)} \le C \|f\|_{L^1(\mathbb{T}^2)}\,.$$

Our main result here is much softer, but a) has a control of degrees of ϕ_i, ψ_i by the degree of f, b) does not rely on a difficult technique of [5]. Instead it uses some softer approaches, but the ones which may have an independent interest. Here is our main result.

Theorem 2.1.

$$Q = \sum_{i,j=0}^{N} Q_{ij} z^i z^j,$$

then $z_2^N Q$ (or $z_1^N Q$) can be written as

$$\sum_{i=1}^{\infty} p_i q_i,$$

frequencies of p_i, q_i being in $[0, N] \times [0, N]$, and

$$\sum_{i=1}^{\infty} \|p_i\|_{L^2(\mathbb{T}^2)} \, \|q_i\|_{L^2(\mathbb{T}^2)} \leq A\|Q\|_{L^1(\mathbb{T}^2)}. \tag{0.6}$$

Theorem contains the control of the degrees but involves the shift of the original data. By proving it here we want to explore the relations of the circle of problems related to the weak factorization: Peller's lemma, extension of bounded finite Toeplitz matrices, Riesz-Fejér theorem in several variables. We believe that one can prove the result without the shift, but the proof will be heavily based on very non-elementary technique of [5]. Our approach is quite elementary, it is based on Riesz-Fejér theorem in several variables. We plan to write down "no shift, degree control" result elsewhere.

1. Function theory proof

Theorem 1.1. $Q = Q_0 + \cdots + Q_n z^n$ can be written as $\sum_{i=1}^{4} p_i q_i$, $\deg p_i$, $\deg q_i \leq \frac{n}{2}$,

$$\|p_i\|_2^2, \|q_i\|_2^2 \leq A\|Q\|_1. \tag{1.1}$$

Remark. I heard this statement from Sergei Shimorin, but not the proof, and then it was easy to restore the proof.

Proof of Theorem 1.1. Start with Q and consider $Q_1(z) = z^{-n/2}Q(z)$, $Q_2(z) = \operatorname{Re} Q_1(z)$, $Q_3 = \operatorname{Im} Q_1$. Notice that their L^1 norms are controlled by $\|Q\|_1$.

Lemma 1.2. Given a real trigonometric polynomial $q = a_{-m}z^{-m} + \cdots + a_0 + \cdots + a_m z^m$, there exists a polynomial p of degree $[sm]$ ($s \in (0, 1]$) such that

1) $p > |q|$ on $\mathbb{T}$;

2) $\int_{\mathbb{T}} p \leq e^2 \int_{\mathbb{T}} |q|$.

Proof. Consider analytic polynomial $h(z) = z^m q(z)$.

Consider it in the disc $\left(1+\dfrac{1}{m+1}\right)\mathbb{D}$.

$$\int_{(1+\frac{1}{m+1})\mathbb{T}} |h| \le e^2 \int_{\mathbb{T}} |h|. \tag{1.2}$$

In fact, $h(z)/z^{2m}$ is analytic outside of the unit disc. So just by mean value theorem

$$\int_{(1+\frac{1}{m+1})\mathbb{T}} |h(z)/z^{2m}| \le \int_{\mathbb{T}} |h(z)/z^{2m}| = \int_{\mathbb{T}} |h(z)|$$

But $\min_{|z|=1+\frac{1}{m+1}} \dfrac{1}{|z|^{2m}} \ge e^{-2}$. So (1.2) is proved.

Using (1.2), consider $H(z) =$ harmonic extension of $|h|\Big|\left(1+\dfrac{1}{m+1}\right)\mathbb{T}$ into $\left(1+\dfrac{1}{m+1}\right)\mathbb{D}$. As $|h(z)|$ is subharmonic there, we get

$$|h(z)| \le H(z), \ z \in \mathbb{T}. \tag{1.3}$$

But by (1.2)

$$\int_{\mathbb{T}} H(z)dm \le \int_{(1+\frac{1}{m+1})\mathbb{T}} |h| \le e^2 \int_{\mathbb{T}} |h|. \tag{1.4}$$

So $H \mid \mathbb{T}$ is what we wanted in 1) and 2) except that $H \mid \mathbb{T}$ is not a polynomial. This is easy to amend. Consider

$$p = K_{sm} * H,$$

where K_{sm} is a Fejér kernel of order sm. Then, from (1.4) one obtains (recall that $h = (q(z)z^m)$)

$$\int_{\mathbb{T}} p \le e^2 \int |h| = e^2 \int |q|. \tag{1.5}$$

On the other hand, p is the averaging of H with Fejér kernel which is $\ge \dfrac{a}{m}\chi_{[-\frac{1}{m},\frac{1}{m}]}$. And on each interval of length $\dfrac{2}{m}$ on the circle $\mathbb{T}$ our harmonic function H is not changing too much:

$$\exists\, A \text{ (abs. const.) } : \frac{\max_I H}{\min_I H} \le A \tag{1.6}$$

for any I, $|I| = \dfrac{2}{m}$, $I \subset \mathbb{T}$. This is because H is a positive harmonic function in $\left(1+\dfrac{1}{m+1}\right)\mathbb{D}$ and we consider it $\dfrac{1}{m}$ distance to the boundary. Now (1.6), (1.3) and $p = K_{sm} * H$ give

$$p(z) \ge a_0|h(z)|. \tag{1.7}$$

And (1.5) and (1.7) prove the lemma. □

Recall that $Q_1(z) = z^{-n/2}Q(z)$, $Q_2(z) = \text{Re}\, Q_1(z)$, $Q_3 = \text{Im}\, Q_1$. Use the lemma for $q = Q_2$, and then for $q = Q_3$. Then $(s = 1,\ m = \frac{n}{2})$

$$\begin{array}{ccc} Q_2 & = & p_2 - (p_2 - Q_2), \qquad Q_3 = p_3 - (p_3 - Q_3) \\ & & \| \qquad\qquad\qquad\qquad \| \\ & & p_2 - p_4 \qquad\qquad\qquad p_3 - p_5 \end{array}$$

Of course, $p_2, p_3, p_4, p_5 \geq 0$. So Fejér-Riesz theorem decomposes them

$$p_i(z) = \alpha_i(z)^* \alpha_i(z), \quad i = 2, 3, 4, 5.$$

$$\|\alpha_i\|_2 = \|p_i\|_1^{1/2} \leq A\Big(\int |Q_2|\Big)^{1/2}$$

$$\leq A\Big(\int_{\mathbb{T}} |Q|\Big)^{1/2}.$$

Degrees of analytic polynomials α_i are at most $m = \frac{n}{2}$.
Finally,

$$\begin{aligned} z^{-n/2}Q(z) &= Q_1(z) = Q_2 + iQ_3 = p_2 - p_4 + ip_3 - ip_5 \\ &= \alpha_2^*\alpha_2 - \alpha_4^*\alpha_4 + i\alpha_3^*\alpha_3 - i\alpha_5^*\alpha_5. \end{aligned}$$

Multiplying α_i^* by $z^{n/2}$ we get new *analytic* polynomials. So Q is decomposed exactly as Theorem 1.1 states. □

Theorem 2.1 below (2 variables) is proved in exactly the same way using operator Fejér-Riesz theorem. Lemma 1.2 holds for 2 variables, and the proof is verbatim the same.

2. Function theory proof. Two variables

Theorem 2.1.

$$Q = \sum_{i,j=0}^{N} Q_{ij} z^i z^j,$$

then $z_2^N Q$ (or $z_1^N Q$) can be written as

$$\sum_{i=1}^{\infty} p_i q_i,$$

frequencies of p_i, q_i being in $[0, N] \times [0, N]$, and

$$\sum_{i=1}^{\infty} \|p_i\|_{L^2(\mathbb{T}^2)} \, \|q_i\|_{L^2(\mathbb{T}^2)} \leq A \|Q\|_{L^1(\mathbb{T}^2)}. \tag{2.1}$$

Proof of Theorem 2.1. We will use Lemma 2.1.1 for two variables (the complete analog of Lemma 1.2 for one variable) combined with an idea from M. Dritschel [3]. First the lemma.

Lemma 2.1.1. Given a real trigonometric polynomial q with frequencies lying in $[-m,m]\times[-m,m]$,there exists a positive polynomial p with frequencies lying in $[-sm,sm]\times[-sm,sm]$ such that

1) $p>|q|$ on $\mathbb{T}^2$;

2) $\int_{\mathbb{T}^2} p \le C\int_{\mathbb{T}^2}|q|$.

Here C is an absolute constant.

Proof. Put $h(z,\zeta):=z^m\zeta^m p(z,\zeta)$. The function is analytic in $\mathbb{C}^2$. Fix any z. Then Lemma 1.2 shows that

$$\int_{(1+\frac{1}{m})\mathbb{T}} |h(z,\zeta)||d\zeta| \le e^2\int_{\mathbb{T}}|h(z,\zeta)||d\zeta|\,. \tag{2.2}$$

The resulting function is called $H_1(z,\zeta)$. It is the Poisson extension of $h((1+\frac{1}{m})e^{i\theta},\zeta)$ to $(1+\frac{1}{m})\mathbb{D}\times\{\zeta\}$. It is harmonic with respect to $z\in(1+\frac{1}{m})\mathbb{D}$. It is defined in $D_m:=(1+\frac{1}{m})\mathbb{D}$. This Poisson averaging of (2.2) immediately gives

$$\int_{(1+\frac{1}{m})\mathbb{T}} H_1(z,\zeta)|d\zeta| \le e^2\int_{\mathbb{T}}H_1(z,\zeta)|d\zeta|\,,\forall z\in D_m\,. \tag{2.3}$$

Now for each $z\in D_m$ let $H_{12}(z,\zeta)$ be the Poisson extension of $H_1(z,(1+\frac{1}{m})e^{i\theta})$ into the disc $\{z\}\times(1+\frac{1}{m})\mathbb{D}$. This gives

$$\int_{\mathbb{T}} H_{12}(z,\zeta)|d\zeta| = \int_{(1+\frac{1}{m})\mathbb{T}} H_1(z,\zeta)|d\zeta| \le e^2\int_{\mathbb{T}}H_1(z,\zeta)|d\zeta|\,,\forall z\in D_m\,. \tag{2.4}$$

We used (2.3) in the last inequality. Now consider (2.4) for all $z\in\mathbb{T}$ and integrate over $z\in\mathbb{T}$ (with respect to Lebesgue measure). Then we get from (2.4)

$$\int_{\mathbb{T}}\int_{\mathbb{T}} H_{12}(z,\zeta)|d\zeta||dz| \le e^2\int_{\mathbb{T}}\int_{\mathbb{T}}H_1(z,\zeta)|dz||d\zeta|\,. \tag{2.5}$$

Now we first integrate H_1 in the z variable. And let us use the fact that it is harmonic in D_m in the first variable for every fixed ζ (in $\mathbb{T}$).

Then

$$\int_{\mathbb{T}} H_1(z,\zeta)|dz| = \int_{(1+\frac{1}{m})\mathbb{D}}|h(z,\zeta)||dz|\,,\forall\zeta\in\mathbb{T}\,. \tag{2.6}$$

Combine (2.6) and (2.5). Then

$$\int_{\mathbb{T}}\int_{\mathbb{T}} H_{12}(z,\zeta)|d\zeta||dz| \le e^2\int_{\mathbb{T}}\int_{(1+\frac{1}{m})\mathbb{D}}|h(z,\zeta)||dz||d\zeta|\,. \tag{2.7}$$

But symmetrically to (2.2) we have

$$\int_{(1+\frac{1}{m})\mathbb{T}} |h(z,\zeta)||dz| \le e^2\int_{\mathbb{T}}|h(z,\zeta)||dz|\,. \tag{2.8}$$

Combining (2.8) and (2.7) we get

$$\int_{\mathbb{T}}\int_{\mathbb{T}} H_{12}(z,\zeta)|d\zeta||dz| \le e^4 \int_{\mathbb{T}}\int_{\mathbb{T}} |h(z,\zeta)||dz||d\zeta|\,. \tag{2.9}$$

Notice that h is subharmonic in z, and H_1 is harmonic and has the same boundary values on $(1+\frac{1}{m})\mathbb{T}$. So for all $z \in D_m$ and all ζ $H_1(z,\zeta) \ge |h(z,\zeta)|$. As the averaging of subharmonic functions $\zeta \to |h(z,\zeta)|$ the function $\zeta \to H_1(z,\zeta)$ is subharmonic in ζ for each fixed $z \in D_m$. Notice that $H_{12}(z,\zeta)$ is harmonic in ζ and has the same boundary values on $(1+\frac{1}{m})\mathbb{T}$. So for all $x \in D_m, \zeta \in D_m$

$$H_{12}(z,\zeta) \ge H_1(z,\zeta) \ge |h(z,\zeta)|\,. \tag{2.10}$$

Now (2.10) and (2.9) almost give us what we want. Only function H_{12} is not a polynomial function. But it is harmonic and *positive* in each variable in $(1+\frac{1}{m})\mathbb{D}$ for every fixed value of the second variable in $\mathbb{T}$. Then Harnack's inequality implies

$$aH_{12}(z,\zeta) \le H_{12}(z',\zeta') \le AH_{12}(z,\zeta)\,, \forall z,\zeta,z',\zeta' \in \mathbb{T}\,, |z-z'| \le \frac{1}{m}, |\zeta-\zeta'| \le \frac{1}{m}\,. \tag{2.11}$$

Consider the Fejér kernel $K := K_{sm}$ of order $[sm]$ in one variable. Exactly as in Lemma 1.2 property (2.11) gives that $q(z,\zeta) := \int_{\mathbb{T}}\int_{\mathbb{T}} K(z-z')K(\zeta-\zeta')H_{12}(z',\zeta')|dz||d\zeta| \ge a\,H_{12}(z,\zeta)$ for all $z,\zeta \in \mathbb{T}$ and for a positive absolute constant a. This q is a positive trigonometric polynomial with frequencies in $[-sm,sm]\times[-sm,sm]$. The last inequality and (2.11) imply that $q \ge a\,|h| = a\,|p|$ on $\mathbb{T}^2$. On the other hand (2.9) implies that $\int_{\mathbb{T}^2} q \le e^4 \int_{\mathbb{T}^2} |p|$. Lemma 2.1.1 is completely proved.

We continue the proof of Theorem 2.1. Our polynomial $Q(z_1,z_2)$ has all frequencies in $[0,N]^2$. Let us shift it (as in the one-dimensional case, but more). Consider $Q_1(z_1,z_2) = z_1^{-N}z_2^{-N}Q(z_1,z_2)$. Now it has frequencies in $[-N,0]^2$. Let

$$Q_2 = \mathrm{Re}\, Q_1, \qquad Q_3 = \mathrm{Im}\, Q_1.$$

Using Lemma 1.2 we can find $p_2, p_4 := p_2 - Q_3,\ p_3,\ p_5 := p_3 - Q_3$ which have frequencies in $[-N,N]^2$, which are all nonnegative polynomials and which satisfy

$$\int_{\mathbb{T}^2} p_i \le A \int_{\mathbb{T}^2} |Q|, \quad i = 2,3,4,5. \tag{2.12}$$

Now it seems to be the time to use operator Fejér-Riesz theorem. But actually it is not. Fix p_i, and call it $\widetilde{P}$. It can be written as block Toeplitz matrix of Toeplitz matrices:

$$\widetilde{P} = \begin{pmatrix} \cdot & \widetilde{P}_{-1} & \widetilde{P}_0 & \widetilde{P}_1 & \cdot & \\ & \cdot & \widetilde{P}_{-1} & \widetilde{P}_0 & \widetilde{P}_1 & \cdot \\ & & \cdot & \cdot & \cdot & \\ & & & \cdot & \cdot & \cdot \end{pmatrix}$$

$\widetilde{P}_i$ is the Toeplitz matrix in which coefficient $\widetilde{P}_{i,j}$ of polynomial $\widetilde{P}$ stands on jth diagonal. It is, of course, a nonnegative infinite matrix, as $\widetilde{P}$ is a nonnegative polynomial.

We use now the idea from [3]. Let us compress each block $\widetilde{P}_i$ to $N \times N$ square block. After that let us divide each block by N. The resulting infinite block Toeplitz matrix whose blocks are $N \times N$ finite Toeplitz matrices is denoted by $P^{(N)}$. Its blocks are $P_i^{(N)}$:

$$P^{(N)} = \begin{pmatrix} \cdot & P_{-1}^{(N)} & P_0^{(N)} & P_1^{(N)} & & \cdot & \\ & \cdot & P_1^{(N)} & P_0^{(N)} & P_1^{(N)} & \cdot & \\ & & \cdot & \cdot & \cdot & & \\ & & & \cdot & \cdot & \cdot & \end{pmatrix}$$

Consider operator polynomial of one variable

$$P^{(N)}(z) = \sum_{i=-N}^{N} P_i^{(N)} z^i$$

(notice that $\widetilde{P}_i$, and so, $P_i^{(N)} = 0$ if $|i| > N$). It is positive definite operator polynomial. We apply operator version of Fejér-Riesz theorem (M. Rosenblum): (F_k are $N \times N$ matrices)

$$P^{(N)}(z) = \left(\sum_{k=0}^{N} F_k z^k\right)^* \left(\sum_{k=0}^{n} F_k z^k\right) =: F(z)^* \, F(z).$$

Then, of course,

$$P_k^{(N)} = \sum_{i=0}^{N} F_i^* \, F_{k+i}, \qquad \text{then} \qquad P_{k;s,t}^{(N)} = \sum_{i=0}^{N} F_{i;s}^* \, F_{k+i;t},$$

where $F_{k+i;t}$ is the t-th column of $N \times N$ matrix F_{k+i} counted from the left. If we are on the j-th diagonal of $P_k^{(N)}$, we have on it $N - j$ times $\frac{1}{N} \, P_{k,j}$ (recall that $P_{k,j}$ are coefficients of original polynomial $\widetilde{P}$, (this polynomial is, say, p_2 (or p_i, i=2,3,4,5)).

Summing up along the j-th diagonal one gets

$$\begin{aligned} \frac{N-j}{N} P_{k,j} &= \sum_{i=0}^{N} \sum_{s=0}^{N-1-|j|} F_{i;s}^* \, F_{k+i;s+j}, & j \geq 0 \\ \frac{N-|j|}{N} P_{k,j} &= \textstyle\sum_{i=0}^{N} \sum_{s=|j|}^{N-1} F_{i;s}^* \, F_{k+i;s+j}, & j < 0 \end{aligned} \tag{2.13}$$

Consider vector polynomial of two variables (coefficients are s-th columns of i-th matrix F)

$$F(z_1, z_1) = \sum_{i=0}^{N} \sum_{s=0}^{N-1} F_{i;s} z_1^i z_2^s.$$

Now (2.13) shows that

$$F(z)^*\ F(z) \quad \text{has coefficient} \quad \frac{N-|j|}{N}\, P_{k,j} \quad \text{in front of} \quad z_1^k\ z_2^j. \tag{2.14}$$

If not for this $\frac{N-|j|}{N}$ coefficient, we would finish the proof of factorization in 2 variables – because the rest would be exactly as in 1 variable.

What we got is the following:

$$Q_1 = z_1^{-N}\ z_2^{-N} Q = Q_2 + iQ_3 = p_2 - p_4 + i(p_3 - p_5),$$

where p_i has frequencies in $[-N, N]^2$, and for $\ell = 2, 3, 4, 5$

$$K_N(z_2) *\ p_\ell = \sum_{m=0}^{N} \left(F^{(m)}(z_1, z_2) \right)^* F^{(m)}(z_1, z_2), \tag{2.15}$$

where $F^{(m)}(z_1, z_2) := \sum_{i=0}^{N} \sum_{s=0}^{N-1} F_{i;s,m}\ z_1^i z_2^s$, where $F_{i;s,m}$ is (m, s)-th element of F_i.

Now just by (2.15)

$$\sum_{m=0}^{N} \|F^{(m)}\|^2_{L^2(\mathbb{T}^2)} \le \|p_\ell\|_{L^1(\mathbb{T}^2)}. \tag{2.16}$$

A problem which awaits us is the existence of a "cut" function K_N in (2.15). Analogous place for one variable did not have any "cut" function (see section 1).

Using (2.15) four times (for $\ell = 2, 3, 4, 5$) we get that

$$K_N(z_2) * (z_1^{-N} z_2^{-N} Q) = \sum \alpha_m^* \alpha_m, \tag{2.17}$$

where frequencies of α_m lie in $[0, N]^2$ and $\sum \|\alpha_m\|_2^2 \le A\|Q\|_1$.
Multiply (2.17) by $z_1^N z_2^N$. Let $\beta_m := z_1^N z_2^N \alpha^*$. These are analytic polynomials with frequencies in $[0, N]^2$. Also

$$\sum \|\beta_m\|_2\ \|\alpha_m\|_2 = \sum \|\alpha_m\|^2 \le A\|Q\|_1. \tag{2.18}$$

And (2.17) becomes

$$V_0^N(z_2) * Q(z_1, z_2) = \sum_{m=0}^{N} \beta_m(z_1, z_2)\ \alpha_m(z_1, z_2). \tag{2.19}$$

Here $V_0^N(z_2)*$ multiplies coefficients $Q_{k,j}$ of Q on $|j|/N$, $j = 0, \dots, N$. And (2.18) is satisfied. The moral is that if we would have this equality without Fejér like cut-off by V we would be done with weak factorization of analytic polynomials of 2 variables. In fact, it looks like the symmetry allows us to get rid of V.
Namely, it looks like we can write a symmetric equality

$$V_N^0(z_2) * Q(z_1, z_2) = \sum_{m=0}^{N} \gamma_m(z_1, z_2)\ \delta_m(z_1, z_2). \tag{2.20}$$

With (2.18) being satisfied for γ, δ instead of β, α. Here $V_N^0(z_2)*$ multiplies coefficients $Q_{k,j}$ of Q by $\frac{N-|j|}{N}$, $j = 0, \ldots, N$. Add (2.19), (2.20). As $(V_0^N + V_N^0)(z_2)^*$ multiply coefficients by $j/N + (N-j)/N = 1$ we get the desired decomposition of $Q(z_1, z_2)$. Unfortunately, (2.19) is, of course, correct (it is just (2.17), and we have proved it already), but (2.20) needs a correction. Let us see why. Let us try to prove this "symmetric" formula.

Consider $\widetilde{Q}(z_1, z_2) := z_2^N Q\left(z_1, \frac{1}{z_2}\right)$. Apply (2.19) to $\widetilde{Q}$:

$$V_0^N * z_2^N Q\left(z_1, \frac{1}{z_2}\right) = V_0^N * \widetilde{Q}(z_1, z_2) = \sum_{m=0}^{N} \tau_m(z_1, z_2)\, \eta_m(z_1, z_2)$$

with frequencies of τ_m, η_m in $[0, N]^2$ and with (2.18) satisfied. This is of course true.

But now we want polynomial Q. Notice that $Q = \widetilde{\widetilde{Q}}$ and

$$\begin{aligned} \widetilde{V_0^N(z_2) * \widetilde{Q}} &= V_N^0(z_2) * Q \widetilde{\left(\sum_{m=0}^{N} \tau_m\, \eta_m\right)} \\ &= z_2^N \sum_{m=0}^{N} \tau_m \left(z_1, \frac{1}{z_2}\right) \eta_m \left(z_1, \frac{1}{z_2}\right) \\ &= z_2^{-N} \sum_{m=0}^{N} \left(z_2^N \tau_m \left(z_1, \frac{1}{z_2}\right)\right) \left(z_2^N \eta_m \left(z_1, \frac{1}{z_2}\right)\right). \end{aligned}$$

Denote $\gamma_m = z_2^N \tau_m \left(z_1, \frac{1}{z_2}\right)$, $\delta_m = z_2^N \eta_m \left(z_1, \frac{1}{z_2}\right)$. Then γ_m, δ_m are analytic polynomials with frequencies in $[0, N]^2$ and

$$V_N^0(z_2) * Q = z_2^{-N} \sum_{m=0}^{N} \gamma_m(z_1, z_2)\, \delta_m(z_1, z_2). \tag{2.21}$$

This is what we get instead of (2.20).

Adding this to (2.19), one gets

$$Q = \sum \alpha_m \beta_m + z_2^{-N} \sum \gamma_m \delta_m.$$

The estimates on norms are okay (see (2.18)):

$$\sum \|\alpha_m\|_2\, \|\beta\|_2 + \sum \|\gamma_m\|_2\, \|\delta_m\|_2 \leq A \|Q\|_1. \tag{2.22}$$

But we get only factorization of $z_2^N Q(z_1, z_2)$, as we see. Theorem 2.1 is proved. □

Question: Is it possible to modify the proof to avoid the shift? The paper by J. Geronimo and H. Woerdeman [4] may give some hope.

3. Operator approach in one variable. Peller's lemma

This approach seems to be very promising for generalizations to higher dimensions, but it will not give exactly what we need ((0.1), (0.2)). If it would give (0.1), (0.2) we would be able to carry out the same proof to more than one variable automatically. So this "proof" is the most promising even being incomplete. We include it here because it seems to be so promising. It is also suited for answering a (more difficult formally) Question 2 (see the Introduction). In the remark after Lemma 3.1 we indicate what is the drawback of the operator approach, and why we cannot generalize it immediately to two variables. However the proof of Lemma 3.1 below seems to us to have an independent interest, and we decided to include it here.

Let us consider $k_n := z^{n/2}\, K_{\frac{n}{2}+1}$ - a shifted Fejér kernel, with Fourier coefficients zero outside $[0, n]$, and "triangle"-like over $[0, n]$ (n is even here).
Also let us consider $v_{\alpha,n}$ - a shifted de la Vallée-Poussin kernel. Fix α in $(0, 1/4)$, consider Fejér kernel K_m with $m = \frac{n}{2} + 1 - [\alpha n]$, then denote $k_{m,n} = z^{n/2} K_m$. Put $c_\alpha := \frac{2}{n+2}[\alpha n]$, and $v_{\alpha,n} := \frac{k_n - k_{m,n}}{c_\alpha} + k_{m,n}$. Notice $\hat{v}_{\alpha,n}$ is supported by $[0, n]$ and is trapezoid-like over $[0, n]$. In particular $\hat{v}_{\alpha,n}$ is identically 1 on interval of integers $([\alpha n], n - [\alpha n])$.

Notice that $\|k_n\|_1 = 1,\ \|v_{\alpha,n}\|_1 \leq A(\alpha) \leq \dfrac{A}{\alpha}$.

Lemma 3.1

(1) Given $Q = Q_0 + \cdots + Q_n z^n$ one can find $\Phi(z) = \Phi_0 + \cdots + \Phi_{n/2} z^{n/2}$, $\Psi(z) = \Psi_0 + \cdots + \Psi_{n/2} z^{n/2}$ such that $\Phi_i, \Psi_j \in \mathbb{C}^{n/2+1}$, and $Q * k_n = \Phi^T \Psi$ and (3.1) holds.

(2) Also fixing $\alpha = \dfrac{m}{n} \in (0, \dfrac{1}{4})$ one can find Φ, Ψ of the same kind such that $Q * v_{\alpha,n} = \Phi^T \Psi$, and $\leq \dfrac{A}{\alpha}$. we also have estimate

$$\Big(\sum \|\Phi_i\|^2_{\mathbb{C}^{\frac{n}{2}+1}}\Big)^{1/2} \Big(\sum \|\Psi_i\|^2_{\mathbb{C}^{\frac{n}{2}+1}}\Big)^{1/2} =$$

$$= \|\Phi\|_{S_2}\, \|\Psi\|_{S_2} \leq \frac{A}{\alpha} \|Q\|_{L^1(\mathbb{T})}. \tag{3.1}$$

Remark. This lemma presents an operator proof of (0.1) and (3.1) factorization. Unfortunately, one has to modify $\widehat{Q}$ by multiplying it on $\hat{k}_n$ or $\hat{v}_{\alpha,n}$. To get rid of this difficulty would mean to have a simple factorization proof in 2 variables.

Proof. Consider the following $\left(\frac{n}{2}+1\right) \times \left(\frac{n}{2}+1\right)$ matrix

$$A := \frac{1}{\frac{n}{2}+1} \begin{bmatrix} Q_0 & Q_1 & Q_2 & \cdots & Q_{n/2} \\ Q_1 & Q_2 & \cdots & Q_{n/2} & Q_{n/2+1} \\ Q_2 & \cdots & Q_{n/2} & Q_{n/2+1} & Q_{n/2+2} \\ \multicolumn{5}{c}{\cdots\cdots\cdots\cdots\cdots\cdots\cdots\cdots} \\ Q_{n/2} & \cdots & Qn-2 & Q_{n-1} & Q_n \end{bmatrix}$$

Below S_1 and S_2 denote trace class and Hilbert-Schmidt classes respectively. Note that its trace norm

$$\|A\|_{S_1} \le \|Q\|_1. \tag{3.2}$$

This is Peller's lemma, see [7]. Le us recall the proof. Fix $t \in [0, 2\pi)$ and consider two vectors $x_t = (1, e^{-it}, \dots, e^{-i\frac{n}{2}t})^T$, $y_t = \overline{Q(e^{it})}\,\overline{x}_t$. Consider rank one operator $a_t := (\cdot, y_t)x_t$ with matrix coefficients $a_t^{k,\ell} = e^{-ikt}e^{-i\ell t}\,Q(e^{it}) = e^{-i(k+\ell)t}\,Q(e^{it})$. Then ($m(t)$ is normalized Lebesgue measure on $[0, 2\pi)$) we have

1) $A = \dfrac{1}{\frac{n}{2}+1} \int a_t \, dm(t)$,

2) $\|a_t\|_{S_1} \le |Q(e^{it})|\, \|x_t\|_2^2 = \left(1 + \frac{n}{2}\right) |Q(e^{it})|$.

Both claims are obvious. Their combination gives

$$\|A\|_{S_1} \le \int |Q(e^{it})| \, dm(t) = \|Q\|_{L^1(\mathbb{T})},$$

and (3.2) is proved.

Having proved (3.2) let us use the polar decomposition of $A = V(A^*A)^{\frac{1}{4}} \cdot (A^*A)^{\frac{1}{4}}$. Consider $\Phi := V(A^*A)^{\frac{1}{4}}$, $\Psi = (A^*A)^{\frac{1}{4}}$. These are two matrices of $\left(\frac{n}{2}+1\right) \times \left(\frac{n}{2}+1\right)$ size, and we denote columns of Ψ counted from left to right as $\Psi_0, \dots, \Psi_{n/2}$. We denote the rows of Φ counted from top to bottom as $\Phi_0, \dots, \Phi_{n/2}$. Then

$$\sum_{0 \le k',\ \ell' \le \frac{n}{2}, k'+\ell'=k+\ell} \langle \Phi_{k'}, \Psi_{\ell'} \rangle_{\mathbb{C}^{\frac{n}{2}+1}} = \frac{k+\ell+1}{\frac{n}{2}+1}\, Q_{k+\ell},\ k+\ell \le \frac{n}{2}, \tag{3.3}$$

$$\sum_{0 \le k',\ \ell' \le \frac{n}{2},\ k'+\ell'=k+\ell} \langle \Phi_{k'}, \Psi_{\ell'} \rangle_{\mathbb{C}^{\frac{n}{2}+1}} = \frac{n+1-(k+\ell)}{\frac{n}{2}+1}\, Q_{k+\ell},\ \frac{n}{2} \le k+\ell \le n, \tag{3.4}$$

But numbers in the right-hand side of (3.3), (3.4) are $\widehat{Q * k_n}(k+\ell)$. In other words, if we denote $\Phi(z) = \sum_{i=0}^{\frac{n}{2}} \Phi_i^T z^i$, $\Psi(z) = \sum_{i=0}^{\frac{n}{2}} \Psi_i z^i$, then (3.3), (3.4) give

$$\Phi^T(z)\Psi(z) = Q(z) * k_n\,. \tag{3.5}$$

We also have estimate

$$\left(\sum \|\Phi_i\|^2_{\mathbb{C}^{\frac{n}{2}+1}}\right)^{1/2} \left(\sum \|\Psi_i\|^2_{\mathbb{C}^{\frac{n}{2}+1}}\right)^{1/2} =$$

$$= \|\Phi\|_{S_2}\, \|\Psi\|_{S_2} = \|A\|_{S_1} \le \|Q\|_{L^1(\mathbb{T})}. \tag{3.6}$$

For the second part of the lemma we want to make the same trick with $Q * v_{\alpha,n}$ instead of $Q * k_n$. This is easy. Fix $m = \frac{n}{2} + 1 - [\alpha n]$. Consider the middle square of size $m \times m$ of the matrix A. Multiply it by $(n/2+1)/m$ (this is bounded by constant 4 by the choice of $\alpha \in (0, 1/4)$). Its entries are $\frac{1}{m} Q_{[\alpha n]}, \dots, \frac{1}{m} Q_{n/2}, \dots, \frac{1}{m} Q_{n-[\alpha n]}$ staying on corresponding diagonals. Consider a new matrix A_m of size $\left(\frac{n}{2}+1\right) \times$

$\left(\frac{n}{2}+1\right)$ which has this square in its middle and zeros everywhere else. This matrix has the same S_1-norm estimate as our A in the first part of the proof (we just consider a square "cut-off" of A, which corresponds to considering PAP with orthogonal projection P, such operation may only diminish S_1 norm):

$$\|A_m\|_{S_1} \leq 4\,\|Q\|_{L^1(\mathbb{T})} \tag{3.7}$$

And so $A_m = \Phi_m \Psi_m$, where Φ_m, Ψ_m are $\left(\frac{n}{2}+1\right) \times \left(\frac{n}{2}+1\right)$ matrices with this estimate:

$$\|\Phi_m\|_{S_2} \|\Psi_m\|_{S_2} \leq 4\,\|Q\|_{L^1(\mathbb{T})}\,. \tag{3.8}$$

For the product of corresponding vector polynomials we have

$$\Phi_m^T(z) \cdot \Psi_m(z) = Q(z) * k_{n,m}, \tag{3.9}$$

where $\widehat{k}_{n,m}$ has been introduces above. In fact, multiplying the polynomial $\Phi_m^T(z)$ (whose coefficients are exactly rows of the matrix Φ_m) and the polynomial $\Psi_m(z)$ (whose coefficients are exactly the columns of Ψ_m), we get in front of $z^{n/2}$ the sum of scalar products of i-th row and j-th column over all $i, j, i+j = n/2$. On the other hand, $A_m = \Phi_m \Psi_m$ shows that the abovementioned sum of scalar products is exactly the sum of elements of A_m stying on the "second main diagonal" (that is on places (i,j), $i = 0, \ldots, n/2; j = 0, \ldots, n/2$, $i + j = n/2$). But all such elements are either zeros or $\frac{1}{m} Q_{n/2}$, and there are exactly m terms of the second kind. So the sum is $Q_{n/2}$. We checked that in (3.9) the coefficients in front of $z^{n/2}$ are the same for the right and left parts. We can repeat the same argument for coefficients in front of z^s, $s = 0, \ldots, n$. For example for $s = \frac{n}{2} - 1$, we get the sum of the elements standing on the corresponding diagonal (one above the "second main diagonal"), and this is $m - 1$ times $\frac{1}{m} Q_{n/2-1}$. This is exactly the coefficient in front of the $z^{\frac{n}{2}-1}$ in the right part of (3.9). Continuing in the same way we prove (3.9). But $v_{\alpha,n} = \frac{k_n - k_{m,n}}{c_\alpha} + k_{m,n}$.

So, using (3.5), (3.9)

$$\begin{aligned} Q * v_{\alpha,n} &= \left(\frac{1}{c_\alpha}(k_n - k_{n,m}) + k_{n,m}\right) * Q = \\ &= \frac{1}{c_\alpha}\left(\Phi^T(z)\Psi(z) - \Phi_m^T(z)\Psi_m(z)\right) + \Phi_m^T(z)\Psi_m(z) =: \tilde{\Phi}^T(z)\tilde{\Psi}(z) \end{aligned}$$

with estimate

$$\begin{aligned} &\left(\sum_{i=0}^{n/2} \|\tilde{\Phi}_i\|^2_{\mathbb{C}^{\frac{n}{2}+1}}\right)^{1/2} \left(\sum_{i=0}^{n/2} \|\tilde{\Psi}_i\|^2_{\mathbb{C}^{\frac{n}{2}+1}}\right)^{1/2} = \\ &= \frac{2}{c_\alpha}\left(\|\Phi\|^2_{S_2} + \|\Phi_m\|^2_{S_2}\right)^{1/2} \left(\|\Psi\|^2_{S_2} + \|\Psi_m\|^2_{S_2}\right)^{1/2}. \end{aligned} \tag{3.10}$$

But, our construction gives more than (3.1). It gives $\|\Phi\|_{S_2} = \|\Psi\|_{S_2} \leq C\,\|A\|^{1/2}_{S_1} \leq C\,\|Q\|^{1/2}_1$. Here C is an absolute constant. Having this in mind and, that $c_\alpha \approx \alpha$, we continue (3.10):

$$\left(\sum_{i=0}^{n/2} \|\tilde{\Phi}_i\|^2_{\mathbb{C}^{\frac{n}{2}+1}}\right)^{1/2} \left(\sum_{i=0}^{n/2} \|\tilde{\Psi}_i\|^2_{\mathbb{C}^{\frac{n}{2}+1}}\right)^{1/2} = \frac{C}{\alpha}\, \|Q\|_1, \tag{3.11}$$

and Lemma 3.1 is completely proved. □

Lemma 3.2. Let $\beta > 0$ be given. Every polynomial $P(z) = P_{\frac{n}{2}} z^{N/2} + \cdots + P_n z^N$ can be represented as $\Phi^T(z)\Psi(z)$, with $\Phi_i, \Psi_j \in \mathbb{C}^M$, Φ, Ψ being polynomials of degrees $\frac{N}{2} + \beta N$ in such a way that

$$\|\Phi\|_2 \|\Psi\|_2 \leq A(\beta)\|P\|_1. \tag{3.12}$$

□

Proof. This is obvious from Lemma 1.1. In fact, we can consider P as a polynomial Q of degree $(1+2\beta)N$. Then

$$Q * v_{\alpha,(1+2\beta)N} = P.$$

It is enough to take $\alpha = \frac{2\beta}{1+2\beta}$. Then (3.12) follows with (3.11) and $A(\beta) \leq \frac{A}{\beta}$. □

Remark. This β in the denominator is our enemy. Suppose we can prove Lemma 3.2 with $\beta = 0$ and $A(0)$ finite. Then we would decompose Q in Lemma 1.1, instead of $Q * v_{\alpha,n}$.

In fact, consider any polynomial $Q(z) = Q_0 + \cdots + Q_n z^n$. Consider $\tilde{Q}(z) = z^{2n} Q(\frac{1}{z}) = Q_n z^n + \cdots + Q_0 z^{2n}$. Apply Lemma 3.2 with $N = 2n$ and $\beta \geq 0$ (only $\beta = 0$ will work correctly, precisely the one which is not given by Lemma 3.2 as it is now).

Then

$$Q_n z^n + \cdots + Q_0 z^{2n} = \Phi^T(z)\Psi(z), \tag{3.13}$$

Φ, Ψ are polynomials of degree $\frac{N}{2} + \beta N = n + 2\beta n = (1+2\beta)n \quad (\beta \geq 0)$ with estimate (3.12).

Write (3.13) as follows:

$$z^{2n}\left(Q_n \frac{1}{z^n} + \cdots + Q_0 \frac{1}{z^{2n}}\right) = z^{2n}\Phi^T\left(\frac{1}{z}\right)\Psi\left(\frac{1}{z}\right) =$$

$$= \left(z^n \Phi\left(\frac{1}{z}\right)\right)^T z^n \Psi\left(\frac{1}{z}\right). \tag{3.14}$$

Then we can denote $z^n\Phi\left(\frac{1}{z}\right)$ by $\Phi_1(z)$, $z^n\Psi\left(\frac{1}{z}\right)$ by $\Psi_1(z)$. Notice that frequencies of Φ_1, Ψ_1 lie in $[-2\beta n, n]$. Therefore, if $\beta = 0$ would be usable, we would get from (3.14)

$$Q(z) = \Psi_1^T(z)\Psi_1(z), \tag{3.15}$$

where Φ_1, Ψ_1 are analytic polynomials (with vector coefficient) of degree n (as Q). And with estimate

$$\|\Phi_1\|_2\|\Psi_1\|_2 \leq A\|Q\|_1. \tag{3.16}$$

This would give us (0.1) and (3.1) – only polynomials with vector coefficients factorizing Q are of degree $n = \deg Q$, not $n/2$ as we originally wanted. But this is good enough.

Unfortunately, we do not know how to prove Lemma 3.2 with $\beta = 0$ and estimate

$$\|\Phi\|_2\|\Psi\|_2 \leq A\|P\|_1. \tag{3.17}$$

Remark. Let us say this again: would we know how to do that, then, by virtue of operator proof we would be able to repeat this in 2 variables and then continue the same way to get (3.15), (3.16) in $z = (z_1, z_2)$ case.

So, we are ready to formulate the problem:

Problem II. $z = (z_1, z_2)$. $P(z)$ is a polynomial with frequencies in

$$\left[\frac{N}{2}, N\right] \times \left[\frac{N}{2}, N\right].$$

Can it be written as

$$P(z) = \sum_m \varphi_m(z)\psi_m(z), \tag{3.18}$$

where φ_m, ψ_m are polynomials with frequencies in

$$\left[0, \frac{N}{2}\right] \times \left[0, \frac{N}{2}\right]$$

and such that the estimate

$$\sum_m \|\varphi_m\|_2\|\psi_m\|_2 \leq A\|P\|_{L^1(\mathbb{T}^2)} \tag{3.19}$$

holds.

Remark. We can do this with

$$\left[0, \frac{N}{2} + \beta N\right] \times \left[0, \frac{N}{2} + \beta N\right]$$

frequency support for φ_m, ψ_m. And $A(\beta)$ in (3.19) grows to ∞ when $\beta \to 0$.

References

[1] A. Böttcher, S. Grudsky, *Fejér means and norms of large Toeplitz matrices*, Preprint No. 331, 2002, de Investigación y de Est. Avanz., Dept. Math.

[2] M. Bakonyi, D. Timotin, *On the extension problem for polynomials*, Bull. London Math. Soc., 33 (2001), No. 5, 599–605.

[3] M. Dritschel, *On factorization of trigonometric polynomials*, Preprint, Dept. of Math., Univ. of Newcastle on Tyne, 2002, pp. 1–30.

[4] J.S Geronimo, H. J. Woederman, *Positive extensions and Riesz-Fejér factorization for two-variables trigonometric polynomials*, Preprint, Georgia Inst. of Technology, 2002, pp. 1–70.

[5] S. Ferguson, M. Lacey, *Characterization of product BMO in the bidisk.* To appear in Acta Math.

[6] Y. Farforofskaya, L. Nikolskaya, *Toeplitz-Schur multipliers,* Preprint, Univ. of Bordeaux, 2002, pp. 1–16.

[7] V.V. Peller, *Hankel operators of class S_p and their applications (rational approximation, Gaussian processes, the problem of majorization of operators)*, Math. USSR Sb., 41 (1982), 443–479.

Alexander Volberg
Department of Mathematics
Michigan State University
East Lansing, MI 48824, USA
e-mail: volberg@math.msu.edu

Operator Theory:
Advances and Applications, Vol. 149, 587–595

The "Action" Variable is not an Invariant for the Uniqueness in the Inverse Scattering Problem

P. Yuditskii

Abstract. We give a simple example of non-uniqueness in the inverse scattering for Jacobi matrices: roughly speaking S-matrix is analytic. Then, multiplying a reflection coefficient by an inner function, we repair this matrix in such a way that it does uniquely determine a Jacobi matrix of Szegö class; on the other hand the transmission coefficient remains the same. This implies the statement given in the title.

1. Jacobi matrices of Szegö class. Direct scattering – Bernstein-Szegö type theorem

As it is well known in the theory of completely integrable systems the absolute values of a reflection coefficient have played the role of the "action" variables and the arguments of this function have meaning of the "angle" variables (in this case we think on the Toda lattice as on an integrable system) [5]. Combining results of our previous work [6] with [3] (see also [4]), we give a wide set of examples where two reflection coefficients, having the same absolute values, possess completely different properties: the first one uniquely determines a Jacobi matrix of Szeqö class and the second one does not. Note that the proof of the main theorem in [3] (and, therefore, our result) essentially uses the analysis of [2] on regularization of so called Arov-singular matrix functions [1].

Let J be a Jacobi matrix defining a bounded self-adjoint operator on $l^2(\mathbb{Z})$:

$$Je_n = p_n e_{n-1} + q_n e_n + p_{n+1} e_{n+1}, \quad n \in \mathbb{Z}, \tag{1}$$

where $\{e_n\}$ is the standard basis in $l^2(\mathbb{Z})$, $p_n > 0$. The resolvent matrix-function is defined by the relation

$$R(z) = R(z, J) = \mathcal{E}^*(J - z)^{-1}\mathcal{E}, \tag{2}$$

Received by the editors October 20, 2002; revised February 5, 2003. Submitted by L. Rodman.
Mathematics Subject Classification (2000). Primary: 47A40. Secondary: 58F07.
Key words and phrases. inverse scattering problem, Jacobi matrix, Bernstein-Szegö type theorem.

where $\mathcal{E} : \mathbb{C}^2 \to l^2(\mathbb{Z})$ is such that

$$\mathcal{E}\begin{bmatrix} c_{-1} \\ c_0 \end{bmatrix} = e_{-1}c_{-1} + e_0 c_0.$$

This matrix-function possesses an integral representation

$$R(z) = \int \frac{d\sigma}{x - z} \tag{3}$$

with a 2×2 matrix-measure having a compact support on $\mathbb{R}$. J is unitarily equivalent to the multiplication operator by an independent variable on

$$L^2_{d\sigma} = \left\{ f = \begin{bmatrix} f_{-1}(x) \\ f_0(x) \end{bmatrix} : \int f^* \, d\sigma \, f < \infty \right\}.$$

The spectrum of J is called absolutely continuous if the measure $d\sigma$ is absolutely continuous with respect to the Lebesgue measure on the real axis,

$$d\sigma(x) = \rho(x)\, dx. \tag{4}$$

It is natural to ask how properties of coefficients of J are reflected on its spectral properties. One is especially interested in J's "close" to the "free" matrix J_0 with constant coefficients, $p_n = 1$, $q_n = 0$ (so called Chebyshev matrix).

Let us mention that J_0 has the following functional representation, besides the general one mentioned above. The resolvent set of J_0 is the domain $\bar{\mathbb{C}} \setminus [-2, 2]$. Let $z(\zeta) : \mathbb{D} \to \bar{\mathbb{C}} \setminus [-2, 2]$ be a uniformization of this domain, $z(\zeta) = 1/\zeta + \zeta$. With respect to the standard basis $\{t^n\}_{n \in \mathbb{Z}}$ in

$$L^2 = \{f(t) : \int_{\mathbb{T}} |f|^2 \, dm\},$$

the matrix of the operator of multiplication by $z(t)$, $t \in \mathbb{T}$, is the Jacobi matrix J_0, since $z(t)t^n = t^{n-1} + t^{n+1}$.

We say that J with absolutely continuous spectrum $[-2, 2]$ is of Szegö class if its spectral density (4) satisfies

$$\log \det \rho(z(t)) \in L^1. \tag{5}$$

Theorem 1.1. *Let J be of Szegö class. Then*

$$p_n \to 1, \; q_n \to 0, \quad n \to \pm\infty. \tag{6}$$

Moreover, there exist generalized eigenvectors

$$\begin{aligned} p_n e^+(n-1,t) + q_n e^+(n,t) + p_{n+1} e^+(n+1,t) &= z(t) e^+(n,t) \\ p_n e^-(-n,t) + q_n e^-(-n-1,t) + p_{n+1} e^-(-n-2,t) &= z(t) e^-(-n-1,t) \end{aligned} \tag{7}$$

such that the following asymptotics hold true

$$\begin{aligned} s(t)e^{\pm}(n,t) &= s(t)t^n + o(1), \quad n \to +\infty \\ s(t)e^{\pm}(n,t) &= t^n + s_{\mp}(t)t^{-n-1} + o(1), \quad n \to -\infty \end{aligned} \tag{8}$$

in L^2.

Thus the eigenvectors of J behave asymptotically as the eigenvectors of J_0 (later we make more precise statement). Note that generalized eigenvectors of J are infinite columns with entry $e^{\pm}(n,t)$ on the n-th position.

The matrix formed by the coefficients of (8)

$$S(t) = \begin{bmatrix} s_- & s \\ s & s_+ \end{bmatrix}(t), \qquad t \in \mathbb{T}, \tag{9}$$

is called the scattering matrix of J. It is unitary-valued, possesses the symmetry property $S^*(\bar{t}) = S(t)$ and the following analytic property: $s(t)$ is an outer function.

In what follows every matrix-function of the form (9) with the above listed properties is called a scattering matrix. Of course we have a good reason for this, since *with every matrix $S(t)$ of this kind one can associate a Jacobi matrix J whose scattering matrix (associated to J according to Theorem 1.1) is the initial matrix-function $S(t)$*. However, $S(t)$, generally speaking, does not determine J uniquely.

To clarify all above statements we need some notation and definitions. First of all for a given functions $s_\pm$ we define the metric

$$\begin{aligned} ||f||^2_{s_\pm} =& \frac{1}{2}\left\langle \begin{bmatrix} 1 & \overline{s_\pm(t)} \\ s_\pm(t) & 1 \end{bmatrix} \begin{bmatrix} f(t) \\ \bar{t}f(\bar{t}) \end{bmatrix}, \begin{bmatrix} f(t) \\ \bar{t}f(\bar{t}) \end{bmatrix} \right\rangle \\ =& \langle f(t) + \bar{t}(s_\pm f)(\bar{t}), f(t)\rangle, \quad f \in L^2, \end{aligned}$$

and we denote by $L^2_{dm,s_\pm}$ or $L^2_{s_\pm}$ (for shortness) the closure of L^2 with respect to this new metric.

The following relations set a unitary map from $L^2_{s_+}$ to $L^2_{s_-}$:

$$\begin{aligned} \begin{bmatrix} sf^+ \\ sf^- \end{bmatrix}(t) &= \begin{bmatrix} s & 0 \\ s_+ & 1 \end{bmatrix}(t) \begin{bmatrix} f^+(t) \\ \bar{t}f^+(\bar{t}) \end{bmatrix} \\ &= \begin{bmatrix} 1 & s_- \\ 0 & s \end{bmatrix}(t) \begin{bmatrix} \bar{t}f^-(\bar{t}) \\ f^-(t) \end{bmatrix}. \end{aligned} \tag{10}$$

Moreover, in this case,

$$||f^+||^2_{s_+} = ||f^-||^2_{s_-} = \frac{1}{2}\{||sf^+||^2 + ||sf^-||^2\}.$$

It is worth to give a scalar variant of relations between $f^+ \in L^2_{s_+}$ and $f^- \in L^2_{s_-}$:

$$s(t)f^{\mp}(t) = \bar{t}f^{\pm}(\bar{t}) + s_\pm(t)f^{\pm}(t).$$

Theorem 1.2. *J is a Jacobi matrix of Szegö class with the spectrum $E = [-2,2]$ if and only if J possesses the scattering representation, i.e.: there exists a unique matrix-function $S(t)$ of the form (9) (with the listed properties) and a unique pair of Fourier transforms*

$$\mathcal{F}^\pm : l^2(\mathbb{Z}) \to L^2_{s_\pm}, \quad (\mathcal{F}^\pm Jf)(t) = z(t)(\mathcal{F}^\pm f)(t), \tag{11}$$

determining each other by the relations

$$s(t)(\mathcal{F}^\pm f)(t) = \bar{t}(\mathcal{F}^\mp f)(\bar{t}) + s_\mp(t)(\mathcal{F}^\mp f)(t), \tag{12}$$

and having the following analytic properties

$$s\mathcal{F}^{\pm}(l^2(\mathbb{Z}_{\pm})) \subset H^2, \tag{13}$$

and asymptotic properties

$$e^{\pm}(n,t) = t^n + o(1) \quad in\ L^2_{s\pm}, \quad n \to +\infty, \tag{14}$$

where

$$e^+(n,t) = (\mathcal{F}^+ e_n)(t), \quad e^-(n,t) = (\mathcal{F}^- e_{-n-1})(t),$$

with $\{e_n\}$ *being the standard basis in* $l^2(\mathbb{Z})$.

We point out that asymptotic relations (8) and (14) are equivalent, moreover (14) directly shows that the eigenvectors of J_0 asymptotically are the eigenvectors of J.

2. Uniqueness and Completeness

Before we proceed with the uniqueness theorem, we show how to construct at least one J with the given scattering matrix $S(t)$.

Consider the space

$$H^2_{s_+} = \operatorname{clos}_{L^2_{s_+}} H^2,$$

and introduce the Hankel operator $\mathcal{H}_{s_+} : H^2 \to H^2$,

$$\mathcal{H}_{s_+} f = P_+ \bar{t}(s_+ f)(\bar{t}), \quad f \in H^2,$$

where P_+ is the Riesz projection from L^2 onto H^2. This operator determines the metric in $H^2_{s_+}$:

$$\begin{aligned} \|f\|^2_{s_+} &= \langle f(t) + \bar{t}(s_+ f)(\bar{t}), f(t)\rangle \\ &= \langle (I + \mathcal{H}_{s_+}) f, f\rangle, \quad \forall f \in H^2. \end{aligned}$$

Theorem 2.1. *Let* $S(t)$ *be a scattering matrix, i.e., the matrix of the form* (9) *with listed properties. Then the space* $H^2_{s_+}$ *is a space of holomorphic functions with a reproducing kernel. Moreover, the reproducing vector* k_{s_+}:

$$\langle f, k_{s_+}\rangle = f(0), \quad \forall f \in H^2_{s_+},$$

is of the form

$$k_{s_+} = (I + \mathcal{H}_{s_+})^{[-1]} \mathbf{1} := \lim_{\epsilon \to 0+} (\epsilon + I + \mathcal{H}_{s_+})^{-1} \mathbf{1} \quad in\ L^2_{s_+}. \tag{15}$$

Put $K_{s_+}(t) = k_{s_+}(t)/\sqrt{k_{s_+}(0)}$. *Then the system of functions* $\{t^n K_{s_+ t^{2n}}(t)\}_{n\in\mathbb{Z}}$ *forms an orthonormal basis in* $L^2_{s_+}$. *With respect to this basis, the multiplication operator by* $z(t)$ *is a Jacobi matrix* $J = J[s_+]$ *of Szegö class. Moreover, the initial* $S(t)$ *serves as the scattering matrix-function, associated with given* J *by Theorem 1.2, and the relations*

$$\mathcal{F}^+(e_n) = t^n K_{s_+ t^{2n}}(t)$$

determine uniquely corresponding Fourier transforms.

Let us fix the notation $J[s_+]$ for the Jacobi matrix associated with $S(t)$ by Theorem 2.1. On the other hand, the system of functions $\{t^n K_{s_- t^{2n}}(t)\}_{n\in\mathbb{Z}}$ forms an orthonormal basis in $L^2_{s_-}$, and we can define a Jacobi matrix $\tilde{J} = J[s_-]$ by the relation

$$z(t)\tilde{e}^+(n,t) = \tilde{p}_n\tilde{e}^+(n-1,t) + \tilde{q}_n\tilde{e}^+(n,t) + \tilde{p}_{n+1}\tilde{e}^+(n+1,t),$$

where $\{\tilde{e}^+(n,t)\}$ is the dual system to the system $\{t^n K_{s_- t^{2n}}(t)\}$ (see (12)), i.e.:

$$s(t)\tilde{e}^+(-n-1,t) := \bar{t}^{n+1} K_{s_- t^{2n}}(\bar{t}) + s_-(t) t^n K_{s_- t^{2n}}(t).$$

None guarantees that operators $J[s_+]$ and $J[s_-]$ are the same (see beginning of the next section). However, if $J[s_+] = J[s_-]$, then the uniqueness theorem takes place.

Theorem 2.2. *A scattering matrix $S(t)$ determines a Jacobi matrix J of Szegö class in a unique way if and only if the following relations take place*

$$s(0)K_{s_\pm}(0)K_{s_\mp t^{-2}}(0) = 1. \tag{16}$$

Of course it is hard to check identities, especially using computer simulation, but, in fact, (16) has a specific approximating meaning.

Let us return to the matrix $J[s_-]$. According to (13) the space $\mathcal{F}^+(l^2(\mathbb{Z}_+))$ is a subspace of $L^2_{s_+}$ consisting of holomorphic in $\mathbb{D}$ functions. In the given case we can even specify this space in the form

$$\mathcal{F}^+(l^2(\mathbb{Z}_+)) = \hat{H}^2_{s_+} := \{f \in L^2_{s_+} : sf \in H^2\}. \tag{17}$$

Every function f from $H^2_{s_+}$ possesses the property $sf \in H^2$, but this means only inclusion:

$$\hat{H}^2_{s_+} \supset H^2_{s_+}. \tag{18}$$

The meaning of (16) is that every function from $\hat{H}^2_{s_+}$ can be approximated by functions from H^2 in $L^2_{s_+}$-norm (H^2 is dense in $\hat{H}^2_{s_+}$). In fact, it is enough to prove that we can approximate just two functions $\tilde{e}^+(0,t)$ and $\tilde{e}^+(1,t)$ from $\hat{H}^2_{s_+}$. This would guarantee (16), completeness and the uniqueness theorem.

3. The result

Our first remark is almost evident. Let us pick an analytic scattering matrix (all entries are in H^∞). For example,

$$S(t) = \begin{bmatrix} \frac{1+\Delta}{2} & \frac{1-\Delta}{2} \\ \frac{1-\Delta}{2} & \frac{1+\Delta}{2} \end{bmatrix}(t),$$

where $\Delta(t) = \overline{\Delta(\bar{t})}$ is a symmetric inner function. Then the analyticity of s_+ implies that $\mathcal{H}_{s_+} = 0$, $H^2_{s_+} = H^2$ and $t^n K_{s_+ t^{2n}}(t) = t^n$, $n \geq 0$. Thus,

$$J[s_+]|l^2(\mathbb{Z}_+(1)) = J_0|l^2(\mathbb{Z}_+(1)),$$

where $\mathbb{Z}_+(m) = \{n \in \mathbb{Z} : n \geq m\}$. The analyticity of s_-, in its turn, implies that

$$J[s_-]|l^2(\mathbb{Z}_-(-1)) = J_0|l^2(\mathbb{Z}_-(-1)),$$

$\mathbb{Z}_-(m) = \{n \in \mathbb{Z} : n \leq m-1\}$.

Thus, the uniqueness would imply that J associated to given $S(t)$ coincides with J_0 up to three coefficients: p_0, q_0, q_{-1}. For fun we give an exact formula for $S(t)$ related to this case:

$$S(t) = \frac{1}{\phi(t)} \begin{bmatrix} \psi_*(t) & p_0(1-t^2) \\ p_0(1-t^2) & \psi(t) \end{bmatrix}, \tag{19}$$

where $\phi(t) = (1-q_0t)(1-q_{-1}t) - p_0^2t^2$, $\psi(t) = (1-q_0t)(q_{-1}-t) + p_0^2t^2$ and $\psi_*(t) = t^2\overline{\psi(t)}$. Note that this matrix has no bound states ($s(t)$ is holomorphic in $\mathbb{D}$) if

$$\left\| \begin{bmatrix} q_{-1} & p_0 \\ p_0 & q_0 \end{bmatrix} \right\| \leq 1.$$

Therefore, once the scattering matrix is not a rational function of degree two, then, for sure, the holomorphic $S(t)$ does not determine J uniquely.

The main result of this note is as follows.

Theorem 3.1. *Given $S(t)$ from H^∞, one can find an inner function $\Phi(t) = \overline{\Phi(\bar{t})}$ such that*

$$S_\Phi(t) = \begin{bmatrix} s_-\Phi & s \\ s & s_+\overline{\Phi} \end{bmatrix}(t), \qquad t \in \mathbb{T},$$

determines a Jacobi matrix of Szegö class uniquely.

Let us point out that we did not change the transmission coefficient s. Moreover the new reflection coefficient $s_-\Phi$ is still analytic one, so, the corresponding Jacobi matrix J coincides with J_0, when both are restricted on $l^2(\mathbb{Z}_-(-1))$.

Theorem 3.1 is the direct consequence of the main result of [3] (especially Lemma 3.4), see also [4], and the following two lemmas.

Lemma 3.2. *Let the reflection coefficient $s_- \in H^\infty$, and $s_-(\zeta) = a\zeta + \dots$, i.e.: $s_-(0) = 0$. Then (16) holds if*

$$\begin{bmatrix} 1 & s_- \\ 0 & s \end{bmatrix} \begin{bmatrix} 1 & \bar{t} \\ \bar{t} & 1 \end{bmatrix} = \lim_{n\to\infty} \begin{bmatrix} s & 0 \\ s_+ & 1 \end{bmatrix} \begin{bmatrix} f_n & \bar{t}g_n \\ \bar{t}f_n(\bar{t}) & g_n(\bar{t}) \end{bmatrix} \tag{20}$$

with a suitable choice of sequences of functions f_n, g_n from H^2, that is, $\frac{1+s_-/t}{s} \in H^2_{s_+}$ and $\frac{1+ts_-}{s} \in H^2_{s_+t^{-2}}$.

Proof. Since s_- is holomorphic, we have $\mathcal{H}_{s_-} = 0$. Moreover, even $\mathcal{H}_{s_-t^{-2}}\mathbf{1} = a\mathbf{1}$, since $s_-(0) = 0$. Therefore,

$$K_{s_-} = \mathbf{1} \quad \text{and} \quad K_{s_-t^{-2}} = \frac{\mathbf{1}}{\sqrt{1+a}}. \tag{21}$$

We work with the second vector. Because of

$$\left\langle \begin{bmatrix} s & 0 \\ s_+ & 1 \end{bmatrix} \begin{bmatrix} f^+(t) \\ \bar{t}f^+(\bar{t}) \end{bmatrix}, \begin{bmatrix} 1 & s_- \\ 0 & s \end{bmatrix} \begin{bmatrix} 1 \\ \bar{t} \end{bmatrix} \right\rangle = \left\langle \begin{bmatrix} s & 0 \\ 0 & \bar{s} \end{bmatrix} \begin{bmatrix} f^+(t) \\ \bar{t}f^+(\bar{t}) \end{bmatrix}, \begin{bmatrix} 1 \\ \bar{t} \end{bmatrix} \right\rangle \\ = 2(sf^+)(0), \quad f^+ \in H^2, \tag{22}$$

and (20), the vector $\frac{1+s_-/t}{s\sqrt{1+a}}$ belongs to H^2_{s+} and it is collinear with the reproducing kernel of this space. Since its norm is one, we get

$$K_{s_+}(\zeta) = \frac{1 + s_-(\zeta)/\zeta}{s(\zeta)\sqrt{1+a}}.$$

Putting here $\zeta = 0$, we get $s(0)K_{s_+}(0)K_{s_-t^{-2}}(0) = 1$. The proof of the second identity is very similar. □

Lemma 3.3. *Let the following approximation hold true*

$$\begin{bmatrix} \bar{t}s_- \\ \frac{s-s(0)}{t} \end{bmatrix} = \lim_{n\to\infty} \begin{bmatrix} su_n \\ P_+s_+u_n \end{bmatrix}, \qquad u_n \in H^2 \tag{23}$$

(we still assume that $s_-(0) = 0$). Then (20) is also true.

Proof. We put

$$\bar{t}f_n(\bar{t}) = s(0)\bar{t} + \bar{t}u_n(\bar{t}) - P_-s_+u_n \tag{24}$$

and $g_n = f_n - (1-t^2)u_n$. In this case

$$\begin{bmatrix} f_n & \bar{t}g_n \\ \bar{t}f_n(\bar{t}) & g_n(\bar{t}) \end{bmatrix} = \begin{bmatrix} f_n - u_n & tu_n \\ \bar{t}u_n(\bar{t}) & f(\bar{t}) - u(\bar{t}) \end{bmatrix} \begin{bmatrix} 1 & \bar{t} \\ \bar{t} & 1 \end{bmatrix}.$$

In fact, we are going to show that

$$\begin{bmatrix} 1 & s_-\bar{t} \\ 0 & s\bar{t} \end{bmatrix} = \lim_{n\to\infty} \begin{bmatrix} s & 0 \\ s_+ & 1 \end{bmatrix} \begin{bmatrix} f_n - u_n & u_n \\ \bar{t}u_n(\bar{t}) & \bar{t}\{f_n(\bar{t}) - u_n(\bar{t})\} \end{bmatrix}. \tag{25}$$

Let us note that the equality in the second column of (25) is just a direct consequence of (23) and (24). Therefore, we need to check the equality in the first column. In the first entry of this column we have

$$\begin{aligned} \lim_{n\to\infty} s(f_n - u_n) &= \lim_{n\to\infty} \{sf_n - s_-\bar{t}\} \\ &= \lim_{n\to\infty} \{ss(0) + su_n - sP_+\bar{t}s_+(\bar{t})u_n(\bar{t}) - s_-\bar{t}\} \\ &= \lim_{n\to\infty} \{ss(0) + s_-\bar{t} - s\overline{s_+}\bar{t}u_n(\bar{t}) + sP_-\bar{t}s_+(\bar{t})u_n(\bar{t}) - s_-\bar{t}\}. \end{aligned}$$

Using (23) and the unitary property of S-matrix, we get

$$\begin{aligned} \lim_{n\to\infty} s(f_n - u_n) = &\lim_{n\to\infty} \{ss(0) + s_-\bar{s}\bar{t}u_n(\bar{t}) + sP_-\bar{t}s_+(\bar{t})u_n(\bar{t})\} \\ =&\{ss(0) + s_-\overline{s_-} + s(\bar{s} - s(0))\} \\ =&\{|s_-|^2 + |s|^2\} = 1. \end{aligned}$$

For the second entry we have

$$\begin{aligned}\lim_{n\to\infty}\{s_+(f_n-u_n)+\bar t u_n(\bar t)\} &= \lim_{n\to\infty}\{s_+s(0)-s_+P_+\overline{s_+}\bar t u_n(\bar t)+\bar t u_n(\bar t)\}\\ &= \lim_{n\to\infty}\{s_+s(0)+s_+P_-\overline{s_+}\bar t u_n(\bar t)+(1-|s_+|^2)\bar t u_n(\bar t)\}\\ &= \lim_{n\to\infty}\{s_+s(0)+s_+(\bar s-s(0))+s\bar s\bar t u_n(\bar t)\}\\ &= s_+s(0)+s_+(\bar s-s(0))+s\overline{s_-}=0. \qquad \square\end{aligned}$$

Lemma 3.4. *Given $S(t)$ from H^∞, one can find an inner function $\Phi(t)=\overline{\Phi(\bar t)}$, $\Phi(0)=0$, such that*

$$\begin{bmatrix}\frac{s_-\Phi}{t}\\ \frac{s-s(0)}{t}\end{bmatrix}=\lim_{n\to\infty}\begin{bmatrix}sh_n\\ P_+s_+\bar\Phi h_n\end{bmatrix},\qquad h_n\in H^2. \tag{26}$$

A naive proof of the lemma looks as follows. We put

$$h=\frac{s_-\Phi/t}{s}, \tag{27}$$

then

$$s_+\frac{s_-\Phi/t}{s}\bar\Phi=\frac{-|s_+|^2}{\bar s}\frac{1}{t}=\frac{|s|^2-1}{\bar s}\frac{1}{t}=\frac{1}{t}s-\frac{1}{t}\frac{1}{\bar s}.$$

Since $\frac{1}{t}\frac{1}{\bar s}$ is anti-analytic, applying P_+, we get $\frac{s-s(0)}{t}$ in the second component. However the function h in (27), in fact, does not belong H^2 (as well as the anti-analytic function $\frac{1}{t}\frac{1}{\bar s}$ does not belong to H^2_-). So, actually, we need in a nontrivial regularization of this procedure. To this end see [3] where the proof was given with all details.

Acknowledgment

The author is thankful to A. Kheifets and the referee for helpful remarks.

References

[1] D. Arov, *Regular and singular J-inner matrix functions and corresponding extrapolation problems.* (Russian) Funktsional. Anal. i Prilozhen. **22** (1988), no. 1, 57–59; translation in Funct. Anal. Appl. **22** (1988), no. 1, 46–48.

[2] V. Katsnelson, *Left and right Blaschke-Potapov products and Arov-singular matrix-valued functions*, Integral Equations Operator Theory **13** (1990), no. 6, 836–848.

[3] A. Kheifets, *On regularization of γ-generating pairs*, J. Funct. Anal. **130** (1995), no. 2, 310–333.

[4] A. Kheifets, *Nehari's interpolation problem and exposed points of the unit ball in the Hardy space H^1*, Proceedings of the Ashkelon Workshop on Complex Function Theory (1996), 145–151, Israel Math. Conf. Proc., **11**, Bar-Ilan Univ., Ramat Gan, 1997.

[5] G. Teschl, *Jacobi operators and completely integrable nonlinear lattices* Mathematical Surveys and Monographs, **72**. American Mathematical Society, Providence, RI, 2000

[6] A. Volberg and P. Yuditskii, *On the inverse scattering problem for Jacobi matrices with the spectrum on an interval, a finite system of intervals or a Cantor set of positive length*, Comm. Math. Phys. **226** (2002), 567–605.

P. Yuditskii
Department of Mathematics
Michigan State University
East Lansing, MI 48824, USA
e-mail: yuditski@math.msu.edu

MIX
Papier aus verantwortungsvollen Quellen
Paper from responsible sources
FSC® C105338

If you have any concerns about our products,
you can contact us on
ProductSafety@springernature.com

In case Publisher is established outside the EU,
the EU authorized representative is:
Springer Nature Customer Service Center GmbH
Europaplatz 3, 69115 Heidelberg, Germany

Printed by Libri Plureos GmbH
in Hamburg, Germany